AF360726

# Placental Related Disorders of Pregnancy

# Placental Related Disorders of Pregnancy

Editors

**Hiten D. Mistry**
**Eun Lee**

MDPI • Basel • Beijing • Wuhan • Barcelona • Belgrade • Manchester • Tokyo • Cluj • Tianjin

*Editors*
Hiten D. Mistry
Department of Women and
Children's Health, School of
Life Course and Population
Health Sciences
King's College London
London
United Kingdom

Eun Lee
Department of Obstetrics and
Gynecology
Virginia Commonwealth
University Health System
Richmond, VA
United States

*Editorial Office*
MDPI
St. Alban-Anlage 66
4052 Basel, Switzerland

This is a reprint of articles from the Special Issue published online in the open access journal *International Journal of Molecular Sciences* (ISSN 1422-0067) (available at: www.mdpi.com/journal/ijms/special_issues/Placental_Pregnancy).

For citation purposes, cite each article independently as indicated on the article page online and as indicated below:

LastName, A.A.; LastName, B.B.; LastName, C.C. Article Title. *Journal Name* **Year**, *Volume Number*, Page Range.

ISBN 978-3-0365-5900-1 (Hbk)
ISBN 978-3-0365-5899-8 (PDF)

Cover image courtesy of Hiten D. Mistry

# Contents

# About the Editors

**Hiten D. Mistry**

Dr. Hiten Mistry is the Senior Research Fellow, Department of Women and Children's Health, School of Life Course and Population Health Sciences, King's College London, working on a range of global women's health studies. Dr. Mistry received a first-class BSc (Hons) in biochemistry with industrial experience from the University of Manchester Institution of Science and Technology (UMIST). He trained as a PhD scientist in Professor Fiona Broughton Pipkin's lab, University of Nottingham, followed by 6 years as a Postdoctoral Research Associate with Professor Lucilla Poston at King's College London and Personal Fellowships at the Universities of Bern (European Renal Association-European Dialysis and Transplantation Association Long-term Fellowship) and Nottingham (British Heart Foundation, Intermediate Basic Science Research Fellowship). Dr. Mistry's main research interests are the pathophysiology of the hypertensive diseases of pregnancy and nutrition in pregnancy. He has published widely in the field, including peer-reviewed journals, expert reviews, book chapters, Guest Editor for Special Issues, and conference proceedings. He is also on the Editorial board for *Placenta*; Associate Editor for *Frontiers in Cardiovascular and Smooth Muscle Pharmacology*; Review Editor for *Frontiers in Reproduction*, *Frontiers in Developmental Endocrinology* and *Frontiers in Integrative Physiology*; and Topical Advisory Panel Member for *IJMS*.

**Eun Lee**

Dr. Eun Lee is the Assistant Professor, Department of Microbiology and Immunology, School of Medicine, Virginia Commonwealth University, Richmond, Virginia, USA. Her laboratory's research interest is in immunological contributions to the development of pre-eclampsia, fetal growth restriction, and repeated spontaneous miscarriage, focusing on placenta development during early pregnancy. Dr. Lee received her PhD from the University of Cincinnati and special research associate training at Harvard University, followed by additional training as a Research Postdoctoral Fellow at Emory University, Department of Transplant Surgery. She has published widely in the field, including peer-reviewed journals and expert reviews, and has served as Guest Editor for Special Issues and conference proceedings. She is also on the editorial board for *Frontiers in Cellular Biology*.

International Journal of
*Molecular Sciences*

*Editorial*

# Placental Related Disorders of Pregnancy

Eun D. Lee [1] and Hiten D. Mistry [2,*]

[1] Department of Microbiology and Immunology, School of Medicine, Massey Cancer Center, Virginia Commonwealth University, Richmond, VA 23298, USA; eun.lee@vcuhealth.org

[2] Department of Women and Children's Health, School of Life Course Sciences, King's College London, London SE5 9NU, UK

* Correspondence: hiten.mistry@kcl.ac.uk

We are pleased to present this Special Issue of *International Journal of Molecular Sciences*, entitled 'Placental Related Disorders of Pregnancy'. The placenta is a unique organ, produced outside the embryo and connected by a cord of vessels, and is formed as a result of various degrees of interactions between fetal and maternal tissues within the pregnant uterus. The placenta fulfils a variety of functions, which are completed by several different organs in adult life. Unlike the relatively stable mature adult organs, the placenta is programmed to complete very different functions during development. Thus, the placenta can be described as a constantly evolving organ. Its major role is the homeostasis of a protected environment for the undisturbed growth and development of an embryo/fetus.

Placental-related disorders of pregnancy are almost unique to the human species and affect around a third of human pregnancies. Many of these disorders result in increased maternal and fetal mortality and morbidity and can have life-long health implications for both the mother and her child. Recent changes in human lifestyle, such as delayed childbirth and hypercaloric diets, may have increased the global incidence of placental-related disorders over recent decades.

This Special Issue is a compilation of 21 research manuscripts and reviews, covering all aspects of placentation, with a particular focus on those related to placental function and disorders of pregnancy. The manuscripts cover aspects of placental physiology, biochemistry and molecular biology, and clinical and animal models are also included in this excellent Special Issue.

This collection contains some excellent reviews. The first review covers the homeostasis of the cytokine interleukin-15 (IL-15) in healthy pregnancy, providing up-to-date mechanisms of the action of IL-15 at the maternal–fetal interface [1]. A fascinating review by Anthony Carter covers why human placentation is so unique, with in-depth details on placentation in different animals to wonderfully illustrate this [2]. This is followed by a comprehensive review covering the important condition of gestational diabetes and the contribution of the placenta in the associated immunoendocrine dysregulation [3]. Finally, a very topical and informative overview highlighting the role of the placenta and the use of low-dose aspirin in the prevention of pre-eclampsia [4,5]. In addition to the reviews, our collection also contains several novel studies covering pre-eclampsia [6–8]; fetal growth restriction [7,9–11]; calcium signaling [12]; placental oxidative stress, nutrition, senescence and apoptosis [6,9,13–15]; sexual dimorphism [16–18], intrahepatic cholestasis [19]; placental vascular modelling [20]; and placental villous explant culture models [21].

This Special Issue presents placental research using a range of established and state-of-the-art techniques showcasing novel and up-to-date data to enhance and facilitate our understanding of placentation as well as mechanisms that result in associated adverse pregnancy outcomes, as well as longer-term risks of complications.

**Funding:** This research received no external funding.

**Conflicts of Interest:** The authors declare no conflict of interest.

**Citation:** Lee, E.D.; Mistry, H.D. Placental Related Disorders of Pregnancy. *Int. J. Mol. Sci.* **2022**, *23*, 3519. https://doi.org/10.3390/ijms23073519

Received: 14 March 2022
Accepted: 22 March 2022
Published: 24 March 2022

**Publisher's Note:** MDPI stays neutral with regard to jurisdictional claims in published maps and institutional affiliations.

## References

1.  Gordon, S.M. Interleukin-15 in Outcomes of Pregnancy. *Int. J. Mol. Sci.* **2021**, *22*, 11094. [CrossRef]
2.  Carter, A.M. Unique Aspects of Human Placentation. *Int. J. Mol. Sci.* **2021**, *22*, 8099. [CrossRef]
3.  Olmos-Ortiz, A.; Flores-Espinosa, P.; Diaz, L.; Velazquez, P.; Ramirez-Isarraraz, C.; Zaga-Clavellina, V. Immunoendocrine Dysregulation during Gestational Diabetes Mellitus: The Central Role of the Placenta. *Int. J. Mol. Sci.* **2021**, *22*, 8087. [CrossRef]
4.  Walsh, S.W.; Strauss, J.F., 3rd. The Road to Low-Dose Aspirin Therapy for the Prevention of Preeclampsia Began with the Placenta. *Int. J. Mol. Sci.* **2021**, *22*, 6085. [CrossRef]
5.  Sun, Y.; Tan, L.; Neuman, R.I.; Broekhuizen, M.; Schoenmakers, S.; Lu, X.; Danser, A.H. Megalin, Proton Pump Inhibitors and the Renin-Angiotensin System in Healthy and Pre-Eclamptic Placentas. *Int. J. Mol. Sci.* **2021**, *22*, 7407. [CrossRef]
6.  Scaife, P.J.; Simpson, A.; Kurlak, L.O.; Briggs, L.V.; Gardner, D.S.; Broughton Pipkin, F.; Jones, C.J.; Mistry, H.D. Increased Placental Cell Senescence and Oxidative Stress in Women with Pre-Eclampsia and Normotensive Post-Term Pregnancies. *Int. J. Mol. Sci.* **2021**, *22*, 7295. [CrossRef]
7.  Mundal, S.B.; Rakner, J.J.; Silva, G.B.; Gierman, L.M.; Ausdal, M.; Basnet, P.; Elschot, M.; Bakke, S.S.; Ostrop, J.; Thomsen, L.C.V.; et al. Divergent regulation of decidual oxidative-stress response by NRF2 and KEAP1 in preeclampsia with and without fetal growth restriction. *Int. J. Mol. Sci.* **2022**, *23*, 1966. [CrossRef]
8.  Sammar, M.; Siwetz, M.; Meiri, H.; Sharabi-Nov, A.; Altevogt, P.; Huppertz, B. Reduced Placental CD24 in Preterm Preeclampsia Is an Indicator for a Failure of Immune Tolerance. *Int. J. Mol. Sci.* **2021**, *22*, 8045. [CrossRef]
9.  Tanner, A.R.; Lynch, C.S.; Kennedy, V.C.; Ali, A.; Winger, Q.A.; Rozance, P.J.; Anthony, R.V. CSH RNA Interference Reduces Global Nutrient Uptake and Umbilical Blood Flow Resulting in Intrauterine Growth Restriction. *Int. J. Mol. Sci.* **2021**, *22*, 8150. [CrossRef]
10. Murphy, C.N.; Walker, S.P.; MacDonald, T.M.; Keenan, E.; Hannan, N.J.; Wlodek, M.E.; Myers, J.; Briffa, J.F.; Romano, T.; Roddy Mitchell, A.; et al. Elevated Circulating and Placental SPINT2 Is Associated with Placental Dysfunction. *Int. J. Mol. Sci.* **2021**, *22*, 7467. [CrossRef]
11. Stojanovska, V.; Shah, A.; Woidacki, K.; Fischer, F.; Bauer, M.; Lindquist, J.A.; Mertens, P.R.; Zenclussen, A.C. YB-1 Is Altered in Pregnancy-Associated Disorders and Affects Trophoblast in Vitro Properties via Alternation of Multiple Molecular Traits. *Int. J. Mol. Sci.* **2021**, *22*, 7226. [CrossRef]
12. Fecher-Trost, C.; Wolske, K.; Wesely, C.; Lohr, H.; Klawitter, D.S.; Weissgerber, P.; Gradhand, E.; Burren, C.P.; Mason, A.E.; Winter, M.; et al. Mutations That Affect the Surface Expression of TRPV6 Are Associated with the Upregulation of Serine Proteases in the Placenta of an Infant. *Int. J. Mol. Sci.* **2021**, *22*, 12694. [CrossRef]
13. Kohan-Ghadr, H.R.; Armistead, B.; Berg, M.; Drewlo, S. Irisin Protects the Human Placenta from Oxidative Stress and Apoptosis via Activation of the Akt Signaling Pathway. *Int. J. Mol. Sci.* **2021**, *22*, 11229. [CrossRef]
14. Lospinoso, K.; Dozmorov, M.; El Fawal, N.; Raghu, R.; Chae, W.J.; Lee, E.D. Overexpression of ERAP2N in Human Trophoblast Cells Promotes Cell Death. *Int. J. Mol. Sci.* **2021**, *22*, 8585. [CrossRef]
15. Steinhauser, C.B.; Lambo, C.A.; Askelson, K.; Burns, G.W.; Behura, S.K.; Spencer, T.E.; Bazer, F.W.; Satterfield, M.C. Placental Transcriptome Adaptations to Maternal Nutrient Restriction in Sheep. *Int. J. Mol. Sci.* **2021**, *22*, 7654. [CrossRef]
16. Bucher, M.; Kadam, L.; Ahuna, K.; Myatt, L. Differences in Glycolysis and Mitochondrial Respiration between Cytotrophoblast and Syncytiotrophoblast In-Vitro: Evidence for Sexual Dimorphism. *Int. J. Mol. Sci.* **2021**, *22*, 10875. [CrossRef]
17. Lien, Y.C.; Zhang, Z.; Cheng, Y.; Polyak, E.; Sillers, L.; Falk, M.J.; Ischiropoulos, H.; Parry, S.; Simmons, R.A. Human Placental Transcriptome Reveals Critical Alterations in Inflammation and Energy Metabolism with Fetal Sex Differences in Spontaneous Preterm Birth. *Int. J. Mol. Sci.* **2021**, *22*, 7899. [CrossRef]
18. Garcia-Martin, I.; Penketh, R.J.; Garay, S.M.; Jones, R.E.; Grimstead, J.W.; Baird, D.M.; John, R.M. Symptoms of Prenatal Depression Associated with Shorter Telomeres in Female Placenta. *Int. J. Mol. Sci.* **2021**, *22*, 7458. [CrossRef]
19. Ontsouka, E.; Epstein, A.; Kallol, S.; Zaugg, J.; Baumann, M.; Schneider, H.; Albrecht, C. Placental Expression of Bile Acid Transporters in Intrahepatic Cholestasis of Pregnancy. *Int. J. Mol. Sci.* **2021**, *22*, 10434. [CrossRef]
20. Courtney, J.A.; Wilson, R.L.; Cnota, J.; Jones, H.N. Conditional Mutation of Hand1 in the Mouse Placenta Disrupts Placental Vascular Development Resulting in Fetal Loss in Both Early and Late Pregnancy. *Int. J. Mol. Sci.* **2021**, *22*, 9532. [CrossRef]
21. Kupper, N.; Pritz, E.; Siwetz, M.; Guettler, J.; Huppertz, B. Placental Villous Explant Culture 2.0: Flow Culture Allows Studies Closer to the In Vivo Situation. *Int. J. Mol. Sci.* **2021**, *22*, 7464. [CrossRef]

International Journal of
*Molecular Sciences*

MDPI

*Review*

# Unique Aspects of Human Placentation

Anthony M. Carter

Cardiovascular and Renal Research, Institute of Molecular Medicine, University of Southern Denmark,
DK-5230 Odense, Denmark; acarter@health.sdu.dk

**Abstract:** Human placentation differs from that of other mammals. A suite of characteristics is shared with haplorrhine primates, including early development of the embryonic membranes and placental hormones such as chorionic gonadotrophin and placental lactogen. A comparable architecture of the intervillous space is found only in Old World monkeys and apes. The routes of trophoblast invasion and the precise role of extravillous trophoblast in uterine artery transformation is similar in chimpanzee and gorilla. Extended parental care is shared with the great apes, and though human babies are rather helpless at birth, they are well developed (precocial) in other respects. Primates and rodents last shared a common ancestor in the Cretaceous period, and their placentation has evolved independently for some 80 million years. This is reflected in many aspects of their placentation. Some apparent resemblances such as interstitial implantation and placental lactogens are the result of convergent evolution. For rodent models such as the mouse, the differences are compounded by short gestations leading to the delivery of poorly developed (altricial) young.

**Keywords:** decidual reaction; fetal membranes; placental hormones; primates; uterine spiral artery; uterine NK cell

**Citation:** Carter, A.M. Unique Aspects of Human Placentation. *Int. J. Mol. Sci.* **2021**, 22, 8099. https://doi.org/10.3390/ijms22158099

Academic Editor: Gernot Desoye

Received: 30 June 2021
Accepted: 26 July 2021
Published: 28 July 2021

**Publisher's Note:** MDPI stays neutral with regard to jurisdictional claims in published maps and institutional affiliations.



## 1. Introduction

Adverse pregnancy outcomes can often be linked to defects in placentation [1]. Ethical considerations preclude detailed exploration of the underlying mechanisms. Unfortunately, there are also limitations to what can be learned from animal models. The mouse (*Mus musculus*) and other murine rodents have exceedingly short gestations. Whilst they may be informative about early events, such as the differentiation of cell lineages, they are unsatisfactory for modelling the events of third-trimester human pregnancy [2]. In addition, there are important differences between rodent and human in placentation and the disposition of fetal membranes such as the yolk sac. The objective of this review is to discuss these unique features of human placentation, to define their appearance during the evolution of primates and to contrast them with rodents.

Primary functions of the placenta are gas exchange and the transfer of substrates from mother to fetus. The underlying mechanisms are similar across mammals. Thus the sheep is an excellent model for studying the oxygen supply to the fetus despite structural differences between human and ovine placentation [3]. Similarly, glucose transfer by facilitated diffusion uses the same set of transporters across species [4]. These topics will not be further explored. The interactions between the trophoblast and the maternal immune system are manifold, and a full reckoning cannot be made here. The section on placental immunology therefore focuses mainly on the uterine natural killer (uNK) cells and their ligands. Information on other immune cells, including macrophages, T-cells and innate lymphoid cells, should be sought elsewhere [5,6].

*Mammalian Evolution and Phylogeny*

The uniqueness of human placentation can best be assessed in the evolutionary framework provided by phylogenetics. I have striven to keep terminology to a minimum, yet some context is needed, especially for primates. In a broader perspective, eutherian mammals can be sorted into four major clades (Figure 1A). Here we shall deal mainly with one of those clades (Euarchontoglires) and its two subdivisions (Figure 1B). Euarchonta comprises primates, tree shrews and colugos. Glires comprises rodents and lagomorphs. The split between Euarchonta and Glires is estimated to have occurred in the Cretaceous period some 80 million years ago [7]. Therefore, it is not surprising that placentation in the mouse and other rodent models differs in significant respects from human placentation [2].

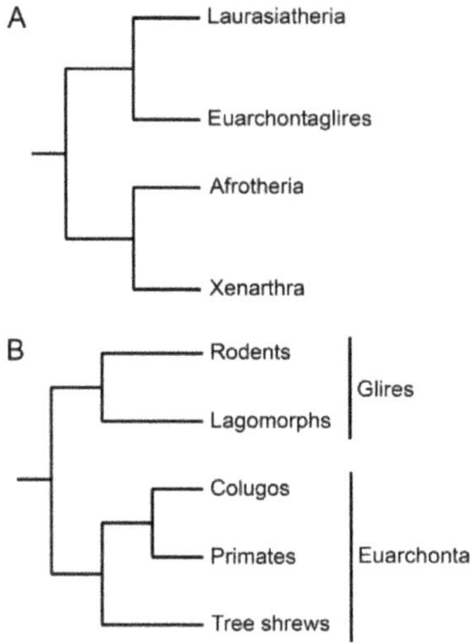


**Figure 1.** The mammalian tree. (**A**) The four major clades of eutherians [8]. (**B**) The orders of Euarchontoglires [9]. Note the separation of Glires (including rodents) from Euarchonta (including primates). There are alternative interpretations of the root of the tree and the position of tree shrews. Reprinted with permission from [2] © 2021 Society for Reproduction and Fertility.

The primate order has two major subdivisions: Strepsirrhini and Haplorrhini (Figure 2). The former comprises lemurs and lorises with placentation that differs radically from that of humans [10]. In contrast, Haplorrhini, to which our species belongs, was defined by commonalities in fetal membrane development [11]. It includes tarsiers (Tarsiiformes), New World monkeys (Platyrrhini), Old World monkeys and apes (together Catarrhini). Gibbons and great apes (orang-utans, gorillas, bonobo, chimpanzees, and man) comprise the superfamily Hominoidea.

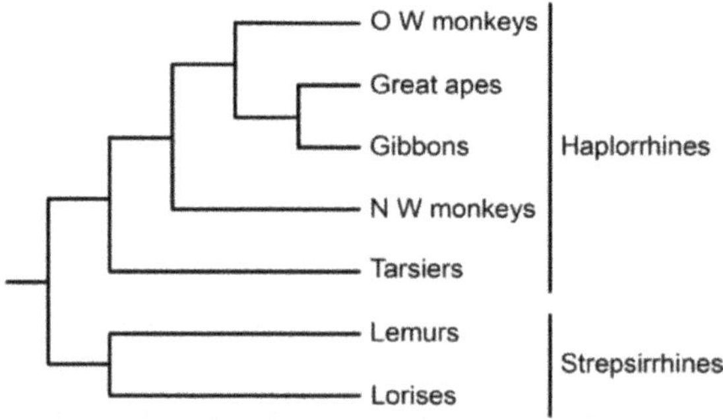


**Figure 2.** Classification of primates [12]. Strepsirrhines and haplorrhines are suborders, tarsiers are regarded as an infraorder, whilst the other clades shown are superfamilies. OW, Old World; NW, New World. Reprinted with permission from [2] © 2021 Society for Reproduction and Fertility.

## 2. Early Development

### 2.1. Interstitial Implantation

In most primates, implantation of the blastocyst is superficial. In macaques and baboons, for example, the trophoblast invades the endometrium to establish a placenta, but the developing embryo remains in the uterine cavity. In contrast, the human blastocyst is pulled into the endometrium, which closes above it so that it is completely embedded by the 12th day [13]. The placental bed is underlain by the basal decidua, and the developing embryo is covered by the capsular decidua. Interstitial implantation is a feature shared with the great apes and the gibbons [14,15]. It does occur in rodents, but the process is not identical and has been independently evolved.

Initial penetration of the endometrium is achieved by syncytiotrophoblast [16]. This is formed by fusion of cellular trophoblast to form a multicellular syncytium. The process depends in large part upon syncytins, which are proteins encoded by endogenous retroviral envelope genes that have been incorporated in the genome and exapted to promote cell fusion in the placenta [17]. Humans have two syncytin genes acquired at different timepoints. Whereas *Syncytin-2* occurs in all haplorrhine primates, *Syncytin-1* is found only in apes [17]. Syncytin genes occur in a wide range of mammals, and each represents a separate gene capture [17]. However, it has been argued that the capture of retroviral envelope genes was a prerequisite for the evolution of invasive placentation in mammals [18].

### 2.2. Initial Decidual Reaction

The maternal response to implantation is the decidual reaction, which involves the transformation of fibroblast-like endometrial stromal cells into polygonal decidual stromal cells [19]. The decidual reaction once was thought to be absent or atypical in elephants and carnivores [20], yet recent work shows it to be a characteristic feature of eutherian mammals [19]. However, the decidual reaction is transient in some species, such as the nine-banded armadillo (*Dasypus novemcinctus*), and has been lost in many with non-invasive placentation such as cattle (*Bos taurus*) [21]. It is now thought that the decidual reaction evolved from an inflammatory response that is present in marsupials, where it imposes a limit on the length of gestation [22,23]. Several of the genes involved in this response have been downregulated in eutherians, while genes beneficial to implantation have been upregulated [24].

In humans and many primates, as well as rodents, decidual stromal cells persist throughout gestation and have acquired an additional role in pregnancy maintenance [19]. These novel functions arose in the lineage of the large clade Euarchontoglires [19].

### 2.3. Early Differentiation of Mesoderm and Secondary Yolk Sac

One of the first fetal membranes to form in mammals is a bilaminar yolk sac comprising an outer layer of trophoblast and an inner lining of the extraembryonic endoderm. It may later acquire blood vessels and function as a choriovitelline placenta. In humans, however, the primary yolk sac is short-lived due to precocious differentiation of the extraembryonic mesoderm, which intrudes between the endoderm and trophoblast (Figure 3). This leads to formation of the secondary yolk sac, which consists of mesoderm and endoderm and becomes a free floating structure within the exocoelomic cavity [25]. Despite lack of contact with maternal tissues, the secondary yolk sac plays an important role in nutrient supply to the first trimester embryo [26]. Precocious development of the extraembryonic mesoderm is a defining feature of haplorrhine primates [11]. Recent work comparing gene expression in the common marmoset (*Callithrix jacchus*), rhesus macaque (*Macaca mulatta*) and human suggests extraembryonic mesoderm is derived in part from the extraembryonic endoderm [27,28].

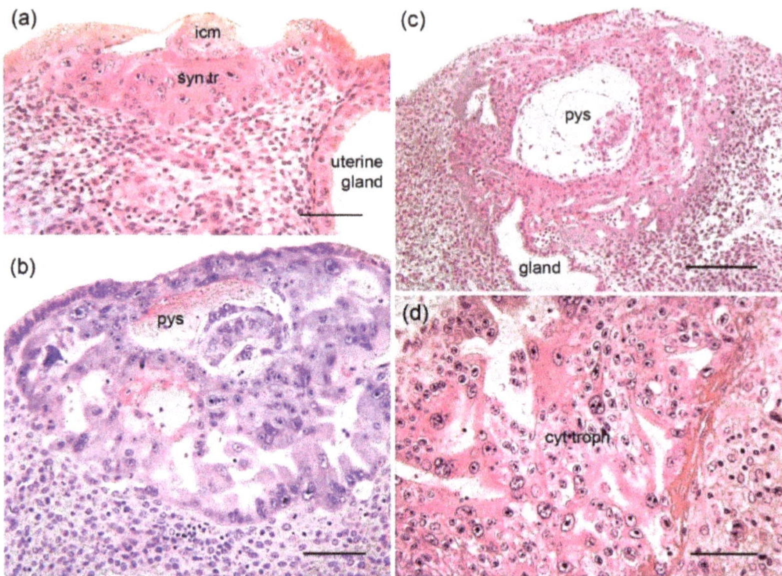

**Figure 3.** Early differentiation of mesoderm in the human embryo. (**a**) Trophoblastic plate stage (Carnegie Stage 5a) showing the inner cell mass (icm). The cavity of the blastocyst is collapsed. The pad of trophoblast has different sized nuclei and both cellular and syncytial trophoblast (syn tr) (Carnegie Embryo #8020). Scale bar, 70 μm. (**b**) Early lacunar stage (Carnegie Stage 5b) Some maternal blood has leaked into the primary yolk sac (pys). Note the irregular shape of the lacunae on the right, which appear to be formed from expanding clefts (Carnegie Embryo #8004). Scale bar, 90 μm. (**c**) Lacunar stage (Carnegie Stage 5c). Note the anastomotic lacunae within the syncytiotrophoblast. In the area between the trophoblast and the already partially constricted primary yolk sac (pys), there are mesenchymal cells (extraembryonic mesoderm) (Carnegie Embryo #7699). Scale bar, 176 μm. (**d**) Predecessors of the primary villi (Carnegie Stage 6). The cytotrophoblast (cyt troph) is accumulating in the partitions between lacunae, initiating the formation of primary villi (Carnegie Embryo #9260). Scale bar = 70 μm. Reprinted with permission from Carter, Enders and Pijnenborg [16] © The Authors. Published by the Royal Society. All rights reserved.

Rodents pursue an entirely different course resulting in an inverted yolk sac with an outward-facing layer of endoderm that persists throughout pregnancy [29]. There are similarities but also marked differences in gene expression and regulatory pathways between the mouse and primates [28].

*2.4. Allantoic Stalk*

The chorioallantoic placenta is formed by the fusion of the allantois with the chorion (trophoblast and extraembryonic mesoderm). In most mammals, the allantois also encloses a fluid-filled space. Indeed, a medium to large allantoic sac is the ancestral state for eutherians [30], and it forms a prominent structure in some species. As an example, cattle have 6–9 litres of allantoic fluid against 2.5 litres of amniotic fluid [31]. In contrast, the human allantois develops as a small diverticulum, and the connection between embryo and placenta, which carries the blood vessels, is the allantoic stalk. This is yet another shared feature that defines haplorrhine primates [11,28]. An allantoic sac is absent in rodents, but this may be due to convergent evolution as it is found in their sister group, the lagomorphs (e.g., rabbit *Oryctolagus cuniculus*) [29].

### 3. Placentation

#### 3.1. Haemochorial Placentation

Invasive placentation is the basal condition in eutherians [30,32]. Some strepsirrhine primates (lemurs and lorises) have epitheliochorial placentation, but this is widely regarded as a derived trait [10]. In haemochorial placentas, the trophoblast is in direct contact with maternal blood. Early in human gestation the interhaemal barrier includes two layers of trophoblast [33], but the cytotrophoblast layer (Langhan's layer) later becomes discontinuous. The interhaemal barrier then comprises syncytiotrophoblast, a thin layer of connective tissue, and the fetal capillary endothelium (Figure 4). Therefore, human placenta is classified as haemomonochorial [34]. Mouse and rat (*Rattus norvegicus*) have three layers of trophoblast, rabbits two, and guinea pigs (*Cavia porcellus*) one, but the significance of these differences should not be overstated [2].

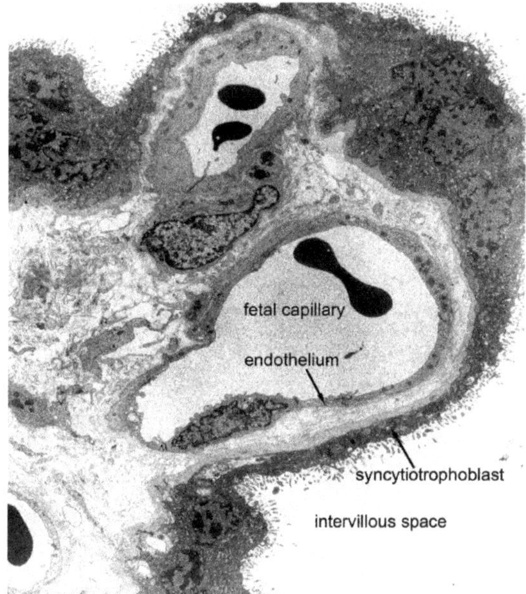


**Figure 4.** The interhaemal barrier of the human placenta is classified as haemomonochorial. The intervillous space is separated from blood in the fetal capillary by syncytiotrophoblast and fetal capillary endothelium with their basal membranes. A very thin layer of connective tissue cytoplasm is interposed between the two basal membranes. Courtesy of Dr. Allen C. Enders. Reproduced with permission from [35] Copyright © American Physiological Society.

#### 3.2. Villous rather Than Labyrinthine Placentation

Of greater consequence is the internal structure of the placenta. Most haemochorial placentas are labyrinthine and organised so that maternal blood channels are arranged in parallel with fetal capillaries. Maternal and fetal blood flow in opposite directions allowing for efficient countercurrent exchange [36,37]. This is also the case in the mouse [38]. In human placenta, on the other hand, the terminal villi are suspended in the intervillous space, which is supplied with blood by the spiral arteries of the basal plate. This is a pattern shared with Old World monkeys and apes [39]. An intermediate form of placentation, where the villi remain connected by bridges of trophoblast (trabeculae), is found in tarsiers and New World monkeys. The evolution of a villous placenta from a labyrinthine one negates the benefit conferred by countercurrent exchange. However, opening up the maternal component allows for a larger volume flow of blood and, thus, greater oxygen delivery, and this likely outweighs the loss of countercurrent exchange [4].

### *3.3. Uterine Spiral Artery Transformation*

A key feature of human placentation is the transformation of the uterine spiral arteries to wide vessels with low resistance to flow. An initial phase involves vacuolation and focal loss of endothelial cells and loosening of the smooth muscle layer. It appears to be dependent on cytokines secreted by uNK cells and macrophages [40–42]. The second phase is associated with invasion of the endometrium and vessel walls by extravillous trophoblast. This leads to complete loss of endothelium and disruption of the smooth muscle with the greatly widened vessels eventually being lined by trophoblast embedded in a fibrinoid layer [41].

With advancing pregnancy there is also dilatation of the radial arteries, arcuate arteries and uterine arteries, none of which are invaded by trophoblast. This is most likely due to stimulation by oestrogens and nitric oxide-mediated flow-dilation signals [43].

### *3.4. Trophoblast Invasion by Interstitial and Intravascular Routes*

In human pregnancy, the trophoblast invades by two routes [16]. Firstly, it migrates from the basal plate into the lumina of the uterine spiral arteries against the direction of flow (the intravascular route). Secondly, the trophoblast differentiating from the anchoring villi migrates through the decidua towards the blood vessels (the interstitial route). In a healthy pregnancy, trophoblast invasion extends through the relatively shallow endometrium to the inner third of the myometrium. Shallower invasion leads to inadequate transformation of the spiral arteries, thereby limiting the blood supply to the intervillous space, and is causally associated with fetal growth restriction and preeclampsia [1,44]. Trophoblasts that invade by the interstitial route undergo endoreduplication [45] and go no deeper than the inner myometrium, where they are found as multinucleate giant cells [46].

Our studies in chimpanzee and gorilla suggest that the depth of trophoblast invasion and spiral artery transformation resembles the human condition [47,48]. In Old World monkeys, however, there is rapid invasion of spiral arteries by the intravascular route but none by the interstitial route; indeed, there is a sharp border between the cytotrophoblastic shell and decidua [49,50]. This also appears to be the case in gibbons, suggesting that invasion by the interstitial route evolved in the lineage of the great apes [51]. In Old World monkeys, intravascular trophoblast is confined largely to the endometrial segments of the spiral arteries [49]. The situation in New World monkeys and tarsiers is not sufficiently known [16].

Rodent models do not readily conform to any of these features. The depth of trophoblast invasion in rodents varies but can extend to the mesometrial arteries, as in the guinea pig [52]. In the mouse, trophoblast glycogen cells migrate to the decidua but do not invade its vessels [53]. Instead, trophoblasts of the giant cell lineage migrate to and line the spiral arteries [54]. However, this does not occur until vascular remodelling is complete [5,55]. Indeed, rodents are unsatisfactory models of trophoblast invasion. Thus, deeper penetration of the arteries was found in a rat model of preeclampsia [56], which is the opposite of the shallower invasion typical of preeclampsia in human pregnancy [1].

## 4. Immunology of Decidua and Trophoblast

The placenta is a semi-allograft, yet it is not rejected by the maternal immune system. This immunological paradox was framed by Sir Peter Medawar [57] and remains a pivotal question in reproductive immunology [6]. As many as 70% of leukocytes in the uterus are uNK cells [58]. Their properties differ from those of peripheral natural killer cells and include lower cytotoxic activity [42]. Their putative role in the early stages of uterine artery transformation, alluded to above, may be explained by the secretion of cytokines, growth factors and proteases [40,42]. Here, we are concerned with their interplay with invasive extravillous trophoblast.

*Interplay of uNK Cell Receptors and HLA Antigens*

Human trophoblast does not express the major histocompatibility antigens (MHC) Class I, which include human leukocyte antigens (HLA) A and B. Instead, the surface of trophoblast presents HLA-C, which exhibits a high degree of allelic polymorphism, as well as HLA-E and HLA-G. There is no equivalent to HLA-C in monkeys or gibbons [59,60], but an invariant form appears in orangutans [61]. A later gene duplication yielded the two epitopes C1 and C2, which are found in chimpanzees, gorillas and humans [62].

HLA-C1 and -C2 are the principal ligands for the killer immunoglobulin-like receptors (KIRs) on uNK cells. KIRs likewise exhibit a high degree of allelic polymorphism, and importantly, there are inhibitory and excitatory variants [63]. Thus, numerous combinations are possible of HLA-C presented by the trophoblast (with one allele being paternal in origin) and KIRs expressed by maternal uNK cells. This can affect pregnancy outcome. When the trophoblast expresses HLA-C2 and the uNK cells express only inhibitory receptors, the combination is associated with a higher incidence of recurrent abortion, fetal growth restriction and preeclampsia [64,65]. KIR genes have evolved along separate pathways in great apes and human [66]. Indeed, it has been proposed that the emergence of the HLA-C2 epitope in apes is causally linked to the advent of preeclampsia. This is difficult to prove as reports of eclampsia in great apes are largely anecdotal (see [51]). In any case, the evolution of KIRs has pursued different paths in nonhuman primates [67].

HLA-G is expressed exclusively on trophoblast and has been implicated in maternal immune tolerance [68,69]. *MHC-G* is expressed in the great apes [70] but is a pseudo-gene in baboons (*Papio* spp.), rhesus macaque (*Macaca mulatta*), cynomolgus macaque (*M. fascicularis*) and vervet monkey (*Chlorocebus aethiops*), where its function is assumed by a new gene *MHC-AG* [60,71]. Receptors for HLA-G and MHA-AG are expressed by uNK cells and include KIR2DL4 [60].

The uterus of rodents is also rich in uNK cells, and they appear to be essential for the transformation of vessels analogous to human spiral arteries [72,73]. However, the principal receptors on rodent uNK cells belong to the lectin-like family (Ly49) [74], so rodents are not useful for exploring interactions between KIRs and HLA antigens. Rodents do not have MHC-G, although *HLA-G* expression has been achieved in transgenic mice [75].

## 5. Endocrinology of the Placenta

Placental hormones are secreted to the maternal circulation and adapt maternal physiology to meet the requirements of pregnancy and subsequent lactation [76]. Many of the peptides made by human trophoblast are unique to the primate lineage. Placental lactogens occur in human and rodents but have arisen through convergent evolution and serve different functions. Pregnancy maintenance depends on progesterone secretion, but an unusual feature of human pregnancy is that parturition occurs without a fall in plasma progesterone.

### 5.1. Chorionic Gonadotrophins

Human chorionic gonadotrophin (hCG) is responsible for early pregnancy maintenance. It evolved through duplication of the gene encoding the β-subunit of luteinising hormone. This occurred in the lineage of haplorrhine primates followed by further duplications so that many primates have multiple genes and pseudogenes [77]. A chorionic gonadotrophin was convergently evolved in the lineage of equids [78].

### 5.2. Placental Lactogens and Growth Hormones

Human placenta also secretes a placental lactogen, hPL, previously known as chorionic somatomammotropin. The genes that code for hPL are derived from the growth hormone gene [79]. A third gene in the cluster codes for placental growth hormone, which supplants pituitary growth hormone in the latter part of pregnancy [80]. All haplorrhine primates have placentally expressed genes related to growth hormone but there is great variation especially between New World and Old World monkeys [81,82], which may have attained placental expression separately [83].

The placental lactogens of muroid rodents, PL1 and PL2, are responsible for the maintenance of the corpus luteum [84]. In contrast to primates, they were derived by duplication from the prolactin gene together with a range of other cytokines [84]. Thus, placental lactogens of primates and rodents differ both in origin and function.

### 5.3. Progesterone and Its Receptors

Pregnancy maintenance in mammals requires the presence of progesterone secreted from the corpus luteum or placenta [85]. In many species, parturition is initiated through a fall in circulating progesterone, so-called progesterone withdrawal. In humans, where the placenta synthesises progesterone from maternal cholesterol, secretion is maintained right up to the start of labour [86]. This contra-intuitive finding led to the concept of a "functional" progesterone withdrawal for which the favoured explanation focusses on progesterone receptors (PR) in the myometrium. Of the two major isoforms, PR-B is the stronger trans-activator of progesterone-responsive genes, and PR-A acts as a trans-suppressor of PR-B's effect. They are equally expressed in the myometrium throughout gestation. However, parturition is associated with a change in the PR-A to PR-B ratio due to increased expression of PR-A [87]. This switch contributes to myometrial activation via the activator protein-1 (AP-1) pathway [88]. Interestingly, there is evidence for adaptive evolution of the progesterone receptor gene (*PGR*) in the human lineage [89,90]. Of note, PRs are also expressed in human decidua, and a recent hypothesis points to a decline in decidual PR expression as a possible factor in the initiation of parturition [91].

Plasma progesterone levels increase before parturition in the rhesus macaque [92], and there is evidence of rapid evolution of *PGR* in catarrhine primates [90].

In mouse and rat, the corpus luteum is the sole source of progesterone, and plasma concentrations fall precipitously before parturition. These models are of limited value in understanding human parturition [93]. The situation is different in hystricomorph rodents: the placenta is a major source of progesterone in the guinea pig, and there is no change in circulating progesterone prior to parturition [86]. Therefore, it is considered a more appropriate model for parturition research [93].

## 6. Pregnancy Duration and Newborn State

An undeniably unique feature of human reproduction is that newborn babies are helpless and entirely dependent on parental care [94]. They differ to some degree from other haplorrhine primates, although a long childhood is a general feature of great apes (Table 1).

**Table 1.** Precocity and parental care in selected primates. Apart from tarsier and marmoset, data are from field observations on free-living populations. Gestation lengths are approximate and based on few observations.

| Clade | Species | Common Name | Length of Gestation | Parental Care | References |
|---|---|---|---|---|---|
| Tarsiers | *Cephalopacus bancanus* | Western tarsier | 178 days | Nutritional and social independence by 60 days | [95] |
| New World monkeys | *Callithrix jacchus* | Common marmoset | 143–144 days | Independent movement by 3 weeks; weaning by 3 months | [96,97] |
| Old World monkeys | *Papio cynocephalus* | Yellow baboon | 178 ± 6 days | Milk supplemented early with plant foods; fully weaned after about a year; carried for 8 months | [98–100] |

**Table 1.** *Cont.*

| Clade | Species | Common Name | Length of Gestation | Parental Care | References |
|---|---|---|---|---|---|
| Lesser apes | *Symphalangus syndactylus* | Siamang | 230–235 days | Partial weaning at 6 months; travel independently by 1 year | [101] |
| Great apes | *Pongo pygmaeus* | Bornean orangutan | 275 days | Partial weaning by 11 months; fully independent at 7–10 years | [102,103] |
| | *Gorilla beringei* | Eastern gorilla | 255 days | Weaning at 3–4 years | [103–105] |
| | *Pan troglodytes* | Chimpanzee | 196–260 days | Weaning at 10 months; dependent on mother for 5 years | [103,106,107] |

Mammals tend to adopt one of two contrasting strategies [108]. In the first, a short gestation with a large litter leads to the birth of poorly developed or altricial offspring. In the second, a long gestation with a small litter (usually singleton) ends with the birth of well-developed offspring with open eyes and ears, a coat of hair, and some degree of independence. Human babies have most of the attributes of precocial offspring but are helpless at birth and require parental care for several years.

The human pelvis has been remodelled to enable bipedal walking. Therefore, there has been a trade-off between prenatal brain development, i.e., the size of the fetal head, and the diameter of the birth canal [109]. As a result, the volume of the brain at birth is about a quarter of adult size compared to 40% in the chimpanzee [110]. Indeed, in humans, the fetal pattern of brain growth continues for a year after birth [94]. Consequently, the fontanelles separating the bones of the skull do not close until 18 months to 2 years after birth [111]. The development of most other organs is as complete at birth as in other primates, all of which deliver precocial young.

In contrast, rodents such as mouse and rat deliver truly altricial young with closed eyes, naked skin, and incomplete development of major organs, such as the kidneys [2]. The short gestation means that there is no period equivalent to the third trimester of human pregnancy when obstetric complications are most evident. Differences in gestation length are even reflected in placental function, the different role of placental lactogens in rodents and primates being a case in point.

## 7. Discussion

### 7.1. Placental Evolution

Placental characters shared with all eutherian mammals are invasive placentation and the decidual reaction (Table 2). Persistence of decidua into late gestation is common to the major clade Euarchontoglires, which includes rodents as well as primates. Many characteristics are shared with the primate suborder Haplorrhini, including features of the fetal membranes that Hubrecht used to justify classing tarsiers with monkeys and apes [11]. Characters shared with Old World monkeys include villous placentation with an intervillous space and some aspects of trophoblast invasion. However, implantation is superficial in all primates except gibbons and great apes, and trophoblast invasion by the interstitial route is shared only with the great apes. This leads to the evolutionary timeline shown in Table 2.

**Table 2.** Characteristics of human placentation and their estimated appearance during evolution. Branching points (Mya, million years ago) are estimates based on molecular data [7,112].

| Character | Taxonomic Clade | Branching Point | Geological Period or Epoch | Comments |
|---|---|---|---|---|
| Invasive placentation | Eutheria | 98.5 Mya | Late Cretaceous | |
| Decidual reaction | Eutheria | 98.5 Mya | Late Cretaceous | An inflammatory response in marsupials |
| Persistence of decidual stromal cells | Euarchontoglires (includes rodents and primates) | 91.8 Mya | Late Cretaceous | |
| Precocious extraembryonic mesoderm | Haplorrhini | 44.8 Mya | Middle Eocene | |
| Secondary yolk sac | Haplorrhini | 44.8 Mya | Middle Eocene | |
| Allantoic stalk | Haplorrhini | 44.8 Mya | Middle Eocene | Many mammals have an allantoic sac |
| Haemomonochorial placentation | Haplorrhini | 44.8 Mya | Middle Eocene | |
| Syncytin-2 *env* gene | Haplorrhini | 44.8 Mya | Middle Eocene | |
| Chorionic gonadotropin | Haplorrhini | 44.8 Mya | Middle Eocene | |
| Placental lactogens and growth hormone | Haplorrhini | 44.8 Mya | Middle Eocene | Vary between primate lineages |
| Trophoblast invasion by intravascular route | Old World monkeys and apes | 29.8 Mya | Oligocene | |
| Villous placentation with an intervillous space | Old World monkeys and apes | 29.8 Mya | Oligocene | Trabecular placentation in tarsiers and NW monkeys |
| Interstitial implantation | Lesser and greater apes | 20.2 Mya | Early Miocene | |
| Syncytin-1 *env* gene | Lesser and greater apes | 20.2 Mya | Early Miocene | |
| Trophoblast invasion by interstitial route | Great apes | 15.1 Mya | Middle Miocene | |
| HLA-C | Great apes | 15.1 Mya | Middle Miocene | |

*7.2. Pregnancy Complications*

Comparatively little is known about pregnancy complications in nonhuman primates. Preeclampsia may occur in great apes, but the evidence is thin, although in one case supported by a renal biopsy [113]. There is, however, much to be said for the argument that deep trophoblast invasion, especially by the interstitial route, can be linked to the emergence of preeclampsia in the great apes [51].

Many changes reminiscent of preeclampsia could be replicated in a baboon model by uterine artery ligation [114], but these may merely reflect responses to reduced oxygen delivery. Hypertension can develop spontaneously in the vervet monkey, even when not pregnant [115], and gestational hypertension in the closely related patas monkey (*Erythrocebus patas*) was accompanied by preeclampsia-like symptoms [116].

Fetal growth restriction is another focus of obstetric research. It occurs in New World monkeys that regularly bear twins or triplets, and the effects on birth weight and neonatal outcomes are currently under investigation [117]. Indeed, the common marmoset (*Callithrix jacchus*) is a promising model for pregnancy research [2].

## 8. Conclusions

The many unique features of human pregnancy and placentation pose problems in planning and interpreting animal experiments. Two factors are involved. The first is phylogenetic distance. Quite a few features are shared with haplorrhine primates, making them the models of choice. Baboons and macaques share additional features such as endovascular trophoblast and spiral artery transformation as well as a true intervillous space. On the other hand, maintenance of breeding colonies is costly. Therefore, it is worth considering the common marmoset for which caging and feeding costs are much lower [118]. As mentioned in the introduction, primates and rodents last shared a common ancestor in the Cretaceous period, so it is not surprising that placental evolution has pursued different paths. There has, for example, been convergent evolution of placental lactogens to serve different purposes.

A second factor compounds the problem with rodent models. This is the difference in reproductive strategies. The short generation times of mice and rats make them ideal laboratory animals. Unfortunately, the same qualities render them unsatisfactory for pregnancy research [119,120]. The major obstetric syndromes become manifest in the third trimester, but there is no equivalent period in the mouse. The newborn mouse is truly altricial, with much of organ development occurring in the postnatal period. I have been at pains to stress, as argued by Martin [111], that human babies are precocial in almost all aspects save brain development; they are not altricial. Alternative rodent models are the spiny mouse (*Acomys cahirinus*) and guinea pig, both of which deliver precocial young [2].

**Funding:** This research received no external funding.

**Data Availability Statement:** No new data were created or analyzed in this study. Data sharing is not applicable to this article.

**Conflicts of Interest:** The author declares no conflict of interest.

# References

1. Brosens, I.; Pijnenborg, R.; Vercruysse, L.; Romero, R. The "Great Obstetrical Syndromes" are associated with disorders of deep placentation. *Am. J. Obstet. Gynecol.* **2011**, *204*, 193–201. [CrossRef]
2. Carter, A.M. Animal models of human pregnancy and placentation: Alternatives to the mouse. *Reproduction* **2020**, *160*, R129–R143. [CrossRef]
3. Carter, A.M. Animal models of human placentation—A review. *Placenta* **2007**, *28*, S41–S47. [CrossRef]
4. Carter, A.M. Evolution of Placental Function in Mammals: The Molecular Basis of Gas and Nutrient Transfer, Hormone Secretion, and Immune Responses. *Physiol. Rev.* **2012**, *92*, 1543–1576. [CrossRef]
5. Huhn, O.; Zhao, X.; Esposito, L.; Moffett, A.; Colucci, F.; Sharkey, A.M. How Do Uterine Natural Killer and Innate Lymphoid Cells Contribute to Successful Pregnancy? *Front. Immunol.* **2021**, *12*, 607669. [CrossRef] [PubMed]
6. Prabhudas, M.; Bonney, E.; Caron, K.; Dey, S.; Erlebacher, A.; Fazleabas, A.; Fisher, S.; Golos, T.; Matzuk, M.; McCune, J.M.; et al. Immune mechanisms at the maternal-fetal interface: Perspectives and challenges. *Nat. Immunol.* **2015**, *16*, 328–334. [CrossRef] [PubMed]
7. Meredith, R.W.; Janečka, J.E.; Gatesy, J.; Ryder, O.A.; Fisher, C.A.; Teeling, E.C.; Goodbla, A.; Eizirik, E.; Simão, T.L.L.; Stadler, T.; et al. Impacts of the Cretaceous Terrestrial Revolution and KPg Extinction on Mammal Diversification. *Science* **2011**, *334*, 521–524. [CrossRef] [PubMed]
8. Murphy, W.J.; Pringle, T.H.; Crider, T.A.; Springer, M.S.; Miller, W. Using genomic data to unravel the root of the placental mammal phylogeny. *Genome Res.* **2007**, *17*, 413–421. [CrossRef]
9. Janečka, J.E.; Miller, W.; Pringle, T.H.; Wiens, F.; Zitzmann, A.; Helgen, K.M.; Springer, M.S.; Murphy, W.J. Molecular and Genomic Data Identify the Closest Living Relative of Primates. *Science* **2007**, *318*, 792–794. [CrossRef] [PubMed]
10. Carter, A.M.; Enders, A.C. The Evolution of Epitheliochorial Placentation. *Annu. Rev. Anim. Biosci.* **2013**, *1*, 443–467. [CrossRef]
11. Hubrecht, A.A.W. Relations of tarsius to the lemurs and apes. *Science* **1897**, *5*, 550–551. [CrossRef] [PubMed]
12. Perelman, P.; Johnson, W.; Roos, C.; Seuánez, H.N.; Horvath, J.E.; Moreira, M.A.M.; Kessing, B.; Pontius, J.; Roelke, M.; Rumpler, Y.; et al. A Molecular Phylogeny of Living Primates. *PLoS Genet.* **2011**, *7*, e1001342. [CrossRef] [PubMed]
13. James, J.; Carter, A.; Chamley, L. Human placentation from nidation to 5 weeks of gestation. Part I: What do we know about formative placental development following implantation? *Placenta* **2012**, *33*, 327–334. [CrossRef] [PubMed]
14. Hill, J.P., II. Croonian lecture—The developmental history of the primates. *Philos. Trans. R. Soc. Lond. Ser. B Contain. Pap. A Biol. Character* **1932**, *221*, 45–178. [CrossRef]
15. Selenka, E. Entwickelung des Gibbon (Hylobates und Siamanga). *Stud. Über Enwickelungsgeschichte Tiere* **1899**, *8*, 163–208.
16. Carter, A.M.; Enders, A.C.; Pijnenborg, R. The role of invasive trophoblast in implantation and placentation of primates. *Philos. Trans. R. Soc. B Biol. Sci.* **2015**, *370*, 20140070. [CrossRef]
17. Dupressoir, A.; Lavialle, C.; Heidmann, T. From ancestral infectious retroviruses to bona fide cellular genes: Role of the captured syncytins in placentation. *Placenta* **2012**, *33*, 663–671. [CrossRef]
18. Lavialle, C.; Cornelis, G.; Dupressoir, A.; Esnault, C.; Heidmann, O.; Vernochet, C.; Heidmann, T. Paleovirology of 'syncytins', retroviral env genes exapted for a role in placentation. *Philos. Trans. R. Soc. B Biol. Sci.* **2013**, *368*, 20120507. [CrossRef]
19. Chavan, A.R.; Bhullar, B.-A.S.; Wagner, G.P. What was the ancestral function of decidual stromal cells? A model for the evolution of eutherian pregnancy. *Placenta* **2016**, *40*, 40–51. [CrossRef]

20. Mossman, H.W. *Vertebrate Fetal Membranes: Comparative Ontogeny and Morphology*; Evolution; Phylogenetic Significance; Basic Functions; Research Opportunities; Rutgers University Press: New Brunswick, NJ, USA, 1987.

21. Kin, K.; Maziarz, J.; Chavan, A.R.; Kamat, M.; Vasudevan, S.; Birt, A.; Wagner, G.P. The transcriptomic evolution of mammalian pregnancy: Gene expression innovations in endometrial stromal fibroblasts. *Genome Biol. Evol.* **2016**, *8*, 2459–2473. [CrossRef]

22. Griffith, O.W.; Chavan, A.R.; Pavlicev, M.; Protopapas, S.; Callahan, R.; Maziarz, J.; Wagner, G.P. Endometrial recognition of pregnancy occurs in the grey short-tailed opossum (*Monodelphis domestica*). *Proc. Biol. Sci.* **2019**, *286*, 20190691. [CrossRef]

23. Griffith, O.W.; Chavan, A.R.; Protopapas, S.; Maziarz, J.; Romero, R.; Wagner, G.P. Embryo implantation evolved from an ancestral inflammatory attachment reaction. *Proc. Natl. Acad. Sci. USA* **2017**, *114*, E6566–E6575. [CrossRef]

24. Erkenbrack, E.M.; Maziarz, J.D.; Griffith, O.; Liang, C.; Chavan, A.R.; Nnamani, M.C.; Wagner, G.P. The mammalian decidual cell evolved from a cellular stress response. *PLoS Biol.* **2018**, *16*, e2005594. [CrossRef]

25. Enders, A.C.; King, B.F. Development of the human yolk sac. In *The Human Yolk Sac and Yolk Sac Tumors*; Nogales, F.F., Ed.; Springer: Berlin, Germany, 1993; pp. 33–47.

26. Burton, G.J.; Cindrova-Davies, T.; Turco, M.Y. Review: Histotrophic nutrition and the placental-endometrial dialogue during human early pregnancy. *Placenta* **2020**, *102*, 21–26. [CrossRef]

27. Boroviak, T.; Stirparo, G.; Dietmann, S.; Hernando-Herraez, I.; Mohammed, H.; Reik, W.; Smith, A.; Sasaki, E.; Nichols, J.; Bertone, P. Single cell transcriptome analysis of human, marmoset and mouse embryos reveals common and divergent features of preimplantation development. *Development* **2018**, *145*, 27. [CrossRef] [PubMed]

28. Ross, C.; Boroviak, T.E. Origin and function of the yolk sac in primate embryogenesis. *Nat. Commun.* **2020**, *11*, 1–14. [CrossRef]

29. Carter, A. IFPA Senior Award Lecture: Mammalian fetal membranes. *Placenta* **2016**, *48*, S21–S30. [CrossRef] [PubMed]

30. Mess, A.; Carter, A.M. Evolutionary transformations of fetal membrane characters in Eutheria with special reference to Afrotheria. *J. Exp. Zool. Part B Mol. Dev. Evol.* **2006**, *306B*, 140–163. [CrossRef] [PubMed]

31. Bongso, T.A.; Basrur, P.K. Foetal fluids in cattle. *Can. Vet. J.* **1976**, *17*, 38–41. [PubMed]

32. Wildman, D.E.; Chen, C.; Erez, O.; Grossman, L.I.; Goodman, M.; Romero, R. Evolution of the mammalian placenta revealed by phylogenetic analysis. *Proc. Natl. Acad. Sci. USA* **2006**, *103*, 3203–3208. [CrossRef] [PubMed]

33. Jones, C.; Jauniaux, E. Ultrastructure of the materno-embryonic interface in the first trimester of pregnancy. *Micron* **1995**, *26*, 145–173. [CrossRef]

34. Enders, A.C. A comparative study of the fine structure of the trophoblast in several hemochorial placentas. *Am. J. Anat.* **1965**, *116*, 29–67. [CrossRef]

35. Carter, A.M. Placental Gas Exchange and the Oxygen Supply to the Fetus. *Compr. Physiol.* **2015**, *5*, 1381–1403. [CrossRef] [PubMed]

36. Mossman, H.W. The rabbit placenta and the problem of placental transmission. *Am. J. Anat.* **1926**, *37*, 433–497. [CrossRef]

37. Metcalfe, J.; Bartels, H.; Moll, W. Gas exchange in the pregnant uterus. *Physiol. Rev.* **1967**, *47*, 782–838. [CrossRef]

38. Adamson, S.L.; Lu, Y.; Whiteley, K.J.; Holmyard, D.; Hemberger, M.; Pfarrer, C.; Cross, J.C. Interactions between trophoblast cells and the maternal and fetal circulation in the mouse placenta. *Dev. Biol.* **2002**, *250*, 358–373. [CrossRef] [PubMed]

39. Ramsey, E.M.; Harris, J.W.S. Comparison of utero-placental vasculature and circulation in the rhesus monkey and man. *Contrib. Embryol. Carnegie Inst.* **1966**, *38*, 61–70.

40. Harris, L. IFPA Gabor Than Award lecture: Transformation of the spiral arteries in human pregnancy: Key events in the remodelling timeline. *Placenta* **2011**, *32*, S154–S158. [CrossRef] [PubMed]

41. Pijnenborg, R.; Vercruysse, L.; Hanssens, M. The Uterine Spiral Arteries in Human Pregnancy: Facts and Controversies. *Placenta* **2006**, *27*, 939–958. [CrossRef] [PubMed]

42. Lash, G.E.; Bulmer, J.N. Do uterine natural killer (uNK) cells contribute to female reproductive disorders? *J. Reprod. Immunol.* **2011**, *88*, 156–164. [CrossRef]

43. Burton, G.; Woods, A.; Jauniaux, E.; Kingdom, J. Rheological and Physiological Consequences of Conversion of the Maternal Spiral Arteries for Uteroplacental Blood Flow during Human Pregnancy. *Placenta* **2009**, *30*, 473–482. [CrossRef]

44. Pankiewicz, K.; Fijałkowska, A.; Issat, T.; Maciejewski, T. Insight into the Key Points of Preeclampsia Pathophysiology: Uterine Artery Remodeling and the Role of MicroRNAs. *Int. J. Mol. Sci.* **2021**, *22*, 3132. [CrossRef] [PubMed]

45. Zybina, T.G.; Frank, H.-G.; Biesterfeld, S.; Kaufmann, P. Genome multiplication of extravillous trophoblast cells in human placenta in the course of differentiation and invasion into endometrium and myometrium. II. *Mech. Polyploidization Tsitologiya* **2004**, *46*, 640–648.

46. Pijnenborg, R.; Bland, J.M.; Robertson, W.B.; Dixon, G.; Brosens, I. The pattern of interstitial trophoblastic invasion of the myometrium in early human pregnancy. *Placenta* **1981**, *2*, 303–316. [CrossRef]

47. Pijnenborg, R.; Vercruysse, L.; Carter, A.M. Deep trophoblast invasion and spiral artery remodelling in the placental bed of the lowland gorilla. *Placenta* **2011**, *32*, 586–591. [CrossRef] [PubMed]

48. Pijnenborg, R.; Vercruysse, L.; Carter, A.M. Deep trophoblast invasion and spiral artery remodelling in the placental bed of the chimpanzee. *Placenta* **2011**, *32*, 400–408. [CrossRef] [PubMed]

49. Pijnenborg, R.; D'Hooghe, T.; Vercruysse, L.; Bambra, C. Evaluation of trophoblast invasion in placental bed biopsies of the baboon, with immunohistochemical localisation of cytokeratin, fibronectin, and laminin. *J. Med. Primatol.* **1996**, *25*, 272–281. [CrossRef] [PubMed]

50. Blankenship, T.N.; Enders, A.C.; King, B.F. Trophoblastic invasion and the development of uteroplacental arteries in the macaque: Immunohistochemical localization of cytokeratins, desmin, type IV collagen, laminin, and fibronectin. *Cell Tissue Res.* **1993**, *272*, 227–236. [CrossRef]

51. Carter, A.M. Comparative studies of placentation and immunology in non-human primates suggest a scenario for the evolution of deep trophoblast invasion and an explanation for human pregnancy disorders. *Reproduction* **2011**, *141*, 391–396. [CrossRef]

52. Verkeste, C.; Slangen, B.; Daemen, M.; Van Straaten, H.; Kohnen, G.; Kaufmann, P.; Peeters, L. The extent of trophoblast invasion in the preplacental vasculature of the guinea-pig. *Placenta* **1998**, *19*, 49–54. [CrossRef]

53. Redline, R.W.; Lu, C.Y. Localization of fetal major histocompatibility complex antigens and maternal leukocytes in murine placenta. Implications for maternal-fetal immunological relationship. *Lab. Investig.* **1989**, *61*, 27–36. [PubMed]

54. Hu, D.; Cross, J.C. Development and function of trophoblast giant cells in the rodent placenta. *Int. J. Dev. Biol.* **2010**, *54*, 341–354. [CrossRef] [PubMed]

55. Ain, R.; Canham, L.N.; Soares, M.J. Gestation stage-dependent intrauterine trophoblast cell invasion in the rat and mouse: Novel endocrine phenotype and regulation. *Dev. Biol.* **2003**, *260*, 176–190. [CrossRef]

56. Geusens, N.; Hering, L.; Verlohren, S.; Luyten, C.; Drijkoningen, K.; Taube, M.; Vercruysse, L.; Hanssens, M.; Dechend, R.; Pijnenborg, R. Changes in endovascular trophoblast invasion and spiral artery remodelling at term in a transgenic preeclamptic rat model. *Placenta* **2010**, *31*, 320–326. [CrossRef] [PubMed]

57. Medawar, P.B. Some immunological and endocrinological problems raised by the evolution of viviparity in vertebrates. *Symp. Soc. Exp. Biol.* **1953**, *7*, 320–338.

58. Lash, G.E.; Robson, S.C.; Bulmer, J.N. Review: Functional role of uterine natural killer (uNK) cells in human early pregnancy decidua. *Placenta* **2010**, *31*, S87–S92. [CrossRef]

59. Abi-Rached, L.; Kuhl, H.; Roos, C.; Ten Hallers, B.; Zhu, B.; Carbone, L.; Walter, L. A small, variable, and irregular killer cell Ig-like receptor locus accompanies the absence of MHC-C and MHC-G in gibbons. *J. Immunol.* **2010**, *184*, 1379–1391. [CrossRef]

60. Golos, T.G.; Bondarenko, G.I.; Dambaeva, S.V.; Breburda, E.E.; Durning, M. On the role of placental Major Histocompatibility Complex and decidual leukocytes in implantation and pregnancy success using non-human primate models. *Int. J. Dev. Biol.* **2010**, *54*, 431–443. [CrossRef]

61. Thomson, G.; Adams, E.J.; Parham, P. Evidence for an HLA-C -like locus in the orangutan *Pongo pygmaeus*. *Immunogenetics* **1999**, *49*, 865–871. [CrossRef]

62. Aguilar, A.M.O.; Guethlein, L.A.; Adams, E.J.; Abi-Rached, L.; Moesta, A.; Parham, P. Coevolution of Killer Cell Ig-Like Receptors with HLA-C To Become the Major Variable Regulators of Human NK Cells. *J. Immunol.* **2010**, *185*, 4238–4251. [CrossRef]

63. Penman, B.S.; Moffett, A.; Chazara, O.; Gupta, S.; Parham, P. Reproduction, infection and killer-cell immunoglobulin-like receptor haplotype evolution. *Immunogenetics* **2016**, *68*, 755–764. [CrossRef]

64. Hiby, S.E.; Apps, R.; Sharkey, A.; Farrell, L.E.; Gardner, L.; Mulder, A.; Claas, F.H.; Walker, J.; Redman, C.C.; Morgan, L.; et al. Maternal activating KIRs protect against human reproductive failure mediated by fetal HLA-C2. *J. Clin. Investig.* **2010**, *120*, 4102–4110. [CrossRef]

65. Hiby, S.E.; Walker, J.; O'Shaughnessy, K.M.; Redman, C.W.; Carrington, M.; Trowsdale, J.; Moffett, A. Combinations of Maternal KIR and Fetal HLA-C Genes Influence the Risk of Preeclampsia and Reproductive Success. *J. Exp. Med.* **2004**, *200*, 957–965. [CrossRef]

66. Khakoo, S.; Rajalingam, R.; Shum, B.P.; Weidenbach, K.; Flodin, L.; Muir, D.G.; Canavez, F.; Cooper, S.L.; Valiante, N.M.; Lanier, L.L.; et al. Rapid Evolution of NK Cell Receptor Systems Demonstrated by Comparison of Chimpanzees and Humans. *Immunity* **2000**, *12*, 687–698. [CrossRef]

67. Wroblewski, E.E.; Parham, P.; Guethlein, L.A. Two to Tango: Co-evolution of Hominid Natural Killer Cell Receptors and MHC. *Front. Immunol.* **2019**, *10*, 177. [CrossRef]

68. Le Bouteiller, P. HLA-G in the human placenta: Expression and potential functions. *Biochem. Soc. Trans.* **2000**, *28*, 208–212. [CrossRef]

69. Hunt, J.S.; Langat, D.L. HLA-G: A human pregnancy-related immunomodulator. *Curr. Opin. Pharmacol.* **2009**, *9*, 462–469. [CrossRef]

70. Castro, M.J.; Morales, P.; Fernández-Soria, V.; Suarez, B.; Recio, M.J.; Alvarez, M.; Arnaiz-Villena, A. Allelic diversity at the primate MHC-G locus: Exon 3 bears stop codons in all Cercopithecinae sequences. *Immunogenetics* **1996**, *43*, 327–336. [CrossRef] [PubMed]

71. Bondarenko, G.I.; Dambaeva, S.V.; Grendell, R.L.; Hughes, A.L.; Durning, M.; Garthwaite, M.A.; Golos, T.G. Characterization of cynomolgus and vervet monkey placental MHC class I expression: Diversity of the nonhuman primate AG locus. *Immunogenetics* **2009**, *61*, 431–442. [CrossRef]

72. Croy, B.A.; Luross, J.A.; Guimond, M.J.; Hunt, J.S. Uterine natural killer cells: Insights into lineage relationships and functions from studies of pregnancies in mutant and transgenic mice. *Nat. Immun.* **1996**, *15*, 22–33.

73. Burke, S.; Barrette, V.F.; Gravel, J.; Carter, A.L.I.; Hatta, K.; Zhang, J.; Chen, Z.; Leno-Durán, E.; Bianco, J.; Leonard, S.; et al. Uterine NK Cells, Spiral Artery Modification and the Regulation of Blood Pressure During Mouse Pregnancy. *Am. J. Reprod. Immunol.* **2010**, *63*, 472–481. [CrossRef]

74. Croy, B.; Esadeg, S.; Chantakru, S.; Heuvel, M.V.D.; Paffaro, V.A.; He, H.; Black, G.P.; Ashkar, A.; Kiso, Y.; Zhang, J. Update on pathways regulating the activation of uterine Natural Killer cells, their interactions with decidual spiral arteries and homing of their precursors to the uterus. *J. Reprod. Immunol.* **2003**, *59*, 175–191. [CrossRef]
75. Nguyen-Lefebvre, A.T.; Ajith, A.; Portik-Dobos, V.; Horuzsko, D.D.; Mulloy, L.L.; Horuzsko, A. Mouse models for studies of HLA-G functions in basic science and pre-clinical research. *Hum. Immunol.* **2016**, *77*, 711–719. [CrossRef] [PubMed]
76. Napso, T.; Yong, H.E.J.; Lopez-Tello, J.; Sferruzzi-Perri, A. The Role of Placental Hormones in Mediating Maternal Adaptations to Support Pregnancy and Lactation. *Front. Physiol.* **2018**, *9*, 1091. [CrossRef] [PubMed]
77. Maston, G.A.; Ruvolo, M. Chorionic gonadotropin has a recent origin within primates and an evolutionary history of selection. *Mol. Biol. Evol.* **2002**, *19*, 320–335. [CrossRef] [PubMed]
78. Chopineau, M.; Stewart, F.; Allen, W.R. Cloning and analysis of the cDNA encoding the horse and donkey luteinizing hormone beta-subunits. *Gene* **1995**, *160*, 253–256. [CrossRef]
79. Chen, E.Y.; Liao, Y.-C.; Smith, D.H.; Barrera-Saldaña, H.A.; Gelinas, R.E.; Seeburg, P.H. The human growth hormone locus: Nucleotide sequence, biology, and evolution. *Genomics* **1989**, *4*, 479–497. [CrossRef]
80. Frankenne, F.; Closset, J.; Gomez, F.; Scippo, M.L.; Smal, J.; Hennen, G. The Physiology of Growth Hormones (GHs) in Pregnant Women and Partial Characterization of the Placental GH Variant. *J. Clin. Endocrinol. Metab.* **1988**, *66*, 1171–1180. [CrossRef] [PubMed]
81. de Mendoza, A.R.; Escobedo, D.E.; Dávila, I.M.; Saldaña, H. Expansion and divergence of the GH locus between spider monkey and chimpanzee. *Gene* **2004**, *336*, 185–193. [CrossRef]
82. Wallis, O.C.; Wallis, M. Evolution of growth hormone in primates: The GH gene clusters of the New World monkeys marmoset (*Callithrix jacchus*) and white-fronted capuchin (*Cebus albifrons*). *J. Mol. Evol.* **2006**, *63*, 591–601. [CrossRef] [PubMed]
83. Papper, Z.; Jameson, N.M.; Romero, R.; Weckle, A.L.; Mittal, P.; Benirschke, K.; Santolaya-Forgas, J.; Uddin, M.; Haig, D.; Goodman, M.; et al. Ancient origin of placental expression in the growth hormone genes of anthropoid primates. *Proc. Natl. Acad. Sci. USA* **2009**, *106*, 17083–17088. [CrossRef] [PubMed]
84. Soares, M.J. The prolactin and growth hormone families: Pregnancy-specific hormones/cytokines at the maternal-fetal interface. Reprod. *Biol. Endocrinol.* **2004**, *2*, 51. [CrossRef] [PubMed]
85. Csapo, A. Progesterone block. *Am. J. Anat.* **1956**, *98*, 273–291. [CrossRef]
86. Thorburn, G.D.; Challis, J.R.G.; Robinson, J.S. *Endocrine Control of Parturition*; Springer Science and Business Media LLC: Tokyo, Japan, 1977; pp. 653–732.
87. Merlino, A.A.; Welsh, T.N.; Tan, H.; Yi, L.J.; Cannon, V.; Mercer, B.M.; Mesiano, S. Nuclear Progesterone Receptors in the Human Pregnancy Myometrium: Evidence that Parturition Involves Functional Progesterone Withdrawal Mediated by Increased Expression of Progesterone Receptor-A. *J. Clin. Endocrinol. Metab.* **2007**, *92*, 1927–1933. [CrossRef] [PubMed]
88. Shynlova, O.; Nadeem, L.; Zhang, J.; Dunk, C.; Lye, S. Myometrial activation: Novel concepts underlying labor. *Placenta* **2020**, *92*, 28–36. [CrossRef]
89. Marinić, M.; Lynch, V.J. Relaxed constraint and functional divergence of the progesterone receptor (PGR) in the human stem-lineage. *PLoS Genet.* **2020**, *16*, e1008666. [CrossRef] [PubMed]
90. Chen, C.; Opazo, J.C.; Erez, O.; Uddin, M.; Santolaya-Forgas, J.; Goodman, M.; Grossman, L.I.; Romero, R.; Wildman, D.E. The human progesterone receptor shows evidence of adaptive evolution associated with its ability to act as a transcription factor. *Mol. Phylogenet. Evol.* **2008**, *47*, 637–649. [CrossRef] [PubMed]
91. Blanks, A.; Brosens, J. Progesterone Action in the Myometrium and Decidua in Preterm Birth. *Facts Views Vis. ObGyn* **2012**, *4*, 188–194.
92. Challis, J.R.; John Davies, I.; Benirschke, K.; Hendrickx, A.G.; Ryan, K.J. The concentrations of progesterone, estrone and estradiol-17 beta in the peripheral plasma of the rhesus monkey during the final third of gestation, and after the induction of abortion with PGF 2 alpha. *Endocrinology* **1974**, *95*, 547–553. [CrossRef]
93. Mitchell, B.F.; Taggart, M.J. Are animal models relevant to key aspects of human parturition? *Am. J. Physiol. Integr. Comp. Physiol.* **2009**, *297*, R525–R545. [CrossRef]
94. Martin, R.D. The evolution of human reproduction: A primatological perspective. *Am. J. Phys. Anthr.* **2007**, *134*, 59–84. [CrossRef]
95. Roberts, M. Growth, development, and parental care in the western tarsier (*Tarsius bancanus*) in captivity: Evidence for a "slow" life-history and nonmonogamous mating system. *Int. J. Primatol.* **1994**, *15*, 1–28. [CrossRef]
96. Abbott, D.H.; Barnett, D.K.; Colman, R.J.; Yamamoto, M.E.; Schultz-Darken, N.J. Aspects of common marmoset basic biology and life history important for biomedical research. *Comp. Med.* **2003**, *53*, 339–350.
97. Tardif, S.D.; Smucny, D.A.; Abbott, D.H.; Mansfield, K.; Schultz-Darken, N.; Yamamoto, M.E. Reproduction in captive common marmosets (*Callithrix jacchus*). *Comp. Med.* **2003**, *53*, 364–368.
98. Altmann, J.; Samuels, A. Costs of maternal care: Infant-carrying in baboons. *Behav. Ecol. Sociobiol.* **1992**, *29*, 391–398. [CrossRef]
99. Rhine, R.J.; Norton, G.W.; Wynn, G.M.; Wynn, R.D. Weaning of free-ranging infant baboons (*Papio cynocephalus*) as indicated by one-zero and instantaneous sampling of feeding. *Int. J. Primatol.* **1985**, *6*, 491–499. [CrossRef]
100. Gesquiere, L.R.; Altmann, J.; Archie, E.A.; Alberts, S.C. Interbirth intervals in wild baboons: Environmental predictors and hormonal correlates. *Am. J. Phys. Anthr.* **2018**, *166*, 107–126. [CrossRef] [PubMed]
101. Lappan, S. Patterns of Infant Care in Wild Siamangs (*Symphalangus syndactylus*) in Southern Sumatra. In *The Gibbons*; Springer Science and Business Media LLC: Tokyo, Japan, 2009; pp. 327–345.

102. McConkey, K. Bornean Orangutan (*Pongo pygmaeus*). In *World Atlas of Great Apes and Their Conservation*; Caldecott, J.O., Miles, L., Eds.; University of California Press, in Association with UNEP-WCMC: Berkeley, CA, USA, 2005; pp. 161–183.
103. Tullner, W.W. Comparative Aspects of Primate Chorionic Gonadotropins. In *Reproductive Biology of the Primates*; Luckett, W.P., Ed.; Karger: Basel, Switzerland, 1974; pp. 235–257.
104. Ferriss, S.; Robbins, M.M.; Williamson, E.A. Eastern gorilla (*Gorilla beringei*). In *World Atlas of Great Apes and Their Conservation*; Caldecott, J.O., Miles, L., Eds.; University of California Press, in Association with UNEP-WCMC: Berkeley, CA, USA, 2005; pp. 129–152.
105. Canington, S.L. *Gorilla beringei* (Primates: Hominidae). *Mamm. Species* **2018**, *967*, 119–133. [CrossRef]
106. Inskipp, T. Chimpanzee (*Pan troglodytes*). In *World Atlas of Great Apes and their Conservation*; Caldecott, J.O., Miles, L., Eds.; University of California Press, in Association with UNEP-WCMC: Berkeley, CA, USA, 2005; pp. 53–81.
107. Jones, C.; Jones, C.A.; Knox Jones, J.; Wilson, D.E. Pan troglodytes. *Mamm. Species* **1996**, *529*, 1–9. [CrossRef]
108. Martin, R.D.; MacLarnon, A. Gestation period, neonatal size and maternal investment in placental mammals. *Nature* **1985**, *313*, 220–223. [CrossRef]
109. Trevathan, W. Primate pelvic anatomy and implications for birth. *Philos. Trans. R. Soc. B Biol. Sci.* **2015**, *370*, 20140065. [CrossRef] [PubMed]
110. Sakai, T.; Hirata, S.; Fuwa, K.; Sugama, K.; Kusunoki, K.; Makishima, H.; Eguchi, T.; Yamada, S.; Ogihara, N.; Takeshita, H. Fetal brain development in chimpanzees versus humans. *Curr. Biol.* **2012**, *22*, R791–R792. [CrossRef]
111. Martin, R.D. *How We Do It: The Evolution and Future of Human Reproduction, xii*; Basic Books: New York, NY, USA, 2013; p. 304.
112. Chatterjee, H.J.; Ho, S.Y.W.; Barnes, I.; Groves, C. Estimating the phylogeny and divergence times of primates using a supermatrix approach. *BMC Evol. Biol.* **2009**, *9*, 259. [CrossRef] [PubMed]
113. Stout, C.; Lemmon, W. Glomerular capillary endothelial swelling in a pregnant chimpanzee. *Am. J. Obstet. Gynecol.* **1969**, *105*, 212–215. [CrossRef]
114. Makris, A.; Thornton, C.; Thompson, J.; Thomson, S.; Martin, R.; Ogle, R.; Waugh, R.; McKenzie, P.; Kirwan, P.; Hennessy, A. Uteroplacental ischemia results in proteinuric hypertension and elevated sFLT-1. *Kidney Int.* **2007**, *71*, 977–984. [CrossRef] [PubMed]
115. Rhoads, M.K.; Goleva, S.B.; Beierwaltes, W.H.; Osborn, J.L. Renal vascular and glomerular pathologies associated with spontaneous hypertension in the nonhuman primate Chlorocebus aethiops sabaeus. *Am. J. Physiol. Integr. Comp. Physiol.* **2017**, *313*, R211–R218. [CrossRef]
116. Palmer, A.E.; London, W.T.; Sly, D.L.; Rice, J.M. Spontaneous preeclamptic toxemia of pregnancy in the patas monkey (*Erythrocebus patas*). *Lab. Anim. Sci.* **1979**, *29*, 102–106. [PubMed]
117. Rutherford, J.N.; Tardif, S.D. Placental efficiency and intrauterine resource allocation strategies in the common marmoset pregnancy. *Am. J. Phys. Anthr.* **2008**, *137*, 60–68. [CrossRef]
118. Mansfield, K. Marmoset models commonly used in biomedical research. *Comp. Med.* **2003**, *53*, 383–392.
119. Schmidt, A.; Prieto, D.M.M.; Pastuschek, J.; Fröhlich, K.; Markert, U.R. Only humans have human placentas: Molecular differences between mice and humans. *J. Reprod. Immunol.* **2015**, *108*, 65–71. [CrossRef]
120. Malassine, A.; Frendo, J.-L.; Evain-Brion, D. A comparison of placental development and endocrine functions between the human and mouse model. *Hum. Reprod. Updat* **2003**, *9*, 531–539. [CrossRef] [PubMed]

International Journal of
*Molecular Sciences*

*Review*

# Interleukin-15 in Outcomes of Pregnancy

**Scott M. Gordon** [1,2]

1   Division of Neonatology, Children's Hospital of Philadelphia, Philadelphia, PA 19104, USA;
    gordons1@chop.edu
2   Department of Pediatrics, Perelman School of Medicine, University of Pennsylvania,
    Philadelphia, PA 19104, USA

**Abstract:** Interleukin-15 (IL-15) is a pleiotropic cytokine that classically acts to support the development, maintenance, and function of killer lymphocytes. IL-15 is abundant in the uterus prior to and during pregnancy, but it is subject to tight spatial and temporal regulation. Both mouse models and human studies suggest that homeostasis of IL-15 is essential for healthy pregnancy. Dysregulation of IL-15 is associated with adverse outcomes of pregnancy. Herein, we review producers of IL-15 and responders to IL-15, including non-traditional responders in the maternal uterus and fetal placenta. We also review regulation of IL-15 at the maternal–fetal interface and propose mechanisms of action of IL-15 to facilitate additional study of this critical cytokine in the context of pregnancy.

**Keywords:** Interleukin-15; CD122; pregnancy; inflammation; placenta; natural killer cells; macrophages; trophoblast

**Citation:** Gordon, S.M.
Interleukin-15 in Outcomes of
Pregnancy. *Int. J. Mol. Sci.* **2021**, *22*,
11094. https://doi.org/10.3390/
ijms222011094

Academic Editors: Hiten D Mistry
and Eun Lee

Received: 30 September 2021
Accepted: 11 October 2021
Published: 14 October 2021

**Publisher's Note:** MDPI stays neutral
with regard to jurisdictional claims in
published maps and institutional affiliations.



## 1. Introduction

The regulation of pro- and anti-inflammatory signals is a critically important aspect of pregnancy, from implantation of the embryo to delivery of a newborn. As in countless other contexts, immune homeostasis must be maintained at the maternal–fetal interface to ensure the health of the mother and fetus. Both inappropriate immune activation and inappropriate immune quiescence are associated with adverse outcomes of pregnancy.

Multiple cytokines and chemokines regulate the unique composition and function of immune cells at the maternal–fetal interface. Interleukin-15 (IL-15) is a pleotropic cytokine that classically supports development, maintenance, and activity of killer lymphocytes [1–5]. Populations of natural killer (NK) cells, NKT cells, and memory CD8+ T cells are severely impaired in the absence of IL-15 in mice. Prior to pregnancy and early in gestation, NK cells are evident in the uterus in staggering concentrations relative to other lymphoid and non-lymphoid organs. While not all functions for NK cells in pregnancy are known, it has become clear that NK cells play key roles in healthy and complicated gestations [6]. Given the abundance and importance of NK cells in pregnancy, detailed study of IL-15 is essential.

The signaling components of the IL-15 receptor complex have been well described [7,8]. In brief, the IL-15 receptor complex is heterotrimeric, consisting of an $\alpha$ chain (CD215), a $\beta$ chain (CD122), and the common $\gamma$ chain ($\gamma$c/CD132) (Figure 1). IL-15R$\alpha$ binds IL-15 with high affinity and presents it to cells expressing high levels of CD122 and $\gamma$c. Presentation of IL-15 may occur in cis, with the IL-15/IL-15R$\alpha$ complex on the same cell as CD122/$\gamma$c, or presentation of IL-15 may occur in trans, with IL-15/IL-15R$\alpha$ on one cell and CD122/$\gamma$c on another [5,9–11].

Of note, CD122/$\gamma$c also represents the intermediate-affinity IL-2 receptor complex. Despite the fact that IL-2 and IL-15 share key components of their respective receptor complexes, we focus here on IL-15. IL-2 is completely absent from the maternal–fetal interface during steady-state pregnancy in mice. In a study of cytokine dynamics during mouse pregnancy, whole uterine tissue, plus placenta later in gestation, was subjected to qRT-PCR at nearly every day of gestation [12]. No *Il2* transcript was detected. In a large, publicly available dataset of single-cell RNA sequencing (scRNAseq) of hematopoietic and non-hematopoietic cells during the first trimester of human pregnancy, *IL2* transcript is also absent from the decidua and placenta [13].

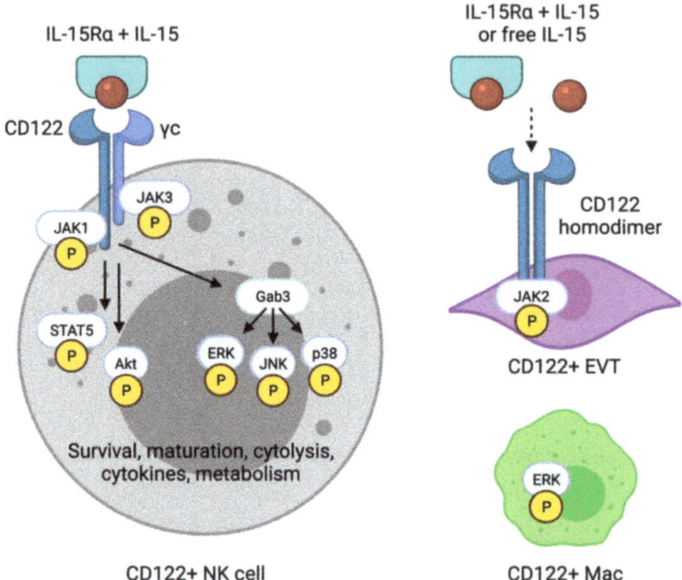



**Figure 1.** Components of IL-15 receptors and signaling cascades in IL-15-responsive cell types at the maternal–fetal interface. NK cells express CD122 and the common gamma chain (γc) and activate JAK-STAT, Akt, and MAP kinases in response to IL-15 presented in the context of IL-15Rα. IL-15 drives numerous functions in killer lymphocytes, including classical and uterine NK cells. Emerging IL-15-responsive cell types, such as CD122+ macrophages (CD122+ Mac) and CD122+ extravillous trophoblasts (EVT), may express alternative forms of the IL-15 receptor and activate non-classical signaling cascades downstream of IL-15. Created with Biorender.com, accessed 3 October 2021.

## 2. Spatial and Temporal Regulation of IL-15 in the Uterus

Like many other inflammatory cytokines, IL-15 is regulated at multiple levels to maintain homeostasis. Human *IL15* transcript exists as two isoforms, one with a long signal peptide sequence (LSP) and one with a short signal peptide sequence (SSP) [14,15]. When translated, the resultant IL-15 proteins are differentially localized in the cell and differentially secreted [15]. Soluble, free IL-15 is evident in the serum of mice in the steady state, as is small but nonzero amounts of IL-15 complexed with IL-15Rα [16]. Both isoforms of IL-15 are stabilized by IL-15Rα and exhibit greatly enhanced bioactivity on target cells when bound to IL-15Rα in vitro and in vivo [15,17–19]. The IL-15/IL-15Rα complex may remain bound to the cell membrane to spatially restrict the activity of IL-15 only to target cells directly contacting the IL-15-presenting cell. Instead, IL-15/IL-15Rα may be cleaved by a variety of mechanisms from the cell surface to act on more distant target cells [20].

Both *Il15* and *Il15ra* transcripts in mice have been detected by qRT-PCR in bulk uterus alone in early gestation or in bulk uterus and placenta later in gestation [12]. *Il15* is induced during the preimplantation period, by embryonic day 3 (E3), and peaks during mid-gestation, from E8 to E10 (Figure 2). It declines thereafter to a lower but constant level through E18.5. With *Il15*, *Il15ra* is also strongly induced from E8 to E10, but it is nearly undetectable in bulk tissue at every other gestational day. These were semi-quantitative and neither completely rule out low expression of *Il15ra* nor rule out the persistence of

IL-15Rα protein even after production of new transcripts ceases. However, uterine NK (uNK) cells in mice, identified by the binding of *Dolichus biflores* agglutinin (DBA), expand and contract with the same kinetics as the *Il15ra* transcript. In a kinetic, flow cytometric study of conventional NK1.1+DX5+NK cells and DBA+ uNK cells, uNK cells are rare prior to pregnancy and at E6.5, expand at least 10-fold and peak at E9.5, and return nearly to pre-pregnancy levels by E13.5 [21]. These data support a model in which IL-15 is induced just prior to implantation of the embryo in mice, peaks at mid-gestation around the time of spiral artery remodeling, and declines but persists to term. Uterine NK cells may then differentiate and expand in response to IL-15/IL-15Rα, but not to IL-15 alone (Figure 1).

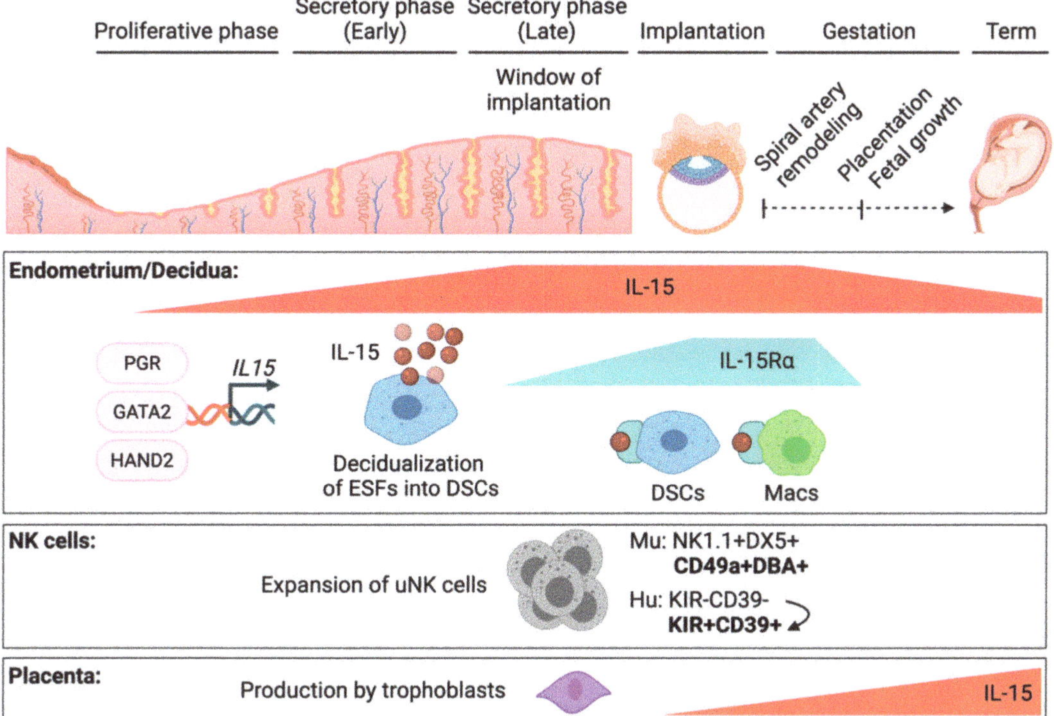


**Figure 2.** Spatial and temporal regulation of IL-15 during the menstrual cycle and during gestation has multiple effects during pregnancy. IL-15 is produced and presented by decidual stromal cells (DSCs) and macrophages (Macs) in the uterus. *IL15* is induced during the late secretory phase, when progesterone is dominant. Transcription of *IL15* is induced by the progesterone receptor (PGR) and GATA2. HAND2 binds the *IL15* promoter directly and induces *IL15* in DSCs but represses it in endometrial stromal fibroblasts, precursors to DSCs. Drawing on murine data, IL-15Rα peaks post-implantation, at the time of uterine spiral artery remodeling. The presumed abundance of IL-15/IL-15Rα complexes during that time expands murine DBA+ uterine NK cells and drives differentiation of KIR+CD39+ NK cells in human decidua. In the placenta, trophoblasts can produce IL-15 early in gestation at low levels but produce IL-15 maximally late in gestation during spontaneous labor. Created with Biorender.com, accessed 3 October 2021.

In the human uterus, the first studies to characterize IL-15 at the maternal–fetal interface did so with semi-quantitative RT-PCR and immunohistochemistry (IHC) [22,23]. With endometrial tissues collected after hysterectomy for leiomyomas or carcinoma in situ of the cervix, work by Kitaya found both the SSP and LSP isoforms of *IL15* transcript evident at low levels during the proliferative phase [22]. Both SSP and LSP *IL15* were induced during the secretory phase, and in first trimester elective terminations of 7–11 weeks gestation, *IL15* transcripts remained elevated (Figure 2). Data by Verma agree with low levels of *IL15* in proliferative phase endometrium [23]. While Verma found minimal to no induction of *IL15* transcript in secretory phase endometrium, at odds with the data by Kitaya, the highest levels of *IL15* transcript were found in pregnant decidua. Different findings between these two studies may reflect differences in when, exactly, tissues were obtained during the secretory phase. In recent studies of human endometrial tissue, exact dating was possible during proliferative and secretory phases [24–26]. With such granular data, it was shown that the *IL15* transcript is induced only during the second half of the secretory phase, which correlates to the window of implantation, or the point in the menstrual cycle, during which the uterus is most ready to receive an embryo (Figure 1) [24–27].

In the study by Kitaya, IHC for IL-15 protein in the proliferative phase endometrium showed staining in endometrial glandular epithelial cells, with very weak stromal cell staining [22]. IL-15 staining increased throughout the stroma and was most prominent in perivascular cells around spiral arteries in the secretory phase endometrium (Figure 2). In first-trimester decidua, IL-15 was found throughout the stroma, as well as in vascular endothelial cells. Newer data confirm the induction of *IL15* in stromal cells of the secretory phase endometrium by RNA in situ hybridization, as well as IL-15 staining by IHC in pregnant decidua [24,25]. The data by Verma add that the *IL15* transcript was abundant in decidual macrophages enriched by adherence to tissue culture plastic [23]. While these data agree with scRNAseq data showing decidual macrophages at the first-trimester maternal–fetal interface express *IL15* [13], decidual stromal cells are also adherent and strongly express *IL15* (as discussed below), perhaps biasing detection of IL-15 transcript in this assay.

By flow cytometry, surface IL-15, presumably bound by IL-15R$\alpha$, was detected on CD14+ and CD14− cells (Figure 2) [22]. CD14+ cells in the uterus most likely represent monocytes, macrophages, and/or monocyte-derived dendritic cells. The identity of the CD14− cells as hematopoietic or non-hematopoietic was not revealed, as flow data were not gated on CD45+ cells. As most decidual stromal cells express a large amount of *IL15* and *IL15RA* [13], we presume that these CD14− cells presenting IL-15 are decidual stromal cells. While Kitaya showed that IL-15 was not detected on the surface of CD56$^{bright}$ NK cells in first-trimester decidua [22], Verma reported the *IL15RA* transcript in CD56$^{bright}$ cells purified by magnetic cell separation [23]. While relatively few NK cells in first-trimester human decidua express *IL15* and/or *IL15RA* by scRNAseq [13], NK cells can express IL-15R$\alpha$ and signal in cis in other contexts [11].

### 3. Hormonal and Transcriptional Regulation of IL-15 in the Uterus

Taken together, the data discussed above support that the uterus expresses a modest amount of IL-15 until the late secretory phase, after the endometrial lining has remodeled, or decidualized [27–29], in preparation for an embryo (Figure 2). The uterus remains rich in IL-15 through the first trimester of pregnancy. IL-15 levels then wane and are accompanied by a withdrawal of IL-15R$\alpha$, which would limit trans-presentation of IL-15 and could lead to contraction of the killer lymphocyte compartment. Contributions by reproductive hormones may shed additional light on the regulation of IL-15 at the maternal–fetal interface.

Estrogen is far in excess of progesterone until the beginning of the secretory phase, when progesterone levels begin to rise and the process of decidualization begins [27–29]. The timeframe during which levels of progesterone become dominant represents the window of implantation. One interpretation of the data presented thus far is that an estrogen-dominant hormonal environment restricts expression of IL-15. Another interpretation is that progesterone needs to reach a critical level before it can drive the expression of IL-15. Alternatively, progesterone may initiate a cascade of events that culminates indirectly, over days, in the induction of IL-15. In other words, expression of IL-15 may be a distal feature of the decidualization program (Figure 2).

Kitaya cultured matched samples of bulk decidual cells for 5 days in medium alone or in medium plus progesterone [22]. Supernatants contained inconsistently higher concentrations of IL-15 after 5 days of exposure to progesterone. Effects of estrogen were not tested. Verma found that cultures enriched for decidual macrophages produced over twofold less IL-15 by ELISA after exposure to progesterone and prostaglandin E2 (PGE2) [23]. This same medium containing progesterone and PGE2 stimulated decidual stromal cells to produce about 50% more IL-15 by ELISA. These data suggest that hormonal control of IL-15 production is complex and may be cell type-specific.

More recently, two groups have examined the role of the transcription factor Heart and neural crest derivatives-expressed transcript 2 (HAND2) in regulating IL-15. Shindoh studied the effects of HAND2 on expression of IL15 in Vimentin+ primary endometrial stromal cells (ESCs) cultured for 12 days in estrogen and progesterone to decidualize them [30]. Knockdown of HAND2 with siRNA resulted in loss of multiple decidualization-associated genes, including IL-15 (Figure 2). More recently, the same group showed coordinate induction of HAND2 transcript and protein, as well as IL15 transcript and protein, in secretory phase endometrium [24]. Of note, not every stromal cell expressed IL-15 protein. IL-15 expressers appeared to be evenly distributed throughout the stroma. These data suggest that certain stromal cells may be specialized in their ability to express IL-15. Instead, IL-15 expression may be stochastic.

To answer whether HAND2 had a direct effect on IL-15 or a secondary effect by virtue of controlling decidualization, Murata found a HAND2 binding motif conserved between mice, macaques and humans at position -1628 to -1622 relative to the IL15 transcriptional start site (TSS) [24]. By CHIP-qPCR, HAND2 was found to bind this site of the IL15 promoter. Next, ESCs were decidualized in culture and transfected with luciferase constructs containing the *IL15* promoter region with the HAND2 motif intact or mutant versions of the *IL15* promoter region lacking the intact HAND2 binding motif. Co-transfection of these constructs with a HAND2 expression vector drove expression of luciferase only in the presence of the intact promoter region.

Similar to findings by Murata, Marinić recently found that levels of *HAND2* and *IL15* transcript were directly correlated in preparation for and during pregnancy [25]. They added a complete kinetic of levels of both *HAND2* and *IL15* over the course of human gestation, showing induction during the mid-to-late secretory phase into the first trimester of pregnancy and showing a relative decrease in levels of *HAND2* and *IL15* transcripts in the basal plate, or the maternal side, of the placenta as gestation approached term (Figure 2). Broadly speaking, these human data agree with the kinetics of murine *Il15* during gestation, but expression of IL-15Rα was not examined. In contrast to prior findings [24,30], however, Marinić found that HAND2 negatively regulated *IL15* using siRNA knockdown in immortalized endometrial stromal fibroblasts (ESFs) [25]. Marinić used the hTERT-immortalized cell line CRL-4003, derived from mid-secretory ESFs, and performed siRNA knockdown of HAND2 for 48hrs [25,31]. Murata and Shindoh used bulk Vimentin+ primary ESCs, which likely contain ESFs and more differentiated DSCs [32], and performed siRNA knockdown in the presence of decidualization hormones for 12 days. So, the discrepant results are likely due to differences in cell type, timing, and differences in culture conditions.

Altogether, these data suggest that HAND2 directly binds the *IL15* promoter and regulates expression of IL15 (Figure 2). HAND2 clearly can bind a motif proximal to the *IL15* TSS [24]. Whether HAND2 promotes or represses transcription may depend on cell type/state, with activation seen in decidualized primary ESCs [24] and repression seen in undifferentiated, immortalized ESFs [25]. It is possible, though it remains to be formally shown, that HAND2 also directs expression of *IL15* by binding to more distant enhancers. In silico analyses show that HAND2 binding motifs exist beyond the *IL15* promoter, within distant enhancers that contact the *IL15* promoter [25].

Additional factors that may drive expression of *IL15* were identified by analysis of publicly available siRNA knockdown data in primary ESFs differentiated to DSCs with cAMP and progesterone. Consistent with the notion that IL-15 is part of a decidualization program (Figure 2), *IL15* was induced by differentiation of ESFs into DSCs with cAMP and progesterone [25]. Knockdown of *PGR*, encoding the progesterone receptor, diminished the *IL15* transcript approximately twofold, consistent with data that the progesterone receptor antagonist asoprisnil suppresses *IL15* [26]. Knockdown of GATA Binding Protein 2 (GATA2) diminished the *IL15* transcript over fourfold [25]. Of potential interest is that GATA2 has been implicated in differentiation and maintenance of CD56$^{bright}$ NK cells in humans, the precursors of more mature CD56$^{dim}$ NK cells and the predominant phenotype of human uNK cells [33]. While GATA2 mutations appear to cause NK cell-intrinsic defects, it was not formally shown whether these inborn errors of immunity also impact the IL-15 axis, with a secondary effect on NK cell differentiation and/or survival. In the context of myelopoiesis and lymphopoiesis in the steady state, expression of IL-15 using a fluorescent reporter was demonstrated in hematopoietic progenitors and myeloid cells, the transcriptional regulators of which remain unknown [34]. The same group showed that upregulation of IL-15 occurred after viral infection and that induction of IL-15 was dependent on the type I interferon (IFN) receptor IFNAR [35]. Type I IFNs are present in the uterus at low levels in the steady state and contribute to host defense [36], but whether IFNs play a role in decidualization and/or induction of IL-15 in the uterus has not been examined.

## 4. Spatial and Temporal Regulation of IL-15 in the Placenta

While IL-15 is rich in the maternal uterus during early gestation, the developing fetal placenta is relatively poor in IL-15 early in gestation. Minimal *IL15* transcript was found in flow-sorted, first-trimester epidermal growth factor receptor-positive (EGFR+) villous trophoblasts (VTs) and erbB2+ extravillous trophoblasts (EVTs) [23], consistent with scRNAseq data mentioned previously [13]. Toth shows minimal expression of IL-15 in multiple cells of the normal first-trimester placenta by IHC, including EVT and syncytiotrophoblast [37]. Agarwal measured IL-15 by ELISA and *IL15* by semi-quantitative RT-PCR in cultured human placental explants [38]. While these were bulk explants containing a heterogenous population of cells, production of IL-15 increased twofold from first-trimester explants to full-term explants from spontaneous vaginal deliveries. Full-term explants from elective cesarean sections in the absence of labor produced an intermediate amount of IL-15. Placental macrophages isolated at term and cultured for 24 h produce an undetectable amount of IL-15 by ELISA [39]. These limited data suggest that the human fetal placenta is capable of making progressively more IL-15 over the course of gestation, with a peak during labor (Figure 2). Trophoblasts, not macrophages, appear to be responsible for the production of IL-15 in the placenta.

## 5. Dysregulation of IL-15 in Adverse Outcomes of Pregnancy

Multiple lines of evidence in mice and humans suggest that either excess or loss of IL-15 leads to adverse outcomes of pregnancy. First, a genetic deficiency of IL-15 in mice leads to fetal growth restriction, morphologically abnormal deciduae, and impaired remodeling of spiral arteries [40,41], a pathologic hallmark of preeclampsia. IL-15 knockout dams were also shown to have a modestly but consistently elevated rate of fetal resorption during pregnancy, compared to IL-15-sufficient dams [42]. However, IL-15-deficient dams mated to IL-15-deficient sires still have a normal length of gestation and deliver a normal number of viable pups [40,41]. Both groups to report this phenotype demonstrate an absence of DBA+ uNK cells, consistent with the notion that IL-15/IL-15Rα drives differentiation of uterine-phenotype NK cells, as discussed above (Figure 2). Of note, DBA-negative NK cells in the uterus were not examined and may support gestation in important ways. More broadly, IL-15-independent NK cells in the uterus have not been examined. It has been shown by two different groups that interleukin-12 can support the development and expansion of NK cells independently of γc cytokines [43,44]. It is also important to appreciate that relatively minor adverse outcomes of pregnancy in IL-15-deficient mice and in other NK cell-deficient mice were observed in the setting of syngeneic matings [40,41,45,46]. Despite the hardiness of syngeneic mouse pregnancy, fetal growth restriction and preeclampsia are of major clinical significance in humans. In allogeneic matings, *Nfil3*/E4BP4-deficient mice that have a severe reduction in NK cells also exhibit substantial fetal loss [47].

Overabundance of IL-15 in mouse models is also associated with poor outcomes of pregnancy. IL-15 permits fetal loss in a model of lipopolysaccharide (LPS)-mediated abortion, in which low-dose LPS is given to pregnant dams post-implantation at E7.5 [42]. In other words, IL-15-deficient dams are completely resistant to fetal loss in this system. Whether IL-15 permits inflammation-mediated fetal loss by activating NK cells was not formally shown. With an anti-NK1.1 antibody known to deplete NK1.1+ NK cells in vivo, though, the authors demonstrated partial rescue of fetal loss in response to LPS. The mechanism of this remains unclear, as successful depletion of NK cells in the uterus was not shown, and NK1.1-DBA+ uNK cells should not be depleted using this method [21].

Another example of unrestrained IL-15 is in the BPH/5 mouse, a model of spontaneous hypertension and preeclampsia [48–50]. Compared to control animals, pregnancies in BPH/5 dams are characterized by implantation defects, placental abnormalities, fetal growth restriction, and fetal loss. Despite having abnormal, delayed decidualization, BPH/5 implantation sites exhibit inappropriately elevated IL-15 transcript and protein early in gestation, during the peri-implantation period and through at least E7.5 [48]. Thus, decidualization and IL-15 are uncoupled in this model. Levels of IL-15 normalize in BPH/5 implantation sites by E10.5, when IL-15 peaks in normal mouse pregnancy. Paradoxically, however, DBA+ uNK cells are decreased in BPH/5 mice by flow cytometry, NK cell-related transcripts, including *Ncr1*, and by immunofluorescence of E7.5 implantation sites. It is worth noting that uNK cells are rare early in gestation [21], and the authors' raw data confirm this [48]. At the same time, the number of CD122+TCRβ- lymphocytes, presumably conventional DBA- NK cells, appears to be severely reduced by flow cytometry in the BPH/5 mouse. Treatment of pregnant WT mice with recombinant IL-15 precomplexed with IL-15Rα also reduced DBA+ NK cells- and *Ncr1* transcript in early implantation sites. Administration of the cyclooxygenase 2 (COX2) inhibitor celecoxib at E6.5 reduced resorption of fetuses, improved fetal-placental weights, and quickly (by E7.5) normalized IL-15 and NK cells back to WT levels. These data support that IL-15 must be tightly regulated in a spatial and temporal manner during pregnancy and that the relationship between uNK cells and IL-15 is complex and not fully understood.

One caveat to applying studies of IL-15 in mice is that mouse and human cells differentially depend on, and respond to, IL-15. A single 10 µg intraperitoneal dose of an antibody raised against murine IL-15 results in near-total elimination of classical splenic DX5+NK1.1+ NK cells in less than 5 days [51]. Blood NK cells also sharply declined in cynomolgus macaques administered 5 weekly doses of intravenous or subcutaneous anti-IL-15 antibody by 2 weeks post-treatment through 20 weeks post-treatment. Despite high levels of circulating IL-15-blocking antibody in healthy volunteers, neither CD56$^{bright}$ nor CD56$^{dim}$ NK cells in the blood declined in number during the study. These data suggest that, in contrast to murine and non-human primate NK cells, human NK cells do not depend on IL-15. While they may not critically depend on IL-15, human NK cells are at least partially dependent on Janus kinase 3 (JAK3), which directly associates with γc. A safety and efficacy study in kidney transplant recipients showed that a JAK3 inhibitor (that also modestly inhibits JAK2) substantially decreases the absolute number of circulating NK cells [52]. Investigations of IL-15-dependent and -independent mechanisms of uNK cell development and function may help to reconcile these differences between mouse and human studies and refine our understanding of roles for IL-15 in human pregnancy.

Further study of IL-15 is essential, as IL-15 is consistently dysregulated in the setting of adverse outcomes of pregnancy in humans. Placental explants from term pregnancies affected by severe preeclampsia produce fourfold less IL-15 in culture than explants prepared from healthy term pregnancies [38]. In the uterus of women with preeclampsia, *IL15* transcript was upregulated tenfold in DSCs but only twofold in ESFs [25]. IL-15 was modestly elevated in the serum of women with preeclampsia at term [53]. In the uterus of women with implantation failure, *IL15* transcript was downregulated over twofold and trended toward marginal downregulation in women with recurrent spontaneous abortion. IL-15 protein by IHC in the placenta is elevated in the setting of miscarriage and further elevated in the setting of recurrent miscarriage [37]. These data underscore that adverse outcomes of pregnancy are associated with disturbed homeostasis of IL-15 and that IL-15 is regulated (and dysregulated) in a cell type-specific manner. Whether abnormal homeostasis of IL-15 is a cause or effect of adverse outcomes of pregnancy, and whether targeting IL-15 might reverse such conditions and restore maternal–fetal health, remain to be investigated.

## 6. Traditional and Non-Traditional Responders to IL-15

Effects of IL-15 on NK cells, as well as signaling and metabolic cascades activated in response to IL-15, have been extensively studied and are the subject of several comprehensive reviews [54–56]. In brief, IL-15 supports NK cells in many ways, including the development, maintenance, cytolysis, antibody-dependent cell-mediated cytotoxicity, and production of cytokines, such as IFNγ (Figure 1). Like so many other inflammatory cytokines, IL-15 also induces counter-regulatory mechanisms to avoid overwhelming IL-15-mediated inflammation. For instance, IL-15 given continuously to human NK cells in vitro results in enhanced proliferation but decreased cytolysis, decreased production of IFNγ, decreased killing of liquid tumors in vivo, and increased cell death, all due (in part) to impaired fatty acid oxidation [57]. The membrane-bound A disintegrin and metalloprotease 17 (ADAM17), activated downstream of IL-15 signaling, restrains proliferation of NK cells in another example of negative feedback [58]. These data may help explain the findings of Sones, in which inappropriately high levels of IL-15 in preeclamptic mice are associated with reduced numbers of uNK cells [48].

Recent studies have also shed new light on how uNK interpret signals from IL-15 to carry out numerous functions essential to pregnancy, such as the release of angiogenic factors and growth factors, driving spiral artery remodeling, and shaping trophoblast invasion [59–61]. Sliz showed that the scaffolding protein GRB2-associated binding protein 3 (Gab3) was required for expansion of NK cells in response to both IL-15 (and IL-2) through CD122/γc [62]. Gab3-deficient NK cells stimulated with IL-15 exhibited major defects in phosphorylation of the mitogen-activated protein (MAP) kinases extracellular signal-related kinase (ERK), p38, and c-Jun N-terminal kinase (JNK) but phosphorylated signal transducer and activator of transcription 5 (STAT5) and Akt normally (Figure 1). Consistent with impaired IL-15 responsivity, NK cells in the uterus are reduced in the absence of Gab3, although only conventional NK1.1+ NK cells, not DBA+ NK cells, were affected. Regardless of the subset affected, abnormal spiral artery remodeling and trophoblast invasion were observed in Gab3-deficient dams, again supporting the importance of IL-15 in healthy pregnancy.

Recent human data show that a subset of decidual NK cells lacking expression of killer cell immunoglobulin-like receptors (KIRs) and CD39 proliferate strongly in response to IL-15 [63]. KIR-CD39- NK cells stimulated with IL-15 then acquire expression of KIRs and CD39, a phenotype previously identified as NK1 and enriched for transcripts encoding the cytolytic mediators perforin and granzymes by scRNAseq [13]. These data indicate that IL-15 supports the differentiation and function of NK cells in the uterus. These data also provide evidence that responsiveness to IL-15 among subsets in the uterus is heterogeneous. Marinić demonstrated that culture of primary NK cells in IL-15 or in conditioned medium from ESFs promotes migration of NK cells into Matrigel [25]. Knockdown of *IL15* in ESFs with siRNA or treatment of conditioned medium with anti-IL-15 antibody led to a marginal decrease in NK cell migration, raising the possibility that not all migration by NK cells in response to conditioned ESF medium is due to IL-15. Future work directed at how uNK subsets differentially integrate signals from IL-15 will be useful to better understand IL-15-mediated adverse outcomes of pregnancy, as discussed above.

While killer lymphocytes are the canonical responders to IL-15 by virtue of high expression of CD122 and γc, several other cells expressing IL-15 receptor components have been identified at the maternal–fetal interface. We recently showed that a subset of macrophages (Macs) in the uterus of mice and humans expresses CD122 [64]. CD122+Macs are enriched for interferon-stimulated transcripts and respond to type I and II IFNs by inducing CD122 in a dose-dependent manner. These Macs respond biochemically to stimulation with the IL-15/IL-15Rα complex by phosphorylating ERK and respond functionally by enhancing the release of proinflammatory cytokines (Figure 1). Unique tools to understand roles for CD122 on Macs have been developed and are being studied in our laboratory to better understand how IL-15-responsive macrophages impact pregnancy.

In addition to Macs, trophoblasts and trophoblastic cell lines have been found to express CD122 and respond to IL-15. Derived from choriocarcinomas, the cell lines JEG-3, BeWo, and JAR all expressed the *IL2RB* transcript by RT-PCR, but protein expression was not examined [65]. IL-15 increases the invasion of JEG-3 cells into Matrigel in a dose-dependent fashion at 1 and 10 ng/mL of recombinant IL-15 [66]. At 10 ng/mL of IL-15, JEG3 invasion increased by a factor of two, relative to untreated JEG3 cells. IL-15 does not affect proliferation of JEG-3 cells at any dose, as measured by activity of mitochondrial dehydrogenases. Similar data were obtained recently using HTR8 cells, which are immortalized first-trimester trophoblasts [25]. Finally, a modest increase in secretion of matrix metalloproteinase 1 (MMP1), but not MMP2 or MMP9, by JEG3 cells was shown by ELISA in response to culture in the presence of IL-15 [66]. Interpretation of these data derived from JEG3 cells is complicated by how distantly related JEG3 cells and primary EVTs are at a global transcriptional level, with almost 1200 genes significantly different by microarray between the two cell types [67].

In primary cells obtained from 8 to 12 weeks gestation, Apps compared the transcriptomes of flow-sorted EGFR+ VTs and HLA-G+ EVTs cultured for 12 h on fibronectin. While this method of obtaining EVT may be criticized, cross-referencing this dataset with that of the recently published scRNAseq data from the first-trimester maternal–fetal interface [13] confirmed that the vast majority of genes associated with EVT or VT in the Apps dataset were indeed associated specifically with EVT or VT (data not shown). *IL2RB* was highly enriched in EVTs [67]. Strong extracellular staining of CD122 protein was evident by flow cytometry on EVTs (Figure 1), with weak but non-zero expression by VTs. Expression of CD122 was also seen by IHC in a column of EVTs outside of a villous, with minimal staining in the villous itself.

Interestingly, γc is not detected in EVT or VT (Figure 1) [13,67]. It has been established in cell lines that CD122 can homodimerize [68,69]. Ferrag showed in CHO cells that chimeric receptors, containing the intracellular domain of CD122 paired with unique extracellular domains, signaled through JAK2 and STAT5 [68]. Co-transfection of COS-7 cells with differentially tagged human CD122, followed by co-immunoprecipitation, showed that CD122 can homodimerize [69]. CD122 prefers to heterodimerize with γc, if γc is available. With fluorescence resonance energy transfer (FRET), Pillet showed that the N-terminal (extracellular) domain of CD122 was absolutely required for homodimerization, while the intracellular domains were dispensable for homodimerization. With flow cytometry using tagged IL-2, CD122 homodimers bound IL-2 that could be blocked by an anti-CD122 antibody. Finally, cells expressing CD122 homodimers bound IL-2 with a similar, intermediate affinity as cells expressing heterodimers of CD122 and γc. Altogether, these data suggest that CD122 homodimers bind IL-2, but it is unclear how these homodimers interact with IL-15 in the presence or absence of IL-15Rα. Further, it remains unknown how IL-15 affects the function of trophoblasts, especially EVTs that invade into an IL-15-rich decidua.

Expression of CD122 on trophoblasts is unusual in several ways. Cohen found a long terminal repeat (LTR) with endogenous retroviral promoter elements 25kb upstream of the native promoter for *IL2RB* [70]. This LTR contains a transcriptional start site (TSS) and can splice into the normal ATG residing in exon 2 of *IL2RB*. RT-PCR for the LTR-containing transcript containing the retroelement showed that this alternative transcript was abundant in placenta (nearly 90% of transcripts) and rare in spleen and other organs, including peripheral blood mononuclear cells (PBMCs). Using a luciferase assay, a construct containing a portion of the LTR upstream of the splice donor site drove luciferase expression strongly in the choriocarcinoma cell line JEG-3. In vitro methylation of the luciferase construct eliminated luciferase expression under control of the LTR. Supporting the notion that epigenetic factors unique to trophoblasts allowed the LTR to drive expression of *IL2RB*, 9 C followed by G dinucleotide (CpG) sites within the LTR were identified and found to be consistently methylated in PBMCs but methylated less frequently in placenta and enriched trophoblasts. Inhibition of histone deacetylases to improve accessibility of the LTR also enhanced expression of *IL2RB* in cell lines. At the protein level, extracts of bulk placental villi were probed with two C-terminal-specific antibodies by Western blot. Full-length 75 kDa CD122 was not detected with these antibodies, only a 37 kDa fragment that may correspond to a cleaved fragment of CD122, something that has been previously demonstrated in leukemic cell lines [71]. Sequencing was not carried out to confirm the identity of this fragment, nor was subcellular localization of CD122 to confirm that it was membrane bound and not localized elsewhere. Further, the lack of full-length CD122 is at odds with flow cytometry data demonstrating clear expression of CD122 on trophoblasts [67]. If CD122 is translated and cleaved subsequently in trophoblasts, the N- and C-terminal fragments may have distinct and novel functions that have yet to be explored.

## 7. Conclusions

IL-15 is abundant and tightly regulated in the uterus during pregnancy. Several lines of evidence suggest that homeostasis of IL-15 must be maintained to ensure the health of the mother and fetus. Given widespread interest in IL-15 as a mediator of anti-tumor immunity in preclinical studies, numerous biological agents have been synthesized to mimic or inhibit the activity of IL-15 on target cells [51,72]. IL-15 agonists succeed in expanding killer lymphocytes but fail to reject tumors in humans when administered alone. Thought to be due to compensatory induction of anti-inflammatory signaling by IL-15, these observations may be of great interest in the context of reproduction, which depends on appropriate inflammation that maintains tolerance to the fetus. With a deeper understanding of the spatial and temporal regulation of IL-15, as well as its effects on classical and novel cell types, we can better detect dysregulation of IL-15 and associated immunopathology during gestation. We may then modulate IL-15 signaling in the uterus to optimize outcomes of pregnancy.

**Funding:** This research was funded by the National Institute of Allergy and Infectious Diseases (NIAID) grant 1K08AI151265-01A1 and Children's Hospital of Philadelphia.

**Informed Consent Statement:** Not applicable.

**Conflicts of Interest:** The author declares no conflict of interest.

## Abbreviations

| | |
|---|---|
| ADAM17 | A disintegrin and metalloprotease 17 |
| cAMP | cyclic adenosine monophosphate |
| CD | cluster of differentiation |
| COX2 | cyclooxygenase 2 |
| CpG | C followed by G dinucleotide |
| DBA | Dolichus biflores agglutinin |
| DSC | decidual stromal cell |
| E | embryonic day |
| ELISA | enzyme-linked immunosorbent assay |
| EGFR | epidermal growth factor receptor |
| ERK | extracellular signal-related kinase |

| | |
|---|---|
| ESF | endometrial stromal fibroblast |
| EVT | extravillous trophoblast |
| FRET | fluorescence resonance energy transfer |
| $\gamma$c | common gamma chain |
| Gab3 | GRB2-associated binding protein 3 |
| GATA2 | GATA Binding Protein 2 |
| HAND2 | Heart and neural crest derivatives-expressed transcript 2 |
| IFN | interferon |
| IHC | immunohistochemistry |
| IL-2 | Interleukin-2 |
| IL-15 | Interleukin-15 |
| JAK | Janus kinase |
| JNK | c-Jun N-terminal kinase |
| KIR | killer cell immunoglobulin-like receptor |
| LPS | lipopolysaccharide |
| LSP | long signal peptide |
| LTR | long terminal repeat |
| MMP | matrix metalloproteinase |
| NK cell | natural killer cell |
| NKT cell | natural killer T cell |
| PBMCs | peripheral blood mononuclear cells |
| PGE2 | prostaglandin E2 |
| PGR | progesterone receptor |
| siRNA | short-interfering RNA |
| RT-PCR | reverse-transcription polymerase chain reaction |
| scRNAseq | single-cell RNA sequencing |
| SSP | short signal peptide |
| STAT | signal transducer and activator of transcription |
| TSS | transcriptional start site |
| VT | villous trophoblast |
| uNK cell | uterine natural killer cell |

## References


1.  Carson, W.E.; Giri, J.G.; Lindemann, M.J.; Linett, M.L.; Ahdieh, M.; Paxton, R.; Anderson, D.; Eisenmann, J.; Grabstein, K.; Caligiuri, M.A. Interleukin (IL) 15 Is a Novel Cytokine That Activates Human Natural Killer Cells via Components of the IL-2 Receptor. *J. Exp. Med.* **1994**, *180*, 1395–1403. [CrossRef]
2.  Puzanov, I.J.; Bennett, M.; Kumar, V. IL-15 Can Substitute for the Marrow Microenvironment in the Differentiation of Natural Killer Cells. *J. Immunol.* **1996**, *157*, 4282–4285. [PubMed]
3.  Kennedy, M.K.; Glaccum, M.; Brown, S.N.; Butz, E.A.; Viney, J.L.; Embers, M.; Matsuki, N.; Charrier, K.; Sedger, L.; Willis, C.R.; et al. Reversible Defects in Natural Killer and Memory CD8 T Cell Lineages in Interleukin 15-Deficient Mice. *J. Exp. Med.* **2000**, *191*, 771–780. [CrossRef]
4.  Becker, T.C.; Wherry, E.J.; Boone, D.; Murali-Krishna, K.; Antia, R.; Ma, A.; Ahmed, R. Interleukin 15 Is Required for Proliferative Renewal of Virus-Specific Memory CD8 T Cells. *J. Exp. Med.* **2002**, *195*, 1541–1548. [CrossRef] [PubMed]
5.  Koka, R.; Burkett, P.R.; Chien, M.; Chai, S.; Chan, F.; Lodolce, J.P.; Boone, D.L.; Ma, A. Interleukin (IL)-15R$\alpha$–Deficient Natural Killer Cells Survive in Normal but Not IL-15R$\alpha$–Deficient Mice. *J. Exp. Med.* **2003**, *197*, 977–984. [CrossRef]
6.  Kanter, J.; Mani, S.; Gordon, S.; Mainigi, M. Uterine Natural Killer Cell Biology and Role in Early Pregnancy Establishment and Outcomes. *F&S Rev.* **2021**. [CrossRef]
7.  Budagian, V.; Bulanova, E.; Paus, R.; Bulfone-Paus, S. IL-15/IL-15 Receptor Biology: A Guided Tour through an Expanding Universe. *Cytokine Growth Factor Rev.* **2006**, *17*, 259–280. [CrossRef]
8.  Mishra, A.; Sullivan, L.; Caligiuri, M.A. Molecular Pathways: Interleukin-15 Signaling in Health and in Cancer. *Clin. Cancer Res.* **2014**, *20*, 2044–2050. [CrossRef]
9.  Dubois, S.; Mariner, J.; Waldmann, T.A.; Tagaya, Y. IL-15Ralpha Recycles and Presents IL-15 In Trans to Neighboring Cells. *Immunity* **2002**, *17*, 537–547. [CrossRef]
10. Huntington, N.D.; Legrand, N.; Alves, N.L.; Jaron, B.; Weijer, K.; Plet, A.; Corcuff, E.; Mortier, E.; Jacques, Y.; Spits, H.; et al. IL-15 Trans-Presentation Promotes Human NK Cell Development and Differentiation in Vivo. *J. Exp. Med.* **2009**, *206*, 25–34. [CrossRef]
11. Zanoni, I.; Spreafico, R.; Bodio, C.; Gioia, M.D.; Cigni, C.; Broggi, A.; Gorletta, T.; Caccia, M.; Chirico, G.; Sironi, L.; et al. IL-15 Cis Presentation Is Required for Optimal NK Cell Activation in Lipopolysaccharide-Mediated Inflammatory Conditions. *Cell Rep.* **2013**, *4*, 1235–1249. [CrossRef]

12. Ye, W.; Zheng, L.M.; Young, J.D.; Liu, C.C. The Involvement of Interleukin (IL)-15 in Regulating the Differentiation of Granulated Metrial Gland Cells in Mouse Pregnant Uterus. *J. Exp. Med.* **1996**, *184*, 2405–2410. [CrossRef]
13. Vento-Tormo, R.; Efremova, M.; Botting, R.A.; Turco, M.Y.; Vento-Tormo, M.; Meyer, K.B.; Park, J.-E.; Stephenson, E.; Polański, K.; Goncalves, A.; et al. Single-Cell Reconstruction of the Early Maternal–Fetal Interface in Humans. *Nature* **2018**, *563*, 347–353. [CrossRef]
14. Tagaya, Y.; Bamford, R.N.; DeFilippis, A.P.; Waldmann, T.A. IL-15: A Pleiotropic Cytokine with Diverse Receptor/Signaling Pathways Whose Expression Is Controlled at Multiple Levels. *Immunity* **1996**, *4*, 329–336. [CrossRef]
15. Bergamaschi, C.; Jalah, R.; Kulkarni, V.; Rosati, M.; Zhang, G.-M.; Alicea, C.; Zolotukhin, A.S.; Felber, B.K.; Pavlakis, G.N. Secretion and Biological Activity of Short Signal Peptide IL-15 Is Chaperoned by IL-15 Receptor Alpha In Vivo. *J. Immunol.* **2009**, *183*, 3064–3072. [CrossRef]
16. Anderson, B.G.; Quinn, L.S. Free IL-15 Is More Abundant Than IL-15 Complexed with Soluble IL-15 Receptor-α in Murine Serum: Implications for the Mechanism of IL-15 Secretion. *Endocrinology* **2016**, *157*, 1315–1320. [CrossRef]
17. Bergamaschi, C.; Rosati, M.; Jalah, R.; Valentin, A.; Kulkarni, V.; Alicea, C.; Zhang, G.-M.; Patel, V.; Felber, B.K.; Pavlakis, G.N. Intracellular Interaction of Interleukin-15 with Its Receptor α during Production Leads to Mutual Stabilization and Increased Bioactivity*. *J. Biol. Chem.* **2008**, *283*, 4189–4199. [CrossRef] [PubMed]
18. Rubinstein, M.P.; Kovar, M.; Purton, J.F.; Cho, J.-H.; Boyman, O.; Surh, C.D.; Sprent, J. Converting IL-15 to a Superagonist by Binding to Soluble IL-15Rα. *Proc. Natl. Acad. Sci. USA* **2006**, *103*, 9166–9171. [CrossRef] [PubMed]
19. Dubois, S.; Patel, H.J.; Zhang, M.; Waldmann, T.A.; Muller, J.R. Preassociation of IL-15 with IL-15Rα-IgG1-Fc Enhances Its Activity on Proliferation of NK and CD8$^+$/CD44$^{high}$ T Cells and Its Antitumor Action. *J. Immunol.* **2008**, *180*, 2099–2106. [CrossRef] [PubMed]
20. Anthony, S.M.; Howard, M.E.; Hailemichael, Y.; Overwijk, W.W.; Schluns, K.S. Soluble Interleukin-15 Complexes Are Generated In Vivo by Type I Interferon Dependent and Independent Pathways. *PLoS ONE* **2015**, *10*, e0120274. [CrossRef]
21. Yadi, H.; Burke, S.; Madeja, Z.; Hemberger, M.; Moffett, A.; Colucci, F. Unique Receptor Repertoire in Mouse Uterine NK Cells. *J. Immunol.* **2008**, *181*, 6140–6147. [CrossRef]
22. Kitaya, K.; Yasuda, J.; Yagi, I.; Tada, Y.; Fushiki, S.; Honjo, H. IL-15 Expression at Human Endometrium and Decidua. *Biol. Reprod.* **2000**, *63*, 683–687. [CrossRef]
23. Verma, S.; Hiby, S.E.; Loke, Y.W.; King, A. Human Decidual Natural Killer Cells Express the Receptor for and Respond to the Cytokine Interleukin 15. *Biol. Reprod.* **2000**, *62*, 959–968. [CrossRef]
24. Murata, H.; Tanaka, S.; Tsuzuki-Nakao, T.; Kido, T.; Kakita-Kobayashi, M.; Kida, N.; Hisamatsu, Y.; Tsubokura, H.; Hashimoto, Y.; Kitada, M.; et al. The Transcription Factor HAND2 Up-Regulates Transcription of the IL15 Gene in Human Endometrial Stromal Cells. *J. Biol. Chem.* **2020**, *295*, 9596–9605. [CrossRef]
25. Marinić, M.; Mika, K.; Chigurupati, S.; Lynch, V.J. Evolutionary Transcriptomics Implicates HAND2 in the Origins of Implantation and Regulation of Gestation Length. *eLife* **2021**, *10*, e61257. [CrossRef]
26. Wilkens, J.; Male, V.; Ghazal, P.; Forster, T.; Gibson, D.A.; Williams, A.R.W.; Brito-Mutunayagam, S.L.; Craigon, M.; Lourenco, P.; Cameron, I.T.; et al. Uterine NK Cells Regulate Endometrial Bleeding in Women and Are Suppressed by the Progesterone Receptor Modulator Asoprisnil. *J. Immunol.* **2013**, *191*, 2226–2235. [CrossRef]
27. Ochoa-Bernal, M.A.; Fazleabas, A.T. Physiologic Events of Embryo Implantation and Decidualization in Human and Non-Human Primates. *Int. J. Mol. Sci.* **2020**, *21*, 1973. [CrossRef]
28. Ng, S.-W.; Norwitz, G.A.; Pavlicev, M.; Tilburgs, T.; Simón, C.; Norwitz, E.R. Endometrial Decidualization: The Primary Driver of Pregnancy Health. *Int. J. Mol. Sci.* **2020**, *21*, 4092. [CrossRef]
29. Okada, H.; Tsuzuki, T.; Murata, H. Decidualization of the Human Endometrium. *Reprod. Med. Biol.* **2018**, *17*, 220–227. [CrossRef]
30. Shindoh, H.; Okada, H.; Tsuzuki, T.; Nishigaki, A.; Kanzaki, H. Requirement of Heart and Neural Crest Derivatives–Expressed Transcript 2 during Decidualization of Human Endometrial Stromal Cells in Vitro. *Fertil. Steril.* **2014**, *101*, 1781–1790. [CrossRef]
31. Krikun, G.; Mor, G.; Alvero, A.; Guller, S.; Schatz, F.; Sapi, E.; Rahman, M.; Caze, R.; Qumsiyeh, M.; Lockwood, C.J. A Novel Immortalized Human Endometrial Stromal Cell Line with Normal Progestational Response. *Endocrinology* **2004**, *145*, 2291–2296. [CrossRef]
32. Rytkönen, K.T.; Erkenbrack, E.M.; Poutanen, M.; Elo, L.L.; Pavlicev, M.; Wagner, G.P. Decidualization of Human Endometrial Stromal Fibroblasts Is a Multiphasic Process Involving Distinct Transcriptional Programs. *Reprod. Sci.* **2019**, *26*, 323–336. [CrossRef] [PubMed]
33. Mace, E.M.; Hsu, A.P.; Monaco-Shawver, L.; Makedonas, G.; Rosen, J.B.; Dropulic, L.; Cohen, J.I.; Frenkel, E.P.; Bagwell, J.C.; Sullivan, J.L.; et al. Mutations in GATA2 Cause Human NK Cell Deficiency with Specific Loss of the CD56bright Subset. *Blood* **2013**, *121*, 2669–2677. [CrossRef] [PubMed]
34. Colpitts, S.L.; Stonier, S.W.; Stoklasek, T.A.; Root, S.H.; Aguila, H.L.; Schluns, K.S.; Lefrancois, L. Transcriptional Regulation of IL-15 Expression during Hematopoiesis. *J. Immunol.* **2013**, *191*, 3017–3024. [CrossRef] [PubMed]
35. Colpitts, S.L.; Stoklasek, T.A.; Plumlee, C.R.; Obar, J.J.; Guo, C.; Lefrançois, L. Cutting Edge: The Role of IFN-α Receptor and MyD88 Signaling in Induction of IL-15 Expression In Vivo. *J. Immunol.* **2012**, *188*, 2483–2487. [CrossRef] [PubMed]
36. Fung, K.Y.; Mangan, N.E.; Cumming, H.; Horvat, J.C.; Mayall, J.R.; Stifter, S.A.; Weerd, N.D.; Roisman, L.C.; Rossjohn, J.; Robertson, S.A.; et al. Interferon-ε Protects the Female Reproductive Tract from Viral and Bacterial Infection. *Science* **2013**, *339*, 1088–1092. [CrossRef]

37. Toth, B.; Haufe, T.; Scholz, C.; Kuhn, C.; Friese, K.; Karamouti, M.; Makrigiannakis, A.; Jeschke, U. Placental Interleukin-15 Expression in Recurrent Miscarriage. *Am. J. Reprod. Immunol.* **2010**, *64*, 402–410. [CrossRef] [PubMed]
38. Agarwal, R.; Loganath, A.; Roy, A.C.; Wong, Y.C.; Ng, S.C. Expression Profiles of Interleukin-15 in Early and Late Gestational Human Placenta and in Pre-Eclamptic Placenta. *Mol. Hum. Reprod.* **2001**, *7*, 97–101. [CrossRef]
39. Pavlov, O.V.; Selutin, A.V.; Pavlova, O.M.; Selkov, S.A. Two Patterns of Cytokine Production by Placental Macrophages. *Placenta* **2020**, *91*, 1–10. [CrossRef]
40. Barber, E.M.; Pollard, J.W. The Uterine NK Cell Population Requires IL-15 but These Cells Are Not Required for Pregnancy nor the Resolution of a Listeria Monocytogenes Infection. *J. Immunol.* **2003**, *171*, 37–46. [CrossRef]
41. Ashkar, A.A.; Black, G.P.; Wei, Q.; He, H.; Liang, L.; Head, J.R.; Croy, B.A. Assessment of Requirements for IL-15 and IFN Regulatory Factors in Uterine NK Cell Differentiation and Function During Pregnancy. *J. Immunol.* **2003**, *171*, 2937–2944. [CrossRef] [PubMed]
42. Lee, A.J.; Kandiah, N.; Karimi, K.; Clark, D.A.; Ashkar, A.A. Interleukin-15 Is Required for Maximal Lipopolysaccharide-Induced Abortion. *J. Leukoc. Biol.* **2013**, *93*, 905–912. [CrossRef] [PubMed]
43. Sun, J.C.; Ma, A.; Lanier, L.L. Cutting Edge: IL-15-Independent NK Cell Response to Mouse Cytomegalovirus Infection. *J. Immunol.* **2009**, *183*, 2911–2914. [CrossRef] [PubMed]
44. Ohs, I.; van den Broek, M.; Nussbaum, K.; Münz, C.; Arnold, S.J.; Quezada, S.A.; Tugues, S.; Becher, B. Interleukin-12 Bypasses Common Gamma-Chain Signalling in Emergency Natural Killer Cell Lymphopoiesis. *Nat. Commun.* **2016**, *7*, 13708. [CrossRef]
45. Boulenouar, S.; Doisne, J.-M.; Sferruzzi-Perri, A.; Gaynor, L.M.; Kieckbusch, J.; Balmas, E.; Yung, H.W.; Javadzadeh, S.; Volmer, L.; Hawkes, D.A.; et al. The Residual Innate Lymphoid Cells in NFIL3-Deficient Mice Support Suboptimal Maternal Adaptations to Pregnancy. *Front. Immunol.* **2016**, *7*, 43. [CrossRef]
46. Fu, B.; Zhou, Y.; Ni, X.; Tong, X.; Xu, X.; Dong, Z.; Sun, R.; Tian, Z.; Wei, H. Natural Killer Cells Promote Fetal Development through the Secretion of Growth-Promoting Factors. *Immunity* **2017**, *47*, 1100–1113. [CrossRef]
47. Fu, B.; Li, X.; Sun, R.; Tong, X.; Ling, B. Natural Killer Cells Promote Immune Tolerance by Regulating Inflammatory TH17 Cells at the Human Maternal–Fetal Interface. *Proc. Natl. Acad. Sci. USA* **2013**, *110*, E231–E240. [CrossRef] [PubMed]
48. Sones, J.L.; Cha, J.; Woods, A.K.; Bartos, A.; Heyward, C.Y.; Lob, H.E.; Isroff, C.E.; Butler, S.D.; Shapiro, S.E.; Dey, S.K.; et al. Decidual Cox2 Inhibition Improves Fetal and Maternal Outcomes in a Preeclampsia-like Mouse Model. *JCI Insight* **2016**, *1*, e75351. [CrossRef]
49. Dokras, A.; Hoffmann, D.S.; Eastvold, J.S.; Kienzle, M.F.; Gruman, L.M.; Kirby, P.A.; Weiss, R.M.; Davisson, R.L. Severe Feto-Placental Abnormalities Precede the Onset of Hypertension and Proteinuria in a Mouse Model of Preeclampsia. *Biol. Reprod.* **2006**, *75*, 899–907. [CrossRef] [PubMed]
50. Davisson, R.L.; Hoffmann, D.S.; Butz, G.M.; Aldape, G.; Schlager, G.; Merrill, D.C.; Sethi, S.; Weiss, R.M.; Bates, J.N. Discovery of a Spontaneous Genetic Mouse Model of Preeclampsia. *Hypertension* **2002**, *39*, 337–342. [CrossRef]
51. Lebrec, H.; Horner, M.J.; Gorski, K.S.; Tsuji, W.; Xia, D.; Pan, W.J.; Means, G.; Pietz, G.; Li, N.; Retter, M.; et al. Homeostasis of Human NK Cells Is Not IL-15 Dependent. *J. Immunol.* **2013**, *191*, 5551–5558. [CrossRef]
52. Busque, S.; Leventhal, J.; Brennan, D.C.; Steinberg, S.; Klintmalm, G.; Shah, T.; Mulgaonkar, S.; Bromberg, J.S.; Vincenti, F.; Hariharan, S.; et al. Calcineurin-Inhibitor-Free Immunosuppression Based on the JAK Inhibitor CP-690,550: A Pilot Study in De Novo Kidney Allograft Recipients. *Am. J. Transpl.* **2009**, *9*, 1936–1945. [CrossRef]
53. Bachmayer, N.; Hamad, R.R.; Liszka, L.; Bremme, K.; Sverremark-Ekström, E. Aberrant Uterine Natural Killer (NK)-Cell Expression and Altered Placental and Serum Levels of the NK-Cell Promoting Cytokine Interleukin-12 in Pre-Eclampsia. *Am. J. Reprod. Immunol.* **2006**, *56*, 292–301. [CrossRef]
54. Rautela, J.; Huntington, N.D. IL-15 Signaling in NK Cell Cancer Immunotherapy. *Curr. Opin. Immunol.* **2017**, *44*, 1–6. [CrossRef]
55. Gotthardt, D.; Trifinopoulos, J.; Sexl, V.; Putz, E.M. JAK/STAT Cytokine Signaling at the Crossroad of NK Cell Development and Maturation. *Front. Immunol.* **2019**, *10*, 2590. [CrossRef] [PubMed]
56. Pfefferle, A.; Jacobs, B.; Haroun-Izquierdo, A.; Kveberg, L.; Sohlberg, E.; Malmberg, K.-J. Deciphering Natural Killer Cell Homeostasis. *Front. Immunol.* **2020**, *11*, 812. [CrossRef] [PubMed]
57. Felices, M.; Lenvik, A.J.; McElmurry, R.; Chu, S.; Hinderlie, P.; Bendzick, L.; Geller, M.A.; Tolar, J.; Blazar, B.R.; Miller, J.S. Continuous Treatment with IL-15 Exhausts Human NK Cells via a Metabolic Defect. *JCI Insight* **2018**, *3*, e96219. [CrossRef] [PubMed]
58. Mishra, H.K.; Dixon, K.J.; Pore, N.; Felices, M.; Miller, J.S.; Walcheck, B. Activation of ADAM17 by IL-15 Limits Human NK Cell Proliferation. *Front. Immunol.* **2021**, *12*, 711621. [CrossRef] [PubMed]
59. Hanna, J.; Goldman-Wohl, D.; Hamani, Y.; Avraham, I.; Greenfield, C.; Natanson-Yaron, S.; Prus, D.; Cohen-Daniel, L.; Arnon, T.I.; Manaster, I.; et al. Decidual NK Cells Regulate Key Developmental Processes at the Human Fetal-Maternal Interface. *Nat. Med.* **2006**, *12*, 1065–1074. [CrossRef] [PubMed]
60. Gaynor, L.M.; Colucci, F. Uterine Natural Killer Cells: Functional Distinctions and Influence on Pregnancy in Humans and Mice. *Front. Immunol.* **2017**, *8*, 467. [CrossRef]
61. Wang, F.; Qualls, A.E.; Marques-Fernandez, L.; Colucci, F. Biology and Pathology of the Uterine Microenvironment and Its Natural Killer Cells. *Cell. Mol. Immunol.* **2021**, *18*, 2101–2113. [CrossRef]

62. Sliz, A.; Locker, K.C.S.; Lampe, K.; Godarova, A.; Plas, D.R.; Janssen, E.M.; Jones, H.; Herr, A.B.; Hoebe, K. Gab3 Is Required for IL-2- and IL-15-Induced NK Cell Expansion and Limits Trophoblast Invasion during Pregnancy. *Sci. Immunol.* **2019**, *4*, eaav3866. [CrossRef]
63. Strunz, B.; Bister, J.; Jönsson, H.; Filipovic, I.; Crona-Guterstam, Y.; Kvedaraite, E.; Sleiers, N.; Dumitrescu, B.; Brännström, M.; Lentini, A.; et al. Continuous Human Uterine NK Cell Differentiation in Response to Endometrial Regeneration and Pregnancy. *Sci. Immunol.* **2021**, *6*, eabb7800. [CrossRef]
64. Gordon, S.M.; Nishiguchi, M.A.; Chase, J.M.; Mani, S.; Mainigi, M.A.; Behrens, E.M. IFNs Drive Development of Novel IL-15-Responsive Macrophages. *J. Immunol.* **2020**, *205*, 1113–1124. [CrossRef]
65. Hamai, Y.; Fujii, T.; Yamashita, T.; Miki, A.; Hyodo, H.; Kozuma, S.; Geraghty, D.E.; Taketani, Y. The Expression of Human Leukocyte Antigen-G on Trophoblasts Abolishes the Growth-Suppressing Effect of Interleukin-2 towards Them. *Am. J. Reprod. Immunol.* **1999**, *41*, 153–158. [CrossRef] [PubMed]
66. Zygmunt, M.; Hahn, D.; Kiesenbauer, N.; Münstedt, K.; Lang, U. Invasion of Cytotrophoblastic (JEG-3) Cells Is Up-Regulated by Interleukin-15 In Vitro. *Am. J. Reprod. Immunol.* **1998**, *40*, 326–331. [CrossRef]
67. Apps, R.; Sharkey, A.; Gardner, L.; Male, V.; Trotter, M.; Miller, N.; North, R.; Founds, S.; Moffett, A. Genome-Wide Expression Profile of First Trimester Villous and Extravillous Human Trophoblast Cells. *Placenta* **2011**, *32*, 33–43. [CrossRef] [PubMed]
68. Ferrag, F.; Pezet, A.; Chiarenza, A.; Buteau, H.; Nelson, B.H.; Goffin, V.; Kelly, P.A. Homodimerization of IL-2 Receptor β Chain Is Necessary and Sufficient to Activate Jak2 and Dowstream Signaling Pathways. *FEBS Lett.* **1998**, *421*, 32–36. [CrossRef]
69. Pillet, A.-H.; Juffroy, O.; Mazard-Pasquier, V.; Moreau, J.-L.; Gesbert, F.; Chastagner, P.; Colle, J.-H.; Thèze, J.; Rose, T. Human IL-Rbeta Chains Form IL-2 Binding Homodimers. *Eur. Cytokine Netw.* **2008**, *19*, 49–59. [CrossRef]
70. Cohen, C.J.; Rebollo, R.; Babovic, S.; Dai, E.L.; Robinson, W.P.; Mager, D.L. Placenta-Specific Expression of the Interleukin-2 (IL-2) Receptor β Subunit from an Endogenous Retroviral Promoter. *J. Biol. Chem.* **2011**, *286*, 35543–35552. [CrossRef] [PubMed]
71. De Oca, P.M.B.; Malardé, V.; Proust, R.; Dautry-Varsat, A.; Gesbert, F. Ectodomain Shedding of Interleukin-2 Receptor β and Generation of an Intracellular Functional Fragment*. *J. Biol. Chem.* **2010**, *285*, 22050–22058. [CrossRef] [PubMed]
72. Waldmann, T.A.; Dubois, S.; Miljkovic, M.D.; Conlon, K.C. IL-15 in the Combination Immunotherapy of Cancer. *Front. Immunol.* **2020**, *11*, 868. [CrossRef] [PubMed]

International Journal of
*Molecular Sciences*

# The Road to Low-Dose Aspirin Therapy for the Prevention of Preeclampsia Began with the Placenta

Scott W. Walsh * and Jerome F. Strauss III 

Department of Obstetrics and Gynecology, Virginia Commonwealth University, Richmond, VA 23298, USA; jerome.strauss@vcuhealth.org
* Correspondence: scott.walsh@vcuhealth.org

**Abstract:** The road to low-dose aspirin therapy for the prevention of preeclampsia began in the 1980s with the discovery that there was increased thromboxane and decreased prostacyclin production in placentas of preeclamptic women. At the time, low-dose aspirin therapy was being used to prevent recurrent myocardial infarction and other thrombotic events based on its ability to selectively inhibit thromboxane synthesis without affecting prostacyclin synthesis. With the discovery that thromboxane was increased in preeclamptic women, it was reasonable to evaluate whether low-dose aspirin would be effective for preeclampsia prevention. The first clinical trials were very promising, but then two large multi-center trials dampened enthusiasm until meta-analysis studies showed aspirin was effective, but with caveats. Low-dose aspirin was most effective when started <16 weeks of gestation and at doses >100 mg/day. It was effective in reducing preterm preeclampsia, but not term preeclampsia, and patient compliance and patient weight were important variables. Despite the effectiveness of low-dose aspirin therapy in correcting the placental imbalance between thromboxane and prostacyclin and reducing oxidative stress, some aspirin-treated women still develop preeclampsia. Alterations in placental sphingolipids and hydroxyeicosatetraenoic acids not affected by aspirin, but with biologic actions that could cause preeclampsia, may explain treatment failures. Consideration should be given to aspirin's effect on neutrophils and pregnancy-specific expression of protease-activated receptor 1, as well as additional mechanisms of action to prevent preeclampsia.

**Keywords:** low-dose aspirin; preeclampsia; placenta; eicosanoids; sphingolipids; thromboxane; prostacyclin; isoprostanes; neutrophils; protease-activated receptor 1

**Citation:** Walsh, S.W.; Strauss, J.F., III The Road to Low-Dose Aspirin Therapy for the Prevention of Preeclampsia Began with the Placenta. *Int. J. Mol. Sci.* **2021**, *22*, 6985. https://doi.org/10.3390/ijms22136985

Academic Editors: Hiten D. Mistry and Eun Lee

Received: 31 May 2021
Accepted: 23 June 2021
Published: 29 June 2021

**Publisher's Note:** MDPI stays neutral with regard to jurisdictional claims in published maps and institutional affiliations.



## 1. Introduction

The rationale for low-dose aspirin therapy began in the 1970s with the discovery of thromboxane and prostacyclin [1,2]. Thromboxane is a potent vasoconstrictor and platelet aggregating agent, whereas prostacyclin is a potent vasodilator and inhibitor of platelet aggregation. Both are synthesized from arachidonic acid by action of cyclooxygenase to generate prostaglandin H2, which is then converted by thromboxane synthase to thromboxane or by prostacyclin synthase to prostacyclin.

In the 1980s, low-dose aspirin was being used to prevent recurrent myocardial infarction and other thrombotic events based on its ability to selectively inhibit thromboxane synthesis without affecting prostacyclin synthesis [3–6]. The reason this is possible is because the synthesis of thromboxane and prostacyclin is compartmentalized in different cell types. In the systemic circulation, thromboxane is produced by platelets. Platelets do not have nuclei and so cannot regenerate cyclooxygenase when it is inhibited. Therefore, the synthesis of thromboxane is inhibited for the life span of the platelets. Prostacyclin is produced by endothelial cells. Endothelial cells do have nuclei and can regenerate cyclooxygenase, so prostacyclin production is minimally affected by low-dose aspirin.

## 2. Low-Dose Aspirin for the Prevention of Preeclampsia

Preeclampsia only occurs in the presence of the placenta or placental tissue. Once the placenta is delivered, symptoms clear. Therefore, the placenta is key to understanding preeclampsia, but treatment must correct placental, as well as maternal, abnormalities. In the early 1980s, the placental imbalance between thromboxane and prostacyclin was discovered. The first reports described a decrease in prostacyclin production. Several groups reported a deficiency in prostacyclin in umbilical arteries, uterine vessels, and placental veins in women with preeclampsia [7–9]. In 1985, we demonstrated that the reduction in placental prostacyclin was associated with a significant increase in placental production of thromboxane (Figure 1). Normal placentas produced equal amounts of thromboxane and prostacyclin, but in preeclampsia the placenta produced 7 times as much thromboxane as prostacyclin [10]. Other studies later confirmed increased placental production of thromboxane in preeclampsia [11–16], with the increase linked to increased phospholipase $A_2$ [15], increased cyclooxygenase-2 [16], and increased thromboxane synthase [14] in trophoblast cells.

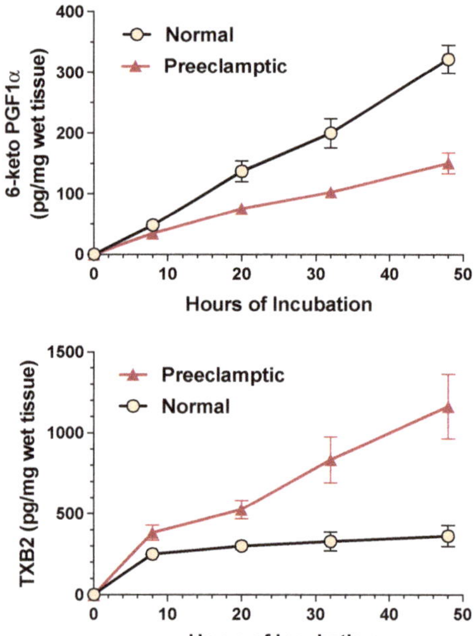


**Figure 1.** Production of prostacyclin and thromboxane in normal and preeclamptic placentas.

With the discovery that there was an imbalance between thromboxane and prostacyclin in preeclampsia, it was reasonable to evaluate whether low-dose aspirin would be effective for preeclampsia prevention. The first clinical trial was published in 1986 by Wallenburg et al. [17]. It was a randomized, placebo-controlled, double-blind trial using 60 mg/day of aspirin. Forty-six normotensive women at 28 weeks' gestation were judged to be at risk for preeclampsia by increased blood pressure response to infused angiotensin II. Twelve of 23 women taking placebo developed preeclampsia, whereas only 2 of 21 women on aspirin developed preeclampsia. The incidence of preeclampsia was decreased 83% by low-dose aspirin (Figure 2). The investigators concluded that low-dose aspirin may correct the thromboxane/prostacyclin imbalance.

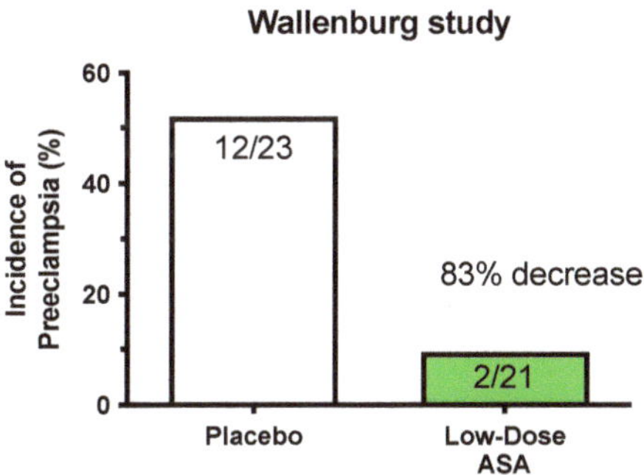


**Figure 2.** Significant reduction in preeclampsia.

A plethora of clinical trials followed, reporting varying degrees of effectiveness of aspirin treatment. Two large multicenter intent-to-treat studies were conducted in nulliparous pregnant women given 60 mg/day of aspirin by the NICHD Maternal-Fetal Medicine Unit Network and the Collaborative Low-dose Aspirin Study in Pregnancy (CLASP) trials [18,19]. Only modest decreases in the incidence of preeclampsia were found (Figure 3). The MFM Unit Network study reported no improvement in perinatal morbidity and an increased risk of placental abruption. Interest in low-dose aspirin declined after the MFM Network Unit and CLASP studies due to concerns about placental abruption and small beneficial effect of aspirin.

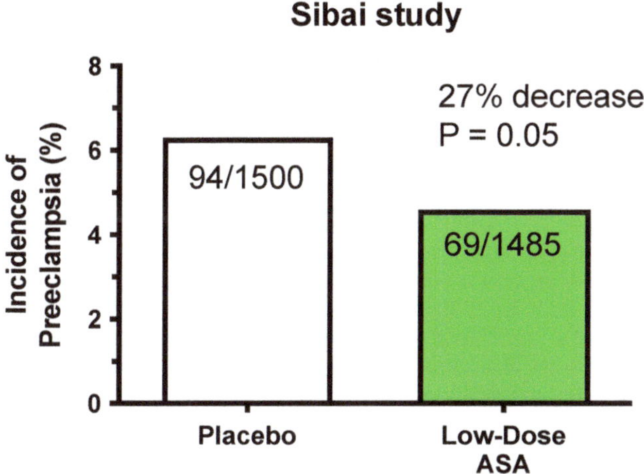


**Figure 3.** Modest reduction in preeclampsia.

However, there were problems with these studies. Regarding placental abruption, only one MFM Network Unit reported this, and abruption was found only on pathologic examination. None were clinically significant, and no other studies previous or since have found an increase in placental abruption due to low-dose aspirin therapy [20]. Another problem was that both the MFM Network and CLASP studies recommended patients use acetaminophen for pain relief. Acetaminophen selectively inhibits prostacyclin without affecting thromboxane [21,22], so the effect of low-dose aspirin to correct the thromboxane/prostacyclin imbalance was compromised. Another major problem was these were intent-to-treat studies. Compliance with low-dose aspirin was not taken into consideration [23]. No drug will work if the patient does not take it.

Hauth et al. reanalyzed the MFM Network data based on compliance [24,25]. They found that women who were more than 75% compliant in taking their aspirin had a significant decrease in the incidence of preeclampsia, from 5.7% to 2.7%, as well as significant decreases in the incidence of low birth weight, preterm birth, and adverse pregnancy outcomes (Figure 4). Unfortunately, these data were only published in abstract form and did not gain recognition.

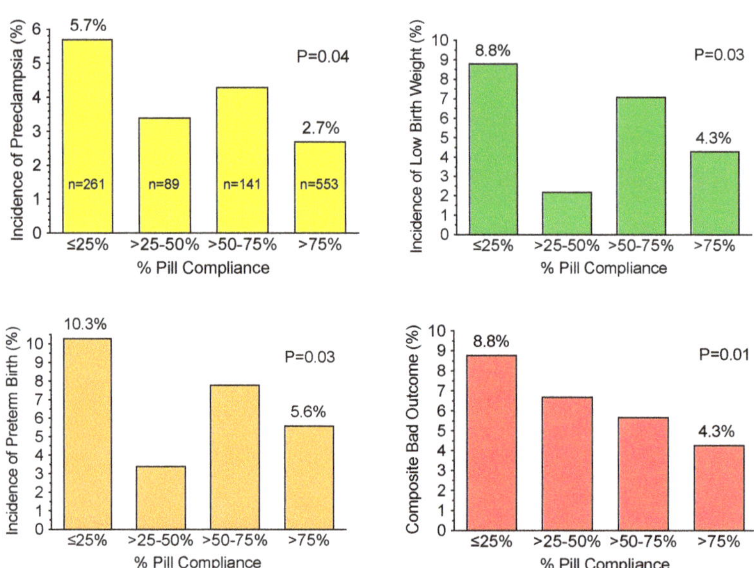

**Figure 4.** Importance of compliance for low-dose aspirin therapy.

In 2007, the first meta-analysis of low-dose aspirin trials was published by Askie et al., who found that in almost all studies low-dose aspirin reduced the incidence of preeclampsia [26]. Additional meta-analysis studies followed, reinforcing the effectiveness of aspirin. Bujold et al. found that aspirin was more effective when started before 16 weeks [27]. Roberge et al. reported that low-dose aspirin was effective in preventing preterm preeclampsia, but not term preeclampsia [28,29]. These investigators also considered the dose of aspirin. They found that studies that used a dose of aspirin $\geq$100 mg were more effective in reducing preeclampsia than studies that used a dose <100 mg [29], and Seidler et al. reported a dose response effect for aspirin when comparing studies using $\leq$81 mg/day to those using >81 mg and up to 150 mg/day [30]. Another study reported that aspirin delays the development of preeclampsia, suggesting this may partly explain why aspirin is more effective in preventing preterm preeclampsia than term preeclampsia because women who would have developed preterm preeclampsia had symptoms delayed to term [31]. The influence of obesity is another factor to consider. A dose of 60 mg/day may have been sufficient in the 1980s when the first clinical trials were started, but since then the United States and other countries have experienced an obesity epidemic. Most study subjects are now overweight or obese, which may explain why meta-analysis studies find that higher doses of aspirin are more effective [29,30,32,33].

Overall, the meta-analysis studies demonstrated that low-dose aspirin not only decreases the incidence of preeclampsia, but also preterm birth < 37 weeks, perinatal death, IUGR, and pregnancies with serious adverse outcomes. In 2013 and 2018, the American College of Obstetrics and Gynecology recommended low-dose aspirin therapy for women at risk of preeclampsia, and it is now the standard of care [34–36].

Consideration should be given to the possibility that the effectiveness of low-dose aspirin could be improved by supplementation with L-arginine, the substrate for nitric oxide synthase. Nitric oxide, like prostacyclin, is a potent vasodilator, so supplementation to increase its production would be beneficial. Supplementation with L-arginine significantly reduced the incidence of preeclampsia in a population at high risk for preeclampsia [37], and a recent study showed favorable effects of L-arginine supplementation in conjunction with low-dose aspirin to improve perinatal outcomes, blood pressure values, and uterine pulsatile index [38].

Another consideration is the finding that low-dose aspirin is most effective when started before 16 weeks gestation. This raises the importance of identifying accurate predictive biomarkers for preeclampsia risk to be used in conjunction with maternal characteristics and medical history, so at-risk women can be identified early in their pregnancy and immediately put on low-dose aspirin therapy.

## 3. Does Low-Dose Aspirin Affect the Placenta?

The actions of low-dose aspirin are generally attributed to selective inhibition of maternal platelet thromboxane; however, beneficial effects must extend to the placenta, which is a major source of eicosanoids. Indeed, preeclampsia only occurs in the presence of placental tissue, and the preeclamptic placenta is characterized by increased thromboxane, decreased prostacyclin, and oxidative stress. Does low-dose aspirin affect the placenta to correct the thromboxane/prostacyclin imbalance and oxidative stress?

As part of the NICHD Human Placental Project, we undertook a comprehensive evaluation of placental lipids in women with normal pregnancy (NP) and women at risk for preeclampsia who were prescribed aspirin [39]. We found the placenta is a rich source of eicosanoids. We measured 30 eicosanoids in numerous different classes of cyclooxygenase and non-cyclooxygenase metabolites. Ten of these were abnormal in women with severe preterm preeclampsia (SPE). Interestingly, thromboxane (TXB2) was not increased, and prostacyclin (6-keto PGF1a) was not decreased (Figure 5), so the imbalance was not present. However, prostaglandins PGE and PGF were decreased, indicating maternal ingestion of aspirin did affect placental cyclooxygenase. These findings suggest low-dose aspirin therapy corrects the thromboxane/prostacyclin imbalance in the placenta.

Correction of the placental imbalance is possible because thromboxane and prostacyclin are compartmentalized within the placenta (Figure 6). Thromboxane is produced by the trophoblast cells on the maternal side of the placenta, whereas prostacyclin is produced by the placental vasculature on the fetal side [40–42]. This allows for selective inhibition of thromboxane because as aspirin enters the maternal intervillous space and starts to cross the placenta, its concentrations are highest in the trophoblast cells to selectively inhibit cyclooxygenase associated with thromboxane production. As aspirin crosses the placenta, its concentration gradually declines according to Fick's second law of diffusion, sparing prostacyclin production by the endothelial cells of the placental vasculature. Only 34% of aspirin from the maternal side crosses to the fetal side [43]. In vitro studies demonstrated that low-dose aspirin preferentially inhibits placental thromboxane while sparing prostacyclin [43–45].

We also found evidence that maternal ingestion of aspirin attenuated placental oxidative stress. Two of the most abundant isoprostanes, 8-isoprostane (8-iso PGF2a) and 5-isoprostane (5-iPF2a), which are significantly elevated in placentas of preeclamptic women [46,47], were not elevated in our study of women who developed preeclampsia while on aspirin therapy (Figure 5) [39]. Isoprostanes are accurate markers of endogenous lipid peroxidation. They are prostaglandin-like products formed in vivo by free-radical catalyzed non-enzymatic peroxidation of arachidonic acid [48–50]. The finding that two of the most abundant isoprostanes were not elevated in preeclampsia is significant because the placental imbalance between thromboxane and prostacyclin is driven by oxidative stress [25,51]. This may explain why the imbalance was not present.

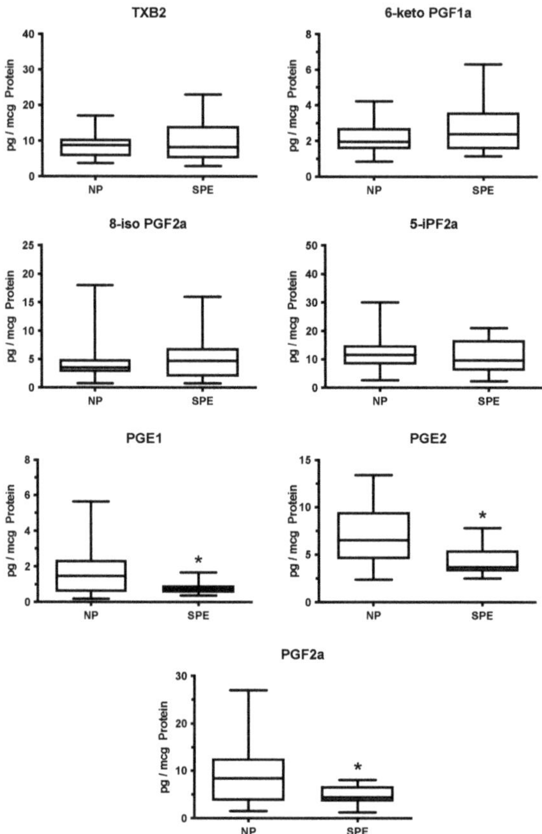

**Figure 5.** Placental production of cyclooxygenase metabolites and isoprostanes in women with severe preeclampsia receiving low-dose aspirin, * $p < 0.05$.

The fact that placental isoprostanes did not increase in women taking low-dose aspirin could be due to an indirect effect of cyclooxygenase inhibition. Cyclooxygenase generates reactive oxygen species (ROS) [52], so inhibition of cyclooxygenase could have removed the source of free radicals to generate isoprostanes from arachidonic acid (Figure 7). This idea is consistent with our previous reports that low-dose aspirin inhibits lipid peroxides along with thromboxane in the maternal circulation and in the placenta [43–45,53]. This action of aspirin could explain the correction of the thromboxane/prostacyclin imbalance because aspirin removed the driving force.

Despite aspirin therapy, some women develop preeclampsia. Low-dose aspirin reduces the risk, but it does not prevent the disease in all women. Significant elevations in levels of placental hydroxyeicosatetraenoic acids (HETEs) and sphingolipids with biologic actions that could cause preeclampsia could explain why.

HETEs are lipoxygenase metabolites of arachidonic acid, and they are, therefore, not affected by low-dose aspirin. The placenta produced four HETEs, two of which, 15-HETE and 20-HETE, were significantly elevated in women who delivered preterm with severe preeclampsia (Figure 8) [39]. Both of these HETEs cause inflammation [54–61], and placental pathologic features of preterm preeclampsia are consistent with chronic inflammation [62]. In addition, 20-HETE promotes hypertension, vasoconstriction, and vascular dysfunction [59–61]. Intrauterine production of 20-HETE by the placenta could contribute to reduced uterine blood flow and placental vasoconstriction in preeclampsia, and placental release into the maternal circulation could contribute to maternal hypertension. In this regard, 20-HETE enhances vascular reactivity to angiotensin II.

## Compartmentalization Allows Selective Inhibition

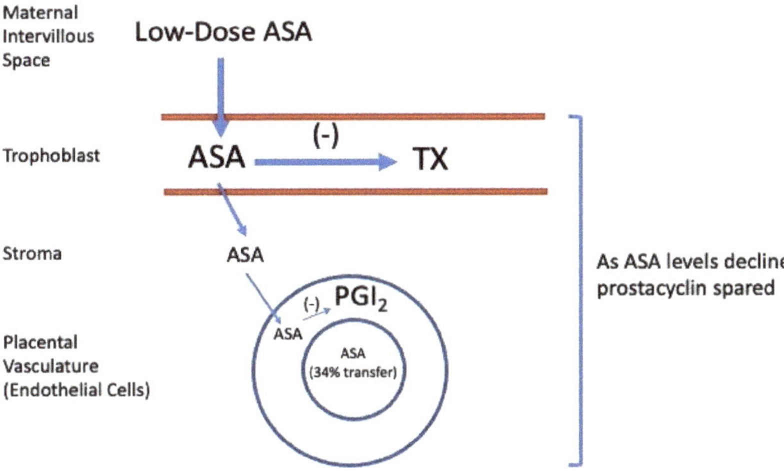


**Figure 6.** Mechanism for selective inhibition of placental thromboxane (TX) by low-dose aspirin (ASA). Prostacyclin (PGI2).

## Inhibition of Isoprostanes

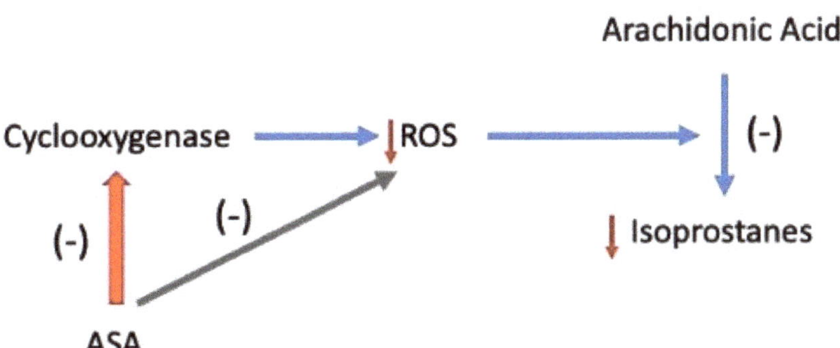


**Figure 7.** Mechanism for inhibition of isoprostanes by aspirin (ASA).

Sphingolipids are major constituents of the cell membrane and are involved in cell signaling (Figure 9). They are long-chain fatty acids of various carbon chain lengths that contain a backbone of sphingosine. Sphingolipids include sphingomyelin, ceramide, sphingosine, and sphingosine-1-phosphate. They are involved in inflammatory signaling pathways and implicated in cardiovascular disease [63–68]. They are not cyclooxygenase metabolites, and so, are not affected by aspirin. The placenta produced 42 sphingolipids, 5 of which were abnormal in women with severe preeclampsia [39]. All sphingolipids that were abnormal were significantly increased compared to normal pregnancy, including major C:18 forms. D-e-$C_{18:0}$ ceramide, D-e-$C_{18:0}$ sphingomyelin, D-e-sphingosine-1-phosphate (S1P), and D-e-sphinganine-1-phosphate were increased 2-fold to over 4-fold in placentas of women with severe preeclampsia compared to placentas of women with a normal pregnancy (Figure 10).

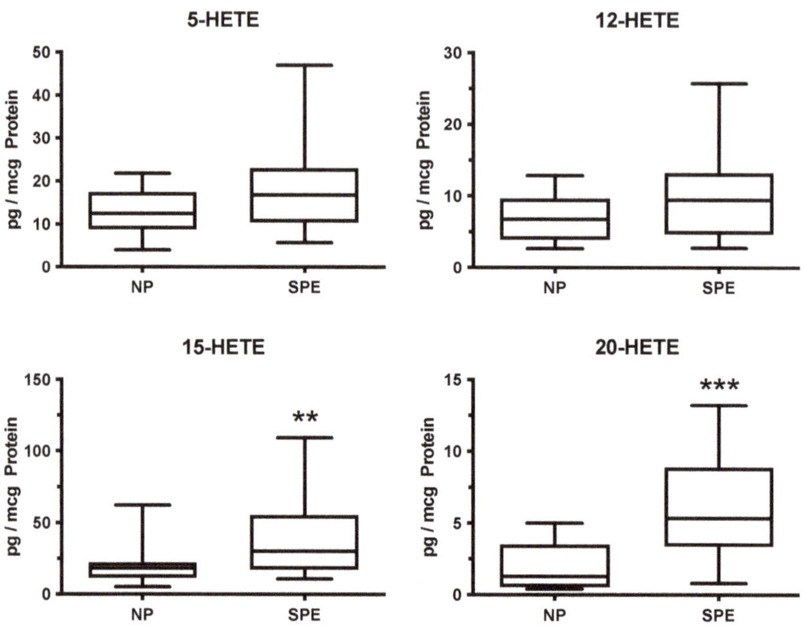


**Figure 8.** Increases in HETEs related to the development of preeclampsia, ** $p < 0.01$, *** $p < 0.001$.

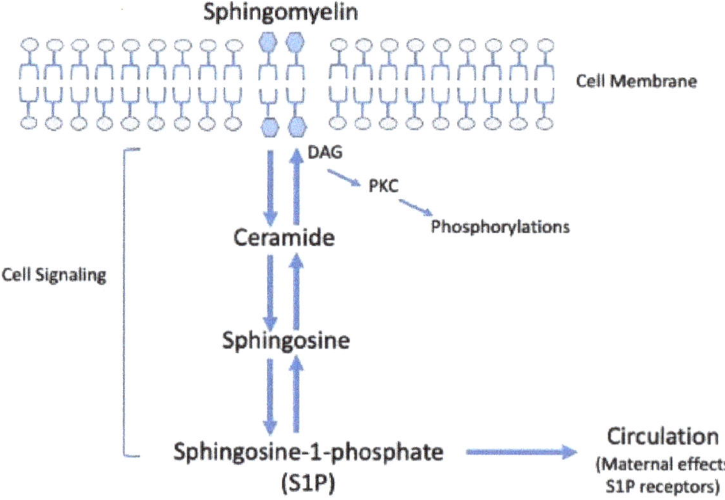


**Figure 9.** Interconversion of sphingolipids. (DAG, diacylglycerol; PKC, protein kinase C).

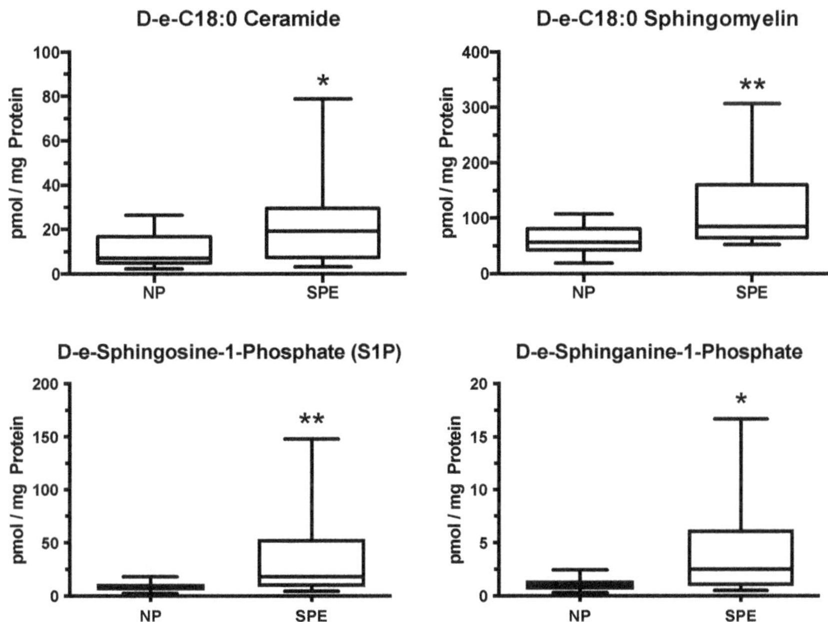



**Figure 10.** Increases in sphingolipids related to the development of preeclampsia, * $p < 0.05$, ** $p < 0.01$.

Abnormal placental sphingolipid production may contribute to several features of preeclampsia. For example, ceramide induces apoptosis, which may contribute to placental cell death in preeclampsia [69], and S1P inhibits extravillous trophoblast migration [70], and so may contribute to failure of extravillous trophoblasts to effectively remodel the spiral arteries in preeclampsia. S1P is also involved in inflammation, vascular permeability, and the immune response. S1P is an intracellular second messenger, but it is also a blood-borne lipid mediator, and as such, has extracellular actions by binding to S1P receptors. Placental secretion of S1P could be responsible for abnormalities in the maternal circulation. Very little information is available about sphingolipids in pregnancy, but maternal levels of ceramide and S1P have been reported to be elevated in preeclampsia and linked to a placental source [71,72].

### 4. Other Considerations Involving Neutrophils and Pregnancy-Specific Expression of Protease-Activated Receptor 1

Normal pregnancy is characterized by leukocytosis caused by proliferation of neutrophils in the 2nd and 3rd trimesters. The number of neutrophils increases 2.5-fold by 30 weeks of gestation in normal pregnancy [73], and the number increases further in preeclampsia [74]. Neutrophils are usually thought of as part of the innate immune system and the first line of defense against infection. A role for neutrophils in non-infectious disease has not been widely considered, but accumulating evidence indicates a role for neutrophils in "sterile" inflammatory diseases [75].

For neutrophils to manifest their inflammatory effects, they need to infiltrate tissue, and in women with preeclampsia there is extensive neutrophil infiltration into the maternal systemic blood vessels (Figure 11) [76–79]. In preeclamptic women, 80–90% of vessels in subcutaneous and omental fat are infiltrated and, although all classes of leukocytes are activated [80,81], vascular infiltration is restricted to neutrophils [77,78]. Neutrophil infiltration is associated with a significant increase in inflammatory markers, e.g., interleukin-8 (IL-8), intercellular adhesion molecule-1 (ICAM-1), cyclooxygenase-2 (COX-2), nuclear factor-kappa B (NF-κB), thromboxane synthase (TBXAS1), and myeloperoxidase (MPO) [76,79,82,83]. The finding of neutrophil infiltration provides a basis for a new way of thinking about vascular dysfunction in preeclampsia. It does not discount the potential role of plasma factors but adds a new dimension to the understanding of the underlying mechanisms of the vascular phenotype.

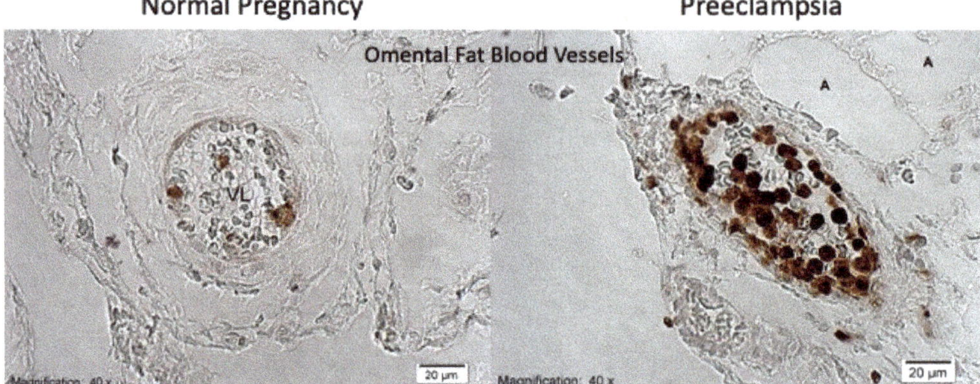


**Figure 11.** Neutrophils (brown) in omental fat arteries.

*4.1. Pregnancy-Specific Expression of PAR-1*

Protease-activated receptor 1 (PAR-1), originally known as thrombin receptor, is activated by serine proteases, such as thrombin, neutrophil elastase, and matrix metalloproteinase-1 (MMP-1) [84–86]. Activation leads to downstream signaling mechanisms that include the RhoA kinase (ROCK) phosphorylation pathway. ROCK is a recognized mediator of enhanced vascular reactivity, and also regulates the shape and movement of cells. There is pregnancy-specific expression of PAR-1. Wang et al. showed that PAR-1 is expressed on neutrophils, but only during pregnancy [87,88]. This suggests that something associated with the placenta is causing the expression of PAR-1 on circulating neutrophils.

Figure 12 shows omental fat vessels of preeclamptic and normal pregnant women immunostained for PAR-1. In preeclampsia, PAR-1 is expressed in endothelial cells (EC), vascular smooth muscle (VSM), and in neutrophils flattened and adherent to the endothelium, infiltrated into the vessel, and present in the lumen of the vessel. In normal pregnancy, weak staining is present in the endothelium and neutrophils in the vessel lumen. There is an 8-fold increase in gene and protein expression of PAR-1 in blood vessels of women with preeclampsia [89].

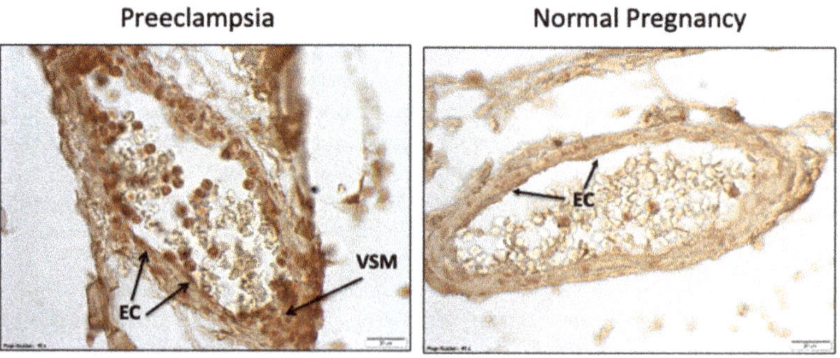


**Figure 12.** Expression of PAR-1 in omental vessels.

*4.2. PAR-1 Mediates Neutrophil Inflammatory Response in Pregnancy*

Proteases, such as MMP-1, neutrophil elastase, and thrombin, are elevated in women with preeclampsia [90]. The expression of PAR-1 on neutrophils is specific to pregnancy, so its activation by elevated proteases in preeclampsia activates an inflammatory mechanism unique to pregnancy. In normal pregnancy, it makes sense that the expression of inflammatory genes would be silenced. A mechanism for this could be DNA methylation to mask binding sites for inflammatory transcription factors, such a NF-κB. However, if the methylation marks were erased, it would open these sites, possibly leading to increased gene expression. One mechanism for erasing methylation marks involves the recently discovered TET proteins (ten-eleven translocation proteins, aka tet methylcytosine dioxygenases). TET proteins regulate gene expression by enzymatic de-methylation of DNA. They catalyze the conversion of 5-methycytosine (5-mC) to 5-hydroxy-methylcytosine (5-hmC) [91–94], which is further oxidized and then removed by the DNA base excision repair enzyme, thymine-DNA glycosylase, and replaced with unmodified cytosine [95]. TET enzymes were first discovered in 2009 [93], and little is known about their regulation or role in disease. TET2 is the main TET protein expressed in leukocytes, and its activation has been shown to play an essential role in regulating hematopoietic differentiation, which proceeds in mature cells without cell division normally during emigration from the circulation into tissue [96–98].

*4.3. Proteases Activate Neutrophil TET2 and NF-κB to Mediate Inflammatory Response*

Protease activation of PAR-1 causes translocation of TET2 from the cytosol into the nucleus in neutrophils obtained from pregnant women as evidenced by immunofluorescence and confocal microscopy (Figure 13) [90]. TET2 (green) is localized to the cytosol in control cells of normal pregnant women. Protease treatment with MMP-1 or elastase results in translocation of TET2 into the nucleus (location identified by DAPI blue) in as early as 15 min, which is consistent for proteins containing a nuclear localization signal (NLS). Nuclear translocation of TET2 coincides with activation of NF-κB. Similar to TET2, protease stimulation of pregnancy neutrophils causes translocation of the p65 subunit of NF-kB (red) from the cytosol to the nucleus. Inhibition of PAR-1, as well as inhibition of ROCK, prevents protease-induced translocation of TET2 and p65 into the nucleus (Figure 13). Inhibition of PAR-1 or ROCK also inhibits inflammatory response as measured by the production of IL-8 and TXB2, which are regulated by NF-κB. Protease treatment of neutrophils from normal pregnant women significantly increases IL-8 and TXB2, demonstrating that proteases stimulate inflammatory response, but when cells are pretreated with PAR-1 or ROCK inhibitors, protease-induced increases in IL-8 and TXB2 are prevented.

Expression and activation of neutrophil TET2 are increased in preeclampsia. Immunohistochemical staining reveals significantly more staining in omental vessels of preeclamptic women than in omental vessels of normal pregnant women (Figure 14) [90]. In preeclampsia, almost 90% of vessels stain for TET2 with neutrophils infiltrated into the vessel wall, as compared to only 16% of vessels in normal pregnancy with staining. When neutrophils are present in normal vessels, they are usually in the lumen of the vessel. High magnification images reveal dark staining of the polymorphonuclear nuclei of neutrophils in preeclampsia (Panel D), as opposed to diffuse staining in normal pregnancy (Panel C). Nuclear staining suggests TET2 is active in preeclampsia, and activation involves translocation from the cytosol to the nucleus just as observed for TET2 translocation induced by protease activation of PAR-1. Staining for TET2 in preeclamptic vessels mirrors the staining for PAR-1 with staining present in endothelium and vascular smooth muscle (VSM), as well as in neutrophils [90]. This close relationship between PAR-1 and TET2 likely has important implications for vascular inflammation in preeclampsia.

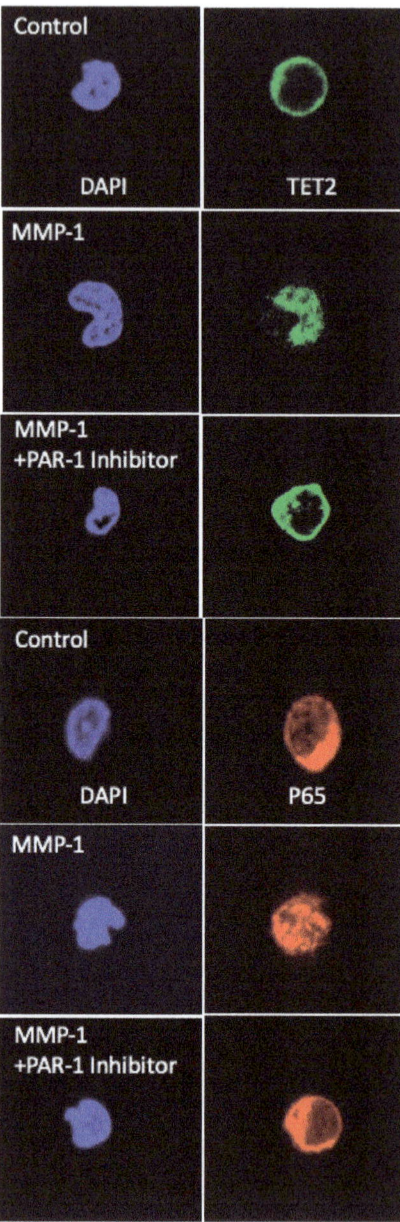


**Figure 13.** Confocal images of TET2 and p65 immuno-fluorescence staining in neutrophils of normal pregnant women. Nuclear localization induced by protease treatment was prevented by inhibition of PAR-1. Images were taken with a Zeiss LSM 700 using x63 lens and then cropped.

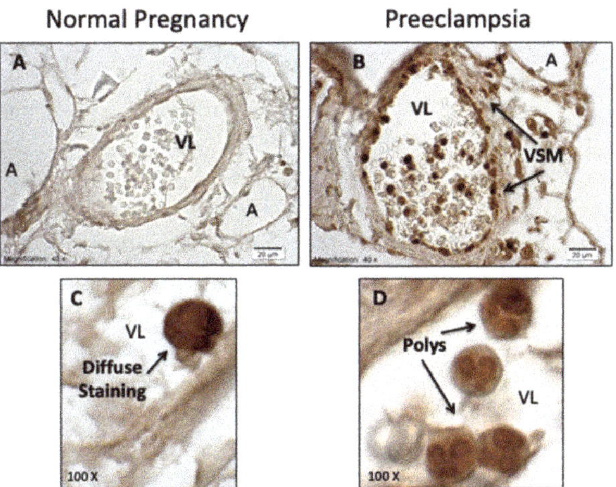



**Figure 14.** TET2 in omental vessels (**A–D**).

Figure 15 summarizes the molecular mechanisms for protease activation of pregnancy neutrophils. In normal pregnancy, circulating proteases are not elevated, TET2 and NF-κB are localized to the cytosol, and inflammatory genes are not expressed. In preeclampsia, circulating proteases are elevated and activate neutrophils due to their pregnancy- specific expression of PAR-1. Activation of PAR-1 results in the movement of TET2 and NF-κB from the cytosol to the nucleus and the expression of inflammatory genes. The PAR-1 pathway involves ROCK phosphorylation because inhibition of either PAR-1 or ROCK blocks the movement of TET2 and NF-κB from the cytosol to the nucleus and the inflammatory response. To summarize, elevated levels of proteases in the maternal circulation of preeclamptic women activate neutrophils due to their pregnancy-specific expression of PAR-1. PAR-1 activates ROCK, which phosphorylates TET2 and NF-κB, causing their translocation from the cytosol to the nucleus. The fact that TET2 translocation into the nucleus coincides with movement of NF-κB implicates epigenetic mechanisms and suggests that TET2 may enzymatically de-methylate DNA, opening up transcription factor binding sites for NF-κB, resulting in the expression of inflammatory genes.

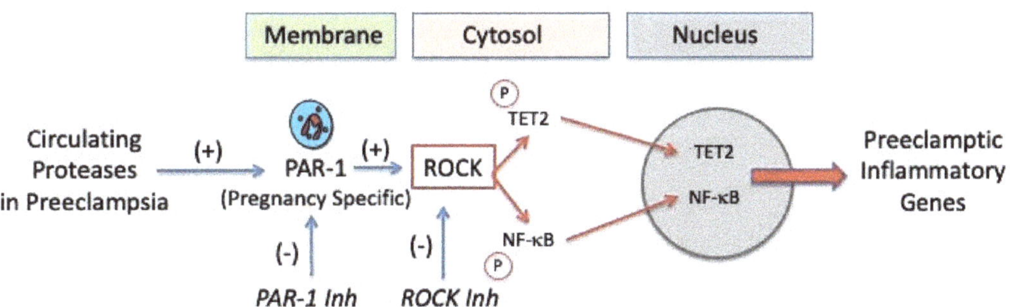



**Figure 15.** Molecular mechanism for protease activation of pregnancy neutrophils.

### 4.4. Expression of PAR-1 in the Placenta

Several studies show PAR-1 is expressed in the placenta [99–102], which is a tissue specific to pregnancy and dysfunctional in preeclampsia. Figure 16 shows staining for PAR-1 in a placental villus. PAR-1 is expressed in the syncytiotrophoblast cells, which are directly bathed by maternal blood. PAR-1 is also present in macrophages of the villous core.

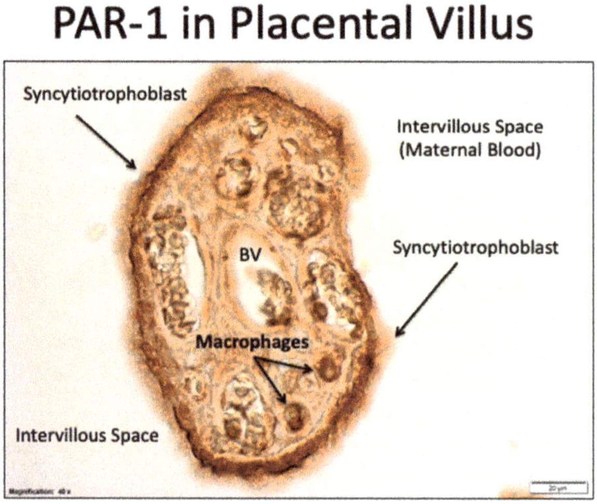

## PAR-1 in Placental Villus



**Figure 16.** Expression of PAR-1 in syncytiotrophoblasts and macrophages in the placenta (dark brown staining).

There is evidence that PAR-1 mediates placental dysfunction in preeclampsia. Because PAR-1 is expressed in the syncytiotrophoblast, elevated levels of proteases in the intervillous space could activate PAR-1, leading to placental dysfunction. For example, protease stimulation of trophoblast PAR-1 causes increased release of the angiogenic factor, sFlt [100,103], by activation of placental NADPH oxidase to generate reactive oxygen species [99]. Activation of NADPH oxidase via PAR-1 could be responsible for placental oxidative stress, which drives the imbalance of increased thromboxane and decreased prostacyclin production [25].

A protease activating mechanism of neutrophil and placental PAR-1 could explain why preeclampsia only occurs in pregnant women, and a protease feed-forward scenario could explain why clinical symptoms progressively worsen. Protease activation of PAR-1 could explain other features of preeclampsia. For example, because neutrophils have a limited life span of about 8 days, their rapid turnover would explain why maternal symptoms clear shortly after delivery because new neutrophils not expressing PAR-1 enter the circulation. Some women develop preeclampsia in the immediate post-partum period. Labor is recognized to be an inflammatory process, and even in normal term labor, there is extensive infiltration of neutrophils into maternal systemic vasculature [104]. Women who develop post-partum preeclampsia might have been on the verge of developing preeclampsia, and then neutrophil infiltration with labor pushed them over the edge.

### 4.5. Central Role for PAR-1 in the Clinical Manifestations of Preeclampsia

Protease activation of PAR-1 may play a central role in the pathology of preeclampsia (Figure 17). Protease activation is involved in the neutrophil TET2 inflammatory response, neutrophil activation, and enhanced vascular reactivity [90,105]. Activation of PAR-1 may explain other pathologic features as well because PAR-1 mediates coagulation abnormalities, platelet aggregation, and thromboxane generation. Protease activation of endothelial PAR-1 activates NF-κB, upregulates cell adhesion molecules (ICAM-1), triggers production of neutrophil chemokines (IL-8), and increases permeability of the endothelium to trigger edema formation [106–112]. PAR-1 may explain the elevation in angiogenic factors because trophoblast and decidual production of sFlt is stimulated by protease activation of

PAR-1 [100,103]. Placental oxidative stress may be explained by protease stimulation of trophoblast PAR-1, which activates NADPH oxidase to generate reactive oxygen species, resulting in the release of sFlt [99]. Activation of NADPH oxidase could also explain the placental imbalance of increased thromboxane and decreased prostacyclin characteristic of preeclampsia because oxidative stress drives this imbalance [25]. The effect of aspirin on PAR-1 signaling should be evaluated. If aspirin interferes with downstream signaling of PAR-1, this would be another action to account for its beneficial effects.

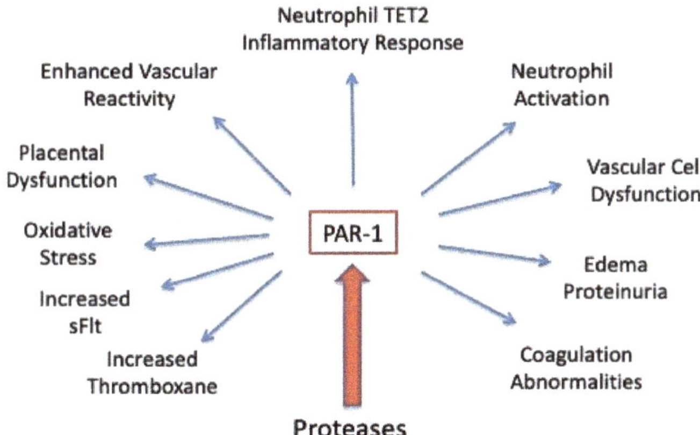


**Figure 17.** Clinical manifestations of preeclampsia that can be explained by protease activation of PAR-1.

### 4.6. Placental Activation of Neutrophils

Although all classes of leukocytes are activated in the circulation of women with preeclampsia [80,81,113–115], only neutrophils extensively infiltrate maternal blood vessels [76–78]. The extensive infiltration of activated neutrophils into blood vessels of women with preeclampsia [76,78,79] could explain systemic vascular inflammation and why multiple organs are affected. The question arises as to how neutrophils are activated. The placenta would seem to be a source for the activator because preeclampsia only occurs in the presence of placental tissue. Lipid peroxides are potent activators of leukocytes [116–118], and the human placenta produces lipid peroxides and secretes them into the maternal circulation [13,42,46,119]. In women with preeclampsia, placental production of lipid peroxides is significantly higher than in women with normal pregnancy [13,42,46]. Therefore, it is plausible that activation occurs as neutrophils circulate through the intervillous space and are exposed to lipid peroxides released by the placenta [51,120,121].

### 4.7. Inhibition of Neutrophils and Treatment of Preeclampsia with Aspirin

Low-dose aspirin is currently standard of care for the prevention of preeclampsia in high-risk populations. Low-dose aspirin selectively inhibits maternal platelet thromboxane production without affecting prostacyclin production and, as shown above, it appears to also selectively inhibit placental thromboxane production, as well as placental oxidative stress. However, maternal platelets and placental trophoblasts may not be the only aspirin targets. Neutrophils may also be a target. The expression of cyclooxygenase-2 is increased in neutrophils of preeclamptic women [79,122], and aspirin inhibits neutrophil production of thromboxane, as well as the generation of reactive oxygen species [117,118]. Neutrophils could be a major source of thromboxane and oxidative stress due to the marked increase in their numbers during pregnancy. Aspirin treatment might also reduce the infiltration of neutrophils into the maternal blood vessels. Future studies are necessary to address the various mechanisms by which low-dose aspirin is able to reduce the incidence of preeclampsia.

Low-dose aspirin is currently being used to prevent preeclampsia in women at risk, but given its effectiveness, consideration should be given to the use of aspirin in treating women with preeclampsia. Aspirin was contraindicated for use in pregnancy due to concern that it might reduce amniotic fluid volume or cause closure of the ductus arteriosus. However, this concern may be unwarranted because only 30% of an aspirin dose crosses from the maternal to the fetal side of the placenta [43], and more importantly, the Collaborative Perinatal Project in the 1970s involving over 40,000 pregnant women and their offspring, over 24,000 of whom took aspirin during their pregnancy, 1500 of whom were heavily exposed, found no harmful effects of aspirin use on the neonates [123].

## 5. Mysterious Beneficial Effects of Low-Dose Aspirin—Is Cyclooxygenase Involved?

The known mechanism of aspirin is to inhibit cyclooxygenase enzymes, the constitutive COX-1 and the inducible COX-2. However, reports are appearing that aspirin also affects non-cyclooxygenase products. For example, placental soluble fms-like tyrosine kinase 1 (sFlt-1) is elevated in the circulation of women with preeclampsia and implicated in preeclampsia pathology [124,125]. sFlit-1 is not a cyclooxygenase product, but low-dose aspirin reduces hypoxia-induced sFlt-1 release by cytotrophoblast cells in vitro [126,127]. Hypoxia causes oxidative stress and the induction of COX-2, so inhibition of sFlt-1 may be related to aspirin's ability to decrease ROS generated by COX-2 (Figure 7).

Aspirin has favorable effects through alterations in phosphoproteins, transcription factors, and microRNAs implicated in placental apoptosis and trophoblast migration [128–131]. Aspirin facilitates trophoblast invasion by regulating a family of microRNAs that inhibit trophoblast invasion [131]. Thus, aspirin may augment extracellular trophoblast remodeling of the spiral arteries, which is deficient in preeclampsia. COX-2 is elevated in the process of apoptosis [129], so aspirin may decrease placental apoptosis by inhibiting COX-2.

Another puzzling effect of low-dose aspirin is on the regulation of placenta-derived exosomes. Exosomes are lipid bilayer nano-vesicles released by many cells and are present in blood [132]. Their lipid makeup reflects their tissue of origin. Placental exosomes can be specifically identified because they contain microRNAs of the chromosome 19 miRNA cluster that are highly and exclusively expressed by the placenta throughout pregnancy [133–136]. The placenta releases exosomes throughout pregnancy into maternal blood and placental exosomes are higher in women with preeclampsia and may contribute to endothelial dysfunction [137]. Aspirin has been shown to inhibit exosome formation and shedding by platelets, erythrocytes, monocytes, and vascular smooth muscle cells, and it has been suggested that low-dose aspirin may have a similar beneficial effect on placental exosome shedding and content during pregnancy [137].

Most of the studies demonstrating beneficial effects of low-dose aspirin on non-cyclooxygenase products were conducted in vitro or with animal models of preeclampsia. It remains to be shown that these effects occur in pregnant women in vivo. However, these studies expose how much there is to learn about how low-dose aspirin achieves its protective effect. Does aspirin affect non-cyclooxygenase pathways, or are cyclooxygenase metabolites involved in regulating these other pathways? Future studies are needed to address these issues.

## 6. Conclusions

In summary, low-dose aspirin therapy for the prevention of preeclampsia began with the discovery of an imbalance of thromboxane and prostacyclin production by placentas of women with preeclampsia. Although the benefits of low-dose aspirin are generally attributed to inhibition of maternal platelet thromboxane, they must extend to the placenta. Maternal low-dose aspirin appears to attenuate placental oxidative stress and correct the thromboxane/prostacyclin imbalance. Abnormalities in eicosanoids and sphingolipids not affected by low-dose aspirin may explain why some aspirin-treated women develop preeclampsia.

Meta-analysis studies provide new considerations for low-dose aspirin therapy beyond those currently recommended by the American College of Obstetrics and Gynecology. These include the following: (1) a higher dose of aspirin of 150 mg/day (or 2 baby aspirin/day) is more effective, (2) aspirin should be started before 16 weeks of gestation, (3) obese women might need a higher dose, (4) low-dose aspirin is most effective in preventing preterm preeclampsia, and (5) compliance is very important and should be emphasized to the patient.

Neutrophils and the pregnancy-specific expression of PAR-1 also play significant roles in preeclampsia. Proteases are elevated in women with preeclampsia and protease activation of PAR-1 on neutrophils and placental trophoblasts can explain major clinical manifestations of preeclampsia. Additional mechanisms of action of aspirin to prevent preeclampsia should be explored, and consideration should be given to using a standard dose of aspirin and possible supplementation with L-arginine for treatment of women with preeclampsia.

**Author Contributions:** S.W.W.—conceptualization, writing—original draft preparation. J.F.S.III—writing—review and editing. All authors have read and agreed to the published version of the manuscript.

**Funding:** This research was funded by Eunice Kennedy Shriver National Institute of Child Health & Human Development (NICHD) grants 5R01 HD088386 (SWW), 1U01 HD087198 (SWW).

**Institutional Review Board Statement:** The study was conducted according to the guidelines of the Declaration of Helsinki, and approved by the Office of Research Subjects Protection of Virginia Commonwealth University (HM20009145, approved 6/10/2021; HM20005160, 10/27/2020).

**Informed Consent Statement:** Informed consent was obtained from all subjects involved in the study.

**Data Availability Statement:** No new data were created or analyzed in this study. Data sharing is not applicable to this article.

**Conflicts of Interest:** The authors declare no conflict of interest.

## References

1. Moncada, S.; Gryglewski, R.J.; Bunting, S.; Vane, J.R. A lipid peroxide inhibits the enzyme in blood vessel microsomes that generates from prostaglandin endoperoxides the substance (prostaglandin X) which prevents platelet aggregation. *Prostaglandins* **1976**, *12*, 715–737. [CrossRef]
2. Moncada, S.; Vane, J.R. Pharmacology and endogenous roles of prostaglandin endoperoxides, thromboxane A2, and prostacyclin. *Pharmacol. Rev.* **1979**, *30*, 293–331.
3. Bunting, S.; Moncada, S.; Vane, J.R. The prostacyclin–thromboxane A2 balance: Pathophysiological and therapeutic implications. *Br. Med. Bull.* **1983**, *39*, 271–276. [CrossRef]
4. Lewis, H.D., Jr.; Davis, J.W.; Archibald, D.G.; Steinke, W.E.; Smitherman, T.C.; Doherty, J.E., 3rd; Schnaper, H.W.; LeWinter, M.M.; Linares, E.; Pouget, J.M.; et al. Protective effects of aspirin against acute myocardial infarction and death in men with unstable angina. Results of a Veterans Administration Cooperative Study. *N. Engl. J. Med.* **1983**, *309*, 396–403. [CrossRef] [PubMed]
5. Marcus, A.J. Aspirin as an Antithrombotic Medication. *N. Engl. J. Med.* **1983**, *309*, 1515–1517. [CrossRef]
6. Salzman, E.W. Aspirin to Prevent Arterial Thrombosis. *N. Engl. J. Med.* **1982**, *307*, 113–115. [CrossRef]
7. Bussolino, F.; Benedetto, C.; Massobrio, M.; Camussi, G. Maternal vascular prostacyclin activity in pre-eclampsia. *Lancet* **1980**, *316*, 702. [CrossRef]
8. Downing, I.; Shepherd, G.L.; Lewis, P.J. Reduced prostacyclin production in pre-eclampsia. *Lancet* **1980**, *316*, 1374. [CrossRef]
9. Remuzzi, G.; Marchesi, D.; Zoja, C.; Muratore, D.; Mecca, G.; Misiani, R.; Rossi, E.; Barbato, M.; Capetta, P.; Donati, M.B.; et al. Reduced umbilical and placental vascular prostacyclin in severe preeclampsia. *Prostaglandins* **1980**, *20*, 105–110. [CrossRef]
10. Walsh, S.W. Preeclampsia: An imbalance in placental prostacyclin and thromboxane production. *Am. J. Obstet. Gynecol.* **1985**, *152*, 335–340. [CrossRef]
11. Cervar, M.; Kainer, F.; Jones, C.J.; Desoye, G. Altered release of endothelin-1,2 and thromboxane B2 from trophoblastic cells in pre-eclampsia. *Eur. J. Clin. Investig.* **1996**, *26*, 30–37. [CrossRef]
12. Ding, Z.Q.; Rowe, J.; Sinosich, M.J.; Saunders, D.M.; Gallery, E.D.M. In-vitro secretion of prostanoids by placental villous cytotrophoblasts in pre-eclampsia. *Placenta* **1996**, *17*, 407–411. [CrossRef]
13. Walsh, S.W.; Wang, Y.; Jesse, R. Placental Production of Lipid Peroxides, Thromboxane, and Prostacyclin in Preeclampsia. *Hypertens. Pregnancy* **1996**, *15*, 101–111. [CrossRef]

14. Woodworth, S.H.; Li, X.; Lei, Z.M.; Rao, C.V.; Yussman, M.A.; Spinnato, J.A., 2nd; Yokoyama, C.; Tanabe, T.; Ullrich, V. Eicosanoid biosynthetic enzymes in placental and decidual tissues from preeclamptic pregnancies: Increased expression of thromboxane-A2 synthase gene. *J. Clin. Endocrinol. Metab.* **1994**, *78*, 1225–1231.

15. Bowen, R.S.; Zhang, Y.; Gu, Y.; Lewis, D.F.; Wang, Y. Increased phospholipase A2 and thromboxane but not prostacyclin production by placental trophoblast cells from normal and preeclamptic pregnancies cultured under hypoxia condition. *Placenta* **2005**, *26*, 402–409. [CrossRef]

16. Johnson, R.D.; Sadovsky, Y.; Graham, C.; Anteby, E.Y.; Polakoski, K.L.; Huang, X.; Nelson, D.M. The Expression and Activity of Prostaglandin H Synthase-2 Is Enhanced in Trophoblast from Women with Preeclampsia. *J. Clin. Endocrinol. Metab.* **1997**, *82*, 3059–3062. [CrossRef]

17. Wallenburg, H.C.S.; Makovitz, J.W.; Dekker, G.A.; Rotmans, P. Low-dose aspirin prevents pregnancy-induced hypertension and pre-eclampsia in angiotensin-sensitive primigravidae. *Lancet* **1986**, *327*, 1–3. [CrossRef]

18. Beroyz, G.; Casale, R.; Farreiros, A.; Palermo, M.; Margulies, M.; Voto, L.; Fabregues, G.; Ramalingam, R.; Davies, T.; Bryce, R. CLASP: A randomised trial of low-dose aspirin for the prevention and treatment of pre-eclampsia among 9364 pregnant women. CLASP (Collaborative Low-dose Aspirin Study in Pregnancy) Collaborative Group. *Lancet* **1994**, *343*, 619–629.

19. Sibai, B.M.; Caritis, S.N.; Thom, E.; Klebanoff, M.; McNellis, D.; Rocco, L.; Paul, R.H.; Romero, R.; Witter, F.; Rosen, M.; et al. Prevention of preeclampsia with low-dose aspirin in healthy, nulliparous pregnant women. The National Institute of Child Health and Human Development Network of Maternal-Fetal Medicine Units. *N. Engl. J. Med.* **1993**, *329*, 1213–1218. [CrossRef]

20. Roberge, S.; Bujold, E.; Nicolaides, K.H. Meta-analysis on the effect of aspirin use for prevention of preeclampsia on placental abruption and antepartum hemorrhage. *Am. J. Obstet. Gynecol.* **2018**, *218*, 483–489. [CrossRef] [PubMed]

21. Grèen, K.; Drvota, V.; Vesterqvist, O. Pronounced reduction of in vivo prostacyclin synthesis in humans by acetaminophen (paracetamol). *Prostaglandins* **1989**, *37*, 311–315. [CrossRef]

22. O'Brien, W.F.; Krammer, J.; O'Leary, T.D.; Mastrogiannis, D.S. The effect of acetaminophen on prostacyclin production in pregnant women. *Am. J. Obstet. Gynecol.* **1993**, *168*, 1164–1169. [CrossRef]

23. Wright, D.; Poon, L.C.; Rolnik, D.L.; Syngelaki, A.; Delgado, J.L.; Vojtassakova, D.; de Alvarado, M.; Kapeti, E.; Rehal, A.; Pazos, A.; et al. Aspirin for Evidence-Based Preeclampsia Prevention trial: Influence of compliance on beneficial effect of aspirin in prevention of preterm preeclampsia. *Am. J. Obstet. Gynecol.* **2017**, *217*, 685.e1–685.e5. [CrossRef]

24. Hauth, J.; Owen, J.; Thom, E.; Goldenberg, R.; Sibai, B.; McNellis, D. Prevention of preeclampsia with low-dose aspirin in nulliparas: A compliance analysis [Abstract #56]. *J. Soc. Gynecol. Investig.* **1996**, *3*, 86A.

25. Walsh, S.W. Eicosanoids in preeclampsia. *Prostaglandins Leukot. Essent. Fat. Acids* **2004**, *70*, 223–232. [CrossRef] [PubMed]

26. Askie, L.M.; Duley, L.; Henderson-Smart, D.J.; Stewart, L.A. Antiplatelet agents for prevention of pre-eclampsia: A meta-analysis of individual patient data. *Lancet* **2007**, *369*, 1791–1798. [CrossRef]

27. Bujold, E.; Roberge, S.; Lacasse, Y.; Bureau, M.; Audibert, F.; Marcoux, S.; Forest, J.-C.; Giguère, Y. Prevention of preeclampsia and intrauterine growth restriction with aspirin started in early pregnancy: A meta-analysis. *Obstet. Gynecol.* **2010**, *116*, 402–414. [CrossRef]

28. Roberge, S.; Villa, P.; Nicolaides, K.; Giguère, Y.; Vainio, M.; Bakthi, A.; Ebrashy, A.; Bujold, E. Early Administration of Low-Dose Aspirin for the Prevention of Preterm and Term Preeclampsia: A Systematic Review and Meta-Analysis. *Fetal Diagn. Ther.* **2012**, *31*, 141–146. [CrossRef]

29. Roberge, S.; Bujold, E.; Nicolaides, K.H. Aspirin for the prevention of preterm and term preeclampsia: Systematic review and metaanalysis. *Am. J. Obstet. Gynecol.* **2018**, *218*, 287–293.e1. [CrossRef]

30. Seidler, A.L.; Askie, L.; Ray, J.G. Optimal aspirin dosing for preeclampsia prevention. *Am. J. Obstet. Gynecol.* **2018**, *219*, 117–118. [CrossRef] [PubMed]

31. Wright, D.; Nicolaides, K.H. Aspirin delays the development of preeclampsia. *Am. J. Obstet. Gynecol.* **2019**, *220*, 580.e1–580.e6. [CrossRef]

32. Finneran, M.M.; Gonzalez-Brown, V.M.; Smith, D.D.; Landon, M.B.; Rood, K.M. Obesity and laboratory aspirin resistance in high-risk pregnant women treated with low-dose aspirin. *Am. J. Obstet. Gynecol.* **2019**, *220*, 385.e1–385.e6. [CrossRef] [PubMed]

33. Rolnik, D.L.; Wright, D.; Poon, L.C.; O'Gorman, N.N.; Syngelaki, A.A.; Matallana, C.C.D.P.; Akolekar, R.R.; Cicero, S.S.; Janga, D.D.; Singh, M.M.; et al. Aspirin versus Placebo in Pregnancies at High Risk for Preterm Preeclampsia. *N. Engl. J. Med.* **2017**, *377*, 613–622. [CrossRef]

34. American College of Obstetricians and Gynecologists. Hypertension in pregnancy. Report of the American College of Obstetricians and Gynecologists' Task Force on Hypertension in Pregnancy. *Obstet. Gynecol.* **2013**, *122*, 1122–1131. [CrossRef]

35. American College of Obstetricians and Gynecologists. ACOG Committee Opinion No. 743: Low-Dose Aspirin Use during Pregnancy. *Obstet. Gynecol.* **2018**, *132*, e44–e52. [CrossRef] [PubMed]

36. Henderson, J.T.; Whitlock, E.P.; O'Connor, E.; Senger, C.A.; Thompson, J.H.; Rowland, M.G. Low-Dose Aspirin for Prevention of Morbidity and Mortality From Preeclampsia: A Systematic Evidence Review for the U.S. Preventive Services Task Force. *Ann. Intern. Med.* **2014**, *160*, 695–703. [CrossRef] [PubMed]

37. Vadillo-Ortega, F.; Perichart-Perera, O.; Espino, S.; Vergara, M.A.A.; Ibarra-Gonzalez, I.; Ahued, R.; Godines, M.; Parry, S.; Macones, G.; Strauss, J.F. Effect of supplementation during pregnancy with L-arginine and antioxidant vitamins in medical food on pre-eclampsia in high risk population: Randomised controlled trial. *BMJ* **2011**, *342*, d2901. [CrossRef]

38. Monari, F.; Menichini, D.; Pignatti, L.; Basile, L.; Facchinetti, F.; Neri, I. Effect of L-Arginine supplementation in pregnant women with chronic hypertension and previous placenta vascular disorders receiving Aspirin prophylaxis: A randomized control trial. *Minerva Obstet. Gynecol.* **2021**. [CrossRef]

39. Walsh, S.W.; Reep, D.T.; Alam, S.M.K.; Washington, S.L.; Al Dulaimi, M.; Lee, S.M.; Springel, E.H.; Strauss, J.F.; Stephenson, D.J.; Chalfant, C.E. Placental Production of Eicosanoids and Sphingolipids in Women Who Developed Preeclampsia on Low-Dose Aspirin. *Reprod. Sci.* **2020**, *27*, 2158–2169. [CrossRef]

40. Nelson, D.M.; Walsh, S.W. Thromboxane and Prostacyclin Production by Different Compartments of the Human Placental Villus. *J. Clin. Endocrinol. Metab.* **1989**, *68*, 676–683. [CrossRef] [PubMed]

41. Thorp, J.A.; Walsh, S.W.; Brath, P.C. Low-dose aspirin inhibits thromboxane, but not prostacyclin, production by human placental arteries. *Am. J. Obstet. Gynecol.* **1988**, *159*, 1381–1384. [CrossRef]

42. Walsh, S.W.; Wang, Y. Trophoblast and placental villous core production of lipid peroxides, thromboxane, and prostacyclin in preeclampsia. *J. Clin. Endocrinol. Metab.* **1995**, *80*, 1888–1893. [CrossRef] [PubMed]

43. Walsh, S.W.; Wang, Y. Maternal Perfusion with Low-Dose Aspirin Preferentially Inhibits Placental Thromboxane While Sparing Prostacyclin. *Hypertens. Pregnancy* **1998**, *17*, 203–215. [CrossRef]

44. Nelson, D.M.; Walsh, S.W. Aspirin differentially affects thromboxane and prostacyclin production by trophoblast and villous core compartments of human placental villi. *Am. J. Obstet. Gynecol.* **1989**, *161*, 1593–1598. [CrossRef]

45. Wang, Y.; Walsh, S.W. Aspirin inhibits both lipid peroxides and thromboxane in preeclamptic placentas. *Free Radic. Biol. Med.* **1995**, *18*, 585–591. [CrossRef]

46. Walsh, S.W.; Vaughan, J.E.; Wang, Y.; Roberts, L.J., II. Placental isoprostane is significantly increased in preeclampsia. *FASEB J.* **2000**, *14*, 1289–1296. [CrossRef] [PubMed]

47. Bilodeau, J.-F. Review: Maternal and placental antioxidant response to preeclampsia—Impact on vasoactive eicosanoids. *Placenta* **2014**, *35*, S32–S38. [CrossRef]

48. Morrow, J.D.; Roberts, L., II. The isoprostanes. Current knowledge and directions for future research. *Biochem. Pharmacol.* **1996**, *51*, 1–9. [CrossRef]

49. Roberts, L.J., II; Morrow, J.D. The isoprostanes: Novel markers of lipid peroxidation and potential mediators of oxidant injury. In *Advances in Prostaglandin, Thromboxane, and Leukotriene Research*; Samuelsson, B., Ramwell, P.W., Paoletti, R., Folco, G., Granstrom, E., Nicosia, S., Eds.; Raven Press: New York, NY, USA, 1995; pp. 219–224.

50. Morrow, J.D.; Hill, K.E.; Burk, R.F.; Nammour, T.M.; Badr, K.F.; Roberts, L.J. A series of prostaglandin F2-like compounds are produced in vivo in humans by a non-cyclooxygenase, free radical-catalyzed mechanism. *Proc. Natl. Acad. Sci. USA* **1990**, *87*, 9383–9387. [CrossRef]

51. Walsh, S. Maternal-Placental Interactions of Oxidative Stress and Antioxidants in Preeclampsia. *Semin. Reprod. Endocrinol.* **1998**, *16*, 93–104. [CrossRef]

52. Kukreja, R.C.; Kontos, H.A.; Hess, M.L.; Ellis, E.F. PGH synthase and lipoxygenase generate superoxide in the presence of NADH or NADPH. *Circ. Res.* **1986**, *59*, 612–619. [CrossRef]

53. Walsh, S.W.; Wang, Y.; Kay, H.H.; McCoy, M.C. Low-dose aspirin inhibits lipid peroxides and thromboxane but not prostacyclin in pregnant women. *Am. J. Obstet. Gynecol.* **1992**, *167*, 926–930. [CrossRef]

54. Powell, W.S.; Rokach, J. Biosynthesis, biological effects, and receptors of hydroxyeicosatetraenoic acids (HETEs) and oxoeicosatetraenoic acids (oxo-ETEs) derived from arachidonic acid. *Biochim. Biophys. Acta* **2015**, *1851*, 340–355. [CrossRef]

55. Li, J.; Rao, J.; Liu, Y.; Cao, Y.; Zhang, Y.; Zhang, Q.; Zhu, D. 15-Lipoxygenase Promotes Chronic Hypoxia–Induced Pulmonary Artery Inflammation via Positive Interaction With Nuclear Factor-κB. *Arter. Thromb. Vasc. Biol.* **2013**, *33*, 971–979. [CrossRef]

56. Stenson, W.F.; Parker, C.W. Leukotrienes. *Adv. Intern. Med.* **1984**, *30*, 175–199.

57. Rubin, P.; Mollison, K.W. Pharmacotherapy of diseases mediated by 5-lipoxygenase pathway eicosanoids. *Prostaglandins Other Lipid Mediat.* **2007**, *83*, 188–197. [CrossRef]

58. Goetzl, E.J.; Goldman, D.W.; Naccache, P.H.; Sha'Afi, R.I.; Pickett, W.C. Mediation of leukocyte components of inflammatory reactions by lipoxygenase products of arachidonic acid. *Adv. Prostaglandin Thromboxane Leukot. Res.* **1982**, *9*, 273–282. [PubMed]

59. Hoopes, S.L.; Garcia, V.; Edin, M.L.; Schwartzman, M.L.; Zeldin, D.C. Vascular actions of 20-HETE. *Prostaglandins Other Lipid Mediat.* **2015**, *120*, 9–16. [CrossRef] [PubMed]

60. Waldman, M.; Peterson, S.J.; Arad, M.; Hochhauser, E. The role of 20-HETE in cardiovascular diseases and its risk factors. *Prostaglandins Other Lipid Mediat.* **2016**, *125*, 108–117. [CrossRef] [PubMed]

61. Garcia, V.; Schwartzman, M.L. Recent developments on the vascular effects of 20-hydroxyeicosatetraenoic acid. *Curr. Opin. Nephrol. Hypertens.* **2017**, *26*, 74–82. [CrossRef]

62. Salafia, C.M.; Pezzullo, J.C.; López-Zeno, J.; Simmens, S.; Minior, V.K.; Vintzileos, A.M. Placental pathologic features of preterm preeclampsia. *Am. J. Obstet. Gynecol.* **1995**, *173*, 1097–1105. [CrossRef]

63. Mihanfar, A.; Nejabati, H.R.; Fattahi, A.; Latifi, Z.; Pezeshkian, M.; Afrasiabi, A.; Safaie, N.; Jodati, A.R.; Nouri, M. The role of sphingosine 1 phosphate in coronary artery disease and ischemia reperfusion injury. *J. Cell. Physiol.* **2019**, *234*, 2083–2094. [CrossRef]

64. Mao, C. Ceramidases: Regulators of cellular responses mediated by ceramide, sphingosine, and sphingosine-1-phosphate. *Biochim. Biophys. Acta* **2008**, *1781*, 424–434. [CrossRef]

65. Zhang, H.; Desai, N.N.; Olivera, A.; Seki, T.; Brooker, G.; Spiegel, S. Sphingosine-1-phosphate, a novel lipid, involved in cellular proliferation. *J. Cell Biol.* **1991**, *114*, 155–167. [CrossRef]
66. Norris, G.H.; Blesso, C.N. Dietary and Endogenous Sphingolipid Metabolism in Chronic Inflammation. *Nutrients* **2017**, *9*, 1180. [CrossRef]
67. Coant, N.; Sakamoto, W.; Mao, C.; Hannun, Y.A. Ceramidases, roles in sphingolipid metabolism and in health and disease. *Adv. Biol. Regul.* **2017**, *63*, 122–131. [CrossRef] [PubMed]
68. Kerage, D.; Brindley, D.; Hemmings, D. Review: Novel insights into the regulation of vascular tone by sphingosine 1-phosphate. *Placenta* **2014**, *35*, S86–S92. [CrossRef]
69. Huppertz, B.; Kadyrov, M.; Kingdom, J.C. Apoptosis and its role in the trophoblast. *Am. J. Obstet. Gynecol.* **2006**, *195*, 29–39. [CrossRef]
70. Westwood, M.; Al-Saghir, K.; Finn-Sell, S.; Tan, C.; Cowley, E.; Berneau, S.; Adlam, D.; Johnstone, E.D. Vitamin D attenuates sphingosine-1-phosphate (S1P)-mediated inhibition of extravillous trophoblast migration. *Placenta* **2017**, *60*, 1–8. [CrossRef] [PubMed]
71. Charkiewicz, K.; Goscik, J.; Blachnio-Zabielska, A.; Raba, G.; Sakowicz, A.; Kalinka, J.; Chabowski, A.; Laudanski, P. Sphingolipids as a new factor in the pathomechanism of preeclampsia—Mass spectrometry analysis. *PLoS ONE* **2017**, *12*, e0177601. [CrossRef] [PubMed]
72. Melland-Smith, M.; Ermini, L.; Chauvin, S.; Craig-Barnes, H.; Tagliaferro, A.; Todros, T.; Post, M.; Caniggia, I. Disruption of sphingolipid metabolism augments ceramide-induced autophagy in preeclampsia. *Autophagy* **2015**, *11*, 653–669. [CrossRef]
73. Van Nieuwenhoven, A.L.V.; Bouman, A.; Moes, H.; Heineman, M.J.; de Leij, L.F.; Santema, J.; Faas, M.M. Cytokine production in natural killer cells and lymphocytes in pregnant women compared with women in the follicular phase of the ovarian cycle. *Fertil. Steril.* **2002**, *77*, 1032–1037. [CrossRef]
74. Lurie, S.; Frenkel, E.; Tuvbin, Y. Comparison of the Differential Distribution of Leukocytes in Preeclampsia versus Uncomplicated Pregnancy. *Gynecol. Obstet. Investig.* **1998**, *45*, 229–231. [CrossRef] [PubMed]
75. Luster, A.D. Chemokines—Chemotactic Cytokines That Mediate Inflammation. *N. Engl. J. Med.* **1998**, *338*, 436–445. [CrossRef] [PubMed]
76. Leik, C.E.; Walsh, S.W. Neutrophils Infiltrate Resistance-Sized Vessels of Subcutaneous Fat in Women with Preeclampsia. *Hypertension* **2004**, *44*, 72–77. [CrossRef]
77. Cadden, K.A.; Walsh, S.W. Neutrophils, but Not Lymphocytes or Monocytes, Infiltrate Maternal Systemic Vasculature in Women with Preeclampsia. *Hypertens. Pregnancy* **2008**, *27*, 396–405. [CrossRef] [PubMed]
78. Mishra, N.; Nugent, W.H.; Mahavadi, S.; Walsh, S.W. Mechanisms of Enhanced Vascular Reactivity in Preeclampsia. *Hypertension* **2011**, *58*, 867–873. [CrossRef]
79. Shah, T.J.; Walsh, S.W. Activation of NF-kappaB and expression of COX-2 in association with neutrophil infiltration in systemic vascular tissue of women with preeclampsia. *Am. J. Obstet. Gynecol.* **2007**, *196*, 48.e1–48.e8. [CrossRef]
80. Gervasi, M.-T.; Chaiworapongsa, T.; Pacora, P.; Naccasha, N.; Yoon, B.H.; Maymon, E.; Romero, R. Phenotypic and metabolic characteristics of monocytes and granulocytes in preeclampsia. *Am. J. Obstet. Gynecol.* **2001**, *185*, 792–797. [CrossRef]
81. Sacks, G.P.; Studena, K.; Sargent, I.L.; Redman, C.W. Normal pregnancy and preeclampsia both produce inflammatory changes in peripheral blood leukocytes akin to those of sepsis. *Am. J. Obstet. Gynecol.* **1998**, *179*, 80–86. [CrossRef]
82. Mousa, A.A.; Strauss, J.F., 3rd; Walsh, S.W. Reduced Methylation of the Thromboxane Synthase Gene Is Correlated with Its Increased Vascular Expression in Preeclampsia. *Hypertension* **2012**, *59*, 1249–1255. [CrossRef]
83. Shukla, J.; Walsh, S.W. Neutrophil Release of Myeloperoxidase in Systemic Vasculature of Obese Women May Put Them at Risk for Preeclampsia. *Reprod. Sci.* **2015**, *22*, 300–307. [CrossRef] [PubMed]
84. Ahn, H.-S.; Chackalamannil, S.; Boykow, G.; Graziano, M.; Foster, C. Development of Proteinase-Activated Receptor 1 Antagonists as Therapeutic Agents for Thrombosis, Restenosis and Inflammatory Diseases. *Curr. Pharm. Des.* **2003**, *9*, 2349–2365. [CrossRef]
85. Trivedi, V.; Boire, A.; Tchernychev, B.; Kaneider, N.C.; Leger, A.J.; O'Callaghan, K.; Covic, L.; Kuliopulos, A. Platelet Matrix Metalloprotease-1 Mediates Thrombogenesis by Activating PAR1 at a Cryptic Ligand Site. *Cell* **2009**, *137*, 332–343. [CrossRef] [PubMed]
86. Suzuki, T.; Moraes, T.J.; Vachon, E.; Ginzberg, H.H.; Huang, T.-T.; Matthay, M.A.; Hollenberg, M.D.; Marshall, J.; McCulloch, C.A.G.; Abreu, M.T.H.; et al. Proteinase-Activated Receptor-1 Mediates Elastase-Induced Apoptosis of Human Lung Epithelial Cells. *Am. J. Respir. Cell Mol. Biol.* **2005**, *33*, 231–247. [CrossRef] [PubMed]
87. Wang, Y.; Gu, Y.; Lucas, M.J. Expression of thrombin receptors in endothelial cells and neutrophils from normal and preeclamptic pregnancies. *J. Clin. Endocrinol. Metab.* **2002**, *87*, 3728–3734. [CrossRef]
88. Shpacovitch, V.; Feld, M.; Hollenberg, M.D.; Luger, T.A.; Steinhoff, M. Role of protease-activated receptors in inflammatory responses, innate and adaptive immunity. *J. Leukoc. Biol.* **2008**, *83*, 1309–1322. [CrossRef]
89. Estrada-Gutierrez, G.; Cappello, R.E.; Mishra, N.; Romero, R.; Strauss, J.F., 3rd; Walsh, S.W. Increased Expression of Matrix Metalloproteinase-1 in Systemic Vessels of Preeclamptic Women: A Critical Mediator of Vascular Dysfunction. *Am. J. Pathol.* **2011**, *178*, 451–460. [CrossRef]
90. Walsh, S.W.; Nugent, W.H.; Al Dulaimi, M.; Washington, S.L.; Dacha, P.; Strauss, J.F., 3rd. Proteases Activate Pregnancy Neutrophils by a Protease-Activated Receptor 1 Pathway: Epigenetic Implications for Preeclampsia. *Reprod. Sci.* **2020**, *27*, 2115–2127. [CrossRef]

91. He, Y.-F.; Li, B.-Z.; Li, Z.; Liu, P.; Wang, Y.; Tang, Q.; Ding, J.; Jia, Y.; Chen, Z.; Li, L.; et al. Tet-Mediated Formation of 5-Carboxylcytosine and Its Excision by TDG in Mammalian DNA. *Science* **2011**, *333*, 1303–1307. [CrossRef]
92. Ito, S.; Shen, L.; Dai, Q.; Wu, S.C.; Collins, L.B.; Swenberg, J.A.; He, C.; Zhang, Y. Tet Proteins Can Convert 5-Methylcytosine to 5-Formylcytosine and 5-Carboxylcytosine. *Science* **2011**, *333*, 1300–1303. [CrossRef]
93. Tahiliani, M.; Koh, K.P.; Shen, Y.; Pastor, W.A.; Bandukwala, H.; Brudno, Y.; Agarwal, S.; Iyer, L.M.; Liu, D.R.; Aravind, L.; et al. Conversion of 5-Methylcytosine to 5-Hydroxymethylcytosine in Mammalian DNA by MLL Partner TET1. *Science* **2009**, *324*, 930–935. [CrossRef] [PubMed]
94. Yin, R.; Mao, S.-Q.; Zhao, B.; Chong, Z.; Yang, Y.; Zhao, C.; Zhang, D.; Huang, H.; Gao, J.; Li, Z.; et al. Ascorbic Acid Enhances Tet-Mediated 5-Methylcytosine Oxidation and Promotes DNA Demethylation in Mammals. *J. Am. Chem. Soc.* **2013**, *135*, 10396–10403. [CrossRef] [PubMed]
95. Nabel, C.S.; Kohli, R.M. Molecular biology. Demystifying DNA demethylation. *Science* **2011**, *333*, 1229–1230. [CrossRef] [PubMed]
96. Klug, M.; Schmidhofer, S.; Gebhard, C.; Andreesen, R.; Rehli, M. 5-Hydroxymethylcytosine is an essential intermediate of active DNA demethylation processes in primary human monocytes. *Genome Biol.* **2013**, *14*, R46. [CrossRef] [PubMed]
97. Wu, H.; Zhang, Y. Mechanisms and functions of Tet protein-mediated 5-methylcytosine oxidation. *Genes Dev.* **2011**, *25*, 2436–2452. [CrossRef]
98. Klug, M.; Heinz, S.; Gebhard, C.; Schwarzfischer, L.; Krause, S.W.; Andreesen, R.; Rehli, M. Active DNA demethylation in human postmitotic cells correlates with activating histone modifications, but not transcription levels. *Genome Biol.* **2010**, *11*, R63. [CrossRef]
99. Huang, Q.-T.; Chen, J.-H.; Hang, L.-L.; Liu, S.-S.; Zhong, M. Activation of PAR-1/NADPH Oxidase/ROS Signaling Pathways is Crucial for the Thrombin-Induced sFlt-1 Production in Extravillous Trophoblasts: Possible Involvement in the Pathogenesis of Preeclampsia. *Cell. Physiol. Biochem.* **2015**, *35*, 1654–1662. [CrossRef]
100. Zhao, Y.; Koga, K.; Osuga, Y.; Nagai, M.; Izumi, G.; Takamura, M.; Harada, M.; Hirota, Y.; Yoshino, O.; Taketani, Y. Thrombin enhances soluble Fms-like tyrosine kinase 1 expression in trophoblasts; possible involvement in the pathogenesis of preeclampsia. *Fertil. Steril.* **2012**, *98*, 917–921. [CrossRef]
101. Erez, O.; Romero, R.; Kim, S.-S.; Kim, J.-S.; Kim, Y.M.; Wildman, D.; Than, N.G.; Mazaki-Tovi, S.; Gotsch, F.; Pineles, B.; et al. Overexpression of the thrombin receptor (PAR-1) in the placenta in preeclampsia: A mechanism for the intersection of coagulation and inflammation. *J. Matern. Fetal Neonatal Med.* **2008**, *21*, 345–355. [CrossRef]
102. Even-Ram, S.C.; Grisaru-Granovsky, S.; Pruss, D.; Maoz, M.; Salah, Z.; Yong-Jun, Y.; Bar-Shavit, R. The pattern of expression of protease-activated receptors (PARs) during early trophoblast development. *J. Pathol.* **2003**, *200*, 47–52. [CrossRef]
103. Lockwood, C.J.; Toti, P.; Arcuri, F.; Norwitz, E.; Funai, E.F.; Huang, S.-T.J.; Buchwalder, L.F.; Krikun, G.; Schatz, F. Thrombin Regulates Soluble fms-Like Tyrosine Kinase-1 (sFlt-1) Expression in First Trimester Decidua: Implications for Preeclampsia. *Am. J. Pathol.* **2007**, *170*, 1398–1405. [CrossRef]
104. Leik, C.E.; Walsh, S.W. Systemic Activation and Vascular Infiltration of Neutrophils With Term Labor. *J. Soc. Gynecol. Investig.* **2006**, *13*, 425–429. [CrossRef]
105. Nugent, W.H.; Mishra, N.; Strauss, J.F., 3rd; Walsh, S.W. Matrix Metalloproteinase 1 Causes Vasoconstriction and Enhances Vessel Reactivity to Angiotensin II via Protease-Activated Receptor 1. *Reprod. Sci.* **2016**, *23*, 542–548. [CrossRef]
106. Henriksen, R.A.; Samokhin, G.P.; Tracy, P.B. Thrombin-induced thromboxane synthesis by human platelets. Properties of anion binding exosite I-independent receptor. *Arter. Thromb. Vasc. Biol.* **1997**, *17*, 3519–3526. [CrossRef] [PubMed]
107. Austin, K.M.; Covic, L.; Kuliopulos, A. Matrix metalloproteases and PAR1 activation. *Blood* **2013**, *121*, 431–439. [CrossRef] [PubMed]
108. Coughlin, S.R. Thrombin signalling and protease-activated receptors. *Nature* **2000**, *407*, 258–264. [CrossRef] [PubMed]
109. Coughlin, S.R. Protease-activated receptors in hemostasis, thrombosis and vascular biology. *J. Thromb. Haemost.* **2005**, *3*, 1800–1814. [CrossRef] [PubMed]
110. Rahman, A.; True, A.L.; Anwar, K.N.; Ye, R.D.; Voyno-Yasenetskaya, T.A.; Malik, A.B. Galpha(q) and Gbetagamma regulate PAR-1 signaling of thrombin-induced NF-kappaB activation and ICAM-1 transcription in endothelial cells. *Circ. Res.* **2002**, *91*, 398–405. [CrossRef] [PubMed]
111. Macfarlane, S.R.; Seatter, M.J.; Kanke, T.; Hunter, G.D.; Plevin, R. Proteinase-activated receptors. *Pharmacol. Rev.* **2001**, *53*, 245–282.
112. Goerge, T.; Barg, A.; Schnaeker, E.-M.; Pöppelmann, B.; Shpacovitch, V.; Rattenholl, A.; Maaser, C.; Luger, T.A.; Steinhoff, M.; Schneider, S.W. Tumor-Derived Matrix Metalloproteinase-1 Targets Endothelial Proteinase-Activated Receptor 1 Promoting Endothelial Cell Activation. *Cancer Res.* **2006**, *66*, 7766–7774. [CrossRef]
113. Barden, A.; Graham, D.; Beilin, L.J.; Ritchie, J.; Baker, R.; Walters, B.N.; Michael, C.A. Neutrophil CD11B Expression and Neutrophil Activation in Pre-Eclampsia. *Clin. Sci.* **1997**, *92*, 37–44. [CrossRef] [PubMed]
114. Greer, I.A.; Haddad, N.G.; Dawes, J.; Johnstone, F.D.; Calder, A.A. Neutrophil activation in pregnancy-induced hypertension. *BJOG Int. J. Obstet. Gynaecol.* **1989**, *96*, 978–982. [CrossRef] [PubMed]
115. Tsukimori, K.; Maeda, H.; Ishida, K.; Nagata, H.; Koyanagi, T.; Nakano, H. The superoxide generation of neutrophils in normal and preeclamptic pregnancies. *Obstet. Gynecol.* **1993**, *81*, 536–540.
116. Görög, P. Activation of human blood monocytes by oxidized polyunsaturated fatty acids: A possible mechanism for the generation of lipid peroxides in the circulation. *Int. J. Exp. Pathol.* **1991**, *72*, 227–237. [PubMed]

117. Vaughan, J.E.; Walsh, S.W. Neutrophils from pregnant women produce thromboxane and tumor necrosis factor-α in response to linoleic acid and oxidative stress. *Am. J. Obstet. Gynecol.* **2005**, *193*, 830–835. [CrossRef]
118. Vaughan, J.E.; Walsh, S.W.; Ford, G.D. Thromboxane mediates neutrophil superoxide production in pregnancy. *Am. J. Obstet. Gynecol.* **2006**, *195*, 1415–1420. [CrossRef]
119. Walsh, S.W.; Wang, Y. Secretion of lipid peroxides by the human placenta. *Am. J. Obstet. Gynecol.* **1993**, *169*, 1462–1466. [CrossRef]
120. Mellembakken, J.R.; Aukrust, P.; Olafsen, M.K.; Ueland, T.; Hestdal, K.; Videm, V. Activation of leukocytes during the uteroplacental passage in preeclampsia. *Hypertension* **2002**, *39*, 155–160. [CrossRef]
121. Walsh, S.W. Lipid Peroxidation in Pregnancy. *Hypertens. Pregnancy* **1994**, *13*, 1–32. [CrossRef]
122. Bachawaty, T.; Washington, S.L.; Walsh, S.W. Neutrophil Expression of Cyclooxygenase 2 in Preeclampsia. *Reprod. Sci.* **2010**, *17*, 465–470. [CrossRef] [PubMed]
123. Shapiro, S.; Siskind, V.; Monson, R.; Heinonen, O.; Kaufman, D.; Slone, D. Perinatal mortality and birth-weight in relation to aspirin taken during pregnancy. *Lancet* **1976**, *307*, 1375–1376. [CrossRef]
124. Maynard, S.E.; Min, J.-Y.; Merchan, J.; Lim, K.-H.; Li, J.; Mondal, S.; Libermann, T.; Morgan, J.P.; Sellke, F.W.; Stillman, I.E.; et al. Excess placental soluble fms-like tyrosine kinase 1 (sFlt1) may contribute to endothelial dysfunction, hypertension, and proteinuria in preeclampsia. *J. Clin. Investig.* **2003**, *111*, 649–658. [CrossRef] [PubMed]
125. Widmer, M.; Villar, J.; Benigni, A.; Conde-Agudelo, A.; Karumanchi, S.A.; Lindheimer, M. Mapping the theories of preeclampsia and the role of angiogenic factors: A systematic review. *Obstet. Gynecol.* **2007**, *109*, 168–180. [CrossRef]
126. Lin, L.; Li, G.; Zhang, W.; Wang, Y.; Yang, H. Low-dose aspirin reduces hypoxia-induced sFlt1 release via the JNK/AP-1 pathway in human trophoblast and endothelial cells. *J. Cell. Physiol.* **2019**, *234*, 18928–18941. [CrossRef] [PubMed]
127. Su, M.-T.; Wang, C.-Y.; Tsai, P.-Y.; Chen, T.-Y.; Tsai, H.-L.; Kuo, P.-L. Aspirin enhances trophoblast invasion and represses soluble fms-like tyrosine kinase 1 production: A putative mechanism for preventing preeclampsia. *J. Hypertens.* **2019**, *37*, 2461–2469. [CrossRef]
128. Ducat, A.; Vargas, A.; Doridot, L.; Bagattin, A.; Lerner, J.; Vilotte, J.-L.; Buffat, C.; Pontoglio, M.; Miralles, F.; Vaiman, D. Low-dose aspirin protective effects are correlated with deregulation of HNF factor expression in the preeclamptic placentas from mice and humans. *Cell Death Discov.* **2019**, *5*, 94. [CrossRef]
129. He, Y.; Chen, L.; Liu, C.; Han, Y.; Liang, C.; Xie, Q.; Zhou, J.; Cheng, Z. Aspirin modulates STOX1 expression and reverses STOX1-induced insufficient proliferation and migration of trophoblast cells. *Pregnancy Hypertens.* **2020**, *19*, 170–176. [CrossRef]
130. Huai, J.; Li, G.-L.; Lin, L.; Ma, J.-M.; Yang, H.-X. Phosphoproteomics reveals the apoptotic regulation of aspirin in the placenta of preeclampsia-like mice. *Am. J. Transl. Res.* **2020**, *12*, 3361–3375.
131. Su, M.-T.; Tsai, P.-Y.; Wang, C.-Y.; Tsai, H.-L.; Kuo, P.-L. Aspirin facilitates trophoblast invasion and epithelial-mesenchymal transition by regulating the miR-200-ZEB1 axis in preeclampsia. *Biomed. Pharmacother.* **2021**, *139*, 111591. [CrossRef]
132. Rani, S.; O'Brien, K.; Kelleher, F.C.; Corcoran, C.; Germano, S.; Radomski, M.W.; Crown, J.; O'Driscoll, L. Isolation of Exosomes for Subsequent mRNA, MicroRNA, and Protein Profiling. *Methods Mol. Biol.* **2011**, *784*, 181–195. [CrossRef] [PubMed]
133. Donker, R.B.; Mouillet, J.F.; Chu, T.; Hubel, C.A.; Stolz, D.B.; Morelli, A.E.; Sadovsky, Y. The expression profile of C19MC microRNAs in primary human trophoblast cells and exosomes. *Mol. Hum. Reprod.* **2012**, *18*, 417–424. [CrossRef] [PubMed]
134. Mouillet, J.-F.; Ouyang, Y.; Bayer, A.; Coyne, C.B.; Sadovsky, Y. The role of trophoblastic microRNAs in placental viral infection. *Int. J. Dev. Biol.* **2014**, *58*, 281–289. [CrossRef] [PubMed]
135. Ouyang, Y.; Mouillet, J.-F.; Coyne, C.; Sadovsky, Y. Review: Placenta-specific microRNAs in exosomes—Good things come in nano-packages. *Placenta* **2014**, *35*, S69–S73. [CrossRef] [PubMed]
136. Tong, M.; Chamley, L.W. Placental Extracellular Vesicles and Feto-Maternal Communication. *Cold Spring Harb. Perspect. Med.* **2015**, *5*, a023028. [CrossRef]
137. Dutta, S.; Kumar, S.; Hyett, J.; Salomon, C. Molecular Targets of Aspirin and Prevention of Preeclampsia and Their Potential Association with Circulating Extracellular Vesicles during Pregnancy. *Int. J. Mol. Sci.* **2019**, *20*, 4370. [CrossRef]

International Journal of
*Molecular Sciences*

**MDPI**

*Article*

# Divergent Regulation of Decidual Oxidative-Stress Response by NRF2 and KEAP1 in Preeclampsia with and without Fetal Growth Restriction

Siv Boon Mundal [1,2], Johanne Johnsen Rakner [1], Gabriela Brettas Silva [1,3], Lobke Marijn Gierman [1,3], Marie Austdal [1,3,4], Purusotam Basnet [2,5], Mattijs Elschot [6,7], Siril Skaret Bakke [1], Jenny Ostrop [1], Liv Cecilie Vestrheim Thomsen [8,9], Eric Keith Moses [10], Ganesh Acharya [2,11], Line Bjørge [8,9] and Ann-Charlotte Iversen [1,3,*]

1 Centre of Molecular Inflammation Research (CEMIR), Department of Clinical and Molecular Medicine, Norwegian University of Science and Technology (NTNU), 7491 Trondheim, Norway; siv.boon@gmail.com (S.B.M.); johanne.j.rakner@ntnu.no (J.J.R.); gabrielabrettas@gmail.com (G.B.S.); lobke.gierman@ntnu.no (L.M.G.); marie.austdal@sus.no (M.A.); siril.s.bakke@ntnu.no (S.S.B.); jenny.ostrop@uib.no (J.O.)
2 Women's Health and Perinatology Research Group, Department of Clinical Medicine, UiT—The Arctic University of Norway, 9037 Tromsø, Norway; purusotam.basnet@uit.no (P.B.); ganesh.acharya@ki.se (G.A.)
3 Department of Gynecology and Obstetrics, St. Olavs Hospital, Trondheim University Hospital, 7030 Trondheim, Norway
4 Department of Research, Stavanger University Hospital, 4068 Stavanger, Norway
5 Department of Obstetrics and Gynecology, University Hospital of Northern Norway, 9037 Tromsø, Norway
6 Department of Circulation and Medical Imaging, Norwegian University of Science and Technology (NTNU), 7491 Trondheim, Norway; mattijs.elschot@ntnu.no
7 Department of Radiology and Nuclear Medicine, St. Olavs Hospital, Trondheim University Hospital, 7030 Trondheim, Norway
8 Department of Gynecology and Obstetrics, Haukeland University Hospital, 5058 Bergen, Norway; liv.vestrheim@uib.no (L.C.V.T.); line.bjorge@uib.no (L.B.)
9 Centre for Cancer Biomarkers CCBIO, Department of Clinical Science, University of Bergen, 5021 Bergen, Norway
10 Menzies Institute for Medical Research, University of Tasmania, Hobart, TAS 7000, Australia; eric.moses@utas.edu.au
11 Department of Clinical Science, Division of Obstetrics and Gynecology, Intervention and Technology, Karolinska Institutet, 141 86 Stockholm, Sweden
* Correspondence: ann-charlotte.iversen@ntnu.no; Tel.: +47-93283877

**Citation:** Mundal, S.B.; Rakner, J.J.; Silva, G.B.; Gierman, L.M.; Austdal, M.; Basnet, P.; Elschot, M.; Bakke, S.S.; Ostrop, J.; Thomsen, L.C.V.; et al. Divergent Regulation of Decidual Oxidative-Stress Response by NRF2 and KEAP1 in Preeclampsia with and without Fetal Growth Restriction. *Int. J. Mol. Sci.* **2022**, *23*, 1966. https://doi.org/10.3390/ijms23041966

Academic Editors: Hiten D Mistry and Eun Lee

Received: 2 December 2021
Accepted: 7 February 2022
Published: 10 February 2022

**Publisher's Note:** MDPI stays neutral with regard to jurisdictional claims in published maps and institutional affiliations.



**Abstract:** Utero-placental development in pregnancy depends on direct maternal–fetal interaction in the uterine wall decidua. Abnormal uterine vascular remodeling preceding placental oxidative stress and placental dysfunction are associated with preeclampsia and fetal growth restriction (FGR). Oxidative stress is counteracted by antioxidants and oxidative repair mechanisms regulated by the transcription factor nuclear factor erythroid 2-related factor 2 (NRF2). We aimed to determine the decidual regulation of the oxidative-stress response by NRF2 and its negative regulator Kelch-like ECH-associated protein 1 (KEAP1) in normal pregnancies and preeclamptic pregnancies with and without FGR. Decidual tissue from 145 pregnancies at delivery was assessed for oxidative stress, non-enzymatic antioxidant capacity, cellular NRF2- and KEAP1-protein expression, and NRF2-regulated transcriptional activation. Preeclampsia combined with FGR was associated with an increased oxidative-stress level and NRF2-regulated gene expression in the decidua, while decidual NRF2- and KEAP1-protein expression was unaffected. Although preeclampsia with normal fetal growth also showed increased decidual oxidative stress, NRF2-regulated gene expression was reduced, and KEAP1-protein expression was increased in areas of high trophoblast density. The trophoblast-dependent KEAP1-protein expression in preeclampsia with normal fetal growth indicates control of decidual oxidative stress by maternal–fetal interaction and underscores the importance of discriminating between preeclampsia with and without FGR.

**Keywords:** antioxidant capacity; decidua; fetal growth restriction; KEAP1; NRF2; oxidative stress; preeclampsia; trophoblast

---

## 1. Introduction

Preeclampsia is an inflammatory multisystem syndrome affecting 2–5% of pregnancies, and it is often further complicated by fetal growth restriction (FGR) [1–3]. Both disorders originate from placental dysfunction due to inadequate blood supply via decidual spiral arteries to the placenta. The decidua basalis develops from the endometrium at the blastocyst implantation site and is composed of spiral arteries, decidualized endometrial cells, maternal immune cells, fetal extravillous trophoblasts, fibrinoid layers and endometrial glands [4]. Close interactions between fetal extravillous trophoblasts and maternal cells are crucial for optimal remodeling of the decidual spiral arteries during pregnancy. Impaired spiral-artery remodeling causes malperfusion of the placenta and leads to placental oxidative-, endoplasmic reticulum (ER)- and inflammatory-stress culminating in established placental dysfunction. The dysfunctional placenta releases stress signals and anti-angiogenic factors to the maternal circulation, leading to endothelial dysfunction and, eventually, to the clinical signs of preeclampsia [1,2].

Reactive oxygen species (ROS) are constantly generated within cells as metabolic by-products, and low-to-moderate levels of ROS are physiological [5,6]. When the production of ROS overwhelms the tissue antioxidant defenses, oxidative stress occurs and causes cellular damage. ROS may provoke damage to multiple cellular organelles and ultimately be detrimental [7]. Nuclear factor erythroid 2-related factor 2 (NRF2) is the master regulator of cellular oxidative-stress responses and initiates the transcription of antioxidants and protective genes for oxidative repair and detoxification [8–10]. During oxidative stress, NRF2 is phosphorylated and translocates to the nucleus [11], binds the antioxidant response element (ARE), and initiates the transcription of antioxidants and protective genes [9]. The NRF2-binding protein Kelch-like ECH-associated protein 1 (KEAP1) is a negative regulator of NRF2 that ensures homeostasis by turning off the NRF2 transcriptional activity when not required, through a continuous degradation of the NRF2 protein [9]. KEAP1 inhibits NRF2 both by binding cytosolic NRF2 and preventing its translocation to the nucleus, and by removing nuclear NRF2 from the ARE [12–14]. In the case of oxidative stress, NRF2-activating electrophilic molecules may modify cysteine residues in KEAP1 and impair its function [15] or other mechanisms may disrupt KEAP1–NRF2 protein–protein interactions [16].

Although decidual oxidative stress plays an essential role in the establishment of normal placental function [5,17], excessive oxidative stress in the placenta, decidua and maternal circulation is central to the development of preeclampsia and FGR [5,18–20]. Placental NRF2 expression has been shown to be both upregulated [21,22] and down-regulated [23] in preeclampsia. Decidual NRF2 is expressed by extravillous trophoblasts and maternal cells such as decidual stroma cells, myometrial cells and leukocytes, and is upregulated in preeclampsia [24,25]. The NRF2 expression and cellular distribution have not been determined in the decidua for different preeclampsia subgroups.

We previously performed decidual transcriptional profiling where the "NRF2-mediated oxidative stress response pathway" was shown to be dysregulated in preeclampsia [26]. A more comprehensive assessment of the decidual oxidative-stress response at the protein level is needed in order to improve our understanding of how it affects the maternal–fetal interaction and pregnancy outcomes. We aimed to determine the role of NRF2- and KEAP1-regulated oxidative-stress responses in decidual maternal and fetal cells in normal pregnancies and preeclamptic pregnancies with and without FGR.

## 2. Results

### *2.1. Characteristics of the Study Population*

A total of 145 pregnant women were recruited to the study. As expected, preeclamptic pregnancies, both with and without FGR, were associated with elevated blood pressure, lower gestational age at delivery, reduced birth weight and placental weight, and increased occurrence in first pregnancies (Table 1). Preeclamptic pregnancies complicated by FGR had lower birth weight and placental weight, placental weight ratio, and gestational age at delivery compared to preeclamptic pregnancies without FGR.

**Table 1.** Clinical characteristics of included pregnancies.

| | Normal Pregnancies (n = 70) | PE without FGR (n = 28) | PE with FGR (n = 47) |
|---|---|---|---|
| **Baseline characteristics** | | | |
| Maternal age, years | 31.8 ($\pm$5.1) | 30.5 ($\pm$4.9) | 30.9 ($\pm$5.5) |
| BMI, kg/m$^2$ * | 25.2 ($\pm$4.2) | 26.5 ($\pm$6.6) | 27.2 ($\pm$5.5) [†] |
| Primipara, n (%) | 12 (17) | 17 (61) [‡] | 26 (55) [‡] |
| **Characteristics at delivery** | | | |
| Systolic BP, mmHg [§] | 121.1 ($\pm$12.8) | 165.1 ($\pm$21.2) [‡] | 161.6 ($\pm$19.8) [‡] |
| Diastolic BP, mmHg [§] | 73.4 ($\pm$8.6) | 102.0 ($\pm$12.4) [‡] | 99.6 ($\pm$8.9) [‡] |
| Gestational age, weeks | 38.5 ($\pm$0.9) | 33.1 ($\pm$3.9) [‡] | 30.1 ($\pm$3) [‡#] |
| Severe PE [‖], n (%) | n.a. | 23 (82) | 34 (72) |
| Early onset PE (<34 weeks), n (%) | n.a. | 15 (54) | 40 (85) [†] |
| Placental weight, g | 620 (193) | 450 (238) [‡] | 275 (140) [‡#] |
| Fetal birth weight, g | 3621 ($\pm$474) | 2208 ($\pm$955) [‡] | 1182 ($\pm$456) [‡] |
| Placental weight ratio ** | 1.04 (0.35) | 0.90 (0.35) [‡] | 0.60 (0.27) [‡#] |

BMI, body mass index; BP, blood pressure; FGR, fetal growth restriction, n.a., not applicable; PE, preeclampsia. Continuous variables, listed as means ($\pm$standard deviations) or medians (interquartile ranges), were assessed for differences between groups by one-way ANOVA with Tukey's post hoc test or Kruskal–Wallis with Dunn's post hoc test. Categorical variables, listed as numbers (percentages in parentheses), were assessed for differences between groups by Fisher's exact test. * Maternal BMI at first antenatal care visit (first trimester). Information is missing for three women (2%). † Significantly different when comparing PE without FGR and PE with FGR to normal pregnancies, $p < 0.05$. ‡ Significantly different when comparing PE without FGR and PE with FGR to normal pregnancies, $p < 0.001$. § BP from last antenatal care visit or before the cesarean section. Information is missing for four women (3%). # Significantly different when comparing PE with FGR to PE without FGR, $p < 0.001$. ‖ PE was classified as severe if one or more severe clinical features were present [27,28]. ** Calculated as observed/expected placental weight according to gestational age and sex in a Norwegian normogram [29]. Information is missing from one woman (0.6%).

### *2.2. Non-Enzymatic Antioxidant Capacity and Oxidative-Stress Levels in the Decidua*

### 2.2.1. Non-Enzymatic Antioxidant Capacity in the Decidua

Frozen decidual samples were available from 126 of the 145 pregnancies. Ten of these were excluded due to insufficient amounts or poor tissue quality, and one was excluded as an outlier. No differences in decidual non-enzymatic antioxidant capacity between preeclampsia with or without FGR when compared to normal pregnancies were observed (Figure 1A).

### 2.2.2. Decidual Oxidative-Stress Levels

Of 126 available frozen decidual samples, 15 were excluded due to insufficient amount or poor tissue quality and one as an outlier. Significantly higher oxidative-stress levels were detected in preeclampsia both with and without FGR compared to normal pregnancies ($p < 0.001$) (Figure 1B).

### *2.3. NRF2-Regulated Transcriptional Activation*

Eighty-six of the 145 included pregnancies had been analyzed in a previous decidual microarray transcriptional data set [26] and were reanalyzed for the current study. The transcriptional data were available for 51 normal pregnancies, 11 preeclamptic pregnancies without FGR, and 24 preeclamptic pregnancies with FGR. The comparison between preeclampsia with and without FGR showed that preeclampsia without FGR was associated with a reduced overall expression of NRF2-regulated transcripts ($p = 0.048$) and the transcripts for "antioxidant proteins" ($p = 0.02$) and the "chaperone and stress response proteins" ($p < 0.05$) (Table 2). In contrast, the comparison between preeclampsia with FGR and normal pregnancies showed that higher expression of transcripts for "chaperone and stress response proteins" was associated with preeclampsia with FGR ($p < 0.001$) (Table 2). The decidual expression of the main "antioxidant protein" heme oxygenase 1 (HO-1) (Supplementary Table S1) was significantly reduced in preeclampsia without FGR compared to preeclampsia with FGR (Table 2). The expression of the HO-1 protein was assessed by reanalyzing the data from a previous study [30]. HO-1 was expressed in most cell types in the decidua (Supplementary Figure S1), and the expression level was significantly higher in preeclampsia overall compared to normal pregnancies ($p = 0.01$) (Figure S2A). A comparison of the overall decidual HO-1 protein expression between preeclampsia with and without FGR showed no differences, but included only three cases of preeclampsia without FGR (Supplementary Figure S2B).

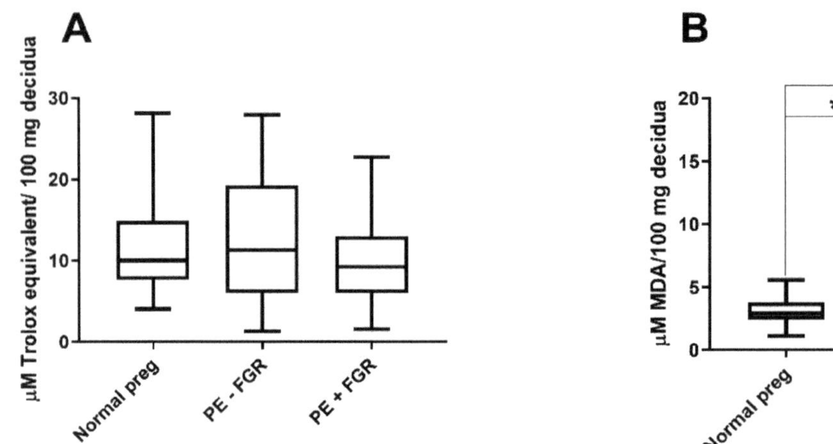


**Figure 1.** Non-enzymatic antioxidant capacity and oxidative-stress levels in decidua. Decidual non-enzymatic antioxidant capacity expressed as µM Trolox equivalent levels per 100 mg decidual tissue (**A**). Decidual oxidative-stress levels assessed by malondialdehyde (MDA) concentrations per 100 mg decidual tissue (**B**). Comparisons between normal pregnancies (preg) and preeclampsia (PE) with or without fetal growth restriction (FGR) (**A**,**B**). Boxes with medians extend from the 25th to 75th percentiles. ** $p < 0.001$.

**Table 2.** Gene-set enrichment analysis of NRF2-regulated functional gene sets.

| Comparison between Diagnostic Groups | | Gene-Set Description | No. of Transcr. | ES | NES | *p*-Value |
|---|---|---|---|---|---|---|
| PE − FGR | Normal preg | Antioxidant proteins | 18 | −0.47 | −1.39 | 0.08 |
| PE − FGR | Normal preg | Phase I and II metabolizing enzymes | 48 | −0.25 | −0.85 | 0.70 |
| PE − FGR | Normal preg | Chaperone and stress response proteins | 43 | −0.23 | −0.82 | 0.76 |
| PE − FGR | Normal preg | Phase III detoxifying proteins | 4 | 0.31 | 0.59 | 0.90 |
| PE − FGR | Normal preg | Ubiquitination/proteasomal degradation | 5 | 0.37 | 0.74 | 0.77 |
| PE − FGR | Normal preg | All five gene sets | 118 | −0.24 | −1.01 | 0.44 |

**Table 2.** *Cont.*

| Comparison between Diagnostic Groups | | Gene-Set Description | No. of Transcr. | ES | NES | *p*-Value |
|---|---|---|---|---|---|---|
| PE + FGR | Normal preg | Antioxidant proteins | 18 | 0.46 | 1.38 | 0.10 |
| PE + FGR | Normal preg | Phase I and II metabolizing enzymes | 48 | −0.33 | −1.23 | 0.24 |
| PE + FGR | Normal preg | Chaperone and stress response proteins | 43 | 0.45 | 1.73 | <0.001 |
| PE + FGR | Normal preg | Phase III detoxifying proteins | 4 | 0.55 | 1.07 | 0.45 |
| PE + FGR | Normal preg | Ubiquitination/proteasomal degradation | 5 | 0.71 | 1.43 | 0.07 |
| PE + FGR | Normal preg | All five gene sets | 118 | 0.32 | 1.32 | 0.08 |
| PE − FGR | PE + FGR | Antioxidant proteins | 18 | −0.60 | −1.66 | 0.02 |
| PE − FGR | PE + FGR | Phase I and II metabolizing enzymes | 48 | 0.31 | 0.99 | 0.55 |
| PE − FGR | PE + FGR | Chaperone and stress response proteins | 43 | −0.44 | −1.59 | <0.05 |
| PE − FGR | PE + FGR | Phase III detoxifying proteins | 4 | −0.60 | −1.06 | 0.45 |
| PE − FGR | PE + FGR | Ubiquitination/proteasomal degradation | 5 | −0.71 | −1.43 | 0.07 |
| PE − FGR | PE + FGR | All five gene sets | 118 | −0.36 | −1.41 | <0.05 |

ES, enrichment score [31]; FGR, fetal growth restriction; NES, normalized enrichment score [31]; PE, preeclampsia; preg, pregnancy.

### 2.4. Cellular Quantitative Decidual NRF2 and KEAP1 Expression

The NRF2- and KEAP1-protein expression in the decidua was assessed in 88 and 82 pregnancies, respectively. In the decidua, extravillous trophoblasts were observed both clustered together surrounded by fibrinoid tissue and as single cells in close proximity to maternal decidual stroma cells, leukocytes, and macrophages (Figure 2). The interstitial clusters of extravillous trophoblasts contained defined multinucleated trophoblast giant cells (Figure 3).

The decidual expression of NRF2 and KEAP1 was detected in both fetal cells and in maternal tissue lacking trophoblasts (Figure 2). NRF2 and KEAP1 were strongly expressed by fetal extravillous trophoblasts, multinucleated trophoblast giant cells, maternal decidual stroma cells and leukocytes, while muscle cells/myometrial cells showed a weaker expression (Figure 2). The decidual NRF2 and KEAP1 expression was localized to the nucleus and cytoplasm and comparably distributed in normal and preeclamptic pregnancies (Figure 3). The nuclear localization of NRF2 was confirmed by the staining of the phosphorylated form of NRF2 (Supplementary Figure S3). The amount and density of decidual trophoblasts did not differ between normal and preeclamptic pregnancies (data not shown).

The decidual NRF2 and KEAP1 expression was quantified in decidual trophoblasts and in maternal tissue without trophoblasts (Figure 4 and Supplementary Table S2), using a novel automated protein-quantification method (Figure 5). The trophoblast-dependent NRF2 expression tended to be higher in preeclamptic pregnancies without FGR than in normal pregnancies and preeclamptic pregnancies with FGR ($p = 0.10$ and $p = 0.09$, respectively) (Figure 4A and Supplementary Table S2). No difference in NRF2-protein expression was detected in maternal tissue without the presence of trophoblasts (Figure 4A). The decidual KEAP1-expression pattern differed from the pattern for NRF2 by being more strongly located to extravillous trophoblast-rich areas, and decidual KEAP1 in trophoblast-rich areas showed increased expression in preeclampsia without FGR compared to both normal pregnancies and preeclampsia with FGR ($p = 0.049$ and $p = 0.02$, respectively) (Figure 4B). The KEAP1-protein expression in maternal tissue (Figure 4B) and the overall KEAP1 expression (protein and mRNA) in decidual tissue did not differ between groups (data not shown).

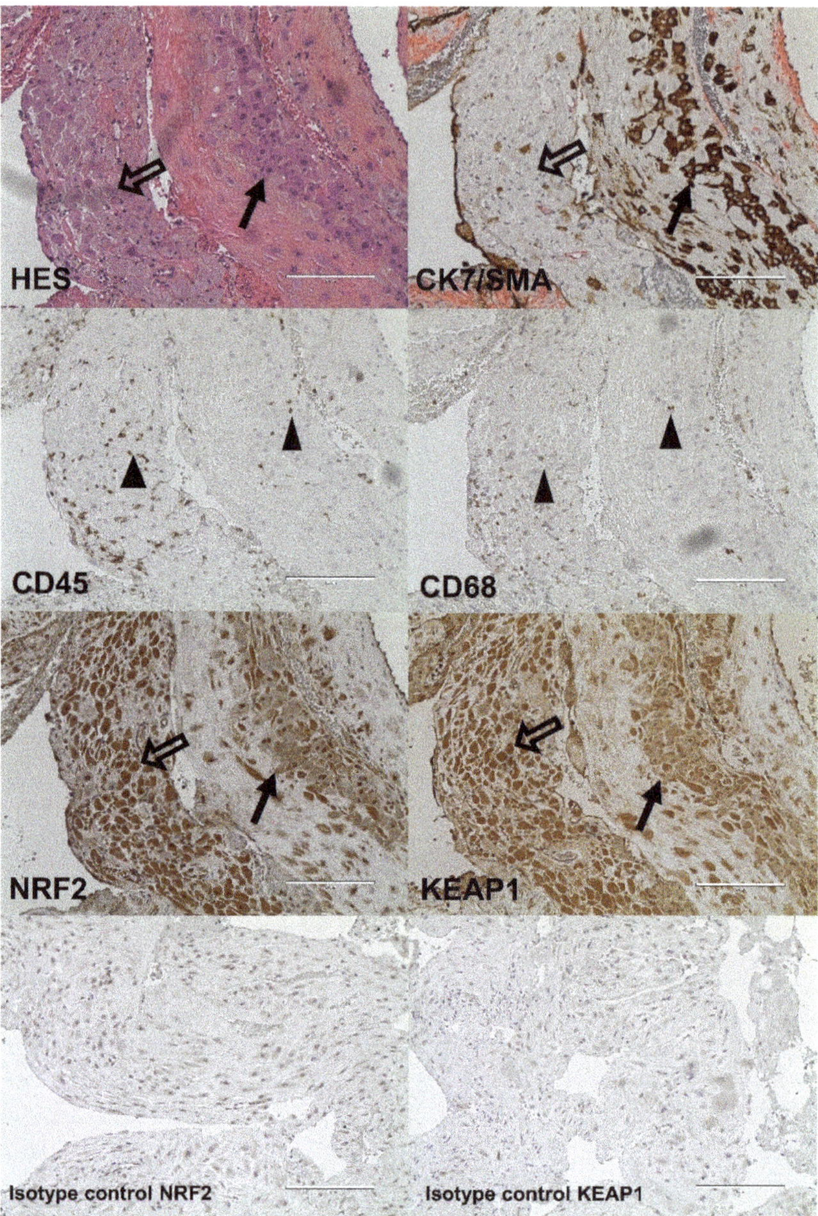



**Figure 2.** Decidual tissue from a preeclamptic pregnancy without fetal growth restriction stained as indicated with hematoxylin and eosin (HES) and markers for trophoblasts (cytokeratin 7 (CK7)), leukocytes (CD45), macrophages (CD68), nuclear factor erythroid 2-related factor 2 (NRF2), Kelch-like ECH-associated protein 1 (KEAP1), or negative isotype controls (20× magnification). CK7 was counterstained with smooth muscle actin (SMA). Black arrows indicate trophoblasts, transparent arrows indicate maternal decidual stroma cells, and black triangles indicate maternal leukocytes. Scale bar: 200 μM.

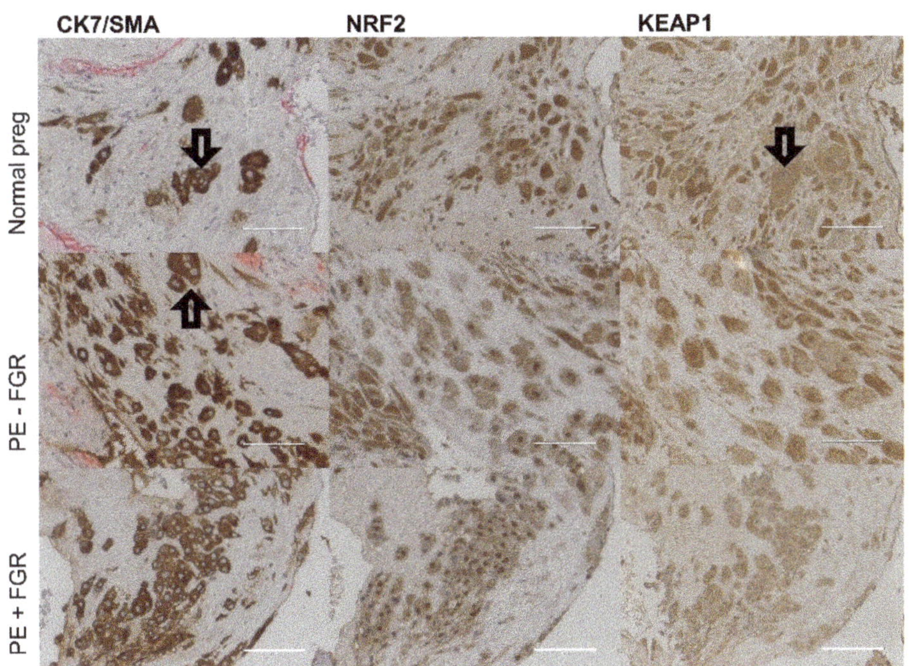


**Figure 3.** Decidual expression of trophoblasts (cytokeratin 7 (CK7)), nuclear factor erythroid 2-related factor 2 (NRF2), and Kelch-like ECH-associated protein 1 (KEAP1) in normal pregnancy, preeclampsia (PE) without fetal growth restriction (FGR), and PE with FGR (40× magnification). CK7 was counterstained with smooth muscle actin (SMA). The transparent arrows indicate examples of trophoblast giant cells. Scale bar: 100 μM.

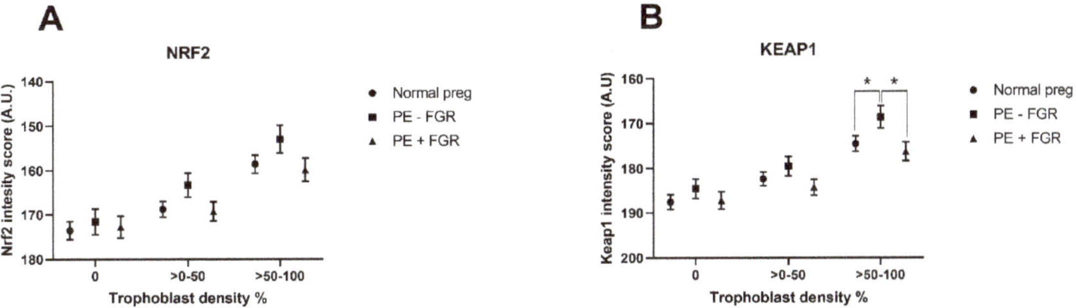


**Figure 4.** Decidual protein expression levels of nuclear factor erythroid 2-related factor 2 (NRF2) and Kelch-like ECH-associated protein 1 (KEAP1) in maternal cells and fetal trophoblasts, and maternal tissue without trophoblasts (0% trophoblast density). The expression of NRF2 (**A**) and KEAP1 (**B**) related to trophoblast density was compared between normal pregnancies (preg) and preeclampsia (PE) with or without fetal growth restriction (FGR). Expression levels are shown as estimated means with standard errors of means. * $p < 0.05$.

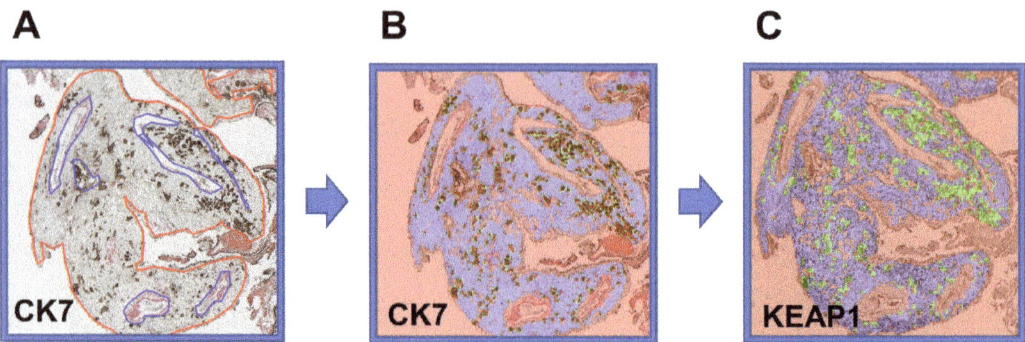


**Figure 5.** Automated identification of trophoblasts and maternal tissue in decidua for cellular quantification of nuclear factor erythroid 2-related factor 2 (NRF2)- and Kelch-like ECH-associated protein 1 (KEAP1)-protein expression using a custom MATLAB script. Tissue scans of slides stained with the trophoblast marker CK7 were used to determine included (red) and excluded (blue) tissue areas (**A**). Positive CK7 staining identified trophoblast-rich areas (green patches), and lack of CK7 identified maternal-tissue areas (blue patches) (**B**). A mask of patches was created for each decidua based on CK7-positive staining and used to relate NRF2- and KEAP1-expression levels to trophoblasts and maternal tissue in the spatially aligned NRF2 or KEAP1 tissue scans (**C**).

## 3. Discussion

This study shows the protein expression of the oxidative-stress regulator NRF2 and its inhibitor KEAP1 by both extravillous trophoblasts and maternal cells in the decidua. By using a novel automated image-based quantification of protein expression, we identified a distinct decidual NRF2-regulated oxidative-stress response in preeclampsia with normal fetal growth. This response was characterized by increased trophoblast-dependent KEAP1 expression and corresponding inhibition of the expression of NRF2-regulated antioxidant-response genes, and with a further corresponding increase in the decidual oxidative-stress levels. Preeclamptic pregnancies with FGR showed increased NRF2-regulated stress-response genes and oxidative stress, but a maternal or fetal effect on decidual NRF2- or KEAP1-protein expression was not identified in those pregnancies.

Our findings are in accordance with other studies that have shown decidual NRF2 expression in extravillous trophoblasts and maternal cells such as decidual stroma cells, myometrial cells, and leukocytes [24,25]. However, while Kweider et al. described exclusive cytoplasmic expression of NRF2 in extravillous trophoblasts [24], our findings show a clear nuclear expression of NRF2 in several cell types, paralleled by transcriptional functionality. This discrepancy may be due to the use of different antibodies. In smaller study populations, the decidual expression of NRF2 has been shown to be upregulated in trophoblasts in early-onset preeclampsia with FGR [24] and downregulated in isolated FGR [25]. To our knowledge, decidual KEAP1-protein expression has not previously been reported, and the presented cytoplasmic and nuclear expression of KEAP1 is supported by its role in sequestering cytoplasmic NRF2 through an active Crm1/exportin-dependent nuclear-transport mechanism [32]. We identified FGR-associated regulation of the oxidative-stress response in preeclampsia, as evidenced by increased decidual trophoblast-dependent KEAP1-protein expression in preeclampsia without FGR. Similar upregulation of Keap1 mRNA could not be detected by microarray, probably since it was performed on decidual tissue consisting of both maternal and fetal cells.

A microarray meta-analysis of the placenta has shown transcriptional downregulation of Nrf2 and Keap1 in preeclampsia [33]. Both reduced [23] and increased [21,22] placental nuclear NRF2-protein expression have been reported in preeclampsia. Likewise, studies assessing NRF2-regulated enzyme activity and the protein expression of NRF2 targets have reported both increases and reductions associated with preeclampsia [25,34–37]. Our novel identification of a diverging regulation of NRF2-regulated oxidative-stress responses in preeclampsia subgroups provides a probable explanation for the lack of consistency in existing data that have not distinguished between preeclampsia subgroups. The NRF2-regulated genes were assessed as overall response pathways in this study. Further gene or protein expression analysis of specific NRF2 target genes would strengthen the findings. Newer technologies such as single-cell RNA sequencing may provide a better approach to understanding this heterogenous tissue consisting of both maternal and fetal cells [38,39].

The increased NRF2-mediated gene activation in preeclampsia with FGR presented here was not supported by the decidual transcriptomics profiling by Tong et al., but they included other subgroups of preeclampsia (i.e., early onset and late onset) and had only three individuals per group [38,40]. The distinct NRF2-mediated regulation correlates, however, with our previous findings of increased decidual ER-stress responses associated with preeclampsia with FGR [41]. Oxidative and ER stress are closely linked and activate NRF2 [42]. ER stress results in the accumulation of misfolded proteins. The increased NRF2-regulated "chaperone and stress response proteins" genes in preeclampsia with FGR may indicate that the accumulation of unfolded proteins in the decidua is a central challenge in this preeclampsia subgroup. The downstream effect of placental ER stress and responses is implicated in FGR, as it may lead to perturbations of post-translational modifications and reduced translation of proteins, resulting in a small and insufficient placenta [42].

The decidual expression of KEAP1 was selectively increased in preeclampsia without FGR, potentially resulting in a reduced NRF2/KEAP1 ratio, supporting a net inhibition of NRF2 activation as an explanation for the corresponding reduction in the NRF2-regulated gene expression. This KEAP1 inhibition by fetal cells was associated with improved fetal outcomes, suggesting that decidual KEAP1 represents a fetal protective mechanism in preeclampsia; this is a novel observation in humans. Our findings are partly supported in a study of mice with pregnancy-associated hypertension where KEAP1 knockdown resulted in reduced fetal weight, and reduced placental angiogenesis stimulated by increased ROS production was suggested as a causative mechanism [43]. In further agreement, NRF2 knockdown has been shown to improve maternal and fetal outcomes in similar murine studies [44,45]. Accepting somewhat higher levels of oxidative stress to improve the placental vascularization in pregnancies with vascular dysfunction, such as in preeclampsia, may be a beneficial fetal compromise. Stronger KEAP1 inhibition may allow for a moderate increase in ROS, stimulating beneficial and immune-suppressive T-cell activity in the decidual microenvironment [6,46].

Chronic NRF2 activation may affect the renin–angiotensin system [47–49]. A prolonged activation of NRF2 has been related to hypertension, kidney injury, and cardiac maladaptation, and these adverse effects were reverted by NRF2 inhibition in murine diabetes [48] and cardiac-pressure-overload models [47]. Prolonged NRF2 activation results in an increased angiotensin II/angiotensin 1-7 ratio, facilitating development of hypertension [48]. The same perturbation of this ratio is observed in maternal plasma and urine in preeclampsia [50], and may be linked to chronic NRF2 activation from oxidative stress. Increased decidual expression of angiotensin II has been reported in preeclampsia [51]. The decidual angiotensin II could act on adjacent fetal chorionic villi, where angiotensin II receptor type 1 expression is increased in preeclampsia, and thereby induce vasoconstriction that impairs fetal blood supply [51–53]. Our study supports the implications of these previous studies, that chronic stress-induced decidual NRF2 activation might be a detrimental step in the development of preeclampsia potentially resulting in FGR through its effect on the renin–angiotensin system. Increased KEAP1 expression in preeclampsia without FGR, as observed here, potentially counteracts these adverse effects of NRF2 activation.

This study substantiates a divergent regulation of the decidual NRF2-mediated oxidative-stress response in preeclampsia with and without FGR, and suggests that the inhibitor KEAP1 is an important regulator. The complex role of NRF2-regulated oxidative stress in pregnancy warrants a more detailed characterization of the NRF2 system locally at the maternal–fetal interface throughout gestation, in relation to maternal–fetal outcomes. The unbiased automated cellular quantification method developed for our study allows for such follow-up studies. The trophoblast-dependent downregulation of NRF2-mediated oxidative-stress responses identifies a role for decidual maternal–fetal interaction in the regulation of oxidative-stress responses in pregnancy.

## 4. Materials and Methods

### 4.1. Study Participants and Decidual Biopsies

The Preeclampsia Study includes healthy and preeclamptic singleton pregnancies delivered by caesarean section (CS) in the absence of labor at St. Olavs and Haukeland University Hospitals between 2002 and 2012. Pregnant women diagnosed with preeclampsia with or without FGR were included as cases. Healthy normotensive pregnant women with no previous history of preeclampsia or FGR were included as normal pregnant controls. For the current study, women with immunosuppressive medications, pre-existing hypertension, or gestational diabetes mellitus were excluded. Preeclampsia was defined as persistent hypertension exceeding 140/90 mmHg plus proteinuria $\geq$0.3 g/24 h or $\geq$+1 by dipstick after 20 weeks of gestation. FGR was diagnosed by serial ultrasound measurements showing reduced intrauterine growth, or by birth weights < the 5th percentile of Norwegian reference curves [54] combined with clinically and sonographically suspected FGR and/or postpartum defined placental pathology.

Decidua basalis tissue was collected by vacuum suction of the placental bed during caesarean section [18,55]. None of the women showed signs of labor prior to the caesarean section. The samples were placed in RNAlater or in 10% neutral-buffered formalin and paraffin embedded or snap frozen in liquid nitrogen within 30 min of collection.

The Norwegian Regional Committee for Medical and Health Research Ethics approved the study (REC no. 2012/1040), and written informed consent was obtained from each participant.

### 4.2. Non-Enzymatic Antioxidant-Capacity Assay

The non-enzymatic antioxidant capacity was measured as the 3-ethylbenzothiazoline-6-sulphonic acid (ABTS)-radical-scavenging activity [56]. In brief, ABTS radicals were generated by mixing 2 mL each of ABTS (7.4 mM, #MAK187, Sigma-Aldrich, St. Louis, MO, USA, Total Antioxidant Assay KIT) and potassium peroxodisulfate (2.6 mM, #60489, Sigma-Aldrich, St. Louis, MO, USA). Decidual tissue lysates were prepared at 4 °C by homogenizing tissue samples in assay buffer from the Total Antioxidant Assay KIT (#MAK187, Sigma-Aldrich, St. Louis, MO, USA) (50 mg/250 µL) with a probe sonicator (10 s with 4 cycles/s) and centrifuged (12,000× $g$ for 15 min) before collecting the supernatant. Reactions were carried out by incubating 190 µL of ABTS-radical solution and 10 µL of decidual lysate for 30 min. The green color of the ABTS radicals scavenged by decidual-lysate antioxidants was measured spectrophotometrically at 731 nm. The vitamin E equivalent Trolox (Sigma-Aldrich, St. Louis, MO, USA, Total Antioxidant Assay KIT) was used as a standard and quantified as µM Trolox/100 mg of decidua, representing the non-enzymatic antioxidant capacity.

### 4.3. Measuring Oxidative-Stress Levels by a Malondialdehyde (MDA) Assay

The total decidual MDA content analyzed with the Lipid Peroxidation (MDA) Assay Kit (#MAK085, Sigma-Aldrich, St. Louis, MO, USA) was used as a measurement of the decidual oxidative-stress levels [57]. In brief, a mixture of 100 µL of decidual extracts (10 µg of tissue/300 µL of extracting buffer) and 300 µL of thiobarbituric acid (TBA, #MAK085, Sigma-Aldrich Assay Kit, St. Louis, MO, USA) solution was incubated at 95 °C for 60 min. Of the reaction mixture, 150 µL was analyzed spectrophotometrically in duplicate at 532 nm. The decidual MDA level was estimated using the MDA standard provided with the kit.

*4.4. NRF2-Regulated Transcriptional Activation*

Decidual microarray transcriptional data from the pregnancies included in this study were published previously [26], and were preprocessed in Sequential Oligogenic Linkage Analysis Routines (SOLAR) [26,58], in accordance with the Minimum Information About a Microarray Experiment (MIAME) guidelines [59]. The data were submitted to ArrayExpress (www.ebi.ac.uk/arrayexpress/ (accessed on 4 May 2014)) under accession no. E-TABM-682. For the current study, downstream targets of NRF2 in "the NRF2-mediated oxidative-stress response pathway" were divided into five functional gene sets identified by Ingenuity Pathway Analysis (QIAGEN Inc., Germantown, MD, USA); 1, "antioxidant proteins"; 2, "phase I and II metabolizing enzymes"; 3, "chaperone and stress response proteins"; 4, "phase III detoxifying proteins"; and 5, "ubiquitination and proteasomal degradation" (Supplementary Table S1). Gene-set enrichment analysis (100 permutations) was run on the five gene sets in the Partek Genomics Suite 6.6 [31].

*4.5. Immunohistochemistry*

Parallel decidual tissue sections (3 μm) were pre-treated in PT link (#PT101, Dako, Glostrup, Denmark) using target retrieval solution (#K8004, Dako, Glostrup, Denmark) at 97 °C for 20 min, and next treated with peroxidase blocking solution (#K4007 or #K5361, Dako, Glostrup, Denmark). The tissue sections were incubated with primary antibodies for KEAP1 (1:150, #10503-2-AP, Proteintech, Rosemount, IL, USA, room temperature for 40 min); NRF2 (1:200, #PA1828, Bosterbio, Pleasanton, CA, USA, overnight at 4 °C); pNRF2 (1:300, #ab76026, Abcam, Cambridge, UK, room temperature for 40 min); cytokeratin 7 (CK7) (1:300, #M7018, Dako, Glostrup, Denmark, room temperature for 45 min); CD45 (1:150, #M0701, Dako, Glostrup, Denmark, room temperature for 40 min); or CD68 (1:6000, #M0718, Dako, Glostrup, Denmark, room temperature for 40 min). All the sections were incubated for 30 min with HRP-labeled polymer (#K4007, Dako, Glostrup, Denmark) and for 10 min with DAB+ as a chromogen (1:50, #K4007 or #K5361, Dako, Glostrup, Denmark). The CK7 sections were double-stained with smooth muscle actin antibodies (1:300, #M0851, Dako, Glostrup, Denmark) with the EnVision G|2 Doublestain System Rabbit/Mouse (DAB+/Permanent Red) Kit system (#K5361, Dako, Glostrup, Denmark). The staining was performed using an Autostainer Plus (#S3800, Dako, Glostrup, Denmark) for KEAP1 and CK7, and manually for NRF2. The sections were counterstained with hematoxylin. Negative isotype controls for KEAP1 and NRF2 were included (1:67, Rabbit IgG #NBP2-24891, Novus, St. Charles, MO, USA, and 1:240, CD3 #A0452, Dako, Glostrup, Denmark). Additional routine staining with hematoxylin (75290, Chemi-Teknik, Oslo, Norway), erythrosine 239 (720-0179, VWR, Radnor, PA, USA), and saffron (75100, Chemi-Teknik, Oslo, Norway) (HES) was performed using a Sakura Tissue-Tek © Prisma Stainer™ (Sakura Finetek, Alphen aan den Rijn, the Netherlands). The reanalysis of HO-1 expression from a previous study [30] is included in the Supplementary Materials.

*4.6. Automated Quantification of Protein Expression*

Parallel sections of decidual tissue stained for NRF2, KEAP1, and CK7 by immuno-histochemistry were used for protein quantification. Tissue scans were obtained with the EVOS™ FL Auto Imaging System (Thermo Fisher Scientific, Waltham, MA, USA) using 20× magnification and defined microscope settings. To ensure a representative analysis, each tissue scan consisted of 9 to 100 bright-field TIFF images (2048 × 1536 pixels) per sample section depending on the amount of available tissue. A custom ImageJ script was used for the background correction and stitching of images [60–62]. The large tissue scans were further analyzed using custom MATLAB scripts (MathWorks, Natick, MA, USA, version 2017a) developed for the identification and automatic quantification of staining intensity [63–65], with the examiner blinded to the pregnancy outcome. Regions of the decidua with muscle cells, villous placental tissue, blood vessels, endometrial glands, and poor morphology were excluded by manually defining regions of disinterest (Figure 5A). A mask of patches (50 × 50 pixels, 662 μm × 662 μm) defining trophoblasts and maternal tissue without trophoblasts was created for each decidua based on CK7-positive staining (Figure 5B). The created masks were used to relate NRF2- and KEAP1-expression intensity to trophoblasts and maternal tissue in the spatially aligned NRF2 and KEAP1 tissue scans (Figure 5C). Trophoblasts were automatically counted, and the trophoblast density was calculated as the total number of trophoblasts divided by the total area of tissue ($mm^2$). The average NRF2- and KEAP1-intensity values were grouped according to trophoblast density; as maternal tissue (0% trophoblasts), low trophoblast density (>0–50%), and high trophoblast density (>50%). The overall decidual KEAP1-expression intensity was calculated as the average value of all the positive patches. The decidual protein-expression-intensity values were measured as gray-level-intensity values ranging from 0 (the absence of color, black) to 255 (the presence of all colors, white) after conversion from RGB to grayscale images. The staining intensity is, therefore, inversely proportional to the protein-expression level.

### 4.7. Statistical Methods

The statistical analyses were performed using the SPSS v. 25, GraphPad Prism 7.03, and Partek Genomic Suite 6.6 software. For the clinical data, one-way ANOVA or Kruskal–Wallis tests with Tukey's or Dunn's test, respectively, were used for comparisons of continuous variables, while Fisher's exact test was applied for categorical variables. For the non-enzymatic antioxidant capacity and oxidative-stress levels, outliers were detected with the Robust regression and Outlier removal (ROUT) method in GraphPad Prism 7.03, and the Kruskal–Wallis with Dunn's test was used for comparisons.

The mRNA level of Keap1 from the decidual microarray data set was compared between normal pregnancies and pregnancies with preeclampsia with or without FGR by the Kruskal–Wallis test.

For the immunohistochemistry data, the amount and density of decidual trophoblasts were compared between the study groups by one-way ANOVA with Tukey's test for pairwise comparisons. To compare the NRF2- and KEAP1-protein expression in maternal tissue (defined by 0% trophoblast density) and the overall decidual KEAP1-protein expression between the study groups, a linear regression model with the recruitment location and study group as additional covariates was used. To compare the NRF2 and KEAP1 expression in trophoblast-containing tissue (>0–50 or >50% trophoblasts), a linear mixed model with the recruitment location and trophoblast density interval as fixed-effects variables was used. Within-subject correlations were accounted for by including a subject-specific random intercept. The significance level was set to 0.05.

**Supplementary Materials:** Supporting information can be found at https://www.mdpi.com/article/10.3390/ijms23041966/s1.

**Author Contributions:** Conceptualization, S.B.M., G.A., L.B. and A.-C.I.; methodology, S.B.M., J.J.R., G.B.S., L.M.G., P.B., M.A., L.C.V.T. and E.K.M.; software, M.E. and J.O.; formal analysis, S.B.M., J.J.R. and S.S.B.; writing—original draft preparation, S.B.M., L.B. and A.-C.I.; writing—review and editing, S.B.M., J.J.R., G.B.S., L.M.G., M.A., P.B., M.E., S.S.B., J.O., L.C.V.T., E.K.M., G.A., L.B. and A.-C.I.; supervision, G.A., L.B. and A.-C.I. All authors have read and agreed to the published version of the manuscript.

**Funding:** This research was funded by the Research Council of Norway through its Centers of Excellence funding scheme, project numbers 223255 (CEMIR) and 223250 (CCBIO), and by the Liaison Committee between the Norwegian University of Science and Technology and The Central Norway Regional Health Authority. The Cellular and Molecular Imaging Core Facility (CMIC) at NTNU is funded by the Faculty of Medicine and Health Science at NTNU and the Central Norway Regional Health Authority.

**Institutional Review Board Statement:** The study was conducted according to the guidelines of the Declaration of Helsinki and approved by the Norwegian Regional Committee for Medical and Health Research Ethics (REC no. 2012/1040; 19 December 2012).

**Informed Consent Statement:** Informed consent was obtained from all the subjects involved in the study.

**Data Availability Statement:** The data presented in this study are available on request from the corresponding author. The data are not publicly available due to privacy.

**Acknowledgments:** The immunohistochemistry was performed by Ingunn Nervik at the Cellular and Molecular Imaging Core Facility (CMIC) at NTNU. Thanks go to Turid Follestad for statistical support.

**Conflicts of Interest:** The authors declare no conflict of interest.

## Abbreviations

| | |
|---|---|
| ABTS | 3-ethylbenzothiazoline-6-sulphonic acid |
| ARE | Antioxidant response element |
| CK7 | Cytokeratin 7 |
| ER | Endoplasmic reticulum |
| FGR | Fetal growth restriction |
| KEAP1 | Kelch-like ECH-associated protein 1 |
| MDA | Malondialdehyde |
| HO-1 | Heme oxygenase 1 |
| NRF2 | Nuclear factor erythroid 2-related factor 2 |
| ROS | Reactive oxygen species |


## References

1. Burton, G.J.; Jauniaux, E. Pathophysiology of placental-derived fetal growth restriction. *Am. J. Obs. Gynecol.* **2018**, *218*, S745–S761. [CrossRef]
2. Chaiworapongsa, T.; Chaemsaithong, P.; Yeo, L.; Romero, R. Pre-eclampsia part 1: Current understanding of its pathophysiology. *Nat. Rev. Nephrol.* **2014**, *10*, 466–480. [CrossRef]
3. Burton, G.J.; Redman, C.W.; Roberts, J.M.; Moffett, A. Pre-eclampsia: Pathophysiology and clinical implications. *BMJ* **2019**, *366*, l2381. [CrossRef]
4. Benirschke, K.; Burton, G.J.; Baergen, R.N. Nonvillous Parts and Trophoblast Invasion. In *Pathology of the Human Placenta*; Springer: Berlin/Heidelberg, Germany, 2012; pp. 157–240.
5. Burton, G.J.; Jauniaux, E. Oxidative stress. *Best Pr. Res. Clin. Obs. Gynaecol.* **2011**, *25*, 287–299. [CrossRef]
6. Nathan, C.; Cunningham-Bussel, A. Beyond oxidative stress: An immunologist's guide to reactive oxygen species. *Nat. Rev. Immunol.* **2013**, *13*, 349–361. [CrossRef]
7. Auten, R.L.; Davis, J.M. Oxygen toxicity and reactive oxygen species: The devil is in the details. *Pediatr. Res.* **2009**, *66*, 121–127. [CrossRef]
8. Dayalan Naidu, S.; Kostov, R.V.; Dinkova-Kostova, A.T. Transcription factors Hsf1 and Nrf2 engage in crosstalk for cytoprotection. *Trends Pharm. Sci.* **2015**, *36*, 6–14. [CrossRef] [PubMed]
9. Bryan, H.K.; Olayanju, A.; Goldring, C.E.; Park, B.K. The Nrf2 cell defence pathway: Keap1-dependent and -independent mechanisms of regulation. *Biochem. Pharm.* **2013**, *85*, 705–717. [CrossRef]
10. Ma, Q. Role of nrf2 in oxidative stress and toxicity. *Annu. Rev. Pharm. Toxicol.* **2013**, *53*, 401–426. [CrossRef] [PubMed]
11. Liu, T.; Lv, Y.F.; Zhao, J.L.; You, Q.D.; Jiang, Z.Y. Regulation of Nrf2 by phosphorylation: Consequences for biological function and therapeutic implications. *Free. Radic. Biol. Med.* **2021**, *168*, 129–141. [CrossRef] [PubMed]
12. Sun, Z.; Wu, T.; Zhao, F.; Lau, A.; Birch, C.M.; Zhang, D.D. KPNA6 (Importin {alpha}7)-mediated nuclear import of Keap1 represses the Nrf2-dependent antioxidant response. *Mol. Cell Biol.* **2011**, *31*, 1800–1811. [CrossRef] [PubMed]
13. Cuadrado, A.; Manda, G.; Hassan, A.; Alcaraz, M.J.; Barbas, C.; Daiber, A.; Ghezzi, P.; Leon, R.; Lopez, M.G.; Oliva, B.; et al. Transcription Factor NRF2 as a Therapeutic Target for Chronic Diseases: A Systems Medicine Approach. *Pharm. Rev.* **2018**, *70*, 348–383. [CrossRef] [PubMed]
14. Sun, Z.; Zhang, S.; Chan, J.Y.; Zhang, D.D. Keap1 controls postinduction repression of the Nrf2-mediated antioxidant response by escorting nuclear export of Nrf2. *Mol. Cell Biol.* **2007**, *27*, 6334–6349. [CrossRef]
15. Szczesny-Malysiak, E.; Stojak, M.; Campagna, R.; Grosicki, M.; Jamrozik, M.; Kaczara, P.; Chlopicki, S. Bardoxolone Methyl Displays Detrimental Effects on Endothelial Bioenergetics, Suppresses Endothelial ET-1 Release, and Increases Endothelial Permeability in Human Microvascular Endothelium. *Oxid. Med. Cell Longev.* **2020**, *2020*, 4678252. [CrossRef]
16. Lee, S.; Hu, L. Nrf2 activation through the inhibition of Keap1-Nrf2 protein-protein interaction. *Med. Chem. Res.* **2020**, *29*, 846–867. [CrossRef] [PubMed]
17. Brosens, I.; Pijnenborg, R.; Vercruysse, L.; Romero, R. The "Great Obstetrical Syndromes" are associated with disorders of deep placentation. *Am. J. Obs. Gynecol.* **2011**, *204*, 193–201. [CrossRef] [PubMed]

18. Staff, A.C.; Ranheim, T.; Khoury, J.; Henriksen, T. Increased contents of phospholipids, cholesterol, and lipid peroxides in decidua basalis in women with preeclampsia. *Am. J. Obs. Gynecol.* **1999**, *180*, 587–592. [CrossRef]

19. Zusterzeel, P.L.; Rutten, H.; Roelofs, H.M.; Peters, W.H.; Steegers, E.A. Protein carbonyls in decidua and placenta of pre-eclamptic women as markers for oxidative stress. *Placenta* **2001**, *22*, 213–219. [CrossRef]

20. Staff, A.C.; Halvorsen, B.; Ranheim, T.; Henriksen, T. Elevated level of free 8-iso-prostaglandin F2alpha in the decidua basalis of women with preeclampsia. *Am. J. Obs. Gynecol.* **1999**, *181*, 1211–1215. [CrossRef]

21. Wruck, C.J.; Huppertz, B.; Bose, P.; Brandenburg, L.O.; Pufe, T.; Kadyrov, M. Role of a fetal defence mechanism against oxidative stress in the aetiology of preeclampsia. *Histopathology* **2009**, *55*, 102–106. [CrossRef]

22. Feng, H.; Wang, L.; Zhang, G.; Zhang, Z.; Guo, W. Oxidative stress activated by Keap-1/Nrf2 signaling pathway in pathogenesis of preeclampsia. *Int. J. Clin. Exp. Pathol.* **2020**, *13*, 382–392.

23. Chigusa, Y.; Tatsumi, K.; Kondoh, E.; Fujita, K.; Nishimura, F.; Mogami, H.; Konishi, I. Decreased lectin-like oxidized LDL receptor 1 (LOX-1) and low Nrf2 activation in placenta are involved in preeclampsia. *J. Clin. Endocrinol. Metab.* **2012**, *97*, E1862–E1870. [CrossRef]

24. Kweider, N.; Huppertz, B.; Wruck, C.J.; Beckmann, R.; Rath, W.; Pufe, T.; Kadyrov, M. A role for Nrf2 in redox signalling of the invasive extravillous trophoblast in severe early onset IUGR associated with preeclampsia. *PLoS ONE* **2012**, *7*, e47055. [CrossRef] [PubMed]

25. Acar, N.; Soylu, H.; Edizer, I.; Ozbey, O.; Er, H.; Akkoyunlu, G.; Gemici, B.; Ustunel, I. Expression of nuclear factor erythroid 2-related factor 2 (Nrf2) and peroxiredoxin 6 (Prdx6) proteins in healthy and pathologic placentas of human and rat. *Acta Histochem.* **2014**, *116*, 1289–1300. [CrossRef] [PubMed]

26. Loset, M.; Mundal, S.B.; Johnson, M.P.; Fenstad, M.H.; Freed, K.A.; Lian, I.A.; Eide, I.P.; Bjorge, L.; Blangero, J.; Moses, E.K.; et al. A transcriptional profile of the decidua in preeclampsia. *Am. J. Obs. Gynecol.* **2011**, *204*, 84.e1–84.e27. [CrossRef]

27. Staff, A.; Henriksen, T.; Langsætre, E.; Magnussen, E.; Michelsen, T. Hypertensive svangerskapskomplikasjoner og eklampsi. In *Den Norske Legeforening*; Legeforeningen: Oslo, Norway, 2014.

28. Sibai, B.; Dekker, G.; Kupferminc, M. Pre-eclampsia. *Lancet* **2005**, *365*, 785–799. [CrossRef]

29. Thompson, J.M.; Irgens, L.M.; Skjaerven, R.; Rasmussen, S. Placenta weight percentile curves for singleton deliveries. *BJOG* **2007**, *114*, 715–720. [CrossRef]

30. Eide, I.P.; Isaksen, C.V.; Salvesen, K.A.; Langaas, M.; Schonberg, S.A.; Austgulen, R. Decidual expression and maternal serum levels of heme oxygenase 1 are increased in pre-eclampsia. *Acta Obs. Gynecol. Scand.* **2008**, *87*, 272–279. [CrossRef]

31. Subramanian, A.; Tamayo, P.; Mootha, V.K.; Mukherjee, S.; Ebert, B.L.; Gillette, M.A.; Paulovich, A.; Pomeroy, S.L.; Golub, T.R.; Lander, E.S.; et al. Gene set enrichment analysis: A knowledge-based approach for interpreting genome-wide expression profiles. *Proc. Natl. Acad. Sci. USA* **2005**, *102*, 15545–15550. [CrossRef]

32. Velichkova, M.; Hasson, T. Keap1 regulates the oxidation-sensitive shuttling of Nrf2 into and out of the nucleus via a Crm1-dependent nuclear export mechanism. *Mol. Cell Biol.* **2005**, *25*, 4501–4513. [CrossRef]

33. van Uitert, M.; Moerland, P.D.; Enquobahrie, D.A.; Laivuori, H.; van der Post, J.A.; Ris-Stalpers, C.; Afink, G.B. Meta-Analysis of Placental Transcriptome Data Identifies a Novel Molecular Pathway Related to Preeclampsia. *PLoS ONE* **2015**, *10*, e0132468.

34. Madazli, R.; Benian, A.; Aydin, S.; Uzun, H.; Tolun, N. The plasma and placental levels of malondialdehyde, glutathione and superoxide dismutase in pre-eclampsia. *J. Obs. Gynaecol.* **2002**, *22*, 477–480. [CrossRef]

35. Mutlu-Turkoglu, U.; Ademoglu, E.; Ibrahimoglu, L.; Aykac-Toker, G.; Uysal, M. Imbalance between lipid peroxidation and antioxidant status in preeclampsia. *Gynecol. Obs. Inves.* **1998**, *46*, 37–40. [CrossRef] [PubMed]

36. Wang, Y.; Walsh, S.W. Antioxidant activities and mRNA expression of superoxide dismutase, catalase, and glutathione peroxidase in normal and preeclamptic placentas. *J. Soc. Gynecol. Investig.* **1996**, *3*, 179–184. [CrossRef]

37. Wang, Y.; Walsh, S.W. Increased superoxide generation is associated with decreased superoxide dismutase activity and mRNA expression in placental trophoblast cells in pre-eclampsia. *Placenta* **2001**, *22*, 206–212. [CrossRef]

38. Tong, J.; Niu, Y.; Chen, Z.J.; Zhang, C. Comparison of the transcriptional profile in the decidua of early-onset and late-onset pre-eclampsia. *J. Obs. Gynaecol. Res.* **2020**, *46*, 1055–1066. [CrossRef] [PubMed]

39. Huang, J.; Li, Q.; Peng, Q.; Xie, Y.; Wang, W.; Pei, C.; Zhao, Y.; Liu, R.; Huang, L.; Li, T.; et al. Single-cell RNA sequencing reveals heterogeneity and differential expression of decidual tissues during the peripartum period. *Cell Prolif.* **2021**, *54*, e12967. [CrossRef] [PubMed]

40. Tong, J.; Zhao, W.; Lv, H.; Li, W.P.; Chen, Z.J.; Zhang, C. Transcriptomic Profiling in Human Decidua of Severe Preeclampsia Detected by RNA Sequencing. *J. Cell Biochem.* **2018**, *119*, 607–615. [CrossRef]

41. Lian, I.A.; Loset, M.; Mundal, S.B.; Fenstad, M.H.; Johnson, M.P.; Eide, I.P.; Bjorge, L.; Freed, K.A.; Moses, E.K.; Austgulen, R. Increased endoplasmic reticulum stress in decidual tissue from pregnancies complicated by fetal growth restriction with and without pre-eclampsia. *Placenta* **2011**, *32*, 823–829. [CrossRef]

42. Burton, G.J.; Yung, H.W.; Murray, A.J. Mitochondrial—Endoplasmic reticulum interactions in the trophoblast: Stress and senescence. *Placenta* **2017**, *52*, 146–155.

43. Nezu, M.; Souma, T.; Yu, L.; Sekine, H.; Takahashi, N.; Wei, A.Z.; Ito, S.; Fukamizu, A.; Zsengeller, Z.K.; Nakamura, T.; et al. Nrf2 inactivation enhances placental angiogenesis in a preeclampsia mouse model and improves maternal and fetal outcomes. *Sci. Signal.* **2017**, *10*. [CrossRef] [PubMed]

44. Li, L.; Li, H.; Xue, J.; Chen, P.; Zhou, Q.; Zhang, C. Nanoparticle-Mediated Simultaneous Downregulation of Placental Nrf2 and sFlt1 Improves Maternal and Fetal Outcomes in a Preeclampsia Mouse Model. *ACS Biomater. Sci. Eng.* **2020**, *6*, 5866–5873. [CrossRef] [PubMed]
45. Zhang, Y.; Liang, B.; Meng, F.; Li, H. Effects of Nrf-2 expression in trophoblast cells and vascular endothelial cells in preeclampsia. *Am. J. Transl. Res.* **2021**, *13*, 1006–1021.
46. Yarosz, E.L.; Chang, C.H. The Role of Reactive Oxygen Species in Regulating T Cell-mediated Immunity and Disease. *Immune Netw.* **2018**, *18*, e14. [CrossRef]
47. Qin, Q.; Qu, C.; Niu, T.; Zang, H.; Qi, L.; Lyu, L.; Wang, X.; Nagarkatti, M.; Nagarkatti, P.; Janicki, J.S.; et al. Nrf2-Mediated Cardiac Maladaptive Remodeling and Dysfunction in a Setting of Autophagy Insufficiency. *Hypertension* **2016**, *67*, 107–117. [CrossRef]
48. Zhao, S.; Ghosh, A.; Lo, C.S.; Chenier, I.; Scholey, J.W.; Filep, J.G.; Ingelfinger, J.R.; Zhang, S.L.; Chan, J.S.D. Nrf2 Deficiency Upregulates Intrarenal Angiotensin-Converting Enzyme-2 and Angiotensin 1-7 Receptor Expression and Attenuates Hypertension and Nephropathy in Diabetic Mice. *Endocrinology* **2018**, *159*, 836–852. [CrossRef] [PubMed]
49. Abdo, S.; Shi, Y.; Otoukesh, A.; Ghosh, A.; Lo, C.S.; Chenier, I.; Filep, J.G.; Ingelfinger, J.R.; Zhang, S.L.; Chan, J.S. Catalase overexpression prevents nuclear factor erythroid 2-related factor 2 stimulation of renal angiotensinogen gene expression, hypertension, and kidney injury in diabetic mice. *Diabetes* **2014**, *63*, 3483–3496. [CrossRef] [PubMed]
50. Spaan, J.J.; Brown, M.A. Renin-angiotensin system in pre-eclampsia: Everything old is new again. *Obs. Med.* **2012**, *5*, 147–153. [CrossRef]
51. Anton, L.; Merrill, D.C.; Neves, L.A.; Diz, D.I.; Corthorn, J.; Valdes, G.; Stovall, K.; Gallagher, P.E.; Moorefield, C.; Gruver, C.; et al. The uterine placental bed Renin-Angiotensin system in normal and preeclamptic pregnancy. *Endocrinology* **2009**, *150*, 4316–4325. [CrossRef] [PubMed]
52. Anton, L.; Merrill, D.C.; Neves, L.A.A.; Stovall, K.; Gallagher, P.E.; Diz, D.I.; Moorefield, C.; Gruver, C.; Ferrario, C.M.; Brosnihan, K.B. Activation of local chorionic villi angiotensin II levels but not angiotensin (1-7) in preeclampsia. *Hypertension* **2008**, *51*, 1066–1072. [CrossRef]
53. Mistry, H.D.; Kurlak, L.O.; Broughton Pipkin, F. The placental renin-angiotensin system and oxidative stress in pre-eclampsia. *Placenta* **2013**, *34*, 182–186. [CrossRef] [PubMed]
54. Johnsen, S.L.; Rasmussen, S.; Wilsgaard, T.; Sollien, R.; Kiserud, T. Longitudinal reference ranges for estimated fetal weight. *Acta Obs. Gynecol. Scand.* **2006**, *85*, 286–297. [CrossRef] [PubMed]
55. Harsem, N.K.; Staff, A.C.; He, L.; Roald, B. The decidual suction method: A new way of collecting decidual tissue for functional and morphological studies. *Acta Obs. Gynecol. Scand.* **2004**, *83*, 724–730. [CrossRef] [PubMed]
56. Joraholmen, M.W.; Skalko-Basnet, N.; Acharya, G.; Basnet, P. Resveratrol-loaded liposomes for topical treatment of the vaginal inflammation and infections. *Eur. J. Pharm. Sci.* **2015**, *79*, 112–121. [CrossRef] [PubMed]
57. Songstad, N.T.; Kaspersen, K.H.; Hafstad, A.D.; Basnet, P.; Ytrehus, K.; Acharya, G. Effects of High Intensity Interval Training on Pregnant Rats, and the Placenta, Heart and Liver of Their Fetuses. *PLoS ONE* **2015**, *10*, e0143095. [CrossRef]
58. Almasy, L.; Blangero, J. Multipoint quantitative-trait linkage analysis in general pedigrees. *Am. J. Hum. Genet.* **1998**, *62*, 1198–1211. [CrossRef]
59. Brazma, A.; Hingamp, P.; Quackenbush, J.; Sherlock, G.; Spellman, P.; Stoeckert, C.; Aach, J.; Ansorge, W.; Ball, C.A.; Causton, H.C.; et al. Minimum information about a microarray experiment (MIAME)—Toward standards for microarray data. *Nat. Genet.* **2001**, *29*, 365–371. [CrossRef]
60. Schindelin, J.; Arganda-Carreras, I.; Frise, E.; Kaynig, V.; Longair, M.; Pietzsch, T.; Preibisch, S.; Rueden, C.; Saalfeld, S.; Schmid, B.; et al. Fiji: An open-source platform for biological-image analysis. *Nat. Methods* **2012**, *9*, 676–682. [CrossRef]
61. Rueden, C.T.; Schindelin, J.; Hiner, M.C.; DeZonia, B.E.; Walter, A.E.; Arena, E.T.; Eliceiri, K.W. ImageJ2: ImageJ for the next generation of scientific image data. *BMC Bioinform.* **2017**, *18*, 529. [CrossRef]
62. Preibisch, S.; Saalfeld, S.; Tomancak, P. Globally optimal stitching of tiled 3D microscopic image acquisitions. *Bioinformatics* **2009**, *25*, 1463–1465. [CrossRef]
63. Gierman, L.M.; Silva, G.B.; Pervaiz, Z.; Rakner, J.J.; Mundal, S.B.; Thaning, A.J.; Nervik, I.; Elschot, M.; Mathew, S.; Thomsen, L.C.V.; et al. TLR3 expression by maternal and fetal cells at the maternal-fetal interface in normal and preeclamptic pregnancies. *J. Leukoc. Biol.* **2020**, *109*, 173–183. [CrossRef] [PubMed]
64. Silva, G.B.; Gierman, L.M.; Rakner, J.J.; Stødle, G.S.; Mundal, S.B.; Thaning, A.J.; Sporsheim, B.; Elschot, M.; Collett, K.; Bjørge, L.; et al. Cholesterol Crystals and NLRP3 Mediated Inflammation in the Uterine Wall Decidua in Normal and Preeclamptic Pregnancies. *Front. Immunol.* **2020**, *11*, 564712. [CrossRef] [PubMed]
65. Rakner, J.J.; Silva, G.B.; Mundal, S.B.; Thaning, A.J.; Elschot, M.; Ostrop, J.; Thomsen, L.C.V.; Bjørge, L.; Gierman, L.M.; Iversen, A.C. Decidual and placental NOD1 is associated with inflammation in normal and preeclamptic pregnancies. *Placenta* **2021**, *105*, 23–31. [CrossRef] [PubMed]

International Journal of
*Molecular Sciences*

*Article*

# Mutations That Affect the Surface Expression of TRPV6 Are Associated with the Upregulation of Serine Proteases in the Placenta of an Infant

Claudia Fecher-Trost [1], Karin Wolske [1], Christine Wesely [1], Heidi Löhr [1], Daniel S. Klawitter [1], Petra Weissgerber [1,2], Elise Gradhand [3], Christine P. Burren [4], Anna E. Mason [5], Manuel Winter [1] and Ulrich Wissenbach [1,*]

[1] Experimental and Clinical Pharmacology and Toxicology, Center for Molecular Signaling (PZMS), Saarland University, Buildings 61.4 and 46, 66421 Homburg, Germany; Claudia.Fecher-Trost@uks.eu (C.F.-T.); Karin.Wolske@uks.eu (K.W.); Christine.Wesely@uks.eu (C.W.); Heidi.Loehr@uks.eu (H.L.); s8daklaw@teams.uni-saarland.de (D.S.K.); Petra.Weissgerber@uks.eu (P.W.); Manuel.Winter@uks.eu (M.W.)

[2] Transgenic Technologies, Center for Molecular Signaling (PZMS), Saarland University, Building 61.4, 66421 Homburg, Germany

[3] Kinder- und Perinatalpathologie Dr. Senckenberg, Institut für Pathologie Universitätsklinikum Frankfurt/Main Theodor-Stern-Kai 7, 60590 Frankfurt, Germany; Elise.Gradhand@kgu.de

[4] Department of Translational Health Sciences, Bristol Medical School, University of Bristol, University Hospitals Bristol and Weston NHS Foundation Trust, Upper Maudlin St, Bristol BS2 8BJ, UK; Christine.Burren@uhbw.nhs.uk

[5] Histopathology Department, Aneurin Bevan University Health Board, Royal Gwent Hospital, Cardiff NP20 2UB, UK; Anna.Mason@wales.nhs.uk

[*] Correspondence: Ulrich.Wissenbach@uks.eu

**Citation:** Fecher-Trost, C.; Wolske, K.; Wesely, C.; Löhr, H.; Klawitter, D.S.; Weissgerber, P.; Gradhand, E.; Burren, C.P.; Mason, A.E.; Winter, M.; et al. Mutations That Affect the Surface Expression of TRPV6 Are Associated with the Upregulation of Serine Proteases in the Placenta of an Infant. *Int. J. Mol. Sci.* **2021**, 22, 12694. https://doi.org/10.3390/ijms222312694

Academic Editor: Hiten D Mistry

Received: 5 November 2021
Accepted: 22 November 2021
Published: 24 November 2021

**Publisher's Note:** MDPI stays neutral with regard to jurisdictional claims in published maps and institutional affiliations.



**Abstract:** Recently, we reported a case of an infant with neonatal severe under-mineralizing skeletal dysplasia caused by mutations within both alleles of the *TRPV6* gene. One mutation results in an in frame stop codon ($R_{510}$stop) that leads to a truncated, nonfunctional TRPV6 channel, and the second in a point mutation ($G_{660}R$) that, surprisingly, does not affect the $Ca^{2+}$ permeability of TRPV6. We mimicked the subunit composition of the unaffected heterozygous parent and child by coexpressing the TRPV6 $G_{660}R$ and $R_{510}$stop mutants and combinations with wild type TRPV6. We show that both the $G_{660}R$ and $R_{510}$stop mutant subunits are expressed and result in decreased calcium uptake, which is the result of the reduced abundancy of functional TRPV6 channels within the plasma membrane. We compared the proteomic profiles of a healthy placenta with that of the diseased infant and detected, exclusively in the latter two proteases, HTRA1 and cathepsin G. Our results implicate that the combination of the two mutant TRPV6 subunits, which are expressed in the placenta of the diseased child, is responsible for the decreased calcium uptake, which could explain the skeletal dysplasia. In addition, placental calcium deficiency also appears to be associated with an increase in the expression of proteases.

**Keywords:** TRPV6; placenta; calcium transport; skeletal dysplasia; serine proteases; subunit assembly; transient receptor potential

## 1. Introduction

TRPV6 is a $Ca^{2+}$ selective ion channel which shows a very restricted expression pattern. Human TRPV6 is expressed in a few glands, including acinar salivary and lacrimal glands, in parts of the small intestine, and in the trophoblast layer of the placenta [1–3]. In addition, TRPV6 is overexpressed in a number of malignancies, namely, prostate, mammary ovarial and endometrial cancer [1,2,4–10]. In the human population, two TRPV6 alleles, TRPV6a and TRPV6b, exist, leading to a coupled polymorphism with three distinct amino acid exchanges detected in TRPV6a ($R_{197}V_{418}T_{721}$) and TRPV6b ($C_{197}M_{418}M_{721}$) [1,11,12].

Whether this polymorphism has a functional consequence is not known. In recent publications, the effects of TRPV6 mutations altering the functionality of TRPV6 channels in humans were published [13–17]. Dysfunction of TRPV6 channels leads to transient neonatal hyperparathyroidism (HRPTTN) and is listed in the OMIM database (Online Mendelian Inheritance in Man). We recently described the case of an infant who suffers from neonatal severe under mineralizing skeletal dysplasia due to underlying severe transient hyperparathyroidism. Both the *TRPV6* alleles of the infant showed mutations (13): one mutation leads to an amino acid exchange of glycine 660 to arginine ($G_{660}R$) at the C-terminus of the TRPV6 protein, which is presumed to be localized intracellularly. The second *TRPV6* locus exhibits a mutation which leads to an in frame stop codon replacing an arginine coding triplet by a stop codon, $R_{510}$stop (stop mutant). The TRPV6 protein contains six hydrophobic transmembrane domains, and the pore region of the channel is located between the fifth and the sixth domain [18]. The $R_{510}$stop mutation is placed in the linker sequence between the fourth and the fifth transmembrane domain and results in a truncated protein without a pore region and any detectable $Ca^{2+}$ permeability. We demonstrate that the mutations of the affected child lead to an inadequate channel assembly and, as a consequence, to a reduced insertion of the maternal $G_{660}R$-mutant in combination with the truncated paternal TRPV6-$R_{510}$stop mutant into the plasma membrane.

In addition, we show by mass spectrometry that two serine proteases were only detectable in the placenta of the affected child. In addition, a protease is upregulated in a TRPV6 expressing human trophoblast cell line cultured under a low $Ca^{2+}$ condition.

## 2. Results

### 2.1. Functional Consequence of Mutations within TRPV6 Channel Subunits

An affected child who exhibits mutations within the *TRPV6* gene was recently analysed using whole exome sequencing [13,14]. The child showed a pronounced dysplasia of the skeleton and died after several months. One *TRPV6* allele of the child contained a mutation that leads to a $G_{660}R$ mutation in the very C-terminus of the coding sequence, whereas the second allele contained an in frame stop codon, $R_{510}$stop, which leads to a truncated protein without the pore region of the TRPV6 channel. We focused on the TRPV6 mutations and cloned a number of TRPV6 constructs in the dicistronic pCAGGS-IRES-GFP or IRES-RFP vectors, allowing the expression of TRPV6 independently from the fluorescent proteins. First, we analysed the $G_{660}R$ mutation present in the affected child. We introduced this mutation in the TRPV6 cDNA and expressed the construct in HEK293 cells, measured $Ca^{2+}$ uptake, and compared the result with wild type TRPV6 expressing cells (Figure 1A,B). Surprisingly, the $Ca^{2+}$ uptake is not significantly different compared to wild type TRPV6 expressing cells (Figure 1B). The peak value of the two constructs was not altered. TRPV6 channels consist of four identical subunits and, in the human placenta, both TRPV6 loci are expressed [1,19–21]. Therefore, we mimicked the TRPV6 expression of the nonaffected parents and the affected child by coexpressing wild type TRPV6 and the $G_{660}R$ mutant (maternal genotype), as well as wild type TRPV6 and the $R_{510}$stop mutant (paternal genotype) and $G_{660}R$ and $R_{510}$stop mutant which reflects the affected child (Figure 1D and Supplementary Figure S1). It can be seen that the combination of the expressed mutant TRPV6 variants strongly reduces the $Ca^{2+}$ uptake of expressing cells. The peak value of the combination present in the affected child is 48% of the maternal and 51% of the paternal combination (Figure 1E). The experiment also shows that the reduced $Ca^{2+}$ uptake is not an effect of the amount of functional TRPV6 channels, otherwise, one would expect to also see a reduced $Ca^{2+}$ signal using the paternal combination (TRPV6 WT and $R_{510}$stop mutant) which is not the case. To test that in the coexpressing experiments, both variants were synthesized and we expressed the combinations of constructs cloned in IRES-GFP vectors and in IRES-RFP vectors. Next, we asked if the amount of TRPV6 mutant proteins might be reduced in TRPV6 expressing cells as consequence of an unfolded protein response. Therefore, we expressed all constructs alone or as combinations that reflect the parents and the affected child. It can be seen on Western blots using two different TRPV6 specific antibodies that all constructs are present (Figure 1F). Thus, according to this experiment, there is no evidence that unfolded protein response/degradation occurs in the overexpressing cells. We also transfected the TRPV6 $R_{510}$stop mutant alone and did not detect a higher $Ca^2$ uptake, as seen in cells expressing the empty vector, which shows that the mutation completely abolishes the $Ca^{2+}$ uptake of the mutated TRPV6 protein in expressing cells. It should be mentioned that the construct of the stop mutant contained the full length *TRPV6* cDNA in which the stop triplet was inserted. This experiment confirms the Western blot experiment and shows that in HEK293 cells the in frame $R_{510}$stop codon present in the cDNA of *TRPV6* is not translated and leads to a truncated protein, as expected (~53 kDA). The experiment was performed to exclude that a read-through phenomenon occurs, as described by Li and Zhang [22].

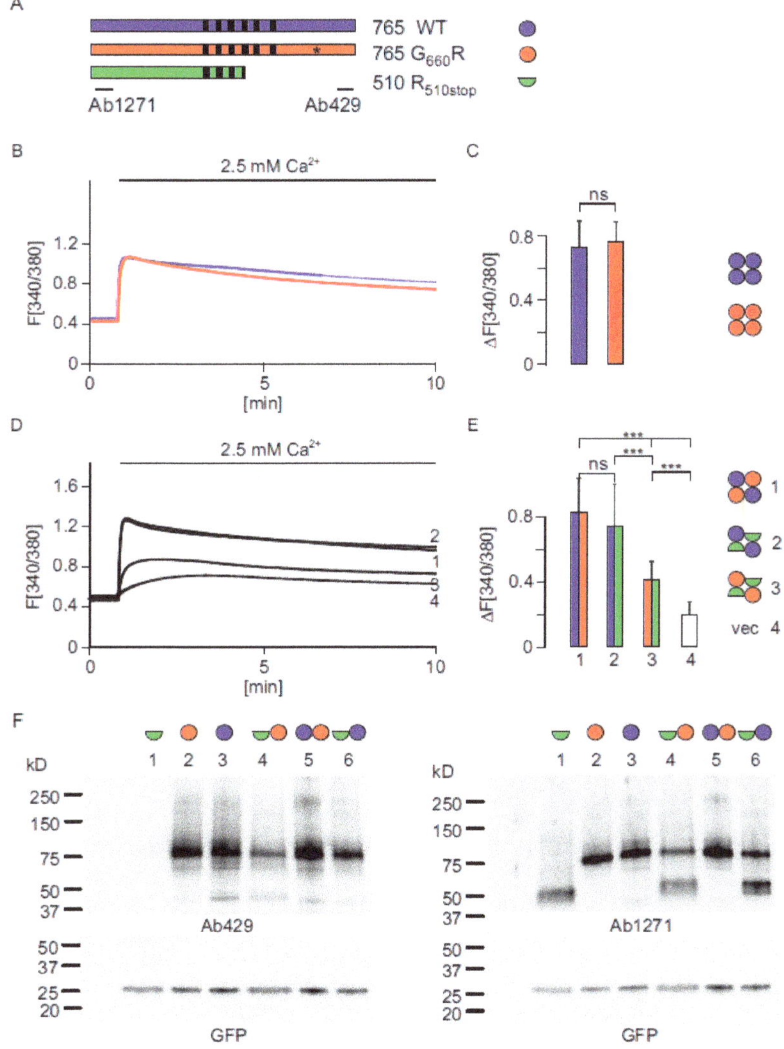

**Figure 1.** TRPV6 activity is reduced in HEK293 cells expressing mutant TRPV6 subunits present in the affected child. (**A**) TRPV6 constructs used for $Ca^{2+}$ imaging and Western blots, TRPV6 WT (blue), TRPV6-$G_{660}$R (red) and TRPV6 $R_{510}$stop

(green). Transmembrane domains (black bars), $G_{660}R$ mutation (*) and binding sites for TRPV6 specific antibodies 1271 and 429 are indicated. (**B,C**) $Ca^{2+}$ imaging of TRPV6 WT (blue, n/N = 132/3) and TRPV6-$G_{660}R$ (red, n/N = 126/3) in HEK293 cells and statistical analysis of the peak values. Circles indicate TRPV6 subunits. (**D,E**) Coexpression of TRPV6 WT-I-GFP and TRPV6-$G_{660}R$-I-RFP, which reflects the maternal TRPV6-genotype (1, blue/red, n/N = 117/3), coexpression of TRPV6 WT-I-RFP and TRPV6-$R_{510}$stop mutant I-GFP, which reflects paternal genotype (2, blue/green, n/N = 72/3), coexpression of TRPV6-$G_{660}R$-I-RFP and TRPV6-$R_{510}$stop mutant I-GFP, which reflects the child (3, red/green, n/N = 87/3), vector control (4, white, n/N = 82/2) and statistical analysis of peak values. n/N = cells/experiments. Asterisks assign significance differences (*** $p < 0.001$, ns = not significant). (**F**) Western blots of cells expressing TRPV6 constructs in HEK293 cells: lane1 TRPV6-$R_{510}$stop mutant (green semicircle), lane2 TRPV6-$G_{660}R$ mutant (red circle), lane 3 TRPV6 WT (blue circle), lane 4 coexpression of TRPV6-$R_{510}$stop and $G_{660}R$ mutants, lane 5 coexpression of TRPV6 WT and $G_{660}R$ mutant, lane 6 coexpression TRPV6 WT and TRPV6-$R_{510}$stop mutant. All TRPV6 variants were expressed as I-GFP constructs. Western blot was probed with antibody 429 (left) and antibody 1271 (right). GFP control below.

### 2.1.1. The $G_{660}R$ Mutation Can Be Rescued by Alanine

Next, we asked if the $G_{660}R$ mutant in combination with the $R_{510}$stop mutant leads to a decreased $Ca^{2+}$ uptake as the result of the positively charged amino acid arginine. If so, is it possible to obtain a rescuing effect by introducing an alanine residue instead ($G_{660}A$ mutation)? We coexpressed the $G_{660}A$ mutant with the $R_{510}$stop mutant and measured $Ca^{2+}$ signals comparable to the combination of wild type/$R_{510}$stop mutant, as present in the father (Figure 2A). This result shows that the $G_{660}A$ mutation rescues the $Ca^{2+}$ uptake. In addition, we replaced the $G_{660}$ residue with another positive charged amino acid, resulting in a $G_{660}K$ mutation (lysine, Figure 2B). This mutation had a similar effect to the $G_{660}R$ mutation, if coexpressed with the truncated $R_{510}$stop mutant. In addition, replacement by a negative charged amino acid, $G_{660}E$, greatly reduced $Ca^{2+}$ uptake (Figure 2C). We also tried to rescue the $G_{660}R$ mutation by introducing several mutations within the interacting sequence of the truncated $R_{510}$stop mutation (Figure 2D described in detail below). The data indicate that, at position 660 of the human TRPV6 sequence, positive as well as negative charged amino acids affect the function of the channel when coexpressed with the truncated TRPV6 $R_{510}$stop mutant. Therefore, one would expect that a noncharged amino acid at corresponding positions is strictly conserved within mammalian TRPV6 proteins and this is, indeed, the case. Although the $G_{660}A$ mutant rescued $Ca^{2+}$ uptake, within all mammalian TRPV6 sequences the $G_{660}$ residue is invariant (Figure 2E).

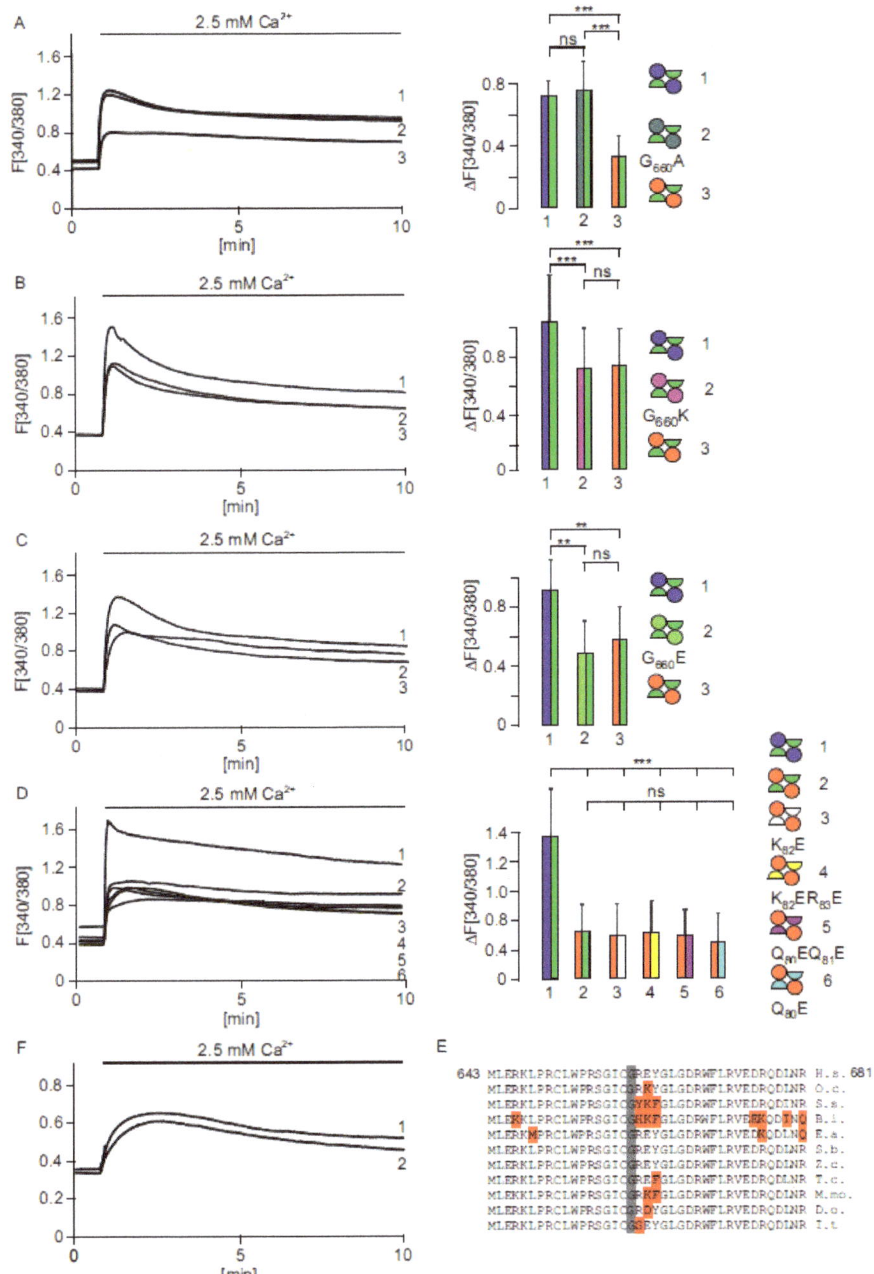



**Figure 2.** A TRPV6-G$_{660}$A mutation rescues the G$_{660}$R mutation. (**A**) Calcium imaging of cells coexpressing TRPV6 WT-I-GFP and TRPV6-R$_{510}$stop-I-RFP (1, blue/green, paternal, n/N = 95/3), TRPV6-R$_{510}$stop mutant I-RFP and TRPV6-G$_{660}$A-I-GFP (2, red/grey, n/N = 97/3) and TRPV6-G$_{660}$R I-GFP and TRPV6-R$_{510}$stop mutant I-RFP (3, green/red, child,

n/N = 91/3) n/N = cells/experiment. (**B**) Coexpression of TRPV6-$R_{510}$stop mutant I-RFP and TRPV6-$G_{660}$K-I-GFP (2, pink/green, n/N = 106/3) compared with the parental combination (1, blue/green, n/N = 79/3) and the child (3, red/green, n/N = 84/3). (**C**) Similar experiment as shown in (**B**), coexpression of TRPV6-$R_{510}$stop mutant I-RFP and TRPV6 $G_{660}$E-I-GFP (2, light green/green n/N = 55/3), compared with the parental combination (1, blue/green n/N = 65/3) and the child (3, red/green n/N = 44/3). (**D**) Coexpression of the TRPV6-$G_{660}$R-I-RFP with several TRPV6-$R_{510}$stop mutants cloned in I-GFP vectors, which, in addition, contain a second mutation within the N-terminal located sequence $QQKR_{83}$. This sequence interacts with the C-terminal sequence in which the $G_{660}$ residue is located. The following mutants were tested: $K_{82}$E (3, red/white, n/N = 50/3), $K_{82}ER_{83}$E (4, red/yellow, n/N = 44/3), $Q_{80}EQ_{81}$E (5, red/magenta, n/N = 65/3), and $Q_{80}$E (6, red/light blue, n/N = 45/3). The mutants were compared with the parental combination (1, blue/green, n/N = 52/3) and the child (2, red/green, n/N = 47/2). Here, n/N = cells/experiments. Asterisks assign significance differences (** $p < 0.01$, *** $p < 0.001$, ns = not significant). (**E**) Alignment of mammalian TRPV6 protein sequences from amino acid 643 to 681. $G_{660}$ is strictly conserved (grey). H.s., Homo sapiens; O.c., Oryctolagus cuniculus; S.s., Sus scrofa; B.i., Bos indicus; E.a., Equus asinus; S.b., Saimiri boliviensis; Z.c., Zalophus californianus; T.c., Tupaia chinensis; M.mo., Monodon Monoceros; D.o., Dipodomys ordii; I.t., Ictidomys tridecemlineatus. (**F**) Expression of artificial TRPV6 construct which contains amino acids 510 to 765 (1, n/N = 117/3) and coexpression with the same construct and the TRPV6-$R_{510}$stop mutant (2, n/N = 70/3).

### 2.1.2. Functional TRPV6 Channels Cannot Be Formed When the Subunits Are Expressed as Two Independent Parts

The TRPV6 stop mutant is characterized by the in frame stop codon which replaces $R_{510}$; thus, the stop mutant corresponds to amino acid 1-509 of the TRPV6 protein. We made a TRPV6 construct in which amino acids 1-509 are not present but $R_{510}$ was replaced by an artificial methionine, resulting in $M_{510}$. This construct contains the amino acids $M_{510}$-to $I_{765}$, which represent the complete C-terminus, including the pore region of the TRPV6 protein. We coexpressed the latter construct with the $R_{510}$stop mutant to test if cells can form functional TRPV6 channels (Figure 2F). We compared the coexpression with the single expression of TRPV6 $M_{510}$-$I_{765}$ but could not find significant differences. This indicates that functional TRPV6 channels cannot be formed from the $R_{510}$stop mutant in combination with the TRPV6 $M_{510}$-$I_{765}$ construct. Next, we analysed the position of the particular $G_{660}$ residue within the structure of the TRPV6 channel [18]. $G_{660}$ is located at the boundary surface of the TRPV6 subunits in a large distance to the pore of the TRPV6 channel. The location implicates an influence of subunit assembly rather than parameters influencing the functionality of the pore directly. We suggest, from the TRPV6 structure, that the $G_{660}$ that is located within the C-terminus of TRPV6 interacts with the N-terminal sequence of the adjacent TRPV6 subunit.

### 2.1.3. The $G_{660}$R Mutation Cannot Be Rescued by Mutations in the N-Terminus of the Interacting Subunit

We identified amino acid residues within the N-terminus of TRPV6, to be considered as interaction partners of the $G_{660}$ using the structural data published by Saotome and coworkers [18]. We identified, as a possible interacting sequence, a $QQKR_{83}$ motif within the N-terminus of TRPV6, with $K_{82}$ being at a distance of about 10.43Å to $G_{660}$. We cloned a number of constructs, introducing one or two negatively charged amino acids in the C-terminus of the $QQKR_{83}$ motif in the $R_{510}$stop mutant, and coexpressed these constructs with the $G_{660}$R mutant to see if a negative charged mutation in the truncated $R_{510}$stop mutant can rescue the effect of the $G_{660}$R mutation. We analysed four mutations within the $QQKR_{83}$ sequence, namely, the mutations $K_{82}$R, $K_{82}ER_{83}$E, $Q_{80}EQ_{81}$E and $Q_{80}$E, which were cloned into the truncated $R_{510}$stop mutant, and coexpressed these constructs with $G_{660}$R mutant and tested if these can rescue the $G_{660}$R mutation present in the full length protein. The expressed combinations of the four mutated truncated constructs showed a decreased $Ca^{2+}$ signal and did not rescue the $G_{660}$R mutation (Figure 2D, Supplementary Figure S1c).

### 2.1.4. The TRPV6-$R_{510}$stop Subunit Interacts with the Full Length TRPV6 Subunit

The previous experiment requires the assumption that the truncated stop mutant can still interact with the $G_{660}R$ mutant. This assumption is supported by the finding that the N-terminal ankyrin repeats which are important for the multimerization of the TRPV6 channel, are also present in the truncated mutant [23]. In addition, we performed a coimmunoprecipitation experiment, which shows that the truncated TRPV6 present in the child can interact with the full length TRPV6 protein (Figure 3A,B). We fused GFP to TRPV6 resulting in TRPV6-$R_{510}$-GFP and TRPV6-GFP. As shown earlier by Hirnet and coworkers, the TRPV6 protein occurs as glycosylated and non-glycosylated protein [24]. The glycosylation site is located in between transmembrane S1 and S2, and is present in the full length TRPV6 as well as in the truncated TRPV6-$R_{510}$-GFP variant. The TRPV6 protein was fished with the TRPV6 specific antibody 429, which binds to the C-terminus of TRPV6, and the co-immunoprecipitate (COIP) was analysed on a Western blot with a GFP antibody. Both proteins, TRPV6-$R_{510}$-GFP and TRPV6-GFP, as well as glycosylated forms, were detected by COIP, which shows that both proteins interact. Another COIP experiment using TRPV6-$G_{660}$R-RFP and TRPV6-$R_{510}$-GFP fusion proteins also shows that the mutant variants present in the affected child can interact (Figure 3C,D). The interaction of both fusion protein was also confirmed by mass spectrometry (Figure 3E).

### 2.1.5. The Amount of the Full Length TRPV6 Channel in the Plasma Membrane Is Reduced

Next, we asked if we can also detect the different TRPV6 subunits in the plasma membrane. We performed a biotinylation experiment, which shows that a small amount of the truncated TRPV6-$R_{510}$stop variant is detectable in the plasma membrane of expressing cells (Supplementary Figure S2A). Furthermore, we expressed the combinations of TRPV6 WT and TRPV6-$R_{510}$stop, as well as the combination of TRPV6-$G_{660}$R and TRPV6-$R_{510}$stop, and performed another biotinylation experiment. Although TRPV6 WT, as well as TRPV6-$G_{660}$R, were present in the plasma membrane, it is clearly visible that the amount of the TRPV6-$G_{660}$R in the plasma membrane is greatly reduced (Supplementary Figure S2B).

### 2.1.6. The $G_{660}K$ Mutation Cannot Be Rescued by Mutating $W_{85}$ of the Interacting Subunit

In the experiment shown in Figure 2D, we identified the N-terminal sequence of TRPV6, which is in close proximity to $G_{660}$. Next, we emulated the effect of the $G_{660}R$ mutation in the affected child. The introduction of the $R_{660}$ residue may lead to a slight shift in the interacting TRPV6 subunit and place the $R_{660}$ residue next to a tryptophan residue, $W_{85}$, of the interacting subunit. To test if an exchange of $W_{85}$ to alanine ($W_{85}A$), arginine ($W_{85}R$) or glutamate ($W_{85}E$) can rescue the $G_{660}R$ mutation, we also coexpressed the latter three mutant constructs as truncated $R_{510}$stop variants with the $G_{660}R$ mutation (Figure 4A,B). The three mutations, $W_{85}A$, $W_{85}R$ and $W_{85}E$, did not rescue the $G_{660}R$ mutation. The data indicate that the $G_{660}$, which is strictly conserved among all mammalian TRPV6 proteins, is important for correct channel function when coexpressed with the TRPV6 $R_{510}$stop mutant. In addition, we compared the paternal TRPV6 combination (TRPV6 and $R_{510}$stop variant) with the combination of the child in the permanent presence of $Ca^{2+}$ ions and measured the basic cytosolic $Ca^{2+}$. It can be seen that the cytosolic $Ca^{2+}$ level is significantly lower in cells mimicking the affected child (Figure 4C).

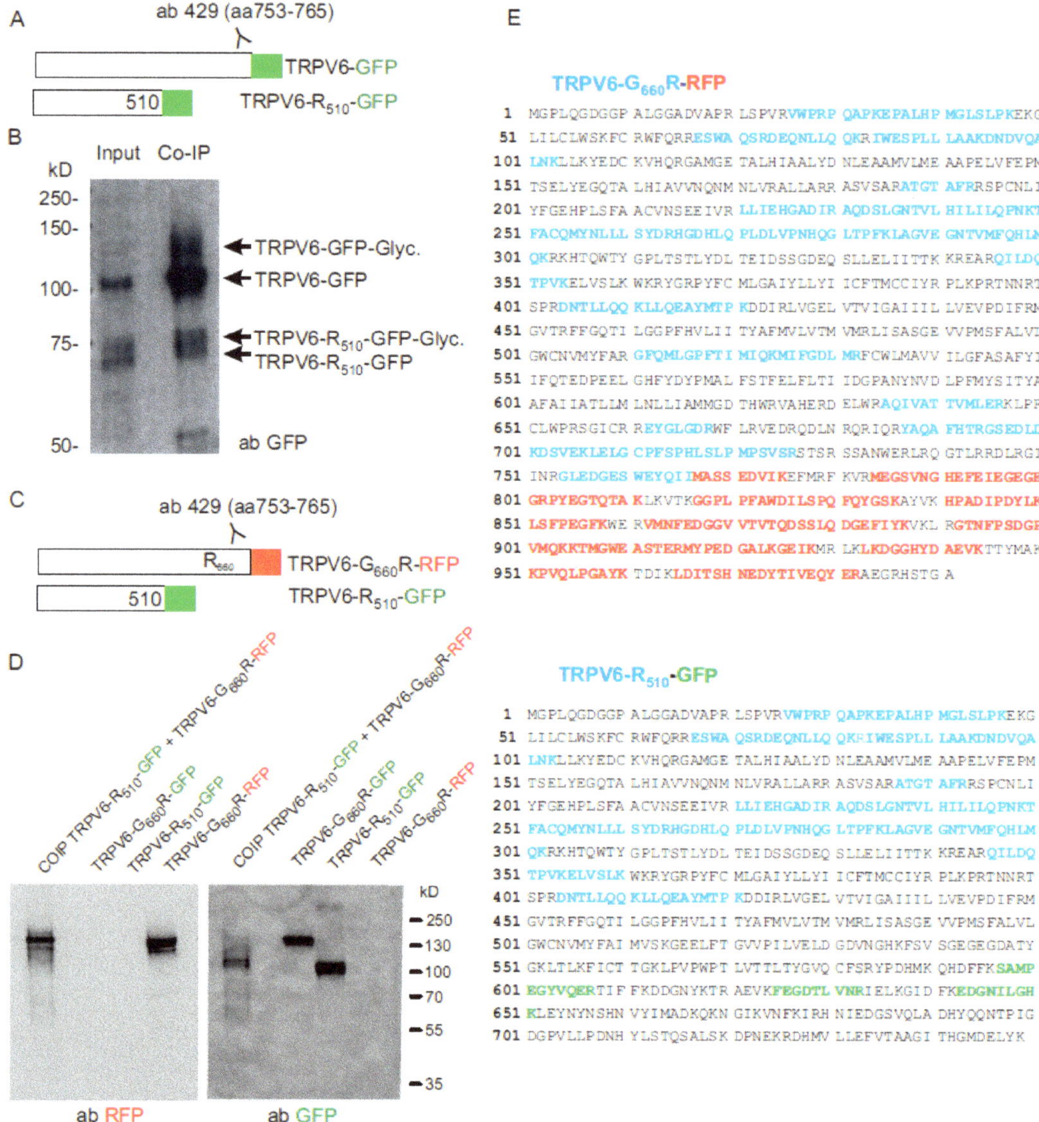

**Figure 3.** (**A**) TRPV6 fused to GFP (TRPV6-GFP) and TRPV6-R510 fused to GFP (TRPV6-R510-GFP, stop codon removed) were cotransfected in HEK293 cells and immunoprecipitated with a C-terminal TRPV6 specific antibody 429 (directed against aa 753-765 of TRPV6). (**B**) Western blot of the input and eluate from co-immunoprecipitaion (COIP) with a GFP antibody. (**C**) TRPV6-G660R fused to mRFP (TRPV6-G660R-RFP) and TRPV6-R510 fused to GFP (TRPV6-R510-GFP) were cotransfected in HEK293 cells. TRPV6-G660R-RFP was immunoprecipitated with TRPV6 antibody 429. (**D**) Detection of fused RFP and GFP tagged TRPV6-G660R and TRPV6-R510 proteins in cell lysates from single transfections and in the eluate obtained after cotransfection/co-immunoprecipitation (COIP). (**E**) Mass spectrometrical identification of TRPV6-G660R-RFP and TRPV6-R510-GFP proteins in the eluate of the COIP (as presented in (**C**,**D**)). Location of tryptic peptides identified by MS/MS fragmentation; TRPV6 (blue), RFP (red) and GFP (green).

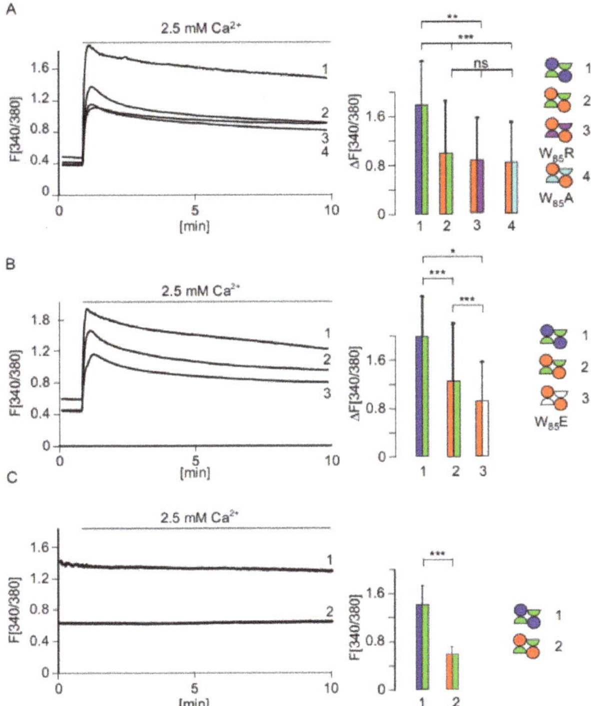

**Figure 4. (A,B)** Coexpression of TRPV6-G660R-I-RFP with several TRPV6-R510stop mutants which, in addition, contain W85 mutations. The tryptophan W85 was mutated to W85R (3, red/magenta, n/N = 47/3), W85A (4, red/light blue, n/N = 92/3), W85E (3 shown in B, red/white, n/N = 98/3) and compared with the TRPV6 combination present in the father (1, blue/green, n/n/N/N = 61/3, 55/3) and the child (2, red/green, n/n/N/N = 60/3, 120/3), respectively. **(C)** Expression of the TRPV6 combination of parental (1, blue/green, n/N = 48/3) and child (Red/green, n/N = 73/3) after loading FURA-2AM and incubation of the cells in the permanent presence of 2.5 mM Ca2+. n/N = cells/experiments. Asterisks assign significance differences (* $p < 0.05$, ** $p < 0.01$, *** $p < 0.001$, ns = not significant).

### 2.2. Comparative Proteome Analysis of Tissue Sections Obtained from a Healthy Placenta and the Placenta of the Affected Child

#### 2.2.1. HTRA1 and Cathepsin G Are Upregulated in the Placenta of the Affected Child

Since the malfunction of the mutated TRPV6 protein changes $Ca^{2+}$ homeostasis in the placenta and leads to hyperparathyroidism, we analysed whether the dysfunction of the channel alters the protein expression profile in the placenta of the affected child. Therefore, we analysed sections of paraformaldehyde embedded placenta tissue from the sick and from a healthy child (with no skeletal dysplasia and hyperparathyroidism). After separation by gel electrophoresis, we analysed the extracted proteins using label free nano LC mass spectrometry. Three independent analyses were performed from each placenta. Using this approach, a total of 740 individual proteins were identified in both placentas (Figure 5A–C), 649 in the placenta of the child and 600 in the control placenta. As we showed previously, TRPV6 is mainly expressed in human syncytiotrophoblasts [1,25]. We used placental alkaline phosphatase (Swissprot: P05187 (PPB1_HUMAN) as a fetal syncytiotrophoblasts marker to evaluate the share of fetal syncytiotrophblast cells present in the tissue sections in both groups [26]. Combining the datasets for both categories (healthy vs. sick placenta), the mean amount of alkaline phosphatase was not different, which indicates that the content of syncytiotrophoblasts in both groups was similar. Next, we performed a label free semiquantitative analysis by counting the total number of peptide spectra belonging to the individual proteins of three datasets from both categories. Doing this, 15 proteins were exclusively or significantly more abundant in the affected placenta, while four proteins were downregulated compared to the healthy placenta (Figure 5A,B). Two proteases, high-temperature requirement A serine peptidase 1 (HTRA1) and cathepsin G, were only identified in the affected placenta. In a previous work, we analysed murine placenta trophoblasts from *Trpv6$^{-/-}$* and wild type mouse, and also found HTRA1 protease being upregulated in the TRPV6 deficient placenta [27].

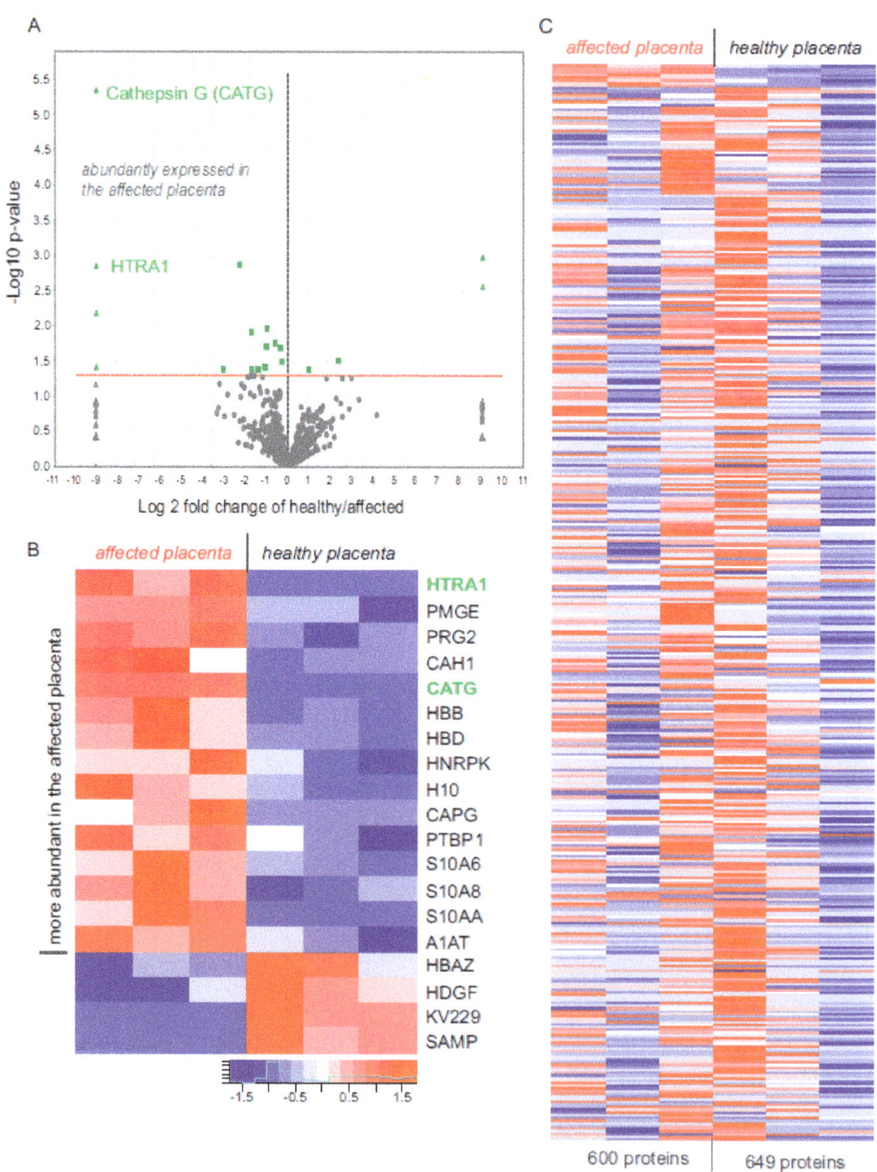


**Figure 5.** Proteome analysis of paraformaldehyde fixed tissue of a healthy placenta and the placenta of the affected child. (**A**) Vulcano-blot of a semiquantitative analysis of differentially expressed proteins identified by mass spectrometry of pooled

tissue sections of a healthy placenta and the placenta of the affected child ($n$ = 3 samples/genotype). Fifteen proteins are upregulated and four proteins downregulated in the placenta of the affected child. Up- and downregulated proteins were identified based on at least 1.3-fold changes in the total spectrum counts, with $p$-values < 0.05 using unpaired two-tailed Student's t-test. Cathepsin G (CATG) and serine protease HTRA1 (green) are only detectable in the affected child. (**B**) Heatmap of Z-scores calculated from the total peptide spectra counts of proteins (UniProt Identifier), which were more abundant in the child (green triangles and squares shown in (**A**)). (**C**) Heatmap of identified proteins on the basis of total spectrum count values (shown as Z-scores) from three independent mass spectrometry samples prepared from placentas from healthy control and affected child ($n$ = 3). In total, 740 proteins were identified by a least two unique peptides/protein of the healthy placenta and the placenta of the affected child.

### 2.2.2. The HTRA4 Protease Is Upregulated in BeWo Cells in the Presence of Low Ca$^{2+}$

Next, we tested if TRPV6 is expressed in a cell line, BeWo, which serves as model for throphoblast cells. After immunoprecipitation, we unambiguously identified several tryptic TRPV6 peptides, covering 15% of the human TRPV6 sequence (SwissProt: Q9H1D0) by mass spectrometry and Western blot (Figure 6A and Supplementary Figure S3C). With the confirmation that TRPV6 is expressed in these human cells, we cultured BeWo cells in the presence of high (0.65 mM) or low Ca$^{2+}$ (0.35 mM), and analysed the protein expression again by mass spectrometry. By this approach, 2337 proteins were detected for both conditions: 2154 proteins were identified in lysates obtained after cultured in low Ca$^{2+}$ and 1890 proteins were identified under high Ca$^{2+}$ conditions. Interestingly, one serine protease, HTRA4, was detected to be five times more abundant in BeWo cells cultured using low Ca$^{2+}$ (Figure 6A–D). Taken together, our Ca imaging experiments and the proteome analysis show that a disturbed assembly of the TRPV6 subunits in the placenta of the sick child leads to a massive reduction in the Ca$^{2+}$ influx and, presumably, to a reduced Ca$^{2+}$ content in trophoblasts, which, in turn, triggers a higher expression of serine proteases. The data may explain the phenotype of the observed child. The results raise the question of if TRPV6 is involved in the syncytialisation of the placenta. BeWo cells were treated with forskolin which induces syncytialisation and we analysed TRPV6 expression by Western blot (Supplementary Figure S3). As an indicator for syncytialisation, we used a zonula occludens (ZO-1) specific antibody. TRPV6 was immunoprecipitated from the BeWo cells and detected using two different TRPV6 specific antibodies. The experiment indicates that TRPV6 expression is very similar in forskolin treated and untreated BeWo cells. On the other hand, we compared murine TRPV6 deficient trophoblasts with wild type trophoblasts and found that the percentage of cells that showed a distinct ZO-1 staining at the cell membrane is very low ( Supplementary Figure S4), meaning that the trophoblast cells are fully syncytialised independent of the *Trpv6* genotype. The result implies that, at least in the murine trophoblasts, TRPV6 is not involved in the syncytialisation.

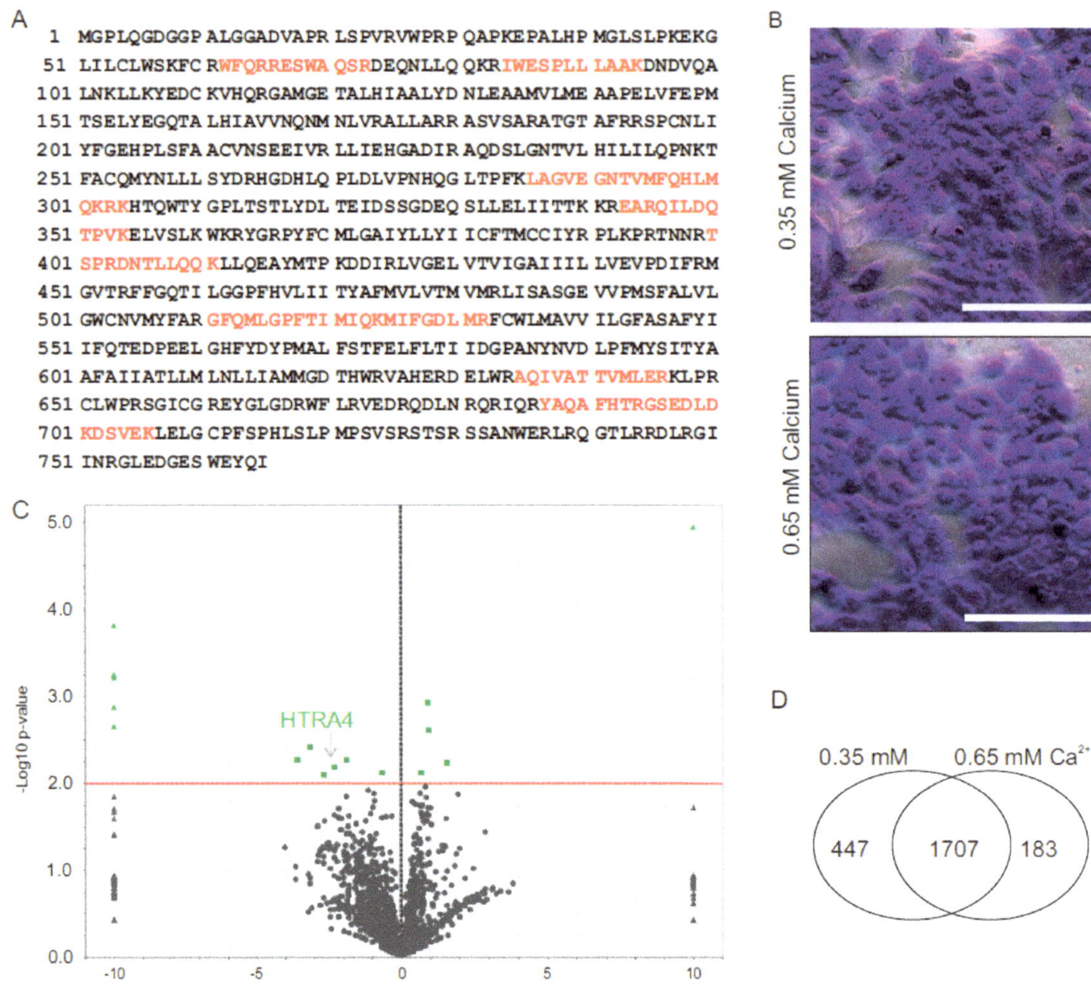

**Figure 6.** Proteome analysis of BeWo cells cultured in the presence of different $Ca^{2+}$ concentrations. (**A**) Tryptic TRPV6 peptides identified by mass spectrometry after immunoprecipitation from BeWo cells (red) with TRPV6 ab 429. (**B**) BeWo cells were cultured in the presence of 0.65 mM (high) or 0.35 mM $Ca^{2+}$ (low). Cells were fixed and stained with eosin/azur, scale bar: 200 μm. (**C**) Total protein identifications in BeWo cells cultured in the presence of high or low $Ca^{2+}$. (**D**) Vulcano blot shows semi quantitative analysis of differentially detected proteins identified in BeWo cell lysates. Proteome analysis of the abundance of peptide spectra detected in BeWo cells in the presence of high or low $Ca^{2+}$. Up- and downregulated proteins were identified based on at least 2-fold changes in the total peptide spectra abundance detected in three independent experiments with a *p*-value < 0.01, calculated using the unpaired two-tailed Student's *t* test. Serine protease HTRA4 is more abundant in BeWo lysates cultured in 0.35 mM $Ca^{2+}$.

## 3. Discussion

The human TRPV6 protein is expressed in a few tissues, e.g., pancreatic acini and the trophoblast layer of the placenta. Several recent studies describe new born children who suffer from hyperparathyroidism with undermineralized bones [13–17]. However, the underlying cause of the disease seems to be the TRPV6 gene. Most of the identified TRPV6 mutations when cloned and overexpressed affect the function of the channel.

Thus, typically, the $Ca^{2+}$ influx via TRPV6 channels is greatly reduced. Interestingly, the parents of the described children were healthy at their time of birth, which indicates that one dysfunctional TRPV6 allele does not lead to a dramatic undermineralization of the skeleton.

*3.1. The TRPV6 Mutations of the Affected Child Lead to a Decreased Surface Expression of Functional Channels*

We examined the role of two TRPV6 mutations in an affected child [13], in which one *TRPV6* allele contains a premature stop codon, $R_{510}$stop, whereas the other allele contains, at position 660, a G to R ($G_{660}R$) substitution. Surprisingly, the overexpressed $G_{660}R$ mutation behaves very similarly to the wild type TRPV6 channel, whereas, as expected, the truncated TRPV6 with the premature stop codon, $R_{510}$stop, exhibits no detectable $Ca^{2+}$ conductance. Coexpression studies that mimic the parental TRPV6 combinations, thus, wild type plus $G_{660}R$ or wild type combined with the truncated $R_{510}$stop TRPV6, also did not reveal significantly reduced $Ca^{2+}$ activity. Only the combination of the truncated $R_{510}$stop TRPV6 with the $G_{660}R$ mutation showed a greatly reduced $Ca^{2+}$ activity. The interpretation of the experimental data leads to the following model (Figure 7A–E): The maternal combination, TRPV6 WT allele and $G_{660}R$ mutation, does have a minor effect on the function of the channel. In addition, the paternal combination of the truncated $R_{510}$stop mutation and a WT allele does not lead to reduced TRPV6 activity. Only the combination of the $G_{660}R$ and $R_{510}$stop mutation shows a reduced $Ca^{2+}$ uptake, most likely due to a disturbed subunit assembly and reduced surface expression. Interestingly, the particular glycine residue, $G_{660}$, is conserved among all mammalian TRP-proteins, which emphasizes its indispensable importance for the correct function of the subunits. However, this becomes evident when the mutation was coexpressed together with the truncated TRPV6 $R_{510}$stop mutant. The next related TRP protein, TRPV5, contains a glycine residue corresponding to the G660 of the human TRPV6 sequence, which is conserved in mammalian TRPV5 proteins. In addition, the interacting sequence within the N-terminus of TRPV6 is also conserved in TRPV5 proteins. We suggest that the G660 residue is critical for the interaction of N and C-termini of TRPV6 subunits. A mutation of this residue disturbs subunit assembly/membrane trafficking, and cannot be rescued by the tested mutations using the truncated TRPV6 R510stop constructs. The data provide the molecular basis of why these mutations lead to a pronounced undermineralization and, as a consequence, to the dysplasia of the skeleton of the affected child [13]. A recent publication shows that a small percentage of patients with nonalcohol dependent pancreatitis contain mutations that affect the TRPV6 gene [28]. All of the patients who were examined contained only one defective TRPV6 allele, which, when overexpressed, showed reduced $Ca^{2+}$ activity [16]. In addition, mutations were found that lead to a dysfunctional channel, in which the closing/inactivation behaviour of the TRPV6 channel is affected. These combinations are thought to be potentially toxic as the result of $Ca^{2+}$ overload. Suzuki and coworkers published a patient who contained the combination of a TRPV6 with reduced $Ca^{2+}$ and one with an enhanced $Ca^{2+}$ conductance [15]. This combination also leads to a reduced mineralisation of the skeleton in this patient. In addition, the overexpression of TRPV6 transcripts seems to be critical. Thus, overexpression of *TRPV6* transcripts is associated with several malignancies, which include cancer derived from prostate and breast. The present work shows that the $G_{660}R$ mutation, which does not alter the function of the channel when expressed as homomultimer, is critical for the mineralization of the bones when coexpressed in combination with the $R_{510}$stop mutation. We suggest that these mutations affect the assembly of the subunits and lead to the reduced surface expression of functional TRPV6 channels. It can be seen in Supplementary Figure S2 that, in the plasma membrane fraction of the paternal combination, the amount of TRPV6 WT is higher compared to the combination which reflects the affected child. Thus, the amount of TRPV6-$G_{660}R$ seems to be greatly decreased. On the other hand, in both cotransfections (paternal and child), occurs an additional protein that is detectable with the N-terminal ab 1271 but not with the C-terminal ab 429. We speculate that this protein is a C-terminal breakdown product of the TRPV6 channel. Most notably, the amount of the breakdown product is higher in the plasma membrane fraction of the affected child. The data are conclusive that the occurrence of a lower amount of TRPV6 channels results in the decreased calcium uptake of expressing cells, and may explain the undermineralization of the skeleton of the affected child.

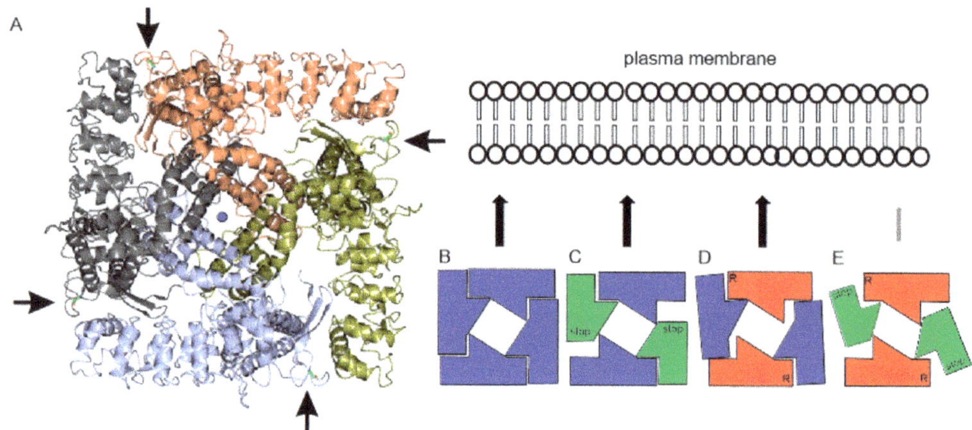


**Figure 7.** Putative assembly of TRPV6 subunits. (**A**) Crystal structure of rat TRPV6 (PDB ID: 5IWK, Saotome et al., 2016). View from the top. $G_{660}$ is located at the border of the TRPV6 subunits (arrows). The pore of the channel is shown in the middle, a $Ca^{2+}$ ion is indicated (blue). (**B**) Assembly of TRPV6 wild type subunits (blue). (**C**) Assembly of wild type and $R_{510}$stop TRPV6 subunits, as present in the father (blue/green). (**D**) Assembly of wild type and the $G_{660}R$ TRPV6 subunit, as present in the mother (blue/red). (**E**) Assembly of truncated $R_{510}$stop and $G_{660}R$ TRPV6 subunits, as present in the affected child (red/green). $Ca^{2+}$ uptake is only decreased in the child because the abundance of functional TRPV6 channels in the plasma membrane is greatly reduced (indicated by grey arrow).

*3.2. Loss of TRPV6 Function Is Associated with the Upregulation of Placental Proteases*

In addition, we show that the loss of TRPV6 is associated with changes in the expression level of a few proteases. We compared the protein profiles of a healthy placenta with the placenta of the affected child and found that two proteases, HTRA1 and cathepsin G, were only detectable in the placenta of the affected child. A link between cathepsin G and $Ca^{2+}$ transport has been reported by Peterson and coworkers [29]. The authors show that, in cultures of endothelial/epithelial cells, cathepsin G affects the formation of intercellular gaps and thereby increases the permeability for $Ca^{2+}$ ions. The underlying mechanism is not known but the expression of cathepsin G in the placenta of the sick child may act as a compensation mechanism to increase the $Ca^{2+}$ uptake/transfer to the foetus, which is reduced as a consequence of the mutations in the TRPV6 channel.

Additionally, the serine protease HTRA1was exclusively detected in the placenta of the affected child. HTRA1 was also more abundant in murine *Trpv6*-deficient placenta trophoblasts [27]. In line with this observation, murine *Trpv6* deficient placenta contains a higher activity of HTRA1, which leads to enhanced degradation of extracellular matrix proteins, such as fibronectin. This also applies to the human trophoblast derived cell line BeWo, in which TRPV6 is endogenously expressed. In the presence of low $Ca^{2+}$ in the medium, which mimics the loss of TRPV6 channels, BeWo cells overexpress the protease HTRA4, which is closely related to the HTRA1 protease. The HTRA1 protease is preferentially expressed in human placenta during the third trimester, is localized in syncytiotrophoblasts and cytotrophoblast intracellularly and is also detectable extracellularly [30]. Both proteases, HTRA1 and HTRA4, are elevated in the placenta of preeclampsia patients [31]. In this study, the overexpression of HTRA1 and HTRA4 in the human trophoblast cell line HTR-8 reduces cell migration and, under hypoxic cell culture conditions, HTRA1 and HTRA4 expression increases. In line with results reported for HTR-8 cells, we detected a 5-fold increase in HTRA4 expression in the presence of low $Ca^{2+}$ concentration in the medium, which obviously induces a comparable stress in BeWo cells.

In addition, in pancreatic acini, which also express TRPV6, there is a link between protease activity and dysfunctional TRPV6 channels. A subpopulation of patients with chronic pancreatitis shows mutations in one *TRPV6* allele [28]. The same study also shows that *Trpv6* knock in mice, which express non $Ca^{2+}$ permeable channels, are more sensitive to treatment with cerulein. After cerulein treatment, the serum level of a few pancreatic enzymes, namely, amylase and lipase, are accelerated in this mouse model. A link between the onset of pancreatitis and the intracellular $Ca^{2+}$ concentration of rodent acinus cells was observed earlier [32]. Taken together, we demonstrate that the mutations in the *TRPV6* gene of the affected child lead to reduced $Ca^{2+}$ uptake in heterologous expression system, which may explain the skeletal dysplasia observed in the new born. In addition, the $Ca^{2+}$ deficiency seems to be connected to enhanced expression of proteases in the placenta of the affected child.

### 4. Materials and Methods

#### 4.1. Cloning

Mutations were introduced using Fusion polymerase (NEB, Ipswich, MA, USA) and a plasmid which contains the *TRPV6* cDNA cloned in pCDNA3. Mutations in the C-terminal part replacing $G_{660}$ were subcloned as BstEII and MfeI fragments in TRPV6-pCAGGS-IRES-eGFP and TRPV6-pCAGGS-IRES-mRFP. To generate an in frame stop codon replacing $R_{510}$, we used a similar strategy. We subcloned mutations within the $QQKR_{8\,3}$ as well as the $W_{85}$ present in the N-terminus of TRPV6 as PshAI-BestEII fragments in the appropriate vectors.

#### 4.2. Immunoprecipitation of TRPV6 from Transfected HEK293 and BeWo Cells

BeWo cells were grown to 90% confluence in cell culture dishes ($282cm^2$ each) and harvested using 10 mL PBS and a cell scraper, cells were sedimented ($1400\times g$, 5 min) and washed with PBS. The cell pellet was resuspended in 1 mL RIPA buffer (150 mM NaCl, 50 mM Tris HCl, pH 8.0, 5 mM EDTA, 1% Nonidet P40, 0.1% SDS, 0.5% Na-deoxycholate, pH 7.4), supplemented with proteinase inhibitors (Roche, Mannheim, Germany). Cell solution was sheared ten times (27G gauge needle) on ice and then incubated for 30 min at 4 °C on a shaker. After centrifugation at $100,000\times g$ at 4 °C for 45 min, the supernatant containing the solubilized proteins was incubated for 16 h at 4 °C in the presence of 10 µg anti-TRPV6 antibody 429 (directed against the c-terminus) coupled to 50µL of Dynabeads™ Protein G (Invitrogen, Schwerte, Germany). The beads were collected using a magnetic rack, washed three times with 1 mL RIPA buffer and were eluted with 50 µL denaturing sample buffer (final concentration: 60 mM Tris HCl, pH 6.8, 4% SDS, 10% glycerol including 0.72 M beta-mercaptoethanol). The elution was incubated for 20 min at 60 °C and analysed by mass spectrometry. To co-immunoprecipitate TRPV6 from HEK293 cells two dishes were cotransfected with WT TRPV6 and truncated TRPV6 ($R_{510}$stop mutant) GFP fusion constructs.

The stop mutant was fused to GFP to discriminate this protein from IgGs on the Western blot. Harvested cells were solved in 1.5 mL lysis buffer (TBS including 1% digitonin and protease inhibitors). After $10\times$ shearing, the cells were incubated for 1 h at 4 °C on a shaker device and centrifuged at $100,000\times g$ for 45 min at 4 °C. The cell lysate was incubated for 2 h at 4 °C with antibody 429 coupled to magnetic protein A/G beads (ThermoFisher, Karlsruhe, Germany). Antibody 429 is directed against the C-terminus of TRPV6 and binds only WT TRPV6 but not the $R_{510}$stop mutant. The beads were washed 3 times using 1ml of 0.1% digitonin buffer including protease inhibitors, denatured in 60 µL sample buffer, subjected to SDS-PAGE electrophoresis following Western blot procedure as described below. The Western blot was incubated with the GFP antibody.

### 4.3. Surface Biotinylation and Western Blot

The method has already been described (Fecher-Trost et al., 2013). A 75cm$^2$ flask with confluent COS cells was transfected with the appropriate constructs and cultured for 24 h, washed twice with ice-cold phosphate buffered saline (137 mM NaCl, 2.7 mM KCl, 10 mM Na$_2$HPO$_4$, 2 mM KH$_2$PO$_4$) containing 1 mM MgCl$_2$ and 0.5 mM CaCl$_2$ (PBSB, pH, 8.0), and incubated in the presence of NHS-LC-biotin freshly diluted in PBSB at 0.5 mg/mL for 30 min at 4 °C. The reaction was stopped by washing twice with PBSB containing 0.1% ($w/v$) bovine serum albumin and once with PBS, pH 7.4. Cells were harvested from the flasks by shaking in PBS supplemented with 2 mM EDTA. The cells were centrifuged at 1000× $g$ at 4 °C for 5 min and resuspended in ice cold lysis buffer (PBS containing 1% Triton X-100, 1 mm EDTA, and a mixture of protease inhibitors). Cell lysates were rotated at 4 °C for 30 min to solubilize proteins; after centrifugation at 1000× $g$ and 4 °C for 5 min, the amount of protein was determined using BCA (ThermoFisher, Waltham, MA, USA), and the protein solution (~1 mg) was added to 100 µL of avidin–agarose beads pre-equilibrated in lysis buffer and incubated at 4 °C for 2 h. The biotin–avidin–agarose complexes were washed 4 times with lysis buffer supplemented with 0.25 mM NaCl. Biotinylated proteins were eluted in 100 µL of 2-times denaturing electrophoresis sample buffer and incubated at 60 °C for 30 min before SDS-polyacrylamide gel electrophoresis (SDS-PAGE) on 8% Bolt gels, (Invitrogen, Carlsbad, CA, USA) in a Bolt buffer system. Cell lysate input corresponds to protein samples taken before adding avidin–agarose beads. The proteins were electrophoretically separated, blotted, and probed with TRPV6 N-terminal ab 1271 and C-terminal ab 429, respectively. The endoplasmic reticulum protein calnexin was used as a control.

### 4.4. Antibodies

The following in house generated anti-TRPV6 antibodies were used: Polyclonal antibody 1271 and 429 directed against N- and C-terminus of human TRPV6 [24,25], respectively and monoclonal YFP/GFP antibody [25]. All antibodies were affinity purified before use. Commercial ZO1 antibody was from Invitrogen.

### 4.5. Calcium Imaging

Intracellular live cell Ca$^{2+}$-imaging experiments were performed using a Polychrome V and CCD camera (TILL Imago)-based imaging system from TILL Photonics (Martinsried, Germany) with a Zeiss Axiovert S100 fluorescence microscope equipped with a Zeiss Fluar 20×/0.75 objective. Data acquisition was accomplished with the imaging Live Acquisition software (TILL Photonics). Data were analysed using the Offline analysis software (TILL Photonics). Cells were incubated in media supplemented with 4 µm Ca$^{2+}$-sensitive fluorescent dye Fura-2-AM (Molecular probes, Eugene, USA) for 30 min in the dark at room temperature and washed 4 times with nominally Ca$^{2+}$ free external solution (140 mM NaCl, 5 mM KCl, 1 mM MgCl$_2$, 10 mM HEPES, 10 mM glucose, adjusted to pH 7.2 with NaOH) to remove excess Fura-2-AM. The Fura-2-AM loaded cells, growing on 2.5 cm glass coverslips, were transferred to a bath chamber containing nominally Ca$^{2+}$ free solution, and Fura-2 fluorescence emission was monitored at >510 nm after excitation at 340 and 380 nm for 30 ms each at a rate of 1 Hz for 600 s. Cells were marked, and the ratio of the background-corrected Fura-2 fluorescence at 340 and 380 nm (F340/F380) were plotted versus time. After reaching a stable F340/F380 base line, 2.5 mM CaCl$_2$ was added to the bath solution, and cytosolic Ca$^{2+}$-signals were measured.

### 4.6. Cell Culture and Transfection of HEK293 Cells

HEK293 cells were grown in culture dishes (3 cm diameter) with poly l-lysine-coated glass coverslips (diameter 2.5 cm) until 80% confluence and then transiently transfected with 2.5 µg of appropriate cDNA constructs in 5 mL of Lipofectamin 3000 (ThermoFisher, Karlsruhe, Germany). For Fura-2-AM measurements, cells were transfected with TRPV6 constructs cloned in pcAGGS-IRES-GFP or IRES-mRFP vectors. Cotransfection was carried out with a combination of the appropriate constructs cloned in vectors with green and red fluorescent proteins (1.25 μg each). Coverslips with transfected cells were used for $Ca^{2+}$ imaging experiments 24 h after transfection.

### 4.7. Modelling $G_{660}R$ Mutation

We used the RCSB PDB [33] software and the rat TRPV6 structure (PDB ID: 5IWK) to identify amino acids in close proximity to the glycine residue, which corresponds to $G_{660}$ in the human TRPV6 sequence [18]. It was concluded that the C-terminal part of wild type TRPV6 interacts with the amino acid sequence $QQKR_{83}$ present within the N-terminus of the interacting TRPV6 subunit. In addition, we analysed the TRPV6 structure in which the $R_{660}$ mutation was modelled. The $R_{660}$ residue comes in close proximity to the tryptophan residue $W_{85}$ present in the N-terminus of the interacting subunit.

### 4.8. Protein Extraction from Paraformaldehyde/Formalin Fixed Placenta Tissue Sections

For protein extraction, two to three unstained slides of 4% formalin fixed placenta tissue (3μm) were scraped from glass microscope slices with a scalpel and resuspended in 80 μL protein extraction buffer (60 mM Tris, pH 6.8, 1.2 M glycerol, 0.78 M β-mercaptoethanol, 70 mM SDS, 10 mM arginine). Samples were consecutively incubated for 15 min on ice, 20 min at 100 °C and for 1 h at 80 °C. Protein extracts were centrifuged for 15 min at $14,000 \times g$ (4 °C) and the supernatant was transferred to a fresh vial. 60 μL per sample was loaded on a 4–12% BoltTM gel (ThermoFisher, Waltham, MA, USA) and electrophoresed. The gel was fixed and stained with a colloidal Coomassie, 16 gel bands/lane were isolated and digested using trypsin as described before [25].

### 4.9. BeWo Cell Culture

Human BeWo cells (ATCC® CCL¬98™) were cultured in 3.5 cm diameter cell culture dishes (Corning, Tewksbury, USA), with culture medium (F-12 Nut Mix with 2 mM glutamax (ThermoFisher, Karlsruhe, Germany), 10% FKS (Corning, Tewksbury, MA, USA) and 1% penicillin/streptomycin (Sigma-Aldrich-Merck, Darmstadt, Germany). The cells grew in the presence of 5% $CO_2$ at 37 °C. Cells were trypsinized, seeded in fresh cell culture dishes and cells grew until 40% of confluency was reached. The medium was changed after 24 h to medium containing 0.65 mM $Ca^{2+}$ or to medium containing 0.35 mM $Ca^{2+}$ including EDTA. Calcium concentration was determined using a Dri-Chem NX500i System (FujiFilm, Japan). The amount of EDTA was calculated using the WEBMAXCSTANDARD7/3/2009 software (https://somapp.ucdmc.ucdavis.edu/pharmacology/bers/maxchelator/webmaxc/webmaxcS.htm retrieved on 31 March 2021). Cells were cultured for additional 48h in either 0.65 mM $Ca^{2+}$ or 0.35 mM $Ca^{2+}$ medium, washed 3 times with PBS and removed from the dish with a cell scraper (Corning, Tewksbury, MA, USA) and resuspended in denaturing sample buffer. To prepare a sample for one mass spectrometry experiment, two cell culture dishes of the same medium condition were pooled. To induce syncytialisation of BeWo cells 200,000 cells were cultured in a flask (Falcon, 75 $cm^2$) and medium was supplemented with 0.3% DMSO or 20 μM forskolin/0,3% DMSO for 48 h. Cells were stained with azur/eosin or with a ZO1-antibody.

### 4.10. Preparation of BeWo Cell Lysates for Proteome Analysis

20 μL of BeWo cell lysates grown in 0.65 mM $Ca^{2+}$ or 0.35 mM $Ca^{2+}$ medium were separated on a NuPAGE® 4–12% gradient gels (ThermoFisher, Karlsruhe, Germany). Proteins were fixed in the presence of 40% ethanol and 10% acetic acid and visualized with colloidal Coomassie stain (20% ($v/v$) methanol, 10% ($v/v$) phosphoric acid, 10% ($w/v$) ammonium sulfate, and 0.12% ($w/v$) Coomassie G-250). Fourteen gel pieces were cut/sampled and trypsin digested as described [25].

### 4.11. Primary Mouse Trophoblast Cell Culture

Trophoblasts from *WT* and *Trpv6$^{-/-}$* placentae were isolated at E13.5 as previously described by Winter et al., 2020 [27]. 200,000 cells were seeded in a 3.5 cm dish on four uncoated glass coverslips (Orsa$^{tec}$, 1.2 cm) and incubated with 2 mL medium (DEMEM, Gibco®, 10% FCS, 100 U/mL penicillin, 100 µg/mL streptomycin) at 37 °C and 5% $CO_2$. After 24 h the medium was changed and cells were cultured for additional 4 days followed by ZO1-antibody staining.

### 4.12. Zona Occludens 1 (ZO-1) Antibody Staining of Trophoblast and BeWo Cells

Coverslips with BeWo or mouse trophoblast cells were removed and washed with PBS. Fixation and permeabilization was performed by incubating the cells with methanol (−20 °C) on ice for 20 min. Cells were washed 4 times with PBS and blocking buffer (3% BSA, 1% normal goat serum and 0.1% Triton X100 in PBS) was added for 1 h at RT. After washing 4 times with PBS, cells were incubated over night at 4 °C in a buffer (1% BSA in PBS) containing a ZO-1 specific primary antibody (1:1000, Invitrogen). Cells were washed with PBS 4 times and incubated with an Alexa-fluor 488 anti-rabbit antibody (Invitrogen) and DAPI (2 µg/mL) in PBS. Cells were washed with PBS 4 times and mounted using Immu-Mount$^{TM}$ (ThermoSCIENTIFIC). Stained BeWo cells were analyzed by using an Imager.M2 microscope (Zeiss) obtaining a Axiocam MRm (Zeiss). Trophoblast cells were analyzed using a Slightscanner (Axio Scan.Z1, Zeiss). The whole set of pictures was processed with the software Imaris (Oxford Instruments) and the number of nuclei were detected automatically. ZO-1 positive cells were counted manually and the percentage of ZO-1 positive cells was calculated.

### 4.13. Mass Spectrometric Measurement (Nano-LC–MS/MS)

Six µL of tryptic digested peptides derived from each gel piece (BeWo cell lysate, IPs from BeWo cells or human placenta lysates) were analysed by nano LC–ESI–MS/MS analysis using the set up (Ultimate 3000 RSLC nano system equipped with an Ultimate3000 RS autosampler coupled to an LTQ Orbitrap Velos Pro, (ThermoFisher, Dreieich, Germany)). Peptides were trapped on a C18 trap column (75 µm × 2 cm, Acclaim PepMap100C18, 3 µm, Dionex) and separated on a reversed phase column (nano viper Acclaim PepMap capillary column, C18; 2 µm; 75 µm × 50 cm, Dionex) at a flow rate of 200 nL/min with buffers A (water and 0.1% formic acid) and B (90% acetonitrile and 0.1% formic acid) using a 94 min gradient (BeWo cell lysates) or 120 min gradient (human placenta lysates). The effluent of the chromatography was sprayed into the mass spectrometer through a coated emitter (PicoTipEmitter, 30 µm, New Objective, Woburn, MA, USA) and ionized at 2.2 kV. MS spectra were acquired in a data dependent mode. For the collision induced dissociation (CID) MS/MS top10 method, full scan MS spectra (m/z 300–1700) were acquired in the Orbitrap analyser using a target value of $10^6$. Peptide ions with charge states >2 were fragmented in the high-pressure linear ion trap by low-energy CID with normalized collision energy of 35%.

### 4.14. Raw Mass Spectrometrical Data Analysis

The fragmented tryptic peptides were identified using the MASCOT algorithm and TF Proteome Discoverer 1.4 software (ThermoFisher, Waltham, MA, USA). Peptides were matched to tandem mass spectra by Mascot version 2.4.0 by searching of a SwissProt database (version2018_03, number of protein sequences, taxonomy human: 20,387). Data were analysed as described by Winter and co-workers [27].

**Supplementary Materials:** All data are available online at https://www.mdpi.com/article/10.339 0/ijms222312694/s1.

**Author Contributions:** Conceptualization, U.W. and C.F.-T.; methodology, U.W., K.W., C.F.-T., D.S.K., H.L., M.W. and C.W.; resources, E.G., data analysis, U.W., C.F.-T. and D.S.K.; data interpretation, U.W., C.F.-T., D.S.K. and P.W.; writing—original draft preparation, C.F.-T. and U.W.; writing—review and editing, C.F.-T., U.W., P.W., D.S.K., E.G., A.E.M., C.P.B., H.L., C.W., M.W. and K.W. All authors have read and agreed to the published version of the manuscript.

**Funding:** This research was funded by the Deutsche Forschungsgemeinschaft (DFG) FE 629/2-1 (CFT), WE4866/1-1 (PW), SFB894/P2 (PW), and HOMFOR (UW, CFT). We acknowledge the support of the Deutsche Forschungsgemeinschaft (DFG, German Research Foundation) and Saarland University within the funding program Open Access Publishing.

**Institutional Review Board Statement:** Ethics approval and consent to participate: The parents provided approval and written, informed consent for a limited postmortem (chest, abdomen, ribs and long bones). We confirm that the postmortem was undertaken in accordance with the policies and procedures of North Bristol NHS Trust, which includes consent for research procedures, approved by the North Bristol NHS Trust Research and Ethics Committee (United Kingdom).

**Informed Consent Statement:** The parents of the patient provided written, informed consent for research and publication of patient information.

**Data Availability Statement:** Data is contained within the article or supplementary material.

**Acknowledgments:** We thank Francis Glaser, Marie Petry, Armin Weber, and Martin Simon-Thomas for excellent technical assistance.

**Conflicts of Interest:** The authors declare no conflict of interest. The funders had no role in the design of the study; in the collection, analyses, or interpretation of data; in the writing of the manuscript, or in the decision to publish the results.

## References

1. Wissenbach, U.; Niemeyer, B.A.; Fixemer, T.; Schneidewind, A.; Trost, C.; Cavalie, A.; Reus, K.; Meese, E.; Bonkhoff, H.; Flockerzi, V. Expression of CaT-like, a novel calcium-selective channel, correlates with the malignancy of prostate cancer. *J. Biol. Chem.* **2001**, *276*, 19461–19468. [CrossRef]
2. Peng, J.B.; Zhuang, L.; Berger, U.V.; Adam, R.M.; Williams, B.J.; Brown, E.M.; Hediger, M.A.; Freeman, M.R. CaT1 expression correlates with tumor grade in prostate cancer. *Biochem. Biophys. Res. Commun.* **2001**, *282*, 729–734. [CrossRef]
3. Bödding, M.; Wissenbach, U.; Flockerzi, V. Characterisation of TRPM8 as a pharmacophore receptor. *Cell Calcium* **2007**, *42*, 618–628. [CrossRef] [PubMed]
4. Fixemer, T.; Wissenbach, U.; Flockerzi, V.; Bonkhoff, H. Expression of the $Ca^{2+}$-selective cation channel TRPV6 in human prostate cancer: A novel prognostic marker for tumor progression. *Oncogene* **2003**, *22*, 7858–7861. [CrossRef]
5. Wissenbach, U.; Niemeyer, B.; Himmerkus, N.; Fixemer, T.; Bonkhoff, H.; Flockerzi, V. TRPV6 and prostate cancer: Cancer growth beyond the prostate correlates with increased TRPV6 Ca2+ channel expression. *Biochem. Biophys. Res. Commun.* **2004**, *322*, 1359–1363. [CrossRef]
6. Bolanz, K.A.; Hediger, M.A.; Landowski, C.P. The role of TRPV6 in breast carcinogenesis. *Mol. Cancer Ther.* **2008**, *7*, 271–279. [CrossRef]
7. Bolanz, K.A.; Kovacs, G.G.; Landowski, C.P.; Hediger, M.A. Tamoxifen inhibits TRPV6 activity via estrogen receptor-independent pathways in TRPV6-expressing MCF-7 breast cancer cells. *Mol. Cancer Res. MCR* **2009**, *7*, 2000–2010. [CrossRef]
8. Bowen, C.V.; DeBay, D.; Ewart, H.S.; Gallant, P.; Gormley, S.; Ilenchuk, T.T.; Iqbal, U.; Lutes, T.; Martina, M.; Mealing, G.; et al. In vivo detection of human TRPV6-rich tumors with anti-cancer peptides derived from soricidin. *PLoS ONE* **2013**, *8*, e58866. [CrossRef] [PubMed]
9. Xue, H.; Wang, Y.; MacCormack, T.J.; Lutes, T.; Rice, C.; Davey, M.; Dugourd, D.; Ilenchuk, T.T.; Stewart, J.M. Inhibition of Transient Receptor Potential Vanilloid 6 channel, elevated in human ovarian cancers, reduces tumour growth in a xenograft model. *J. Cancer* **2018**, *9*, 3196–3207. [CrossRef]
10. Stewart, J.M. TRPV6 as A Target for Cancer Therapy. *J. Cancer* **2020**, *11*, 374–387. [CrossRef] [PubMed]
11. Kessler, T.; Wissenbach, U.; Grobholz, R.; Flockerzi, V. TRPV6 alleles do not influence prostate cancer progression. *BMC Cancer* **2009**, *9*, 380. [CrossRef]
12. Akey, J.M.; Swanson, W.J.; Madeoy, J.; Eberle, M.; Shriver, M.D. TRPV6 exhibits unusual patterns of polymorphism and divergence in worldwide populations. *Hum. Mol. Genet.* **2006**, *15*, 2106–2113. [CrossRef]
13. Burren, C.P.; Caswell, R.; Castle, B.; Welch, C.R.; Hilliard, T.N.; Smithson, S.F.; Ellard, S. TRPV6 compound heterozygous variants result in impaired placental calcium transport and severe undermineralization and dysplasia of the fetal skeleton. *Am. J. Med. Genet. Part. A* **2018**, *176*, 1950–1955. [CrossRef]
14. Mason, A.E.; Grier, D.; Smithson, S.F.; Burren, C.P.; Gradhand, E. Post-mortem histology in transient receptor potential cation channel subfamily V member 6 (TRPV6) under-mineralising skeletal dysplasia suggests postnatal skeletal recovery: A case report. *BMC Med. Genet.* **2020**, *21*, 64. [CrossRef]

15. Suzuki, Y.; Chitayat, D.; Sawada, H.; Deardorff, M.A.; McLaughlin, H.M.; Begtrup, A.; Millar, K.; Harrington, J.; Chong, K.; Roifman, M.; et al. TRPV6 Variants Interfere with Maternal-Fetal Calcium Transport through the Placenta and Cause Transient Neonatal Hyperparathyroidism. *Am. J. Hum. Genet.* **2018**, *102*, 1104–1114. [CrossRef]
16. Yamashita, S.; Mizumoto, H.; Sawada, H.; Suzuki, Y.; Hata, D. TRPV6 Gene Mutation in a Dizygous Twin With Transient Neonatal Hyperparathyroidism. *J. Endocr. Soc.* **2019**, *3*, 602–606. [CrossRef] [PubMed]
17. Almidani, E.; Elsidawi, W.; Almohamedi, A.; Bin Ahmed, I.; Alfadhel, A. Case Report of Transient Neonatal Hyperparathyroidism: Medically Free Mother. *Cureus* **2020**, *12*, e7000. [CrossRef] [PubMed]
18. Saotome, K.; Singh, A.K.; Yelshanskaya, M.V.; Sobolevsky, A.I. Crystal structure of the epithelial calcium channel TRPV6. *Nature* **2016**, *534*, 506–511. [CrossRef]
19. Hoenderop, J.G.; Voets, T.; Hoefs, S.; Weidema, F.; Prenen, J.; Nilius, B.; Bindels, R.J. Homo- and heterotetrameric architecture of the epithelial Ca2+ channels TRPV5 and TRPV6. *EMBO J.* **2003**, *22*, 776–785. [CrossRef]
20. Hellwig, N.; Albrecht, N.; Harteneck, C.; Schultz, G.; Schaefer, M. Homo- and heteromeric assembly of TRPV channel subunits. *J. Cell Sci.* **2005**, *118*, 917–928. [CrossRef]
21. Schaefer, M. Homo- and heteromeric assembly of TRP channel subunits. *Pflug. Arch. Eur. J. Physiol.* **2005**, *451*, 35–42. [CrossRef] [PubMed]
22. Li, C.; Zhang, J. Stop-codon read-through arises largely from molecular errors and is generally nonadaptive. *PLoS Genet.* **2019**, *15*, e1008141. [CrossRef]
23. Erler, I.; Hirnet, D.; Wissenbach, U.; Flockerzi, V.; Niemeyer, B.A. $Ca^{2+}$-selective transient receptor potential V channel architecture and function require a specific ankyrin repeat. *J. Biol. Chem.* **2004**, *279*, 34456–34463. [CrossRef]
24. Hirnet, D.; Olausson, J.; Fecher-Trost, C.; Bödding, M.; Nastainczyk, W.; Wissenbach, U.; Flockerzi, V.; Freichel, M. The TRPV6 gene, cDNA and protein. *Cell Calcium* **2003**, *33*, 509–518. [CrossRef]
25. Fecher-Trost, C.; Wissenbach, U.; Beck, A.; Schalkowsky, P.; Stoerger, C.; Doerr, J.; Dembek, A.; Simon-Thomas, M.; Weber, A.; Wollenberg, P.; et al. The in vivo TRPV6 protein starts at a non-AUG triplet, decoded as methionine, upstream of canonical initiation at AUG. *J. Biol. Chem.* **2013**, *288*, 16629–16644. [CrossRef]
26. Leitner, K.; Szlauer, R.; Ellinger, I.; Ellinger, A.; Zimmer, K.P.; Fuchs, R. Placental alkaline phosphatase expression at the apical and basal plasma membrane in term villous trophoblasts. *J. Histochem. Cytochem. Off. J. Histochem. Soc.* **2001**, *49*, 1155–1164. [CrossRef] [PubMed]
27. Winter, M.; Weissgerber, P.; Klein, K.; Lux, F.; Yildiz, D.; Wissenbach, U.; Philipp, S.E.; Meyer, M.R.; Flockerzi, V.; Fecher-Trost, C. Transient Receptor Potential Vanilloid 6 (TRPV6) Proteins Control the Extracellular Matrix Structure of the Placental Labyrinth. *Int. J. Mol. Sci.* **2020**, *21*, 9674. [CrossRef]
28. Masamune, A.; Kotani, H.; Sorgel, F.L.; Chen, J.M.; Hamada, S.; Sakaguchi, R.; Masson, E.; Nakano, E.; Kakuta, Y.; Niihori, T.; et al. Variants That Affect Function of Calcium Channel TRPV6 Are Associated with Early-Onset Chronic Pancreatitis. *Gastroenterology* **2020**, *158*, 1626–1641. [CrossRef] [PubMed]
29. Peterson, M.W.; Gruenhaupt, D.; Shasby, D.M. Neutrophil cathepsin G increases calcium flux and inositol polyphosphate production in cultured endothelial cells. *J. Immunol.* **1989**, *143*, 609–616. [PubMed]
30. De Luca, A.; De Falco, M.; Fedele, V.; Cobellis, L.; Mastrogiacomo, A.; Laforgia, V.; Tuduce, I.L.; Campioni, M.; Giraldi, D.; Paggi, M.G.; et al. The serine protease HtrA1 is upregulated in the human placenta during pregnancy. *J. Histochem. Cytochem. Off. J. Histochem. Soc.* **2004**, *52*, 885–892. [CrossRef]
31. Liu, C.; Xing, F.; He, Y.; Zong, S.; Luo, C.; Li, C.; Duan, T.; Wang, K.; Zhou, Q. Elevated HTRA1 and HTRA4 in severe preeclampsia and their roles in trophoblast functions. *Mol. Med. Rep.* **2018**, *18*, 2937–2944. [CrossRef] [PubMed]
32. Krüger, B.; Albrecht, E.; Lerch, M.M. The role of intracellular calcium signaling in premature protease activation and the onset of pancreatitis. *Am. J. Pathol.* **2000**, *157*, 43–50. [CrossRef]
33. Berman, H.M.; Westbrook, J.; Feng, Z.; Gilliland, G.; Bhat, T.N.; Weissig, H.; Shindyalov, I.N.; Bourne, P.E. The Protein Data Bank. *Nucleic Acids Res.* **2000**, *28*, 235–242. [CrossRef] [PubMed]

International Journal of
*Molecular Sciences*

*Article*

# Irisin Protects the Human Placenta from Oxidative Stress and Apoptosis via Activation of the Akt Signaling Pathway

Hamid-Reza Kohan-Ghadr [†], Brooke Armistead [†], Mikaela Berg and Sascha Drewlo *

Department of Obstetrics, Gynecology and Reproductive Biology, College of Human Medicine, Michigan State University, Grand Rapids, MI 49503, USA; kohangha@msu.edu (H.-R.K.-G.); armiste9@msu.edu (B.A.); bergmik1@msu.edu (M.B.)
* Correspondence: sdrewlo@msu.edu; Tel.: +1-616-234-2754
† Designates authors with equal contribution.

**Abstract:** Irisin is a newly discovered exercise-mediated polypeptide hormone. Irisin levels increase during pregnancy however, women with preeclampsia (PE) have significantly lower levels of Irisin compared to women of healthy pregnancies. Even though many studies suggest a role of Irisin in pregnancy, its function in the human placenta is unclear. In the current study, we aimed to understand key roles of Irisin through its ability to protect against apoptosis is the preeclamptic placenta and in ex vivo and in vitro models of hypoxia/re-oxygenation (H/R) injury. Our studies show that Irisin prevents cell death by reducing pro-apoptotic signaling cascades, reducing cleavage of PARP to induce DNA repair pathways and reducing activity of Caspase 3. Irisin caused an increase in the levels of anti-apoptotic BCL2 to pro-apoptotic BAX and reduced ROS levels in an in vitro model of placental ischemia. Furthermore, we show that Irisin treatment acts through the Akt signaling pathway to prevent apoptosis and enhance cell survival. Our findings provide a novel understanding for the anti-apoptotic and pro-survival properties of Irisin in the human placenta under pathological conditions. This work yields new insights into placental development and disease and points towards intervention strategies for placental insufficiencies, such as PE, by protecting and maintaining placental function through inhibiting hypoxic ischemia-induced apoptosis.

**Keywords:** Irisin; placenta; trophoblast; preeclampsia; apoptosis; oxidative stress

**Citation:** Kohan-Ghadr, H.-R.; Armistead, B.; Berg, M.; Drewlo, S. Irisin Protects the Human Placenta from Oxidative Stress and Apoptosis via Activation of the Akt Signaling Pathway. *Int. J. Mol. Sci.* **2021**, *22*, 11229. https://doi.org/10.3390/ijms222011229

Academic Editors: Hiten D. Mistry and Eun Lee

Received: 7 September 2021
Accepted: 14 October 2021
Published: 18 October 2021

**Publisher's Note:** MDPI stays neutral with regard to jurisdictional claims in published maps and institutional affiliations.



## 1. Introduction

Preeclampsia (PE) is a hypertensive disorder of pregnancy described by systemic endothelial damage in the mother. In its severe form, PE establishes clinically as early as 20 weeks of gestation. It often necessitates preterm delivery and presents a significant risk to the immediate and long-term well-being of the baby, furthermore, causing vast neonatal intensive care unit costs [1,2]. The etiology of PE is thought to originate from the placenta since, at present, the only available treatment is the removal of the placenta, requiring delivery of the baby.

In healthy pregnancy, the extra-villous trophoblast (EVT) cells of the placenta invade from the anchoring villi into the uterine wall and participate in spiral artery remodeling to establish blood supply of the placenta [3,4]. This results in a low pressure/high flow blood delivery, which maintains a steady perfusion of the placental villi and its exchange function. The hemodynamic adaptations are disturbed in severe PE pregnancies due to low EVT cell invasion and reduced spiral artery remodeling causing decreased oxygen availability for the placental villi [5]. This pathological process is accompanied with increased hypoxic and oxidative stress often resulting in significant placental apoptosis [6,7].

In the mother, endothelial dysfunction is a major contribution to maternal hypertension causing damage to the kidneys and resulting in proteinuria and renal failure [5,8]. If untreated, severe forms of PE can advance to include the hepatic and coagulation systems and damage the brain. These disease phenotypes are largely attributed to abnormal placental function resulting in an excess release of anti-angiogenic proteins secreted by the placenta [9]. Abnormal placental protein secretion such as soluble fms-like tyrosine kinase 1 (sFLT1), endoglin and others can damage the maternal endothelium and is commonly observed in PE [5].

Irisin is among many peptides that are upregulated during pregnancy and function to regulate energy homeostasis across gestation [10]. Irisin was first described in 2012 as a myokine polypeptide secreted from skeletal muscle that regulates glucose and lipid metabolism in adipose tissues in response to exercise [11]. Irisin is a secreted form of the Fibronectin Type III Domain Containing 5 (FNDC5) transmembrane protein. Studies showed that FNDC5/Irisin is expressed in adipose tissue, cardiomyocytes, the brain and other parts of the body [12,13]. Recently the interest in the molecular actions of Irisin revealed that it is not only involved in energy storage and sensing, but it has a variety of molecular functions including differentiation, inflammation, oxidative stress in different systems [14–18]. The inhibitory effects of Irisin on cell apoptosis was shown to occur by modulation of pro-apoptotic markers such as the BCL2 associated agonist of death (BAD), BCL2 associated X apoptosis regulator (BAX), Caspase-9 and Caspase-3 and anti-apoptotic proteins such as BCL-2 and BCL-XL [16,19,20].

The Akt signaling pathway plays an important role in regulating cell proliferation, cell migration and apoptosis inhibition [21–23]. Akt activation has been shown to regulate trophoblastic cell migration and invasion [24]. In the human placenta, Akt regulates trophoblast invasion through the upregulation of matrix metalloproteinase 9 (MMP-9) and tissue inhibitor of metalloproteinase-1 (TIMP-1) [25]. Under oxidative stress and hypoxia conditions, Akt is inactivated in trophoblast cells and in primary cytotrophoblast cells via Hypoxia inducible factor-1α (HIF-1α) upregulation, which subsequently triggers Glial cell missing 1 (GCM1) degradation that, in turn, inhibits migration and invasion of trophoblast cells [26–30]. Inactivation of the Akt pathway in human PE placentas suggests its possible contribution in PE pathophysiology and progression [31]. In addition to the crucial role in trophoblast differentiation, Akt activation has also cytoprotective effect in trophoblasts [32,33].

Irisin was shown to be reduced in the circulation of women with preeclampsia [34,35], which has led us to question its importance human reproduction and disease. Previously, we identified that Irisin modulates trophoblast differentiation through activation of the AMPK pathway [36]. However, the role of Irisin in placental function and pathophysiology, especially in pregnancy complications such as PE, is still not fully known. In our study, we hypothesized that Irisin can protect from apoptosis in the human placenta through activation of the Akt signaling pathway. To test this hypothesis, we employed Irisin treatment in preeclamptic placentas and in ex vivo and in vitro placental models of ischemic-reperfusion injury and evaluated Irisin's effect on Akt activation and downstream apoptotic pathways.

## 2. Results

### 2.1. Irisin Rescues Villous Trophoblast Cells from Apoptosis in the Preeclamptic Term Placenta

The preeclamptic placenta is characterized by increased apoptosis believed to be caused by the prolonged hypoxic-ischemic stress caused by intermittent reperfusion injury during pregnancy. To test the effect of Irisin on apoptosis in the preeclamptic placenta, tissues were dissected and cultured overnight with 10 and 50 nM of Irisin and collected for protein analysis and embedded for staining. Placental sections were stained for late-stage apoptosis. DNA-strand breaks were identified by a fluorescein-based TUNEL (terminal deoxynucleotidyl transferase [TdT] dUTP nick-end labeling) assay. DNA strand breaks are thus identified by the green fluorescence as shown Figure 1A. We observed a significant reduction of apoptotic cells in the villous trophoblast cells of preeclamptic placental explants exposed to different concentrations of Irisin (Figure 1A). Downstream apoptosis markers were analyzed by western blots. There was a significant 50% and 70% increase in the ratio for the anti-apoptotic BCL2: pro-apoptotic BAX protein levels in the PE tissues treated with 10 or 50 nm Irisin, respectively, supporting the anti-apoptotic effect of Irisin in the preeclamptic placenta (n = 9, Figure 1B).

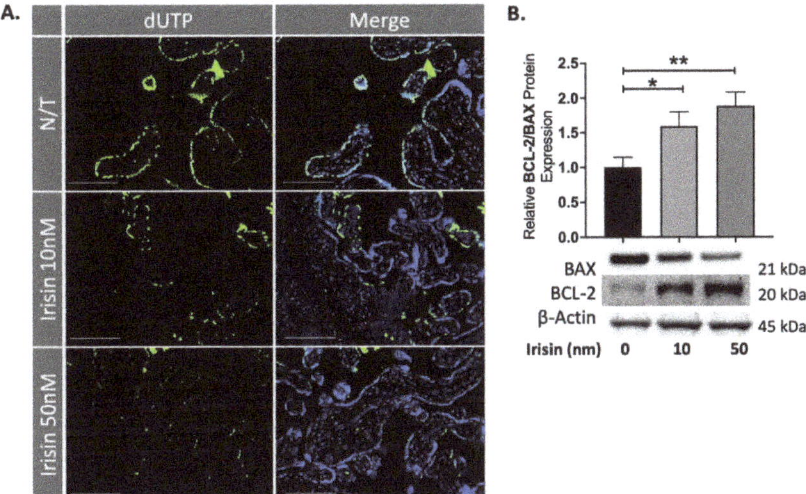



**Figure 1.** Irisin reduces apoptosis in the preeclamptic placenta. (**A**) A TUNEL assay shows a significant reduction of double-stranded DNA breaks, as measured by dUTP fluorescent molecules, when treated with Irisin. (**B**) 10 nm and 50 nm Irisin treatment in the preeclamptic placenta increased the anti-apoptotic BCL-2: pro-apoptotic BAX protein levels (n = 8). (Relative protein expression was determined by normalization to β-actin, a one-way analysis of variance and subsequent Tukey's post hoc test to analyze differences between cohorts; * $p < 0.05$, ** $p < 0.01$. Bar plots are presented as mean ± SEM. Scale bar = 100 μm. Green staining indicates DNA strand breaks; Blue staining is for Dapi indicating the cell nucleus. dUTP, deoxyUridine TriPhosphate; PE, Preeclampsia).

### 2.2. Irisin Significantly Decreases Apoptosis and Improves Cell Survival in 1st Trimester Human Placental Explants Stressed by Hypoxia/Re-Oxygenation

First trimester placental explants were cultured in hypoxia/re-oxygenation (H/R) conditions (24-hours 1% $O_2$ followed by 5-hours 8% $O_2$), to imitate the intermittent reperfusion injury observed in the preeclamptic placenta and treated with 10 or 50 nM Irisin. Cell death was quantified by TUNEL assay. There was a significant decrease in apoptotic cells in response to Irisin treatment compared to the no treatment control (n = 4, Figure 2A,B). Early-stage apoptosis was measured by using proximity ligation assay (PLA) to investigate the interaction of Apoptotic protease activating factor-1 (APAF-1) and Cytochrome C at single molecule resolution. The binding of Cytochrome C and APAF-1 is a key step in the initiation of apoptosis to permit the formation of apoptosome complexes [37]. Irisin treatment resulted in a visual reduction of APAF1 and Cytochrome C co-localization during H/R (Figure 2C).

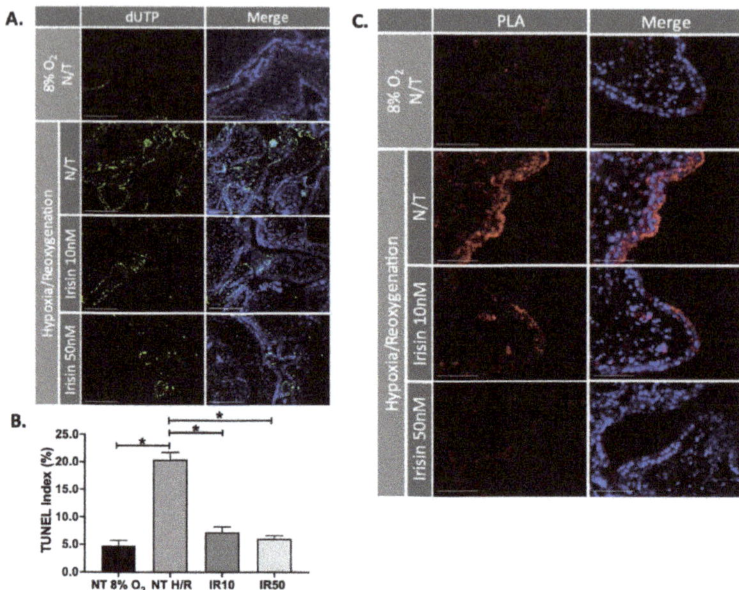

**Figure 2.** Irisin reduces apoptosis and improves cell survival in the ischemic 1st trimester human placenta. (**A,B**) Irisin significantly reduced the number of apoptotic cells during H/R in the first trimester placenta, as identified by a reduction of green fluorescent molecules in the TUNEL assay (n = 4). (**C**) The interaction between APAF1 and Cytochrome C (the early sign for initiation of apoptosis) indicated by red fluorescent molecules, was significantly diminished when H/R stressed first trimester placental explants were exposed to Irisin. The molecular interaction was studied by immunofluorescence in situ PLA, a novel technique that enables visualization of molecular proximity at single molecule resolution. (Significant changes in TUNEL index were measured by a one-way analysis of variance and subsequent Tukey's post hoc test to analyze differences between cohorts; * $p < 0.05$. Scale bar = 100 µM. Green fluorescent staining indicates DNA strand breaks (**A**); Blue fluorescent staining is for Dapi indicating the cell nucleus (**A,C**); Red fluorescent staining indicates APAF1 and Cytochrome C interaction (**C**). Bar plots are presented as mean ± SEM; N/T, non-treatment; H/R, hypoxia/re-oxygenation; PLA, Proximity Ligation Assay; dUTP, deoxyUridine TriPhosphate).

### 2.3. Anti-Apoptotic Effect of Irisin in 1st Trimester Placenta Explants Coincides with Akt Activation

To evaluate how Irisin can reduce apoptosis in the ischemic first trimester placenta, we investigated expression of various proteins in the apoptotic pathways. Irisin treatment of first trimester placental explants exposed to H/R resulted in a 25–40% decrease in gene expression of the pro-apoptotic BAX, while gene expression of anti-apoptotic BCL2A1 did not significantly change (n = 3, Figure 3A). Irisin further showed a 65–105% increase in the ratio for the anti-apoptotic BCL2: pro-apoptotic BAX protein levels in the first trimester explants (n = 3, Figure 3B). We additionally investigated the levels of Poly (ADP-ribose) polymerase (PARP) cleavage in the first trimester placenta after H/R injury. PARP functions to detect DNA damage and provide base excision repair and during apoptosis [38], and its cleavage by caspases prevents the ability to undergo DNA repair. We observed a significant reduction in cleaved PARP (cPARP) compared to total PARP (tPARP) levels when treated with 50 nm Irisin (n = 3, Figure 3C). Further, we observed an increase in the ratio of phosphorylated Akt: total Akt protein levels from Irisin treatment during H/R (n = 3, Figure 3D).

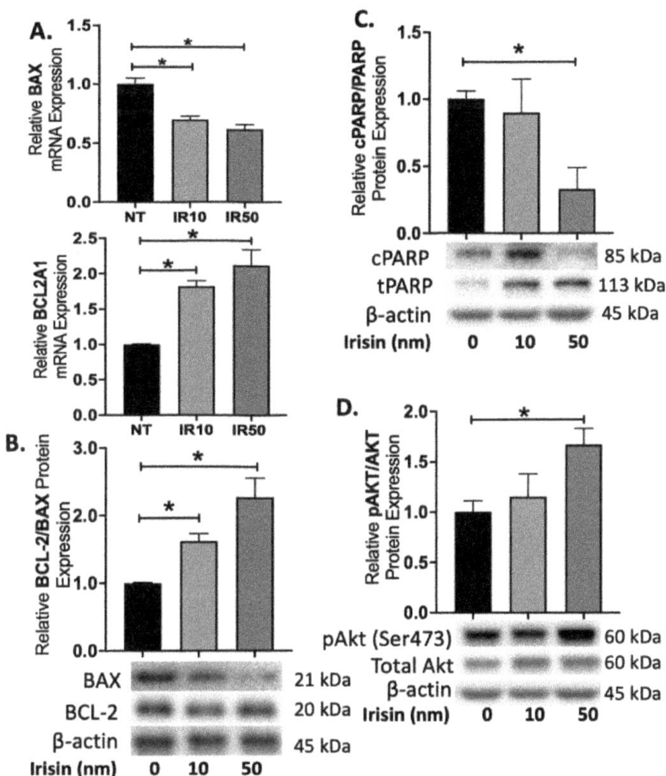

**Figure 3.** Irisin activates Akt pathway to protect against apoptosis during first trimester oxidative stress. (**A**) The gene expression of pro-apoptotic BAX decreased in presence of Irisin however the gene expression of anti-apoptotic BCL2A1 did not change with Irisin treatment (n = 3). (**B**) Protein expression for the BCL2: BAX ratio were significantly increased with Irisin treatment during H/R (n = 3). (**C**) cPARP was significantly reduced in the first trimester placenta treated with Irisin during H/R (n = 3). (**D**) Treatment of 50 nm of Irisin led to a significant increase in the ratio for phosphorylated Akt: total Akt protein levels in the first trimester placenta during H/R (n = 3). (Relative protein expression was determined by normalization to β-actin, followed by a one-way analysis of variance and subsequent Tukey's post hoc test to analyze differences between cohort; * $p < 0.05$. Bar plots are presented as mean ± SEM. H/R, Hypoxia/Re-oxygenation, PARP, Poly (ADP) Polymerase; cPARP, cleaved PARP; tPARP, total PARP).

### 2.4. Treatment IRISIN with Rescued Ischemic Injury in JEG-3 Cells

In addition to our ex vivo experiments, we used the choriocarcinoma JEG-3 cell line to validate our findings in H/R conditions. PLA showed that H/R induced APAF1/Cytochrome C interaction (apoptosis) in the cell-based model, which was significantly reduced in response to Irisin treatment (n = 3, Figure 4A). Caspase 3 assists in apoptotic pathways by cleaving several cellular proteins [39]. We observed a reduction of Caspase 3 activity in the JEG-3 cells exposed to H/R after treatment with Irisin (n = 3, Figure 4B).

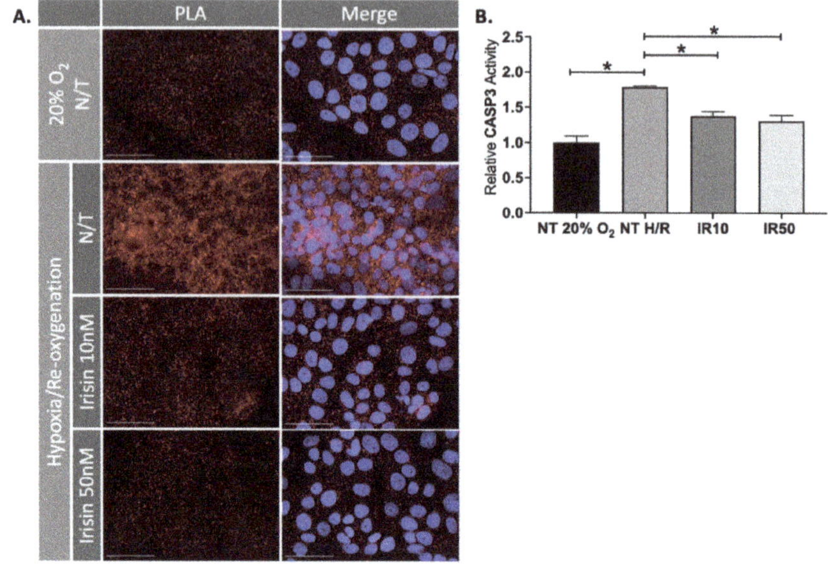

**Figure 4.** Irisin reduced apoptosis during ischemic injury in JEG-3 cells. (**A**) Proximity ligation assay (PLA) showed that the elevated APAF1/Cytochrome C interaction (apoptosis), as shown by red fluorescent molecules, under H/R condition was antagonized by Irisin in JEG-3 cells, similar to our observation in human 1st trimester explants. (**B**) This process coincided with the inhibition of Caspase 3 activity as confirmed by specific fluorogenic substrates Ac-DEVD-AMC (n = 3). (Significant changes in Caspase 3 activity were measured by a one-way analysis of variance and subsequent Tukey's post hoc test to analyze differences between cohorts; * $p < 0.05$. Scale bar = 100 μM. Blue fluorescent staining is for Dapi indicating the cell nucleus; Red fluorescent staining indicates APAF1 and Cytochrome C interaction. Bar plots are presented as mean ± SEM. N/T, non-treatment; PLA, Proximity Ligation Assay).

## 2.5. Perifosine, a Specific Akt Antagonist, Inhibited Anti-Apoptotic Effect of Irisin in JEG-3 Cells

To investigate the potential role of Akt signalling in cellular response to Irisin, JEG-3 cells were incubated with 10 or 50 nm Irisin and within 5-min we observed a significant increase in the ratio for the phosphorylated Akt: total Akt protein expression levels (n = 3, Figure 5A). We incubated JEG-3 cells with different concentration of Perifosine (10–100 μM), a specific Akt inhibitor. Perifosine significantly reduced the ratio of phosphorylated Akt: total Akt protein expression (n = 3, Figure 5B). Treatment with both 50 nm Irisin and 50 μM Perifosine resulted in a significant reduction of phosphorylated Akt: total Akt protein expression levels (n = 3, Figure 5C). To further support the notion of Irisin's apoptotic effects, we identified a significant reduction of reactive oxygen species (ROS) during H/R which was successfully blocked by Perifosine, showing that the antioxidant effect of Irisin is an Akt-dependent phenomenon (Figure 5D).

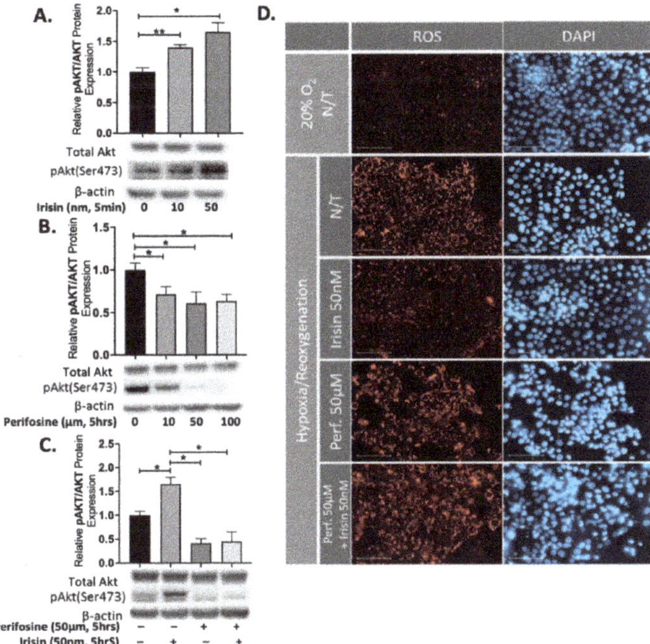



**Figure 5.** Perifosine inhibited the anti-apoptotic effect of Irisin in JEG-3 cells. (**A**) Irisin treatment increased the ratio of phosphorylated Akt: total Akt protein expression within 5-minutes of treatment (n = 3). (**B**) The ratio of phosphorylated: total Akt protein expression was significantly reduced after 5 h of increasing amounts of Perifosine (n = 3). (**C**) The promoting effect of Irisin on Akt phosphorylation was attenuated in presence of Perifosine (n = 3). (**D**) The rescue effect of Irisin on intracellular generation of ROS, shown by red fluorescence, in JEG-3 cells was blocked by Perifosine during H/R. (Relative protein expression was determined by normalization to β-actin, followed a one-way analysis of variance and subsequent Tukey's post hoc test to analyze differences between cohorts; * $p < 0.05$, ** $p < 0.01$. Bar plots are presented as mean ± SEM. Scale bar = 100 μM. Blue fluorescent staining is for Dapi indicating the cell nucleus; Red fluorescent staining indicates ROS presence in the cells. N/T, non-treatment, H/R, Hypoxia/Re-oxygenation; ROS, Reactive Oxygen Species).

## 3. Discussion

Placental disorders such as preeclampsia often involve ischemic reperfusion injury that largely contributes to increased cell death in the placenta. In this study we provided evidence of the anti-apoptotic role for Irisin in the human preeclamptic placenta and in the human first trimester placenta and in the choriocarcinoma JEG-3 cell model during pathological conditions.

Irisin is a secreted myokine that is involved in energy metabolism. Irisin has a wide range of effects in the cell involving changes in inflammation, oxidative stress and apoptosis [16]. Exercise upregulates Irisin secretion which acts on fat tissue to induce tissue browning for the purpose of energy mobilization. Therefore, many studies have been focused on the roles of Irisin in the treatment of metabolic disorders such as obesity and type 2 diabetes, and less is known of Irisin's roles during pregnancy.

Our study is one of the first to report anti-apoptotic features of Irisin treatment in the placenta. Evidence to support this is shown by the reduction of apoptotic cells and by an increase in the ratio of anti-apoptotic BCL2: pro-apoptotic BAX protein levels in response to Irisin-treated preeclamptic placentas and in ex vivo and in vitro model systems during ischemic reperfusion injury. The reduction of cleaved PARP induced by Irisin provides the cells an opportunity to undergo repair of DNA strand breaks in the first trimester placenta during ischemic conditions. Caspase 3 is a crucial member of apoptotic pathways by acting to cleave several cellular proteins [39] and its reduction in activity by Irisin further convinces the role for Irisin in promoting anti-apoptotic pathways. Furthermore, Caspase and BAX reduction has been described in lung injury models [40] and vascular diseases [41] in response to irisin treatment, which correlates with the findings of our study.

The increase in BAX activity during oxidative stress compromises mitochondrial membrane integrity to release Cytochrome C into the cytoplasm [42] where it is free to interact with apoptotic protease-activating factor-1 (Apaf-1) [43]. The binding of Cytochrome C with Apaf-1 initiates apoptosis signaling cascades [43]. Remarkably, we observe a dose-dependent decrease in the interaction between Cytochrome C and Apaf-1 with increasing Irisin treatment, which provides greater support to our claims that Irisin protects against apoptosis.

The Akt signaling pathway is an important intracellular signal transduction pathway with a key role in the regulation of apoptosis and cell survival [44]. Akt phosphorylation acts on several proteins including the Foxo family members, YAP, BAD and Caspase 9 to inhibit apoptosis [43]. We observed significant increases in the ratio of phosphorylated Akt: total Akt protein levels from Irisin treatment in our ischemic models. Blocking Akt activation with the Akt specific inhibitor, Perifosine, abolished the anti-apoptotic effects of Irisin, which suggests that Irisin likely acts through the Akt pathway to protect against apoptosis. Furthermore, our results correlate with additional reports in the literature. Specifically, a study by Li et al. assessed Irisin's effect on endothelial function in apolipoprotein E-Null diabetic mice and found that Irisin treatment protected the endothelium through activation of AMPK and Akt pathways [41].

The exact signaling mechanisms on how Irisin exhibits its action are still under investigation. While we have shown that Irisin activates Akt phosphorylation at the serine 473 residue, more studies investigating members upstream and downstream of Akt are needed to further explore this pathway. A putative receptor for Irisin, $\alpha V/\beta 5$ integrin has been recently identified [45,46], which was initially characterized in the human placenta making it a prime target for Irisin signaling [47]. $\alpha V/\beta 5$ integrin has a variety of downstream targets and although it is likely Irisin will initiate these same $\alpha V/\beta 5$ integrin signaling pathways in the placenta, future studies will need to validate this process. While the placenta is known to express the full length FNDC5 protein, it is unclear if the placenta is a significant source of its secreted form, Irisin. Overall, the many physiological roles of Irisin are still under investigation and here we were able to show that Irisin has cytoprotective capabilities in the human preeclamptic placenta and models of placental disease. Future research should focus on the various sources of Irisin in pregnancy and the effects on the placenta to better understand its roles in maternal and feto-placental metabolism and disease.

## 4. Materials and Methods

### 4.1. Placental Tissue Collection

First trimester (10–12 weeks of gestation) placental tissues (n = 4) were obtained with written informed consent from healthy pregnant women undergoing elective termination of pregnancy. Term placental samples were obtained either by the Research Centre for Women's and Infants' Health BioBank program of Hutzel Women's Hospital in Detroit, MI or by Women's Health Center at Spectrum Hospital in Grand Rapids, MI, USA. The Institutional Review Board of Wayne State University and Michigan State University and approved all consent forms and protocols used in this study, which abide by the National Institutes of Health (NIH) research guidelines. Specimens were collected from pregnancies complicated by preeclampsia (n = 8; gestational age = 31–37 weeks) and were delivered either by Cesarean section or vaginal birth. Inclusion criteria for preeclampsia was in accordance with current guidelines including blood pressure >140/90 mm Hg on 2 occasions longer than 6 h apart, evidence of end-organ damage including proteinuria, with or without fetal growth restriction [48].

The collected tissues were washed and transported to the laboratory in ice-cold Hank's balanced salt solution and were processed within a maximum of 2 h after collection. Upon arrival, tissues were snap-frozen in liquid nitrogen for further analysis. For ex vivo modeling, individual clusters of villous trees were dissected under a stereomicroscope and cultured in 1 mL of Dulbecco's modified Eagle's medium/Ham's F-12 nutrient mixture (DMEM/F-12; 1:1; Life Technologies; Grand Island, NY, USA) containing 10% fetal bovine serum (FBS; Life Technologies) and 1% Gibco™ antibiotic-antimycotic. The explants were maintained overnight at either 8% $O_2$ or Hypoxia/Reoxygenation (H/R) with 5% $CO_2$ at 37 °C. During hypoxic exposures, the gas mixtures were balanced with $N_2$. HR conditions were performed by culturing the first trimester placental explants in 1% $O_2$ overnight followed by culture at 8% $O_2$ and replacement of fresh medium and incubation for 5 h. For the experiments, placental explants were cultured in a medium supplemented with low (10 nM) or high (50 nM) physiological doses of active recombinant Irisin (Enzo Life Sciences, Farmingdale, NY, USA) [34,49].

### 4.2. Human Trophoblast Cell Culture

The human choriocarcinoma cell line JEG-3 was purchased from the American Type Culture Collection (ATCC). Cells were cultured in Dulbecco Modified Eagle Medium (DMEM) and Ham F12 (1:1 DMEM/F12) medium (Invitrogen, Waltham, MA, USA) containing 10% FBS and 1% Antibiotic-Antimycotic (Gibco, Amarillo, TX, USA) in a humidified incubator at 5% $CO_2$. The H/R (hypoxia/reoxygenation) was performed as 1.5% $O_2$ overnight followed by replacement with fresh medium equilibrated at 20% $O_2$ and incubation for 5 h. To examine the effect of irisin on trophoblast differentiation, the cells were treated with 10 nM or 50 nM of recombinant Irisin (Enzo Life Sciences, Farmingdale, NY, USA). JEG-3 cells were also treated with 10 μm, 50 μm and 100 μm of Perifosine (Selleckchem, Houston, TX, USA) to inhibit Akt phosphorylation.

### 4.3. Protein Extraction and Immunoblotting

Protein extraction from tissues (20–30 mg) was performed as previously described before [50]. Protein concentration was determined with BCA™ protein assay reagent (Thermo Fisher Scientific, Rockford, IL, USA) according to the manufacturer's instructions. Equal protein amounts (35 μg) were denatured (8 min, 95 °C) in Laemmli sample buffer (Bio-Rad Laboratories; Hercules, CA, USA) and separated using sodium dodecyl sulfate-polyacrylamide gel electrophoresis, with subsequent semi-dry transfer (Trans-Blot®; Bio-Rad Laboratories) to a polyvinylidene difluoride membrane. The membranes were blocked with 5% nonfat dry milk in 1× Tris-buffered saline containing 0.05% Tween-20 and were incubated overnight at 4 °C with anti-BAX (1:1000; Cell Signaling Technology, Danvers, MA, USA), anti-BCL-2α (1:1000; Cell Signaling Technology, Danvers, MA, USA), anti-cleaved PARP (Cell Signaling Technology, Danvers, MA, USA), anti-PARP (Cell Signaling Technology, Danvers, MA, USA), anti-pan-Akt (1:1000; Cell Signaling Technology, Danvers, MA, USA) and anti-phospho-Akt (Ser473) (Cell Signaling Technology, Danvers, MA, USA) primary antibodies. Subsequently, membranes were incubated with horseradish peroxidase-conjugated secondary antibodies for 1 h at room temperature and were developed with Western Lightning® ECL Pro (PerkinElmer, Waltham, MA, USA). Signals were visualized using a ChemiDoc™ Imaging System (Bio-Rad Laboratories) and Image Lab Version 5.1 software (Bio-Rad Laboratories). Densities of immunoreactive bands were measured as arbitrary units by the ImageJ software (NIH, Bethesda, MD, USA). Protein levels were normalized to a housekeeping protein (β-actin; 1:4000; Cell Signaling Technology, Danvers, MA, USA).

### 4.4. Cell Death Assay

A total of 5 placentas were used for histological evaluation and quantification of apoptosis in paraffin-embedded section. A similar number of villi were dissected from all five regions mixed and then randomly picked from the pool for each experiment to avoid sampling bias. Each dissected explant contained approximately 8–10 mg tissue, four to five villi, or cultured cells. All treatments were performed in triplicates. DNA-strand breaks were detected by TUNEL (terminal deoxynucleotidyl transferase [TdT] dUTP nick-end labeling), using a fluorescein-based in situ cell death detection kit (Roche Applied Science, Indianapolis, IN, USA), according to the manufacturer's instructions. Nuclei were counterstained with DAPI (EMD Biosciences, Billerica, MA, USA). Sections were imaged with a Nikon Eclipse 90i epifluorescence microscope (Nikon Inc., Melville, NY, USA). The apoptotic cells (TUNEL-positive nuclei) were counted at 40xX from four random fields on each section from three samples for each treatment, along with the total number of nuclei (DAPI-labeled) to calculate the percentage of TUNEL/DAPI-labeled nuclei (TUNEL index). Additional sections subjected to treatment without TdT were assessed as negative controls.

### 4.5. Proximity Ligation Assay (PLA)

Proximity ligation assay (PLA) was performed in situ using a Duolink In Situ Red Starter Kit for Mouse/Rabbit (Sigma, St. Louis, MO, USA), according to the manufacturer's instructions. Briefly, first trimester explants were fixed in 4% paraformaldehyde, imbedded in paraffin block and sectioned onto slides for staining, followed by standard dewaxing and rehydrating conditions. JEG-3 cells were fixed and permeabilized. Cells or tissues and incubated overnight at 4 °C with primary anti-Cytochrome C and anti-APAF1 antibodies (Abcam) in pre-blocking buffer (0.05% Triton X-100 in PBS, pH 7.4). A negative control was prepared by incubating cells/tissues in blocking solution without primary antibodies. Cells/tissues were washed and incubated with rabbit plus and mouse minus PLA probes for 60 min at 37 °C. After washing, the ligation-ligase mixture was added and cells were incubated for 30 min at 37 °C, followed by an amplification step that generates a rolling circle DNA. Hoechst 33342 was used to stain nuclei. The fluorescently labeled oligonucleotides were visualized by a Nikon Eclipse 90i epifluorescence microscope (Nikon Inc.).

### 4.6. Caspase Activity Assay

The activation of caspase 3 was determined using a fluorometric substrate, Ac-DEVD-AMC (Enzo Life Sciences). JEG-3 cells were seeded at a density of 70,000 cells/well in a 6-well plate. After treatment, a nondenaturing lysis buffer was added to extract cellular protein, as described previously [51]. Total protein (35 μg) and 40 μL of substrate were added to 50 μL of reaction buffer (1% NP-40, 10% glycerol in TBS). The mixture was incubated at 37 °C for 3 h, and the fluorescence intensity was quantified using a microplate reader.

### 4.7. Fluorometric and Quantitative Evaluation of ROS Generation

To observe the basic changes of intracellular ROS, a Cellular Reactive Oxygen Species Detection Assay Kit (Abcam, Cambridge, MA, USA) was utilized as instructed by the manufacturer. JEG-3 cells were seeded in an eight-chamber slide at 10,000 cells/chamber. After a 48-h incubation, cells were washed with PBS and preloaded with $1\times$ ROS probes for 45 min at 37 °C. Cells were then washed with PBS. Cells were examined under a Nikon Eclipse 90i epifluorescence microscope (Nikon Inc.) with appropriate filters.

### 4.8. Statistical Analysis

All statistical analysis was performed with GraphPad Prism 7.0 software. Raw mRNA and protein expressions were normalized to respective housekeeping genes or protein. All experiments were performed at least three times. A one-way analysis of variance and subsequent Tukey's post hoc test was performed to analyze differences between cohorts. An effect was considered significant when $p < 0.05$ and is indicated with (*) on each graph. For arbitrary units, results were calculated relative to non-treatment (N/T) controls (set as 1) and presented as mean $\pm$ standard error of the mean (SEM).

**Author Contributions:** All authors listed contributed significantly to this manuscript. Specific roles include conceptualization, H.-R.K.-G. and S.D.; methodology, H.-R.K.-G., B.A. and M.B.; formal analysis, H.-R.K.-G. and B.A.; writing—original draft preparation, H.-R.K.-G. and S.D.; writing—review and editing, B.A.; supervision, H.-R.K.-G. and S.D.; funding acquisition, S.D. All authors have read and agreed to the published version of the manuscript.

**Funding:** This research was supported by the Department of Obstetrics, Gynecology and Reproductive Biology at Michigan State University College of Human Medicine and the March of Dimes Foundation awarded to S.D. Research support for B.A.'s work in this publication was provided by the Eunice Kennedy Shriver National Institute of Child Health and Human Development of the National Institutes of Health under Award Number T32HD087166 awarded to Michigan State University AgBio Research. The content is solely the responsibility of the authors and does not necessarily represent the official views of the National Institutes of Health Support.

**Institutional Review Board Statement:** The Institutional Review Boards of Wayne State University and Michigan State University approved all consent forms and protocols used in this study, which abide by the National Institutes of Health research guidelines.

**Informed Consent Statement:** Informed consent was obtained from all subjects involved in the study.

**Data Availability Statement:** Not applicable.

**Acknowledgments:** We thank the women who have graciously donated their tissues for this study. We also thank the Spectrum Health Accelerator of Research Excellence (SHARE) biorepository (Spectrum Health, Grand Rapids, MI, USA) for their work in collecting and providing de-identified biospecimens for this study.

**Conflicts of Interest:** The authors declare no conflict of interest.

## References

1.  Toal, M.; Chan, C.; Fallah, S.; Alkazaleh, F.; Chaddha, V.; Windrim, R.C.; Kingdom, J.C. Usefulness of a placental profile in high-risk pregnancies. *Am. J. Obs. Gynecol.* **2007**, *196*, 363.e1–363.e7. [CrossRef] [PubMed]
2.  Sibai, B.M. Evaluation and management of severe preeclampsia before 34 weeks' gestation. *Am. J. Obs. Gynecol.* **2011**, *205*, 191–198. [CrossRef] [PubMed]
3.  Armistead, B.; Kadam, L.; Drewlo, S.; Kohan-Ghadr, H.-R. The Role of NFκB in Healthy and Preeclamptic Placenta: Trophoblasts in the Spotlight. *Int. J. Mol. Sci.* **2020**, *21*, 1775. [CrossRef] [PubMed]
4.  Knöfler, M.; Haider, S.; Saleh, L.; Pollheimer, J.; Gamage, T.K.J.B.; James, J. Human placenta and trophoblast development: Key molecular mechanisms and model systems. *Cell. Mol. Life Sci.* **2019**, *76*, 3479–3496. [CrossRef] [PubMed]
5.  Rana, S.; Lemoine, E.; Granger, J.P.; Karumanchi, S.A. Preeclampsia: Pathophysiology, Challenges, and Perspectives. *Circ. Res.* **2019**, *124*, 1094–1112. [CrossRef] [PubMed]
6.  Can, M.; Guven, B.; Bektas, S.; Arikan, I. Oxidative stress and apoptosis in preeclampsia. *Tissue Cell* **2014**, *46*, 477–481. [CrossRef] [PubMed]
7.  Tong, W.; Giussani, D.A. Preeclampsia link to gestational hypoxia. *J. Dev. Orig. Health Dis.* **2019**, *10*, 322–333. [CrossRef] [PubMed]
8.  Armistead, B.; Kadam, L.; Siegwald, E.; McCarthy, F.P.; Kingdom, J.C.; Kohan-Ghadr, H.-R.; Drewlo, S. Induction of the PPARγ (Peroxisome Proliferator-Activated Receptor γ)-GCM1 (Glial Cell Missing 1) Syncytialization Axis Reduces sFLT1 (Soluble fms-Like Tyrosine Kinase 1) in the Preeclamptic Placenta. *Hypertension* **2021**, *78*, 230–240. [CrossRef]
9.  Shahul, S.; Medvedofsky, D.; Wenger, J.B.; Nizamuddin, J.; Brown, S.M.; Bajracharya, S.; Salahuddin, S.; Thadhani, R.; Mueller, A.; Tung, A.; et al. Circulating Antiangiogenic Factors and Myocardial Dysfunction in Hypertensive Disorders of Pregnancy. *Hypertension* **2016**, *67*, 1273–1280. [CrossRef]
10. Armistead, B.; Johnson, E.; Vanderkamp, R.; Kula-Eversole, E.; Kadam, L.; Drewlo, S.; Kohan-Ghadr, H.-R. Placental Regulation of Energy Homeostasis During Human Pregnancy. *Endocrinology* **2020**, *161*, bqaa076. [CrossRef]
11. Boström, P.; Wu, J.; Jedrychowski, M.P.; Korde, A.; Ye, L.; Lo, J.C.; Rasbach, K.A.; Boström, E.A.; Choi, J.H.; Long, J.Z.; et al. A PGC1-α-dependent myokine that drives brown-fat-like development of white fat and thermogenesis. *Nature* **2012**, *481*, 463–468. [CrossRef] [PubMed]
12. Dun, S.; Lyu, R.-M.; Chen, Y.-H.; Chang, J.-K.; Luo, J.; Dun, N. Irisin-immunoreactivity in neural and non-neural cells of the rodent. *Neuroscience* **2013**, *240*, 155–162. [CrossRef]
13. Huh, J.Y.; Panagiotou, G.; Mougios, V.; Brinkoetter, M.; Vamvini, M.T.; Schneider, B.E.; Mantzoros, C.S. FNDC5 and irisin in humans: I. Predictors of circulating concentrations in serum and plasma and II. mRNA expression and circulating concentrations in response to weight loss and exercise. *Metabolism* **2012**, *61*, 1725–1738. [CrossRef]
14. Zhu, J.; Wang, Y.; Cao, Z.; Du, M.; Hao, Y.; Pan, J.; He, H. Irisin promotes cementoblast differentiation via p38 MAPK pathway. *Oral Dis.* **2020**, *26*, 974–982. [CrossRef]

15. Li, H.; Zhang, Y.; Wang, F.; Donelan, W.; Zona, M.C.; Li, S.; Reeves, W.; Ding, Y.; Tang, D.; Yang, L. Effects of irisin on the differentiation and browning of human visceral white adipocytes. *Am. J. Transl. Res.* **2019**, *11*, 7410–7421. [PubMed]
16. Askari, H.; Rajani, S.F.; Poorebrahim, M.; Haghi-Aminjan, H.; Raeis-Abdollahi, E.; Abdollahi, M. A glance at the therapeutic potential of irisin against diseases involving inflammation, oxidative stress, and apoptosis: An introductory review. *Pharm. Res.* **2018**, *129*, 44–55. [CrossRef]
17. Li, Q.; Tan, Y.; Chen, S.; Xiao, X.; Zhang, M.; Wu, Q.; Dong, M. Irisin alleviates LPS-induced liver injury and inflammation through inhibition of NLRP3 inflammasome and NF-κB signaling. *J. Recept. Signal Transduct. Res.* **2021**, *41*, 294–303. [CrossRef]
18. Bi, J.; Zhang, J.; Ren, Y.; Du, Z.; Li, Q.; Wang, Y.; Wei, S.; Yang, L.; Zhang, J.; Liu, C.; et al. Irisin alleviates liver ischemia-reperfusion injury by inhibiting excessive mitochondrial fission, promoting mitochondrial biogenesis and decreasing oxidative stress. *Redox Biol.* **2018**, *20*, 296–306. [CrossRef] [PubMed]
19. Storlino, G.; Colaianni, G.; Sanesi, L.; Lippo, L.; Brunetti, G.; Errede, M.; Colucci, S.; Passeri, G.; Grano, M. Irisin Prevents Disuse-Induced Osteocyte Apoptosis. *J. Bone Min. Res.* **2020**, *35*, 766–775. [CrossRef]
20. Liu, S.; Du, F.; Li, X.; Wang, M.; Duan, R.; Zhang, J.; Wu, Y.; Zhang, Q. Effects and underlying mechanisms of irisin on the proliferation and apoptosis of pancreatic β cells. *PLoS ONE* **2017**, *12*, e0175498. [CrossRef]
21. Carnero, A.; Blanco-Aparicio, C.; Renner, O.; Link, W.; Leal, J.F.M. The PTEN/PI3K/AKT signalling pathway in cancer, therapeutic implications. *Curr. Cancer Drug Targets* **2008**, *8*, 187–198. [CrossRef]
22. Yu, J.S.L.; Cui, W. Proliferation, survival and metabolism: The role of PI3K/AKT/mTOR signalling in pluripotency and cell fate determination. *Development* **2016**, *143*, 3050–3060. [CrossRef]
23. Duan, C.; Bauchat, J.R.; Hsieh, T. Phosphatidylinositol 3-kinase is required for insulin-like growth factor-I-induced vascular smooth muscle cell proliferation and migration. *Circ. Res.* **2000**, *86*, 15–23. [CrossRef]
24. Long, Y.; Jiang, Y.; Zeng, J.; Dang, Y.; Chen, Y.; Lin, J.; Wei, H.; Xia, H.; Long, J.; Luo, C.; et al. The expression and biological function of chemokine CXCL12 and receptor CXCR4/CXCR7 in placenta accreta spectrum disorders. *J. Cell. Mol. Med.* **2020**, *24*, 3167–3182. [CrossRef]
25. Qiu, Q.; Yang, M.; Tsang, B.K.; Gruslin, A. EGF-induced trophoblast secretion of MMP-9 and TIMP-1 involves activation of both PI3K and MAPK signalling pathways. *Reproduction* **2004**, *128*, 355–363. [CrossRef]
26. Chiang, M.H.; Liang, F.-Y.; Chen, C.-P.; Chang, C.-W.; Cheong, M.-L.; Wang, L.-J.; Liang, C.-Y.; Lin, F.-Y.; Chou, C.-C.; Chen, H. Mechanism of hypoxia-induced GCM1 degradation: Implications for the pathogenesis of preeclampsia. *J. Biol. Chem.* **2009**, *284*, 17411–17419. [CrossRef] [PubMed]
27. Martorell, L.; Gentile, M.; Rius, J.; Rodriguez, C.; Crespo, J.; Badimon, L.; Martínez-González, J. The Hypoxia-Inducible Factor 1/NOR-1 Axis Regulates the Survival Response of Endothelial Cells to Hypoxia. *Mol. Cell. Biol.* **2009**, *29*, 5828–5842. [CrossRef]
28. Zhu, X.; Cao, Q.; Li, X.; Wang, Z. Knockdown of TACC3 inhibits trophoblast cell migration and invasion through the PI3K/Akt signaling pathway. *Mol. Med. Rep.* **2016**, *14*, 3437–3442. [CrossRef] [PubMed]
29. Flügel, D.; Görlach, A.; Michiels, C.; Kietzmann, T. Glycogen Synthase Kinase 3 Phosphorylates Hypoxia-Inducible Factor 1α and Mediates Its Destabilization in a VHL-Independent Manner. *Mol. Cell. Biol.* **2007**, *27*, 3253–3265. [CrossRef] [PubMed]
30. Mottet, D.; Dumont, V.; Deccache, Y.; Demazy, C.; Ninane, N.; Raes, M.; Michiels, C. Regulation of hypoxia-inducible factor-1alpha protein level during hypoxic conditions by the phosphatidylinositol 3-kinase/Akt/glycogen synthase kinase 3beta pathway in HepG2 cells. *J. Biol. Chem.* **2003**, *278*, 31277–31285. [CrossRef] [PubMed]
31. Xu, Y.; Sui, L.; Qiu, B.; Yin, X.; Liu, J.; Zhang, X. ANXA4 promotes trophoblast invasion via the PI3K/Akt/eNOS pathway in preeclampsia. *Am. J. Physiol. Physiol.* **2019**, *316*, C481–C491. [CrossRef]
32. Wang, W.; Shi, Y.; Bai, G.; Tang, Y.; Yuan, Y.; Zhang, T.; Li, C. HBxAg suppresses apoptosis of human placental trophoblastic cell lines via activation of the PI3K/Akt pathway. *Cell Biol. Int.* **2016**, *40*, 708–715. [CrossRef] [PubMed]
33. Lim, W.; Yang, C.; Bazer, F.W.; Song, G. Luteolin Inhibits Proliferation and Induces Apoptosis of Human Placental Choriocarcinoma Cells by Blocking the PI3K/AKT Pathway and Regulating Sterol Regulatory Element Binding Protein Activity. *Biol. Reprod.* **2016**, *95*, 82. [CrossRef] [PubMed]
34. Garcés, M.F.; Peralta, J.J.; Ruiz-Linares, C.E.; Lozano, A.R.; Poveda, N.E.; Torres-Sierra, A.L.; Eslava-Schmalbach, J.H.; Alzate, J.P.; Sánchez, Y.; Sanchez, E.; et al. Irisin Levels During Pregnancy and Changes Associated With the Development of Preeclampsia. *J. Clin. Endocrinol. Metab.* **2014**, *99*, 2113–2119. [CrossRef] [PubMed]
35. Zhang, L.-J.; Xie, Q.; Tang, C.-S.; Zhang, A.-H. Expressions of irisin and urotensin II and their relationships with blood pressure in patients with preeclampsia. *Clin. Exp. Hypertens.* **2017**, *39*, 460–467. [CrossRef]
36. Drewlo, S.; Johnson, E.; Kilburn, B.A.; Kadam, L.; Armistead, B.; Kohan-Ghadr, H. Irisin induces trophoblast differentiation via AMPK activation in the human placenta. *J. Cell. Physiol.* **2020**, *235*, 7146–7158. [CrossRef]
37. Shakeri, R.; Kheirollahi, A.; Davoodi, J. Apaf-1: Regulation and function in cell death. *Biochimie* **2017**, *135*, 111–125. [CrossRef] [PubMed]
38. Morales, J.; Li, L.; Fattah, F.J.; Dong, Y.; Bey, E.A.; Patel, M.; Gao, J.; Boothman, D.A. Review of Poly (ADP-ribose) Polymerase (PARP) Mechanisms of Action and Rationale for Targeting in Cancer and Other Diseases. *Crit. Rev. Eukaryot. Gene Expr.* **2014**, *24*, 15–28. [CrossRef] [PubMed]
39. Porter, A.G.; Jänicke, R.U. Emerging roles of caspase-3 in apoptosis. *Cell Death Differ.* **1999**, *6*, 99–104. [CrossRef]
40. Shao, L.; Meng, D.; Yang, F.; Song, H.; Tang, D. Irisin-mediated protective effect on LPS-induced acute lung injury via suppressing inflammation and apoptosis of alveolar epithelial cells. *Biochem. Biophys. Res. Commun.* **2017**, *487*, 194–200. [CrossRef] [PubMed]

41. Lu, J.; Xiang, G.; Liu, M.; Mei, W.; Xiang, L.; Dong, J. Irisin protects against endothelial injury and ameliorates atherosclerosis in apolipoprotein E-Null diabetic mice. *Atherosclerosis* **2015**, *243*, 438–448. [CrossRef] [PubMed]

42. Westphal, D.; Dewson, G.; Czabotar, P.E.; Kluck, R. Molecular biology of Bax and Bak activation and action. *Biochim. Biophys. Acta (BBA)—Bioenerg.* **2011**, *1813*, 521–531. [CrossRef]

43. Abeyrathna, P.; Su, Y. The critical role of Akt in cardiovascular function. *Vasc. Pharm.* **2015**, *74*, 38–48. [CrossRef] [PubMed]

44. Manning, B.D.; Toker, A. AKT/PKB Signaling: Navigating the Network. *Cell* **2017**, *169*, 381–405. [CrossRef] [PubMed]

45. Park, E.J.; Myint, P.K.; Ito, A.; Appiah, M.G.; Darkwah, S.; Kawamoto, E.; Shimaoka, M. Integrin-Ligand Interactions in Inflammation, Cancer, and Metabolic Disease: Insights into the Multifaceted Roles of an Emerging Ligand Irisin. Front. *Cell Dev. Biol.* **2020**, *8*, 588066. [CrossRef] [PubMed]

46. Kim, H.; Wrann, C.D.; Jedrychowski, M.; Vidoni, S.; Kitase, Y.; Nagano, K.; Zhou, C.; Chou, J.; Parkman, V.-J.A.; Novick, S.J.; et al. Irisin Mediates Effects on Bone and Fat via αV Integrin Receptors. *Cell* **2018**, *175*, 1756–1768. [CrossRef] [PubMed]

47. Smith, J.W.; Vestal, D.J.; Irwin, S.V.; Burke, T.A.; Cheresh, D.A. Purification and functional characterization of integrin alpha v beta 5. An adhesion receptor for vitronectin. *J. Biol. Chem.* **1990**, *265*, 11008–11013. [CrossRef]

48. Espinoza, J.; Pettker, C.M.; Hyagriv, S.; Vidaeff, A. ACOG Practice Bulletin No. 202: Gestational Hypertension and Preeclampsia. *Obstet. Gynecol.* **2019**, *133*, e1–e25.

49. Ural, M.; Şahin, S.B.; Tekin, Y.B.; Cüre, M.C.; Sezgin, H. Alteration of maternal serum irisin levels in gestational diabetes mellitus. *Ginekol. Pol.* **2016**, *87*, 395–398. [CrossRef]

50. Becker, J.; Barysch, S.V.; Karaca, S.; Dittner, C.; Hsiao, H.-H.; Diaz, M.B.; Herzig, S.; Urlaub, H.; Melchior, F. Detecting endogenous SUMO targets in mammalian cells and tissues. *Nat. Struct. Mol. Biol.* **2013**, *20*, 525–531. [CrossRef]

51. Marino, J.; Vior, M.C.G.; Furmento, V.A.; Blank, V.C.; Awruch, J.; Roguin, L.P. Lysosomal and mitochondrial permeabilization mediates zinc (II) cationic phthalocyanine phototoxicity. *Int. J. Biochem. Cell Biol.* **2013**, *45*, 2553–2562. [CrossRef] [PubMed]

International Journal of
*Molecular Sciences*

*Article*

# Differences in Glycolysis and Mitochondrial Respiration between Cytotrophoblast and Syncytiotrophoblast In-Vitro: Evidence for Sexual Dimorphism

**Matthew Bucher, Leena Kadam, Kylia Ahuna and Leslie Myatt ***

Department of Obstetrics and Gynecology, Oregon Health and Science University, Portland, OR 97239, USA; mbucher@uoregon.edu (M.B.); kadam@ohsu.edu (L.K.); ahuna@ohsu.edu (K.A.)
* Correspondence: myattl@ohsu.edu; Tel.: +1-(503)-418-2781

**Abstract:** In the placenta the proliferative cytotrophoblast cells fuse into the terminally differentiated syncytiotrophoblast layer which undertakes several energy-intensive functions including nutrient uptake and transfer and hormone synthesis. We used Seahorse glycolytic and mitochondrial stress tests on trophoblast cells isolated at term from women of healthy weight to evaluate if cytotrophoblast (CT) and syncytiotrophoblast (ST) have different bioenergetic strategies, given their different functions. Whereas there are no differences in basal glycolysis, CT have significantly greater glycolytic capacity and reserve than ST. In contrast, ST have significantly higher basal, ATP-coupled and maximal mitochondrial respiration and spare capacity than CT. Consequently, under stress conditions CT can increase energy generation via its higher glycolytic capacity whereas ST can use its higher and more efficient mitochondrial respiration capacity. We have previously shown that with adverse in utero conditions of diabetes and obesity trophoblast respiration is sexually dimorphic. We found no differences in glycolytic parameters between sexes and no difference in mitochondrial respiration parameters other than increases seen upon syncytialization appear to be greater in females. There were differences in metabolic flexibility, i.e., the ability to use glucose, glutamine, or fatty acids, seen upon syncytialization between the sexes with increased flexibility in female trophoblast suggesting a better ability to adapt to changes in nutrient supply.

**Keywords:** placenta; metabolism; glycolysis; mitochondrial respiration; cytotrophoblast; syncytiotrophoblast; placental metabolism; trophoblast glycolysis; trophoblast mitochondrial respiration; sexual dimorphism

**Citation:** Bucher, M.; Kadam, L.; Ahuna, K.; Myatt, L. Differences in Glycolysis and Mitochondrial Respiration between Cytotrophoblast and Syncytiotrophoblast In-Vitro: Evidence for Sexual Dimorphism. *Int. J. Mol. Sci.* **2021**, *22*, 10875. https://doi.org/10.3390/ijms221910875

Academic Editors: Hiten D Mistry and Eun Lee

Received: 2 September 2021
Accepted: 30 September 2021
Published: 8 October 2021

**Publisher's Note:** MDPI stays neutral with regard to jurisdictional claims in published maps and institutional affiliations.



## 1. Introduction

The placenta is a highly specialized fetal organ responsible for supporting growth and development of the fetus in utero. It forms an immune and physical barrier between the mother and fetus and provides metabolic, transport, and endocrine functions [1]. Each of these functions exact a metabolic cost so it comes with no surprise that the placenta has an extraordinarily high metabolic rate, accounting for approximately 40% of the total oxygen consumed by the fetus and placenta combined [2]. The chorionic villi of the placenta, which contain fetal capillaries, are bathed with maternal blood and it is here on its outer surface where oxygen and nutrient uptake and transfer between maternal and fetal circulations occurs across the syncytiotrophoblast (ST) cell layer, a multinuclear syncytium 13 m$^2$ in area at term. The ST also synthesizes and secretes large amounts of peptide and steroid hormones [3]. The ST cell layer is formed from underlying mononucleated villous cytotrophoblast (CT) cells which constantly proliferate in vivo and fuse into the fully differentiated multinucleated ST. Formation of ST can be recapitulated in vitro by culture of isolated CT cells which, although they cannot proliferate in-vitro, spontaneously fuse, and differentiate into ST [4]. As the majority of energy requiring placental functions (nutrient/waste transfer and hormone production) are carried out by the ST layer, it has traditionally been assumed that the ST is more metabolically active than CT cells, although recent studies suggest a more complex metabolic status [5].

Trophoblast cells, like other cells, produce chemical energy in the form of adenosine triphosphate (ATP) mainly via oxidative phosphorylation. Glycolysis, the TCA cycle, and fatty acid oxidation all result in formation of energy-rich NADH and FADH2 which donate their electrons into the electron transport chain (ETC) for shuttling down a chain of protein complexes while protons are pumped out of the mitochondrial matrix into the intermembrane space, creating a proton gradient across the inner mitochondrial membrane. In the final step of oxidative phosphorylation, protons travel down their concentration gradient through complex 5 (ATP synthase) and phosphorylate adenosine diphosphate (ADP), creating ATP. In most cells, breakdown of glucose via glycolysis and formation of acetyl CoA is the primary pathway that provides metabolites for oxidative phosphorylation. However, cells can switch to other metabolites in either the absence of glucose or excess of fatty acids. This ability to switch metabolite substrates depending on nutrient availability is called metabolic flexibility and is a crucial cell survival mechanism when faced with sub-optimal metabolic conditions. We recently showed that in addition to glucose, trophoblast cells can also utilize amino acids, e.g., glutamine, and fatty acids for generation of ATP via the ETC and that the proportions of each used can change with metabolic condition, e.g., obesity or gestational diabetes [6]. Since, the proliferative CT and differentiated ST have different role in terms of transport, metabolism, and steroid and peptide hormone production, we hypothesized that they might differ in their use of fuel sources and metabolic flexibility.

Cytotrophoblast cells share many similarities with cancer cells which proliferate, migrate, and invade tissues to establish a continuous nutrient supply to support the development of a tumor. In-vivo, CT proliferate, migrate, and invade (as extravillous trophoblast) endometrial tissue to establish a nutrient supply but also as villous cytotrophoblast undergo fusion to form ST [7]. Otto Warburg described a phenomenon, the Warburg effect, where cancer cells preferentially utilize glycolysis in the presence of oxygen (aerobic glycolysis) to produce the bulk of their ATP requirement, unlike normal body cells that generate ATP through mitochondrial respiration using metabolites from glycolysis, the TCA cycle and $\beta$-oxidation of fatty acids [8–10]. Based on the similarities between CT cells and cancer cells, we therefore hypothesized that CT might have higher glycolytic function, compared to ST cells.

There is now an overwhelming body of data indicating a sexual dimorphism exists in placental physiology underpinned by a sex-dependent difference in placental gene expression [11–14]. This may be linked to the different fetal growth and survival strategies where male fetuses grow larger than female fetuses but are therefore at a greater risk of suffering from adverse pregnancy outcomes if maternal nutrition and placental function are not optimal [15–17]. We have previously reported maternal obesity, preeclampsia, and gestational diabetes mellitus to be associated with sexually dimorphic effects on energetics and autophagy in the placenta, and have also shown sexual dimorphism in placental antioxidant enzyme activity [6,18–20]. In this study we also investigated if fetal sex had effects on glycolytic and mitochondrial metabolism in CT vs. ST cells from women of a healthy weight.

## 2. Results

### 2.1. Clinical Characteristics

All the women who donated their placentas to this study were chosen because they were of healthy pre-pregnancy (lean) BMI (<25 kg/m$^2$). There were no significant differences in gestational age at delivery, maternal age, pre-pregnancy BMI (Body Mass Index), gestational weight gain, or placental weight between the groups (Table 1). However, there were significant differences in fetal weight and the fetal/placental weight ratio between male vs. female pregnancies with the male being significantly heavier and, hence, with a more "efficient" placenta.

**Table 1.** Clinical characteristics of study participants.

| Fetal Sex | Pre-Pregnancy BMI (kg/m$^2$) | Maternal Age (yrs) | Gestational Age (wks) | Fetal Weight (Grams) | Placental Weight (Grams) | Fetal/ Placental Ratio | Gestational Weight Gain (kg) | Ethnicity (Hispanic, Non-Hispanic) |
|---|---|---|---|---|---|---|---|---|
| Males $n = 8$ | $22.9 \pm 1.6$ | $35.9 \pm 6.7$ | $39.0 \pm 0.5$ | $3612 \pm 257$ | $508 \pm 87.6$ | $7.2 \pm 1.1$ | $15.0 \pm 3.7$ | 0, 8 |
| Females $n = 8$ | $22.3 \pm 1.5$ | $32.1 \pm 4.5$ | $38.6 \pm 1.0$ | $3208 * \pm 400$ | $518 \pm 71.9$ | $6.2 * \pm 0.6$ | $15.1 \pm 4.2$ | 1, 7 |

Data presented as mean $\pm$ SD. Significant differences between male and female groups were determined using the student's *t* test. * $p < 0.05$ male vs. female.

On average, male fetuses are born larger than female fetuses [21], with little differences in placental weight, resulting in a larger fetal to placental weight ratio in males [22]. Our data agrees with these findings (Table 1).

*2.2. Isolated Cytotrophoblast Differentiate into Syncytiotrophoblast in Culture*

Isolating intact ST from the placenta is not feasible as the digestion process destroys the syncytial layer. However, CT can be isolated and in culture will aggregate and fuse to form ST over 96 hrs. Figure 1A shows individual cells positive for cytokeratin-7 confirming identity as single CT at 24 hrs. Over the course of the culture, these undergo fusion to form ST as evidenced by multinucleate structures with positive cytokeratin-7 stain (Figure 1B,C) and E-cadherin stain (Supplemental Figure S1B).

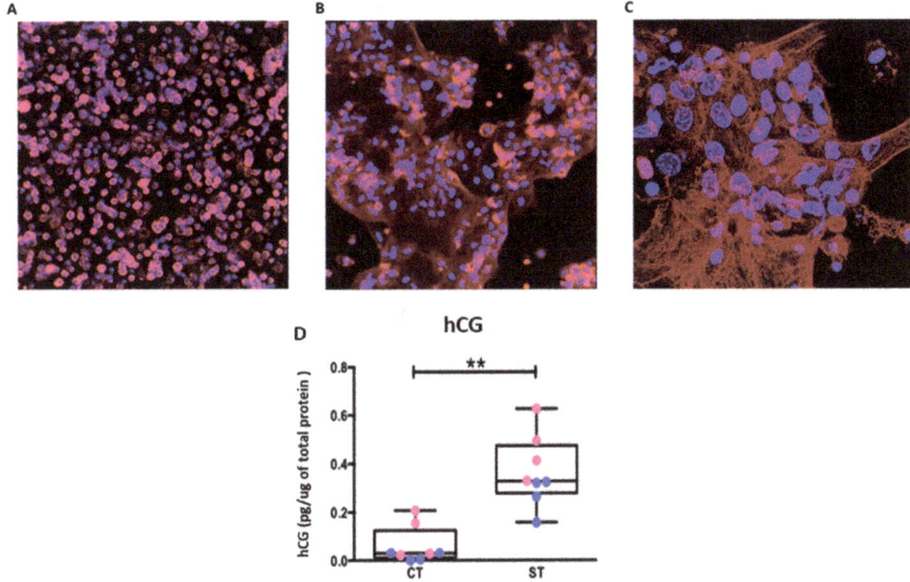


**Figure 1.** Identification of trophoblast cells and their syncytialization. (**A**) Cytotrophoblast at 24 h (20×), (**B**) Syncytiotrophoblast at 96 hrs (20×), and (**C**) Syncytiotrophoblast (63×) stained with cytokeratin 7 (red) and counterstained with Hoechst 33,342 for nuclei (blue). (**D**) Human Chorionic Gonadotropin (hCG) production pg of hormone per µg of cell protein. Data presented as minimum, maximum, median, 25th and 75th quartiles boxes, and whisker plots, $n = 8$, male = blue, female = pink. ** $p < 0.01$, (Wilcoxon test CT vs. ST).

To further verify that our technique of culturing trophoblasts results in ST formation, we measured human chorionic gonadotropin (hCG) production. With data from both fetal sexes combined, ST, as expected had significantly higher hCG production ($p = 0.007$) compared to CT (Figure 2D). With fetal sex separated, ST from both males ($p = 0.01$) and females ($p = 0.02$) have significantly increased hCG production, compared to CT of the same sex (Supplemental Figure S1) however interestingly, the increase in hCG production upon syncytialization appears to be greater in female vs. male trophoblast ($p = 0.02$).

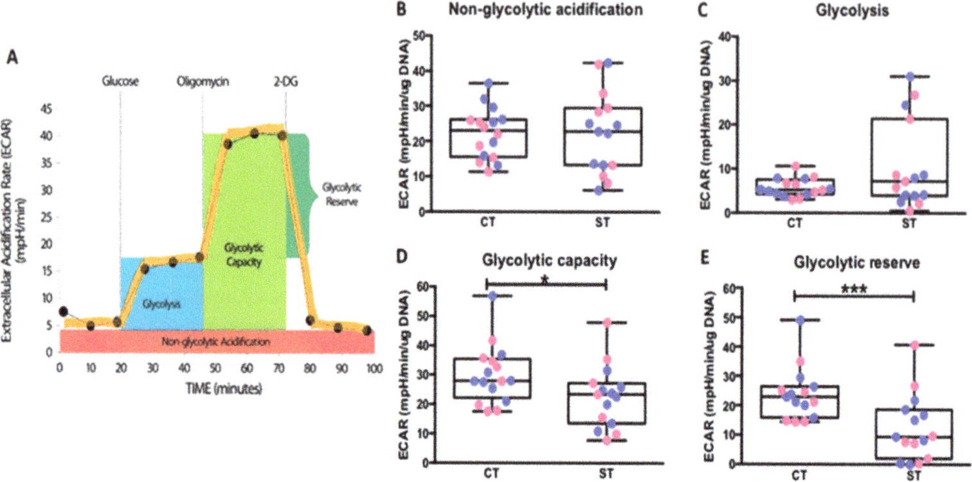

**Figure 2.** Glycolytic function of CT vs. ST analyzed using the glycolysis stress test. (**A**) Graphical representation of the glycolysis stress test, (**B**) non-glycolytic acidification, (**C**) glycolysis, (**D**) glycolytic capacity, and (**E**) glycolytic reserve. Male (blue, $n = 8$) and female (pink, $n = 8$) groups combined. Data presented as minimum, maximum, median, 25th and 75th quartiles boxes, and whisker plots. * $p < 0.05$, *** $p < 0.001$ (Wilcoxon signed-rank test). 2-DG: 2-deoxy-glucose, ECAR: extracellular acidification rate.

### 2.3. Cytotrophoblast Have Higher Glycolytic Capacity and Reserve Capacity

The glycolytic function of CT and ST cells was measured using the glycolysis stress test (Figure 2A). When analyzing with fetal sex combined, no differences were observed in non-glycolytic acidification or rates of glycolysis (Figure 2B,C) suggesting both CT and ST have similar rates of basal glycolysis and basal bioenergetics. However, CT showed significantly higher glycolytic capacity ($p = 0.01$) and glycolytic reserve ($p = 0.0003$) when compared to ST (Figure 2D,E, Supplemental Figure S2G,H). Glycolytic capacity indicates the maximum amount of glycolysis/glucose breakdown the cells can perform acutely, whereas glycolytic reserve (glycolytic capacity−glycolysis rate) is the difference between the basal and maximal glycolytic capacity. The glycolytic reserve thus indicates the cells potential to increase ATP production via glycolysis under stress or other physiologically energy-demanding situations. Our results thus suggest that whereas CT and ST have similar basal rates of glycolysis, CT have higher potential for energy/ATP generation via glycolysis when stressed.

We then separated the data to determine the effects of fetal sex (Supplemental Figure S2). Non-glycolytic acidification and basal glycolysis rate which were not different between CT and ST were also not different between the sexes (Supplemental Figure S2A,B,E,F). Male CT however showed significantly higher glycolytic capacity ($p = 0.04$) when compared to their ST whereas no difference was observed between the female CT and ST. Interestingly, there was no sexually dimorphic effect on glycolytic reserve as male ($p = 0.015$) and female ST ($p = 0.039$) both had significantly lower reserve as compared to their CT, suggesting that under energetically demanding or stressed conditions, both male and female ST have less potential to use glycolysis for ATP production (Supplemental Figure S2C,D).

*2.4. Syncytiotrophoblast Have Higher Mitochondrial Respiration Compared to Cytotrophoblast*

The Mitochondrial stress assay was performed to determine how mitochondrial oxidative respiration and the resultant ATP production change as CT differentiate to ST (Figure 3A). With data from both fetal sexes combined, ST had significantly higher basal respiration (oxygen consumption in the resting state) ($p = 0.003$) and higher ATP-coupled respiration ($p = 0.0008$), suggesting ST are energetically more demanding than CT (Figure 3B,C, Supplemental Figure S3G,H). In addition, the ST also showed significantly higher maximal respiration ($p = 0.0001$) and spare capacity ($p = 0.0001$), suggesting that ST can achieve a higher rate of mitochondrial respiration if needed and have a higher ability to respond to demand when compared to CT (Figure 3D,E). Syncytiotrophoblast also showed significantly higher non-mitochondrial respiration ($p = 0.009$) and proton leak ($p = 0.04$), compared to CT (Figure 3F,G). Proton leak is the amount of oxygen consumption not coupled to ATP production in the mitochondria and has been linked to the levels of reactive oxygen species (ROS) and oxidative stress in the cell [23–25].

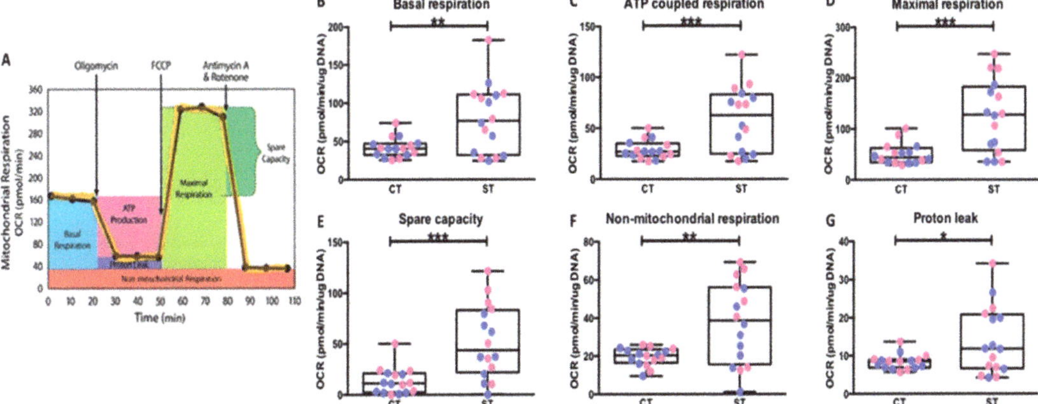

**Figure 3.** Mitochondrial respiration of CT vs. ST analyzed using the mitochondrial stress test. (**A**) Graphical representation of the mitochondrial stress test, (**B**) basal respiration, (**C**) ATP-coupled respiration, (**D**) maximal respiration, (**E**) spare capacity, (**F**) non-mitochondrial respiration, and (**G**) proton leak. Male (blue, $n = 8$) and female (pink, $n = 8$) groups combined. Data presented as minimum, maximum, median, 25th and 75th quartiles boxes, and whisker plots. * $p < 0.05$, ** $p < 0.01$, *** $p < 0.001$, and Wilcoxon signed-rank test. FCCP: Trifluoromethoxy carbonylcyanide phenylhydrazone.

To determine the effect fetal sex has on mitochondrial function, data were analyzed separately for male and female groups (Supplemental Figure S3). Overall, ST from both males and females showed trends similar to that observed in the combined analysis. In both sexes, ST had significantly higher ATP-coupled respiration (M, $p = 0.03$, F, and $p = 0.01$), maximal respiration (M, $p = 0.007$, F, and $p = 0.007$) and spare capacity (M, $p = 0.016$, F, and $p = 0.007$) compared to CT. In females, ST had significantly higher basal respiration ($p = 0.02$) and non-mitochondrial respiration ($p = 0.03$) compared to CT. In males, ST had significantly higher proton leak ($p = 0.03$) compared to CT (Supplemental Figure S3A–F,I,J).

*2.5. Cytotrophoblast and Syncytiotrophoblast Differ in Their Capacity to Respond to Stress*

To more clearly visualize how the metabolic phenotype changes as CT fuse to form ST, basal OCR vs. basal ECAR measurements were plotted against each other (Figure 4A). Both male and female trophoblasts increase glycolysis (ECAR) and oxidative phosphorylation (OCR) upon syncytialization showing the increased energy demands upon fusion into ST. The metabolic potential of CT (Figure 4B) and ST (Figure 4C) in response to stress was visualized by plotting basal OCR/ECAR rates and maximal OCR/ ECAR rates upon addition of FCCP or Oligomycin, respectively, mimicking a physiologically stressed situation.

CT did not show any increase in their oxygen consumption rate (OCR), a direct measure of their ability to generate ATP via oxidative phosphorylation, when stressed by addition of FCCP but did show increased glycolytic function (ECAR, M, $p = 0.02$, F, and $p = 0.008$). This is expected as CT do not have appreciable mitochondrial spare capacity (Figure 3E) but do have appreciable glycolytic reserve (Figure 2E). In contrast, ST (Figure 4C) can increase oxidative phosphorylation (OCR) as well as glycolysis (ECAR) to meet energy demands when presented with a stress (M, $p = 0.11$, F, and $p = 0.05$). This is also expected as ST have significantly higher mitochondrial spare capacity (Figure 3E) and some glycolytic reserve, although less glycolytic reserve than CT (Figure 2E). Taken together, these results suggest that CT and ST have different strategies and capabilities to respond to physiologically demanding/stressful conditions.

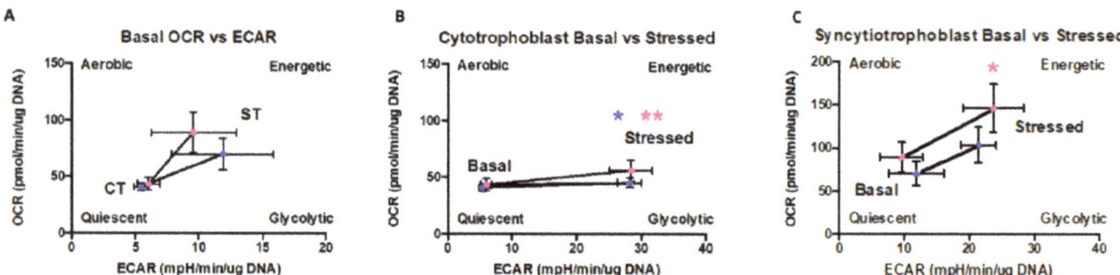


**Figure 4.** Comparison of metabolic phenotypes of CT and ST. (**A**) Metabolic shift of CT to ST and metabolic potential of (**B**) CT and (**C**) ST. Data presented as mean $+/-$ SEM. Male (blue, $n = 8$) and female (pink, $n = 8$) groups separated. * $p < 0.05$, ** $p < 0.01$, Wilcoxon signed-rank test.

*2.6. Syncytiotrophoblast Have a Higher Capacity and Efficiency for Substrate Utilization under Stress*

We examined the impact of different substrates on trophoblast spare respiratory capacity (the ability to respond to increased energy demand or under stress) using an adapted version of the mitochondrial stress test. A combination of two pathway inhibitors (UK5099 (glucose), BPTES (glutamine) or Etomoxir (long-chain fatty acid)) was added, leaving only a single substrate utilization pathway open, before performing the mitochondrial stress test protocol. The impact of availability of just a single substrate—glucose or glutamine or fatty acids—on the ability of cells to increase respiration rate in physiologically stressed conditions (spare/reserve capacity) was then calculated as described in the methods section.

With both sexes combined, CT had a significantly higher capacity to use either glucose ($p = 0.01$) or fatty acids ($p = 0.04$) alone than glutamine alone. However ST were observed to have a significantly higher spare capacity for glucose utilization than glutamine ($p < 0.02$) or fatty acids ($p < 0.05$) when each was present alone which indicates a higher efficiency in glucose utilization over the other two substrates. When compared to CT, ST also had significantly higher spare capacity when glucose ($p = 0.02$) or glutamine ($p = 0.003$) were available alone as substrates (Figure 5A) with no difference apparent between CT and ST spare capacity when only fatty acids were available as substrate. These results suggest that upon syncytialization there is a change in the preference for utilization of the different substrates to generate energy under physiologically stressed conditions.

We then separated the data based on fetal sex to determine if the cells differed in their capacities for substrate utilization under stress (Figure 5B,C). In males, ST showed a trend towards an increase in spare capacity for each substrate over CT, but the values reached significance only for glutamine ($p = 0.04$) (Figure 5B). Interestingly, in females CT had a higher spare capacity (and hence efficiency in utilization) for glucose ($p = 0.03$) over glutamine and a trend towards increased spare capacity for long-chain fatty acids over glutamine ($p = 0.09$). Female ST had significantly higher capacity for use of glucose compared to glutamine ($p = 0.01$) and long-chain fatty acids ($p = 0.01$) (Figure 5C). Thus, it appears that female trophoblast account for most changes in substrate utilization under stress conditions when male and female are combined.

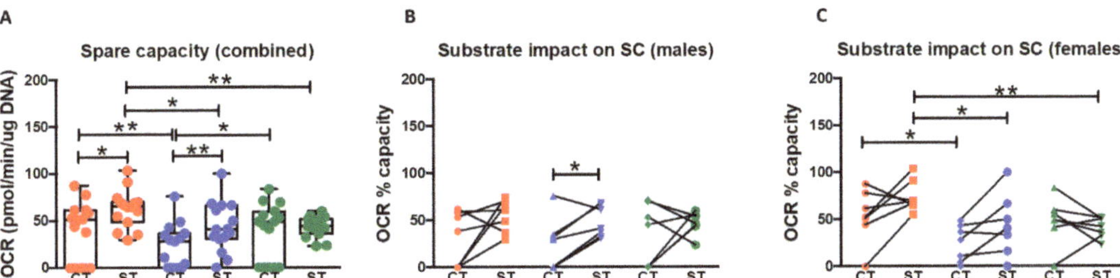

**Figure 5.** Impact of specific substrates on spare capacity of CT vs. ST. (**A**) Impact of glucose (red), glutamine (blue), and long chain fatty acid (green) on spare capacity for each fuel source with fetal sex groups combined (*n* = 16). Combined data presented as minimum, maximum, median, 25th and 75th quartiles boxes, and whisker plots. * *p* < 0.05, ** *p* < 0.01, (Friedman test when comparing substrates or Wilcoxon test when comparing CT vs. ST). (**B**) Substrate impact on spare capacity in males (*n* = 8), and (**C**) females (*n* = 8). Glucose (red), glutamine (blue), and long-chain fatty acid (green). Data plotted as individual values of paired CT and ST from the same sample. * *p* < 0.05, (Friedman test when comparing substrates or the Wilcoxon test when comparing CT vs. ST).

### 2.7. Syncytiotrophoblast Have Lower Mitochondrial Content but Higher Citrate Synthase Activity

To determine if the increased overall mitochondrial respiration observed in ST was a function of increased number of mitochondria, we measured mitochondrial content using the mitochondria specific dye MitoTracker™ (normalized to total DNA amount). Surprisingly, we found the opposite to be true. With data from both fetal sexes combined, CT have significantly greater mitochondrial content compared to ST (*p* = 0.007) (Figure 6A). However, when separated by fetal sex, CT from males (*p* = 0.01) account for the majority of this difference with significantly higher mitochondrial content compared to ST, while females only approached significance (*p* = 0.07) (Supplemental Figure S4A).

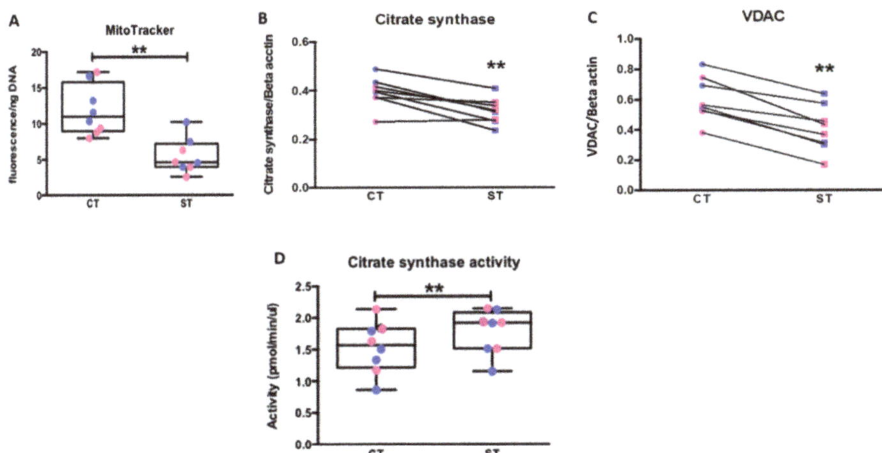

**Figure 6.** Mitochondrial content and activity measurements in cyto- and syncytiotrophoblast. (**A**) MitoTracker™, (**B**) citrate synthase protein, and (**C**) VDAC protein levels. (**D**) Citrate synthase activity (in picomole/min/μL of substrate). Male (blue, *n* = 4) and female (pink, *n* = 4). A, D: Data presented as minimum, maximum, median, 25th and 75th quartiles boxes, and whisker plots. (**B,C**): Data plotted as individual values of paired CT and ST from the same sample Male (blue, *n* = 4) and female (pink, *n* = 4) fetal sex groups combined. ** *p* < 0.01, (Wilcoxon test, CT vs. ST).

To further validate the above observation, we quantified the expression using western blotting of two other mitochondrial markers, citrate synthase, and voltage-dependent anion channel (VDAC) found in the mitochondrial outer membrane. In agreement with the MitoTracker$^{TM}$ data, the ST had lower expression of both citrate synthase ($p = 0.01$) and VDAC ($p = 0.007$) (Figure 6B,C). When the data was separated and analyzed based on fetal sex the decrease in citrate synthase expression upon syncytialization was significant only in male mirroring the change seen with MitoTracker$^{TM}$ whereas VDAC significantly decreased in both male and female trophoblast with syncytialization (Supplemental Figure S4B,C).

We subsequently measured citrate synthase activity as an additional marker for overall mitochondrial activity. Citrate synthase is responsible for catalyzing the first step of the citric acid cycle by combining acetyl-CoA (end product of all three fuel oxidation pathways) with oxaloacetate to generate citrate which then enters the TCA cycle to generate FADH2 and NADH. With data from both sexes combined, ST have significantly higher citrate synthase activity ($p = 0.007$) compared to CT (Figure 6D), however, separation by fetal sex revealed male ($p = 0.008$) ST have significantly increased citrate synthase activity compared to CT, while female ST only approached significance ($p = 0.09$) (Supplemental Figure S4D). Increased citrate synthase activity in ST aligns with our results of increased mitochondrial respiration rate in ST.

### 2.8. Effect of Syncytialization on Mitochondrial Protein Expression

We next investigated if the increased mitochondrial respiration and citrate synthase activity measured in ST corresponded with an increase in the expression of proteins involved in mitochondrial catabolic pathways (outlined in Table 2).

**Table 2.** List of mitochondrial metabolism proteins assessed by western blotting grouped in 3 subgroups (capitalized).

| **ELECTRON TRANSPORT CHAIN COMPLEXES** |
| :---: |
| NADH reductase (Complex I) |
| Succinate dehydrogenase (Complex II) |
| Cytochrome C reductase (Complex III) |
| Cytochrome C oxidase (Complex II) |
| ATP synthase (Complex V) |
| **METABOLITE PROCESSING ENZYMES** |
| Glutamate dehydrogenase, Mitochondrial (GLUD 1/2) |
| Carnitine palmitoyl transferase one alpha (CPT1$\alpha$) |
| Hexokinase 2 |
| Glutaminase |
| Glucose Transporter Type 1(GLUT1) |
| **MITOCHONDRIAL BIOGENESIS** |
| Peroxisome proliferator-activated receptor gamma coactivator 1-alpha (PGC1$\alpha$) |

Surprisingly, we also found that every mitochondrial specific protein we measured significantly decreased in ST compared to CT. As seen in Figure 7, the expression of all five complexes in the respiratory chain, I. NADH dehydrogenase ($p = 0.007$), II. Succinate dehydrogenase ($p = 0.007$), III. Cytochrome C reductase ($p = 0.02$), IV. Cytochrome C oxidase ($p = 0.007$) and V. ATP synthase ($p = 0.01$) significantly decrease in ST compared to CT (Figure 7E–I). Glutaminase and glutamate dehydrogenases (GLUD 1/2) the mitochondrial enzymes that convert glutamine into glutamate, and glutamate to $\alpha$-ketoglutarate, a precursor of the citric acid cycle intermediate, respectively, were also found to be significantly decreased in ST compared to CT ($p = 0.0078$,) (Figure 7J,K). However, no differences were observed in Hexokinase 2, Carnitine palmitoyltransferase one alpha (CPT1$\alpha$), or Glucose Transporter Type 1 (GLUT1) expression (Figure 7L–N). Peroxisome proliferator-activated receptor gamma coactivator 1-alpha (PGC1$\alpha$), which is the master regulator of mitochondrial biogenesis, was also found to be significantly decreased in ST compared to CT ($p = 0.007$) (Figure 7O).

Similar observations were made when data was separated by fetal sex (Supplemental Figure S5). Both male and female ST had significantly decreased protein expression of NADH dehydrogenase (M, $p = 0.006$, F, and $p = 0.001$), Succinate dehydrogenase (M, $p = 0.003$, F, and $p = 0.001$), Cytochrome C oxidase (M, $p = 0.01$, F, and $p = 0.001$), GLUD1/2 (M, $p = 0.01$, F, and $p = 0.008$), Glutaminase (M, $p = 0.002$, F, and $p = 0.02$) and PGC1$\alpha$ (M, $p = 0.03$, F, and $p = 0.0005$) compared to CT of the same fetal sex. Male ST had significantly decreased Glucose transporter 1 ($p = 0.029$) while female ST had significantly decreased ATP synthase ($p = 0.02$) and trended to have decreased Cytochrome C reductase ($p = 0.09$). No differences were seen in CPT1$\alpha$ or Hexokinase 2 across any of the groups.

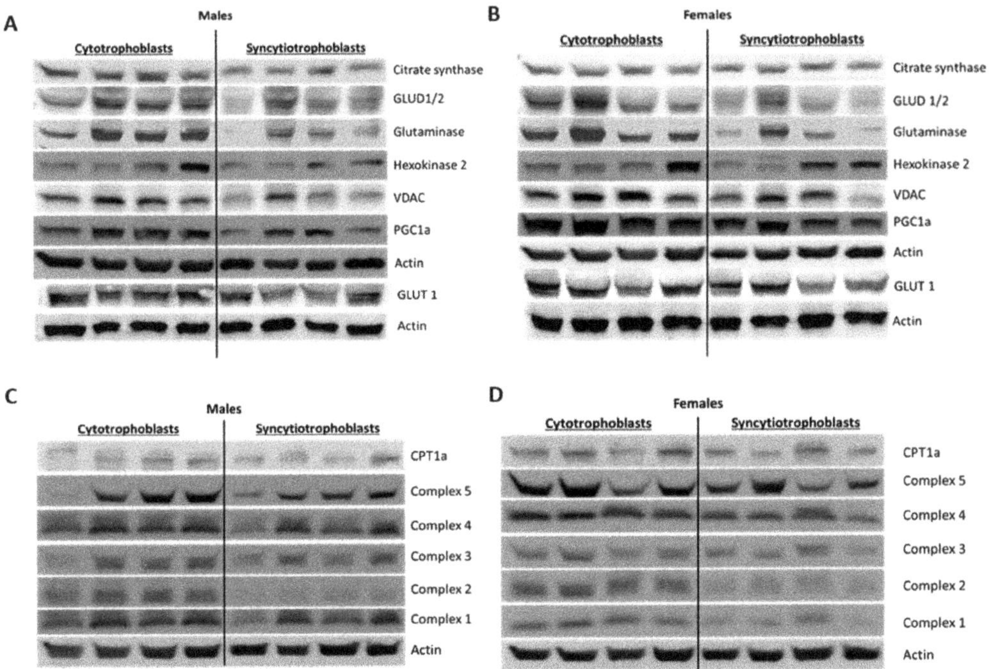


**Figure 7.** *Cont.*

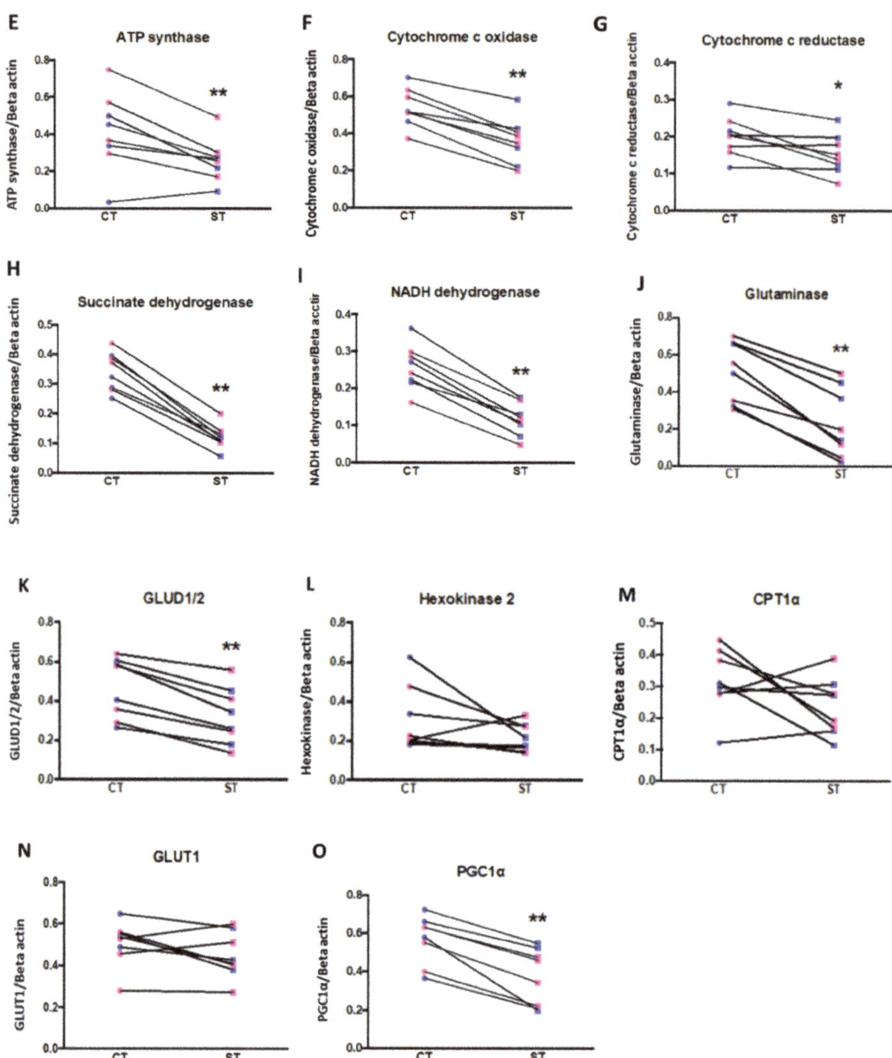

**Figure 7.** Effect of trophoblast differentiation on specific mitochondrial protein expression. Representative western blots (**A–D**) and quantification (**E–O**) of cellular proteins in CT vs. ST. Data plotted as individual values of paired CT and ST from the same sample Male (blue, $n = 4$) and female (pink, $n = 4$) fetal sex groups combined. * $p < 0.05$, ** $p < 0.01$, (Wilcoxon test, CT vs. ST).

## 3. Discussion

Cell differentiation and differentiated functions are highly energy-consuming processes [26]. Several studies have reported significant modifications in cellular bioenergetics as progenitor cells differentiate [27,28]. However, the shifts in mitochondrial and cellular bioenergetic pathways during ST differentiation are not well understood. Additionally, while sexual dimorphism in placental function has been reported, the effect of fetal sex on CT and ST bioenergetics and mitochondrial function has been largely unexplored.

The present study provides several lines of evidence that CT and ST differ significantly in their metabolic phenotypes. CT have equivalent basal glycolysis but a higher glycolytic capacity and reserve than ST whereas ST have a higher mitochondrial respiratory function than CT under both basal conditions and conditions mimicking physiological stress and increased energy demand. ST also appear to utilize glucose and glutamine more efficiently than CT whereas the two cell types show no difference in their ability to use fatty acids to generate energy. Further, both CT and ST show a distinct sexual dimorphism in their energy metabolism with male ST having lower glycolytic capacity and reserve compared to their CT and with female ST having comparable glycolytic capacity, but lower reserve than their CT. On the other hand, both male and female ST have higher mitochondrial respiration (compared to their respective CT) for all parameters except basal respiration which is not different in male ST vs. CT and proton leak which is not different in female ST vs. CT.

In the current study, we used isolated term CT cells cultured for 24 h and 96 h representing progenitor CT cells and syncytialized ST, respectively. Syncytialization over this timeframe was confirmed by staining for the trophoblast marker CK-7 and for nuclear aggregates and measuring hCG secretion as shown in Figure 1. We then assessed glycolytic function and mitochondrial respiration in both CT and ST using the Seahorse assay. The assay measures the rate of depletion of $O_2$ from the media, "oxygen consumption rate" (OCR) and protons released into the media, "extracellular acidification rate" (ECAR) as indicators of mitochondrial oxidative phosphorylation and glycolytic function, respectively.

Although, there was no statistical difference in the basal rate of glycolysis between CT and ST, we observed that CT had a significantly higher glycolytic capacity and reserve capacity than ST (Figure 2). Kolahi et al. previously reported significantly higher basal glycolysis rate in CT but no difference in the glycolytic reserve. However, their study was performed with media containing pyruvate, a product in the glycolysis pathway which upon breakdown releases lactate and proton measured as ECAR in the Seahorse assay. The presence of pyruvate would thus affect the baseline measurements performed in the study and may account for the differences seen in this study. Higher glycolytic capacity and reserve in CT suggests that under physiologically energy demanding conditions, CT but not ST could rapidly increase their glycolytic function to survive. From a bioenergetic perspective, glycolysis is not as efficient as mitochondrial respiration for ATP production with 2 vs. 36 ATP molecules being generated per glucose molecule respectively. However, it is generally accepted that proliferating progenitor cells, such as cytotrophoblast, are glycolytic in nature [29–32] and it is the preferred way to generate ATP in cancer cells described as the Warburg effect [8].

We also observed that the differentiated ST have significantly higher levels of both basal mitochondrial respiration and higher reserve capacity (Figure 3). ST were also more flexible in their fuel dependency and were able to better utilize glucose and glutamine for energy generation under conditions mimicking physiological stress and energy demand (Figure 5). Studies assessing bioenergetics of neuronal, osteogenic, and erythroid differentiation also made similar observations where differentiation was accompanied by, and required, a shift towards mitochondrial respiration [27,28,32,33]. Interestingly, previous studies on mitochondrial function in human placenta have reported observations contradictory to ours. Fisher et al. reported reduced oxygen consumption, mitochondrial respiration, and ATP production in mitochondria of ST vs. CT [34]. However, these studies use intact mitochondria isolated from whole placental tissue by homogenization followed by density gradient purification. This separates bigger mitochondria from small mitochondria which the authors refer to as cyto–mito and syncytio–mito, respectively, based on previous studies that reported large circular mitochondria in cytotrophoblast and small irregular shaped ones in syncytiotrophoblast [35]. As prepared this "cyto–mito" fraction will also contain normal sized mitochondria found in other cell types of the placenta such as stromal and endothelial cells and hence does not only represent cytotrophoblast mitochondrial respiration. Our use of intact cells (individual or syncytialized) better mimics the physiological milieu which may have impact on availability of substrates, membrane potential and proton gradients all crucial for mitochondrial function. Similarly, Kolahi et al. reported reduced OCR in ST contradictory to our observations. However, their study was focused on fatty acid metabolism and assays contained high concentrations of saturated long chain and monounsaturated fatty acids which could account for the observed differences in OCR seen.

The differentiated functions of syncytiotrophoblast means their mitochondria have several functions distinct from those of the proliferative cytotrophoblast, particularly steroidogenesis. Martinez et al. reported that ST mitochondria have significantly increased cytochrome P450 expression an enzyme responsible for catalyzing the first step in steroidogenesis, highlighting the role of syncytiotrophoblast in hormone synthesis [35]. Similarly, Fisher et al. reported increased CYP11A1 expression and increased progesterone production in ST mitochondria. While we did not assess steroidogenesis, we measured citrate synthase activity as a marker for mitochondrial activity. Like the above studies, we found that ST had higher citrate synthase activity, again implying greater mitochondrial function. However, our assessment showed significantly reduced mitochondrial content, as well as decreased protein expression of citrate synthase, VDAC, mitochondrial ETC complexes and other enzymes involved in mitochondrial respiration seemingly counter-intuitive to our results on mitochondrial function. (Figures 6 and 7, Supplementary Figures S4 and S5). We assessed mitochondrial content using the specific dye MitoTracker$^{TM}$ deep red and normalizing it to the nuclear DNA content determined by the Hoechst DNA stain. The dye accumulates in active mitochondria and is used for mitochondrial tracking in live cells. Its correlation to mitochondrial mass is, however, not clearly defined. Complementing our results with additional assays such as quantifying cardiolipin content or ratio of mitochondrial DNA to nuclear DNA might provide a better idea of mitochondrial mass in ST [36].

Several studies have also highlighted how mitochondrial ultrastructure and cristae organization play a critical role in its function (comprehensive review in [37]). Detailed ultrastructure studies using cryo-electron tomography have suggested that ATP synthase dimers preferentially localize in (and even aid in formation of) the curved regions of the cristae, such as the tips, whereas the ETC complexes are in less curved regions, such as the stalks [38–41]. These observations suggest that the cristae structure is finely tuned to support the energetic needs of the respective cells. Increased number of cristae could improve mitochondrial function but substantially reduce the available matrix space for metabolic enzymes [37] which would explain the reduced expression but increased function observed in our study. Recently, Cagiliati et al. showed that cristae structure drives the assembly of respiratory chain super complexes (RCS) (consisting of ATP synthase and ETC enzymes) on their surface and therefore affect the efficiency of mitochondrial respiration [42]. They further reported that mitochondrial fusion protein OPA1 (Optic Atrophy Protein 1) was crucial for cristae organization and structure, assembly of the RCS, and respiratory function. Increased expression of another fusion protein mitochondrial fusion protein-2 (Mfn2) has also been correlated to increased mitochondrial function further emphasizing the correlation between mitochondrial ultrastructure, function and 'mitochondria-shaping' proteins that regulate the organelle's fission and fusion [43]. We propose that a detailed analysis of ST and CT mitochondrial cristae structure and studying expression of mitochondrial shaping proteins might provide further insights into the above results.

An important aim for the study was to assess sexual dimorphism, if any, in placental mitochondrial function. Sexual dimorphism in fetal and placental development as well as placental gene expression has been reported before [14,44]. Male fetuses are known to be bigger and heavier than females with equivalent placental weight as observed in our study [21,22,45] and are therefore considered more efficient, but vulnerable to gestational stressors. Placental responses to environmental stress, such as hyperlipidemia and asthma, are influenced by fetal sex wherein female fetus growth is limited increasing the chances of survival, but male fetuses continue growing normally, increasing their chances of a poor outcome in case of acute exacerbation of the stressors [11,16,46]. We have previously shown that indeed male and female syncytiotrophoblast show differences in metabolic flexibility in use of glucose, glutamine, or fatty acids when they are exposed to different intrauterine environments, i.e., from normal weight, obese, or type A2 gestational diabetes, with male trophoblasts being more severely affected [6,14,47]. To the best of our knowledge, this is the first study assessing sexual dimorphism in basal mitochondrial function and response to stressors as CT from normal pregnancies differentiate to ST. We report that when CT differentiate into ST, they reduce their glycolytic capacity with a more pronounced reduction in male ST. On the other hand, while ST from both sexes have an efficient and higher rate of mitochondrial respiration over their respective CT, this is more pronounced in female ST. The reduced capability of male trophoblasts (ST) to shift to the more efficient mitochondrial oxidation suggests that they might not be equipped at coping with situations that require an increase in energy production. This is further evident in their reduced metabolic flexibility in using either glucose, glutamine, or fatty acids as substrates. Our results thus provide evidence for sexual dimorphism on the cellular, metabolic, and functional level in placental trophoblast. Collectively, our results fortify the notion that male placentas function at near their maximum limit, and if presented with a stress, may not be able to increase energy production and are at a higher risk of suffering from adverse pregnancy outcomes [16,46].

## 4. Materials and Methods

### 4.1. Ethical Approval of the Study

Placentae were collected from the Labor and Delivery Unit at Oregon Health and Science University into a tissue repository under a protocol approved by the Institutional Review Board with informed consent from the patients. Fetal weight was recorded. All tissues and clinical data were de-identified before being made available to the investigative team.

### 4.2. Collection of Placental Tissues

Placentae were collected and weighed immediately following Cesarean section from uncomplicated pregnancies at term in the absence of labor from patients with either a male or a female fetus and a pre-pregnancy BMI in the normal weight range (NW, BMI = 18.5–24.9, $n$ = 8 male, 8 female). Exclusion criteria included overweight or obesity, multifetal gestation, gestational diabetes mellitus, preeclampsia, chronic inflammatory diseases, use of tobacco/illicit drugs, and recent bariatric surgery. Five random samples of tissue (~80 g) were collected from each placenta and placed in PBS to be transported back to the lab.

### 4.3. Primary Cell Isolation and Culture

The chorionic plate and decidua were removed from each randomly isolated placental sample, leaving only villous tissue, which was thoroughly rinsed in PBS to remove excess blood. Primary cytotrophoblast were isolated from villous tissue using a protocol adapted from Eis et al. [48] using trypsin/DNAse digestion followed by density gradient purification. Isolated cytotrophoblast cells were then frozen in freezing media (10% DMSO in FBS) and stored in liquid nitrogen until usage.

Cytotrophoblast cells were rapidly thawed in a 37 °C water bath and immediately diluted in Iscove's modified Dulbecco's medium (25 mM glucose, 4mM glutamine, and 1 mM pyruvate (ThermoFisher Scientific, Waltham, MA, Cat. #12440053) supplemented with 10% FBS and 1% penicillin/streptomycin (complete media) (ThermoFisher Scientific, Cat. #MT35010CV, #15140 respectively). Cells were centrifuged at 1000 × RCF for 10 min and re-suspended in fresh complete media. Trophoblast cells were plated in a 96-well Seahorse plate (100,000 cells/well) in 100µL of complete media for glycolysis and respiration measurements or plated in a 6-well plate (4 million cells/well) in 2 mL complete media for protein expression studies. The following day, additional complete media was added to each well. All studies were performed at two time points—24 hrs (labelled as cytotrophoblast/CT) and 96 hrs to allow fusion and formation of syncytiotrophoblast (ST). ST formation was confirmed by staining the cells for the trophoblast marker Cytokeratin-7.

### 4.4. Immunocytochemistry

CT cells were plated on circular coverslips at a cell density of 1.5 million cells/mL in a volume of 0.3 mL. CT (24 h) and ST (96 h) were fixed in ice-cold methanol for 10 min at $-20\ °C$ and washed three times with cold PBS. Cells were then blocked in 3% BSA diluted in PBS + 0.1% Tween 20 (PBST) for 2 hrs at room temperature. Cytokeratin-7 primary antibody (1:100) (ThermoFisher Scientific, Waltham, MA, Cat. #MA1-06315) was incubated overnight at 4 °C. Following primary antibody incubation, cells were washed three times in PBST and incubated with anti-mouse Alexa fluor 555 secondary antibodies (1:1000) (Thermofisher Scientific, Cat. #A31570) for 3 hrs at room temperature. Cells were then washed three times in PBST followed by Hoechst 33342 (1:10,000) counterstain for 30 s. Cells were washed three more times with PBST and mounted on slides using SlowFade Diamond Antifade Mountant (Thermofisher Scientific, Cat. #S36972). After allowing to set for 24 hr, cover-slips were sealed in place using clear nail polish. Images were captured using a Zeiss LSM 880 confocal microscope and processed using ImageJ Software (Bethesda, Rockville, MD, USA).

### 4.5. Metabolic Analysis and Cellular Bioenergetics Measurements

CT and ST bioenergetics were measured using Seahorse XF Analyzer (Agilent Technologies, Santa Clara, CA, USA) assays following the manufacturer's protocol outlined briefly below. For all assays, 100,000 cells were plated per well in a 96-well Seahorse assay plate.

#### 4.5.1. Mitochondrial Stress Test

This was used to assess mitochondrial function parameters: basal respiration, ATP production-coupled respiration, maximal respiration, spare capacity, and non-mitochondrial respiration using the Seahorse XF Cell Mito Stress Test (Agilent Technologies, Cat # 103010). One hr prior to running the mitochondrial stress test, complete media was exchanged with basal Seahorse media supplemented with glucose, glutamine, and pyruvate to match culture conditions. The cells were then allowed to equilibrate in a non-$CO_2$ 37 °C incubator for 1 hr before the first rate measurement, called 'Basal respiration rate', and is defined as the initial oxygen consumption rate (OCR). This represents the total mitochondrial respiration rate. After measuring the baseline, 75 μL of oligomycin (ATP synthase inhibitor), FCCP (protonophore), and a combination of rotenone (NADH dehydrogenase inhibitor) and antimycin A (cytochrome c reductase inhibitor) solutions were sequentially added to each well at a 1 μM working concentration to determine the ATP coupled respiration, maximum respiration, and non-mitochondrial oxygen consumption rates, respectively. The ATP coupled response is defined the rate of oxygen consumption linked to ATP production and is calculated as the difference between the basal OCR and the OCR after oligomycin injection. Maximal respiratory rate was calculated as the difference between the OCR after uncoupled addition (FCCP) and the lowest OCR reached after oligomycin addition. Spare (reserve) capacity is calculated as the difference between OCR after FCCP and basal respiration and represents the spare metabolic potential thought to guard against stressful conditions (Figure 3A) [49].

#### 4.5.2. Modified Mitochondrial Stress Test

An adapted version of the mitochondrial stress test described above that was used to examine substrate impact on spare capacity by determining the rate of oxidation of a single substrate (glucose, glutamine, or long-chain fatty acids) while the other two substrate pathways are blocked. The pathway inhibitors used were 2 μM UK5099 (inhibitor of glucose oxidation, blocks action of mitochondrial pyruvate carrier (MPC), which converts glucose to pyruvate), 3 μM BPTES (inhibitor of glutamine oxidation, blocks glutaminase

(GSL1), which converts glutamine to glutamate) and 4 µM Etomoxir (inhibitor of long-chain fatty acid oxidation, which blocks carnitine palmitoyltransferase 1 alpha (CPT1α). The cells were treated with either a combination of two pathway inhibitors or a combination of all three pathway inhibitors followed by the mitochondrial stress test ETC inhibitors to calculate the capacity of each pathway using the following formula.

$$\text{Substrate impact on Spare capacity} = 1 - \left[ \frac{\text{No OCR inhibitor} - \text{Two OCR inhibitors}}{\text{No OCR inhibitor} - \text{Three OCR inhibitors}} \right] \times 100$$

### 4.5.3. Glycolysis Stress Test

This was used to assess glycolytic function parameters: glycolysis, glycolytic capacity, glycolytic reserve, and non-glycolytic acidification using the Seahorse XF Glycolysis Stress kit (Agilent Technologies, Cat # 103020). One hr prior to running the glycolysis stress test, the cell culture medium was exchanged with basal Seahorse media supplemented with glutamine (excluding glucose and pyruvate) to match culture conditions. The cells were then allowed to equilibrate in a non-$CO_2$ 37 °C incubator for 1 hr before the first rate measurement called 'Non-glycolytic acidification' and is defined as the extracellular acidification rate (ECAR) that is not attributed to glycolysis. After measuring Non-glycolytic acidification rate, 75 µL of glucose (converted to pyruvate through glycolysis), Oligomycin (ATP synthase inhibitor), and 2-deoxyglucose-glucose (competitive inhibitor of hexokinase, the first enzyme in the glycolysis pathway) solutions were sequentially added to each well at a 10 mM glucose, 1 µM Oligomycin and 50 mM 2-deoxy-glucose working concentration to determine the rate of glycolysis under basal conditions, maximum glycolytic capacity and to confirm the initial ECAR measured is due to glycolysis, respectively. Glycolysis is defined as the glucose-induced increase in ECAR and is calculated by subtracting non-glycolytic acidification from the highest ECAR measurement following the addition of glucose. Maximum glycolytic capacity was calculated as the difference between the highest ECAR measurement during non-glycolytic acidification and the highest ECAR measurement after the addition of Oligomycin. Glycolytic reserve was calculated as the difference between ECAR after glucose and after oligomycin.

Data from all Seahorse assays were normalized to cellular DNA content measured immediately after the assay was finished. Hoescht 33342 dye (Thermofisher Scientific, Cat. #H1399) was added to each well (1:1000 final concentration) and incubated for 30 min at 37 °C with constant shaking. Fluorescence was measured using a plate reader (excitation 350 nm emission 461 nm).

### 4.6. Protein Extraction and Western Blotting

Proteins were extracted from cultured trophoblast cells (after 24 hrs for CT fraction and after 96 hrs for ST fraction). Briefly, media was collected and frozen for ELISA analysis. To isolate protein, cells were washed in PBS followed by lysing in 100 uL RIPA buffer with added protease/phosphatase inhibitors (ThermoFisher Scientific, Cat. #89901 #A32959 respectively). Cells were then scraped, and the cell lysate transferred to a sterile 1.5 mL tube and placed on ice. Cell debris was removed by centrifuging the cell lysate at 1000 × RCF for 10 min at 4 °C and storing the supernatant at −80 °C. Total protein was quantified using the Pierce BCA Protein Assay Kit (ThermoFisher Scientific, Cat. #23225). Approximately 20 µg of protein was separated on 12% sodium dodecyl sulphate-polyacrylamide gel electrophoresis (SDS-PAGE) hand-cast gels for approximately 30 min at 30V followed by 2 hrs at 100 V and transferred for 1 hr at 100 V onto polyvinylidene difluoride (PVDF) membranes using Mini-PROTEAN tetra cell electrophoresis chamber (BioRad, Hercules, CA, Cat. # 1658004). Membranes were blocked in 5% (*w/v*) nonfat milk in TBS + 0.1% Tween 20 (TBST) for 1 hr and incubated with primary antibody overnight at 4 °C. On the next day, membranes were washed three times in TBST for 5 min each and incubated with HRP-conjugated secondary antibodies. Membranes were washed and incubated in Supersignal West Pico Plus ECL Substrate (ThermoFisher Scientific, Cat. #34578) for 5 min and imaged using the GBOX system (Syngene, Frederick, MD, USA). All samples were normalized to β-Actin and analyzed using Genetools software (Syngene).

The following primary antibodies were used for western blotting: Citrate Synthase (Cell Signaling Technology, Danvers, MA, USA, Cat# 14309, RRID:AB_2665545), glutamate dehydrogenase GLUD1/GLUD2 (Abcam, Cambridge, UK, Cat# ab154027), Glutaminase (Abcam, Cat# ab93434, RRID:AB_10561964), Hexokinase 2 (Cell Signaling Technology, Cat# 2867, RRID:AB_2232946), VDAC (Cell Signaling Technology, Cat# 4661, RRID:AB_10557420), PGC1α (Novus Biologicals, Littleton, CO, USA, Cat# NBP1-04676SS, RRID: AB_1522119), CPT1α (Cell Signaling Technology, Cat# 12252, RRID:AB_2797857), OXPHOS (Abcam, Cat# ab110411, RRID:AB_2756818) and β actin (Sigma-Aldrich, St. Louis, MO, USA, Cat# A2228, RRID:AB_476697). The following HRP conjugated secondary antibodies were used: goat anti-rabbit (Cell Signaling Technology, Cat# 7074, RRID:AB_2099233) and horse anti-mouse (Cell Signaling Technology, Cat# 7076, RRID:AB_330924).

### 4.7. Enzyme Linked Immunosorbent (ELISA) Assay

The levels of human chorionic gonadotropin (hCG) hormone were measured in media collected from CT and ST cells using an ELISA based assay (R&D Systems, Minneapolis, MN, Cat. #DY9034-05) following manufacturer instructions. Data were then normalized to cellular protein measured using the Pierce BCA Protein Assay Kit (ThermoFisher Scientific, Cat. #23225).

### 4.8. Citrate Synthase Activity

Citrate synthase activity was measured using the citrate synthase activity kit (Millipore Sigma, St. Louis, MO, USA, Cat. #MAK193) following manufacturer instructions. Briefly, $2 \times 10^6$ cells/well were plated in 12-well tissue-culture plates. At 24 hrs and 96 hrs cells were lysed using 90 µL ice cold CS Assay Buffer. The total protein in the lysate was determined using Pierce BCA Protein Assay Kit (ThermoFisher Scientific, Cat. #23225) and all samples were adjusted to 40 µg of protein/50 µL using the CS assay buffer. 50 µL of the lysate was transferred to a 96-well reaction plate along with the standards supplied in the kit. 50 µL Reaction buffer was added to each well and an initial absorbance was measured at 412 nm. The plate was incubated at 25 °C for a total of 10 min before the final measurement was taken. The CS activity was calculated as $S_a/(\text{Reaction Time}) \times S_v$; where Sa = Amount of GSH (nmole) generated in unknown sample well between $T_{initial}$ and $T_{final}$ from standard curve, Reaction Time = $T_{final} - T_{initial}$ (minutes) and Sv = sample volume (mL) added to well. CS activity is reported as pmole/min/µL = microunit/µL.

### 4.9. Quantitation of Mitochondrial Content

To quantitate mitochondrial number CT cells were plated in a 96-well tissue-culture dish at a cell density of 1 million cells/mL in a volume of 0.1 mL/well for 24 hrs (CT) or 96 hrs (ST). Cells were then incubated with 200 nM MitoTracker™ Deep Red (Thermo Fisher Scientific, Cat. #M22426) diluted in HBSS for 30 min at 37 °C. Cells were washed three times in HBSS and MitoTracker™ fluorescence (excitation 644 nm/emission 665 nm). MitoTracker™ Deep Red specifically stains the mitochondria, and the OD data was normalized to DNA content measured using Quant-it Pico Green dsDNA Reagent (Thermo Fisher Scientific, Cat. #P7581).

### 4.10. Statistical Analysis

Data are reported as box-and-whisker plots (min to max with mean) with individual data points. Data separated by fetal sex are reported as individual symbols and lines. Statistical significance between groups was calculated using the Friedman test, Wilcoxon test or paired t-test where appropriate. * $p < 0.05$, ** $p < 0.01$, and *** $p < 0.001$ are reported as statistically significant. Graphpad Prism was used to perform all statistical analyses and to generate all graphs.

## 5. Conclusions

The current study outlines fundamental differences between CT and ST energy metabolism, their responses to stressful conditions and how these are influenced by fetal sex. The study justifies further research into how exposure to in utero adverse conditions, like diabetes and obesity, might affect placental function and emphasizes the need for understanding these in the context of sexual dimorphism.

**Supplementary Materials:** The following are https://www.mdpi.com/article/10.3390/ijms221910875/s1.

**Author Contributions:** M.B. and L.M. conceived and planned the experiments. M.B., K.A. and L.K. carried out the experiments and analyzed the data. M.B., L.K. and L.M. discussed the results and contributed to the final manuscript. All authors have read and agreed to the published version of the manuscript.

**Funding:** This research was funded by the National Institutes of Health, Grant number HD095610 (LM).

**Institutional Review Board Statement:** The study was conducted according to the guidelines of the Declaration of Helsinki and approved by the Institutional Review Board (or Ethics Committee) of Oregon Health and Science University 00016328 Approved 13 July 2021.

**Informed Consent Statement:** Informed consent was obtained from all subjects involved in the study.

**Acknowledgments:** The authors thank the Labor and Delivery Department at OHSU and the Maternal and Fetal Research Team for coordinating the collection of the placentas. We also thank all the women who participated in this study by donating their placentas, and the Maloyan lab for helping collect and process placentas and isolate trophoblasts.

**Conflicts of Interest:** The authors declare no conflict of interest.

## References

1. Gude, N.M.; Roberts, C.T.; Kalionis, B.; King, R.G. Growth and function of the normal human placenta. *Thromb. Res* **2004**, *114*, 397–407. [CrossRef] [PubMed]
2. Burton, G.J.; Fowden, A.L. The placenta: A multifaceted, transient organ. *Philos. Trans. R. Soc. Lond. B Biol. Sci.* **2015**, *370*, 20140066. [CrossRef] [PubMed]
3. Handwerger, S. New insights into the regulation of human cytotrophoblast cell differentiation. *Mol. Cell. Endocrinol.* **2010**, *323*, 94–104. [CrossRef] [PubMed]
4. Mele, J.; Muralimanoharan, S.; Maloyan, A.; Myatt, L. Impaired mitochondrial function in human placenta with increased maternal adiposity. *Am. J. Physiol. Endocrinol. Metab.* **2014**, *307*, E419–E425. [CrossRef]
5. Kolahi, K.S.; Valent, A.M.; Thornburg, K.L. Cytotrophoblast, Not Syncytiotrophoblast, Dominates Glycolysis and Oxidative Phosphorylation in Human Term Placenta. *Sci. Rep.* **2017**, *7*, 42941. [CrossRef] [PubMed]
6. Wang, Y.; Bucher, M.; Myatt, L. Use of Glucose, Glutamine and Fatty Acids for Trophoblast Respiration in Lean, Obese and Gestational Diabetic Women. *J. Clin. Endocrinol. Metab.* **2019**, *104*, 4178–4187. [CrossRef]
7. Soundararajan, R.; Rao, A.J. Trophoblast 'pseudo-tumorigenesis': Significance and contributory factors. *Reprod. Biol. Endocrinol.* **2004**, *2*, 15. [CrossRef]
8. Potter, M.; Newport, E.; Morten, K.J. The Warburg effect: 80 years on. *Biochem. Soc. Trans.* **2016**, *44*, 1499–1505. [CrossRef]
9. Perazzolo, S.; Hirschmugl, B.; Wadsack, C.; Desoye, G.; Lewis, R.M.; Sengers, B.G. The influence of placental metabolism on fatty acid transfer to the fetus. *J. Lipid. Res.* **2017**, *58*, 443–454. [CrossRef]
10. Cleal, J.K.; Lofthouse, E.M.; Sengers, B.G.; Lewis, R.M. A systems perspective on placental amino acid transport. *J. Physiol.* **2018**, *596*, 5511–5522. [CrossRef]
11. Tarrade, A.; Rousseau-Ralliard, D.; Aubrière, M.-C.; Peynot, N.; Dahirel, M.; Bertrand-Michel, J.; Aguirre-Lavin, T.; Morel, O.; Beaujean, N.; Duranthon, V.; et al. Sexual dimorphism of the feto-placental phenotype in response to a high fat and control maternal diets in a rabbit model. *PLoS ONE* **2013**, *8*, e83458. [CrossRef] [PubMed]
12. Cvitic, S.; Longtine, M.S.; Hackl, H.; Wagner, K.; Nelson, M.D.; Desoye, G.; Hiden, U. The human placental sexome differs between trophoblast epithelium and villous vessel endothelium. *PLoS ONE* **2013**, *8*, e79233. [CrossRef] [PubMed]
13. Gong, S.; Sovio, U.; Aye, I.L.; Gaccioli, F.; Dopierala, J.; Johnson, M.D.; Wood, A.M.; Cook, E.; Jenkins, B.J.; Koulman, A.; et al. Placental polyamine metabolism differs by fetal sex, fetal growth restriction, and preeclampsia. *JCI Insight* **2018**, *3*, e120723. [CrossRef] [PubMed]
14. Sood, R.; Zehnder, J.L.; Druzin, M.L.; Brown, P.O. Gene expression patterns in human placenta. *Proc. Natl. Acad. Sci. USA* **2006**, *103*, 5478–5483. [CrossRef] [PubMed]
15. Eriksson, J.G.; Kajantie, E.; Osmond, C.; Thornburg, K.; Barker, D.J.P. Boys live dangerously in the womb. *Am. J. Hum. Biol.* **2010**, *22*, 330–335. [CrossRef] [PubMed]

16. Clifton, V.L. Review: Sex and the human placenta: Mediating differential strategies of fetal growth and survival. *Placenta* **2010**, *31*, S33–S39. [CrossRef] [PubMed]

17. Di Renzo, G.C.; Rosati, A.; Sarti, R.D.; Cruciani, L.; Cutuli, A.M. Does fetal sex affect pregnancy outcome? *Gend. Med.* **2007**, *4*, 19–30. [CrossRef]

18. Muralimanoharan, S.; Gao, X.; Weintraub, S.; Myatt, L.; Maloyan, A. Sexual dimorphism in activation of placental autophagy in obese women with evidence for fetal programming from a placenta-specific mouse model. *Autophagy* **2016**, *12*, 752–769. [CrossRef] [PubMed]

19. Muralimanoharan, S.; Maloyan, A.; Myatt, L. Evidence of sexual dimorphism in the placental function with severe preeclampsia. *Placenta* **2013**, *34*, 1183–1189. [CrossRef]

20. Evans, L.; Myatt, L. Sexual dimorphism in the effect of maternal obesity on antioxidant defense mechanisms in the human placenta. *Placenta* **2017**, *51*, 64–69. [CrossRef]

21. Alur, P. Sex Differences in Nutrition, Growth, and Metabolism in Preterm Infants. *Front. Pediatr.* **2019**, *7*, 22. [CrossRef] [PubMed]

22. Gabory, A.; Roseboom, T.J.; Moore, T.; Moore, L.G.; Junien, C. Placental contribution to the origins of sexual dimorphism in health and diseases: Sex chromosomes and epigenetics. *Biol. Sex Differ.* **2013**, *4*, 5. [CrossRef] [PubMed]

23. Brookes, P.S. Mitochondrial H(+) leak and ROS generation: An odd couple. *Free. Radic. Biol. Med.* **2005**, *38*, 12–23. [CrossRef] [PubMed]

24. Echtay, K.S.; Roussel, D.; St-Pierre, J.; Jekabsons, M.B.; Cadenas, S.; Stuart, J.A.; Harper, J.A.; Roebuck, S.J.; Morrison, A.; Pickering, S.; et al. Superoxide activates mitochondrial uncoupling proteins. *Nature* **2002**, *415*, 96–99. [CrossRef]

25. Herrero, A.; Barja, G. ADP-regulation of mitochondrial free radical production is different with complex I- or complex II-linked substrates: Implications for the exercise paradox and brain hypermetabolism. *J. Bioenerg. Biomembr.* **1997**, *29*, 241–249. [CrossRef]

26. Paldi, A. What makes the cell differentiate? *Prog. Biophys. Mol. Biol.* **2012**, *110*, 41–43. [CrossRef]

27. Metabolic Enhancement of Glycolysis and Mitochondrial Respiration Are Essential for Neuronal Differentiation. *Cell. Reprogram.* **2020**, *22*, 291–299. [CrossRef]

28. Energy Metabolism During Osteogenic Differentiation: The Role of Akt. *Stem Cells Dev.* **2021**, *30*, 149–162. [CrossRef]

29. Richard, A.; Vallin, E.; Romestaing, C.; Roussel, D.; Gandrillon, O.; Gonin-Giraud, S. Erythroid differentiation displays a peak of energy consumption concomitant with glycolytic metabolism rearrangements. *PLoS ONE* **2019**, *14*, e0221472. [CrossRef]

30. O'Brien, L.C.; Keeney, P.M.; Bennett, J.P., Jr. Differentiation of human neural stem cells into motor neurons stimulates mitochondrial biogenesis and decreases glycolytic flux. *Stem Cells Dev.* **2015**, *24*, 1984–1994. [CrossRef]

31. Hui, S.; Ghergurovich, J.M.; Morscher, R.J.; Jang, C.; Teng, X.; Lu, W.; Esparza, L.A.; Reya, T.; Zhan, L.; Guo, J.Y. Glucose feeds the TCA cycle via circulating lactate. *Nature* **2017**, *551*, 115–118. [CrossRef] [PubMed]

32. Faubert, B.; Li, K.Y.; Cai, L.; Hensley, C.T.; Kim, J.; Zacharias, L.G.; Yang, C.; Do, Q.N.; Doucette, S.; Burguete, D. Lactate metabolism in human lung tumors. *Cell* **2017**, *171*, 358–371.e359. [CrossRef] [PubMed]

33. Agathocleous, M.; Love, N.K.; Randlett, O.; Harris, J.J.; Liu, J.; Murray, A.J.; Harris, W.A. Metabolic differentiation in the embryonic retina. *Nat. Cell Biol.* **2012**, *14*, 859–864. [CrossRef] [PubMed]

34. Fisher, J.J.; McKeating, D.R.; Cuffe, J.S.; Bianco-Miotto, T.; Holland, O.J.; Perkins, A.V. Proteomic Analysis of Placental Mitochondria Following Trophoblast Differentiation. *Front. Physiol.* **2019**, *10*, 1536. [CrossRef]

35. Martínez, F.; Kiriakidou, M.; Strauss, J.F., III. Structural and Functional Changes in Mitochondria Associated with Trophoblast Differentiation: Methods to Isolate Enriched Preparations of Syncytiotrophoblast Mitochondria. *Endocrinology* **1997**, *138*, 2172–2183. [CrossRef]

36. Larsen, S.; Nielsen, J.; Hansen, C.N.; Nielsen, L.B.; Wibrand, F.; Stride, N.; Schroder, H.D.; Boushel, R.; Helge, J.W.; Dela, F.; et al. Biomarkers of mitochondrial content in skeletal muscle of healthy young human subjects. *J. Physiol.* **2012**, *590*, 3349–3360. [CrossRef]

37. Glancy, B.; Kim, Y.; Katti, P.; Willingham, T.B. The Functional Impact of Mitochondrial Structure Across Subcellular Scales. *Front. Physiol.* **2020**, *11*, 541040. [CrossRef] [PubMed]

38. Davies, K.M.; Anselmi, C.; Wittig, I.; Faraldo-Gómez, J.D.; Kühlbrandt, W. Structure of the yeast F1Fo-ATP synthase dimer and its role in shaping the mitochondrial cristae. *Proc. Natl. Acad. Sci. USA* **2012**, *109*, 13602–13607. [CrossRef]

39. Mühleip, A.W.; Joos, F.; Wigge, C.; Frangakis, A.S.; Kühlbrandt, W.; Davies, K.M. Helical arrays of U-shaped ATP synthase dimers form tubular cristae in ciliate mitochondria. *Proc. Natl. Acad. Sci. USA* **2016**, *113*, 8442–8447. [CrossRef]

40. Vogel, F.; Bornhövd, C.; Neupert, W.; Reichert, A.S. Dynamic subcompartmentalization of the mitochondrial inner membrane. *J. Cell Biol.* **2006**, *175*, 237–247. [CrossRef]

41. Ikon, N.; Ryan, R.O. Cardiolipin and mitochondrial cristae organization. *Biochim. Biophys. Acta (BBA)—Biomembr.* **2017**, *1859*, 1156–1163. [CrossRef]

42. Cogliati, S.; Frezza, C.; Soriano, M.E.; Varanita, T.; Quintana-Cabrera, R.; Corrado, M.; Cipolat, S.; Costa, V.; Casarin, A.; Gomes, L.C.; et al. Mitochondrial Cristae Shape Determines Respiratory Chain Supercomplexes Assembly and Respiratory Efficiency. *Cell* **2013**, *155*, 160–171. [CrossRef]

43. Yao, C.H.; Wang, R.; Wang, Y.; Kung, C.P.; Weber, J.D.; Patti, G.J. Mitochondrial fusion supports increased oxidative phosphorylation during cell proliferation. *Elife* **2019**, *8*, e41351. [CrossRef] [PubMed]

44. Mao, J.; Zhang, X.; Sieli, P.T.; Falduto, M.T.; Torres, K.E.; Rosenfeld, C.S. Contrasting effects of different maternal diets on sexually dimorphic gene expression in the murine placenta. *Proc. Natl. Acad. Sci. USA* **2010**, *107*, 5557–5562. [CrossRef] [PubMed]

45. Misra, D.P.; Salafia, C.M.; Miller, R.K.; Charles, A.K. Non-Linear and Gender-Specific Relationships among Placental Growth Measures and The Fetoplacental Weight Ratio. *Placenta* **2009**, *30*, 1052–1057. [CrossRef] [PubMed]
46. Martin, E.; Smeester, L.; Bommarito, P.A.; Grace, M.R.; Boggess, K.; Kuban, K.; Karagas, M.R.; Marsit, C.J.; O'Shea, T.M.; Fry, R.C. Sexual epigenetic dimorphism in the human placenta: Implications for susceptibility during the prenatal period. *Epigenomics* **2017**, *9*, 267–278. [CrossRef] [PubMed]
47. Buckberry, S.; Bianco-Miotto, T.; Bent, S.J.; Dekker, G.A.; Roberts, C.T. Integrative transcriptome meta-analysis reveals widespread sex-biased gene expression at the human fetal-maternal interface. *Mol. Hum. Reprod.* **2014**, *20*, 810–819. [CrossRef] [PubMed]
48. Eis, A.L.; Brockman, D.E.; Pollock, J.S.; Myatt, L. Immunohistochemical localization of endothelial nitric oxide synthase in human villous and extravillous trophoblast populations and expression during syncytiotrophoblast formation in vitro. *Placenta* **1995**, *16*, 113–126. [CrossRef]
49. Dranka, B.P.; Hill, B.G.; Darley-Usmar, V.M. Mitochondrial reserve capacity in endothelial cells: The impact of nitric oxide and reactive oxygen species. *Free. Radic. Biol. Med.* **2010**, *48*, 905–914. [CrossRef] [PubMed]

International Journal of
**Molecular Sciences**

*Communication*

# Placental Expression of Bile Acid Transporters in Intrahepatic Cholestasis of Pregnancy

Edgar Ontsouka [1,†], Alessandra Epstein [1,†], Sampada Kallol [1], Jonas Zaugg [1], Marc Baumann [2], Henning Schneider [2] and Christiane Albrecht [1,*]

1   Institute of Biochemistry and Molecular Medicine, Faculty of Medicine, University of Bern, Bühlstrasse 28, 3012 Bern, Switzerland; edgar.ontsouka@ibmm.unibe.ch (E.O.); alexandra.epstein@students.unibe.ch (A.E.); sampuak@gmail.com (S.K.); Jonas.zaugg@ibmm.unibe.ch (J.Z.)
2   Department of Obstetrics and Gyneacology, University Hospital, Effingerstrasse 102, 3010 Bern, Switzerland; Marc.Baumann@insel.ch (M.B.); henning.schneider@hispeed.ch (H.S.)
*   Correspondence: christiane.albrecht@ibmm.unibe.ch; Tel.: +41-31-684-48-57
†   Contributed equally and share the first authorship.

**Abstract:** Intrahepatic cholestasis of pregnancy (ICP) is a pregnancy-related condition characterized by increased maternal circulating bile acids (BAs) having adverse fetal effects. We investigated whether the human placenta expresses specific regulation patterns to prevent fetal exposition to harmful amounts of BAs during ICP. Using real-time quantitative PCR, we screened placentae from healthy pregnancies ($n$ = 12) and corresponding trophoblast cells ($n$ = 3) for the expression of 21 solute carriers and ATP-binding cassette transporter proteins, all acknowledged as BA- and/or cholestasis-related genes. The placental gene expression pattern was compared between healthy women and ICP patients ($n$ = 12 each). Placental *SLCO3A1* (OATP3A1) gene expression was significantly altered in ICP compared with controls. The other 20 genes, including *SLC10A2* (ASBT) and *EPHX1* (EPOX, mEH) reported for the first time in trophoblasts, were comparably abundant in healthy and ICP placentae. *ABCG5* was undetectable in all placentae. Placental *SLC10A2* (ASBT), *SLCO4A1* (OATP4A1), and *ABCC2* mRNA levels were positively correlated with BA concentrations in ICP. Placental *SLC10A2* (ASBT) mRNA was also correlated with maternal body mass index. We conclude that at the transcriptional level only a limited response of BA transport systems is found under ICP conditions. However, the extent of the transcriptional response may also depend on the severity of the ICP condition and the magnitude by which the maternal BA levels are increased.

**Keywords:** intrahepatic cholestasis of pregnancy; human placenta; bile acids; transporters; pregnancy complications

**Citation:** Ontsouka, E.; Epstein, A.; Kallol, S.; Zaugg, J.; Baumann, M.; Schneider, H.; Albrecht, C. Placental Expression of Bile Acid Transporters in Intrahepatic Cholestasis of Pregnancy. *Int. J. Mol. Sci.* **2021**, 22, 10434. https://doi.org/10.3390/ijms221910434

Academic Editors: Daniel Vaiman and Hiten D Mistry

Received: 12 August 2021
Accepted: 22 September 2021
Published: 28 September 2021

**Publisher's Note:** MDPI stays neutral with regard to jurisdictional claims in published maps and institutional affiliations.

## 1. Introduction

Intrahepatic cholestasis of pregnancy (ICP) is a pregnancy-specific liver disorder affecting women worldwide, characterized by the onset of pruritus and elevation of serum bile acid (BA) concentrations. Abnormal metabolic profile including elevated cholesterolemia and maternal comorbidity, such as gestational diabetes, have been also reported [1–3]. Depending on the severity of the ICP, the occurring fetal adverse outcomes may include spontaneous preterm labor, fetal distress, and stillbirth [4–6]. Numerous factors (e.g., genetic, hormonal, and environmental conditions) are thought to be implicated in the pathogenesis of ICP. Among genetic factors, the mutations in the genes coding familial intrahepatic cholestasis protein-1 (*FIC1*, also named *ATPase phospholipid transporting 8B1 (ATP8B1)*), bile salt excretory protein (*BSEP*, also known as *ATP-binding cassette (ABC) subfamily B member 11 (ABCB11)*), and multi-drug-resistance protein 3 (*MDR3*, also named *ABCB4*) and altered activities of multi-drug-resistance-related protein 2 (*MRP2*, also named *ABCC2*) have been reported. Nonetheless, the underlying mechanism of ICP is not precisely known so far.

There is accumulating evidence linking the toxicity of BAs to adverse fetal and maternal outcomes. Therefore, the BA equilibrium within the maternal–fetal pool, which necessarily depends on a balanced BA transport across the placental barrier, is critical. In healthy pregnancies, the concentration of BAs is higher in the fetal than in the maternal circulation [7]. Therefore, vectorial transfer of BAs across the placenta mainly occurs from fetus to mother [8]. The transport from fetus to trophoblast is primarily mediated by anion/BA exchangers, whereas the transport from trophoblast to mother especially occurs via ABC transporter proteins, comparable to BA uptake and efflux in hepatocytes. In ICP, the transplacental gradient for BAs is reversed. Consequently, the net transport of BAs is directed towards the fetus, rather than being transported to the maternal side [9]. Surprisingly, fetal BAs are raised to a lesser extent than maternal BAs, implying a protective mechanism that limits BA uptake into the fetal circulation and/or enhances ATP-dependent carriers that transport against concentration gradients towards the maternal circulation [10]. Since the placenta is a vital organ, which plays a key role in fetal protection, it could be assumed that it would prevent fetal exposure to greater amounts of endobiotic toxic compounds, such as BAs. Although the exploration of human placental BA transport systems is clinically relevant, experimental studies reporting the placental gene expression of BA transporters and carriers are scarce or even lacking. This is particularly noticeable for transporters such as the solute carrier (SLC) family 10 member 2 (*SLC10A2*, also known as *apical sodium-dependent bile acid transporter (ASBT)*), epoxide hydrolase 1 (*EPHX1*, known in hepatic tissue as epoxide hydrolase *(EPOX)*), solute carrier organic anion transport protein 3A1 *(SLCO3A)*, *ATP8B1*, *ABCB11*, and *ABCB4* in health and disease conditions. An overview of currently known SLC and ABC transporters associated with BA transport is presented in Table 1.

**Table 1.** Overview of currently known genes associated with bile acid transport.

| Membrane Protein Class | Entry Number | Gene Name [1] | Previous Symbols/Aliases [1] | Approved Name [1] |
|---|---|---|---|---|
| Solute carriers | Q14973 (NTCP_HUMAN) | *SLC10A1* | NTCP | Solute carrier family 10 member 1 |
| | P46721 (SO1A2_HUMAN) | *SLCO1A2* | OATP, OATP1A2, OATP-A | Solute carrier organic anion transporter family member 1A2 |
| | Q9Y6L6 (SO1B1_HUMAN) | *SLCO1B1* | SLC21A6/OATP1B1, OATP-C, LST-1 | Solute carrier organic anion transporter family member 1B1 |
| | Q9NPD5 (SO1B3_HUMAN) | *SLCO1B3* | SLC21A8/OATP1B3 OATP8, | Solute carrier organic anion transporter family member 1B3 |
| | O94956 (SO2B1_HUMAN) | *SLCO2B1* | SLC21A9/ OATP2B1, OATP-B | Solute carrier organic anion transporter family member 2B1 |
| | Q9UIG8 (SO3A1_HUMAN) | *SLCO3A1* | SLC21A11/ OATP3A1, OATP-D | Solute carrier organic anion transporter family member 3A1 |
| | Q96BD0 (SO4A1_HUMAN) | *SLCO4A1* | SLC21A12/ OATP4A1, OATP-E | Solute carrier organic anion transporter family member 4A1 |
| | P07099 (HYEP_HUMAN) | *EPHX1* | EPOX/EPHX1/ mEH | Epoxide hydrolase 1 |
| | Q86UW1 (OSTA_HUMAN) | *SLC51A* | OST-$\alpha$ | Organic solute transporter subunit alpha |

**Table 1.** *Cont.*

| Membrane Protein Class | Entry Number | Gene Name [1] | Previous Symbols/Aliases [1] | Approved Name [1] |
|---|---|---|---|---|
| | Q86UW2 (OSTB_HUMAN) | *SLC51B* | OST-β | Organic solute transporter subunit beta |
| | Q12908 (NTCP2_HUMAN) | *SLC10A2* | ISBT/ASBT | Solute carrier family 10 member 2 |
| ABC transporters | O95342 (ABCBB_HUMAN) | *ABCB11* | BSEP, PFIC2/ ABC16 | ATP-binding cassette subfamily B member 11 |
| | P08183 (MDR1_HUMAN) | *ABCB1* | MDR1/P-gp; CD243 | ATP-binding cassette subfamily B member 1 |
| | Q92887 (MRP2_HUMAN) | *ABCC2* | MRP2/CMOAT1 | ATP-binding cassette subfamily C member 2 |
| | Q9UNQ0 (ABCG2_HUMAN) | *ABCG2* | BCRP, MXR, ABCP, CD338 | ATP-binding cassette subfamily G member 2 |
| | P21439 (MDR3_HUMAN) | *ABCB4* | MDR3, PGY3/ MDR2, PFIC-3 | ATP-binding cassette subfamily B member 4 |
| | O43520 (AT8B1_HUMAN) | *ATP8B1* | FIC1, PFIC1/ATPIC, PFIC | ATPase phospholipid transporting 8B1 |
| | Q9H222 (ABCG5_HUMAN) | *ABCG5* | STSL | ATP-binding cassette subfamily G member 5 |
| | P33527 (MRP1_HUMAN) | *ABCC1* | MRP1/GS-X | ATP-binding cassette subfamily C member 1 |
| | O15438 (MRP3_HUMAN) | *ABCC3* | MRP3, MOAT-D, cMOAT2, MLP2 | ATP-binding cassette subfamily C member 3 |
| | O15439 (MRP4_HUMAN) | *ABCC4* | MRP4/CFTR MOAT-B | ATP-binding cassette subfamily C member 4 |

[1] Source: Gene nomenclature committee (https://www.genenames.org/data/genegroup/#!/group/752, accessed on 22 July 2021).

Based on these premises, we carried out the current investigations on placental tissues and trophoblast cells obtained from healthy and ICP pregnancies. We hypothesized that (i) the expression of BA- and cholestasis-related transporters (Table 1) is altered in placentae from ICP patients as a protective response to higher maternal serum BA concentrations, and (ii) correlations exist between placental mRNA levels of BA transport proteins and maternal clinical data. Our objectives were to (i) determine in human placental tissues and trophoblast cells obtained from healthy pregnancies mRNA levels of 21 candidate solute carriers and ABC transporters, whose cellular localization and functional role in BA transport are already well established in hepatocytes and enterocytes; (ii) assess the effect of ICP (i.e., BA "overload") on the placental mRNA levels of BA transport proteins; (iii) examine the association between placentally expressed BA- and cholestasis-related transport proteins and selected maternal clinical parameters as well as baby sex; and (iv) summarize the currently available knowledge on the BA transport machinery in human placenta.

## 2. Results

### 2.1. Study Participants

The clinical characteristics of pregnant women enrolled in ICP and control groups are summarized in Table 2. ICP patients had a comparable body mass index (BMI) to healthy pregnant controls. Maternal circulating BA levels were monitored only in case of serious suspicions of ICP. Thus, corresponding data are available only for women diagnosed to be positive for ICP.

**Table 2.** Maternal clinical parameters.

| Parameters | Controls ($n$ = 12) | ICP ($n$ = 12) |
| --- | --- | --- |
| Maternal age, years | 33.1 ± 4.1 | 30.5 ± 6.7 |
| Gravidity | 2.5 ± 1.4 | 2.5 ± 1.2 |
| Parity | 1.9 ± 0.9 | 1.5 ± 1.2 |
| Gestational age, weeks | 39.2 ± 0.8 | 37.9 ± 1.9 |
| BMI, kg/m$^2$ | 22.1 ± 2.4 | 23.1 ± 6.5 |
| Baby gender (male/female) | 6/6 | 6/6 |
| Bile acid levels, μmol/L | n.a. | 55.5 ± 61.7 |
| De-Ursil® treatment applied | $n$ = 0 | $n$ = 12 |

Data are expressed as mean ± SD. ICP, intrahepatic cholestasis of pregnancy; BMI, body mass index; n.a., not analyzed.

### 2.2. Expression of Selected Solute Carriers and ATP-Dependent Transporter Genes with Affinity for BA- and Cholestasis-Related Molecules in Control Placentae and Trophoblast Cells

The primers listed in Table 3 (see Section 5), used for the amplification of corresponding genes in placental tissues and trophoblast cells, were validated on positive control tissues and cells (liver and hepatocyte cell line). They amplified the expected products in positive controls (data not shown).

**Table 3.** Primers used for gene amplification.

| Gene | Forward Primer (5′-3′) | Reverse Primer (5′-3′) | Accession Number |
| --- | --- | --- | --- |
| *SLC10A1*/NTCP | GGAGGGAACCTGTCCAATGTC | CATGCCAAGGGCACAGAAG | NM_003049.3 |
| *SLCO1A2*/OATP1A2 | CACCACCTTCAGATACAT | GTAGATGACACTTCCTCAA | NM_005630.2 |
| *SLCO1B1*/OATP1B1 | CTTGTATTTAGGTAGTTTGA | CTTAGGAGTTATTCTGATAG | NM_019844.3 |
| *SLCO1B3*/OATP1B3 | ATAGAGCATCACCTGAGA | TCCACGAAGCATATTACC | NM_006446.4 |
| *SLCO2B1*/OATP2B1 | CACGAAGAAGCAGGATGG | CTGGGGAAGACTTTAATGAACT | NM_007256.4 |
| *SLCO3A1*/OATP3A1 | TTGTTGGGCTTCATCCCTCC | CGAAGGATTTGAGCGCGATG | NM_013272.3 |
| *SLCO4A1*/OATP4A1 | GAATACTAGGGGGCATCCCG | ATGGCAAAGAAGAGGACGCC | NM_016354.3 |
| *EPHX1*/EPOX/mEH | CCCAAGGAGTAATCAGAGGGTG | ACATGGCTCCTGTACCTCAG | NM_000120.3 |
| *SLC51A*/OST-α | CAGGTCTCAAGTGATGAA | CTTCGGTAGTACATTCGT | NM_152672.5 |
| *SLC51B*/OST-β | GCTGCTGGAAGAGATGCTTTG | TTTCTTTTCTGCTTGCCTGGATG | NM_178859.3 |
| *SLC10A2*/ASBT | CCTGGTACAGGTGCCGAAC | TGAGCGGGAAGGTGAATACG | NM_000452.2 |
| *ABCB11*/BSEP | GACATGCTTGCGAGGACCTT | GGTTCGTGCACCAGGTAAGAA | NM_003742.2 |
| *ABCB1*/MDR1 | GCCAGAAACAACGCATTGCC | GGGCTTCTTGGACAACCTTTTC | NM_000927.4 |
| *ABCC2*/MRP2 | GATGCACAAAAGGCCTTCACC | GGAAACACTGGCCTGGAGCAT | NM_000392.4 |
| *ABCG2*/BCRP | TGTGTTTATGATGGTCTGTTGGTC | GCTGCAAAGCCGTAAATCCA | NM_001257386.1 |
| *ABCB4*/MDR3 | GGACAGTGCTTCTCGATGGTC | TACAACCCGGCTGTTGTCTC | NM_000443.3 |
| *ATP8B1*/FIC1 | AGCAGTTTAAGAGAGCAGCC | TATGGCGAGCCACATCGTC | NM_005603.4 |
| *ACGG5*/STSL | CCTCTCATCTTTGACCCCCG | CTCACGCGGTGGCTGAC | NM_022436.2 |
| *ABCC1*/MRP1 | TTAAGGTGTTATACAAGAC | GATGAGCAACTTTAAGAT | NM_004996.3 |
| *ABCC3*/MRP3 | GATACGCTCGCCACAGTCC | CAGTTGCCGTGATGTGGCTG | NM_003786.3 |
| *ABCC4*/MRP4 | CCATTGAAGATCTTCCTGG | GGTGTTCAATCTGTGTGC | NM_005845.4 |
| *β-actin* | AACTCCATCATGAAGTGTGACG | GATCCACATCTGCTGGAAGG | NM_001101.5 |
| *YWHAZ* | CCGTTACTTGGCTGAGGTTG | AGTTAAGGGCCAGACCCAGT | NM_145690.3 |
| *GAPDH* | GCTCCTCCTGTTCGACAGTCA | ACCTTCCCCATGGTGTCTGA | NM_002046.7 |
| *Ubiquitin* | TCGCAGCCGGGATTTG | GCATTGTCAAGTGACGATCACA | NM_021009 |

The primers used for amplification in placentae were designed with Beacon (Premier Biosoft, Palo Alto, CA, USA). They were validated for accurateness of amplification on positive tissues using immortalized liver carcinoma (HEPG2) cells. NTCP: sodium (Na)-taurocholate cotransporting polypeptide; OATP: organic anion transport; OST: organic solute transporter; ASBT: apical sodium-dependent bile acid transporter; EPHX1/EPOX/mEH: Epoxide hydrolase 1/microsomal epoxide hydrolase; BSEP: bile salt excretory protein; MDR: multi-drug-resistance protein; FIC1: familial intrahepatic cholestasis protein-1; MRP: multi-drug-resistance-related protein; ASBT: sodium-dependent bile acid transporter; ABC: ATP-binding cassette transporter; SLC: solute carrier protein; SLCO: solute carrier organic anion transporter; ATP8B1: ATPase phospholipid transporting 8B1.

As illustrated in Table 4, in human placental tissues, except *ABCG5*, whose mRNA transcripts were not found, the remaining 20 BA- and cholestasis-related transport genes were detected by qPCR and categorized as either expressed (defined in our studies as Ct values < 35) or only marginally expressed (defined as Ct values > 35).

**Table 4.** Expression of the investigated bile acid solute carriers and ABC transporters in control placentae and primary trophoblast cells.

| Membrane Protein | | | mRNA Transcripts Detectable in | |
|---|---|---|---|---|
| Class | Protein Name | Gene Name | Placental Tissue ($n$ = 12) | Trophoblasts ($n$ = 3) |
| Solute carriers | NTCP | *SLC10A1* | 3/12 | all |
| | OATP1A2 | *SLCO1A2* | all | n.d. |
| | OATP1B1 | *SLCO1B1* | 2/12 | n.d. |
| | OATP1B3 | *SLCO1B3* | 2/12 | n.d. |
| | OATP2B1 | *SLCO2B1* | all | all |
| | OATP3A1 | *SLCO3A1* | all | all |
| | OATP4A1 | *SLCO4A1* | all | all |
| | EPOX/mEH | *EPHX1* | all | all |
| | OST-$\alpha$ | *SLC51A* | 9/12 | all |
| | OST-$\beta$ | *SLC51B* | all | all |
| | ASBT | *SLC10A2* | all | all |
| ABC transporters | BSEP | *ABCB11* | Ct > 35 | all |
| | MDR1 | *ABCB1* | all | all |
| | MRP2 | *ABCC2* | all | all |
| | BCRP | *ABCG2* | all | all |
| | MDR3 | *ABCB4* | all | all |
| | FIC1 | *ATP8B1* | all | all |
| | ABCG5 | *ABCG5* | n.d. | n.d. |
| | MRP1 | *ABCC1* | 5/12 | n.d. |
| | MRP3 | *ABCC3* | all | all |
| | MRP4 | *ABCC4* | 6/12 | all |

NTPC: sodium (Na)-taurocholate cotransporting polypeptide; OATP: organic anion transport; OST: organic solute transporter; ASBT: apical sodium-dependent bile acid transporter; EPHX1: epoxide hydrolase 1; mEH/EPOX: microsomal epoxide hydrolase; BSEP: bile salt excretory protein; MDR: multi-drug-resistance protein; FIC1: familial intrahepatic cholestasis protein-1; MRP: multi-drug-resistance-related protein; ASBT: sodium-dependent bile acid transporter; ABC: ATP-binding cassette transporter; SLC: solute carrier protein; SLCO: solute carrier organic anion transporter; n.d.: not detected. The threshold of gene expression in this study was set at Ct < 35 amplification cycles. Details regarding procedures related to quantitative RT-PCR are given in Section 5.

Among the expressed genes, some exhibited, however, an inconsistent expression pattern. This was the case for *SLC10A1*, *SLC51A*, *SLCO1B1*, and *SLCO1B3* since the corresponding mRNA transcripts were detected only in 3/12, 2/12, 2/12, and 9/12 control specimens, respectively. The same applies for *ABCC1* and *ABCC4* expressions, as they were only detected in 5/12 and 6/12 control tissues, respectively (Table 4). The mRNA expression of *ABCB11* was marginal (Table 4). We did not find any sex-specific expression profiles of the transport proteins tested.

The gene expression profile of the investigated BA- and cholestasis-related transport proteins did not fully correspond between primary trophoblast cells and control placental tissues (Table 4). This is illustrated, for instance, by the dissimilar gene expression of *SLCO1A2* in placental tissue compared with trophoblast cells. In our samples, *ABCG5* expression was detected neither in control placental tissues nor in primary trophoblast cells.

### 2.3. Comparison of Transporter Expression in Patients and Healthy Controls

In a next step, only the 13 BA- and cholestasis-related transport proteins, which were unequivocally detected in all heathy placentae, were compared with ICP placental tissues. Interestingly, we found that solely *SLCO3A1* mRNA expression was differentially expressed ($p$ = 0.0177) in ICP placentae as compared with controls (Figure 1).

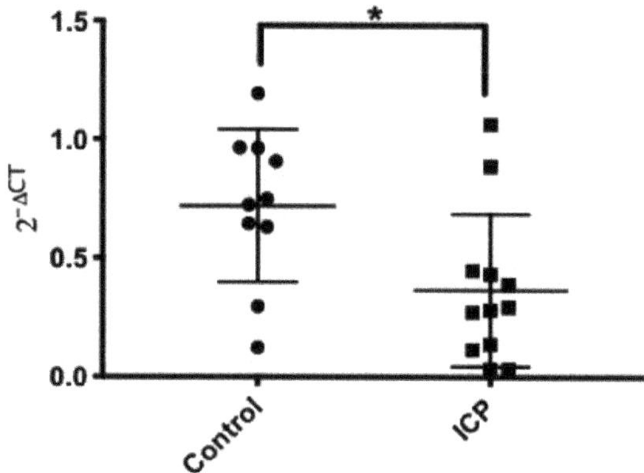


**Figure 1.** Comparative mRNA levels of SLCO3A1 in healthy and ICP placentae. Circle symbols represent healthy control placentae, and square symbols represent intrahepatic cholestasis of pregnancy (ICP). Real-time quantitative PCR and statistical evaluations of data are as described in the Section 5. * $p$ = 0.0177.

The remaining tested genes were unaltered ($p$ > 0.05) by the ICP condition (Table 5). There were no sex-specific expression patterns found.

**Table 5.** Summary of gene expression comparisons between ICP and controls.

| Gene | SLC10A2 | ABCB1 | ABCB4 | ABCG2 | ABCC2 | ABCC3 | ATP8B1 | SLC51A | EPHX1 | SLCO2B1 | SLCO4A1 | SLCO1A2 |
|---|---|---|---|---|---|---|---|---|---|---|---|---|
| $p$-value | 0.99 | 0.86 | 0.73 | 0.84 | 0.37 | 0.47 | 0.78 | 0.33 | 0.37 | 0.71 | 0.43 | 0.44 |

Differences between ICP and controls were evaluated by using unpaired *t*-test. SLC: solute carrier protein; ABC: ATP-binding cassette protein; SLCO: solute carrier organic anion transporter; EPHX1: epoxide hydrolase 1; ATP8B1: ATPase phospholipid transporting 8B1.

### 2.4. Correlation between Placental BA Transport Proteins and Clinical Parameters

We found significant positive relationships between placental *SLC10A2* mRNA levels and maternal BMI values (Table 6), independently of the maternal health status. In patients, *SLC10A2* mRNA levels were correlated with circulating BA levels (Table 6), whereas *SLCO4A1* and *ABCC2* mRNA levels were positively correlated with maternal serum BA concentrations (Table 6).

**Table 6.** Summary of significant correlations.

| | SLC10A2 | SLCO4A1 | ABCC2 |
|---|---|---|---|
| BMI * | R2 = 0.27; $p$ = 0.013 | | |
| Serum bile acids ** | R2 = 0.58; $p$ = 0.004 | R2 = 0.34; $p$ = 0.047 | R2 = 0.70; $p$ = 0.0007 |

Pearson's correlation coefficients are shown. All pregnant women independent of their health status (*) or only patients (**) were included in the analysis. ABC: ATP-binding cassette protein; SLC: solute carrier protein; SLCO: solute carrier organic anion transporter.

## 3. Discussion

### 3.1. Screening of Transporters in Placental Tissues and Trophoblast Cells

The present study describes the mRNA expression profile of important BA solute carriers and ABC transporters in placental tissues/cells and discusses their potential relevance as protective mechanisms preventing fetal exposure to excessive harmful BAs.

One of the main findings of the study is that, for the first time, the gene expression of *SLC10A2* is described in human placental tissue and trophoblast cells. SLC10A2/ASBT is a sodium-dependent transporter that exerts a crucial function in the enterohepatic circulation. It enables, at the apical membrane of enterocytes, the uptake of BAs from the intestinal lumen [11]. Previous findings have established that SLC10A2/ASBT abundance in the intestine is inversely correlated with maternal BA concentrations [12]. This is consistent with a role of this gene in preventing excess absorption of BAs into the portal circulation. By assuming that the (apical) localization of SLC10A2/ASBT is conserved in the trophoblast, the unexpected positive relationship of placental *SLC10A2* mRNA and maternal BA levels found in the present study appears intriguing. This finding does not argue for a role of placental SLC10A2/ASBT as a feto-protective mechanism against the deleterious effect of elevated maternal serum BA concentrations in ICP. Indeed, the herein observed positive correlation would suggest a parallel increase in placental *SLC10A2* expression with augmentation of maternal serum BA concentrations.

Nonetheless, the interpretation of the mentioned relationship requires some caution. Maternal blood samples (used for BA measurements during diagnosis) and placental tissue (for *SLC10A2* mRNA analysis after birth) have not been collected at identical time points. Hence, due to limitations based on our ethical approval, maternal BA levels at delivery (i.e., at the time of placenta tissue collection) could not be monitored. Thus, it is not certain whether the detected placental *SLC10A2* mRNA expression fully reflects its gene expression at the time of blood sampling.

Next, we identified mRNA transcripts of *SLC51B* in human placental tissue and in primary trophoblasts isolated from term control placentae. This finding is valuable since the available literature concerning trophoblast cells has reported so far only the gene expression of SLC51A/OST-$\alpha$ [13]. Considering the identification of *SLC51A* and *SLC51B* mRNA isoforms in the current study, it is likely that SLC51A/OST-$\alpha$ and SLC51B/OST-$\beta$ are important in modulating BA fluxes across the placental barrier, similar to their role in other tissue/cell types [14,15].

An additional new finding in this study is based on the detection of mRNA expression of *EPHX1* (also called EPOX or mEH in hepatic tissue; see Table 1) in trophoblast cells. This result complements investigations by Coller et al., who, studying human placental tissue, described the presence of EPHX1/*EPOX* only in placental blood vessels and Hofbauer cells [16]. The discrepancy between these studies may be explained by the difference of the sensitivity of the methods employed. We applied the highly sensitive real-time quantitative PCR using cDNA from well-characterized isolated trophoblast cells, whereas Coller et al. used an immunostaining technique. Considering that EPHX1/EPOX operates as a sodium-dependent BA transporter in other mammalian cells [17,18], the identification of its mRNA in placental tissues and trophoblast cells may suggest a similar function in the human placenta. However, the lack of a significant correlation between the placental *EPHX1* gene expression and maternal BA concentrations could indicate a minor role of EPHX1/EPOX in controlling BA fluxes across the human placenta.

Considering that pregnant women with ICP are also prone to other metabolic features, especially dyslipidemia [2], *ABCG5* mRNA expression was determined. Surprisingly, we did not detect *ABCG5* mRNA expression, neither in our human placental tissues nor in the isolated trophoblast cells. These data are in contrast to findings in rats, where *Abcg5* and *Abcg8* mRNAs were detected [19]. Nonetheless, we did not find literature data reporting the expression of ABCG5/STSL in human placenta, implying that the placental expression of this membrane protein could be species specific.

### 3.2. Summary of the BA Transport Machinery in Human Placenta

Given the expression profiles of BA transport proteins detected in healthy placental tissues and primary trophoblasts in the current study, we suggest the following scheme summarizing the BA transport machinery in human placental tissue (Figure 2). This overview is based on the assumption that the genes' substrate affinity and polarization patterns are conserved and would therefore reflect findings in enterocytes/hepatocytes [14,15,20]. The fetal liver produces BAs as early as 12 weeks of gestation, which are eliminated as waste products by transporting them across the placenta towards the maternal circulation (Figure 2A). Conversely, maternal-originating BAs are also directed to the fetus through the placenta (Figure 2B,C). The transport proteins involved are located at the plasma membrane of the placental apical (microvillus) and basal layers (Figure 2C). The expressed (Ct value < 35) and consistently (present in all specimens) detected transport proteins in placental tissue and syncytiotrophoblasts are illustrated with symbols, filled in green and red colors, respectively. Faded colors indicate equivocally expressed transport proteins. The indicated localization of transport proteins in the trophoblast (Figure 2C), the substrate affinity, and the directionality of transport are according to existing literature under physiological conditions.

Notably, for a few transport proteins, such as SLCO1A2/OATP1A2 and SLCO1B3/OATP1B3, whose mRNA transcripts were absent in trophoblast cells used in this study, their placental expression is still a matter of controversy. Both absence [21] and presence [13,22,23] have been reported.

### 3.3. Comparison of ICP Versus Controls

Considering recently published data, stratifying the severity of ICP and its relationship to hazard risks of the prevalence of adverse perinatal outcomes [3], patients investigated in this study appeared to be "mildly" affected. The mean concentration of serum BAs was around 56 μmol/L, although a considerable interindividual variation was observed within the studied cohort. Of all 13 analyzed genes, solely the *SLCO3A1* mRNA expression was significantly altered in ICP compared with controls. It cannot be excluded that the clinical severity of ICP has an impact on the maternal and fetal regulation of the studied genes. Thus, depending on the severity of the disease and the cohort size, the expression of the genes might vary. The downregulation of *SLCO3A1* detected in this study per se seems interesting. SLCO3A1/OATP3A1 is an uptake transporter that has transport affinity for BAs, various steroid hormones, and others [21,24]. Among them are also prostaglandins, key regulators of myometrium contraction, which plays an important role, for example, during preterm labor associated with ICP [25].

Next, all subjects enrolled into the ICP cohort were treated with ursodeoxycholic acid (UDCA treatment). Contrary to the finding by Azzaroli et al. [26], who reported the gene-promoting effect of UDCA on *ABCC2* expression, we did not observe alterations in the *ABCC2* expression in this study. We were unable to determine whether UDCA treatment caused changes in the placental *SLCO3A1* mRNA expression or whether the altered expression pattern resulted from the effect of maternal BA. Moreover, further studies aiming to precisely identify the SLCO3A1/OATP3A1 localization (apical versus basal membranes) and functionality (e.g., with placental trophoblast cell-based Transwell® system or ex vivo dual perfusion of the human placenta) are needed. They will help to draw firm conclusions regarding whether *SLCO3A1* gene expression downregulation constitutes a protective measure for the fetus exposed to maternal ICP.

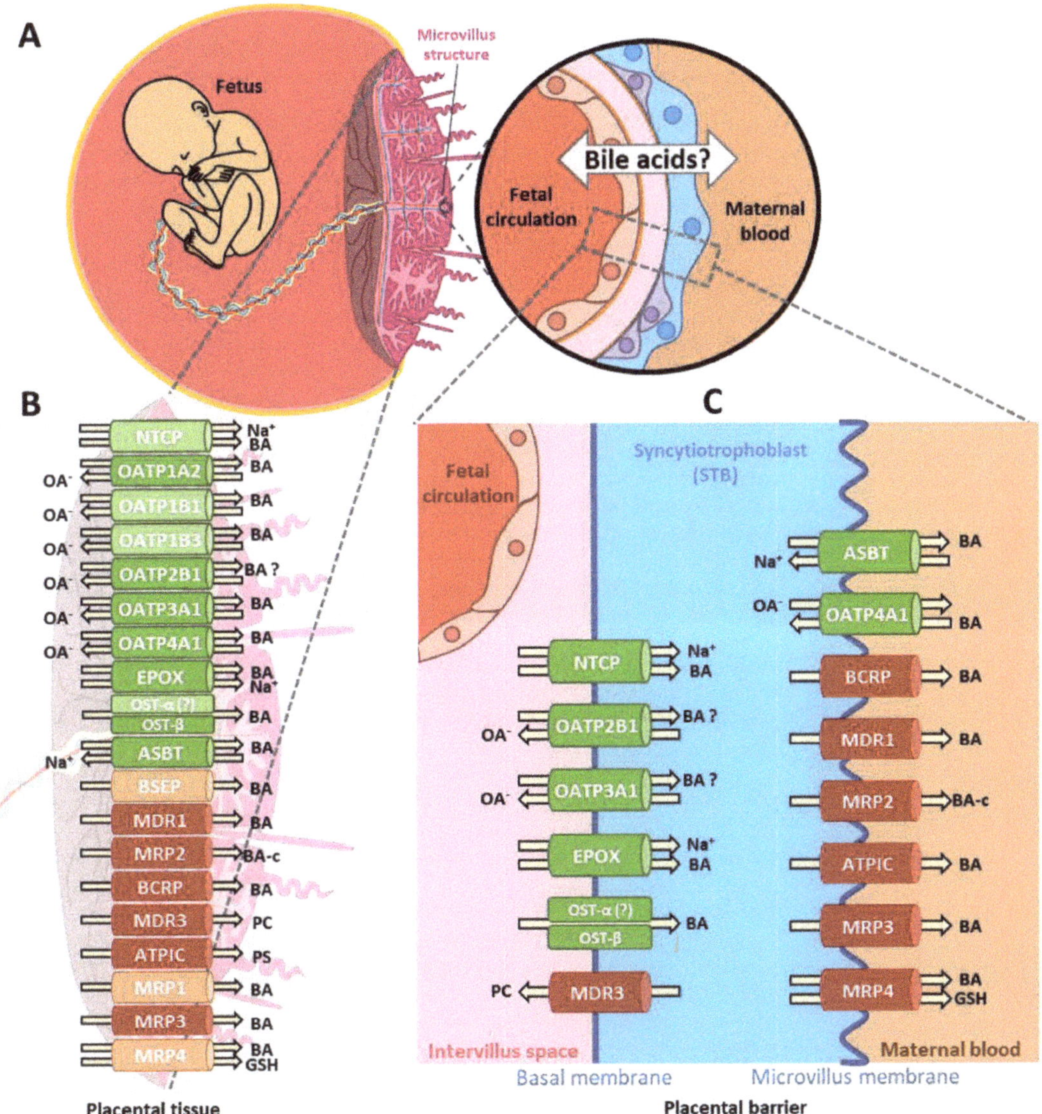



**Figure 2.** Schematic illustration of the bile acid transport machinery in human placenta. Fetal BAs are eliminated by transport across the placenta towards the maternal circulation (**A**). Arrows indicate the direction of substrate and cosubstrate transport exchange by the various transporter proteins across the human placenta (**B**) and across the trophoblast membranes, respectively (**C**), representing the critical part of the placental barrier. Details are described in the text. The gene and protein names of the depicted transporters are listed in Tables 1 and 4. Abbreviations: BA: bile acids. BA-c: bile acid conjugates. OA$^-$: organic anions. PS: phosphatidylserine. PC: phosphatidylcholine. BA (?) indicates uncertainties regarding the BA transport, while OST-α (?) indicates uncertainties regarding expression.

## 4. Conclusions

Data reported in the current study indicate that the human placenta exhibits a limited response at the transcriptional level when the mother suffers from a "moderate" ICP condition. The newly identified gene expression of *SLC10A2* and *EPHX1* in human placenta tissue and trophoblasts was unaltered in ICP, while *SLC10A2* mRNA strongly correlated with both maternal BMI and BA levels. Nonetheless, given the relatively small cohort size of controls and ICP patients in the present study, the reported findings and interpretations may not be generalized unless confirmed in a larger cohort.

## 5. Materials and Methods

### 5.1. Human Placental Tissue and Trophoblast Cells

The study was approved by the ethics institutional review board of the Canton of Bern with an informed consent obtained from each participant prior to giving birth. The study was conducted in accordance with the Declaration of Helsinki. Pregnant women were under obstetrical care at the Department of Obstetrics and Gynecology, University Hospital, Bern, Switzerland. Placentae from healthy controls ($n = 12$) and ICP ($n = 12$) pregnancies were obtained after elective caesarean section or from spontaneous delivery between 2009 and 2011 after having obtained informed consent from the pregnant women. Placentae from healthy pregnancies were used. The clinical characteristics of pregnancies, whose placentae were investigated, are summarized in Table 2.

Upon serious suspicion, pregnant women were diagnosed for ICP following the routinely applied procedure at the University Hospital. The criteria of eligibility for the pregnant women's inclusion to the ICP group include, among others, increased serum BA concentration in combination with pruritus. All ICP women were treated with appropriate doses of De-Ursil® from diagnosis until delivery.

In addition to placental tissues, a cDNA pool of three independent trophoblast cell isolations was also tested. The procedure of trophoblast isolation has been previously described in detail [27].

### 5.2. RNA Extraction and Quantitative RT-PCR

Considering the heterogeneity of the human placenta, we standardized the placental tissue collection procedure for gene expression analysis [28]. Thus, each specimen analyzed was taken from the central area of the placenta. Total RNA was extracted from placental tissues using Trizol® reagent (Thermo Fisher Scientific, Waltham, MA, USA). The concentrations of the extracted total RNA were calculated by measuring absorbance (A) at 260 nm. Purity was assessed by the A260/280 and A260/230 ratios, measured using a NanoDrop ™ 1000 Spectrophotometer (Thermo Fisher Scientific, Waltham, MA, USA). Next, 2 µg of total RNA from each sample was reverse-transcribed using the GoScript ™ Reverse Transcriptase System (Promega, Madison, WI, USA) according to the manufacturer's instructions. Real-time PCR was carried out on the ViiA 7 Real-Time PCR Detection System (Bio-Rad Laboratories Inc., Hercules, CA, USA) using an SYBR® Green PCR master mix detection kit (Promega, Madison, WI, USA). The reaction conditions were as follows: an initial denaturation at 95 °C for 10 min, followed by 40 cycles of 95 °C for 15 s and 60 °C for 60 s, for melting curve temperature was increased from 60 °C to 95 °C with an increment rate of 0.5 °C every 0.05 s. The primers used for PCR amplification of the 21 currently known human BA- and cholestasis-related transport proteins are shown in Table 3. The relative mRNA expression of BA transport proteins was calculated by using the formula $2^{-dCt}$. For each individual sample, the dCt equates the difference between the Ct value of the transport protein of interest and the mean Ct value of the measured references. The latter were β-actin, YWHAZ, GAPDH, and ubiquitin.

### 5.3. Statistical Analysis

The statistical evaluation was performed using GraphPad Prism® (GraphPad Software Inc., San Diego, CA, USA). All data are shown as mean ± SD. The gene expression data were analyzed for normality of distribution. Differences in the placental mRNA expression of targeted BA transport proteins between healthy controls and ICP patients were analyzed with unpaired *t*-test. The correlations of clinical data and gene expression of transport proteins were calculated with Pearson's correlation test. The level of statistical significance was set at $p \leq 0.05$.

**Author Contributions:** Conceptualization, C.A., M.B. and H.S.; methodology, A.E., S.K. and J.Z.; validation, A.E., J.Z., S.K., M.B., E.O. and C.A.; formal analysis, A.E., E.O., J.Z., M.B. and S.K.; writing—original draft preparation, A.E. and E.O.; writing—review and editing, E.O., A.E., S.K., J.Z., M.B., H.S. and C.A.; visualization, E.O., A.E., M.B., S.K. and H.S.; supervision, J.Z., S.K., H.S. and C.A.; project administration, C.A., E.O., M.B. and H.S.; funding acquisition, C.A. All authors have read and agreed to the published version of the manuscript.

**Funding:** The current study was supported by the Swiss National Science Foundation (Grant No. 310030_149958) and NCCR TransCure (Grant No. 51NF40-185544).

**Institutional Review Board Statement:** The study was conducted according to the guidelines of the Declaration of Helsinki, and approved by the ethical committee of the Canton of Bern, Switzerland (approval number 178/03; 26/09/2005).

**Informed Consent Statement:** Informed consent was obtained from all subjects involved in the study.

**Data Availability Statement:** Not applicable.

**Acknowledgments:** The authors wish to express their gratitude to the patients, physicians, and midwives from the Department of Obstetrics and Gynecology, University Hospital, for participating in this study. We are grateful to Michael Luethi for his technical support.

**Conflicts of Interest:** The authors declare no conflict of interest.

## References

1. Dann, A.T.; Kenyon, A.P.; Wierzbicki, A.S.; Seed, P.T.; Shennan, A.H.; Tribe, R.M. Plasma lipid profiles of women with intrahepatic cholestasis of pregnancy. *Obstet. Gynecol.* **2006**, *107*, 106–114. [CrossRef] [PubMed]
2. Martineau, M.G.; Raker, C.; Dixon, P.H.; Chambers, J.; Machirori, M.; King, N.M.; Hooks, M.L.; Manoharan, R.; Chen, K.; Powrie, R.; et al. The metabolic profile of intrahepatic cholestasis of pregnancy is associated with impaired glucose tolerance, dyslipidemia, and increased fetal growth. *Diabetes Care* **2015**, *38*, 243–248. [CrossRef] [PubMed]
3. Ovadia, C.; Seed, P.T.; Sklavounos, A.; Geenes, V.; Di Illio, C.; Chambers, J.; Kohari, K.; Bacq, Y.; Bozkurt, N.; Brun-Furrer, R.; et al. Association of adverse perinatal outcomes of intrahepatic cholestasis of pregnancy with biochemical markers: Results of aggregate and individual patient data meta-analyses. *Lancet* **2019**, *393*, 899–909. [CrossRef]
4. Fisk, N.; Storey, G. Fetal outcome in obsteric cholestasis. *BJOG* **1988**, *95*, 1137–1143. [CrossRef] [PubMed]
5. Williamson, C.; Geenes, V.L. Intrahepatic Cholestasis of Pregnancy. *Obs. Gynecol.* **2014**, *88*, 13–16. [CrossRef] [PubMed]
6. Reid, R.; Ivey, K.J.; Rencoret, R.H.; Storey, B. Fetal complications of obstetric cholestasis. *Br. Med. J.* **1976**, *1*, 870–872. [CrossRef] [PubMed]
7. Colombo, C.; Roda, A.; Roda, E.; Buscaglia, M.; Albert, C.; Agnola, D.; Filippetti, P.; Ronchi, M. Correlation between Fetal and Maternal Serum Bile Acid Concentrations. *Pediatr. Res.* **1984**, *19*, 227–231. [CrossRef] [PubMed]
8. Colombo, C.; Zuliani, G.; Ronchi, M.; Breidenstein, J.; Setchell, K.D.R. Biliary Bile Acid Composition of the Human Fetus in Early Gestation. *Pediatr. Res.* **1987**, *2*, 68–71. [CrossRef]
9. Geenes, V.; Lövgren-Sandblom, A.; Benthin, L.; Lawrence, D.; Chambers, J.; Gurung, V.; Thornton, J.; Chappell, L.; Khan, E.; Dixon, P.; et al. The Reversed Feto-Maternal Bile Acid Gradient in Intrahepatic Cholestasis of Pregnancy Is Corrected by Ursodeoxycholic Acid. *PLoS ONE* **2014**, *9*, e83828. [CrossRef]
10. Laatikainen, T.; Lehtonen, P.; Hesso, A. Fetal sulfated and nonsulfated bile acids in intrahepatic cholestasis of pregnancy. *J. Lab. Clin. Med.* **1978**, *92*, 681810.
11. Hagenbuch, B.; Dawson, P. The sodium bile salt cotransport family SLC10. *Pflügers Arch.* **2004**, *447*, 566–570. [CrossRef]
12. Hruz, P.; Zimmermann, C.; Gutmann, H.; Degen, L.; Beuers, U.; Terracciano, L.; Drewe, J.; Beglinger, C. Adaptive regulation of the ileal apical sodium dependent bile acid transporter (ASBT) in patients with obstructive cholestasis. *Gut* **2006**, *55*, 395–402. [CrossRef]
13. Serrano, M.A.; Macias, R.I.R.; Briz, O.; Monte, M.J.; Blazquez, A.G.; Williamson, C.; Kubitz, R.; Marin, J.J.G. Expression in Human Trophoblast and Choriocarcinoma Cell Lines, BeWo, Jeg-3 and JAr of Genes Involved in the Hepatobiliary-like Excretory Function of the Placenta. *Placenta* **2007**, *28*, 107–117. [CrossRef]
14. Wang, W.; Seward, D.J.; Li, L.; Boyer, J.L.; Ballatori, N. Expression cloning of two genes that together mediate organic solute and steroid transport in the liver of a marine vertebrate. *Proc. Natl. Acad. Sci. USA* **2001**, *98*, 9431–9436. [CrossRef]

15. Dawson, P.A.; Hubbert, M.; Haywood, J.; Craddock, A.L.; Zerangue, N.; Christian, W.V.; Ballatori, N. The Heteromeric Organic Solute Transporter OST-alpha-OST-beta, Is an Ileal Basolateral Bile Acid Transporter. *J. Biol. Chem.* **2005**, *280*, 6960–6968. [CrossRef]

16. Coller, J.K.; Fritz, P.; Zanger, U.M.; Siegle, I.; Eichelbaum, M.; Kroemer, H.K.; Thomas, E.M. Distribution of microsomal epoxide hydrolase in humans: An immunohistochemical study in normal tissues, and benign and malignant tumours. *Histochem. J.* **2001**, *33*, 329–336. [CrossRef] [PubMed]

17. Zhu, Q.; Von Dippe, P.; Xing, W.; Levy, D. Membrane Topology and Cell Surface Targeting of Microsomal. *J. Biol. Chem.* **1999**, *274*, 27898–27904. [CrossRef]

18. Von Dippe, P.; Amoui, M.; Stellwagen, R.H.; Levy, D. The Functional Expression of Sodium-dependent Bile Acid Transport in Madin-Darby Canine Kidney Cells Transfected with the cDNA for Microsomal Epoxide Hydrolase. *J. Biol Chem.* **1996**, *271*, 18176–18180. [CrossRef]

19. Leazer, T.M.; Klaassen, C.D. The presence of xenobiotic transporters in rat placenta. *Drug. Metab. Dispos.* **2003**, *31*, 153–167. [CrossRef]

20. König, J.; Cui, Y.; Nies, A.T.; Keppler, D. Localization and genomic organization of a new hepatocellular organic anion transporting polypeptide. *J. Biol. Chem.* **2000**, *275*, 23161–23168. [CrossRef]

21. Tamai, I.; Nezu, J.I.; Uchino, H.; Sai, Y.; Oku, A.; Shimane, M.; Tsuji, A. Molecular identification and characterization of novel members of the human organic anion transporter (OATP) family. *Biochem. Biophys. Res. Commun.* **2000**, *273*, 251–260. [CrossRef] [PubMed]

22. Patel, P.; Weerasekera, N.; Hitchins, M.; Boyd, C.; Johnston, D.; Williamson, C. Semi Quantitative Expression Analysis of MDR3, OATP-D, OATP-E, NTCP Gene Transcripts in 1st and 3rd Trimester Human Placenta. *Placenta* **2003**, *24*, 39–44. [CrossRef] [PubMed]

23. Wang, H.; Yan, Z.; Dong, M.; Zhu, X.; Wang, H.; Wang, Z. Alteration in placental expression of bile acids transporters OATP1A2, OATP1B1, OATP1B3 in intrahepatic cholestasis of pregnancy. *Arch. Gynecol. Obstet.* **2012**, *285*, 1535–1540. [CrossRef] [PubMed]

24. Huber, R.D.; Gao, B.; Pfändler, M.S.; Zhang-fu, W.; Leuthold, S.; Hagenbuch, B.; Folkers, G.; Meier, P.J.; Stieger, B. Characterization of two splice variants of human organic anion transporting polypeptide 3A1 isolated from human brain. *Am. J. Physiol. Cell Physiol.* **2006**, *292*, 795–806. [CrossRef]

25. Olson, D.M.; Zaragoza, D.B.; Shallow, M.C.; Cook, J.L.; Mitchell, B.F.; Grigsby, P.; Hirst, J. Myometrial Activation and Preterm Labour: Evidence Supporting a Role for the Prostaglandin F Receptor—A Review. *Placenta* **2003**, *17*, A47–A52. [CrossRef]

26. Azzaroli, F.; Mennone, A.; Feletti, V.; Simoni, P.; Baglivo, E.; Montagnani, M.; Rizzo, N.; Pelusi, G.; De Aloysio, D.; Lodato, F.; et al. Modulation of human placental multidrug resistance proteins in cholestasis of pregnancy by ursodeoxycholic acid. *Aliment. Pharmacol. Ther.* **2007**, *26*, 1139–1146. [CrossRef]

27. Kallol, S.; Moser-Haessig, R.; Ontsouka, C.E.; Albrecht, C. Comparative expression patterns of selected membrane transporters in differentiated BeWo and human primary trophoblast cells. *Placenta* **2018**, *72–73*, 48–52. [CrossRef]

28. Huang, X.; Baumann, M.; Nikitina, L.; Wenger, F.; Surbek, D.; Körner, M.; Albrecht, C. RNA degradation differentially affects quantitative mRNA measurements of endogenous reference genes in human placenta. *Placenta* **2013**, *34*, 544–547. [CrossRef]

International Journal of
*Molecular Sciences*

*Article*

# Conditional Mutation of Hand1 in the Mouse Placenta Disrupts Placental Vascular Development Resulting in Fetal Loss in Both Early and Late Pregnancy

Jennifer A. Courtney [1], Rebecca L. Wilson [2,3], James Cnota [4] and Helen N. Jones [2,3,*]

[1] Center for Fetal and Placental Therapy, Cincinnati Children's Hospital Medical Center, Cincinnati, OH 45229, USA; jcourtneymdb@gmail.com
[2] Center for Research in Perinatal Outcomes, College of Medicine, University of Florida, Gainesville, FL 32603, USA; rebecca.wilson@ufl.edu
[3] Department of Physiology and Functional Genomics, College of Medicine, University of Florida, Gainesville, FL 32603, USA
[4] Heart Institute, Cincinnati Children's Hospital Medical Center, Cincinnati, OH 45229, USA; James.Cnota@cchmc.org
[*] Correspondence: jonesh@ufl.edu; Tel.: +1-352-846-1503

**Abstract:** Congenital heart defects (CHD) affect approximately 1% of all live births, and often require complex surgeries at birth. We have previously demonstrated abnormal placental vascularization in human placentas from fetuses diagnosed with CHD. *Hand1* has roles in both heart and placental development and is implicated in CHD development. We utilized two conditionally activated $Hand1^{A126fs/+}$ murine mutant models to investigate the importance of cell-specific *Hand1* on placental development in early ($Nkx2-5^{Cre}$) and late ($Cdh5^{Cre}$) pregnancy. Embryonic lethality occurred in $Nkx2-5^{Cre}/Hand1^{A126fs/+}$ embryos with marked fetal demise occurring after E10.5 due to a failure in placental labyrinth formation and therefore the inability to switch to hemotrophic nutrition or maintain sufficient oxygen transfer to the fetus. Labyrinthine vessels failed to develop appropriately and vessel density was significantly lower by day E12.5. In late pregnancy, the occurrence of $Cdh5^{Cre+};Hand1^{A126fs/+}$ fetuses was reduced from 29% at E12.5 to 20% at E18.5 and remaining fetuses exhibited reduced fetal and placental weights, labyrinth vessel density and placenta angiogenic factor mRNA expression. Our results demonstrate for the first time the necessity of *Hand1* in both establishment and remodeling of the exchange area beyond early pregnancy and in patterning vascularization of the placental labyrinth crucial for maintaining pregnancy and successful fetal growth.

**Keywords:** congenital heart disease; placenta; hand1; vascular development; pregnancy

Citation: Courtney, J.A.; Wilson, R.L.; Cnota, J.; Jones, H.N. Conditional Mutation of Hand1 in the Mouse Placenta Disrupts Placental Vascular Development Resulting in Fetal Loss in Both Early and Late Pregnancy. *Int. J. Mol. Sci.* **2021**, *22*, 9532. https://doi.org/10.3390/ijms22179532

Academic Editors: Hiten D. Mistry and Eun Lee

Received: 6 July 2021
Accepted: 31 August 2021
Published: 2 September 2021

**Publisher's Note:** MDPI stays neutral with regard to jurisdictional claims in published maps and institutional affiliations.



## 1. Introduction

Congenital heart disease (CHD) is the most common birth defect, affecting ~1% of all live births [1]. Babies born with a CHD often undergo corrective surgeries within days of being born, and survival from these surgeries often depends on size at birth [2]. Population studies have demonstrated that pregnancies complicated by CHD carry a higher risk of developing pathologies associated with abnormal placental development and function including growth disturbances [3–5], preeclampsia [6–9], preterm birth [10,11], and stillbirth [12]. The placenta serves as the mediator between mother and developing fetus to provide nutrition and gas exchange, remove fetal wastes, and prevents mixing of maternal and fetal blood [13]. It is unsurprising, while relatively recently demonstrated, that placental development is often affected in pregnancies with CHD as both the heart and placenta are vascular organs that develop concurrently very early in gestation, and shared pathways direct the development of both [14]. Placentas of fetuses with CHD often exhibit changes that disrupt the proper patterning and function of the maternal/fetal exchange area including underdeveloped vasculature and impaired nutrient transport [15,16].

Genome-wide associations studies (GWAS) undertaken previously have identified almost 400 genes associated with CHD [17–19]. However, despite decades of research with a primary focus on genetic etiology, the underlying cause of these CHD remains unknown in the majority of cases [20]. Additionally, modeling studies implicating genes and/or environmental influences responsible for abnormal heart development [21] often overlook the involvement of extraembryonic tissues or circumvent it by using 'cardiac-specific' conditional knockout mouse models. However, this does not truly reflect the situation in cases of CHD where genetic perturbations would occur in all cells/tissues expressing that gene and many of the identified genes are expressed in other cell types in addition to those found in the heart. Importantly, it has been shown that 68% of 103 knockout models of heart development that exhibited embryonic lethality at or after mid-gestation had abnormal placental development not previously investigated [22]. Such outcomes highlight the strong correlation between placental dysmorphology, angiogenesis and heart development.

One example of a gene involved in both cardiac and placental development is *Hand1* [23]. *Hand1* is a basic helix-loop-helix transcription factor found in multiple organ systems during embryogenesis. *Hand1*-null mice are embryonic lethal by E8.5 due to extraembryonic defects in the yolk sac, chorion, allantois, and trophoblast giant cells [23–26]. Lineage tracing verified *Hand1* expression at E9.5 in these tissues [27]. Placental labyrinth vascular endothelium also expressed *Hand1* at E14.5 as well as the endothelium of the umbilical vein [27]. Early pregnancy lethality of the global *Hand1* knockout models increases the difficulty in studying its role on placental development as the developing conceptuses are lost before the placenta has fully formed. Furthermore, the relevance of the disruption to placentation by trophoblast giant cell failure is controversial in its applicability to the development of the human placenta due to species differences. There are a number of conditional knockout mouse models in which gene disruption is isolated within the fetus, which, depending on the Cre-driver, show varying developmental outcomes from mid-gestation lethality to viable offspring with mild phenotypes [28,29]. However, these studies focused solely on the fetus, despite the fact that some of the Cre-drivers function in extra-embryonic cells of the yolk sac and placenta, which may explain the wide disparities in outcomes.

The implications for *Hand1* on heart development have been well characterized [23], however currently, there have been no investigations into the placental contribution to embryonic lethality. In the current study, we aimed to determine the placental contribution to embryonic lethality in the *Hand1*-mutant mice in early and late pregnancy using two conditional activation mouse models. Nkx2-5 is a well-known cardiac development gene, but is also required for yolk sac angiogenesis and is expressed in endothelial and hematopoietic cells within yolk sac mesoderm [30–32]. Previous investigations using the conditional activation knock-in *Hand1$^{SFA126fs}$* mice generated with *Nkx2-5$^{Cre}$* have shown that *Hand1$^{A126fs/+}$* fetuses die at embryonic day 15.5 (E15.5) and display outflow tract abnormalities, thin myocardium and ventricular septal defects [28]. However, the extraembryonic tissue was not investigated. Cadherin 5 (Cdh5) is an endothelial cell specific gene expressed in the placental fetal vessels. Thus, the use of the *Nkx2-5$^{Cre}$* allows for the study of mutating *Hand1* from E8.5 in the yolk sac and labyrinth trophoblast progenitor cells at a time essential for labyrinth establishment and the switch from histiotrophic to hemotrophic nutrition. The *Cdh5$^{Cre}$* model on the other hand, targets the requirements for *Hand1* in placental vascular branching in pregnancy when *Hand1* is expressed in endothelial cells from E12.5.

## 2. Results

### 2.1. Nkx2-5-Cre Is Expressed in Yolk Sac, Trophoblast Cells, and Cardiomyocytes

Conditionally activated *Hand1$^{A126FS/+}$* females were time mated with homozygous *Nkx2-5$^{Cre}$* males to produce litters containing *Nkx2-5$^{Cre}$;Hand1$^{A126FS/+}$* and *Nkx2-5$^{Cre}$;Hand1$^{+/+}$* embryos. In order to verify the efficacy and spaciotemporal expression of the *Nkx2-5$^{Cre}$*, males were first mated with homozygous tdTomato reporting females. In control mice, immunofluorescence and microscopy confirmed trophoblast expression of the Nkx2-5-Cre at E8.5 as well as Hand1 expression in the trophoblast nucleus (Figure 1A); co-localization of Nkx2-5-Cre expression and the trophoblast marker cytokeratin 7 further confirmed trophoblast expression (Figure 1B). Co-expression of Nkx2-5-Cre and Hand1 in yolk sac, labyrinth trophoblast progenitor cells and syncytiotrophoblast was also confirmed at E9.5 (Figure 1C) and E10.5 (Figure 1D). Sinusoidal giant cells did not express Nkx2-5-Cre at E10.5 but did express cytokeratin 7 (Figure 1E). Analysis of the heart at E9.5 showed strong expression for Nkx2-5-Cre, and Hand1 was localized in the cardiac ventricular tissue (Figure 1F).

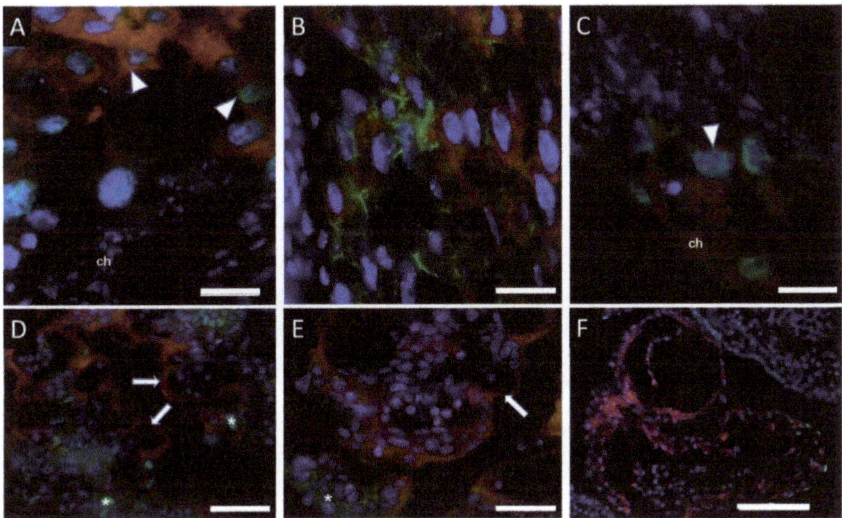

**Figure 1.** Expression of Nkx2-5-Cre and Hand1 or Cytokeratin 7 in the placenta and fetal heart at embryonic day 8.5 (E8.5), E9.5 and E10.5. (**A**) *Nkx2-5$^{Cre}$* (red) is expressed in trophoblast progenitor cells at E8.5, overlapping with Hand1 protein expression, but not in chorion which does express Hand1; (**B**) Colocalization of *Nkx2-5$^{Cre}$* (red) expression with cytokeratin 7 (green) at E8.5; (**C**) At E9.5, *Nkx2-5$^{Cre}$* (red) and Hand1 (green) are co-expressed in both chorion and labyrinth trophoblast progenitor cells; (**D**) Hand1 (green) and *Nkx2-5$^{Cre}$* (red) overlap in syncytiotrophoblasts of the placental labyrinth, whilst sinusoidal giant cells do not express *Nkx2-5$^{Cre}$* at E10.5; (**E**) Colocalization of *Nkx2-5$^{Cre}$* (red) expression with cytokeratin 7 (green) at E10.5. (**F**) Hand1 is co-expressed in a subset of ventricular cardiomyocytes at E9.5. Arrowhead is trophoblast progenitor cell. Arrow is syncytiotrophoblast. Asterisk is sinusoidal giant cell. ch = chorion. Scale bar = 50 μm (**A–E**) and 100 μm (**F**).

### 2.2. Hand1 Disruption under Nkx2-5-Cre Results in Fetal Demise

*Hand1$^{A126FS/+}$* females mated with *Nkx2-5$^{Cre}$* males were sacrificed at E8.5, E9.5, E10.5, E12.5 and E14.5, and fetuses and placentas collected. The expected ratio of *Nkx2-5$^{Cre}$;Hand1$^{+/+}$* to *Nkx2-5$^{Cre}$;Hand1$^{A126fs/+}$* was 50% each. While *Nkx2-5$^{Cre}$;Hand1$^{+/+}$* littermates showed consistent survival at all timepoints, marked fetal demise occurred after E10.5 in *Nkx2-5$^{Cre}$;Hand1$^{A126fs/+}$* fetuses (Figure 2) and none survived to E14.5. *Nkx2-5$^{Cre}$;Hand1$^{A126fs/+}$* fetuses were overrepresented at E10.5 (55%); however, they showed a marked decrease at E12.5 (36%) and no viable fetuses were recovered at E14.5. Resorptions were not genotyped due to a lack of available fetal tissue.

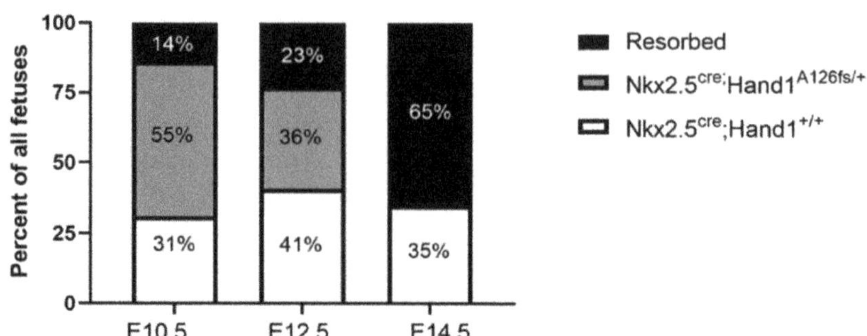


**Figure 2.** Percentages of resorptions (black), *Nkx2-5$^{Cre}$;Hand1$^{A126fs/+}$* (grey) and *Nkx2-5$^{Cre}$; Hand1$^{+/+}$* (white) embryos in litters recovered at embryonic day 10.5 (E10.5), E12.5 and E14.5. From E10.5, embryos with the *Nkx2-5$^{Cre}$;Hand1$^{A126fs/+}$* genotype show increased fetal demise. *Nkx2-5$^{Cre}$;Hand1$^{A126fs/+}$* fetuses were overrepresented at E10.5 (55%); however, they showed a marked decrease at E12.5 (36%) and no viable fetuses were recovered at E14.5. *n* = 4–8 litters.

### 2.3. Early Placental Morphogenesis Is Impaired in the Nkx2-5$^{cre}$;Hand1$^{A126fs/+}$ Implantation Sites

Lack of viable *Nkx2-5$^{Cre}$;Hand1$^{A126fs/+}$* fetuses by E14.5 indicated impaired early placental morphogenesis. Analysis of control litter mate implantation sites at E9.5 revealed invagination and folding of the chorion with scattered Hand1-positive labyrinthine progenitor trophoblasts and a single layer of trophoblast giant cells separating labyrinth from decidua (Figure 3A). Conversely, *Nkx2-5$^{Cre}$;Hand1$^{A126fs/+}$* implantation sites had a disorganized layer of Hand1-positive trophoblast giant cells but lacked Hand1-positive labyrinthine progenitor trophoblasts (Figure 3B). While maternal erythrocytes were present in both genotypes in the developing labyrinthine area, maternal blood spaces were dilated in the conditional activated compared to control littermates and structural integrity of the area disrupted. In addition, yolk sac morphology appeared abnormal in *Nkx2-5$^{Cre}$;Hand1$^{A126fs/+}$* fetuses at E9.5. The yolk sac was adjacent to the chorionic plate in the control littermates (Figure 3C), but was separated from the trophoblast layer in *Nkx2-5$^{Cre}$;Hand1$^{A126fs/+}$* fetuses (Figure 3D).

### 2.4. Nkx2-5$^{Cre}$;Hand1$^{A126fs/+}$ Placentas Exhibit Failed Labyrinthine Vascularization and Syncytiotrophoblast Differentiation

Further analysis to identify trophoblast (cytokeratin 7) and fetal vessels (CD-31) in the *Nkx2-5$^{Cre}$;Hand1$^{A126fs/+}$* placentas at E10.5 and E12.5 showed developmental failure of labyrinthine vasculature. CD-31 positive labyrinthine vessels were visible in the control littermate placentas at E12.5 and E14.5 (Figure 4A,B, respectively). Both syncytium and sinusoidal cytokeratin-7-positive trophoblast giant cells were present in the control labyrinth with clear delineation of fetal vasculature from maternal blood spaces (Figure 4A,B). In contrast, very few labyrinthine blood vessels were observed in the *Nkx2-5$^{Cre}$;Hand1$^{A126fs/+}$* placentas at either timepoint (Figure 4C,D). Labyrinthine vessel density trended lower (*P* = 0.088; Figure 4E) in the conditional activated placentas at E10.5, and by E12.5, vessel counts were significantly lower in *Nkx2-5$^{Cre}$;Hand1$^{A126fs/+}$* labyrinths compared to control (*P* = 0.001; Figure 4F). Additionally, the labyrinths at E10.5 showed disorganization of the trophoblasts and lack of syncytial layers while retaining sinusoidal trophoblast giant cells, demonstrating a lack of differentiation along the syncytiotrophoblast lineage (Figure 4C). By E12.5, *Nkx2-5$^{Cre}$;Hand1A$^{126fs/+}$* placentas had very few remaining cytokeratin-7-positive trophoblasts (Figure 4D). There were no significant sex differences identified in labyrinth morphology, fetal vessel density or survival.

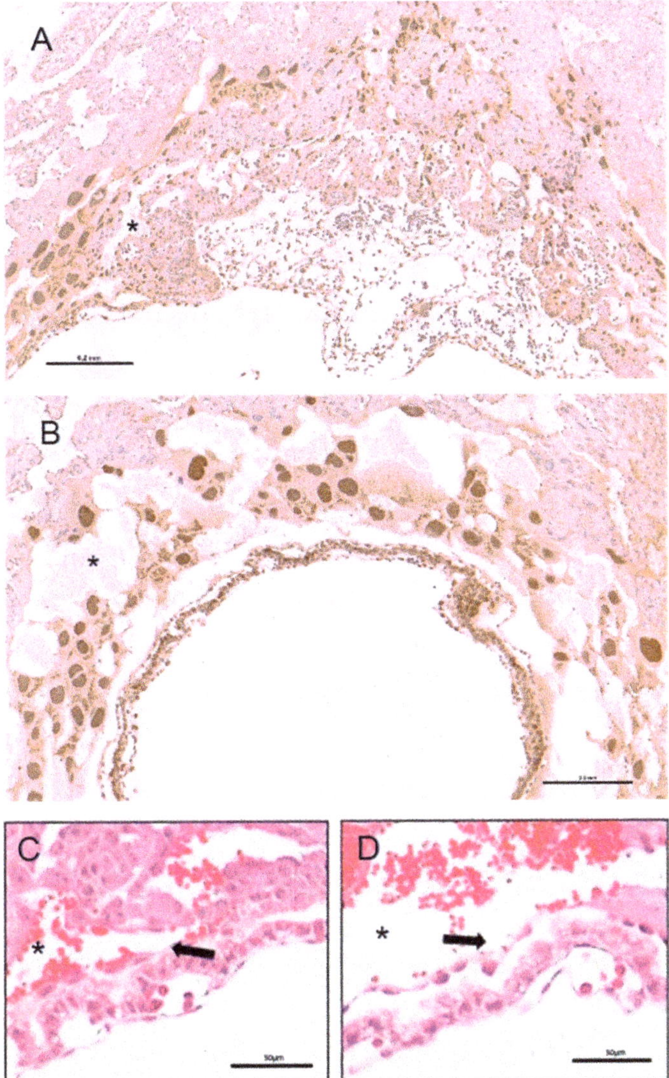

**Figure 3.** Placenta morphogenesis is impaired as early as embryonic day 9.5 (E9.5) in *Nkx2-5^{Cre};Hand1^{A126fs/+}* implantations. (**A**) Control littermate placentas (*Nkx2-5^{Cre};Hand1^{+/+}*) exhibit invagination of the chorion and development of an early labyrinth at E9.5; (**B**) In contrast, *Nkx2-5^{Cre};Hand1^{A126fs/+}* placentas lack a chorionic plate and only have a disorganized layer of trophoblast giant cells surrounding the amniotic cavity; (**C**) High magnification of the yolk sac morphology in the normal littermates show adjacent chorion and yolk sac; (**D**) The *Nkx2-5^{Cre};Hand1^{A126fs/+}* placentas lacked a chorionic layer and yolk sacs were detached from the trophoblast layer separated by large maternal blood spaces. Arrows indicate attachment/detachment of the yolk sac. Asterisk indicate maternal blood spaces. (**A,B**) are immunostained for Hand1; scale bar = 0.2 mm; (**C,D**) are stained with hematoxylin and eosin; scale bar = 50 μm.

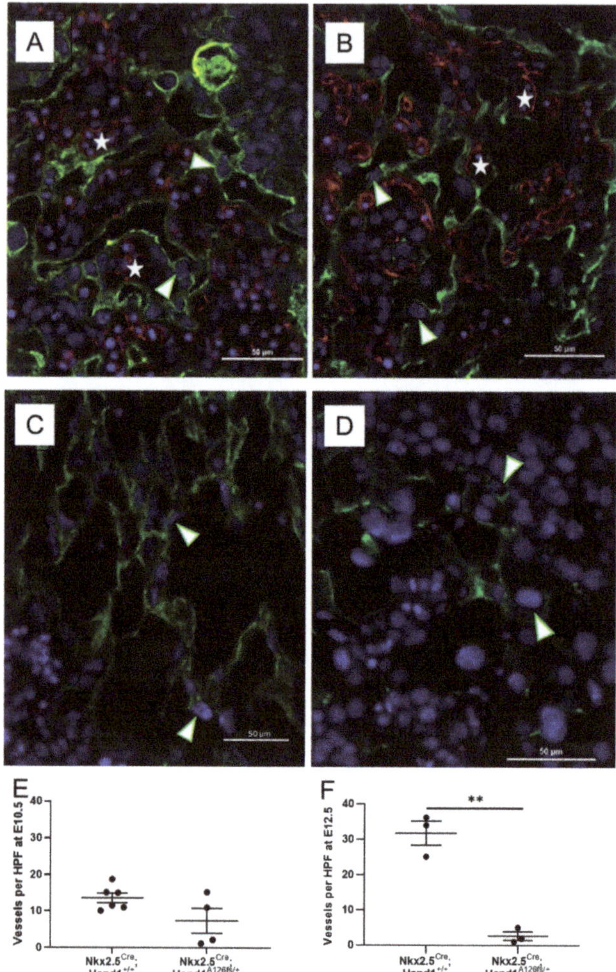

**Figure 4.** Conditional activation of mutant Hand1 in the placenta results in failed labyrinthine vascular development by embryonic day 12.5 (E12.5). (**A,B**) Control (*Nkx2-5$^{Cre}$;Hand1$^{+/+}$*) placentas exhibited normal development of the fetal vasculature with fetal vessels (Red/Star; CD-31 positive) clearly separated from maternal blood space by trophoblast (Green; Cytokeratin-7 positive) at E10.5 (**A**) and E12.5 (**B**); (**C,D**) *Nkx2-5$^{Cre}$;Hand1$^{A126fs/+}$* labyrinths however, had very few fetal labyrinthine blood vessels at both E10.5 (**C**) and E12.5 (**D**) and showed disorganization of the trophoblasts with very few remaining cytokeratin-7-positive trophoblasts remaining at E12.5; (**E**) Whilst not statistically significant, fetal vessel counts at E10.5 confirmed a decreased number of fetal vessels (*P* = 0.088); (**F**) There was significantly reduced by E12.5 (*P* = 0.001). Arrowhead indicates sinusoidal trophoblast giant cells. Nuclei are labelled Blue with DAPI. (**A–D**) are representative images. (**E,F**) individual dots represent average per litter. Data are mean ± SEM. *n* = 3–6 litters. Statistical significance determined using a students *t*-test. ** *P* = 0.001.

## 2.5. Conditional Activation of Mutant Hand1 in Placental Endothelium Resulted in a Reduced Percentage of Cdh5$^{Cre+}$;Hand1$^{A126fs/+}$ Fetuses by E18.5

Having confirmed the requirement for *Hand1* to be expressed in yolk sac and labyrinth trophoblast progenitor cells during early pregnancy on successful development of the labyrinth exchange area, we chose to further assess the requirement for *Hand1* in later pregnancy, examining vascular remodeling and expansion in the *Cdh5$^{Cre+}$;Hand1$^{A126fs/+}$*. Mating the *Cdh5$^{Cre}$* mice with tdTomato reporting mice confirmed Cdh5-Cre and Hand1 expression in the endothelial cells of the placental labyrinth at GD16.5 (Figure 5). Conditional activated *Hand1$^{A126fs/+}$* females were then time mated to hemizygous *Cdh5$^{Cre}$* males to produce *Cdh5$^{Cre+}$;Hand1$^{A126fs/+}$*, *Cdh5$^{Cre+}$;Hand1$^{+/+}$*, *Cdh5$^{Cre-}$;Hand1$^{A126fs/+}$* and *Cdh5$^{Cre-}$;Hand1$^{+/+}$* fetuses. Fetuses and placentas were harvested at E12.5, E16.5 and E18.5. At E12.5, there was a 3% resorption rate which increased to 16% at E16.5 and to 24% at E18.5 (Figure 6). Meanwhile, the percentage of *Cdh5$^{Cre+}$;Hand1$^{A126fs/+}$* fetuses was reduced from 29% at E12.5 to 25% at E16.5 and 20% at E18.5.

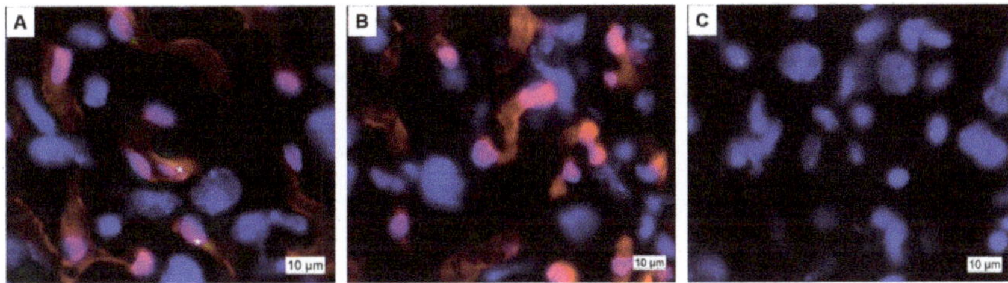


**Figure 5.** Expression of Cdh5-Cre and Hand1 in the placental labyrinth at embryonic day 16.5. (**A**) Cdh5-Cre (red) is expressed in endothelial cells, overlapping with Hand1 protein expression (green); (**B**) Representative negative control for Hand1 immunofluorescence; (**C**) Representative image of tdTomato fluorescence in a *Cdh5$^{Cre-}$* placenta. Asterix indicates fetal capillary.

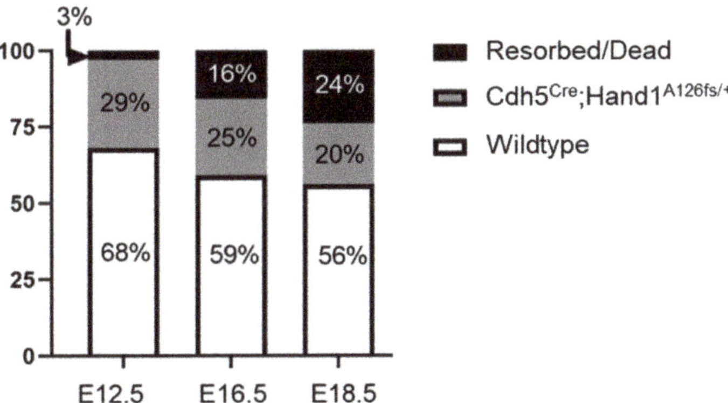


**Figure 6.** Percentage of resorptions (black), *Cdh5$^{Cre+}$;Hand1$^{A126fs/+}$* (grey) and wildtype (white) fetuses in litters recovered at embryonic day 12.5 (E12.5), E16.5 and E18.5. As gestation progressed, the percentage of resorptions/dead fetuses increased while the percentage of *Cdh5$^{Cre}$;Hand1$^{A126fs/+}$* fetuses decreased. Numbers in parentheses are the total number of pups at each timepoint. *n* = 3–8 dams.

### 2.6. Fetal and Placental Weights Were Reduced in Cdh5$^{cre}$; Hand1$^{A126fs/+}$ by E18.5

At E16.5, no difference in fetal weight was found between *Cdh5$^{Cre+}$;Hand1$^{A126fs/+}$* and *Cdh5$^{Cre-}$;Hand1$^{+/+}$* littermates ($P > 0.05$; Figure 7A), however, placental weight of the *Cdh5$^{Cre+}$;Hand1$^{A126fs/+}$* fetuses was significantly lower ($P = 0.01$; Figure 7B). By E18.5, both fetal and placental weight were lower in the *Cdh5$^{Cre+}$;Hand1$^{A126fs/+}$* genotype when compared to their *Cdh5$^{Cre-}$;Hand1$^{+/+}$* littermates ($P < 0.001$ and $P = 0.002$; Figure 7C,D, respectively). There was no difference in fetal or placental weight between males and females in either genotype.

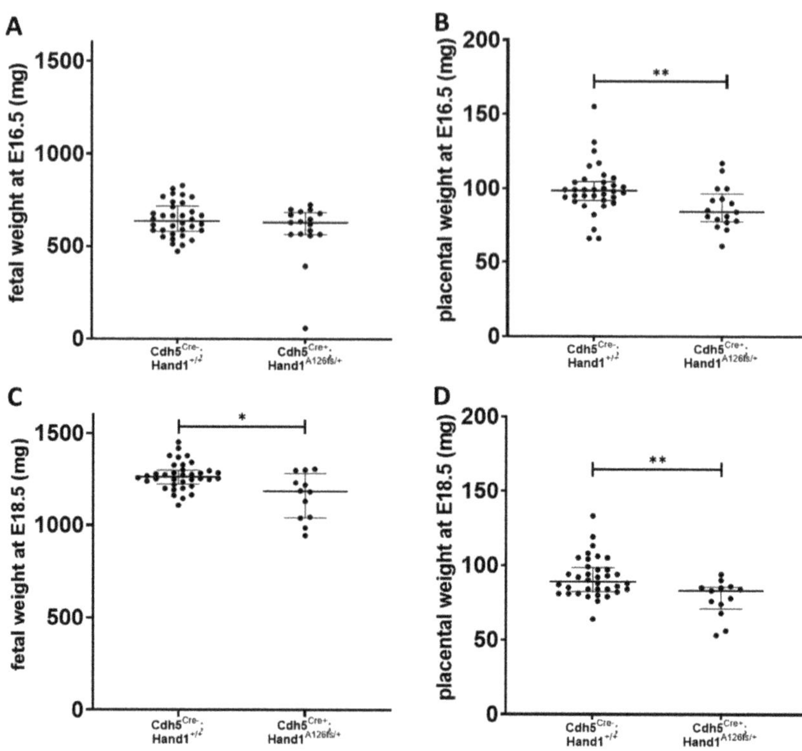

**Figure 7.** Effect of conditionally activating mutant Hand1 in placental fetal endothelium in late pregnancy. (**A**) At embryonic day 16.5 (E16.5) fetal weight was not different between $Cdh5^{Cre+};Hand1^{A126fs/+}$ and control ($Cdh5^{Cre-};Hand1^{+/+}$) littermates; (**B**) However, placental weight at E16.5 was significantly lower in the $Cdh5^{Cre+};Hand1^{A126fs/+}$ fetuses; (**C,D**) By E18.5, both fetal (**C**) and placental (**D**) weight was significantly lower in the $Cdh5^{Cre+};Hand1^{A126fs/+}$ fetuses when compared to the control littermates. Data are median ± interquartile range. Individual dots represent individual fetuses or placentas. $n = 8$ litters per time point. Statistical significance determined using a Mann–Whitney test. * $P < 0.05$; ** $P < 0.01$.

*2.7. Labyrinthine Vessel Density and Angiogenic Factor mRNA Expression Was Significantly Reduced in the Cdh5$^{Cre+/-}$;Hand1$^{A126fs/+}$ Placentas*

Using immunohistochemistry, CD-31 positive fetal vessels in the placental labyrinth were identified and counted to assess placental vascularization at E16.5 and E18.5. There was no difference in in the number of labyrinth vessels in the placenta between $Cdh5^{Cre+};Hand1^{A126fs/+}$ and littermate controls at E16.5 (Figure 8A). However, by E18.5, there was a significant reduction in the number of vessels in the labyrinth of the $Cdh5^{Cre+};Hand1^{A126fs/+}$ fetuses ($P < 0.05$; Figure 8B). Whilst vascular density was not reduced in the $Cdh5^{Cre+};Hand1^{A126fs/+}$ placentas at E16.5, there was a reduction in the mRNA expression of *Angiopoietin 1* (*Angpt1*) and *Angpt1 Receptor* (*Tie2*) in the placentas of the $Cdh5^{Cre+};Hand1^{A126fs/+}$ at E16.5 when compared to littermate controls ($P < 0.001$ and $P = 0.012$; Figure 9A,B, respectively). mRNA expression of *Angpt2*, *Vascular endothelial growth factor α* (*Vegfα*) and *Placenta growth factor* (*Plgf*) were also assessed however, there was no difference between the genotypes (Data not shown).

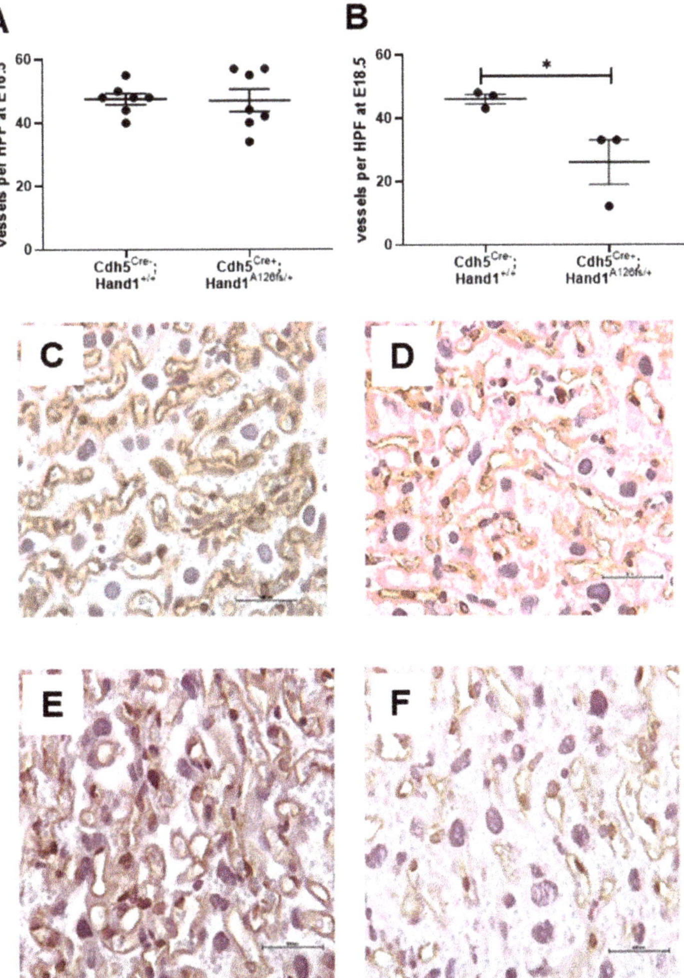

**Figure 8.** Fetal vessel density in the placenta labyrinth at embryonic day 16.5 (E16.5) and E18.5 between wildtype (*Cdh5$^{Cre-}$;Hand1$^{+/+}$*) and *Cdh5$^{Cre+}$;Hand1$^{A126fs/+}$* placentas. (**A**) At E16.5 there was no difference in the number of fetal vessels per high powered field (HPF) between *Cdh5$^{Cre-}$;Hand1$^{+/+}$* and *Cdh5$^{Cre+}$;Hand1$^{A126fs/+}$* placentas; (**B**) At E18.5, the number of fetal vessels per HPF was reduced in the labyrinth of the *Cdh5$^{Cre+}$;Hand1$^{A126fs/+}$* placentas compared to wildtype; (**C,D**) are representative images at E16.5 and E18.5 of a *Cdh5$^{Cre-}$;Hand1$^{+/+}$* placenta, respectively. (**E,F**) are representative images at E16.5 and E18.5 of a *Cdh5$^{Cre+}$;Hand1$^{A126fs/+}$* placenta, respectively. Data are mean $\pm$ SEM; (**A,B**) individual dots represent average per litter. $n$ = 3–4 litters per time point. Statistical significance determined using a students *t*-test. * $P < 0.05$. CD-31 staining (brown) with hematoxylin counterstain. Scale bars = 500 μm.

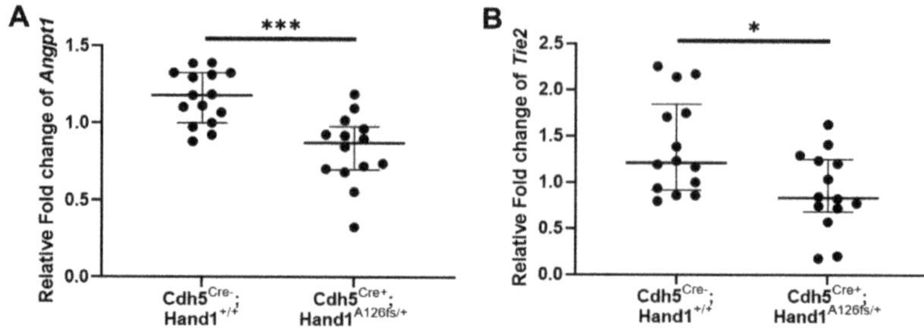

**Figure 9.** Analysis of angiogenic factor *Angiopoietin 1* (*Angpt1*) and *Angpt1 Receptor* (*Tie2*) mRNA expression in wildtype (*Cdh5*$^{Cre-}$;*Hand1*$^{+/+}$) and *Cdh5*$^{Cre+}$;*Hand1*$^{A126fs/+}$ placentas at embryonic day 16.5. mRNA expression of both Angpt1 (**A**) and Tie2 (**B**) was reduced in the *Cdh5*$^{Cre+}$;*Hand1*$^{A126fs/+}$ placentas compared to wildtype. Expression was not different between the fetal sexes for either genotype. Data are median ± interquartile range. Individual dots represent individual placentas. *n* = 3–6 litters per time point. Statistical significance determined using a Mann–Whitney test. * *P* < 0.05, *** *P* < 0.001.

## 3. Discussion

CHDs often require complex surgeries at birth to correct the defect, and one critical predictor of survival is birth weight [2]. In pregnancies complicated by fetal CHD, abnormalities of placental development and function likely contribute to the growth restriction and prematurity that negatively impact clinical outcomes [3–5,10,11]. Nearly 400 genes known to be associated with CHD [17–19]. However, many of the more in-depth studies, aimed at determining the particular contribution of the genes or environment to cardiac defects, overlook the potential disruption to extraembryonic development that is crucial for maintaining the appropriate in utero environment and fetal organ development and growth. In the present study, we show the effect of targeted loss of *Hand1*, a gene known to be crucial in heart development, on placental development throughout gestation. Targeted loss of *Hand1* in chorionic and labyrinthine progenitor trophoblasts at E8.5-9.5 led to abnormal formation of the placental labyrinth from both a trophoblast and endothelial perspective, ultimately contributing to embryonic lethality from E12.5. Interestingly, the loss of *Hand1* in labyrinthine endothelial cells from E12.5 resulted in reduced vascular expansion and remodeling associated with reduced fetal and placental weights, and elevated risk of fetal demise by near-term.

*Hand1* has been shown to be expressed during the differentiation of human trophoblast stem cells to syncytiotrophoblast cells in early pregnancy [33]. Interestingly at term, neither isolated human placental cytotrophoblasts nor villous tissue express HAND1 [34]. However, no studies have identified if vascular remodeling throughout gestation requires HAND1. In our present study, targeted early pregnancy loss of *Hand1* in chorionic and labyrinthine progenitor trophoblasts resulted in embryonic lethality after E12.5. Co-expression of the Nkx2-5 reporter and Hand1 was limited to labyrinthine progenitor trophoblasts at gestation day 8.5 and expanded to include chorion by E9.5 and syncytium by E10.5. It was previously demonstrated that in *Hand1*-null mice, trophoblast stem cells fail to differentiate into trophoblast giant cells (TGCs) and exhibited reduced invasion, failed placentation and fetal demise around E6.5 [35], however in our study, *Hand1* expression was maintained in the TGCs of *Nkx2-5*$^{Cre}$;*Hand1*$^{A126fs/+}$ placentas allowing us to study labyrinth components and development. Our histological analyses indicate that *Nkx2-5*$^{Cre}$-driven disruption to *Hand1* does not significantly impact differentiation of the trophoblast stem cells into TGCs (sinusoidal, canal, and parietal) for establishment of the placenta, but does disrupt labyrinth formation and syncytialization. In mice, at E10-10.5 there is a switch in nutrient and oxygen supply mechanisms whereby support for continued embryonic survival changes from histiotrophic to hemotrophic [36]. In the case of the *Nkx2-5$^{Cre}$;Hand1$^{A126fs/+}$* placentas, failure to form the appropriate layers of the exchange area of the placenta within the labyrinth (syncytium and vasculature) by E10/10.5 ultimately prevented this switch. While subsets of trophoblast cells within the labyrinth were still capable of receiving oxygen and nutrients carried in blood into maternal blood spaces, failure of syncytial layers and inadequate vascularization of the labyrinth resulted in a failure to transfer nutrients and oxygen to the fetal circulation, resulting in fetal demise.

Similar to targeted loss of *Hand1* in early pregnancy, targeted loss in the placental endothelial cells after E12.5 resulted in disrupted placental vascular expansion and re-modeling. Altering *Hand1* expression in endothelium did not appear to cause early fetal loss, as survival at E12.5 was not significantly lower than the spontaneous loss levels in this mouse strain. However, later-gestation fetal demise was significantly higher in the *Cdh5$^{Cre+}$;Hand1$^{A126fs/+}$* fetuses compared to wildtype littermates, pointing to a role for *Hand1* in placental vascular remodeling and expansion of the placental labyrinth. The expansion of the labyrinth and increase in branching of the labyrinthine vasculature is vital to support the exponential fetal growth phase of late pregnancy and it's failure, as we observed from E16.5 in the *Cdh5$^{Cre+}$;Hand1$^{A126fs/+}$*, leads to impaired fetal growth and in some cases late term fetal demise. Despite no differences in placental fetal vessel density at E16.5, we did identify reductions in the expression of *Angpt1* and *Tie2*. *Angpt1* is a growth factor which predominantly acts on endothelial cells, and whose signaling through its receptor *Tie2* is capable of various vascular shaping functions including maintaining vascular integrity, vessel remodeling, cell migration and as anti-inflammatory potential [37]. Further studies, beyond the scope of the present one are required, but reduced expression of *Angpt1* and *Tie2*, as well as reduced labyrinthine vessel density at E18.5 does provide strong evidence that *Hand1* expression in placental endothelial cells is required for the angiogenic remodeling required to support fetal growth in late pregnancy.

Complete knockout of *Hand1* results in embryonic lethality [23–26] and given that deletion of *Hand1* in only cardiomyocytes results in fetal survival to term with only mild cardiac phenotypes [28], this suggests that the embryonic lethality is not solely due to issues with cardiac development. Results of the current study support this hypothesis; that abnormal placental development due to *Hand1* disruption plays a primary role in the fetal demise in the *Hand1*-null mouse model. The role that the placenta may play in con-tributing to, or exacerbating the development of CHD, remains understudied. Many genes previously associated with CHD [17,20] have not been adequately investigated in placental development or function. The Deciphering the Mechanisms of Developmental Disorders (DMDD) mouse screen [22], and our study of placental gene expression in human CHD placenta samples identifying multiple 'heart-specific' pathways [15,38], underscore the importance of understanding the roles of developmental genes shared between placenta and heart. The dual impact of a genetic disruption to the development of embryonic (heart) and extra-embryonic (placenta, yok sac) organs, plus in utero environmental disruptions in oxygen and nutrient supply secondary to abnormal placenta development, may pro-vide mechanisms that underlie early fetal loss, and most importantly those that underlie the high rate of miscarriage in humans associated with CHD directly or in prior/future pregnancy [39].

This study presents clear evidence that placental development is critical in driving fetal growth and survival in the context of CHD. One limitation to the study is that mouse and human placenta structure and development differs substantially [40], hence there are implications for the translatability of the findings to human pregnancy; although the placenta cell types in which *Hand1* was manipulated are present in both mouse and human placenta. Additionally, the manipulation of *Hand1* expression occurred either before or after establishment of the placental labyrinth (E10.5) and as such, future studies are required in order to fully understand the implications at this critical time in placental development.

In conclusion, by assessing placental development in the setting of a previously developed *Hand1* mutation known to result in cardiac defects [28], we have begun to explore the mechanisms that may result in adverse pregnancy outcomes in the setting of

CHD. Disruption of critical stages in placental development and function could contribute to the known clinical observations of stillbirth, fetal growth restriction, and prematurity. Our data highlights the necessity for future research to not only consider the contribution of genetic manipulation to the organ of interest but also the extraembryonic tissue and the in utero environment resulting from its disruption. By doing so, we can then better understand the broad overlap of placental and cardiac development which may ultimately drive novel therapies to improve outcomes for children with CHD.

## 4. Materials and Methods

### 4.1. Animal Procedures

All animal procedures were performed under protocols approved by the Institutional Animal Care and Use Committee of CCHMC (IACUC 2018-0065, 12 November 2018). Mice were housed and maintained under temperature and humidity-controlled environments on a 12:12 h light cycle with water and food provided ad libitum.

### 4.2. Cre Expression and Validation

To verify to efficacy and spaciotemporal expression of the $Nkx2\text{-}5^{IRESCre}$ (Jackson Laboratory; 024637) [31] and $Cdh5^{cre}$ (Jackson Laboratory, 017968), homozygous $Nkx2\text{-}5^{IRESCre}$ males or $Cdh5^{Cre}$ males were crossed with homozygous B6.Cg-$Gt(ROSA)26Sor^{tm14(CAG\text{-}tdTomato)Hze}$/J female mice (tdTomato: Jackson Laboratory; 007914). Fetoplacental units were then collected at E8.5, E9.5 and E10.5 for the $Nkx2\text{-}5^{Cre}$, and E14.5 and E16.5 for the $Cdh5^{Cre}$, and processed for cryopreservation and frozen-sectioning. Placentas and embryos were fixed in 4% paraformaldehyde (PFA) with 2.5% polyvinylpyrrolidone (PVP), in phosphate-buffered saline (PBS) for 4 h at room temperature on a rocker plate. Cryoprotecting was achieved by rinsing the fixed tissue in PBS before placing in a 30% sucrose solution until fully infused, then embedding the tissue in OCT and storing at −80 °C. At time of sectioning, blocks were warmed to −20 °C and 7 μm sections were obtained using a cryotome (Leica, Wetzlar, Germany) before being mounted onto slides for immunofluorescent analysis.

### 4.3. Mice Mating and Genotyping

Conditionally activated $Hand1^{A126FS/+}$ [28] females were time mated with homozygous $Nkx2\text{-}5^{IREScre}$ males by placing the breeding pair together overnight and the following morning was designated GD0.5. The mating strategy produced litters containing $Nkx2\text{-}5^{cre};Hand1^{A126FS/+}$ and $Nkx2\text{-}5^{cre};Hand1^{+/+}$ embryos. For the Cdh5 study, conditional activated $Hand1^{A126fs/+}$ females were time mated to hemizygous $Cdh5^{cre}$ males to produce $Cdh5^{Cre+};Hand1^{A126fs/+}$, $Cdh5^{Cre+};Hand1^{+/+}$, $Cdh5^{Cre-};Hand1^{A126fs/+}$, and $Cdh5^{Cre-};Hand1^{+/+}$ fetuses. Females were sacrificed via carbon dioxide ($CO_2$) asphyxiation and embryos and placentas weighed and collected at E8.5, E9.5, E10.5, E12.5, and E14.5 (Nkx2-5 study) and E12.5, E16.5 and E18.5 (Cdh5 study). At E8.5 and E9.5, implantation sites were collected, fixed and paraffin embedded for histological analysis. At E10.5 to E18.5, placentas were halved with one half fixed for histology and the other flash-frozen in liquid nitrogen and stored at −80 °C for molecular analyses.

Genotyping was performed on all feto-placental units within the litter by removing part of the yolk sac (E8.5, E9.5 and E10.5) or clipping the tail of fetuses (E12.5, E14.5, E16.5 and E18.5). Tissue was digested in 72.75 μL Direct PCR lysis buffer and 2.25 μL Proteinase K for 4 h at 56 °C, vortexed, then denatured for 30 min at 85 °C. PCR was then performed on 1 μg extracted DNA using FastStart PCR Master (Roche) and primers specific to *Hand1* (Supplementary Table S1) under the following conditions: 95 °C for 4 min, 35 cycles of 95 °C × 30 s, 65 °C × 30 s, 72 °C × 1 min, and 72 °C for 7 min. To detect the frameshift mutation at the *Hand1* locus, the *Hand1* PCR product was digested with the restriction enzyme FauI (R0651, New England Biolabs, Ipswich, MA, USA) following standard protocol using 1 μL of restriction enzyme with 1.5 μL of PCR product, 2 μL CutSmart buffer, and 20 μL RNAse-free water and visualized on a 1% agarose gel. Fetal sex determination was performed using PCR following the protocol outlined in [41] with primers provided in Supplementary Table S1.

*4.4. RNA Expression via Reverse Transcription Quantitative Polymerase Chain Reaction(qPCR)*

For RNA extraction and qPCR gene expression, flash-frozen half placentas at E10.5 (Nkx2-5 study) and E16.5 (Cdh5 study) were placed into RLT lysis buffer (Qiagen, Hilden, Germany) and homogenized. RNA was extracted using a RNAEasy Mini kit (Qiagen) following standard protocol, and quantification and quality control assessed using Nanodrop Spectrophotometer (Thermo Fisher, Waltham, MA, USA). From each sample 1 µg of RNA was converted to cDNA using the Applied Biosystems high-capacity cDNA conversion kit per protocol. cDNA was stored at $-20$ °C. mRNA expression levels of *Angpt1*, *Angpt2*, *Vegfα*, *Plgf* and *Tie2* were assayed in 25 µL SYBR Green (Applied Biosystems, Waltham, MA, USA) PCR Master Mix reactions containing 1/40th of the cDNA template and 300 nm/L of forward and reverse primer (Table S1). qPCR was performed in triplicate using the Applied Biosystems StepOne Plus Real-Time PCR System. Expression was normalized using the housekeeping gene *Rps20*. Relative quantification expression levels were calculated by comparative CT method using StepOne software v2.3 (Applied Biosystems).

*4.5. Immunohistochemistry and Immunofluorescence*

For paraffin-embedded implantation sites at E8.5 and E9.5, blocks were systematically sectioned until mid-sagittal placenta tissue could be identified and then 5 µm serial sections were obtained. Paraffin-embedded placenta tissue at E10.5 through to E18.5 were also serial sectioned at 5 µm. To assess gross morphology at E9.5 hematoxylin and eosin staining was performed. Briefly, sections were deparaffinized, rehydrated, placed in hematoxylin for 45 s, washed in running tap water, placed in 80% ethanol, dipped 3 times in eosin followed by dehydration, clearing and mounting in xylene-based mounting medium.

Immunohistochemistry for Hand1 was performed on sections of mouse placenta at E9.5 in the Nkx2-5 study, and for CD-31 at E10.5 and E12.5 in the NKx2-5 study and at E16.5 and E18.5 in the Cdh5 study. Sections were deparaffinized, rehydrated, and antigen retrieval was performed using Target Retrieval Solution (Dako, Santa Clara, CA, USA). Endogenous peroxidase activity was blocked using 3% hydrogen peroxide. Nonspecific binding was blocked with a 10% serum solution corresponding to the secondary antibody host species followed by overnight incubation with the primary antibody (antibodies and dilutions outlined Supplementary Table S1). Sections were then washed and incubated with appropriate biotinylated secondary antibodies. Antibody binding was amplified using the avidin-biotin complex (ABC) kit (Vector Laboratories, Inc., Burlingame, CA, USA) and visualized using the enzyme substrate DAB (Vector Laboratories, Inc.). Sections were counterstained briefly in hematoxylin before dehydration, clearing and mounting. Protein localization was observed by light microscopy (Nikon, Tokyo, Japan). In order to assess the number of fetal vessels within the placental labyrinth, two CD-31 stained sections from each placenta were analyzed. Ten high-powered images at least 50 µm apart across the placental labyrinth were obtained, the number of fetal vessels counted manually and then averaged to obtain a number per high powered field (HPF).

To verify the efficacy and spaciotemporal expression of Nkx2-5 and Cdh5, placenta samples from pregnancies between *Nkx2-5* and *Cdh5* males and tdTomato reporting females were stained for Hand1 or cytokeratin 7 (*Nkx2-5* study only) using immunofluorescence. Sections were treated as described for immunohistochemistry, incubated with an appropriate fluorochrome-conjugated secondary antibody and nuclei counter stained with Dapi. For the Nkx-2-5 study, representative images were obtained using the Nikon Eclipse 80i fluorescent microscope. For the Cdh5 study, representative images were obtained using the Axioscan (Zeiss, Oberkochen, Germany).

To assess placental microstructure in the Nkx2-5 study, double-label immunofluorescence for cytokeratin 7 and CD-31 was performed on sections of mouse placental at E9.5, E10.5 and E12.5. Sections were treated as described for immunohistochemistry until the primary antibody was applied. Following, primary antibody incubation, sections were washed, incubated with appropriate fluorochrome-conjugated secondary antibodies and counterstained with DAPI before being mounted with antifade mounting media. Fluorescence microscopy was performed using the Nikon Eclipse 80i fluorescent microscope.

*4.6. Statistical Analysis and Data Presentation*

Statistical analysis was performed using Prism GraphPad (v8.4.3). Shapiro–Wilk test was used to test for normality in the data. Statistical significance in data which passed the normality test was determined using student's *t*-test and expressed as mean $\pm$ SEM. Statistical significance in data which did not pass the normality test was determined using Mann–Whitney test and expressed as median $\pm$ interquartile range. n number refers to the number of litters. Level of significance (*P*-value) was defined as $P \leq 0.05$.

**Supplementary Materials:** The following are available online at https://www.mdpi.com/article/10.3390/ijms22179532/s1.

**Author Contributions:** Conceptualization, J.A.C. and H.N.J.; methodology, J.A.C. and R.L.W.; formal analysis, J.A.C.; writing—original draft preparation, J.A.C.; writing—review and editing, R.L.W., J.C. and H.N.J.; supervision, J.C. and H.N.J.; funding acquisition, H.N.J. All authors have read and agreed to the published version of the manuscript.

**Funding:** This research was funded by Cincinnati Children's Hospital Medical Center Fetal Research Gift Fund.

**Institutional Review Board Statement:** All animal procedures were performed under protocols approved by the Institutional Animal Care and Use Committee of CCHMC (2018-0065).

**Informed Consent Statement:** Not applicable.

**Data Availability Statement:** Data supporting these results can be provided upon request to corresponding author.

**Conflicts of Interest:** The authors declare no conflict of interest.

# References

1. van der Linde, D.; Konings, E.E.; Slager, M.A.; Witsenburg, M.; Helbing, W.A.; Takkenberg, J.J.; Roos-Hesselink, J.W. Birth prevalence of congenital heart disease worldwide: A systematic review and meta-analysis. *J. Am. Coll. Cardiol.* **2011**, *58*, 2241–2247. [CrossRef] [PubMed]
2. Best, K.E.; Tennant, P.W.; Rankin, J. Survival, by birth weight and gestational age, in individuals with congenital heart disease: A population-based study. *J. Am. Heart Assoc.* **2017**, *6*, e005213. [CrossRef]
3. Cnota, J.F.; Hangge, P.T.; Wang, Y.; Woo, J.G.; Hinton, A.C.; Divanovic, A.A.; Michelfelder, E.C.; Hinton, R.B. Somatic growth trajectory in the fetus with hypoplastic left heart syndrome. *Pediatr. Res.* **2013**, *74*, 284–289. [CrossRef] [PubMed]
4. Puri, K.; Warshak, C.R.; Habli, M.A.; Yuan, A.; Sahay, R.D.; King, E.C.; Divanovic, A.; Cnota, J.F. Fetal somatic growth trajectory differs by type of congenital heart disease. *Pediatr. Res.* **2018**, *83*, 669–676. [CrossRef] [PubMed]
5. Ruiz, A.; Ferrer, Q.; Sanchez, O.; Ribera, I.; Arevalo, S.; Alomar, O.; Mendoza, M.; Cabero, L.; Carreras, E.; Llurba, E. Placenta-related complications in women carrying a foetus with congenital heart disease. *J. Matern.-Fetal Neonatal. Med.* **2016**, *29*, 3271–3275. [CrossRef]
6. Auger, N.; Fraser, W.D.; Healy-Profitos, J.; Arbour, L. Association between Preeclampsia and Congenital Heart Defects. *JAMA* **2015**, *314*, 1588–1598. [CrossRef]
7. Boyd, H.A.; Basit, S.; Behrens, I.; Leirgul, E.; Bundgaard, H.; Wohlfahrt, J.; Melbye, M.; Oyen, N. Association Between Fetal Congenital Heart Defects and Maternal Risk of Hypertensive Disorders of Pregnancy in the Same Pregnancy and Across Pregnancies. *Circulation* **2017**, *136*, 39–48. [CrossRef] [PubMed]
8. Brodwall, K.; Leirgul, E.; Greve, G.; Vollset, S.E.; Holmstrom, H.; Tell, G.S.; Oyen, N. Possible Common Aetiology behind Maternal Preeclampsia and Congenital Heart Defects in the Child: A Cardiovascular Diseases in Norway Project Study. *Paediatr. Perinat. Epidemiol.* **2016**, *30*, 76–85. [CrossRef]
9. Llurba, E.; Sanchez, O.; Ferrer, Q.; Nicolaides, K.H.; Ruiz, A.; Dominguez, C.; Sanchez-de-Toledo, J.; Garcia-Garcia, B.; Soro, G.; Arevalo, S.; et al. Maternal and foetal angiogenic imbalance in congenital heart defects. *Eur. Heart J.* **2014**, *35*, 701–707. [CrossRef]
10. Laas, E.; Lelong, N.; Thieulin, A.C.; Houyel, L.; Bonnet, D.; Ancel, P.Y.; Kayem, G.; Goffinet, F.; Khoshnood, B.; Group, E.S. Preterm birth and congenital heart defects: A population-based study. *Pediatrics* **2012**, *130*, e829–e837. [CrossRef]

11. Tararbit, K.; Lelong, N.; Goffinet, F.; Khoshnood, B.; Group, E.S. Assessing the risk of preterm birth for newborns with congenital heart defects conceived following infertility treatments: A population-based study. *Open Heart* **2018**, *5*, e000836. [CrossRef] [PubMed]
12. Jorgensen, M.; McPherson, E.; Zaleski, C.; Shivaram, P.; Cold, C. Stillbirth: The heart of the matter. *Am. J. Med. Genet. A* **2014**, *164A*, 691–699. [CrossRef]
13. Burton, G.J.; Fowden, A.L. The placenta: A multifaceted, transient organ. *Philos. Trans. R. Soc. Lond. Ser. B Biol. Sci.* **2015**, *370*, 20140066. [CrossRef] [PubMed]
14. Burton, G.J.; Jauniaux, E. Development of the Human Placenta and Fetal Heart: Synergic or Independent? *Front. Physiol.* **2018**, *9*, 373. [CrossRef] [PubMed]
15. Courtney, J.; Troja, W.; Owens, K.J.; Brockway, H.M.; Hinton, A.C.; Hinton, R.B.; Cnota, J.F.; Jones, H.N. Abnormalities of placental development and function are associated with the different fetal growth patterns of hypoplastic left heart syndrome and transposition of the great arteries. *Placenta* **2020**, *101*, 57–65. [CrossRef] [PubMed]
16. Jones, H.N.; Olbrych, S.K.; Smith, K.L.; Cnota, J.F.; Habli, M.; Ramos-Gonzales, O.; Owens, K.J.; Hinton, A.C.; Polzin, W.J.; Muglia, L.J.; et al. Hypoplastic left heart syndrome is associated with structural and vascular placental abnormalities and leptin dysregulation. *Placenta* **2015**, *36*, 1078–1086. [CrossRef]
17. Jin, S.C.; Homsy, J.; Zaidi, S.; Lu, Q.; Morton, S.; DePalma, S.R.; Zeng, X.; Qi, H.; Chang, W.; Sierant, M.C.; et al. Contribution of rare inherited and de novo variants in 2,871 congenital heart disease probands. *Nat. Genet.* **2017**, *49*, 1593–1601. [CrossRef]
18. Homsy, J.; Zaidi, S.; Shen, Y.; Ware, J.S.; Samocha, K.E.; Karczewski, K.J.; DePalma, S.R.; McKean, D.; Wakimoto, H.; Gorham, J.; et al. De novo mutations in congenital heart disease with neurodevelopmental and other congenital anomalies. *Science* **2015**, *350*, 1262–1266. [CrossRef]
19. Sifrim, A.; Hitz, M.P.; Wilsdon, A.; Breckpot, J.; Turki, S.H.; Thienpont, B.; McRae, J.; Fitzgerald, T.W.; Singh, T.; Swaminathan, G.J.; et al. Distinct genetic architectures for syndromic and nonsyndromic congenital heart defects identified by exome sequencing. *Nat. Genet.* **2016**, *48*, 1060–1065. [CrossRef]
20. Zaidi, S.; Brueckner, M. Genetics and Genomics of Congenital Heart Disease. *Circ. Res.* **2017**, *120*, 923–940. [CrossRef] [PubMed]
21. Chapman, G.; Moreau, J.L.M.; Ip, E.; Szot, J.O.; Iyer, K.R.; Shi, H.; Yam, M.X.; O'Reilly, V.C.; Enriquez, A.; Greasby, J.A.; et al. Functional genomics and gene-environment interaction highlight the complexity of congenital heart disease caused by Notch pathway variants. *Hum. Mol. Genet.* **2020**, *29*, 566–579. [CrossRef]
22. Perez-Garcia, V.; Fineberg, E.; Wilson, R.; Murray, A.; Mazzeo, C.I.; Tudor, C.; Sienerth, A.; White, J.K.; Tuck, E.; Ryder, E.J.; et al. Placentation defects are highly prevalent in embryonic lethal mouse mutants. *Nature* **2018**, *555*, 463–468. [CrossRef]
23. Firulli, A.B.; McFadden, D.G.; Lin, Q.; Srivastava, D.; Olson, E.N. Heart and extra-embryonic mesodermal defects in mouse embryos lacking the bHLH transcription factor Hand1. *Nat. Genet.* **1998**, *18*, 266–270. [CrossRef] [PubMed]
24. Cserjesi, P.; Brown, D.; Lyons, G.E.; Olson, E.N. Expression of the novel basic helix-loop-helix gene eHAND in neural crest derivatives and extraembryonic membranes during mouse development. *Dev. Biol.* **1995**, *170*, 664–678. [CrossRef]
25. Riley, P.; Anson-Cartwright, L.; Cross, J.C. The Hand1 bHLH transcription factor is essential for placentation and cardiac morphogenesis. *Nat. Genet.* **1998**, *18*, 271–275. [CrossRef] [PubMed]
26. Morikawa, Y.; Cserjesi, P. Extra-embryonic vasculature development is regulated by the transcription factor HAND1. *Development* **2004**, *131*, 2195–2204. [CrossRef]
27. Barnes, R.M.; Firulli, B.A.; Conway, S.J.; Vincentz, J.W.; Firulli, A.B. Analysis of the Hand1 cell lineage reveals novel contributions to cardiovascular, neural crest, extra-embryonic, and lateral mesoderm derivatives. *Dev. Dyn.* **2010**, *239*, 3086–3097. [CrossRef] [PubMed]
28. Firulli, B.A.; Toolan, K.P.; Harkin, J.; Millar, H.; Pineda, S.; Firulli, A.B. The HAND1 frameshift A126FS mutation does not cause hypoplastic left heart syndrome in mice. *Cardiovasc. Res.* **2017**, *113*, 1732–1742. [CrossRef]
29. Firulli, B.A.; George, R.M.; Harkin, J.; Toolan, K.P.; Gao, H.; Liu, Y.; Zhang, W.; Field, L.J.; Liu, Y.; Shou, W.; et al. HAND1 loss-of-function within the embryonic myocardium reveals survivable congenital cardiac defects and adult heart failure. *Cardiovasc. Res.* **2020**, *116*, 605–618. [CrossRef] [PubMed]
30. Tanaka, M.; Wechsler, S.B.; Lee, I.W.; Yamasaki, N.; Lawitts, J.A.; Izumo, S. Complex modular cis-acting elements regulate expression of the cardiac specifying homeobox gene Csx/Nkx2.5. *Development* **1999**, *126*, 1439–1450. [CrossRef] [PubMed]
31. Stanley, E.G.; Biben, C.; Elefanty, A.; Barnett, L.; Koentgen, F.; Robb, L.; Harvey, R.P. Efficient Cre-mediated deletion in cardiac progenitor cells conferred by a 3′UTR-ires-Cre allele of the homeobox gene Nkx2-5. *Int. J. Dev. Biol.* **2002**, *46*, 431–439. [PubMed]
32. Tanaka, M.; Chen, Z.; Bartunkova, S.; Yamasaki, N.; Izumo, S. The cardiac homeobox gene Csx/Nkx2.5 lies genetically upstream of multiple genes essential for heart development. *Development* **1999**, *126*, 1269–1280. [CrossRef] [PubMed]
33. Yabe, S.; Alexenko, A.P.; Amita, M.; Yang, Y.; Schust, D.J.; Sadovsky, Y.; Ezashi, T.; Roberts, R.M. Comparison of syncytiotrophoblast generated from human embryonic stem cells and from term placentas. *Proc. Natl. Acad. Sci. USA* **2016**, *113*, E2598–E2607. [CrossRef] [PubMed]
34. Knofler, M.; Meinhardt, G.; Bauer, S.; Loregger, T.; Vasicek, R.; Bloor, D.J.; Kimber, S.J.; Husslein, P. Human Hand1 basic helix-loop-helix (bHLH) protein: Extra-embryonic expression pattern, interaction partners and identification of its transcriptional repressor domains. *Biochem. J.* **2002**, *361 Pt 3*, 641–651. [CrossRef]
35. Hemberger, M.; Hughes, M.; Cross, J.C. Trophoblast stem cells differentiate in vitro into invasive trophoblast giant cells. *Dev. Biol.* **2004**, *271*, 362–371. [CrossRef] [PubMed]

36. Hemberger, M.; Hanna, C.W.; Dean, W. Mechanisms of early placental development in mouse and humans. *Nat. Rev. Genet.* **2020**, *21*, 27–43. [CrossRef] [PubMed]
37. Bilimoria, J.; Singh, H. The Angiopoietin ligands and Tie receptors: Potential diagnostic biomarkers of vascular disease. *J. Recept. Signal Transduct. Res.* **2019**, *39*, 187–193. [CrossRef] [PubMed]
38. Courtney, J.A.; Cnota, J.F.; Jones, H.N. The Role of Abnormal Placentation in Congenital Heart Disease; Cause, Correlate, or Consequence? *Front. Physiol.* **2018**, *9*, 1045. [CrossRef]
39. Koerten, M.A.; Niwa, K.; Szatmári, A.; Hajnalka, B.; Ruzsa, Z.; Nagdyman, N.; Niggemeyer, E.; Peters, B.; Schneider, K.T.; Kuschel, B.; et al. Frequency of Miscarriage/Stillbirth and Terminations of Pregnancy Among Women with Congenital Heart Disease in Germany, Hungary and Japan. *Circ. J.* **2016**, *80*, 1846–1851. [CrossRef]
40. Georgiades, P.; Ferguson-Smith, A.C.; Burton, G.J. Comparative developmental anatomy of the murine and human definitive placentae. *Placenta* **2002**, *23*, 3–19. [CrossRef]
41. Wilson, R.L.; Buckberry, S.; Spronk, F.; Laurence, J.A.; Leemaqz, S.; O'Leary, S.; Bianco-Miotto, T.; Du, J.; Anderson, P.H.; Roberts, C.T. Vitamin D Receptor Gene Ablation in the Conceptus Has Limited Effects on Placental Morphology, Function and Pregnancy Outcome. *PLoS ONE* **2015**, *10*, e0131287. [CrossRef] [PubMed]

International Journal of
*Molecular Sciences*

*Article*

# Overexpression of ERAP2N in Human Trophoblast Cells Promotes Cell Death

Kristen Lospinoso [1], Mikhail Dozmorov [2,†], Nadine El Fawal [1,†], Rhea Raghu [3,†], Wook-Jin Chae [1] and Eun D. Lee [1,*]

[1]  Department of Microbiology and Immunology, School of Medicine, Massey Cancer Center, Virginia Commonwealth University, Richmond, VA 23298, USA; lospinosokr@mymail.vcu.edu (K.L.); elfawalna@vcu.edu (N.E.F.); Wook-Jin.Chae@vcuhealth.org (W.-J.C.)

[2]  Department of Biostatics and Pathology, Massey Cancer Center, Virginia Commonwealth University, Richmond, VA 23298, USA; mikhail.dozmorov@vcuhealth.org

[3]  Tenafly H.S., Tenafly, NJ 07670, USA; 23rraghu@tenafly.k12.nj.us

*  Correspondence: eun.lee@vcuhealth.org

†  These authors contributed equally to this work.

**Abstract:** The genes involved in implantation and placentation are tightly regulated to ensure a healthy pregnancy. The endoplasmic reticulum aminopeptidase 2 (ERAP2) gene is associated with preeclampsia (PE). Our studies have determined that an isoform of ERAP2-arginine (N), expressed in trophoblast cells (TC), significantly activates immune cells, and ERAP2N-expressing TCs are preferentially killed by both cytotoxic T lymphocytes (CTLs) and Natural Killer cells (NKCs). To understand the cause of this phenomenon, we surveyed differentially expressed genes (DEGs) between ERAP2N expressing and non-expressing TCs. Our RNAseq data revealed 581 total DEGs between the two groups. 289 genes were up-regulated, and 292 genes were down-regulated. Interestingly, most of the down-regulated genes of significance were pro-survival genes that play a crucial role in cell survival (LDHA, EGLN1, HLA-C, ITGB5, WNT7A, FN1). However, the down-regulation of these genes in ERAP2N-expressing TCs translates into a propensity for cell death. The Kyoto Encyclopedia of Genes and Genomes (KEGG) analysis showed that 64 DEGs were significantly enriched in nine pathways, including "Protein processing in endoplasmic reticulum" and "Antigen processing and presentation", suggesting that the genes may be associated with peptide processes involved in immune recognition during the reproductive cycle.

**Keywords:** ERAP2; trophoblast cells; pregnancy; pre-eclampsia; RNA sequencing; differentially expressed genes

**Citation:** Lospinoso, K.; Dozmorov, M.; Fawal, N.E.; Raghu, R.; Chae, W.-J.; Lee, E.D. Overexpression of ERAP2N in Human Trophoblast Cells Promotes Cell Death. *Int. J. Mol. Sci.* **2021**, 22, 8585. https://doi.org/10.3390/ijms22168585

Academic Editor: Marie-Pierre Piccinni

Received: 24 June 2021
Accepted: 22 July 2021
Published: 10 August 2021

**Publisher's Note:** MDPI stays neutral with regard to jurisdictional claims in published maps and institutional affiliations.



## 1. Introduction

The placenta is the largest organ in the fetus; it begins developing with the implantation of the blastocyst into the uterine wall by the end of the first week after fertilization in humans [1,2]. After the 16-cell stage, the morula begins differentiating into the two major cell types of a blastocyst: an internal cell mass that becomes the embryo and an outer layer of trophoblastic cells (TCs) that mostly become the placenta [1,2]. Through a tightly regulated process, the trophoblasts differentiate into two major cell types: an inner layer of cytotrophoblasts and an outer layer of syncytiotrophoblasts that line the placental villi that invade the endometrium [3]. As fetal trophoblasts invade maternal tissue, immunologic adjustments occur in the decidua to ensure the establishment of an immune environment favorable to fetal growth [4]. In a normal pregnancy, myeloid dendritic cells (DCs) with immature phenotypes secrete anti-inflammatory cytokines and promote fetal tolerance, along with antigen-specific regulatory T (Treg) cells [5]. In addition to their inflammatory function to protect against pathogens, Natural Killer (NK) cells in the decidua reinforce uterine vasculature development, thereby supporting fetal growth demands [6]. The abnormal expression or function of any of these immune modulators is implicated in several maternal and fetal pathologies, including pre-eclampsia (PE) [5].

The responses of immune cells after implantation can be modulated via endoplasmic reticulum aminopeptidase (ERAP)1 and ERAP2 expression in TCs. These are enzymes located in the endoplasmic reticulum whose primary function is to trim peptide antigens for presentation on MHC Class I molecules [6,7]. In addition to their antigen processing function, the ERAP1 and ERAP2 proteins have been associated with the development of autoimmune disorders, eclampsia, and PE, respectively [7–9].

Our group and others have reported that the ERAP2 gene is genetically linked to PE in multiple distinct patient populations; the literature highlights Norwegian, Australian/ New Zealand, and African American patients, in particular [8,9]. During a continued surveillance for genetic links to PE, our study discovered a Chilean population that did not show the genetic link of ERAP2 to PE. The haplotype of ERAP2 found in this specific Chilean population is a novel one, and its role in pregnancy and potential for immunogenicity has yet to be explored [10]. PE is not the immediate focus of our study. However, our goal of gathering data on ERAP2's influences on the intrinsic expression of genes will increase our understanding of its role in immunogenic responses and abnormal placentation in conditions such as PE.

ERAP2 exists in two isoforms based on a single nucleotide polymorphism (SNP), rs2549782. ERAP2N has an asparagine (N) present at the 392 positions. ERAP2K has a lysine residue (K) present at the same position. The polymorphism leads to a conformational change in the catalytic site of the protein [11]. This conformational change results in altered antigen processing and thus may have significant implications for immune modulation. ERAP2N has approximately 165-fold greater activity for hydrophobic amino acids than the ERAP2K protein, potentially increasing the antigen presentation to the host immune system [11,12]. Previous studies have demonstrated that Haplotype A (ERAP2K) and Haplotype B (ERAP2N) are selected for frequencies of 0.44 and 0.56, respectively; however, Haplotype B undergoes differential splicing and nonsense-mediated mRNA decay. As a consequence, Haplotype B is never observed in nature [13]. Notably, there was a complete absence of homozygosity of ERAP2N in the maternal and fetal genetic screening in Chilean populations [10]. This suggests that the expression of ERAP2N is incompatible with life in utero. The explanation for this observation is demonstrated in our previous study. We observed that when ERAP2N TCs are co-cultured with peripheral blood monocytic cells (PBMCs), there is an increased immune activation of the CTL- and NKC-induced apoptosis of these cells in vitro [12,14].

The exact mechanism as to how ERAP2N expression leads to increased apoptosis has yet to be outlined in the literature. Before investigating further, the possible heightened immune response elicited by ERAP2N, we opted first to characterize the RNA-level changes observed with or without ERAP2N-expressing TCs. While the ultimate aim of our research is to apply these findings to live human subjects, culture cells (JEG-3) were utilized in this experiment rather than human trophoblast cells from terminated specimens. This was due to the logistical and ethical difficulties associated with obtaining terminated specimens of Chilean populations that would express ERAP2N, as well as the fact that our experiment is a preliminary attempt to explore the differences, and thus cultured cells are adequate for reaching our immediate goals. It is our hope that, in the future, the application of similar experimental protocols may be performed with human tissue to ensure that the results are compatible with our initial reports in this study. To accomplish this, we performed an unbiased mRNA sequencing analysis of ERAP2N-positive TCs compared to those without ERAP2N. Because we have shown previously that ERAP2N-expressing TCs increased immune cell-induced apoptosis, we highlight RNA expression changes in the category of pro-proliferation/survival or pro-apoptosis/death [12–15]. This study aims to elucidate key gene alterations and whether the gene expression profile may explain the increased apoptotic activity in TCs that express ERAP2N.

## 2. Results

### 2.1. Sample Identification

Total RNA was extracted from three different ERAP2N-positive JEG-3 stable transfectants and three stable transfectants with an empty vector without ERAP2N as a control. The paired-end transcript reads from each of these groups were aligned to the hg38/GRCh38 reference genome using the Spliced Transcripts Alignment to a Reference (STAR) aligner v2.6.1 [16].

### 2.2. An Overview of Differentially Expressed Genes in ERAP2N-Positive vs. ERAP2N-Negative Trophoblast Cells

The volcano plot represents the global overview of genes that were differentially affected by the ERAP2N expression in human TCs compared to the TCs that do not express ERAP2N (Figure 1). The differential expression analysis using the edgeR method using the adjusted p-value cutoff <0.3 identified 581 genes differentially expressed between the two groups [17]. Overall, 289 genes were up-regulated and 292 genes were down-regulated, indicating the balanced shift in gene expression changes induced by ERAP2N overexpression.

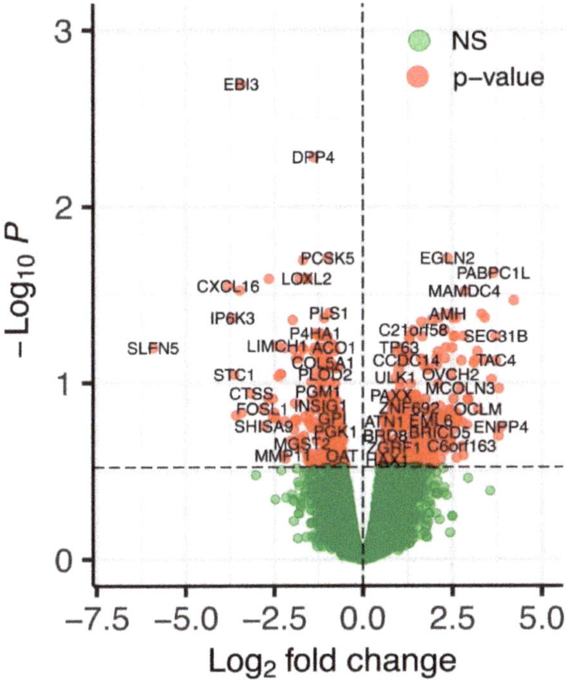


**Figure 1.** A volcano plot of differentially expressed genes in ERAP2N-positive vs. ERAP2N-negative trophoblast cells. X-axis—log2 fold change. Y-axis—(−log10) edgeR-calculated *p*-values. Red points correspond to significantly differentially expressed genes. Green points correspond to genes unchanged (NS: not significant) by ERAP2N expression level.

### 2.3. The Heatmap of the Top 50 Most Significantly Differentially Expressed Genes in ERAP2N-Positive vs. ERAP2N-Negative Trophoblast Cells

The heatmap represents the cluster of the top 50 genes, in which the first three rows show the gene expression in ERAP2N-negative TCs, while the three rows on the right display the gene expression in ERAP2N-positive TCs (Figure 2). Interestingly, the top half of the heatmap cluster genes appears to be mostly involved in pro-apoptotic pathways, while the bottom half centers on those involved in cell survival. In ERAP2N-positive TCs, pro-apoptotic cells are up-regulated compared to ERAP2N-negative control cells, in which these same genes are down-regulated. By contrast, pathways that support cell survival are down-regulated in ERAP2N-positive TCs but up-regulated in those that are ERAP2N-negative.

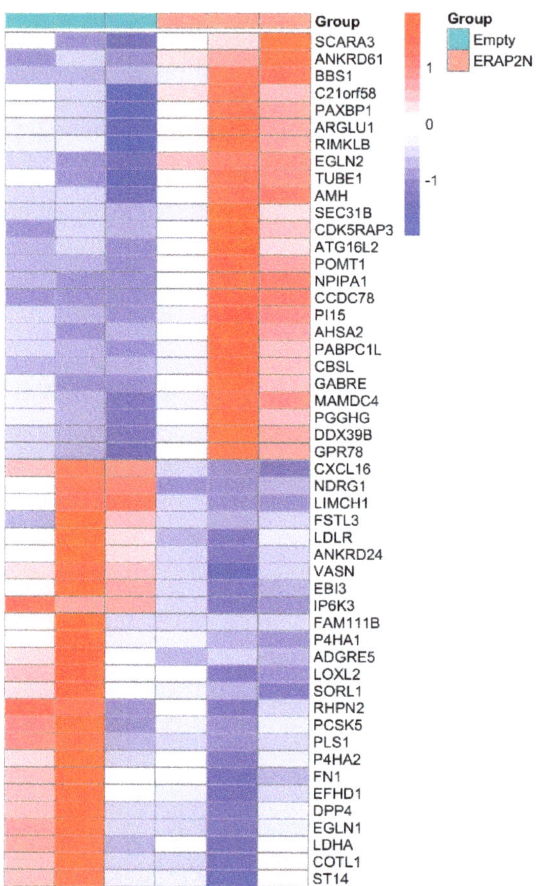

**Figure 2.** A heatmap of the top 50 most significantly differentially expressed genes (DEGs) in ERAP2N-positive vs. ERAP2N-negative trophoblast cells. The first three columns represent the cells transfected with an empty vector (no ERAP2N), and the last three columns represent ERAP2N-expressing trophoblast cells (TCs). Row-scaled log2-transformed TPMs (TPM = transcript count per million) are shown (N = 3/each group).

### 2.4. Functional Enrichment Analysis of KEGG Pathways

Table 1 lists the major KEGG pathways that were significantly enriched with an adjusted *p*-value < 0.3. The entire list of significant pathways at the adjusted *p*-value > 0.3 is in the Supplemental Data (Table S1). The color-coded genes are 12 highly differentially expressed genes (top 50 DEGs) that were also found in these major KEGG pathways which we believe might be significant in determining the cell survival rate with ERAP2N-positive TCs. Specifically, these 12 genes (FN1, LDHA, WNT7A, EGLN1, ITGB5, HLA-C, PPP1R12B, ENO3, CDC42, EGLN2, THBS3, SEC31B) were found to be involved in the following enriched pathways, some genes overlapping in multiple pathways. The rest of top 50 genes divided into cell survival and cell death is listed in the Supplemental Data (Table S2).

**Table 1.** Kyoto Encyclopedia of Genes and Genomes (KEGG) canonical pathways enriched in Differentially Expressed Genes (DEGs) with altered expression in the setting of ERAP2N. The underlined red (up-regulated) and blue (down-regulated) genes are the top 50 differentially expressed genes, appearing in multiple enriched pathways and having known functions that could potentially affect the trophoblast cell's survival or death. The pathways also include genes that are not in the top 50 DEGs (genes in black).

| Pathways | $p$-Value | Adjusted $p$-Value | Genes |
|---|---|---|---|
| Glycolysis/Gluconeogenesis | $6.66 \times 10^{-8}$ | $1.75 \times 10^{-5}$ | GPI; TPI1; PGAM1; ENO1; *ENO3*; HK2; *LDHA*; PKM; PGK1; ALDOC; ALDOA; PGM1; PFKP |
| HIF-1 signaling pathway | $3.30 \times 10^{-5}$ | $3.27 \times 10^{-3}$ | *EGLN1*; *LDHA*; EGLN3; *EGLN2*; TFRC; STAT3; PGK1; SLC2A1; ENO1; ALDOA; *ENO3*; HK2 |
| Fructose and mannose metabolism | $3.73 \times 10^{-5}$ | $3.27 \times 10^{-3}$ | PFKFB4; TPI1; AKR1B1; ALDOC; ALDOA; HK2; PFKP |
| Protein processing in the endoplasmic reticulum | $9.74 \times 10^{-5}$ | $6.41 \times 10^{-3}$ | ERO1A; PDIA3; HSPA5; WFS1; RRBP1; CKAP4; DDOST; PDIA4; HSP90B1; OS9; CALR; P4HB; SEC24D; *SEC31B*; HSPA1B |
| Antigen processing and presentation | $3.85 \times 10^{-4}$ | $1.45 \times 10^{-2}$ | PDIA3; HSPA5; RFX5; *HLA-C*; CALR; CTSS; HSPA1B; CTSB; TAPBP |
| Central carbon metabolism in cancer | $5.68 \times 10^{-4}$ | $1.66 \times 10^{-2}$ | *LDHA*; PKM; PGAM1; IDH1; SLC2A1; SLC16A3; HK2; PFKP |
| Proteoglycans in cancer | $6.05 \times 10^{-3}$ | $1.23 \times 10^{-1}$ | TGFB1; *ITGB5*; FZD7; STAT3; *FN1*; *WNT7A*; ITPR2; *CDC42*; WNT11; SDC1; ITGA5; EZR; *PPP1R12B* |
| ECM-receptor interaction | $9.79 \times 10^{-3}$ | $1.84 \times 10^{-1}$ | *ITGB5*; *FN1*; SDC1; COL9A3; ITGB6; ITGA5; *THBS3* |
| Human papillomavirus infection | $1.68 \times 10^{-2}$ | $2.45 \times 10^{-1}$ | *ITGB5*; FZD7; *FN1*; *HLA-C*; *WNT7A*; *THBS3*; *CDC42*; SLC9A3R1; PKM; WNT11; HEY1; ATP6V0A4; COL9A3; MAML3; ITGB6; ITGA5; TLR |

The list of differentially expressed genes was also analyzed for enrichment in the Kyoto Encyclopedia of Genes and Genomes (KEGG) pathways using ShinyGO (Figure 3). All pathways are indicated through green nodes and are considered "connected" if they share 20% or more genes. The darker the node, the more significantly enriched the gene sets, and the larger the node, the larger the gene sets. Glycolysis/Gluconeogenesis, Protein processing in the endoplasmic reticulum, and the HIF-1 signaling pathway all seem to be the pathways with the most significantly enriched gene sets. Since thicker lines delineate more overlapped genes between the pathways, or nodes, Carbon metabolism, Glycolysis/Gluconeogenesis, and the Biosynthesis of amino acids contain more overlapped genes between the pathways as compared to the other significant pathways. Interestingly, Protein processing in the endoplasmic reticulum shares no overlapped genes with the rest of the pathways.

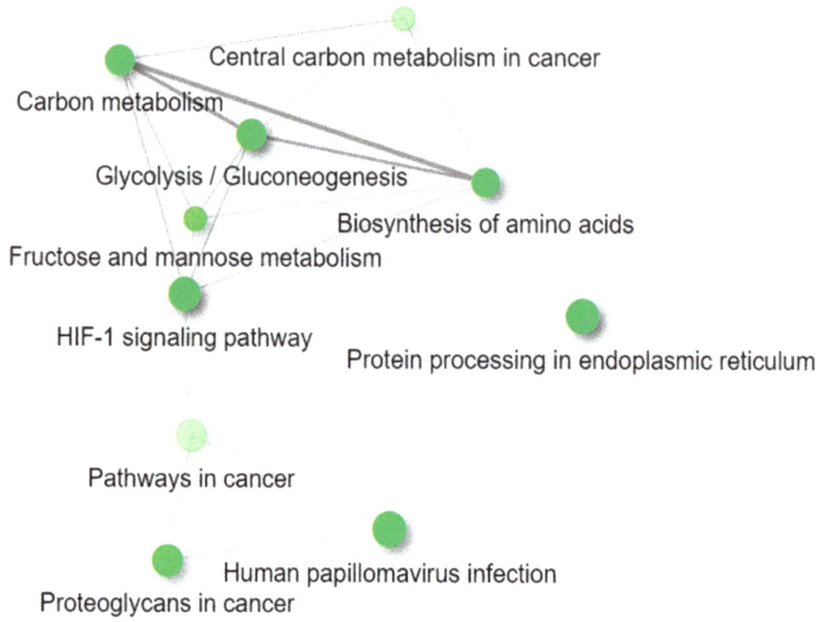

**Figure 3.** Web Diagram of the top 10 most significant KEGG pathways. This diagram represents how the differentially expressed genes listed in the enriched Kyoto Encyclopedia of Genes and Genomes (KEGG) pathways in Table 1 are connected. Two pathways (nodes) are connected if they share 20% or more genes. Darker nodes are more significantly enriched gene sets, and larger nodes represent larger gene sets. Thicker lines represent more overlapped genes between the pathways.

*2.5. Differentially Expressed Genes (DEGs) in ERAP2N-Positive Cells Compared to ERAP2N-Negative Cells*

The S-curve (Figure 4A) is a plot of genes differentially expressed in ERAP2N-positive TCs compared to ERAP2N-negative empty vector TCs, ranked by log2 fold change. The genes plotted on the blue curve are those that are significantly down-regulated, while those on the red curve are significantly up-regulated in ERAP2N-positive TCs, compared to TCs that do not express ERAP2N. LogFC measures how much the gene expression changes between two different conditions. A negative LogFC indicates that the expression of a particular gene is decreasing (i.e., it is being down-regulated) in ERAP2N-positive cells, relative to TCs that are ERAP2N-negative. In the same respect, a positive LogFC indicates that a gene is up-regulated, or that its expression increases in ERAP2N-positive cells compared to cells that do not express ERAP2N. We chose to highlight genes of interest that are relevant in cell survival or cell death. The majority of the genes that are necessary for placental function and cell survival, such as WNT7A, appear to become down-regulated in ERAP2N-positive cells [18–20]. In addition, apoptosis-linked genes such as PPP1R12B and EGLN2 appear to be up-regulated in ERAP2N-positive trophoblasts [21,22].

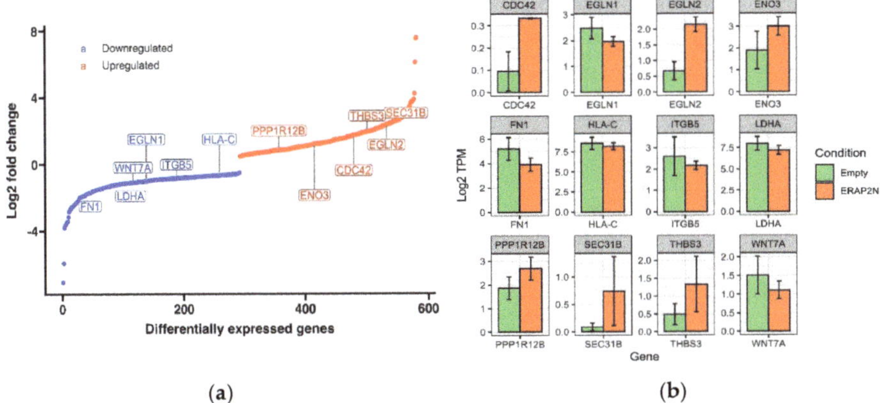

**Figure 4.** (**a**) Visualization of selected differentially expressed genes ranked by log2 fold change in ERAP2N-positive cells compared to ERAP2N-negative cells. The S-curve is a plot of genes differentially expressed in ERAP2N-positive cells compared to ERAP2N-negative cells, ranked by log2 fold change. The blue curve represents genes that are significantly down-regulated in ERAP2N-positive trophoblasts, while the red represents significantly up-regulated genes. (**b**) Bar Graph of Expression Level of Specific Genes from KEGG Pathways. Barplot of 12 differentially expressed genes, comparing the level of expression in ERAP2N-positive trophoblast cells to gene expression in ERAP2N-negative cells. Expression level reported as Log2 TPM (TPM = transcript count per million).

Next, the expression level of each individual gene between the ERAP2N-positive and cells without ERAP2N is directly compared in Figure 4b using the barplot of 12 differentially expressed genes (DEGs). The expression level is reported as Log2 TPM (TPM= transcript count per million). TPM scales the read count of RNA transcripts of a particular gene to the total read count of the sequencing run, thereby measuring the amount of RNA in a sample. The genes represented by the bar plots are a subset of many genes for which the amount of RNA was significantly different in ERAP2N-positive trophoblasts compared to its control. Similarly, PPP1R12B, ENO3, CDC42, THBS3, EGLN2, and SEC31B are up-regulated with ERAP2N, whereas HLA-C, ITGB5, EGLN1, WNT7A, LDHA, and FN1 are down-regulated.

### 2.6. Top 12 Genes in Cellular Fate Expressed in ERAP2N-Positive Cells

Even though the highlighted genes are divided approximately equally between up- and down-expression levels, Figure 5 clearly depicts how these genes determine the fate of ERAP2N-expressing TCs. According to the literature, out of these 12, eight genes (LDHA, EGLN1, HLA-C, ITGB5, WNT7A, FN1, PPP1R12B, and EGLN2) favored cell death depending on their expression level: LDHA, EGLN1, HLA-C, ITGB5, WNT7A, and FN1 are reduced, whereas PPP1R12B and EGLN2 are increased [19–30]. The six genes labeled green (LDHA, EGLN1, HLA-C, ITGB5, WNT7A, FN1) promote cell death due to their main function of maintaining the homeosis that promotes cell survival; however, these genes promote cell death when their expression level is down-regulated in ERAP2N-positive TCs, as indicated with the blue down arrow. The two red genes (PPP1R12A, EGLN2) are pro-cell death when up-regulated, so naturally, in ERAP2N-positive cells, they favor cell death, indicated with the red up arrow [19–30]. Interestingly, ERAP2N-positive TCs still have four green genes that promote cell survival (ENO-3, CDC42, THBS3, SEC31B) when their expressions are up-regulated [31–34]. Nonetheless, the diagram supports our hypothesis that ERAP2N expression in trophoblast cells alters gene expression levels and patterns to favor cell death.

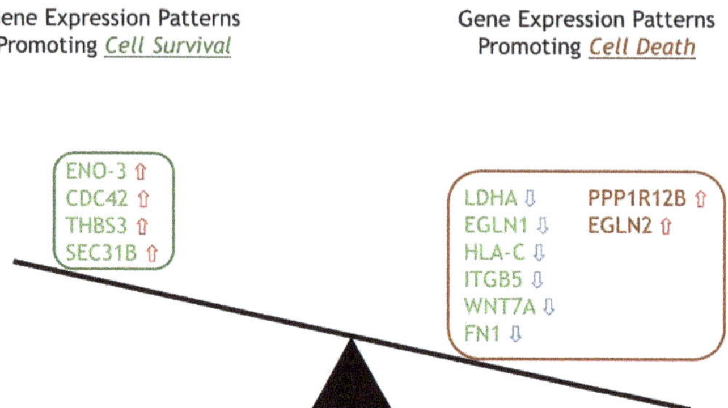


**Figure 5.** A balance diagram depicting significant differentially expressed genes classified into their role in the ERAP2N-positive trophoblast cell fate. Balance diagram of 12 significant differentially expressed genes divided into favoring either a cell survival or cell death role in the setting of ERAP2N. Green represents expression patterns favoring cell survival, and maroon represents expression patterns favoring cell death. Red arrows indicate that the gene was up-regulated, and blue arrows indicate that the gene was down-regulated in the setting of ERAP2N.

The pathway scheme in Figure 6 illustrates the association between ITGB5 and WNT, which explains how these two genes of significance were down-regulated with the presence of ERAP2N promoting cell death instead of cell proliferation/survival. The diagram shows that ITGB5 associates with FAK to activate the Erk cascades that promote transcription. In addition, β-catenin and Wnt contribute to this process through the β-catenin/Wnt signaling pathway. When Wnt binds to the Frizzled receptor, disheveled I (Dvl) is activated, which, through β-catenin, also leads to the Wnt target gene transcription. Thus, both ITGB5 and Wnt promote the proliferation and differentiation of the cell, leading to a pro-survival cell fate.

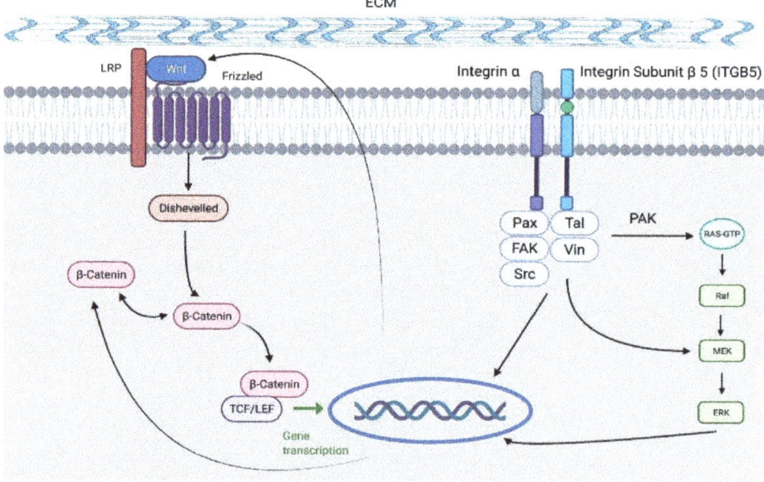


**Figure 6.** Diagram displaying the role of both ITGB5 and WNT in the transcription of genes to promote proliferation and differentiation [15] (Created with BioRender.com).

## 3. Discussion

Trophoblast cells (TCs) of the placenta invade the maternal uterine wall during implantation and are critical in providing adequate signaling to neighboring maternal immune cells to orchestrate a healthy pregnancy [35]. However, we have recently reported that cytotoxic T-lymphocytes (CTLs) and natural killer (NK) cells are significantly activated and increased, preferentially targeting and killing ERAP2N-expressing TCs, potentially disrupting normal pregnancy [16]. This may explain why Chilean populations did not display any ERAP2N homozygosity [10].

To explore the mechanism of ERAP2N and to gain a deeper understanding of the biology of TCs, we assessed changes that may indicate susceptibility to cell death. We successfully identified the gene expression profiles of TCs with or without ERAP2N using RNA-Seq (Quantification) and analyzed the gene expression differences between the two libraries. Furthermore, the KEGG pathways for the DEGs were analyzed.

In our study, a total of 581 genes were significantly differentially expressed between the two groups. A large number of DEGs are associated with the following pathways with an adjusted $p$-value more significant than 0.3: Glycolysis/Gluconeogenesis, HIF-1 signaling pathway, Fructose and mannose metabolism, Protein processing in the endoplasmic reticulum, Antigen processing and presentation, Central carbon metabolism in cancer, Proteoglycans in cancer, ECM-receptor interaction, and Human papillomavirus infection (Table 1). As represented in Figure 2, all these pathways are highly associated with each other, except for the protein processing in the endoplasmic reticulum. Interestingly, genes involved in protein processing in the endoplasmic reticulum and antigen processing and presentation pathways are displayed in significance ($p$ < 0.3, Table 1), supporting ERAP2N's functional role of peptide processing for antigen presentation on HLA class I molecules to potentially modulate immune responses. This is the first reporting of all the genes that are affected by the expression level of ERAP2N.

The top 50 differential gene expression between ERAP2N-expressing TCs and cells that do not express ERAP2N is clearly represented in the heat map. There is a complete expression level reversal of these groups. Observing that cell-survival genes are up-regulated in TCs that do not express ERAP2N while simultaneously being down-regulated in ERAP2N-expressing TCs supports the hypothesis that ERAP2N-expressing TCs are highly prone to cell death prior to immune induced apoptosis, as previously reported [16]. Within these pathways, we have focused on genes that are associated with cell survival when expressed (ENO-3, CDC42, THBS3, SEC31B, LDHA, EGLN2, HLA-C, ITGB5, WNT7A, FN1) and with cell death (PPP1R12B, EGLN2). Interestingly, the current literature determined the function of these genes in the context of cancer and metabolism, which we could speculate given the similarity between cancer and placenta development [19–34]. As shown in Figure 5, the number of genes modulating cell survival are significantly increased compared to the genes promoting cell death. However, most of these homeostatically pro-survival genes are down-regulated in ERAP2N-positive TCs, thereby favoring cell death [19–34]. Even though SEC31B is named apoptosis-linked gene, the gene regulation seems to be complicated. When it is up-regulated, it promotes survival, but when it is down-regulated, it promotes cell death via apoptosis [34]. Interestingly, SEC31B was up-regulated with ERAP2N-expressing TCs. However, our balance diagram demonstrates nicely that the down-regulation of cell survival genes which promote cell death is more significant than the up-regulation of cell death genes. Our RNAseq analysis supports our in vitro data that cell death outcomes were significantly higher with ERAP2N-expressing TCs compared to those that did not express ERAP2N. Among these genes, we focused on human WNT7A because its coordinated expression is essential for the female reproductive tract development [3,36]. To date, most of the available data on WNTs' gene expression in the uterine tissues concern rodents. In mice, Wnt7a is necessary for the appropriate expansion of endometrial glands and the organization of myometrium, as well as for the establishment of the gene expression of further WNT family members (Wnt4, Wnt5a) [3,36]. It has been proposed that WNT7A plays an important role in regulating uterine smooth muscle patterning and maintaining adult uterine function in mice [18]. In humans, high protein expression levels of WNT7A were observed in the cytoplasm and basal plasma membrane of the syncytium, indicating that WNT7A may be produced in syncytiotrophoblast cells and released toward fetal circulation [37]. The same study proposed that WNT7A might be expressed by villous stromal cells, possibly placental macrophages. Our results imply that ERAP2N expression may down-regulate WNT7A-mediated developmental processes, lowering the appropriate expansion of endometrial glands and the organization of the myometrium [18,37]. This translates into trouble during pregnancy. Figure 6 clearly demonstrates the association of ITGB5 and the WNT pathway, where, in ERAP2N-expressing TCs, both are down-regulated, which would promote cell death [15].

Lastly, PDIA3 was listed in the following KEGG pathways: protein processing in the endoplasmic reticulum and antigen processing and presentation pathways. Even though it was not in the top 50 DEGs and was not part of the top 12 genes we have focused on, it must be noted that PDIA3 is down-regulated in ERAP2N-expressing TCs. It was reported that PDIA3 expression was decreased in the TCs from women with PE, and decreased PDIA3 expression induced trophoblast apoptosis and repressed trophoblast proliferation by regulating the MDM2/p53/p21 pathway [38].

A direction for future study is to isolate the chemokines and cytokine signaling responsible for the apoptotic process observed due to ERAP2N expression, as it may help to direct future therapeutic targets and treatments. Running studies to identify the key immunologic cell types involved at each stage of apoptosis may similarly aid in illustrating the mechanism of immune targeting and eventual cell death in ERAP2N-positive TCs. Lastly, genetic screening for the ERAP2N of TCs may determine the success of the pregnancy, and a further evaluation of the genes involved in cell survival can help develop a therapeutic treatment of PE and other pregnancy-related disorders.

Collectively, this study provides further evidence for exploring mechanisms that could expand our knowledge of TCs in cancer and pregnancy, especially ERAP2-associated PE.

## 4. Materials and Methods

### 4.1. Cell Lines and pcDNA Stable Transfection of JEG-3 Cells

The JEG-3 trophoblast cell line, derived from gestational choriocarcinoma, was obtained from ATCC (Manassas, Virginia) and grown in T75 flasks with a MEM medium supplemented with 10% FBS and 1% Pen/Strep antibiotics at 37 °C, 5% CO2 for 24–48 h. For a stable transfection with mammalian expression empty vector pcDNA, 3.1 or pcDNA-ERAP2N, JEG-3 cells were plated at $3.0 \times 10^4$ cells per well with appropriate media. They were allowed to attach overnight at 37 °C before transfection. The pcDNA plasmid with and without an ERAP2N insert was added at a concentration of 200 ng per well using Promega FuGENE® Transfection Reagent (Promega #E2311). Untransfected cells were treated with FuGENE® Reagent alone as a control. After 48 h, the cells were washed once, and the media changed to RPMI-1640 supplemented with 10% FBS(hi). Zeocin at 1 mg/mL concentration was added as a selection agent, and only the cells that survived were propagated. The ERAP2 protein expression level was confirmed by Western blot analysis as previously described [16].

### 4.2. RNA Sample Collection and Preparation

RNA was isolated from a cellular suspension of ERAP2N-positive trophoblast cells or ERAP2N-negative trophoblast cells. The suspended cells were rinsed with a PBS buffer. TRIzol reagent was added (1 mL per $5 \times 10^6$ cells). The cell suspension was aspirated using a syringe to break up any noticeable clumps until the suspension became clear. Before use, samples were stored at −80 °C.

### 4.3. Clustering and RNA Sequencing

According to the manufacturer's instructions, the clustering of the index-coded samples was performed on a cBot Cluster Generation System using a PE Cluster Kit cBot-HS (Illumina). After cluster generation, the library preparations were sequenced on an Illumina platform, and paired-end reads were generated. One microgram of RNA was used for the cDNA library construction at Novogene (Sacramento, CA, USA) using an NEBNext® Ultra II RNA Library Prep Kit for Illumina® (cat NEB #E7775, New England Biolabs, Ipswich, MA, USA) according to the manufacturer's protocol. Briefly, mRNA was enriched using oligo(dT) beads followed by two rounds of purification and fragmented randomly by adding a fragmentation buffer. The first-strand cDNA was synthesized using a random hexamers primer. A custom second-strand synthesis buffer (Illumina), dNTPs, RNase H and DNA polymerase I were added to generate the second strand (ds cDNA). After a series of terminal repairs, poly-adenylation, and sequencing adaptor ligation, the double-stranded cDNA library was completed following the size selection and PCR enrichment. The resulting 250–350 bp insert libraries were quantified using a Qubit 2.0 fluorometer (Thermo Fisher Scientific, Waltham, MA, USA) and quantitative PCR. The size distribution was analyzed using an Agilent 2100 Bioanalyzer (Agilent Technologies, Santa Clara, CA, USA). The qualified libraries were sequenced on an Illumina NovaSeq 6000 Platform (Illumina, San Diego, CA, USA) using a paired-end 150 run (2 × 150 bases). Twenty million paired raw reads were generated from each library.

Paired-end reads were aligned to the hg38/GRCh38 reference genome using the Spliced Transcripts Alignment to a Reference (STAR) aligner v2.6.1 [39]. HTSeq v0.6.1 was used to count the read numbers mapped from each gene. RNA-seq counts were preprocessed and analyzed for differential expression using the edgeR v.3.30.0 [17], R package. *p*-values for differentially expressed genes were corrected using a False Discovery Rate (FDR) multiple testing correction method [40]. A functional enrichment analysis (KEGG) was performed using the enricher R package v.2.1 [41]. Row-median centered log2(TPM + 1) expression profiles for selected genes were visualized using the heatmap package v.1.0.12. All statistical calculations were performed within the R/Bioconductor environment v4.0.0 [42].

## 5. Conclusions

In conclusion, the expression of ERAP2N in trophoblast cells has a high potential to create a hostile uterine environment for fetal development and potentially result in an unsuccessful pregnancy. ERAP2N expression clearly reduces the expression of many cell survival genes and may increase the susceptibility to immune induced apoptosis via the up-regulation of genes involved in cell death, such as PPP1R12B and EGLN2. This is the starting point to define the mechanism of ERAP2N and how each of these significantly affected genes is involved in pregnancy success and failure. This knowledge will be important to develop genetic screening and therapeutic targeting to ensure healthy pregnancies for the mother and the baby.

**Supplementary Materials:** The following are available online at https://www.mdpi.com/article/10.3390/ijms22168585/s1.

**Author Contributions:** Conceptualization: E.D.L.; investigation: N.E.F. and E.D.L.; methodology and formal analysis: M.D.; writing—original draft preparation: K.L., N.E.F., R.R., M.D., W.-J.C. and E.D.L.; writing—review and editing: M.D. and W.-J.C.; data curation: N.E.F., R.R., M.D. and E.D.L.; project administration: E.D.L.; funding acquisition: E.D.L. All authors have read and agreed to the published version of the manuscript.

**Funding:** This publication was [in part] supported by CTSA award No. UL1TR002649 from the National Center for Advancing Translational Sciences. Its contents are solely the responsibility of the authors and do not necessarily represent official views of the National Center for Advancing Translational Sciences or the National Institutes of Health.

**Institutional Review Board Statement:** Not applicable.

**Informed Consent Statement:** Not applicable.

**Data Availability Statement:** Not applicable.

**Acknowledgments:** We would like to thank Maria Teves for editing the manuscript.

**Conflicts of Interest:** The authors declare no conflict of interest. The funders had no role in the design of the study; in the collection, analyses, or interpretation of data; in the writing of the manuscript, or in the decision to publish the results.

## References


1. Burton, G.J.; Jauniaux, E. Development of the human placenta and fetal heart: Synergic or independent? *Front. Physiol.* **2018**, *9*, 1–10. [CrossRef] [PubMed]
2. Human Placenta Project: How Does the Placenta Form? *Placent. Dev. Fertil. Full Term* **2019**, 2–3.
3. Kim, S.-M.; Kim, J.-S. A Review of Mechanisms of Implantation. *Dev. Reprod.* **2017**, *21*, 351–359. [CrossRef] [PubMed]
4. Than, N.G.; Hahn, S.; Rossi, S.W.; Szekeres-Bartho, J. Editorial: Fetal-Maternal Immune Interactions in Pregnancy. *Front. Immunol.* **2019**, *10*, 1–4. [CrossRef]
5. Ehrentraut, S.; Sauss, K.; Neumeister, R.; Luley, L.; Oettel, A.; Fettke, F.; Costa, S.-D.; Langwisch, S.; Zenclussen, A.; Schumacher, A.; et al. Human Miscarriage Is Associated with Dysregulations in Peripheral Blood-Derived Myeloid Dendritic Cell Subsets. *Front. Immunol.* **2019**, *10*, 1–12. [CrossRef] [PubMed]
6. Shmeleva, E.V.; Colucci, F. Maternal natural killer cells at the intersection between reproduction and mucosal immunity. *Mucosal Immunol.* **2021**, 1–15. [CrossRef]
7. Fiorillo, M.T.; van Endert, P.M.; Bouvier, M.; López De Castro, J.A. How ERAP1 and ERAP2 Shape the Peptidomes of Disease-Associated MHC-I Proteins. *Front. Immunol.* **2018**, *9*, 2463.
8. Hill, L.D.; Hilliard, D.D.; York, T.P.; Srinivas, S.; Kusanovic, J.P.; Gomez, R.; Elovitz, M.A.; Romero, R.; Strauss, J.F. Fetal ERAP2 Variation is Associated with Preeclampsia in African Americans in A Case-Control Study. *BMC Med. Genet.* **2011**. Available online: http://www.biomedcentral.com/1471-2350/12/64 (accessed on 1 July 2021). [CrossRef]
9. Johnson, M.P.; Roten, L.T.; Dyer, T.D.; East, C.; Forsmo, S.; Blangero, J.; Brennecke, S.; Austgulen, R.; Moses, E. The ERAP2 gene is associated with preeclampsia in Australian and Norwegian populations. *Hum. Genet.* **2009**, *126*, 655–666. [CrossRef]
10. Vanhille, D.L.; Hill, L.D.; Hilliard, D.D.; Lee, E.D.; Teves, M.E.; Srinivas, S.; Kusanovic, J.P.; Gómez, R.; Stratikos, E.; Elovitz, M.; et al. A novel ERAP2 haplotype structure in a Chilean population: Implications for ERAP2 protein expression and preeclampsia risk. *Mol. Genet. Genom. Med.* **2013**, *1*, 98–107. [CrossRef]
11. Evnouchidou, I.; Birtley, J.; Seregin, S.; Papakyriakou, A.; Zervoudi, E.; Samiotaki, M.; Panayotou, G.; Giastas, P.; Petrakis, O.; Georgiadis, D.; et al. Altered Antigen Processing Induces a Specificity Switch That Leads to in Endoplasmic Reticulum Aminopeptidase 2 A Common Single Nucleotide Polymorphism. *J. Immunol.* **2021**, *189*, 2383–2392. [CrossRef]
12. Lee, E.D. Endoplasmic Reticulum Aminopeptidase 2, a common immunological link to adverse pregnancy outcomes and cancer clearance? *Placenta* **2017**, *56*, 40–43. [CrossRef] [PubMed]
13. Andrés, A.M.; Dennis, M.Y.; Kretzschmar, W.W.; Cannons, J.L.; Lee-Lin, S.-Q. Balancing Selection Maintains a Form of ERAP2 that Undergoes Nonsense-Mediated Decay and Affects Antigen Presentation. *PLoS Genet.* **2010**, *6*, e1001157. [CrossRef] [PubMed]
14. Plácido, A.; Pereira, C.M.F.; Duarte, A.; Candeias, E.; Correia, S.; dos Santos, R.X.C.; Carvalho, C.; Cardoso, S.M.; Oliveira, C.; Moreira, P.; et al. The role of endoplasmic reticulum in amyloid precursor protein processing and trafficking: Implications for Alzheimer's disease. *Biochim. Biophys. Acta Mol. Basis Dis.* **2014**, *1842*, 1444–1453. [CrossRef] [PubMed]
15. Sa, S.; Wong, L.; Mccloskey, K.E. Combinatorial Fibronectin and Laminin Signaling Promote Highly Efficient Cardiac Differentiation of Human Embryonic Stem Cells. *BioRes. Open Access* **2014**, *3*, 150–161. [CrossRef]
16. Warthan, M.D.; Washington, S.L.; Franzese, S.E.; Ramus, R.M.; Kim, K.-R.; York, T.P.; Stratikos, E.; Strauss, J.F.; Lee, E.D. The role of endoplasmic reticulum aminopeptidase 2 in modulating immune detection of choriocarcinoma. *Biol. Reprod.* **2018**, *98*, 309–322. [CrossRef]
17. Robinson, M.D.; Mccarthy, D.J.; Smyth, G.K. edgeR: A Bioconductor package for differential expression analysis of digital gene expression data. *Bioinformatics* **2010**, *26*, 139–140. [CrossRef]
18. Dunlap, K.A.; Filant, J.; Hayashi, K.; Rucker, E.B.; Song, G.; Deng, J.M.; Behringer, R.R.; DeMayo, F.J.; Lydon, J.; Jeong, J.-W.; et al. Postnatal Deletion of Wnt7a Inhibits Uterine Gland Morphogenesis and Compromises Adult Fertility in Mice 1. *Biol. Reprod.* **2011**, *85*, 386–396. [CrossRef]
19. Yoshioka, S.; King, M.L.; Ran, S.; Okuda, H.; Ii, J.A.M.; McAsey, M.E.; Sugino, N.; Brard, L.; Watabe, K.; Hayashi, K. WNT7A Regulates Tumor Growth and Progression in Ovarian Cancer through the WNT/b-Catenin Pathway. *Mol. Cancer Res.* **2012**, *10*, 469–482. [CrossRef] [PubMed]
20. Haseeb, M.; Hassan Pirzada, R.; Ul Ain, Q.; Choi, S. Wnt Signaling in the Regulation of Immune Cell and Cancer Therapeutics. *Cells* **2019**, *8*, 1380. [CrossRef]
21. Ding, C.; Tang, W.; Wu, H.; Fan, X.; Luo, J.; Feng, J.; Wen, K.; Wu, G. The PEAK1-PPP1R12B axis inhibits tumor growth and metastasis by regulating Grb2/PI3K/Akt signalling in colorectal cancer. *Cancer Lett.* **2019**, *442*, 383–395. [CrossRef]
22. Price, C.; Gill, S.; Ho, Z.V.; Davidson, S.M.; Merkel, E.; McFarland, J.M.; Leung, L.; Tang, A.; Kost-Alimova, M.; Tsherniak, A.; et al. Molecular Cell Biology Genome-Wide Interrogation of Human Cancers Identifies EGLN1 Dependency in Clear Cell Ovarian Cancers. *Cancer Res.* **2019**, *79*, 2564–2579. [CrossRef]
23. Cai, X.; Liu, C.; Zhang, T.N.; Zhu, Y.W.; Dong, X.; Xue, P. Down-regulation of FN1 inhibits colorectal carcinogenesis by suppressing proliferation, migration, and invasion. *J. Cell. Biochem.* **2018**, *119*, 4717–4728. [CrossRef] [PubMed]

24. Soikkeli, J.; Podlasz, P.; Yin, M.; Nummela, P.; Jahkola, T.; Virolainen, S.; Krogerus, L.; Heikkilä, P.; von Smitten, K.; Saksela, O.; et al. Metastatic outgrowth encompasses COL-I, FN1, and POSTN up-regulation and assembly to fibrillar networks regulating cell adhesion, migration, and growth. *Am. J. Pathol.* **2010**, *177*, 387–403. [CrossRef] [PubMed]

25. An, J.; Zhang, Y.; He, J.; Zang, Z.; Zhou, Z.; Pei, X.; Zheng, X.; Zhang, W.; Yang, H.; Li, S.; et al. Lactate Dehydrogenase A Promotes the Invasion and Proliferation of Pituitary Adenoma. *Sci. Rep.* **2017**, *7*, 1–12. [CrossRef]

26. Feng, Y.; Xiong, Y.; Qiao, T.; Li, X.; Jia, L.; Han, Y. Lactate dehydrogenase A: A key player in carcinogenesis and potential target in cancer therapy. *Cancer Med.* **2018**, *7*, 6124–6136. [CrossRef]

27. Contini, P.; Ghio, M.; Poggi, A.; Filaci, G.; Induveri, F.; Ferrone, S.; Puppo, F. Soluble HLA-A,-B,-C and-G Molecules Induce Apoptosis in T and NK CD8 + Cells and Inhibit Cytotoxic T Cell Activity through CD8 Ligation. *Eur. J. Immunol.* **2003**, *33*, 125–134. [CrossRef]

28. Hiby, S.E.; Walker, J.; O'Shaughnessy, K.M.; Redman, C.W.; Carrington, M.; Trowsdale, J.; Moffett, A. Combinations of Maternal KIR and Fetal HLA-C Genes Influence the Risk of Preeclampsia and Reproductive Success. *J. Exp. Med.* **2004**, *200*, 957–965. [CrossRef]

29. The Mechanisms of β-Catenin on Keloid Fibroblast Cells Proliferation and Apoptosis. Available online: https://www.semanticscholar.org/paper/The-mechanisms-of-%CE%B2-catenin-on-keloid-fibroblast-Chen-Yu/5387841c19eb188b07740da750976375dabcf5b3 (accessed on 1 July 2021).

30. Zhang, L.Y.; Guo, Q.; Guan, G.F.; Cheng, W.; Cheng, P.; Wu, A.H. Integrin Beta 5 Is a Prognostic Biomarker and Potential Therapeutic Target in Glioblastoma. *Front. Oncol.* **2019**, *9*, 904. [CrossRef] [PubMed]

31. Park, C.; Lee, Y.; Je, S.; Chang, S.; Kim, N.; Jeong, E.; Yoon, S. Molecules and Cells Overexpression and Selective Anticancer Efficacy of ENO3 in STK11 Mutant Lung Cancers. *Mol. Cells* **2019**, *42*, 804–809.

32. Qadir, M.I.; Parveen, A.; Ali, M. Cdc42: Role in Cancer Management. *Chem. Biol. Drug Des.* **2015**, *86*, 432–439. [CrossRef]

33. Dalla-Torre, C.A.; Yoshimoto, M.; Lee, C.H.; Joshua, A.M.; de Toledo, S.R.; Petrilli, A.S.; Andrade, J.A.; Chilton-MacNeill, S.; Zielenska, M.; Squire, J.A.; et al. Effects of THBS3, SPARC and SPP1 Expression on Biological Behavior and Survival in Patients with Osteosarcoma. *BMC Cancer* **2006**, *6*, 1–10. [CrossRef] [PubMed]

34. Hu, H.; Gourguechon, S.; Wang, C.C.; Li, Z. The G1 Cyclin-dependent Kinase CRK1 in Trypanosoma brucei Regulates Anterograde Protein Transport by Phosphorylating the COPII Subunit Sec31. *J. Biol. Chem.* **2016**, *291*, 15527–15539. [CrossRef] [PubMed]

35. Carter, A.M.; Enders, A.C.; Pijnenborg, R. The role of invasive trophoblast in implantation and placentation of primates. *Phil. Trans. R. Soc. B* **2015**, *370*, 20140070. [CrossRef]

36. Kiewisz, J.; Kaczmarek, M.M.; Andronowska, A.; Blitek, A.; Ziecik, A.J. Gene expression of WNTs, β-catenin and E-cadherin during the periimplantation period of pregnancy in pigs—Involvement of steroid hormones. *Theriogenology* **2011**, *76*, 687–699. [CrossRef] [PubMed]

37. Saben, J.; Zhong, Y.; McKelvey, S.; Dajani, N.K.; Andres, A.; Badger, T.M.; Gomez-Acevedo, H.; Shankar, K. A comprehensive analysis of the human placenta transcriptome. *Placenta* **2014**, *35*, 125–131. [CrossRef]

38. Mo, H.Q.; Tian, F.J.; Ma, X.L.; Zhang, Y.C.; Zhang, C.X.; Zeng, W.H.; Zhang, Y.; Lin, Y. PDIA3 regulates trophoblast apoptosis and proliferation in preeclampsia via the MDM2/p53 pathway. *Reproduction* **2020**, *160*, 293–305. [CrossRef]

39. Dobin, A.; Davis, C.A.; Schlesinger, F.; Drenkow, J.; Zaleski, C.; Jha, S. Sequence analysis STAR: Ultrafast universal RNA-seq aligner. *Oxford Acad.* **2013**, *29*, 15–21.

40. Url, S.; Society, R.S. Controlling the False Discovery Rate: A Practical and Powerful Approach to Multiple Testing Yoav Benjamini; Yosef Hochberg. *J. R. Stat. Soc. Ser. B (Methodol.)* **1995**, *57*, 289–300.

41. Chen, E.Y.; Tan, C.M.; Kou, Y.; Duan, Q.; Wang, Z.; Meirelles, G.V.; Clark, N.R.; Ma'Ayan, A. Enrichr: Interactive and Collaborative HTML5 Gene List Enrichment Analysis Tool. 2013. Available online: http://amp.pharm.mssm.edu/Enrichr (accessed on 1 July 2021).

42. Gentleman, R.C.; Carey, V.J.; Bates, D.M.; Bolstad, B.; Dettling, M.; Dudoit, S.; Ellis, B.; Gautier, L.; Ge, Y.; Gentry, J.; et al. Bioconductor: Open software development for computational biology and bioinformatics. *Genome Biol.* **2004**, *5*, 1–16. [CrossRef]

International Journal of
*Molecular Sciences*

# CSH RNA Interference Reduces Global Nutrient Uptake and Umbilical Blood Flow Resulting in Intrauterine Growth Restriction

Amelia R. Tanner [1], Cameron S. Lynch [1], Victoria C. Kennedy [1], Asghar Ali [1], Quinton A. Winger [1], Paul J. Rozance [2] and Russell V. Anthony [1,*]

[1]  College of Veterinary Medicine, Colorado State University, Fort Collins, CO 80523, USA; amelia.tanner@colostate.edu (A.R.T.); cameronlynch553@gmail.com (C.S.L.); Tori.Kennedy@colostate.edu (V.C.K.); asghar.ali20@alumni.colostate.edu (A.A.); Quinton.Winger@colostate.edu (Q.A.W.)
[2]  Department of Pediatrics, University of Colorado Anschutz Medical Campus, Aurora, CO 80045, USA; paul.rozance@cuanschutz.edu
*  Correspondence: russ.anthony@colostate.edu

**Citation:** Tanner, A.R.; Lynch, C.S.; Kennedy, V.C.; Ali, A.; Winger, Q.A.; Rozance, P.J.; Anthony, R.V. CSH RNA Interference Reduces Global Nutrient Uptake and Umbilical Blood Flow Resulting in Intrauterine Growth Restriction. *Int. J. Mol. Sci.* **2021**, *22*, 8150. https://doi.org/ 10.3390/ijms22158150

Academic Editors: Hiten D Mistry and Eun Lee

Received: 17 June 2021
Accepted: 27 July 2021
Published: 29 July 2021

**Publisher's Note:** MDPI stays neutral with regard to jurisdictional claims in published maps and institutional affiliations.

**Abstract:** Deficiency of the placental hormone chorionic somatomammotropin (CSH) can lead to the development of intrauterine growth restriction (IUGR). To gain insight into the physiological consequences of CSH RNA interference (RNAi), the trophectoderm of hatched blastocysts (nine days of gestational age; dGA) was infected with a lentivirus expressing either a scrambled control or CSH-specific shRNA, prior to transfer into synchronized recipient sheep. At 90 dGA, umbilical hemodynamics and fetal measurements were assessed by Doppler ultrasonography. At 120 dGA, pregnancies were fitted with vascular catheters to undergo steady-state metabolic studies with the $^3H_2O$ transplacental diffusion technique at 130 dGA. Nutrient uptake rates were determined and tissues were subsequently harvested at necropsy. CSH RNAi reduced ($p \leq 0.05$) both fetal and uterine weights as well as umbilical blood flow (mL/min). This ultimately resulted in reduced ($p \leq 0.01$) umbilical IGF1 concentrations, as well as reduced umbilical nutrient uptakes ($p \leq 0.05$) in CSH RNAi pregnancies. CSH RNAi also reduced ($p \leq 0.05$) uterine nutrient uptakes as well as uteroplacental glucose utilization. These data suggest that CSH is necessary to facilitate adequate blood flow for the uptake of oxygen, oxidative substrates, and hormones essential to support fetal and uterine growth.

**Keywords:** chorionic somatomammotropin; blood flow; intrauterine growth restriction; nutrient uptake; uterus

## 1. Introduction

Intrauterine growth restriction (IUGR), the second-leading cause of perinatal mortality, results from a failure of the fetus to reach its growth potential in utero, and impacts up to 6% of pregnancies worldwide [1]. This pathology has also been linked epidemiologically to the development of adult-onset diseases such as cardiovascular disease, diabetes, and hypertension [2–5]. While many causes of IUGR remain unknown, deficiency of the placental hormone chorionic somatomammotropin (CSH) has been associated with fetal growth restriction in both humans [6,7] and sheep [8]. This relationship was directly demonstrated by our laboratory, utilizing lentiviral mediated RNA interference (RNAi) in the sheep placenta [9]. Using this model, CSH RNAi resulted in fetal and placental growth restriction in near-term (135 dGA) pregnancies [9], and by the end of the first third (50 dGA) of pregnancy [10]. However, CSH deficiency does not always result in IUGR, a phenomenon which is also documented in human pregnancies [6,11]. We have previously described this normal weight phenotype in response to CSH RNAi, which is characterized by decreased umbilical IGF1 [12], reduced uterine blood flow and increased placental glucose utilization [13]. These perturbations, in spite of normal fetal and placental weights, support the necessity of CSH for maintaining adequate uterine blood flow and regulating placental glucose consumption. Additionally, these CSH dependent effects hint at potential mechanisms that could be responsible for the development of IUGR in more severe CSH RNAi phenotypes. Thus, to better understand the actions of CSH in modulating fetal growth, our objective was to assess the physiological ramifications of CSH RNAi induced IUGR, to ascertain how CSH deficiency leads to the progression of fetal and placental growth restriction. We hypothesized that CSH RNAi induced IUGR results from altered uterine blood flow and nutrient uptake. It was determined that CSH RNAi resulted in decreased uterine and umbilical blood flows, reduced global nutrient transport and altered fetal IGF1 concentrations, driving the development of IUGR.

## 2. Results

All equations used to calculate blood flow, nutrient uptake, nutrient utilization, and nutrient quotients are summarized in Table 1.

**Table 1.** Calculations [1] for blood flow, nutrient uptake, nutrient utilization, and quotients.

| **Blood Flow (Doppler Ultrasound)** | |
| --- | --- |
| Pulsatility index | (PSV–EDV)/Timed-average mean velocity (TAMV) |
| Resistance index | (PSV-EDV)/PSV |
| Umbilical Blood Flow (mL/min) | TAMV $\times$ $(\pi/4)$ $\times$ Artery cross-sectional area $\times$ 60 |
| **Blood Flow ($^3$H$_2$O Tracer)** | |
| $R_{inf}$ $^3$H$_2$O (dpm/min) | Pump rate $\times$ [infusate] |
| $R_{acc(f)}$ (dpm/min) | $\alpha$pl slope $\times$ (0.8 $\times$ fetal weight) |
| $R_{acc(m)}$ (dpm/min) | $R_{acc(f)}$ + [$\alpha$pl slope $\times$ 0.8(uterine weight)] |
| Umbilical Blood Flow (UBF; mL/min) | $(R_{inf} - R_{acc(f)})/([^3H_2O]_{\alpha(WB)} - [^3H_2O]_{\gamma(WB)})$ |
| Umbilical Plasma Flow (UPF; mL/min) | UBF $\times$ [1 − Hct$_{f(avg)}$] |
| Uterine Blood Flow (UtBF; mL/min) | $(R_{inf} - R_{acc(m)})/([^3H_2O]_{V(WB)} - [^3H_2O]_{A(WB)})$ |
| Uterine Plasma Flow (UtPF; mL/min) | UtBF $\times$ [1 − Hct$_{m(avg)}$] |
| **Nutrient Uptake and Utilization Rates** | |
| Umbilical Oxygen Uptake (UOU; mmol/min) | UBF $\times$ $([O_2]_{\gamma(WB)} - [O_2]_{\alpha(WB)})$ |
| Uterine Oxygen Uptake (UtOU; mmol/min) | UtBF $\times$ $([O_2]_{A(WB)} - [O_2]_{V(WB)})$ |
| Uteroplacental Oxygen Utilization (mmol/min) | UtOU − UOU |
| Plasma to WB Glucose Conversion | $[G]_{pl} \times [1 - (0.24 \times Hct)] - (3.3 \times Hct)$ |
| Umbilical Glucose Uptake (UGU; μmol/min) | UBF $\times$ $([G]_{\gamma(WB)} - [G]_{\alpha(WB)})$ |
| Uterine Glucose Uptake (UtGU; μmol/min) | UtBF $\times$ $([G]_{A(WB)} - [G]_{V(WB)})$ |
| Uteroplacental Glucose Utilization (μmol/min) | UtGU − UGU |
| Umbilical Lactate Uptake (μmol/min; ULU) | UBF $\times$ $([L]_{\gamma(pl)} - [L]_{\alpha(pl)})$ |
| Uterine Lactate Secretion (μmol/min; UtLS) | UtBF $\times$ $([L]_{V(pl)} - [L]_{A(pl)})$ |
| Uteroplacental Lactate Production (μmol/min) | ULU + UtLS |
| Umbilical AA Uptake (μmol/min; UAAU) | UPF $\times$ $([AA]_{\gamma(pl)} - [AA]_{\alpha(pl)})$ |
| Uterine AA Uptake (μmol/min; UtAAU) | UtPF $\times$ $([AA]_{A(pl)} - [AA]_{V(pl)})$ |
| Umbilical AA Carbon Uptake (μmol/min; UCU) | (#AA carbons) $\times$ UAAU |
| Uterine AA Carbon Uptake (μmol/min; UtCU) | (#AA carbons) $\times$ UtAAU |
| Umbilical AA Nitrogen Uptake (μmol/min; UNU) | (#AA nitrogens) $\times$ UAAU |
| Uterine AA Nitrogen Uptake (μmol/min; (UtNU) | (#AA nitrogens) $\times$ UtAAU |
| **Fetal Nutrient: Oxygen Quotients** | |
| Glucose:Oxygen (G:O) quotient | $6 \times ([G]_{\gamma(WB)} - [G]_{\alpha(WB)})/([O_2]_{\gamma(WB)} - [O_2]_{\alpha(WB)})$ |
| Lactate:Oxygen (L:O) quotient | $3 \times ([L]_{\gamma(pl)} - [L]_{\alpha(pl)})/([O_2]_{\gamma(WB)} - [O_2]_{\alpha(WB)})$ |
| Amino Acid:Oxygen (AA:O) quotient | $Q \times ([AA]_{\gamma(pl)} - [AA]_{\alpha(pl)})/([O_2]_{\gamma(WB)} - [O_2]_{\alpha(WB)})$ |
| Total Nutrient:Oxygen quotient | G:O quotient + L:O quotient + Total AA:O quotient |

[1] Calculations are derived from Cilvik et al., 2021.

*2.1. 90dGA Doppler Velocimetry*

As assessed by Doppler ultrasound and velocimetry, fetal binocular distance (cm), biparietal circumference (cm), abdominal circumference (cm), femur length (cm), and tibia length (cm) did not differ ($p \geq 0.10$; Table 2) between treatments. Umbilical blood flow (mL/min) was reduced ($p = 0.05$; Figure 1) by 36% in CSH RNAi pregnancies. Umbilical artery cross sectional area (CSA, cm$^2$) and cross-sectional diameter (CSD, cm) both tended ($p \leq 0.10$; Table 2) to be reduced in CSH RNAi pregnancies. Resistance indices (RI), pulsatility indices (PI), systolic/diastolic ratios (S/D ratios), and fetal heart rate (bpm) did not differ between treatments.

**Table 2.** Measures of blood flow and fetal growth as assessed by 90 dGA Doppler velocimetry and 130 dGA $^3$H$_2$0 transplacental diffusion.

| | CON RNAi | CSH RNAi | % Change | *p*-Value |
|---|---|---|---|---|
| **90 dGA Doppler Ultrasound Measurements** | **(*n* = 6)** | **(*n* = 6)** | | |
| Binocular distance, cm | 4.91 ± 0.25 | 4.63 ± 0.19 | 5.70 | 0.48 |
| Biparietal circumference, cm | 16.56 ± 0.63 | 15.11 ± 0.54 | 8.72 | 0.19 |
| Abdominal circumference, cm | 22.03 ± 0.95 | 20.30 ± 1.27 | 7.82 | 0.40 |
| Femur length, cm | 4.24 ± 0.08 | 4.17 ± 0.09 | 1.81 | 0.62 |
| Tibia length, cm | 3.05 ± 0.11 | 3.03 ± 0.04 | 0.75 | 0.87 |
| Pulsatility Index | 1.98 ± 0.16 | 2.04 ± 0.14 | 2.88 | 0.84 |
| Resistance Index | 0.69 ± 0.04 | 0.70 ± 0.02 | 1.12 | 0.89 |
| Systolic:Diastolic | 3.63 ± 0.34 | 3.59 ± 0.28 | 0.97 | 0.95 |
| Fetal heart rate, bpm | 190.48 ± 3.38 | 199.81 ± 8.91 | 4.90 | 0.48 |
| Umbilical artery cross-sectional area, cm$^2$ | 0.24 ± 0.02 | 0.18 ± 0.02 | 25.23 | 0.09 |
| Umbilical artery cross-sectional diameter, cm | 0.55 ± 0.02 | 0.47 ± 0.03 | 15.31 | 0.08 |
| **130 dGA Transplacental Diffusion Blood Flow Measurements** | **(*n* = 4)** | **(*n* = 4)** | | |
| Uterine plasma flow (mL/min) | 1417.74 ± 328.48 | 807.99 ± 208.40 | 43.01 | 0.17 |
| Relative uterine blood flow (mL/min/kg fetus) | 500.07 ± 86.44 | 400.98 ± 59.35 | 19.81 | 0.38 |
| Relative uterine plasma flow (mL/min/kg fetus) | 348.20 ± 61.93 | 274.78 ± 41.11 | 21.09 | 0.36 |
| Uterine blood flow/100 g placenta | 429.05 ± 54.03 | 320.80 ± 59.32 | 25.23 | 0.23 |
| Umbilical plasma flow (mL/min) | 490.84 ± 52.03 | 293.18 ± 63.87 | 40.27 | 0.05 |
| Relative umbilical blood flow (mL/min/kg fetus) | 185.35 ± 8.81 | 155.61 ± 11.26 | 16.04 | 0.08 |
| Relative umbilical plasma flow (mL/min/kg fetus) | 123.17 ± 8.27 | 101.08 ± 12.74 | 17.93 | 0.20 |
| Umbilical blood flow/100 g placenta | 163.27 ± 14.51 | 123.84 ± 13.63 | 24.15 | 0.09 |
| Uterine: umbilical blood flow | 2.68 ± 0.37 | 2.58 ± 0.37 | 3.74 | 0.85 |
| Average umbilical arterial hematocrit | 0.34 ± 0.02 | 0.36 ± 0.04 | 6.25 | 0.65 |
| Average uterine arterial hematocrit | 0.30 ± 0.01 | 0.31 ± 0.02 | 2.94 | 0.64 |

Data are shown as means ± SEM for all pregnancies in each treatment group. CSH, chorionic somatomammotropin; RNAi, RNA interference.

*2.2. 130dGA Uterine and Umbilical Blood Flows*

As calculated by the transplacental diffusion technique, uterine blood (Figure 1) and plasma flows (mL/min; Table 2) in CSH RNAi pregnancies were not statistically different between treatments, nor were relative uterine blood or plasma flow (mL/min/kg fetus or mL/min/100 g placenta) different between treatments (Table 2). While caruncular endothelial nitric oxide synthase (NOS3) was numerically reduced by 38% in CSH RNAi pregnancies, this difference was not statistically significant ($p = 0.38$; Supplemental Figure S1).

Total umbilical (Figure 1) and plasma blood flows (mL/min) were reduced ($p \leq 0.05$; Table 2) by 40% in CSH RNAi pregnancies. Umbilical blood flow relative to fetal or placental weight (mL/min/kg fetus and mL/min/100 g placenta) both tended to be reduced ($p \leq 0.10$) in CSH RNAi pregnancies, but umbilical plasma flow relative to fetal weight was not different. Cotyledonary NOS3 was numerically elevated by 59% ($p = 0.11$; Supplemental Figure S1). Neither the uterine to umbilical blood flow ratio nor uterine and umbilical hematocrits were significantly altered by treatment.

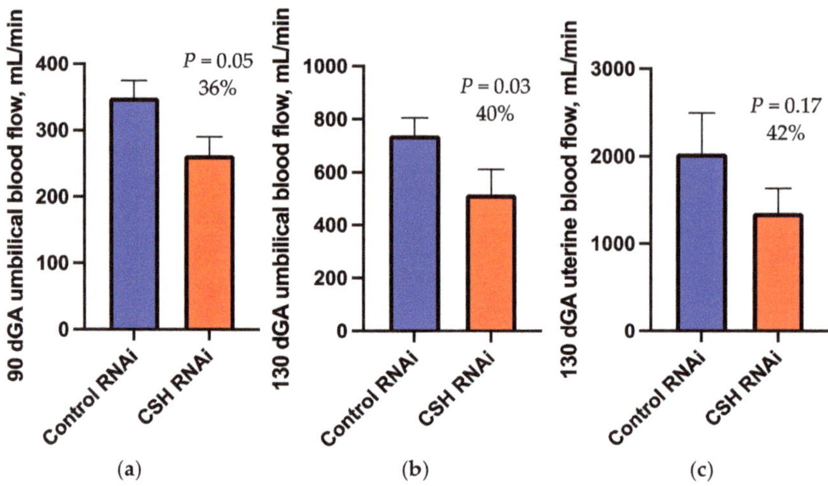



**Figure 1.** Measures of blood flow: (**a**) umbilical blood flow as assessed by Doppler ultrasound at 90 dGA; (**b**) umbilical blood flow as assessed by the transplacental diffusion technique at 130 dGA; (**c**) uterine blood flow as assessed by the transplacental diffusion technique at 130 dGA. Data are shown as means ± SEM for all pregnancies in each treatment group. CSH, chorionic somatomammotropin; RNAi, RNA interference.

### 2.3. Fetal and Uteroplacental Characteristics Near-Term (130 dGA)

Fetal weight was reduced ($p = 0.04$; Figure 2) by 30% by CSH RNAi, but the fetal measurements (Table 3) of crown-rump length (cm) and ponderal index did not differ between treatments. Fetal hindlimb leg length (cm) however, was reduced ($p = 0.02$) in CSH RNAi fetuses. Fetal liver weights tended ($p = 0.10$; Table 3) to be reduced in CSH RNAi pregnancies, with right liver lobe mass tending to be lighter ($p = 0.06$) by 37%. Fetal heart weight was also numerically smaller by 23% ($p = 0.11$) in CSH RNAi fetuses, with left ventricular weights significantly reduced ($p = 0.05$) by 26%. Fetal brain weight did not differ ($p = 0.68$) between treatments, but brain weight relative to fetal weight was increased ($p = 0.03$) by 43% in CSH RNAi fetuses, an indicator of asymmetric fetal growth. Fetal muscle mass including the biceps femoris, soleus, flexor digitorum superficialis (FDS), tibialis anterior (TA), and extensor digitorum longus (EDL) were all reduced ($p \leq 0.05$) in CSH RNAi fetuses.

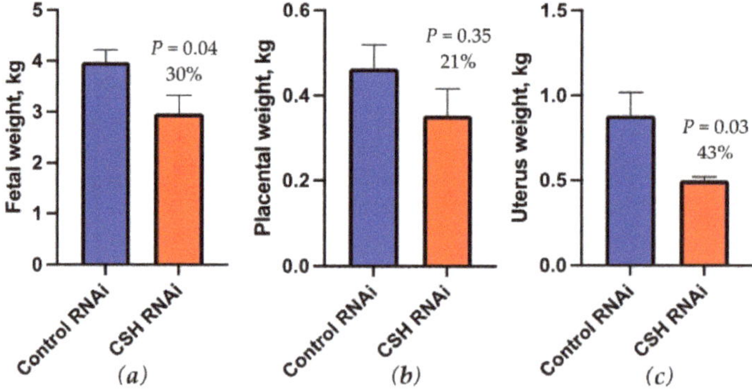



**Figure 2.** Measures of fetal and uteroplacental mass at 130 dGA: (**a**) fetal weight in kg (**b**) placental weight in kg; (**c**) uterus weight in kg. Data are shown as means ± SEM for all pregnancies in each treatment group. CSH, chorionic somatomammotropin; RNAi, RNA interference.

**Table 3.** Fetal weights, growth parameters, and uteroplacental characteristics assessed at necropsy (130 dGA).

|  | CON RNAi | CSH RNAi | % Change | *p*-Value |
|---|---|---|---|---|
|  | (*n* = 4) | (*n* = 4) |  |  |
| Crown-rump length, cm | 49.98 ± 1.67 | 45.73 ± 2.10 | 8.50 | 0.16 |
| Ponderal index | 3.19 ± 0.15 | 2.86 ± 0.24 | 10.48 | 0.28 |
| Lower leg length, cm | 37.25 ± 1.16 | 32.38 ± 1.15 | 13.09 | 0.02 |
| Brain, g | 47.71 ± 1.21 | 46.05 ± 3.72 | 3.48 | 0.69 |
| Brain: fetal weight | 0.0121 ± 0.000 | 0.0172 ± 0.002 | 42.63 | 0.03 |
| Liver, g | 100.33 ± 9.13 | 65.50 ± 15.80 | 34.71 | 0.10 |
| Brain: liver | 0.45 ± 0.03 | 0.79 ± 0.21 | 73.65 | 0.17 |
| Liver: fetal weight | 0.0267 ± 0.001 | 0.0247 ± 0.004 | 7.64 | 0.62 |
| Left liver lobe, g | 27.19 ± 4.53 | 17.74 ± 4.07 | 34.76 | 0.17 |
| Right liver lobe, g | 75.46 ± 3.24 | 47.88 ± 11.71 | 36.54 | 0.06 |
| Heart, g | 22.48 ± 1.79 | 17.28 ± 2.14 | 23.15 | 0.11 |
| Heart: fetal weight | 0.0056 ± 0.000 | 0.0063 ± 0.001 | 12.29 | 0.23 |
| Left ventricle, g | 9.02 ± 0.81 | 6.68 ± 0.48 | 25.91 | 0.05 |
| Right ventricle, g | 4.99 ± 0.32 | 3.81 ± 0.70 | 23.56 | 0.18 |
| Lungs, g | 123.18 ± 8.76 | 101.61 ± 15.44 | 17.51 | 0.27 |
| Lungs: fetal weight | 0.031 ± 0.001 | 0.0364 ± 0.001 | 17.31 | 0.01 |
| Pancreas, g | 3.25 ± 0.33 | 2.40 ± 0.42 | 26.21 | 0.16 |
| Kidneys, g | 20.05 ± 1.21 | 15.61 ± 2.33 | 22.17 | 0.14 |
| Perirenal adipose tissue (PRAT), g | 13.36 ± 1.96 | 10.58 ± 1.23 | 20.80 | 0.27 |
| Spleen, g | 6.48 ± 0.58 | 5.24 ± 1.24 | 19.24 | 0.40 |
| Adrenal glands, g | 0.32 ± 0.02 | 0.34 ± 0.06 | 4.38 | 0.86 |
| Biceps femoris (BF), g | 28.69 ± 1.22 | 17.97 ± 2.19 | 37.35 | 0.01 |
| Soleus, g | 0.35 ± 0.06 | 0.14 ± 0.04 | 60.28 | 0.02 |
| Flexor digitorum superficialis (FDS), g | 3.08 ± 0.27 | 1.93 ± 0.31 | 37.45 | 0.03 |
| Tibialis anterior (TA), g | 3.8 ± 0.35 | 2.37 ± 0.32 | 37.59 | 0.02 |
| Extensor digitorum longus (EDL), g | 1.01 ± 0.10 | 0.59 ± 0.10 | 42.22 | 0.03 |
| Uteroplacental weight, g | 1829.40 ± 136.67 | 1333.65 ± 207.93 | 27.10 | 0.09 |
| Membrane weight, g | 483.55 ± 32.18 | 470.28 ± 112.73 | 2.75 | 0.91 |
| Total placentome, # | 67.25 ± 4.09 | 71.75 ± 10.09 | 6.69 | 0.69 |

Data are shown as means ± SEM for all pregnancies in each treatment group. CSH, chorionic somatomammotropin; RNAi, RNA interference.

Placental weight of CSH RNAi pregnancies was not different from controls (Figure 2), however uterine weights were 43% smaller ($p$ = 0.03; Figure 2) compared with control pregnancies. Uteroplacental weights also tended ($p$ = 0.09; Table 3) to be reduced by 27% in CSH RNAi pregnancies. Fetal membrane weight and placentome number were not significantly altered by CSH RNAi. Cotyledonary CSH was numerically reduced by 36% in CSH RNAi placentae, but did not reach statistical significance ($p$ = 0.17; Supplemental Figure S2).

### 2.4. Blood Gas and Oxygen Uptakes

As calculated by the $^3H_2O$ transplacental diffusion technique, uterine oxygen uptake (mmol/min) was reduced ($p$ = 0.05; Figure 3) by 43% in CSH RNAi pregnancies, but not on a relative basis (mmol/min/kg uterus; Table 4). Umbilical oxygen uptake (mmol/min) was reduced ($p$ = 0.02; Figure 3) by 37% in CSH RNAi pregnancies, with relative (mmol/min/kg fetus) oxygen uptakes also tending ($p$ = 0.07) to be suppressed. Uteroplacental oxygen utilization, both absolute (mmol/min; Figure 3) and relative (mmol/min/kg placenta; Table 4) were numerically lower ($p$ = 0.12) in CSH RNAi pregnancies, but did not reach statistical significance. Uterine artery and vein blood gas and biochemistry measurements were not altered by CSH RNAi, (Supplemental Table S1), but both umbilical artery and vein oxygen content ($O_2$ ct; mmol/L) tended to be reduced ($p \leq 0.10$; Supplemental Table S2).

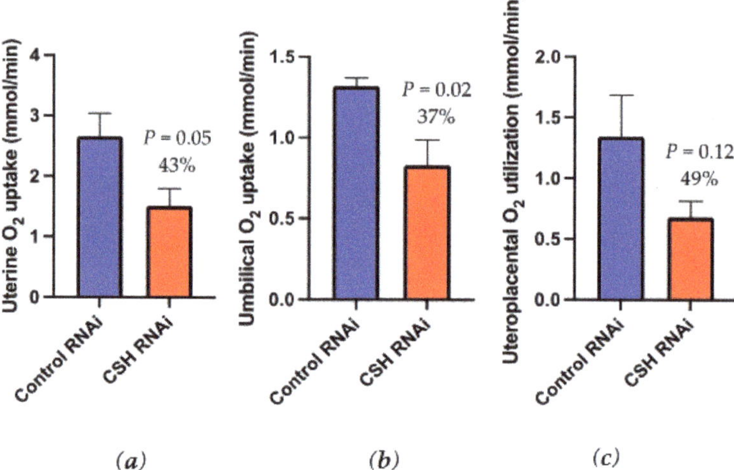

**Figure 3.** Uterine, umbilical, and uteroplacental oxygen uptakes as assessed by $^3H_2O$ transplacental diffusion at 130 dGA: (**a**) uterine oxygen uptakes, mmol/min (**b**) umbilical oxygen uptakes, mmol/min; (**c**) uteroplacental oxygen utilization, mmol/min. Data are shown as means $\pm$ SEM for all pregnancies in each treatment group. CSH, chorionic somatomammotropin; RNAi, RNA interference.

**Table 4.** In vivo measurements of nutrient transfer and uptake based on the $^3H_2O$ transplacental diffusion technique.

| 130 dGA Nutrient Uptakes | CON RNAi | CSH RNAi | % Change | *p*-Value |
|---|---|---|---|---|
| | (*n* = 4) | (*n* = 4) | | |
| Relative umbilical oxygen uptake (mmol/min/kg fetus) | 0.33 ± 0.01 | 0.29 ± 0.02 | 12.25 | 0.07 |
| Relative uterine oxygen uptake (mmol/min/kg uterus) | 3.18 ± 0.54 | 2.96 ± 0.45 | 6.93 | 0.76 |
| Relative uteroplacental oxygen utilization (mmol/min/kg placenta) | 2.78 ± 0.42 | 1.90 ± 0.25 | 31.61 | 0.12 |
| Relative umbilical glucose uptake (μmol/min/kg fetus) | 34.07 ± 2.98 | 24.75 ± 2.70 | 27.36 | 0.06 |
| Glucose transferred per placental weight (μmol/kg/min) | 301.82 ± 41.64 | 199.24 ± 32.59 | 33.99 | 0.10 |
| Relative uterine glucose uptake (μmol/min/kg uterus) | 515.33 ± 85.74 | 461.76 ± 70.98 | 10.39 | 0.65 |
| Relative uteroplacental glucose utilization (μmol/min/kg placenta) | 634.00 ± 59.81 | 472.72 ± 76.49 | 25.44 | 0.15 |
| Umbilical lactate uptake (μmol/min) | 123.16 ± 5.97 | 75.52 ± 16.11 | 38.68 | 0.03 |
| Relative umbilical lactate uptake (μmol/min/kg fetus) | 31.23 ± 1.78 | 26.26 ± 2.37 | 15.91 | 0.14 |
| Uterine lactate secretion (μmol/min) | 134.17 ± 1.78 | −80.41 ± 154.00 | 159.93 | 0.22 |
| Relative uterine lactate secretion (μmol/min/kg uterus) | 301.02 ± 49.60 | −518.48 ± 692.82 | 272.24 | 0.28 |
| Uteroplacental lactate production (μmol/min) | 257.33 ± 16.19 | −4.89 ± 168.14 | 101.90 | 0.17 |
| Relative uteroplacental lactate production (μmol/min/kg placenta) | 576.83 ± 68.83 | −311.58 ± 705.10 | 154.02 | 0.26 |

Data are shown as means $\pm$ SEM for all pregnancies in each treatment group. CSH, chorionic somatomammotropin; RNAi, RNA interference.

### 2.5. Glucose and Lactate Uptakes

Uterine glucose uptake (μmol/min) was reduced ($p$ = 0.04; Figure 4) by 45% in CSH RNAi pregnancies but relative uterine glucose uptake (μmol/min/kg uterus; Table 4) was not changed by treatment. Umbilical glucose uptake (μmol/min; Figure 4) was suppressed ($p$ = 0.02) by 47% in CSH RNAi fetuses, and by 27% ($p$ = 0.06) relative to fetal weight (μmol/min/kg fetus). Uteroplacental glucose utilization (μmol/min) was 44% lower ($p$ = 0.05; Figure 4) in CSH RNAi pregnancies whereas relative uteroplacental glucose utilization (μmol/min/kg placenta) was not statistically impacted ($p$ = 0.15). On a placental weight basis, the quantity of glucose transferred to the fetus (μmol/kg placenta/min) tended ($p$ = 0.10) to be reduced by 34% in CSH RNAi pregnancies. Due to the decrease in glucose uptake and utilization, we assessed the concentrations of the key placental glucose transporters, SLC2A1 and SLC2A3. The placental concentrations of SLC2A1 and SLC2A3 were not altered by treatment (Supplemental Figures S3 and S4).

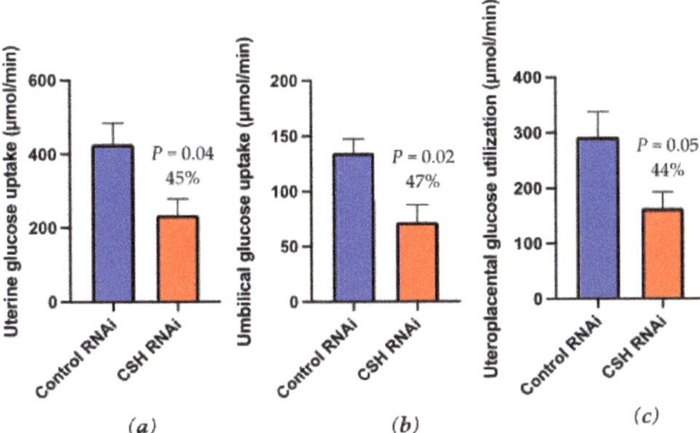


**Figure 4.** Uterine, umbilical, and uteroplacental glucose uptakes as assessed by $^3H_2O$ transplacental diffusion at 130 dGA: (**a**) uterine glucose uptakes, μmol/min (**b**) umbilical glucose uptakes, μmol/min; (**c**) uteroplacental glucose utilization, μmol/min. Data are shown as means ± SEM for all pregnancies in each treatment group. CSH, chorionic somatomammotropin; RNAi, RNA interference.

Uterine absolute (μmol/min; $p = 0.22$) and relative (μmol/min/kg uterus; $p = 0.28$) lactate secretions were not significantly impacted (Table 4) in CSH RNAi pregnancies. In contrast to control pregnancies which had positive uterine lactate secretion (Table 4), CSH RNAi pregnancies had negative uterine lactate secretion which indicates a net uptake of lactate by the uterine tissues. This was also reflected by negative uteroplacental utilization of lactate in CSH RNAi pregnancies (Table 4). In spite of limited uteroplacental lactate utilization, umbilical lactate uptake (μmol/min) was reduced ($p = 0.03$) by 39% in CSH RNAi fetuses.

### 2.6. Amino Acid Uptakes

The uterine uptakes (μmol/min) of alanine, arginine, asparagine, glutamine, histidine, isoleucine, leucine, lysine, ornithine, phenylalanine, serine, threonine, tyrosine, and valine were all reduced ($p \leq 0.05$; Figure 5) in CSH RNAi pregnancies, with the uptakes of both citrulline and methionine also tending ($p \leq 0.10$; Figure 5) to be reduced. The relative uterine uptakes of alanine and lysine (μmol/min/kg uterus) were also reduced ($p \leq 0.05$; Supplemental Table S3) in CSH RNAi pregnancies, while the relative uterine uptake of phenylalanine tended ($p = 0.07$) to be reduced.

Not only were the uterine uptakes of amino acids impaired in CSH RNAi pregnancies, but the umbilical uptakes (μmol/min) of asparagine, leucine, and tyrosine were also reduced, whereas the fetal production of glutamate was decreased ($p \leq 0.05$; Figure 6). Additionally, the umbilical uptakes of glutamine, alanine, and isoleucine also tended ($p \leq 0.10$; Figure 6) to be reduced in CSH RNAi fetuses. The relative fetal production of taurine (μmol/min/kg fetus) tended ($p = 0.08$; Supplemental Table S4) to be increased in CSH RNAi pregnancies whereas the relative umbilical uptake of asparagine tended ($p = 0.09$) to be reduced.

The placenta was also impacted by perturbed amino acid utilization in CSH RNAi pregnancies. The uteroplacental utilization (μmol/min) of glutamate and ornithine was reduced ($p \leq 0.05$; Figure 7) in CSH RNAi pregnancies, whereas the uteroplacental utilization of isoleucine and lysine was negative ($p \leq 0.10$). This response may indicate a lack of utilization to facilitate transfer to the fetus. On a relative basis (μmol/min/kg placenta), the uteroplacental utilization of lysine was negative ($p = 0.03$; Supplemental Table S5), again indicating a lack of utilization to facilitate transfer to the fetus. Additionally, the uteroplacental utilization of glutamate and isoleucine tended ($p \leq 0.10$) to be reduced in CSH RNAi pregnancies.

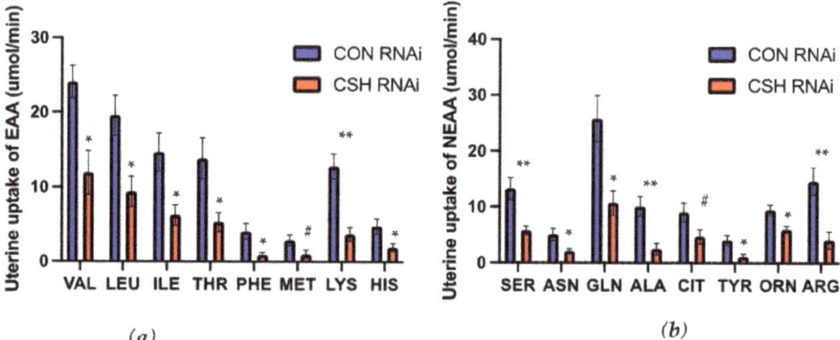



**Figure 5.** Uterine uptakes of essential (EAA) and nonessential (NEAA) assessed by $^3H_2O$ transplacental diffusion at 130 dGA: (**a**) uterine uptakes of essential amino acids, μmol/min (**b**) uterine uptakes of essential amino acids, μmol/min. Data are shown as means ± SEM for all pregnancies in each treatment group. CSH, chorionic somatomammotropin; RNAi, RNA interference. Ala, alanine; Arg, arginine; Asn, asparagine; Cit, citrulline; Gln, glutamine; His, histidine; Ile, isoleucine; Leu, leucine; Lys, lysine; Met, methionine; Orn, ornithine; Phe, phenylalanine; Ser, serine; Thr, threonine; Tyr, tyrosine; Val, valine. ** $p \leq 0.01$, * $p \leq 0.05$, # $p \leq 0.10$, when CSH RNAi pregnancies are compared with control RNAi pregnancies.

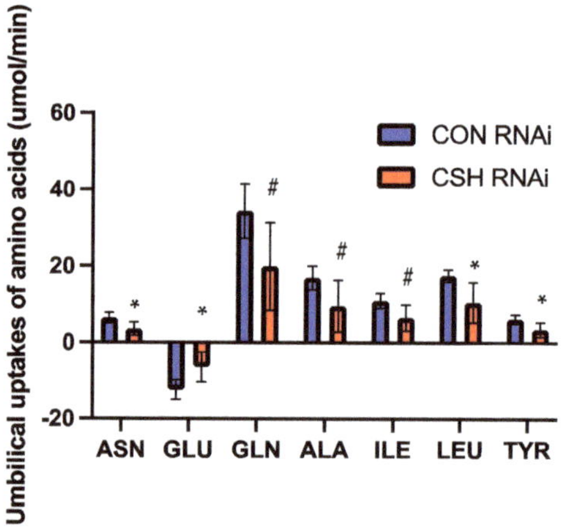



**Figure 6.** Umbilical uptakes (μmol/min) of amino acids as assessed by $^3H_2O$ transplacental diffusion at 130 dGA. Data are shown as means ± SEM for all pregnancies in each treatment group. CSH, chorionic somatomammotropin; RNAi, RNA interference. Ala, alanine; Asn, asparagine; Glu, glutamate; Gln, glutamine; Ile, isoleucine; Leu, leucine; and Tyr, tyrosine. * $p \leq 0.05$, # $p \leq 0.10$, when CSH RNAi pregnancies are compared with control RNAi pregnancies.

### 2.7. Total Nutrient Uptakes

With the transplacental diffusion technique, it is possible to calculate substrate specific carbon supply to both the uterus and the fetus (Table 1). Furthermore, it is possible to calculate the total carbon and nitrogen available for catabolic processes. As summarized in Supplemental Table S6, uterine carbon uptakes from amino acids ($p = 0.004$), glucose ($p = 0.04$), and lactate ($p = 0.22$) were reduced in CSH RNAi pregnancies. This resulted in a 60% reduction ($p = 0.04$) in total carbon available for uptake in CSH RNAi pregnancies. Uterine nitrogen uptake was also reduced ($p = 0.001$) by 61% in CSH RNAi pregnancies.

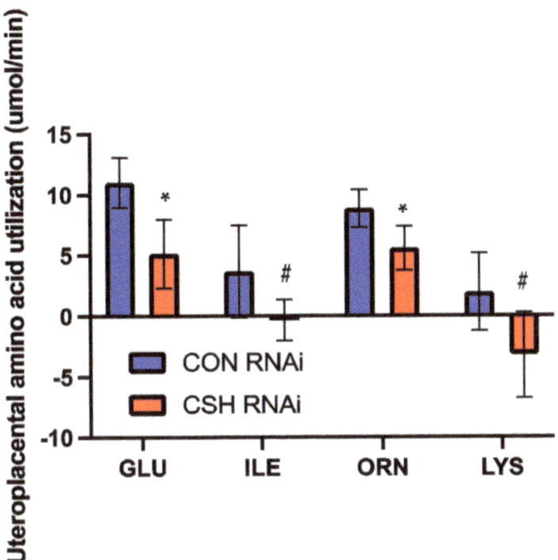



**Figure 7.** Uteroplacental utilization (μmol/min) of amino acids as assessed by $^3H_2O$ transplacental diffusion at 130 dGA. Data are shown as means ± SEM for all pregnancies in each treatment group. CSH, chorionic somatomammotropin; RNAi, RNA interference. Glu, glutamate; Ile, isoleucine; Lys, lysine; and Orn, ornithine. * $p \leq 0.05$, # $p \leq 0.10$, when CSH RNAi pregnancies are compared with control RNAi pregnancies.

Umbilical carbon uptakes from amino acids ($p = 0.13$; Supplemental Table S6), glucose ($p = 0.02$), and lactate ($p = 0.03$) were reduced in CSH RNAi fetuses leading to a 42% reduction ($p = 0.04$) in total carbon uptake. Umbilical nitrogen uptake was numerically lower ($p = 0.13$) in CSH RNAi fetuses. No umbilical nutrient to oxygen quotients differed between treatments (Supplemental Table S6).

### 2.8. 130dGA Hormones

Uterine concentrations of insulin, insulin-like growth factor 1 (IGF1), cortisol, and estradiol-17β were not impacted by RNAi treatment, as summarized in Table 5. In contrast, umbilical artery IGF1 was reduced ($p \leq 0.01$; Table 5) by 48% in CSH RNAi fetuses. Umbilical artery concentrations of insulin, cortisol and estradiol were not statistically impacted by CSH RNAi.

**Table 5.** Concentrations of maternal or fetal hormones.

| 130 dGA Nutrient Uptakes | CON RNAi | CSH RNAi | % Change | *p*-Value |
|---|---|---|---|---|
| **Maternal (Uterine)** | **($n = 4$)** | **($n = 4$)** | | |
| Uterine artery insulin, ng/mL | 0.92 ± 0.17 | 0.77 ± 0.10 | 15.92 | 0.49 |
| Uterine artery IGF1, ng/mL | 297.91 ± 45.18 | 260.24 ± 25.74 | 12.64 | 0.50 |
| Uterine artery cortisol, ng/mL | 126.53 ± 12.95 | 96.27 ± 29.07 | 23.91 | 0.38 |
| Uterine vein estradiol, pg/mL | 7.32 ± 2.34 | 4.42 ± 1.81 | 39.71 | 0.36 |
| **Fetal (Umbilical)** | **($n = 4$)** | **($n = 4$)** | | |
| Umbilical artery insulin, ng/mL | 1.12 ± 0.15 | 0.69 ± 0.23 | 38.63 | 0.16 |
| Umbilical artery IGF1, ng/mL | 140.89 ± 9.30 | 72.85 ± 14.94 | 48.29 | 0.01 |
| Umbilical artery cortisol, ng/mL | 11.98 ± 4.14 | 58.62 ± 29.83 | 389.25 | 0.17 |
| Umbilical vein estradiol, pg/mL | 5.59 ± 1.99 | 6.59 ± 0.75 | 17.99 | 0.65 |

Data are shown as means ± SEM for all pregnancies in each treatment group. CSH, chorionic somatomammotropin; RNAi, RNA interference.

### 3. Discussion

To describe the progression of CSH RNAi induced IUGR, we set out to document the physiological ramifications of CSH RNAi in a cohort of IUGR pregnancies. We hypothesized that CSH RNAi would reduce both fetal and maternal blood flow, resulting in reductions in global nutrient uptake. Unfortunately, due to fetal demise and catheter failures, complete studies were only accomplished on $n = 4$ pregnancies per group, somewhat limiting our statistical power. Additionally, while fetal sex was equally distributed (one female, three males) in each treatment, fetal sex could not be included in our statistical model. Regardless, results from this study directly support this hypothesis.

By using Doppler ultrasonography, we established that by 90 dGA, umbilical blood flow was reduced in CSH RNAi pregnancies. It is somewhat surprising, however, that the decrease in blood flow was not accompanied by a corresponding increase in vascular resistance, a hallmark of placental insufficiency [14,15]. This might suggest that CSH is not acting in a way that reduces placental vascularity, which would have likely resulted in increased vascular resistance, but in a way that modulates the size of major vessels such as the umbilical artery, which was reduced in our current study. While not statistically different ($p = 0.11$), the ~60% increase in cotyledonary NOS3 in the CSH RNAi near term (130 dGA) can perhaps, in part, explain why there was not an increase in vascular resistance. This is supported by an ex vivo study of placental perfusion from human fetal growth-restricted pregnancies, where flow mediated vasodilation was reduced but NOS3 was still elevated [16]. This combination of normal vascular resistance with decreased umbilical blood flow appears to be a unique distinction between CSH RNAi induced IUGR and other forms of placental insufficiency investigated in sheep. For example, in hyperthermia induced IUGR pregnancies, umbilical blood flow is reduced, and vascular resistance is increased [14,17]. Those pregnancies also have reduced umbilical artery NOS3 mRNA concentrations [18].

Near term (130 dGA) uterine and umbilical blood flow was assessed by the $^3H_2O$ transplacental diffusion technique. Total umbilical blood and plasma flows were significantly reduced in CSH RNAi pregnancies, as well as relative umbilical blood flows. This suggests the decrease in umbilical blood flow was somewhat independent of both fetal weight and placental weight. While the reduction (42%) in uterine blood flow in CSH RNAi pregnancies was not statistically different, it agrees with the statistically significant reduction in uterine blood flow reported earlier for CSH RNAi pregnancies [13]. In that study [13], uterine blood flow was reduced by only 24% with no change in uterine or fetal weights. Perhaps one driver of CSH RNAi induced IUGR is decreased uterine blood flow, which precedes a subsequent uterine mass reduction and ultimately prevents the uteroplacental unit from transporting adequate nutrients to sustain fetal growth.

With the noted decreases in blood flow, it is not surprising that fetal weights were reduced in this cohort of CSH RNAi pregnancies. Compared to a previous study that also reported near-term IUGR in CSH RNAi pregnancies, a similar degree of fetal growth restriction was observed with a 30% weight reduction in the current cohort vs. 32% in the Baker et al. [9] study. It appears that CSH RNAi induced IUGR can be classified as asymmetric IUGR as the brain to fetal weight ratio was significantly elevated, a proven indicator of asymmetric fetal growth [19]. This cohort of CSH RNAi pregnancies with IUGR also had reduced lower leg length, which was also observed in CSH RNAi pregnancies with normal fetal weights [13].

Similar to fetal weight reductions, the fetal livers of CSH RNAi pregnancies in this study were comparably reduced (35% vs. 36.5%) as in the Baker et al. [9] study, suggesting the necessity of CSH in preserving liver mass and corresponding liver-mediated IGF1 production. A novel finding reported on this cohort of CSH RNAi pregnancies was proportionately smaller hearts, especially the left ventricles. In a sheep model of experimentally induced placental growth restriction (carunclectomy), left ventricular cardiomyocytes number was reduced in the offspring during adulthood, and was correlated to birth weight [20]. Fetal muscle weights were also significantly reduced in this CSH RNAi IUGR cohort, as has been documented in other sheep models of IUGR [21,22]. In this current study, this effect does appear to be a function of fetal weight reduction, as normalizing each muscle relative to fetal mass removes the differences except for the flexor digitorum superficialis.

In the current cohort, there was a 21% reduction in placental weight, which wasn't statistically different from the control pregnancies. While it can certainly be argued that CSH RNAi did induce metabolic, hemodynamic and growth perturbations to these pregnancies, it is also accurate that this was a less severe placental growth restriction compared to the 52% placental mass reduction in the Baker et al. [9] study. While difficult to truly assess the differences between the two studies, one possible explanation includes variability of robustness in the activity of the RNAi over gestation. If the RNAi is more robustly expressed during key timepoints in placentation, it is possible that the degree of placental mass reduction would be more severe. Based on placental efficiency, the smaller placentas in the Baker et al. study [9] could be argued to be more efficient, as they produced approximately the same degree of fetal growth restriction as the current study. Thus, it is possible that the impacts of CSH RNAi yield consistent degrees of reduced fetal growth in the IUGR phenotype, which could suggest similar placental dysfunction between the two studies.

One truly novel insight into the current cohort provided was the potential impacts of CSH on uterine size, as it was not assessed in the Baker et al. [9] study. There is some evidence during early pregnancy that CSH can impact the uterus. In ovariectomized ewes that have been hormonally manipulated to mimic early pregnancy, CSH infusion acts on the endometrial glands [23]. Through the intrauterine infusion of exogenous recombinant CSH from 16–25 days post estrus, endometrial gland density increased [23]. A similar effect occurred when a separate treatment of growth hormone (GH) was infused over the same timeframe, suggesting that both CSH and GH may act on the early-pregnancy endometrium in a similar fashion [23]. This also may suggest a potential role for CSH, working through the same receptor to support uterine modifications for successful pregnancy outcomes.

The direct vs. indirect impacts of CSH on uterine mass are quite difficult to tease apart. In sheep, pregnancy increases nitric oxide, which has been shown to stimulate uterine blood flow and increase NOS3 in the uterine vascular endothelium [24]. As reported by Tanner et al. [13], CSH RNAi results in similar reductions in both uterine blood flow (24%) and caruncular NOS3 (24%), even in the absence of IUGR or uterine weight reductions. In the present cohort of CSH RNAi induced IUGR pregnancies, similar degrees of reductions in uterine blood flow (42%) and NOS3 (38%) were also observed. It is plausible that this relationship could influence the perfusion of the uterus, partially explaining the reduced uterine weights. This is supported by the global decrease in the uterine uptakes of oxygen, glucose, and amino acids.

Through the application of the $^3H_2O$ transplacental diffusion technique, the uptake of substrates from maternal circulation and their partitioning to uterus, fetus and placenta can be directly assessed. Pregnancies perturbed by IUGR often experience global reductions in the uptakes of all of the key substrates including oxygen, glucose, and amino acids. This study again demonstrated this relationship with global reductions in the uptakes of all three substrates by all aspects of uteroplacental circulation. It is worth noting that the uterine uptakes of substrates (Table 1) really represent the quantity of substrate the entire uteroplacental unit (uterus and placenta) is taking up from maternal circulation. When the umbilical uptake of the substrate is accounted for, the difference between uterine (uteroplacental) and umbilical uptakes represents what the uteroplacental unit is utilizing. A novel but not entirely unexpected finding is the global reduction in oxygen uptake by the CSH RNAi induced IUGR pregnancies. Oxygen uptake is considered a flow limited substrate [25,26] and blood flow was reduced in this cohort of CSH RNAi induced IUGR pregnancies. Other sheep models of placental insufficiency IUGR are also characterized by uteroplacental reductions in oxygen uptakes [19,27]. Kingdom and Kaufmann [28] described several modalities for examining pregnancy hypoxia and their respective etiologies. It appears that CSH deficiency results in placental hypoxia, where maternal blood is adequately oxygenated (as evidenced by no differences in maternal oxygen content, see

Supplemental Table S1) but oxygen cannot adequately diffuse across the placenta [28], as evidenced by reduced fetal uptakes and uteroplacental utilization. In this case, it appears that reduced oxygen uptake is likely originating from reduced blood flow, resulting in a smaller uterine mass which cannot transport as much oxygen from maternal circulation. Functionally, this means both the fetus and placenta had less oxygen available for oxidative metabolism and, not surprisingly, their growth was reduced.

While we have previously demonstrated that CSH RNAi results in reductions to the fraction of glucose transferred to the fetus, even in cases of normal fetal weights [13], the effects on glucose uptake in this cohort of CSH RNAi induced IUGR pregnancies were far more severe. As glucose is considered the primary substrate for fetal oxidative metabolism in both humans and sheep [29], it is not surprising that reductions in glucose uptake and utilization by the uteroplacental unit and the fetus in this cohort led to IUGR. Even when umbilical glucose uptake was normalized to fetal weight, glucose uptake was still reduced, suggesting a direct impact of CSH on umbilical glucose uptake as it is not just a function of fetal mass. CSH RNAi also reduced umbilical lactate uptake, leading to further reductions in carbon sources for the fetus.

Placental glucose transport is dependent on several factors, including the maternofetal glucose gradient, the availability of transporters to shuttle glucose across the placenta, the oxidative needs of the placenta, and placental mass [25,30]. In the current study, there were no significant changes to placental SLC2A1 or SLC2A3 concentrations. This fits with the lack of SLC2A1 changes in the human placentae of IUGR pregnancies [31]. Ultimately, this suggests that the reductions in glucose uptake and utilization by the uteroplacental unit are likely due to decreases in blood flow and uteroplacental mass.

One clear distinction between IUGR and normal weight CSH RNAi phenotypes is the impact on uteroplacental (placental) glucose utilization. In CSH RNAi pregnancies with normal fetal weights, uteroplacental glucose utilization was increased but the fetal uptakes of glucose were not altered [13]. This suggests that with reduced CSH, the placenta requires additional glucose to support appropriate fetal growth and transfer of nutrients to the fetus. This might hint at a potential compensatory mechanism in the normal weight pregnancies that was not sufficient to maintain fetal growth in our CSH RNAi IUGR pregnancies. In the more severe IUGR CSH RNAi phenotype, it is likely that this compensatory mechanism is overridden, possibly being preceded by greater proportional reductions in blood flow leading to a smaller uterus, unable to transfer enough glucose from maternal circulation to support adequate placental function and, therefore, fetal growth.

Amino acids are also considered an important oxidative substrate for fetal development [25]. In the current cohort of CSH RNAi pregnancies with IUGR, the uterine uptakes of eight essential amino acids (including all three branch-chain amino acids) were reduced along with eight nonessential amino acids. This resulted in substantial reductions in carbon (60%) and nitrogen (61%) taken up by the uterus (see Supplemental Table S6). This could contribute to the explanation of why the uteri containing CSH RNAi pregnancies were smaller. In normal weight CSH RNAi pregnancies, only the uterine uptake of taurine and glycine were reduced [13]. It is interesting that neither of these amino acids were statistically reduced in CSH RNAi IUGR pregnancies. Regardless of these variations, it does appear that uterine uptakes of amino acids were more profoundly impacted in cases of CSH RNAi IUGR.

The decreased uterine uptakes of amino acids also resulted in fewer amino acids available for fetal uptake in CSH RNAi pregnancies, with reduced uptakes of both isoleucine and leucine and five non-essential amino acids, including glutamine. Because the concentration of amino acids is usually higher in fetal circulation, the transport of amino acids is dependent on energy availability due to active transport mechanisms necessary to transport against the concentration gradient [32]. In CSH RNAi pregnancies with IUGR, there was reduced placental utilization of both glutamate and ornithine, as well as limited uteroplacental utilization of isoleucine and lysine. In both sheep and humans, there is no net placental transfer of glutamate [33,34]. The fetus takes up glutamine from uterine circulation, and the fetal liver converts it to glutamate to supply the placenta [35]. Thus, the decreased utilization of glutamate by the CSH RNAi uteroplacental unit is likely due to the decreased umbilical uptakes of glutamine, leading to the reduced production of glutamate from by the fetal liver.

One of the key postulated roles for CSH in regulating fetal growth is the stimulation of the insulin-like growth factors (IGFs) [36]. This is supported by suggestive evidence in humans of parallel increases in IGF1 and CSH in maternal circulation [37]. In hypophysectomized rats, ovine CSH was infused which stimulated IGF1 secretion [38]. Furthermore, in pregnant rats and sheep, plasma IGF1 concentrations are maintained after hypophysectomy until delivery of the placenta [36,39]. Based on that evidence, Handwerger hypothesized that CSH stimulated IGF1 in both maternal and fetal circulation [36]. Interestingly, in our CSH RNAi pregnancies [9,12,13], maternal concentrations of IGF1 were not influenced by CSH RNAi, which differs from what was originally hypothesized.

However, our data does support the proposed actions of CSH on IGF1 secretion in the fetus. Handwerger suggested CSH likely stimulated fetal IGF1 and IGF2 production, as evidenced by the treatment of rat embryonic fibroblasts with ovine CSH [40]. We have directly and repeatedly demonstrated that CSH RNAi causes reductions in umbilical IGF1 concentrations, in both the IUGR ([9] and the current cohort) and normal weight phenotypes [12]. Furthermore, CSH is also likely acting in a paracrine fashion on the placenta to impact placental IGFs, as evidenced by CSH RNAi reducing placental mRNA concentrations of IGF1 and IGF2 [10]. This is relevant to the current study, as one of the roles of the IGFs includes the placental transport of glucose and amino acids [41]. Because of the direct effects of CSH on the secretion of IGF1 in fetal circulation, it is not surprising that CSH RNAi results in reduced placental glucose and amino acid transfer, as well as subsequent IUGR.

The IGFs are not the only hormone postulated to be impacted by CSH. CSH has been postulated to increase maternal insulin secretion and impair glucose tolerance [36]. Ex vivo, CSH has been demonstrated to stimulate insulin secretion by isolated pancreatic islet cells [42]. However, our in vivo data do not support this, as in both the CSH RNAi IUGR ([9] and current study) and normal weight phenotypes [12], maternal circulating insulin concentrations are not altered. Furthermore, maternal glucose concentrations are not impacted by CSH RNAi in either the IUGR (current study) or normal weight [13] phenotypes. Maternal glucose and insulin concentrations were also unaltered in a model of CSH infusion into the late gestation sheep [43]. In the fetus, however, CSH does appear to contribute to insulin concentration. In previous [9] and current cohorts of CSH RNAi fetuses with IUGR, umbilical insulin concentrations were reduced by 48% and 39%, respectively. It would be very interesting to follow a cohort of CSH RNAi offspring into adulthood to examine adult glucose homeostasis and insulin sensitivity, as CSH RNAi results in both hypoglycemia and likely hypoinsulinemia. Overall, our in vivo data support the postulated roles of CSH in fetal circulation, but not maternal.

## 4. Materials and Methods

All animal procedures were approved by the Colorado State University Institutional Animal Care and Use Committee (Protocol # 18-7866A), the Institutional Biosafety Committee (18-029B) and the University of Colorado Anschutz Medical Campus Institutional Animal Care and Use Committee (Protocol #00714).

### 4.1. Lentiviral Generation

Lentiviral generations of hLL3.7 tg6 (target 6; CSH RNAi) and hLL3.7 NTS (scramble control/non-targeting sequence; control RNAi) are summarized in Supplemental Table S7 as described previously [9,10,13]. Briefly, both the NTS and tg6 sequences [9,10,13] were cloned into the LL3.7 vector, and all subsequent virus generation and titration were completed in accordance with our previously described procedures [9].

### 4.2. Generation of CSH RNAi Pregnancies

All ewes (Dorper breed composition) were group housed in pens at the Colorado State University Animal Reproduction and Biotechnology Laboratory, and were provided access to hay, trace mineral, and water in order to meet or slightly exceed their National Research Council [44] requirements. All animal procedures were conducted as previously described [9,10,13]. In summary, after synchronization and subsequent breeding, fully expanded and hatched blastocysts were collected by flushing the uteri at 9 days post conception. Each blastocyst was infected with 100,000 transducing units of either NTS/control RNAi or CSH RNAi virus. After infection for approximately 5 hours, each blastocyst was washed and a single blastocyst was transferred surgically [9,10,13] into a synchronized recipient ewe. Each recipient ewe was monitored daily for return to standing estrus, and confirmed pregnant at 50 days of gestational age (dGA) by ultrasound (Mindray Medical Equipment, Mahwah, NJ, USA). Using these methods, 6 Control RNAi and 6 CSH RNAi pregnancies were generated.

### 4.3. Doppler Velocimetry

At 90 dGA, all pregnancies ($n$ = 12) underwent Doppler velocimetry using a Mindray M7 Premium (Mindray) with a convex C5-2s probe (2.1–5.1 MHz) abdominal transducer by a single technician. All ewes were examined supine in a recumbent position. After confirming fetal viability, fetal binocular distance (cm), biparietal circumference (cm), abdominal circumference (cm), femur (cm) and tibia length (cm) were recorded and averaged across 3 independent measurements. Pulse-waved Doppler velocimetry measurements of the umbilical artery were performed with color flow Doppler with an angle of insonation of 30 degrees [14,17]. Doppler indices of perfusion including pulsatility indices (PI; Table 1) and resistance indices (RI; Table 1), as well as a systolic:diastolic ratio (S/D), were recorded on 3 independent sets of 3 cardiac cycles (minimum of 9 cardiac waveforms) and averaged [14,17]. Fetal heart rate was also recorded as a 3 cardiac cycle average. Umbilical blood flow (mL/min) was calculated by time-averaged mean velocity (cm/s) $\times$ $\pi/4$ $\times$ vessel cross-sectional area (cm$^2$) $\times$ 60. One control RNAi ewe had umbilical hemodynamic measurements excluded due to failure to achieve a 30-degree angle of insonation, but all fetal measurements from that animal were included.

### 4.4. Surgical Instrumentation of Fetus and Ewe

At approximately 115 dGA, pregnant recipient ewes were transported to the University of Colorado Anschutz Medical Campus, Perinatal Research Center (Aurora, CO, USA). Animals had access to ad libitum alfalfa pellets (Standlee Hay, Kimberly, ID) and water. All animals (6 = control RNAi and 6 = CSH RNAi pregnancies) underwent surgical placement of fetal and maternal catheters at 126 dGA, to determine blood flow and nutrient flux as previously described [13,19,45–48]. Briefly, the following catheters were placed: fetal descending aorta (representing umbilical artery blood), fetal femoral vein and umbilical vein, maternal femoral artery (representing uterine artery blood), maternal femoral vein, and uterine vein. Due to two fetal demises (one CSH RNAi and one control RNAi) and catheters failing to draw on the study day, a total of four control RNAi and four CSH RNAi animals completed the full study and were included in the final analysis. Fetal sex was balanced between both treatments, with 3 males and 1 female in each treatment group.

### 4.5. Blood Flow Calculations and Tissue Collection

At 130 dGA, uterine and umbilical blood flows were determined by the steady state $^3$H$_2$O transplacental diffusion technique, as summarized previously [13]. Briefly, samples were simultaneously collected from the maternal femoral artery (A), uterine vein (V), umbilical vein ($\gamma$) and fetal descending aorta ($\alpha$) every 20 minutes and averaged across 4 draws for analysis of blood biochemistry, nutrient content, $^3$H$_2$O and hormone concentrations [13].

As described extensively in Table 1, uterine and umbilical blood flows were calculated by the transplacental diffusion technique described previously [13,49,50]. Uterine, umbilical, and uteroplacental utilization of oxygen, glucose, lactate and amino acids were calculated by the transplacental diffusion technique [51] and reported as an average of draws one through four. All other calculations were described extensively in Tanner et al. [13].

Ewes and fetuses were euthanized, and tissues were harvested at 130 dGA as previously described by Tanner et al. [13]. Briefly, after trimming, placentomes were selected from each placenta and separated into cotyledonary (fetal) and caruncular (maternal) components, then snap frozen in liquid $N_2$ and stored at $-80$ °C. Fetal weight and dissected organ weights were recorded, and tissues were snap frozen in liquid nitrogen. Ponderal index and fetal brain:liver weight ratios were calculated, as previously described [13].

### 4.6. Biochemical Analysis of Blood Samples

Whole blood $O_2$ content, hemoglobin $O_2$ saturation ($SO_2$) partial pressure of oxygen ($PO_2$), partial pressure of carbon dioxide ($PCO_2$), pH and hematocrit measurements were analyzed by an ABL 800 Blood Gas analyzer, as previously described [13]. Plasma glucose and lactate were measured by Yellow Springs Instrument 2900 (YSI Incorporated, Yellow Springs, OH, USA), as described previously [13]. Plasma amino acids were measured by HPLC [13]. Maternal (uterine) and fetal (umbilical) concentrations of insulin, IGF1, and cortisol were assessed by enzyme-linked immunosorbent assay (ALPCO Immunoassays 80-INSOV-E01, 22-IGFHU-E01, and 11-CORHU-E01-SLV, respectively), as described previously [50,52,53]. Estradiol was analyzed by radioimmunoassay, as described by Gonzalez-Padilla et al. [54].

### 4.7. Western Blot Analysis

Protein isolation and analysis was conducted in accordance with methods described previously [13]. Cotyledonary or caruncular tissue (100 mg) was lysed in 500 μL of lysis buffer and sonicated on ice. For analysis of cotyledonary CSH, 5 μg of protein was electrophoresed through a 4–15% Tris-Glycine Stain-Free gel (BioRad, Hercules, CA, USA) and transferred via a Trans-Blot Turbo semi-dry transfer system (Bio-Rad) to a 0.20-μm pore nitrocellulose membrane. Total protein per lane was visualized using the ChemiDoc XRS+ chemiluminescence system (BioRad) to use for normalization. To visualize CSH, the blot was incubated in a 1:25,000 dilution (in 5% Non-Fat Dry Milk / 1X Tris-Bis Solution+ 1% Tween) of rabbit α- oPL-S4 [55] for 24 h at 4 °C. After washing the membrane, the membrane was transferred into a 1:5000 dilution (in 5% Non-Fat Dry Milk / 1X Tris-Bis Solution+ 1% Tween) of mouse α-rabbit IgG conjugated to horse radish peroxidase (ab-97051; Abcam, Cambridge, MA, USA). Nitrocellulose membranes were developed using an ECL Western Blotting Detection Reagent chemiluminescent kit (Amersham, Pittsburgh, PA, USA) and imaged using the ChemiDoc XRS+ chemiluminescence system. Densitometry calculations were performed using Image Lab Software (BioRad).

For analysis of caruncular and cotyledonary concentrations of NOS3, 20 μg of each sample was electrophoresed through 4–15% Tris-glycine stain-free gels (Bio-Rad), transferred and analyzed as described for CSH. NOS3 was detected using a 1:2000 dilution of mouse α-NOS3 (BD 610297; BD Biosciences, San Jose, CA, USA) and a 1:5000 dilution of goat α-mouse IgG conjugated to horseradish peroxidase (sc-2005; Santa Cruz Biotechnology Inc., Dallas, TX). For analysis of caruncular and cotyledonary (5 μg/sample) concentrations of SLC2A1 (GLUT1), samples were electrophoresed, transferred, and analyzed as described above. SLC2A1 was detected using a 1:40,000 dilution of rabbit α-SLC2A1 (07–1401; EMD Millipore) and a 1:80,000 dilution of goat α-rabbit IgG conjugated to horseradish peroxidase (ab205718; Abcam). As described above, densitometry of SLC2A1 was normalized on total protein/lane. Analysis of SLC2A3 (GLUT3) was conducted as previously detailed in Tanner et al. [13]. Briefly, 10 μg samples of caruncular or cotyledonary tissue were electrophoresed through NuPAGE 4–12% Bis-Tris Gels (Life Technologies), transferred to nitrocellulose, and the resulting blots were stained with Ponceau S to assess total protein per lane using the ChemiDoc XRS+ (Bio-Rad). Subsequent procedures were as described above, using a 1:1000 dilution of CSU-α-SLC2A3-22, and a 1:5000 dilution of a goat α-rabbit IgG conjugated to horseradish peroxidase (ab97051; Abcam). All densitometry was conducted in accordance with the description above.

### 4.8. Statistical Analysis

Data were analyzed by Student's t-test in GraphPad Prism (8.3.1). Due to the limited number of females in each treatment group ($n = 1$), fetal sex x treatment interactions were not examined. Statistical significance was set at $p \leq 0.05$ and a statistical tendency at $p \leq 0.10$. Data are reported as the mean $\pm$ standard error of the mean (SEM).

### 5. Conclusions

While CSH has been previously demonstrated as necessary for adequate fetal growth, the specific role of CSH in modulating fetal growth has yet to be determined. In this current study, we examined the physiological ramifications of CSH deficiency in cases of IUGR during late gestation. These data suggest that CSH is not only important for uterine blood flow and uteroplacental glucose utilization, but it also facilitates adequate umbilical blood flow necessary for the uptakes of oxygen, oxidative substrates, and hormones necessary to support fetal growth. Additionally, the current study demonstrated a novel role of CSH is supporting uterine growth to facilitate adequate transfer of nutrients.

**Supplementary Materials:** The following are available online at https://www.mdpi.com/article/10.3390/ijms22158150/s1, Table S1: Maternal (uterine) blood gas measurements, Table S2: Fetal (umbilical) blood gas measurements, Table S3: Relative uterine uptake of individual amino acids (μmol/min/kg uterus), Table S4: Relative umbilical uptake of individual amino acids (μmol/min/kg fetus), Table S5: Relative placental uptake of individual amino acids (μmol/min/kg placenta), Table S6: Total nutrient uptakes (μmol/min), Table S7: Scramble control and CSH-targeting shRNA sequences, Figure S1: Densiometric analysis of CSH in cotyledons, Figure S2: Densiometric analysis of NOS3 in cotyledons and caruncles, Figures S3 and S4: Densiometric analysis of SLC2A1 and SLC2A3 in caruncles and cotyledons.

**Author Contributions:** Conceptualization, R.V.A. and P.J.R.; methodology, R.V.A., P.J.R., A.R.T., C.S.L., V.C.K., A.A. and Q.A.W.; validation, A.R.T. and R.V.A.; formal analysis, A.R.T., R.V.A. and P.J.R.; investigation, R.V.A., P.J.R., A.R.T., C.S.L., V.C.K., A.A. and Q.A.W.; resources, A.R.T., R.V.A. and P.J.R.; writing—original draft preparation, A.R.T. and R.V.A.; writing—review and editing, R.V.A., P.J.R., A.R.T., C.S.L., V.C.K., A.A. and Q.A.W.; supervision, R.V.A. and P.J.R.; project administration, R.V.A. and P.J.R.; funding acquisition, R.V.A., P.J.R. and A.R.T. All authors have read and agreed to the published version of the manuscript.

**Funding:** This work was supported by National Institutes of Health Grants HD093701, HD094952, DK088139, and S10OD023553, and Agriculture and Food Research Initiative Grant 2019–67011-29614 from the United States Department of Agriculture.

**Institutional Review Board Statement:** All animal procedures were approved by the Colorado State University Institutional Animal Care and Use Committee (Protocol # 18-7866A; 04/26/2018) and the University of Colorado Anschutz Medical Campus Institutional Animal Care and Use Committee (Protocol #00714; 5/24/2018).

**Informed Consent Statement:** Not applicable.

**Data Availability Statement:** Not applicable.

**Acknowledgments:** The authors thank Bailyn Furrow, David Caprio, Gates Roe, Gregory Harding, Karen Tremble, Larry Toft, Megan Puget, Richard Brandes, and Vince Abushaban for animal care and additional technical support.

**Conflicts of Interest:** The authors declare no conflict of interest. The funders had no role in the design of the study; in the collection, analyses, or interpretation of data; in the writing of the manuscript, or in the decision to publish the results.

# References


1. Gagnon, R. Placental insufficiency and its consequences. *Eur. J. Obs. Gynecol. Reprod. Biol.* **2003**, *110*, S99–S107. [CrossRef]
2. Barker, D.J.; Osmond, C. Low birth weight and hypertension. *BMJ* **1988**, *297*, 134–135. [CrossRef]
3. Barker, D.J.; Bull, A.R.; Osmond, C.; Simmonds, S.J. Fetal and placental size and risk of hypertension in adult life. *BMJ* **1990**, *301*, 259–262. [CrossRef]
4. Barker, D.J. The fetal and infant origins of adult disease. *BMJ* **1990**, *301*, 1111. [CrossRef]
5. Hales, C.N.; Barker, D.J. The thrifty phenotype hypothesis. *Br. Med. Bull.* **2001**, *60*, 5–20. [CrossRef]
6. Daikoku, N.H.; Tyson, J.E.; Graf, C.; Scott, R.; Smith, B.; Johnson, J.W.C.; King, T.M. The relative significance of human placental lactogen in the diagnosis of retarded fetal growth. *Am. J. Obst. Gynecol.* **1979**, *135*, 516–521. [CrossRef]
7. Spellacy, W.N.; Buhi, W.C.; Birk, S.A. Human placental lactogen and intrauterine growth retardation. *Obstet. Gynecol.* **1976**, *47*, 446–448. [PubMed]
8. Lea, R.G.; Wooding, P.; Stewart, I.; Hannah, L.T.; Morton, S.; Wallace, K.; Aitken, R.P.; Milne, J.S.; Regnault, T.R.; Anthony, R.V.; et al. The expression of ovine placental lactogen, StAR and progesterone-associated steroidogenic enzymes in placentae of overnourished growing adolescent ewes. *Reproduction* **2007**, *133*, 785–796. [CrossRef] [PubMed]
9. Baker, C.M.; Goetzmann, L.N.; Cantlon, J.D.; Jeckel, K.M.; Winger, Q.A.; Anthony, R.V. Development of ovine chorionic somatomammotropin hormone-deficient pregnancies. *Am. J. Physiol. Regul. Integr. Comp. Physiol.* **2016**, *310*, R837–R846. [CrossRef]
10. Jeckel, K.M.; Boyarko, A.C.; Bouma, G.J.; Winger, Q.A.; Anthony, R.V. Chorionic somatomammotropin impacts early fetal growth and placental gene expression. *J. Endocrinol.* **2018**, *237*, 301–310. [CrossRef] [PubMed]
11. Rygaard, K.; Revol, A.; Esquivel-Escobedo, D.; Beck, B.L.; Barrera-Saldaña, H.A. Absence of human placental lactogen and placental growth hormone (HGH-V) during pregnancy: PCR analysis of the deletion. *Hum. Genet.* **1998**, *102*, 87–92. [CrossRef]
12. Ali, A.; Swanepoel, C.M.; Winger, Q.A.; Rozance, P.J.; Anthony, R.V. Chorionic somatomammotropin RNA interference alters fetal liver glucose utilization. *J. Endocrinol.* **2020**, *247*, 169–180. [CrossRef]
13. Tanner, A.R.; Lynch, C.S.; Ali, A.; Winger, Q.A.; Rozance, P.J.; Anthony, R.V. Impact of chorionic somatomammotropin RNA interference on uterine blood flow and placental glucose uptake in the absence of intrauterine growth restriction. *Am. J. Physiol. Regul. Integr. Comp. Physiol.* **2021**, *320*, R138–R148. [CrossRef]
14. Galan, H.L.; Anthony, R.V.; Rigano, S.; Parker, T.A.; de Vrijer, B.; Ferrazi, E.; Wilkening, R.B.; Regnault, T.R.H. Fetal hypertension and abnormal Doppler velocimetry in an ovine model of intrauterine growth restriciton. *Am. J. Obstet. Gynecol.* **2005**, *192*, 272–279. [CrossRef] [PubMed]
15. Regnault, T.R.H.; Galan, H.L.; Parker, T.A.; Anthony, R.V. Placental development in normal and compromised pregnancies—A review. *Placenta* **2002**, *16*, S119–S129. [CrossRef] [PubMed]
16. Jones, S.; Bischof, H.; Lang, I.; Desoye, G.; Greenwood, S.L.; Johnstone, E.D.; Wareing, M.; Sibley, C.P.; Brownbill, P. Dysregulated flow-mediated vasodilatation in the human placenta in fetal growth restriction. *J. Physiol.* **2015**, *593*, 3077–3092. [CrossRef] [PubMed]
17. Galan, H.L.; Hussey, M.J.; Chung, M.; Chyu, J.K.; Hobbins, J.C.; Battaglia, F.C. Doppler velocimetry of growth-restricted fetuses in an ovine model of placental insufficiency. *Am. J. Obstet. Gynecol.* **1998**, *178*, 451–456. [CrossRef]
18. Arroyo, J.A.; Anthony, R.V.; Parker, T.A.; Galan, H.L. eNOS, NO, and the activation of ERK and AKT signaling at mid-gestation and near-term in an ovine model of intrauterine growth restriction. *Sys. Biol. Reprod. Med.* **2010**, *56*, 62–73. [CrossRef] [PubMed]
19. Regnault, T.R.H.; de Vrijer, B.; Galan, H.L.; Davidsen, M.L.; Trembler, K.A.; Battaglia, F.C.; Wilkening, R.B.; Anthony, R.V. The relationship between transplacental O2 diffusion and placental expression of PlGF, VEGF and their receptors in a placental insufficiency model of fetal growth restriction. *J. Physiol.* **2003**, *550*, 641–656. [CrossRef]
20. Vranas, S.; Heinemann, G.K.; Liu, H.; De Blasio, M.J.; Owens, J.A.; Gatford, K.L.; Black, M.J. Small size at birth predicts decreased cardiomyocyte number in the adult ovine heart. *J. Dev. Orig. Health Dis.* **2017**, *8*, 618–625. [CrossRef]
21. Chang, E.I.; Rozance, P.J.; Wesolowski, S.R.; Nguyen, L.M.; Shaw, S.C.; Sclafani, R.A.; Bjorkman, K.K.; Peter, A.K.; Hay, W.W.; Brown, L.D. Rates of myogenesis and myofiber numbers are reduced in late gestation IUGR fetal sheep. *J. Endocrinol.* **2019**, *244*, 339–352. [CrossRef]
22. Rozance, P.J.; Zastoupil, L.; Wesolowski, S.R.; Goldstrohm, D.A.; Strahan, B.; Cree-Green, M.; Sheffield-Moore, M.; Meschia, G.; Hay, W.W.; Wilkening, R.B.; et al. Skeletal muscle protein accretion rates and hindlimb growth are reduced in late gestation intrauterine growth-restricted fetal sheep. *J. Physiol.* **2018**, *596*, 67–82. [CrossRef]
23. Spencer, T.E.; Gray, A.; Johnson, G.A.; Taylor, K.M.; Gerler, A.; Gootwine, E.; Ott, T.L.; Bazer, F.W. Effects of recombinant ovine interferon tau, placental lactogen, and growth hormone in the ovine uterus. *Biol. Reprod.* **1999**, *61*, 1409–1418. [CrossRef]
24. Magness, R.R.; Shaw, C.E.; Phernetton, T.M.; Zheng, J.; Bird, I.M. Endothelial vasodilator production by uterine and systemic arteries. II. Pregnancy effects on NO synthase expression. *Am. J. Physiol.* **1997**, *272*, H1730–H1740. [CrossRef] [PubMed]
25. Barry, J.S.; Anthony, R.V. The pregnant sheep as a model for human pregnancy. *Theriogenology* **2008**, *69*, 55–67. [CrossRef]
26. Carter, A.M. Factors affecting gas transfer across the placenta and the oxygen supply to the fetus. *J. Dev. Physiol.* **1989**, *12*, 305–322. [PubMed]
27. Regnault, T.R.H.; de Vrijer, B.; Galan, H.L.; Wilkening, R.B.; Battaglia, F.C.; Meschia, G. Development and mechanisms of fetal hypoxia in severe fetal growth restriction. *Placenta* **2007**, *28*, 714–723. [CrossRef] [PubMed]

28. Kingdom, J.C.P.; Kaufmann, P. Current topic: Oxygen and placental villous development: Origins of fetal hypoxia. *Placenta* **1997**, *18*, 613–621. [CrossRef]
29. Battaglia, F.C.; Meschia, G. *An Introduction to Fetal Physiology*; Academic Press: New York, NY, USA, 1986.
30. Hay, W.W.; Molina, R.A.; DiGiacomo, J.E.; Meschia, G. Model of placental glucose consumption and glucose transfer. *Am. J. Physiol.* **1990**, *258*, R569–R577. [CrossRef]
31. Jansson, T.; Wennergren, M.; Illsley, N.P. Glucose transporter protein expression in human placenta throughout gestation and in intrauterine growth retardation. *J. Clin. Endocrinol. Metab.* **1993**, *77*, 1554–1562. [PubMed]
32. Hay, W.W. Energy and substrate requirements of the placenta and fetus. *Proc. Nutr. Soc.* **1991**, *50*, 321–336. [CrossRef]
33. Lemons, J.A.; Adcock, E.W., III; Jones, M.D.; Naughton, M.A.; Meschia, G.; Battaglia, F.C. Umbilical uptake of amino acids in the unstressed fetal lamb. *J. Clin. Investig.* **1976**, *58*, 1428–1434. [CrossRef]
34. Schneider, H.; Mohlen, K.H.; Dancis, J. Transfer of amino acids across the in vitro perfused human placenta. *Ped. Res.* **1979**, *13*, 236–240. [CrossRef] [PubMed]
35. Battaglia, F.C. Glutamine and Glutamate Exchange between the Fetal Liver and the Placenta. *J. Nutr.* **2000**, *130*, 974S–977S. [CrossRef] [PubMed]
36. Handwerger, S. Clinical counterpoint: The physiology of placental lactogen in human pregnancy. *Endocr. Rev.* **1991**, *12*, 329–336. [CrossRef] [PubMed]
37. Breuer, C.B. Stimulation of DNA synthesis in cartilage of hypophysectomized rats by native and modified placental lactogen and available hormones. *Endocrinology* **1969**, *85*, 989. [CrossRef] [PubMed]
38. Hurley, T.W.; D'Ercole, A.J.; Handwerger, S.; Underwood, L.E.; Fulanetto, R.W.; Fellows, R.E. Ovine placental lactogen induces somatomedin: A possible role in fetal growth. *Endocrinology* **1977**, *101*, 1635. [CrossRef]
39. Daughaday, W.H.; Kapadia, M. Maintenance of serum somatomedin activity in hypophysectomized pregnant rats. *Endocrinology* **1978**, *102*, 1317. [CrossRef]
40. Adams, S.O.; Nissley, S.P.; Handwerger, S.; Rechler, M.M. Developmental patterns of insulin-like growth factor I and II synthesis and regulation in rat fibroblasts. *Nature* **1983**, *302*, 150–153. [CrossRef] [PubMed]
41. Kniss, D.A.; Shubert, P.J.; Zimmerman, P.D.; Landon, M.B.; Gabbe, S.G. Insulinlike growth factors. Their regulation of glucose and amino acid transport in placental trophoblasts isolated from first-trimester chorionic villi. *J. Reprod. Med.* **1994**, *39*, 249–256. [PubMed]
42. Martin, J.W.; Friesen, H.G. Effect of human placental lactogen on the isolated islets of Langerhans in vitro. *Endocrinology* **1969**, *84*, 619. [CrossRef]
43. Oliver, M.H.; Harding, J.E.; Breier, B.H.; Evans, P.C.; Gallaher, B.W.; Gluckman, P.D. The effects of ovine placental lactogen infusion on metabolites, insulin-like growth factors and binding proteins in the fetal sheep. *J. Endocrinol.* **1995**, *144*, 333–338. [CrossRef]
44. National Research Council. *Nutrient Requirements of Small Ruminants: Sheep, Goats, Cervids, and New World Camelids*; The National Academies Press: Washington, DC, USA, 2007.
45. Bonds, D.R.; Anderson, S.; Meschia, G. Transplacental diffusion of ethanol under steady state conditions. *J. Dev. Physiol.* **1980**, *2*, 409–416.
46. Brown, L.D.; Rozance, P.J.; Bruce, J.L.; Friedman, J.E.; Hay, W.W.; Wesolowski, S.R. Limited capacity for glucose oxidation in fetal sheep with intrauterine growth restriction. *Am. J. Physiol. Regul. Integr. Comp. Physiol.* **2015**, *309*, R920–R928. [CrossRef]
47. Hay, W.W.; Sparks, J.W.; Quissell, B.J.; Battaglia, F.C.; Meschia, G. Simultaneous measurements of umbilical uptake, fetal utilization rate, and fetal turnover rate of glucose. *Am. J. Physiol. Endocrinol. Metab.* **1981**, *240*, E662–E668. [CrossRef]
48. Jones, A.K.; Rozance, P.J.; Brown, L.D.; Goldstrohm, D.A.; Hay, W.W.; Limesand, S.W.; Wesolowski, S.R. Sustained hypoxemia in late gestation potentiates hepatic gluconeogenic gene expression but does not activate glucose production in the ovine fetus. *Am. J. Physiol. Endocrinol. Metab.* **2019**, *317*, E1–E10. [CrossRef]
49. Meschia, G.; Cotter, J.R.; Breathnach, C.S.; Barron, D.H. Simultaneous measurement of uterine and umbilical blood flows and oxygen uptake. *Q. J. Exp. Physiol.* **1966**, *52*, 1–18. [CrossRef]
50. Cilvik, S.N.; Wesolowski, S.R.; Anthony, R.V.; Brown, L.D.; Rozance, P.J. Late gestation fetal hyperglucagonaemia impairs placental function and results in diminished fetal protein accretion and decreased fetal growth. *J. Physiol.* **2021**, *599*, 3403–3427. [CrossRef] [PubMed]
51. Meschia, G.; Battaglia, F.C.; Hay, W.W.; Sparks, J.W. Utilization of substrates by the ovine placenta in vivo. *Fed. Proc.* **1980**, *39*, 245–249. [PubMed]
52. Andrews, S.E.; Brown, L.D.; Thorn, S.R.; Limesand, S.W.; Davis, M.; Hay, W.W.; Rozance, P.J. Increased adrenergic signaling is responsible for decreased glucose-stimulated insulin secretion in the chronically hyperinsulinemic ovine fetus. *Endocrinology* **2015**, *156*, 367–376. [CrossRef] [PubMed]
53. Benjamin, J.S.; Culpepper, C.B.; Brown, L.D.; Wesolowski, S.R.; Jonker, S.S.; Davis, M.A.; Limesand, S.W.; Wilkening, R.B.; Hay, W.W.; Rozance, P.J. Chronic anemic hypoxemia attenuates glucose-stimulated insulin secretion in fetal sheep. *Am. J. Physiol. Regul. Integr. Comp. Physiol.* **2017**, *312*, R492–R500. [CrossRef] [PubMed]
54. Gonzalez-Padilla, E.; Wiltbank, J.N.; Niswender, G.D. Puberty in beef heifers. The interrelationship between pituitary, hypothalamic and ovarian hormones. *J. Anim. Sci.* **1975**, *40*, 1091–1104. [CrossRef] [PubMed]
55. Kappes, S.M.; Warren, W.C.; Pratt, S.L.; Liang, R.; Anthony, R.V. Quantification and cellular localization of ovine placental lactogen messenger ribonucleic acid expression during mid- and late gestation. *Endocrinology* **1992**, *131*, 2829–2838. [CrossRef] [PubMed]

International Journal of
**Molecular Sciences**

*Article*

# Reduced Placental CD24 in Preterm Preeclampsia Is an Indicator for a Failure of Immune Tolerance

Marei Sammar [1,*], Monika Siwetz [2], Hamutal Meiri [3], Adi Sharabi-Nov [4], Peter Altevogt [5] and Berthold Huppertz [2]

[1]   Prof. Ephraim Katzir's Department of Biotechnology Engineering, ORT Braude College, 51 Snunit St, Karmiel 2161002, Israel
[2]   Division of Cell Biology, Histology and Embryology, Gottfried Schatz Research Center, Medical University of Graz, Neue Stiftingtalstr. 6/II, 8010 Graz, Austria; monika.siwetz@medunigraz.at (M.S.); berthold.huppertz@medunigraz.at (B.H.)
[3]   Hylabs, Rehovot and TeleMarpe, 21 Beit El St., Tel Aviv 6908742, Israel; hamutal62@hotmail.com
[4]   Ziv Medical Center, Safed, and Tel Hai College, Tel Hai 1220800, Israel; adi_nov@hotmail.com
[5]   Skin Cancer Unit, DKFZ and Department of Dermatology, Venereology and Allergology, University Medical Center Mannheim, Ruprecht-Karl University of Heidelberg, Theodor-Kutzer-Ufer 1–3, 68167 Mannheim, Germany; p.altevogt@dkfz-heidelberg.de
*   Correspondence: sammar@braude.ac.il; Tel.: +972-(04)-9901769; Fax: +972-(04)990171

**Citation:** Sammar, M.; Siwetz, M.; Meiri, H.; Sharabi-Nov, A.; Altevogt, P.; Huppertz, B. Reduced Placental CD24 in Preterm Preeclampsia Is an Indicator for a Failure of Immune Tolerance. *Int. J. Mol. Sci.* **2021**, 22, 8045. https://doi.org/10.3390/ijms22158045

Academic Editors: Hiten D. Mistry and Eun Lee

Received: 6 July 2021
Accepted: 21 July 2021
Published: 28 July 2021

**Publisher's Note:** MDPI stays neutral with regard to jurisdictional claims in published maps and institutional affil-iations.



**Abstract:** Introduction: CD24 is a mucin-like glycoprotein expressed at the surface of hematopoietic and tumor cells and was recently shown to be expressed in the first trimester placenta. As it was postulated as an immune suppressor, CD24 may contribute to maternal immune tolerance to the growing fetus. Preeclampsia (PE), a major pregnancy complication, is linked to reduced immune tolerance. Here, we explored the expression of CD24 in PE placenta in preterm and term cases. Methods: Placentas were derived from first and early second trimester social terminations (N = 43), and third trimester normal term delivery (N = 67), preterm PE (N = 18), and preterm delivery (PTD) (N = 6). CD24 expression was determined by quantitative polymerase chain reaction (qPCR) and Western blotting. A smaller cohort included 3–5 subjects each of term and early PE, and term and preterm delivery controls analyzed by immunohistochemistry. Results: A higher expression (2.27-fold) of CD24 mRNA was determined in the normal term delivery compared to first and early second trimester cases. The mRNA of preterm PE cases was only higher by 1.31-fold compared to first and early second trimester, while in the age-matched PTD group had a fold increase of 5.72, four times higher compared to preterm PE. The delta cycle threshold (ΔCt) of CD24 mRNA expression in the preterm PE group was inversely correlated with gestational age (r = 0.737) and fetal size (r = 0.623), while correlation of any other group with these parameters was negligible. Western blot analysis revealed that the presence of CD24 protein in placental lysate of preterm PE was significantly reduced compared to term delivery controls ($p$ = 0.026). In immunohistochemistry, there was a reduction of CD24 staining in villous trophoblast in preterm PE cases compared to gestational age-matched PTD cases ($p$ = 0.042). Staining of PE cases at term was approximately twice higher compared to preterm PE cases ($p$ = 0.025) but not different from normal term delivery controls. Conclusion: While higher CD24 mRNA expression levels were determined for normal term delivery compared to earlier pregnancy stages, this expression level was found to be lower in preterm PE cases, and could be said to be linked to reduced immune tolerance in preeclampsia.

**Keywords:** CD24; preeclampsia; placenta; cytotrophoblast; syncytiotrophoblast; immunohistochemistry; quantitative polymerase chain reaction (qPCR); immune tolerance

## 1. Introduction

The presence of foreign tissues in a host leads to a strong immune rejection directed to destroy the allo-antigen-expressing tissue. Such a response is not developed in normal pregnancy, although half of the fetal genes come from the father and hence are foreign to the mother. The maternal tolerance to the semi-allogeneic fetus enables a normal fetal development that is achieved due to multiple mechanisms [1,2]. This unique example of immune system suppression to hinder a destructive allo-immune response is supported by several placenta-specific genes including proteins such as the signal transducing CD24 protein [3] and placental protein 13 [4]. Many studies have drawn correlations between immune tolerance during pregnancy and tolerance to malignant tumor growth with respect to proliferation, invasion, and immune modulation [5]. Accordingly, many cell pathways used by tumors were evaluated for their function in normal and complicated placental development and vice versa [3].

The CD24 protein is made of a small core of 27 amino acids, which is attached to the membrane via a glycosylphosphatidylinositol (GP-I) anchor [6,7]. It has many potential glycosylation sites for N- and O-linked carbohydrates, rendering the molecule structurally similar to mucins [8]. P-selectin was identified to bind to CD24 under physiological conditions supporting the adhesion of leukocytes to endothelial cells and platelets [8–10]. Siglec-10 was found to serve as an additional receptor for CD24 in the immune system [11]. The Siglec10–CD24 binding axis was demonstrated to be an important immune checkpoint for the induction of immune tolerance in mouse autoimmune models [12]. Importantly, recombinant CD24 fusion protein bound to the fragment crystallizable (Fc) region of the immunoglobulin (CD24-Fc) was recently demonstrated to be a powerful drug for blocking overshooting immune reactions (cytokine storm) in SARS-2-COVID-19 infections [13].

In a previous study, we suggested that CD24 may provide similar immune-suppressive features to the developing placenta during pregnancy. We showed that CD24 is expressed in glandular epithelial cells of uterine glands and in other decidual cells [14]. The expression was found to be in close vicinity to invasive extravillous trophoblasts [14]. Additionally, as in tumor cells, first trimester CD24 was found to be co-expressed with Siglec 10 [15,16].

Preeclampsia (PE) is one of the major pregnancy complications. Annually, this life-threatening complication affects 10 million pregnant women globally. It is characterized by hypertension and proteinuria and/or organ failure, requiring earlier delivery to save women's lives from seizures and stroke (eclampsia). If developed early, inducing premature delivery is often required to save the mother's life. However, premature delivery may in turn be accompanied by fetal loss and major newborn disabilities [17,18].

The origin of PE, representing a multifactorial disorder, is unknown, but several studies have suggested that it is tightly linked to the loss of maternal immune tolerance to the growing fetus [4,19,20]. When the rejection develops early in pregnancy, it often leads to miscarriage and pregnancy loss [4,21,22]. Overall, immune rejection is implicated as an important process, leading to the clinical features of preeclampsia [19,20,23]. About 70% of all PE cases show clinical symptoms around term (late PE), while about one-third of the cases develop before term (preterm PE, before 37 weeks gestation), of which a minor fraction of cases develops symptoms and is delivered before 34 weeks (early PE) [17,18]. The latter are typically very severe cases and often accompanied by fetal growth restriction (FGR) requiring admission to intensive care units (NICU) due to lower birthweight and many other complications [24–26].

In this study, we explored the expression and distribution of CD24 in placentas from the first and early second trimesters and compared them to cases of the third trimester including cases of normal term delivery controls, preterm delivery (PTD), and preterm ($\leq$37 weeks) PE. A smaller cohort was used to compare the latter third trimester cases also to PE cases at term ($\geq$37 weeks).

## 2. Results

### 2.1. The Cohort for Testing CD24 mRNA Expression

The first trimester and early second trimester placentas were derived from social termination of pregnancy at a mean gestational age of 8.6 weeks (Table 1). The control group of third trimester placentas collected at delivery had a mean gestational week at delivery of 38.3 weeks compared to 30.1 weeks in the preterm PE group and 32.8 weeks in the age-matched PTD group. Blood pressure, mean arterial blood pressure, and proteinuria were significantly higher in the preterm PE group compared to any of the other groups. Birthweight was significantly lower for both the preterm PE and PTD groups compared to term controls. However, in the preterm PE group birthweights were at the lower 10th centile, while all babies of the PTD group were >45% centile according to the gestational age (Table 1).

**Table 1.** Demographic and pregnancy characterization—Israelian cohort.

| Parameter (Mean (Range)) | First and Early Second Trimester (*n* = 43) | Normal Term Delivery (*n* = 67) | Preterm PE (<37 Weeks) (*n* = 18) | PTD (<37 Weeks) (*n* = 6) |
|---|---|---|---|---|
| Gestational age at delivery (weeks) | 8.6 (6.0–18.0) | 38.3 * (37.0–41.0) | 30.1 *# (24.0–36.0) | 32.8 *# (30.0–36.0) |
| Maternal age (years) | 27 (19–39) | 32 (21–42) | 29 (20–47) | 30 (20–41) |
| Ethnicity (Jew, n, %) | 40 (93.0%) | 56 (84.8%) | 13 (72.2%) | 5 (83.3%) |
| Parity | 0.9 (0.6–1.2) | 1.3 (1.1–1.6) | 0.6 # (0.1–1.0) | 1.3 (0.2–2.4) |
| Growth centile (%) | 77 (32–99) | 84 (59–108) | 7 *# (3–10) | 54 *# (45–74) |
| Urine protein (mg/mL) | NA | NA | 3057 # (0–17,000) | 5.0 ω (0–305) |
| Systolic BP (mmHg) | 100 (89–134) | 122 (93–152) | 169 *# (134–220) | 125 ω (115–135) |
| Diastolic BP (mmHg) | 69 (60–82) | 70 (51–87) | 102 *# (81–120) | 69 ω (60–82) |
| MAP (mmHg) | 79 (70–98) | 87 (46–109) | 125 *# (99–147) | 88 ω (79–100) |
| Baby's birthweight (grams) | - | 3428 (2500–4290) | 1228 # (470–2520) | 1858 #ω (1135–2645) |
| Baby's gender (male, n, %) | - | 35 (52%) | 7 (41%) | 4 (67%) |

Values are shown as means and ranges. Mann–Whitney analysis was used to calculate the statistical difference between the groups: significantly different values are marked when *p* < 0.05, as * compared to first trimester, # as compared to term delivery, and ω for preterm PE compared to PTD. NA, not available. No gender and weight were determined in social termination.

## 2.2. Expression of CD24 mRNA in Placental Tissues throughout Pregnancy

Real-time qPCR analysis revealed a 2.27-fold increase of CD24 in normal term delivery cases compared to 5.72-fold increase in the PTD group and 1.31-fold in preterm PE, all compared to the first and early second trimester group (Table 2, Figure 1). There was a large diversity among individuals in any group. The results are thus best described as a tendency towards increased expression when the first and early second trimester expression level is evaluated versus normal term delivery, a much larger tendency for increase when compared to PTD, and a much lower tendency towards decreased expression in preterm PE, especially compared to the age-matched PTD cases.

**Table 2.** Placental CD24 mRNA expression.

| Median (95% CI) | First and Early Second Trimester (*n* = 43) | Normal Term Delivery (*n* = 67) | Preterm PE (*n* = 18) | PTD (*n* = 6) | *p*-Value |
|---|---|---|---|---|---|
| ΔCT | −0.01 (−1.30–0.71) | −1.19 (−1.79–(−0.09)) | −0.37 (−1.36–0.85) | −2.52 (−2.92–(−0.57)) | 0.151 |
| ΔCT/growth centile | −0.001 (−0.015–0.008) | −0.014 (−0.021)–(−0.001) | −0.018 (−0.138–0.157) | −0.044 (−0.054)–(−0.011) | 0.126 |
| Expression fold change ($2^{-\text{Median}\,\Delta\Delta CT}$) | 1.00 | 2.27 (1.05–3.43) | 1.31 (0.55–2.54) | 5.72 (1.47–7.50) | 0.353 |

ΔCT: expression cycle (CD24)-expression cycle (housekeeping gene). ΔΔCT = ΔCT (X) − median ΔCT (first and early second trimester) (95% CI). *p*-values were calculated by the Kruskal–Wallis a-parametric test between all four groups, which showed no statistical differences. Mann–Whitney comparison between any group pair was insignificant.

## Placental Expression of CD24

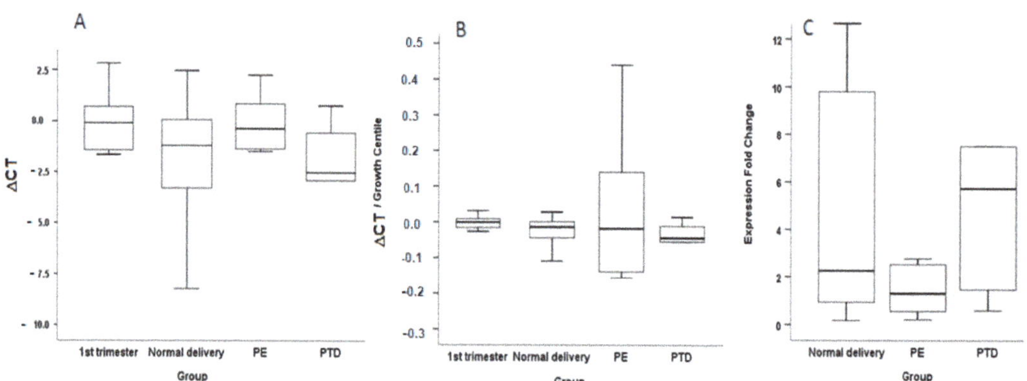

**Figure 1.** Box plot of placental mRNA expression of CD24. (**A**) ΔCT, (**B**) ΔCT versus fetal growth, (**C**) expression fold ($2^{-\text{Median}\Delta\Delta CT}$) change.

### 2.3. Changes in CD24 mRNA Expression in Preterm Preeclampsia

For the preterm PE group, sharp inverse relationships were identified when the relative placental CD24 mRNA expression (ΔCT) was plotted against the fetal growth centile (Figure 2 left), with a correlation coefficient of $R^2 = 0.388$. The ratio of ΔCT standardized to fetal growth (ΔCT/growth centile) plotted against gestational age yielded an $R^2 = 0.545$ (Figure 2 right). These results indicated a direct correlation between CD24 mRNA level, fetal growth, and gestational age. There was no correlation for any of the other groups ($R^2 < 0.1$, Figure 2 left and right). Accordingly, only in cases of preterm PE did the CD24 mRNA expression increase with fetal weight and gestational age, while in all the other groups, including the age-matched PTD cases, no correlation between CD24 expression and fetal growth or gestational age could be identified. The variance of CD24 mRNA expression in the four groups is depicted in Figure 2.

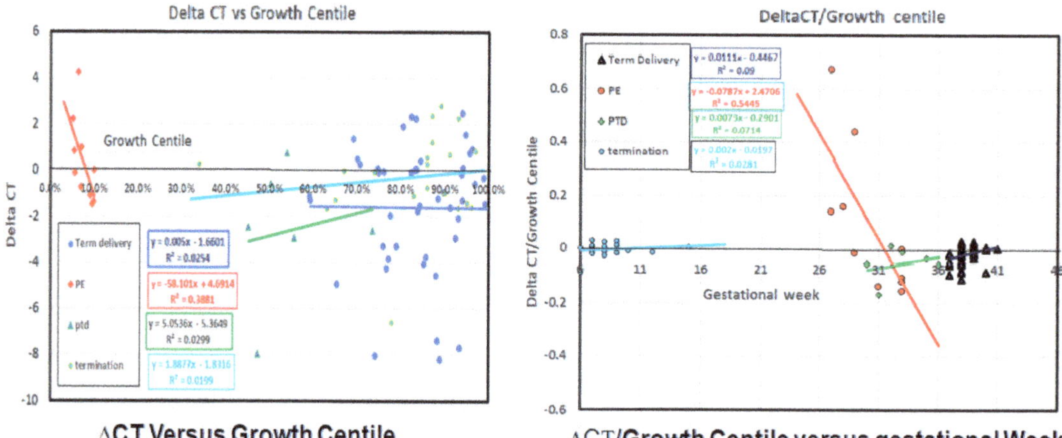

**Figure 2.** CD24 mRNA expression and fetal growth characteristics. **Left**-ΔCT versus growth centile. The mRNA expression assessed via ΔCT is presented for the various clinical outcome groups, indicating high correlation only for the case of the preterm preeclampsia (PE) group. **Right**-ΔCT/growth centile versus gestational week. The ΔCT/growth centile shows a correlation to gestational week at delivery only for the preterm PE cases. PE-preeclampsia, PTD-preterm delivery. Linear regression for each group is shown with regression curve and coefficient listed in the figure.

### 2.4. The Cohort for Testing Placental CD24 Protein Localization

The comparison of CD24 immunostaining in term and early PE became available through a smaller cohort collected at the Medical University of Graz, Austria (Table 3). In this cohort, the early PE cases and the late PE cases had higher blood pressure and proteinuria as anticipated from the definition of this complication (Table 3) compared to their matched preterm and term delivery groups. The early and late PE cases had similar disease severity in terms of elevated blood pressure and proteinuria, and the difference in severity was due to earlier delivery and lower birthweight (Table 3).

**Table 3.** Demographic and pregnancy characterization—Austrian cohort.

| Parameter | Early PE (n = 3) | PTD (n = 3) | Late PE (n = 3) | Term Control (n = 5) | $p$-Value (#) |
|---|---|---|---|---|---|
| Gestational age at delivery (weeks) | 32.3 [b] (31.0–34.0) | 31.3 [b] (29.0–34.0) | 38.7 [ab] (38.0–40.0) [y] | 40.0 [a] (39.0–41.0) [e] | 0.013 |
| Proteinuria (g/24 h) | 2973 [a] (2113–4562) * | 0 [b] (0–0) | 2293 [ab] (2293–2293) * | 0 [b] (0–0) | 0.015 |
| Systolic BP (mmHg) | 171 (150–196) | NA | 176 (148–202) * | 109 (90–111) | 0.319 |
| Diastolic BP (mmHg) | 106 (100–114) | NA | 104 (96–116) * | 80 (75–85) | 0.319 |
| MAP (mmHg) | 128 (117–141) | NA | 128 (117–145) * | 97 (90–9100) | 0.319 |
| Mode of delivery (vaginal, n, %) | 3 (100%) | 3 (100%) | 2 (66.7%) * | 1 (33.3%) | 0.506 |
| Baby's birthweight (gram) | 1267 [b] * | 2030 [ab] (1870–2190) | 2939 [a] (2398–3630) [y] | 3023 [a] (2905–3140) [e] | 0.052 |
| Baby's gender (male, n, %) | 1 (33.3%) | 2 (66.7%) | 2 (66.7%) | NA | 0.836 |
| Growth centile (%) | 12.6 (0.9–21.5) | 65.2 (51.0–79.3) | 51.9 (4.7–90.5) | 39.0 (29.9–48.2) | 0.218 |
| Placental weight (gram) | 350 * | 440 [b] (350–500) | 520 [ab] (450–650) *[y] | 617 [a] (465–720) [e] | 0.041 |
| Placental weight/fetal weight (ratio) | 3.7 [b] (3.3–4.1) * | 5.7 [a] (5.3–6.1) | 5.0 [a] (4.7–5.3) *[y] | 4.7 [ab] (4.5–4.8) | 0.049 |

Values are shown as means and ranges. * $p \leq 0.05$ for Mann–Whitney a-parametric test between early PE to PTD or term PE and term delivery. [y] $p \leq 0.05$ for Mann–Whitney a-parametric test between early to late PE. [e] $p \leq 0.05$ for Mann–Whitney a-parametric test between PTD to term delivery. # Kruskal–Wallis a-parametric test between all four groups. In Kruskal–Wallis analysis, "a" is significantly higher from values of other groups, "b" is significantly lower, "ab" is in between "a" and "b", and "c" is the lowest. NA, not available. $p$-values in bold indicate significant differences.

### 2.5. Pattern of Immunostaining with CD24-Specific Antibodies

Our previous study [14] has shown that in the first trimester, there is clear immuno-labeling for CD24 in the villous and extravillous cytotrophoblasts with a lower staining intensity in villous stroma cells. Interestingly, the protein was not found to be localized in the first trimester syncytiotrophoblast. In this study, cases from preterm delivery as well as normal term and late PE cases showed a varying degree of staining in the syncytiotrophoblast with only minor staining of villous cytotrophoblasts. By contrast, in cases of the early PE group, there was a strong staining of the villous cytotrophoblasts with only minor staining of the syncytiotrophoblast (Figure 3A–D).

Looking at the images of this study in detail, it appears as if in the PTD (Figure 3B1) and normal term delivery group (Figure 3D1) there was a moderate staining of the syncytiotrophoblast (blue arrows) and light staining of villous blood vessels (red arrows, Figure 3B2,D2), with very little staining of the villous cytotrophoblast (green arrow, Figure 3B1,D1). In late PE, staining of the syncytiotrophoblast showed very strong impregnation (blue arrow, Figure 3C1), while blood vessels staining remained very weak (Figure 3C2). By contrast, in the early PE cases, as in first trimester placenta, the villous cytotrophoblasts were stained (green arrow, Figure 3A1), while the syncytiotrophoblast remained mostly negative or pale (blue arrow, Figure 3A1).

The strongest staining for CD24 was found in the villous trophoblast layer of cases with late PE (Figure 3C) compared to moderate staining in PTD (Figure 3B) and normal term cases (Figure 3D). CD24 staining in cases from the early PE group was very low (Figure 3A). Villous stroma and blood vessels showed light staining for CD24, and there was no clear staining pattern (Figure 3A–D, red arrows).

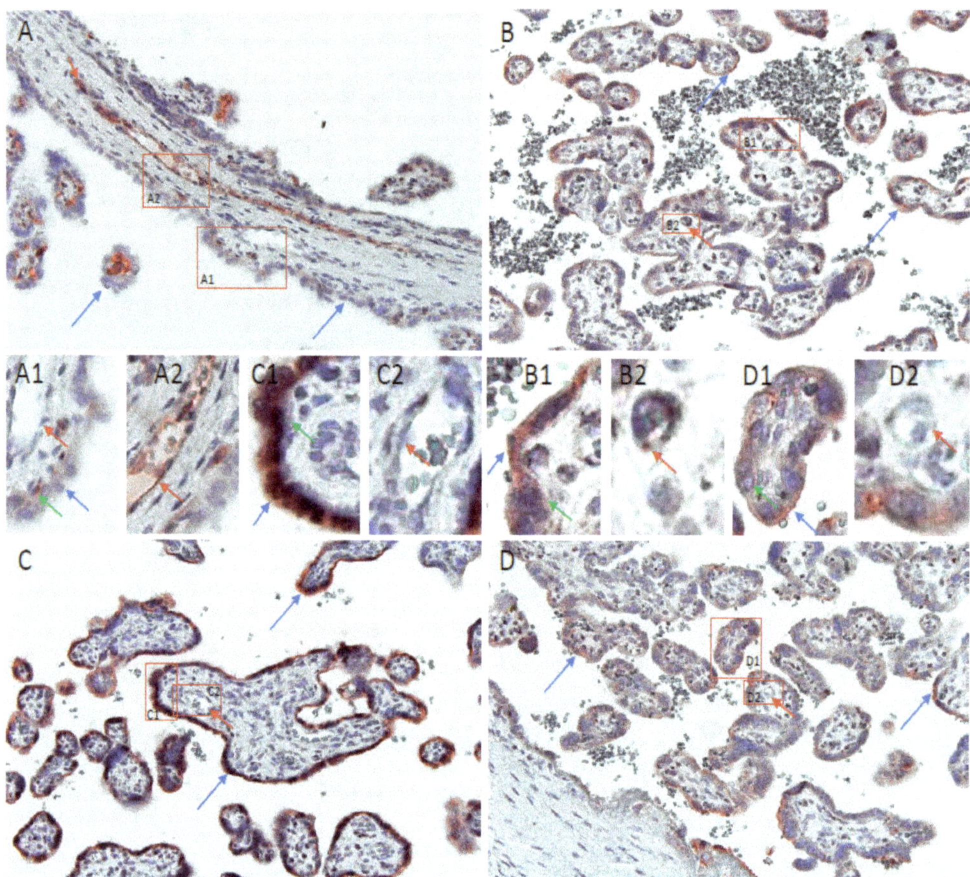

**Figure 3.** Immunohistochemical staining for CD24 in placental tissues. (**A**) early PE; (**B**) preterm delivery; (**C**) late PE; (**D**) term control. Blue arrows point to villous syncytiotrophoblast, green arrows point to villous cytotrophoblasts, and red arrows to villous blood vessels. Original magnification of A to D × 200. The small images are higher magnifications from the four larger images to better show staining of villous syncytiotrophoblast and cytotrophoblast (**A1–D1**), as well as villous blood vessels (**A2–D2**).

### 2.6. Semi-Quantitative Analysis of CD24 Staining

Villous trophoblast. CD24 staining was significantly reduced in early PE cases compared to age-matched preterm delivery cases ($p = 0.042$) (Figure 4A, Table 4a). Staining was mainly found in villous cytotrophoblast but not in the syncytiotrophoblast, which could account for the reduced overall staining intensity in the early PE cases. Staining for CD24 in late PE was nearly two times higher compared to early PE ($p = 0.025$) (Figure 4A, Table 4a). Staining for CD24 in late PE placental tissues was not different from the PTD group or normal term delivery ($p = 0.143$) (Figure 4A, Table 4a,b). The comparison of all unaffected cases (term and preterm delivery combined) to all PE cases (early and late PE cases combined) did not reveal any difference. Thus, it appears that unaffected cases were similar in staining intensity, while in the case of the PE subjects, the lower values in early PE were compensated by the higher values in the late PE group (Figure 4A, Table 4c). Altogether, lower trophoblast staining for the CD24 protein in early PE corresponded to lower CD24 mRNA expression in preterm PE as described in Figure 2 and Table 2.

Stroma and vessels. We did not identify any significant differences for any comparison between the groups (Figure 4B,C, Table 4).

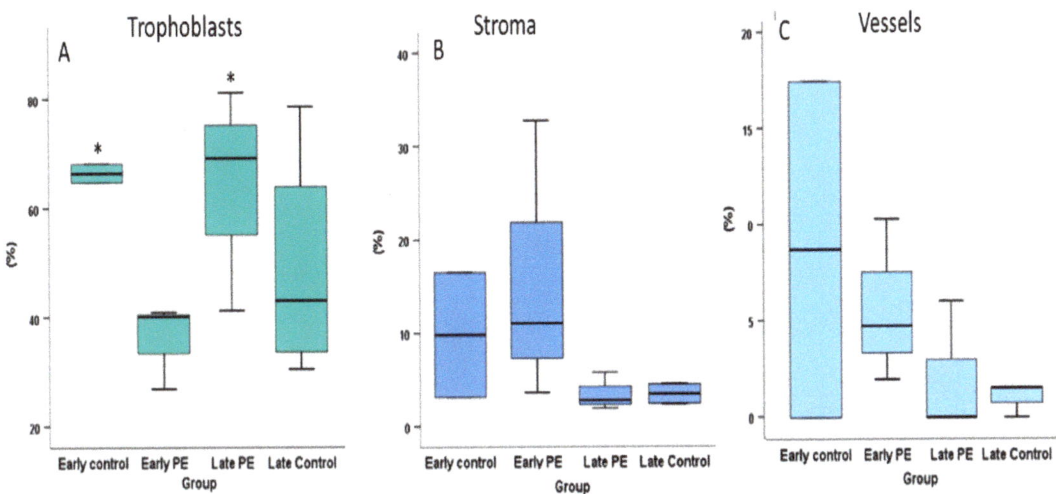

**Figure 4.** Semi-quantitative analysis of CD24 staining. (**A**) Villous trophoblast, (**B**) stroma, (**C**) vessels. (**A**) Villous trophoblast staining in early preeclampsia cases was lower compared to staining in gestational age-matched preterm delivery cases. In comparison, labeling of term delivery controls was not significantly different to late PE cases. Early PE was significantly lower compared to age-matched PTD cases as well as compared to late PE cases. (**B,C**) No significant differences could be detected for villous stroma and placental blood vessels. * $p < 0.05$ by Mann Whitney a-parametric test between the two early and the two term groups.

**Table 4.** Quantitative analysis of CD24 staining.

**(a) Comparison of Trophoblast, Stroma, and Vessel Staining Comparing Each Sub-Pathology to Its Matched Control Group**

| Median [ranges] | Early PE | Early Control | $p$ * | Late PE | Late Control | $p$ * | $p$ # |
|---|---|---|---|---|---|---|---|
| Trophoblast, % | 35.9 (26.8–40.9) | 66.4 (64.7–68.1) | 0.042 | 63.8 (41.2–81.1) | 48.7 (30.5–78.6) | 0.143 | 0.175 |
| Stroma, % | 15.8 (3.6–32.7) | 9.8 (3.1–16.4) | 0.282 | 3.4 (1.9–5.7) | 3.4 (2.3–4.6) | 0.499 | 0.300 |
| Vessels, % | 5.7 (2.0–10.3) | 8.8 (0–17.5) | 0.500 | 2.0 (0–6.1) | 1.2 (0–1.6) | 0.357 | 0.369 |

**(b) Comparison Between the Pathology Groups According to Staining of Trophoblast, Stroma, and Vessels**

| | Early PE | Late PE | $p$ * | Early Control | Late Control | $p$ * |
|---|---|---|---|---|---|---|
| Trophoblast, % | 35.9 (26.8–40.9) | 63.8 (41.2–81.1) | 0.025 | 66.4 (64.7–68.1) | 48.7 (30.5–78.6) | 0.167 |
| Stroma, % | 15.8 (3.6–32.7) | 3.4 (1.9–5.7) | 0.064 | 9.8 (3.1–16.4) | 3.4 (2.3–4.6) | 0.177 |
| Vessels, % | 5.7 (2.0–10.3) | 2.0 (0–6.1) | 0.134 | 8.8 (0–17.5) | 1.2 (0–1.6) | 0.407 |

**(c) Comparison of Trophoblast, Stroma, and Vessel Staining According to All PE versus All Unaffected**

| | All PE | All Unaffected | $p$ |
|---|---|---|---|
| Trophoblast, % | 49.9 (26.8–81.1) | 54.6 (30.5–78.6) | 0.436 |
| Stroma, % | 9.6 (1.9–32.7) | 5.5 (2.3–16.4) | 0.325 |
| Vessels, % | 3.9 (0–10.3) | 3.7 (0–17.5) | 0.257 |

Values are shown as means and ranges. * $p < 0.05$ by Mann–Whitney a-parametric test between two groups. # $p < 0.05$ by Kruskal–Wallis a-parametric test between all 4 groups (part a and b). In C All PE are early and term cases combined, and all unaffected—are term and preterm controls-combined.

*2.7. CD24 Protein Determination in Placental Lysate*

We further determined CD24 protein in lysates of placental tissue derived from preterm preeclamptic and term delivery controls by Western blot analysis. These immunoblots revealed diffuse bands of CD24 with molecular weights ranging between 30 and 60 kDa (Figure 5A). The diffuse bands from the total placental lysate reflected the high degree of CD24 glycosylation. Semi-quantitative analysis by densitometry revealed a significantly reduced CD24 protein expression ($p$ = 0.026) in placentas derived from preterm PE cases compared to term controls (Figure 5B).

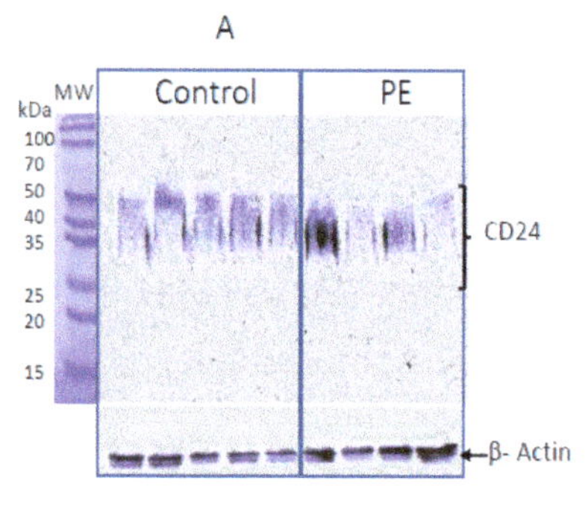


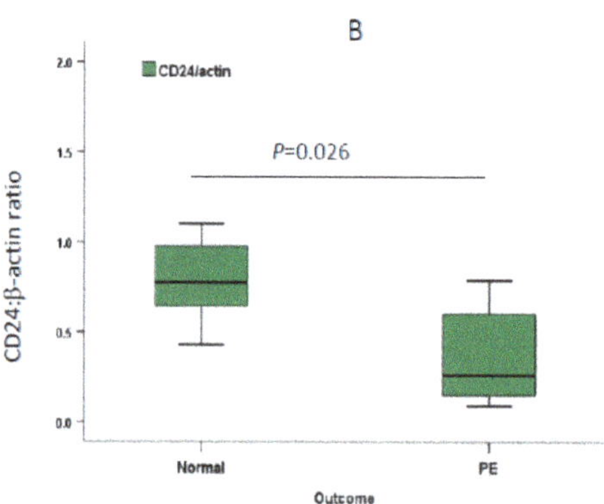


**Figure 5.** Western blot analysis of placental CD24 protein expression. (**A**) Representative Western blot of CD24 protein expression in the placenta. Placental CD24 was analyzed by Western blotting with a specific anti-CD24 antibody. β-Actin was used as housekeeping protein control for loading of equal amounts of proteins on the gel. (**B**) Semi-quantitative densitometric assessment of CD24 and β-actin bands was performed by Image Gauge software. The intensity of the CD24 bands were normalized to β-actin as a housekeeping protein. The analysis included 8 cases of preterm preeclampsia and 10 term control cases.

There was some diversity among the PE cases (Figure 5A), which could not be attributed to any of the following: disease severity (hypertension, proteinuria), maternal or gestational age, or newborn birthweight. Answer: The preterm group was composed of 14 early cases (delivered before 34 weeks) and 4 preterm cases (delivered between 34 weeks and 0 days and 36 weeks and 6 days). All four cases in Figure 5A were early PE cases, and the calculation for Figure 5B was made for the entire group and for the early and preterm subtypes separately. The calculations did not show any differences between the two subtypes.

## 3. Discussion

We have previously reported expression of CD24 in the first trimester placenta [14]. In the present study, we extended the study by showing firstly that the level of CD24 increased from the first and early second trimesters compared to term delivery as assessed by qRT-PCR analysis. Reduced CD24 mRNA expression in the placenta from PE cases has already been reported [27]. Secondly, we compared CD24 mRNA expression of term delivery to preterm PE and the aged-matched PTD group and found that it is lower in preterm PE compared to normal term delivery and especially compared to age-matched PTD. Thirdly, immunohistochemistry revealed reduced expression of CD24 protein in early PE, corresponding to reduced mRNA and protein expression in placental lysates from preterm PE cases. Altogether, our different methods of assessment indicate reduced CD24 expression in cases of early and preterm PE. The fourth finding was the increased level of CD24 protein expression in term PE cases compared to term controls, providing additional evidence that the two types of PE (early and preterm versus term), although similar in many clinical features, may be associated with different underlying pathways as was already suggested by others [18,28,29].

CD24 labeling of the villous syncytiotrophoblast was noted in normal term control specimens. A similar staining pattern was previously identified in a study by McDonald et al. [30]. The presence of CD24 in the syncytiotrophoblast in normal term delivery may be associated with placental maturation. Staining of other regions of the placenta at term delivery was very light. In comparison, we have already shown that during the first trimester, and already in gestational week eight [14], CD24 showed high expression in villous and extravillous cytotrophoblasts. One may ask why there is such a strong labeling at this period in the villous cytotrophoblasts, and especially in endometrial glands and extravillous trophoblasts in the placental bed. It appears that such labeling at that period may indicate the frontier of immune tolerance linked to escape rejection of the invading trophoblasts during the first trimester [4,28]. In comparison, the labeling of the syncytiotrophoblast in the third trimester and the very high level in late PE may indicate the frontier in immune tolerance to be related to this component of the placenta, as was demonstrated with another immune tolerance molecule in this period [31].

Understanding the pathways underlying immune tolerance in pregnancy has important implications as it may allow for the development of new ways to prevent pregnancy complications such as PE and also FGR, as was shown by Guleria et al. [32]. A variety of molecular and cellular processes were described, aiming to explain how immune tolerance is established [33,34]. In previous works, using placental protein 13 (PP13) as a marker [4], we have suggested that the development of early/preterm PE may be linked to a reduced immune tolerance at the feto–maternal interphase. In this respect, we proposed that placental proteins such as PP13 may render the mother immune-tolerant to the invading trophoblasts by inducing apoptosis of the decidual leukocytes, particularly T-cells [35,36]. We suggested that reduced first trimester levels of PP13, either due to primary structure mutation, promotor polymorphism, or reduced mRNA expression, could well be associated with an increased risk to develop PE [37].

This may also explain the difference in CD24 localization between the first trimester and term delivery. Accordingly, during the period of placentation (first trimester), the immune suppression impact of CD24 may be required to escape rejection of the invading trophoblasts [4,29,30]. In comparison, near delivery CD24 labeling of the syncytiotrophoblast might indicate requirements to protect the supply of nutrients and oxygen from the mother to the placenta and the fetus.

In the present study, we evaluated CD24 expression by comparing the protein labeling by immunohistochemistry in early, preterm, and term PE cases. We assumed that altered levels of CD24 could affect immune tolerance via the Siglec10-CD24 binding axis [15,16]. Indeed, we observed that in early/preterm cases of PE, there is a tendency of reduced mRNA and protein expression of CD24, as well as a significantly lower intensity of placental labeling, especially in the villous syncytiotrophoblast of the placenta. This could be an indication for a process of reduced immune tolerance in the context of early/preterm PE. Interestingly, we found that CD24 labeling of the syncytiotrophoblast is increased in PE that are developed around term.

Today, there is no universal agreement as to the differences underlying the pathways leading to the development of PE subtypes, whether term, preterm, or early. The definition is mainly related to the time the clinical situation requires delivery, as the time of onset is usually not clearly defined. The pathological borderline or marker differences between early cases (before 34 weeks) and preterm cases (between 34 weeks and 0 days and 36 weeks and 6 days) are very blurred. By contrast, the differences between the two subgroups of preterm and term PE are quite clear. The main differences between preterm and term PE can be found (1) with regards to blood vessel remodeling as estimating by measuring the blood flow through the uterine arteries and detected by the increased Doppler pulsatility Index (UTPI) and (2) with regards to a larger imbalance between soluble FMF-like tyrosine kinase-1 (sFlt-1) and placental growth factor (PlGF). In addition, aspirin prevention of PE has a relative success (62%) in preterm but not term PE cases [18,28,29,38].

The difference in distribution of CD24 labeling in term PE versus early PE may be linked to an underlying process related to coagulation. In fact, in certain cases of PE around term a strong process of coagulation is identified in the placental villi. Increased P-selectin and CD24 in these cases may be linked to enhanced activation of platelets and endothelial cells linked to blood vessel activation, all of which are part of the process of hypercoagulation that often occurs in term PE cases. This may create another resemblance between PE and tumors [8–10].

In human tumors, CD24 has been introduced as a diagnostic and prognostic marker [39,40]. Strong CD24 expression was linked to a rapid tumor progression in epithelial ovarian cancer, breast cancer, non-small cell lung carcinoma, prostate cancer, and pancreatic cancer [40,41]. Additionally, CD24 expression analysis allowed for the prediction of metastasis in malignant melanoma [41,42].

The role of CD24 in inducing immune tolerance was initially investigated in the immune system but later also in tumors [40–43]. It is interesting to note that the CD24–Siglec-10 binding axis has adverse effects in human cancer [15,16]. Barkal et al. [44] demonstrated that CD24 is a potent anti-phagocytic "don't eat me" signal [43] that is capable of directly protecting cancer cells from attack by Siglec-10-expressing macrophages. Monoclonal antibody blockade of CD24–Siglec-10 signaling robustly enhances clearance of CD24+ tumors, which indicates promise for CD24 blockade in immunotherapy. This is the underlying rational for new drug developments such as EXO-CD24 (N. Arber, Israel) or Merck's CD24-Fc to dampen overshooting immune reactions as observed during the COVID-19-related cytokine storm.

We have focused on villous tissues (the "fetal part" of the placenta) as this tissue represents the largest surface of fetal tissues in direct contact to maternal blood and all the circulating immune cells. We anticipate that in the placenta, there is a potential blockade of CD24, resulting in blocking macrophage attack of placental villi and increased immune tolerance, as is the case in tumors [43,44]. It is tempting to speculate that during the normal course of pregnancy, CD24 blocks macrophages and thereby supports the growth of the placenta. Therefore, CD24 reduction in early PE may be linked to increased rejection of the placenta, and its reduced expression may be associated with early cases of PE and could be relevant to cases also complicated by fetal growth restriction.

Today, only one additional study explored CD24 in the placenta of PE cases [27]. Our study is thus opening a whole new line of research in evaluating the role of this protein the placenta. We hope our manuscript will stimulate other scientists to develop CD24 knockout/knockdown models to further illuminate the role of this protein in the placenta in general and in particularly in immune tolerance during pregnancy.

The main limitations of our present study are as follows: CD24 was not determined at fixed time points, but rather when the patients were admitted to the hospital; however, this reflects clinical reality. A second limitation is that the design of the study was such that we did cross-sectional and not repeated measurements during pregnancy, which have been shown to improve accuracy. A third limitation is that although this is a first study to show reduced CD24 mRNA and protein expression as well as immune staining in preterm PE, the sample size was limited, and additional studies are warranted to add power and strength to our findings. The fourth limitation is that the comparison between term and preterm PE looks promising but requires more diversified analysis of the two PE subtypes. Finally, we did not evaluate the release of this protein from the placenta into the maternal circulation for future use of predictive estimates with simple blood tests. Huppertz et al. [45] have demonstrated the importance of obtaining results of a placental protein from diversified origins (placenta, serum, amniotic fluid) to reach a comprehensive understanding of the actual origin of the measured protein and further verify its role in the placental pathology in PE [45–47].

In conclusion, this is the first study to systematically evaluate placental CD24 expression in the first trimester compared to preterm and term delivery (early and late third trimester). Our study found an increased expression level during pregnancy, and the shift from staining of placental bed towards intensified villous staining. Our study showed decreased CD24 mRNA and protein expression and changes in protein localization in placentas from early and preterm PE cases corresponding to reduced immune tolerance. Higher CD24 staining in PE at term might be linked to hypercoagulation and may indicate a potential difference in the underlying pathways. The role of altered levels of CD24 in pregnancy should be further evaluated in the context of immune tolerance as a marker and potentially also as a new therapy target.

## 4. Materials and Methods

### 4.1. Antibodies

The monoclonal antibody (mAb) clone SWA11 was used for the detection of CD24 [48–50]. This mAb is specific for CD24 and reacts with the leucine–alanine–proline (LAP) motif in the protein core, as shown by peptide inhibition studies [49]. In addition, SWA11 exhibits specific binding to CD24-transfected cells but not to vector control [50]. Anti β-actin (clone 4) was purchased from MP Biomedical (Santa Ana, CA, USA).

### 4.2. Specimens

Two cohorts were used for this study. The first was a cohort of patients who donated their placentas according to the ethics approval of the Institutional Review Committee (IRB) of Bnai Zion Medical Center, Haifa, Israel (#BZ-06-021-972). This cohort included placentas from social terminations in the first and early second trimesters of pregnancy (N = 43), placentas from normal term delivery control (N = 67), and preterm PE (N = 18) and preterm delivery (PTD) (N = 6) patients. The preterm PE patients of this groups can be subdivided into 14 early cases (delivered before 34 weeks) and 4 preterm cases (delivered between 34 weeks and 0 days and 36 weeks and 6 days). However, there was no significant difference between the two subgroups and none of the results pointed to the need to subdivide them for improving group description or clarity of the results.

A small complementary cohort was collected at the Department of Obstetrics and Gynecology of the Medical University of Graz, Austria, after obtaining the approval of the ethics committee of the Medical University of Graz, Austria (# 24-112 ex 11/12). This cohort included placentas from normal term delivery controls (N = 5), late PE (N = 3), early PE (N = 3), and preterm delivery (PTD) (N = 3) patients. All women signed a written informed consent form before providing their specimens. The Israelian cohort was used for PCR and Western blotting, whereas the Austrian cohort was used for immunohistochemistry.

### 4.3. Definition of Pregnancy Complications

First and early second trimester social terminations were approved by the Hospital Pregnancy Termination Committee according to the National Law in Israel. Gestational age was determined according to last menstrual period and verified by measurements of crown–rump length (CRL) [51]. Normal term delivery control placentas were obtained from women delivering a baby at the 75–100 centile at gestational age of 37–41 weeks.

Preeclampsia (PE): In this paper, we used the updated criteria for the definition of preeclampsia as published in June 2020 by the American College of Obstetrics and Gynecology (ACOG) and of the International Society for the Study of Hypertension Disorders of Pregnancy (ISSHP) [18,52]. PE was defined as systolic blood pressure of 140 mm Hg or more or diastolic blood pressure of 90 mm Hg or more on two occasions at least 4 h apart after 20 weeks of gestation in a woman with a previously normal blood pressure. New onset proteinuria was defined as 300 mg or more per 24 h urine collection (or this amount extrapolated from a timed collection) or protein/creatinine ratio of 0.3 mg/dl or more or dipstick reading of 2+ (used only if other quantitative methods were not available) [52,53]. In the absence of proteinuria, new-onset hypertension with the new onset of any of the following were taken into account: (1) thrombocytopenia (platelet count less than $100 \times 10^9$/L), (2) renal insufficiency (serum creatinine concentrations greater than 1.1 mg/dl or doubling of the serum creatinine concentration in the absence of other renal disease), and/or (3) impaired liver function (elevated blood concentrations of liver transaminases to twice the normal concentration) [54–56]. Other symptoms included pulmonary edema, new-onset headache unresponsive to medication, and those not accounted for by alternative diagnoses or visual symptoms [18,52]. Given the new ACOG and ISSHP definition of PE published after the study was completed, we reviewed the database on a patient-by-patient basis to verify that patients included in the PE group according to our hospital clinical guidelines comply with these new ACOG and ISSHP definitions [18,52].

Small for gestational age (SGA) and fetal growth restriction (FGR) were defined according to the criteria of the international society for ultrasound in obstetric gynecology (ISUOG) [24] as the estimated fetal weight < 10th percentile and according to abdominal circumference < 10th percentile (SGA) [24–26]. When SGA was combined with oligohydramnion (AFI < 5 cm) and/or high pulsatility index of the blood flow through the umbilical cord and middle cerebral arteries and the ductus venosus at the highest (>90th percentile) level, it was defined as FGR [24]. Estimation of fetal weight was made with the use of head, body, and femur measurements [57], as adjusted to the Israelian population [58].

Preterm delivery (PTD) was defined as delivery before 37 weeks [59,60] not related to FGR and PE, or placental abruption, contaminated amniotic fluid, or preterm premature rupture of membranes (PPROM).

### 4.4. Immunohistochemistry

Placental samples were formalin-fixed and paraffin-embedded according to standard procedures. Sections (5 μm) were de-paraffinized in xylene and rehydrated in a graded series of alcohol. Epitope retrieval was performed for 7 min at 120 °C using a de-cloaking chamber (Biocare Medical, Pacheco, CA, USA). The procedure was performed with antigen retrieval solution pH 9 (Leica Biosystems, Newcastle, UK).

Sections were immunostained using the UltraVision Detection System HRP Polymer (LabVision, Thermo Fischer Scientific, Kalamazoo, MI, USA) according to the manufacturer's instruction. In brief, slides were incubated with hydrogen peroxide block to quench endogenous peroxidase. After washing with TBS/T (Tris-buffered saline (pH 7.4) containing 0.05% Tween20; Merck, Darmstadt, Germany), non-specific background was blocked by incubation with UltraVision Protein Block. The anti-CD24 antibody, clone SWA 11, was diluted 1:750 in Antibody Diluent (Dako, Carpinteria, CA, USA) and incubated for 45 min at RT.

After washing with TBS/T, Primary Antibody Enhancer, and after another washing step, UltraVision HRP polymer was applied for 20 min. The polymer complex was visualized by incubating the slides with 3-amino-9-ethylcarbazole (AEC) Chromogen Single Solution (Lab Vision) for 10 min. Sections were counterstained with hemalaun and mounted with Kaiser's glycerin gelatin (Merck). Controls were performed by using a mouse IgG negative control antibody (NeoMarkers, Thermo Scientific, Waltham, MA, USA) and revealed no staining.

### 4.5. Semi-Quantitative Analysis of Immunohistochemical Staining

For each case, images of a placental tissue section, stained for CD24, were taken in a stereology workstation using a Leica DM 6000B Microscope with an Olympus DP72 Camera and VIS Visiopharm software (Visiopharm A/S, Hoersholm, Denmark) by applying systematic uniform random sampling. For each slide, 10 images were systematically and randomly selected, representing the whole section. Each of these 10 images was analyzed as follows: A dot grid with $16 \times 12$ dots was placed on each image using the newCAST software (Visiopharm A/S). Each of these dots was assigned to the structure it was lying on top, defined as intervillous space, villous trophoblast, placental blood vessel, or villous stroma. Additionally, the staining intensity was categorized into strong (++), weak (+), or negative (-). Each dot was assigned to the respective combination of categories (tissue type/staining intensity). The sum values of all positive and negative dots of the 10 sections per block was arithmetically calculated and the percentiles of the positive staining were calculated.

### 4.6. Western Blot

Placental tissue processing was performed as described by Sammar et al. [31]. Briefly, placental tissue was homogenized and solubilized in RIPA lysis buffer (1% NP-40, 0.5% sodium deoxycholate, 0.1% SDS, 20 mM Tris/HCl (pH 8.0), 150 mM NaCl) containing a complete set of protease inhibitors (Hoffman la Roche, Switzerland) for 30 min at 2–8 °C. Insoluble material was removed by centrifugation and protein concentration was determined by BCA reagent (Pierce, Rockford, IL, USA). Placental lysate samples were aliquoted and stored at −70 °C until use.

For Western blot analysis, 50 µg of total protein lysates was separated on 12.5% SDS-PAGE and electro-transferred to a nitrocellulose membrane. After blocking free binding sites with 5% non-fat milk in 20 mM Tris/HCl buffer at pH 8.0 supplemented with 150 mM NaCl, membranes were probed with the anti-CD24 mAb SWA11 and anti-β-actin antibodies (0.1 µg/mL) overnight at 4 °C. Bound immune complexes were detected by horseradish peroxidase-conjugated rabbit anti-mouse IgG and developed by ECL detection kit (Biological Industries, Beit Haemek, Israel). β-Actin was used as reference housekeeping protein for equal protein loading and densitometry analysis. Signals were developed by chemiluminescence and were captured by an imager (Bio-Rad, Hercules, CA, USA).

### 4.7. Placental CD24 mRNA Amplification by PCR

Total RNA was isolated from frozen blocks of placental tissues and was reverse transcribed as described before [57]. cDNA synthesis was performed using superscript cDNA synthesis kit (Invitrogen, Carlsbad, CA, USA). Expression of CD24 gene was quantified by TaqMan RT-PCR utilizing the Applied Biosystem StepOne Plus cycler (Applied Biosystems, Austin, TX, USA) and TaqMan Gene Expression Assay with primer and probe sets (Applied Biosystems) for CD24 (Hs02379687_s1). Hypoxanthine-guanine phosphor-ribosyl-transferase (HGPRT) (Hs99999909_m1) was used as housekeeping gene. The relative amount of CD24 was calculated by employing the comparative Ct method (2−ΔΔCt).

### 4.8. Statistics

We used the statistical SPSS version 24 (SPSS Inc., Chicago, IL, USA) for analysis. The differences across groups were calculated by Kruskal–Wallis non-parametric tests (P). Mann–Whitney non-parametric tests were used to compare two groups. Statistical significance was defined when $p < 0.05$. Box-plot graphs provide a visualization of value distribution across quartiles.

The formula for the linear correlation coefficient was calculated according to

$$r_{xy} = \frac{n \sum_{i=1}^{n} x_i y_i - \sum_{i=1}^{n} x_i \sum_{i=1}^{n} y_i}{\sqrt{n \sum_{i=1}^{n} x_i^2 - \left(\sum_{i=1}^{n} x_i\right)^2} \sqrt{n \sum_{i=1}^{n} y_i^2 - \left(\sum_{i=1}^{n} y_i\right)^2}}$$

where $n$ is the number of specimens analyzed, $\Sigma x$ = total of the growth centile or gestational week (first variable value), $\Sigma y$ = total of ΔCT or ΔCT/growth centile (the second variable value), $\Sigma xy$ = sum of the product of first and second values, $\Sigma x^2$ = sum of the squares of the first value, and $\Sigma y^2$ = sum of the squares of the second value.

**Author Contributions:** This study was initiated by M.S. (Marei Sammar) who constructed the study on all fronts and designed its flow. H.M., B.H. and M.S. (Marei Sammar) prepared the clinical study protocols for patient enrolments, obtaining ethics approval and patient signatures on the informed consent forms. M.S. (Monika Siwetz) and B.H. conducted all aspects of immunohistochemistry including block preparation, sectioning, staining, and quantitative analysis. M.S. (Marei Sammar) conducted the experiments on CD24 mRNA and protein expression in the placental tissues. A.S.-N. and H.M. organized the database, and A.S.-N. conducted the statistical analysis. All authors participated in data requisition, management, analysis, and preparing of the manuscript. All authors have read and agreed to the published version of the manuscript.

**Funding:** This work was supported by a travel grant from the DKFZ to S.M. and an ORT Braude Research Exchange Program Grant to S.M.

**Institutional Review Board Statement:** The study was conducted according to the guidelines of the Declaration of Helsinki for human subjects involved in clinical research. Compliance with ethical standards occurred. All the procedures performed with human specimens were conducted in accordance with the ethical standards of the institutional and/or national research committees of the countries involved and complied with the 1964 Declaration of Helsinki and its later amendments or comparable ethical standards. Upon recruitment to the study, all women signed an informed consent form, and the study received approval from the Bnai Zion Medical Center Institutional Review Board (#BZ-06-021-972) and the Medical University of Graz Ethics Committee (# 24-112 ex 11/12).

**Informed Consent Statement:** A written informed consent form was obtained from each pregnant woman involved in the study.

**Data Availability Statement:** The data that support the findings of this study are available from the corresponding author upon reasonable request.

**Conflicts of Interest:** The authors declare no conflict of interest. The funders had no role in the design of the study; in the collection, analyses, or interpretation of data; in the writing of the manuscript; or in the decision to publish the results.

### List of Abbreviations

PE: preeclampsia, PTD: preterm delivery.


## References

1. Hsu, P.; Nanan, R.K. Innate and adaptive immune interactions at the fetal-maternal interface in healthy human pregnancy and pre-eclampsia. *Front. Immunol.* **2014**, *5*, 125. [CrossRef]
2. Lash, G.E. Molecular Cross-Talk at the Feto-Maternal Interface. *Cold Spring Harb. Perspect. Med.* **2015**, *5*, a023010. [CrossRef] [PubMed]
3. Enninga, E.A.; Nevala, W.K.; Holtan, S.G.; Markovic, S.N. Immune Reactivation by Cell-Free Fetal DNA in Healthy Pregnancies Re-Purposed to Target Tumors: Novel Checkpoint Inhibition in Cancer Therapeutics. *Front. Immunol.* **2015**, *6*, 424. [CrossRef]
4. Balogh, A.; Toth, E.; Romero, R.; Parej, K.; Csala, D.; Szenasi, N.L.; Hajdu, I.; Juhasz, K.; Kovacs, A.F.; Meiri, H.; et al. Placental Galectins Are Key Players in Regulating the Maternal Adaptive Immune Response. *Front. Immunol.* **2019**, *10*, 1240. [CrossRef] [PubMed]
5. Holtan, S.G.; Creedon, D.J.; Haluska, P.; Markovic, S.N. Cancer and pregnancy: Parallels in growth, invasion, and immune modulation and implications for cancer therapeutic agents. *Mayo Clin. Proc.* **2009**, *84*, 985–1000. [CrossRef]
6. Kay, R.; Rosten, P.M.; Humphries, R.K. CD24, a signal transducer modulating B cell activation responses, is a very short peptide with a glycosyl phosphatidylinositol membrane anchor. *J. Immunol.* **1991**, *147*, 1412–1416. [PubMed]
7. Rougon, G.; Alterman, L.A.; Dennis, K.; Guo, X.J.; Kinnon, C. The murine heat-stable antigen: A differentiation antigen expressed in both the hematolymphoid and neural cell lineages. *Eur. J. Immunol.* **1991**, *21*, 1397–1402. [CrossRef] [PubMed]
8. Aigner, S.; Sthoeger, Z.M.; Fogel, M.; Weber, E.; Zarn, J.; Ruppert, M.; Zeller, Y.; Vestweber, D.; Stahel, R.; Sammar, M.; et al. CD24, a mucin-type glycoprotein, is a ligand for P-selectin on human tumor cells. *Blood* **1997**, *89*, 3385–3395. [CrossRef] [PubMed]
9. Aigner, S.; Ruppert, M.; Hubbe, M.; Sammar, M.; Sthoeger, Z.; Butcher, E.C.; Vestweber, D.; Altevogt, P. Heat stable antigen (mouse CD24) supports myeloid cell binding to endothelial and platelet P-selectin. *Int. Immunol.* **1995**, *7*, 1557–1565. [CrossRef]
10. Friederichs, J.; Zeller, Y.; Hafezi-Moghadam, A.; Grone, H.J.; Ley, K.; Altevogt, P. The CD24/P-selectin binding pathway initiates lung arrest of human A125 adenocarcinoma cells. *Cancer Res.* **2000**, *60*, 6714–6722. [CrossRef]
11. Chen, G.Y.; Tang, J.; Zheng, P.; Liu, Y. CD24 and Siglec-10 selectively repress tissue damage-induced immune responses. *Science* **2009**, *323*, 1722–1725. [CrossRef]
12. Liu, Y.; Zheng, P. CD24: A genetic checkpoint in T cell homeostasis and autoimmune diseases. *Trends Immunol.* **2007**, *28*, 315–320. [CrossRef] [PubMed]
13. CD24Fc as a Non-antiviral Immunomodulator in COVID-19 Treatment (SAC-COVID). 2020. Available online: https://clinicaltrials.gov/ct2/show/NCT04317040 (accessed on 22 July 2021).
14. Sammar, M.; Siwetz, M.; Meiri, H.; Fleming, V.; Altevogt, P.; Huppertz, B. Expression of CD24 and Siglec-10 in first trimester placenta: Implications for immune tolerance at the fetal-maternal interface. *Histochem. Cell Biol.* **2017**, *147*, 565–574. [CrossRef] [PubMed]

15. Crocker, P.R.; Paulson, J.C.; Varki, A. Siglecs and their roles in the immune system. *Nat. Rev. Immunol.* **2007**, *7*, 255–266. [CrossRef] [PubMed]
16. Pillai, S.; Netravali, I.A.; Cariappa, A.; Mattoo, H. Siglecs and immune regulation. *Ann. Rev. Immunol* **2012**, *30*, 357–392. [CrossRef] [PubMed]
17. Redman, C.W.; Sargent, I.L. Latest advances in understanding preeclampsia. *Science* **2005**, *308*, 1592–1594. [CrossRef] [PubMed]
18. Brown, M.A.; Magee, L.A.; Kenny, L.C.; Karumanchi, S.A.; McCarthy, F.P.; Saito, S.; Hall, D.R.; Warren, C.E.; Adoyi, G.; Ishaku, S.; et al. The hypertensive disorders of pregnancy: ISSHP classification, diagnosis & management recommendations for international practice. *Pregnancy Hypertens.* **2018**, *13*, 291–310. [CrossRef] [PubMed]
19. Perez-Sepulveda, A.; Torres, M.J.; Khoury, M.; Illanes, S.E. Innate immune system and preeclampsia. *Front. Immunol.* **2014**, *5*, 244. [CrossRef]
20. Martinez-Varea, A.; Pellicer, B.; Perales-Marin, A.; Pellicer, A. Relationship between maternal immunological response during pregnancy and onset of preeclampsia. *J. Immunol. Res.* **2014**, *2014*, 210241. [CrossRef]
21. Scott, J.R.; Rote, N.S.; Branch, D.W. Immunologic aspects of recurrent abortion and fetal death. *Obs. Gynecol.* **1987**, *70*, 645–656.
22. Wilczynski, J.R. Fetal rejection: Infertility and immunity. *Expert Rev. Clin. Immunol.* **2007**, *3*, 871–882. [CrossRef] [PubMed]
23. Nguyen, T.A.; Kahn, D.A.; Loewendorf, A.I. Maternal-Fetal rejection reactions are unconstrained in preeclamptic women. *PLoS ONE* **2017**, *12*, e0188250. [CrossRef]
24. Lees, C.C.; Stampalija, T.; Baschat, A.; da Silva Costa, F.; Ferrazzi, E.; Figueras, F.; Hecher, K.; Kingdom, J.; Poon, L.C.; Salomon, L.J.; et al. ISUOG Practice Guidelines: Diagnosis and management of small-for-gestational-age fetus and fetal growth restriction. *Ultrasound Obs. Gynecol.* **2020**, *56*, 298–312. [CrossRef] [PubMed]
25. Khalil, A.; Gordijn, S.J.; Beune, I.M.; Wynia, K.; Ganzevoort, W.; Figueras, F.; Kingdom, J.; Marlow, N.; Papageorghiou, A.T.; Sebire, N.; et al. Essential variables for reporting research studies on fetal growth restriction: A Delphi consensus. *Ultrasound Obs. Gynecol.* **2019**, *53*, 609–614. [CrossRef] [PubMed]
26. Figueras, F.; Gratacos, E. An integrated approach to fetal growth restriction. *Best Pract. Res. Clin. Obs. Gynaecol.* **2017**, *38*, 48–58. [CrossRef]
27. Nagy, B.; Berkes, E.; Rigo, B.; Ban, Z.; Papp, Z.; Hupuczi, P. Under-expression of CD24 in pre-eclamptic placental tissues determined by quantitative real-time RT-PCR. *Fetal Diagn.* **2008**, *23*, 263–266. [CrossRef]
28. Wojtowicz, A.; Zembala-Szczerba, M.; Babczyk, D.; Kolodziejczyk-Pietruszka, M.; Lewaczynska, O.; Huras, H. Early- and Late-Onset Preeclampsia: A Comprehensive Cohort Study of Laboratory and Clinical Findings according to the New ISHHP Criteria. *Int. J. Hypertens.* **2019**, *2019*, 4108271. [CrossRef]
29. Huppertz, B. Placental origins of preeclampsia: Challenging the current hypothesis. *Hypertension* **2008**, *51*, 970–975. [CrossRef] [PubMed]
30. McDonald, E.A.; Wolfe, M.W. The pro-inflammatory role of adiponectin at the maternal-fetal interface. *Am. J. Reprod. Immunol.* **2011**, *66*, 128–136. [CrossRef]
31. Sammar, M.; Nisemblat, S.; Fleischfarb, Z.; Golan, A.; Sadan, O.; Meiri, H.; Huppertz, B.; Gonen, R. Placenta-bound and body fluid PP13 and its mRNA in normal pregnancy compared to preeclampsia, HELLP and preterm delivery. *Placenta* **2011**, *32*, S30–S36. [CrossRef]
32. Guleria, I.; Sayegh, M.H. Maternal acceptance of the fetus: True human tolerance. *J. Immunol.* **2007**, *178*, 3345–3351. [CrossRef] [PubMed]
33. Alijotas-Reig, J.; Llurba, E.; Gris, J.M. Potentiating maternal immune tolerance in pregnancy: A new challenging role for regulatory T cells. *Placenta* **2014**, *35*, 241–248. [CrossRef]
34. La Rocca, C.; Carbone, F.; Longobardi, S.; Matarese, G. The immunology of pregnancy: Regulatory T cells control maternal immune tolerance toward the fetus. *Immunol. Lett.* **2014**, *162*, 41–48. [CrossRef] [PubMed]
35. Kliman, H.J.; Sammar, M.; Grimpel, Y.I.; Lynch, S.K.; Milano, K.M.; Pick, E.; Bejar, J.; Arad, A.; Lee, J.J.; Meiri, H.; et al. Placental protein 13 and decidual zones of necrosis: An immunologic diversion that may be linked to preeclampsia. *Reprod. Sci.* **2012**, *19*, 16–30. [CrossRef] [PubMed]
36. Than, N.G.; Romero, R.; Goodman, M.; Weckle, A.; Xing, J.; Dong, Z.; Xu, Y.; Tarquini, F.; Szilagyi, A.; Gal, P.; et al. A primate subfamily of galectins expressed at the maternal-fetal interface that promote immune cell death. *Proc. Natl. Acad. Sci. USA* **2009**, *106*, 9731–9736. [CrossRef]
37. Sammar, M.; Drobnjak, T.; Mandala, M.; Gizurarson, S.; Huppertz, B.; Meiri, H. Galectin 13 (PP13) Facilitates Remodeling and Structural Stabilization of Maternal Vessels during Pregnancy. *Int. J. Mol. Sci.* **2019**, *20*, 3192. [CrossRef]
38. Rolnik, D.L.; Wright, D.; Poon, L.C.; O'Gorman, N.; Syngelaki, A.; de Paco Matallana, C.; Akolekar, R.; Cicero, S.; Janga, D.; Singh, M.; et al. Aspirin versus Placebo in Pregnancies at High Risk for Preterm Preeclampsia. *N. Engl. J. Med.* **2017**, *377*, 613–622. [CrossRef] [PubMed]
39. Kristiansen, G.; Sammar, M.; Altevogt, P. Tumour biological aspects of CD24, a mucin-like adhesion molecule. *J. Mol. Histol.* **2004**, *35*, 255–262. [CrossRef]
40. Gschaider, M.; Neumann, F.; Peters, B.; Lenz, F.; Cibena, M.; Goiser, M.; Wolf, I.; Wenzel, J.; Mauch, C.; Schreiner, W.; et al. An attempt at a molecular prediction of metastasis in patients with primary cutaneous melanoma. *PLoS ONE* **2012**, *7*, e49865. [CrossRef]

41. Sagiv, E.; Arber, N. The novel oncogene CD24 and its arising role in the carcinogenesis of the GI tract: From research to therapy. *Expert Rev. Gastroenterol. Hepatol.* **2008**, *2*, 125–133. [CrossRef]
42. Fang, X.; Zheng, P.; Tang, J.; Liu, Y. CD24: From A to Z. *Cell. Mol. Immunol.* **2010**, *7*, 100–103. [CrossRef]
43. Barkal, A.A.; Brewer, R.E.; Markovic, M.; Kowarsky, M.; Barkal, S.A.; Zaro, B.W.; Krishnan, V.; Hatakeyama, J.; Dorigo, O.; Barkal, L.J.; et al. CD24 signalling through macrophage Siglec-10 is a target for cancer immunotherapy. *Nature* **2019**, *572*, 392–396. [CrossRef]
44. Bradley, C.A. CD24—A novel 'don't eat me' signal. *Nat. Rev. Cancer* **2019**, *19*, 541. [CrossRef]
45. Huppertz, B.; Meiri, H.; Gizurarson, S.; Osol, G.; Sammar, M. Placental protein 13 (PP13): A new biological target shifting individualized risk assessment to personalized drug design combating pre-eclampsia. *Hum. Reprod. Update* **2013**, *19*, 391–405. [CrossRef]
46. Ehrlich, L.; Hoeller, A.; Golic, M.; Herse, F.; Perschel, F.H.; Henrich, W.; Dechend, R.; Huppertz, B.; Verlohren, S. Increased placental sFlt-1 but unchanged PlGF expression in late-onset preeclampsia. *Hypertens. Pregnancy* **2017**, *36*, 175–185. [CrossRef] [PubMed]
47. Hoeller, A.; Ehrlich, L.; Golic, M.; Herse, F.; Perschel, F.H.; Siwetz, M.; Henrich, W.; Dechend, R.; Huppertz, B.; Verlohren, S. Placental expression of sFlt-1 and PlGF in early preeclampsia vs. early IUGR vs. age-matched healthy pregnancies. *Hypertens. Pregnancy* **2017**, *36*, 151–160. [CrossRef] [PubMed]
48. Ducat, A.; Couderc, B.; Bouter, A.; Biquard, L.; Aouache, R.; Passet, B.; Doridot, L.; Cohen, M.B.; Ribaux, P.; Apicella, C.; et al. Molecular Mechanisms of Trophoblast Dysfunction Mediated by Imbalance between STOX1 Isoforms. *iScience* **2020**, *23*, 101086. [CrossRef] [PubMed]
49. Weber, E.; Lehmann, H.P.; Beck-Sickinger, A.G.; Wawrzynczak, E.J.; Waibel, R.; Folkers, G.; Stahel, R.A. Antibodies to the protein core of the small cell lung cancer workshop antigen cluster-w4 and to the leucocyte workshop antigen CD24 recognize the same short protein sequence leucine-alanine-proline. *Clin. Exp. Immunol.* **1993**, *93*, 279–285. [CrossRef]
50. Kristiansen, G.; Machado, E.; Bretz, N.; Rupp, C.; Winzer, K.J.; Konig, A.K.; Moldenhauer, G.; Marme, F.; Costa, J.; Altevogt, P. Molecular and clinical dissection of CD24 antibody specificity by a comprehensive comparative analysis. *Lab. Investig.* **2010**, *90*, 1102–1116. [CrossRef] [PubMed]
51. Hadlock, F.P.; Shah, Y.P.; Kanon, D.J.; Lindsey, J.V. Fetal crown-rump length: Reevaluation of relation to menstrual age (5–18 weeks) with high-resolution real-time US. *Radiology* **1992**, *182*, 501–505. [CrossRef] [PubMed]
52. Gestational Hypertension and Preeclampsia: ACOG Practice Bulletin, Number 222. *Obs. Gynecol.* **2020**, *135*, e237–e260. [CrossRef] [PubMed]
53. Cassia, M.A.; Daminelli, G.; Zambon, M.; Cardellicchio, M.; Cetin, I.; Gallieni, M. Proteinuria in pregnancy: Clinically driven considerations. *Nephrol. Point Care* **2018**, *4*, 2059300718755622. [CrossRef]
54. Burwick, R.M.; Rincon, M.; Beeraka, S.S.; Gupta, M.; Feinberg, B.B. Evaluation of Hemolysis as a Severe Feature of Preeclampsia. *Hypertension* **2018**, *72*, 460–465. [CrossRef]
55. Ekun, O.A.; Olawumi, O.M.; Makwe, C.C.; Ogidi, N.O. Biochemical Assessment of Renal and Liver Function among Preeclamptics in Lagos Metropolis. *Int. J. Reprod. Med.* **2018**, *2018*, 1594182. [CrossRef]
56. Jodkowska, A.; Martynowicz, H.; Kaczmarek-Wdowiak, B.; Mazur, G. Thrombocytopenia in pregnancy—Pathogenesis and diagnostic approach. *Postepy Hig. Med. Dosw. (Online)* **2015**, *69*, 1215–1221. [CrossRef] [PubMed]
57. Hadlock, F.P.; Harrist, R.B.; Martinez-Poyer, J. In utero analysis of fetal growth: A sonographic weight standard. *Radiology* **1991**, *781*, 129–133. [CrossRef]
58. Dollberg, S.; Haklai, Z.; Mimouni, F.B.; Gorfein, I.; Gordon, E.S. Birth weight standards in the live-born population in Israel. *Isr. Med. Assoc. J.* **2005**, *7*, 311–314. [PubMed]
59. Goldenberg, R.L.; Culhane, J.F.; Iams, J.D.; Romero, R. Epidemiology and causes of preterm birth. *Lancet* **2008**, *371*, 75–84. [CrossRef]
60. Romero, R.; Dey, S.K.; Fisher, S.J. Preterm labor: One syndrome, many causes. *Science* **2014**, *345*, 760–765. [CrossRef] [PubMed]

International Journal of
*Molecular Sciences*

*Article*

# Human Placental Transcriptome Reveals Critical Alterations in Inflammation and Energy Metabolism with Fetal Sex Differences in Spontaneous Preterm Birth

Yu-Chin Lien [1,2], Zhe Zhang [3], Yi Cheng [4], Erzsebet Polyak [4], Laura Sillers [2], Marni J. Falk [4], Harry Ischiropoulos [2], Samuel Parry [1,5] and Rebecca A. Simmons [1,2,*]

1    Center for Research on Reproduction and Women's Health, Department of Obstetrics and Gynecology, Perelman School of Medicine, University of Pennsylvania, Philadelphia, PA 19104, USA; ylien@pennmedicine.upenn.edu (Y.-C.L.); parry@pennmedicine.upenn.edu (S.P.)
2    Division of Neonatology, Department of Pediatrics, Children's Hospital of Philadelphia, Philadelphia, PA 19104, USA; laura.sillers@pennmedicine.upenn.edu (L.S.); ischirop@pennmedicine.upenn.edu (H.I.)
3    Department of Biomedical and Health Informatics, Children's Hospital of Philadelphia, Philadelphia, PA 19104, USA; zhangz@chop.edu
4    Mitochondrial Medicine Frontier Program, Division of Human Genetics, Department of Pediatrics, Children's Hospital of Philadelphia, Philadelphia, PA 19104, USA; chengy1@email.chop.edu (Y.C.); polyake@email.chop.edu (E.P.); falkm@chop.edu (M.J.F.)
5    Department of Obstetrics & Gynecology, University of Pennsylvania Perelman School of Medicine, Philadelphia, PA 19104, USA
*    Correspondence: rsimmons@pennmedicine.upenn.edu

**Citation:** Lien, Y.-C.; Zhang, Z.; Cheng, Y.; Polyak, E.; Sillers, L.; Falk, M.J.; Ischiropoulos, H.; Parry, S.; Simmons, R.A. Human Placental Transcriptome Reveals Critical Alterations in Inflammation and Energy Metabolism with Fetal Sex Differences in Spontaneous Preterm Birth. *Int. J. Mol. Sci.* **2021**, *22*, 7899. https://doi.org/10.3390/ijms22157899

Academic Editors: Hiten D. Mistry and Eun Lee

Received: 16 June 2021
Accepted: 20 July 2021
Published: 23 July 2021

**Publisher's Note:** MDPI stays neutral with regard to jurisdictional claims in published maps and institutional affiliations.

**Abstract:** A well-functioning placenta is crucial for normal gestation and regulates the nutrient, gas, and waste exchanges between the maternal and fetal circulations and is an important endocrine organ producing hormones that regulate both the maternal and fetal physiologies during pregnancy. Placental insufficiency is implicated in spontaneous preterm birth (SPTB). We proposed that deficits in the capacity of the placenta to maintain bioenergetic and metabolic stability during pregnancy may ultimately result in SPTB. To explore our hypothesis, we performed a RNA-seq study in male and female placentas from women with SPTB (<36 weeks gestation) compared to normal pregnancies (≥38 weeks gestation) to assess the alterations in the gene expression profiles. We focused exclusively on Black women (cases and controls), who are at the highest risk of SPTB. Six hundred and seventy differentially expressed genes were identified in male SPTB placentas. Among them, 313 and 357 transcripts were increased and decreased, respectively. In contrast, only 61 differentially expressed genes were identified in female SPTB placenta. The ingenuity pathway analysis showed alterations in the genes and canonical pathways critical for regulating inflammation, oxidative stress, detoxification, mitochondrial function, energy metabolism, and the extracellular matrix. Many upstream regulators and master regulators important for nutrient-sensing and metabolism were also altered in SPTB placentas, including the PI3K complex, TGFB1/SMADs, SMARCA4, TP63, CDKN2A, BRCA1, and NFAT. The transcriptome was integrated with published human placental metabolome to assess the interactions of altered genes and metabolites. Collectively, significant and biologically relevant alterations in the transcriptome were identified in SPTB placentas with fetal sex disparities. Altered energy metabolism, mitochondrial function, inflammation, and detoxification may underly the mechanisms of placental dysfunction in SPTB.

**Keywords:** placenta; transcriptome; spontaneous preterm birth; bioenergetic metabolism; interactome; fetal sex disparity

## 1. Introduction

Preterm birth (delivery before or at 37 weeks of gestation) is the leading cause of morbidity and mortality in newborn infants worldwide [1]. Fifteen million babies are born prematurely annually, resulting in an excess of one million deaths. Infants who survive preterm birth often have serious and lifelong health problems, including lung disease, vision loss, and neurodevelopmental disorders. Spontaneous preterm birth (SPTB) remains a significant and poorly understood perinatal complication. SPTB includes the preterm spontaneous rupture of membranes, cervical insufficiency, and preterm labor. While the exact etiology remains unknown, many factors may contribute to SPTB, including placental dysfunction, abnormal cervical remodeling, uterine distension, vascular disorders, and chorioamnionitis [2,3].

During pregnancy, the placenta facilitates nutrient transport and gas exchange and supports the growth and development of the fetus. It also produces and releases hormones into the maternal and fetal circulation to regulate uterine functions, the maternal metabolism, and fetal growth and development. Therefore, a well-functioning placenta is crucial for normal gestation. Placental dysfunction is associated with preeclampsia and fetal growth restriction. Emerging evidence suggests that placental insufficiency is also associated with a significant proportion of preterm births, especially early preterm births, as well as those complicated by chorioamnionitis [4–6]. The placenta protects the fetus against infections, toxins, xenobiotic molecules, and maternal diseases [7]. The placenta also produces a wide variety of metabolites, many of which are involved in energy production [8,9]. In our previous metabolomic analysis of placenta samples obtained from women with SPTB, we observed a significant elevation in the levels of amino acids, prostaglandins, sphingolipids, lysolipids, and acylcarnitines in SPTB placentas compared to term placentas [10], which suggests an imbalance between the supply capacity and metabolic demands in SPTB placentas.

Fetal sex plays an important role in pregnancy complications and perinatal outcomes. Male fetal sex is a risk factor for preeclampsia and gestational diabetes, as well as presents a higher cardiovascular and metabolic load for the mother [11–14]. A higher incidence of preterm birth is also observed among women carrying male fetuses. Although the underlying mechanism for the increased preterm birth rate of male newborns is still unclear, a potential more proinflammatory intrauterine milieu generated by male placentas may partially contribute to the increased incidences [15–17].

We hypothesized that deficits in the capacity of the placenta to maintain bioenergetic and metabolic stability throughout the course of pregnancy may ultimately result in SPTB. To test this hypothesis, we assessed the transcriptomes in both male and female placenta samples obtained from women with spontaneous preterm deliveries. We also integrated the transcriptomic data with our previously published metabolomic data [10] to assess the interactions of the altered genes and metabolites. Therefore, the aim of this study was to elucidate the underlying mechanisms for placental insufficiency and dysfunction, especially metabolic changes and sex differences, which will lead to a better understanding of the etiology of prematurity and the development of preventative treatments.

## 2. Results

### 2.1. Clinical Characteristics

There were no differences in maternal ages between term and spontaneous preterm birth (SPTB) placenta samples (Table 1). Of note, none of the preterm or term placental samples were from mothers with preeclampsia or gestational diabetes, and none of the women received low-dose aspirin for the prevention of preeclampsia. All of the women contributing placenta samples presented in labor with either a preterm premature rupture of membranes (PPROM), premature rupture of membranes (PROM), or cervical dilation. Thirteen women had preterm labors, and none of the women who labored at term received a betamethasone treatment prior to delivery. Three of the women with preterm labors received 17-hydroxy progesterone and five received vaginal progesterone for prematurity prevention. None of the women with term delivery received supplemental progesterone. A greater percentage of women with preterm labor received antibiotics, with a primary indication of PPROM or Group B—streptococcus (GBS) prophylaxis; however, only one woman with preterm labor was diagnosed with chorioamnionitis. Chronic medications administered to women with preterm deliveries included albuterol and inhaled corticosteroids for asthma in three women, psychotropic medications in two women, and antihypertensive drugs in four women. The chronic medications documented for women with term deliveries included albuterol in one woman and ferrous sulfate medication in one woman.

**Table 1.** Demographics of the study cohort.

| | Preterm_Male | Term_Male | *p*-Value | Preterm_Female | Term_Female | *p*-Value |
| --- | --- | --- | --- | --- | --- | --- |
| | (*n* = 8) | (*n* = 7) | | (*n* = 8) | (*n* = 8) | |
| Gestational age at birth wks, mean ± SD | 29.4 ± 4.3 | 39.7 ± 0.7 | *p* < 0.0001 | 32.1 ± 4.0 | 39.5 ± 0.7 | *p* = 0.0001 |
| Maternal age at delivery yrs, mean ± SD | 28.4 ± 5.5 | 28.6 ± 4.4 | *p* = 0.94 | 26.3 ± 4.1 | 27.9 ± 6.2 | *p* = 0.55 |
| Maternal BMI at first visit Kg/m², mean ± SD | 33.4 ± 13.2 | 25.6 ± 4.4 | *p* = 0.17 | 28.3 ± 5.6 | 28.9 ± 6.9 | *p* = 0.85 |
| Mode of delivery, n (%) Vaginal | 3 (38) | 6 (86) | ND | 8 (100) | 5 (62) | ND |
| C-section | 5 (62) | 1 (14) | | 0 (0) | 3 (38) | |
| Fetal growth restriction, n (%) | 1 (13) | 0 (0) | ND | 0 (0) | 0 (0) | ND |
| Antibiotic administration Yes, *n* (%) | 6 (75) | 0 (0) | ND | 4 (50) | 2 (25) | ND |

ND: not determined.

### 2.2. Global Assessment of Transcriptome Profiles in Placentas

To investigate the genes and novel pathways that are disrupted in placentas from SPTB, the gene expression profiles of placental tissues from eight male preterm cases with a mean gestational age (GA) of 29.4 weeks, seven male term controls with a mean GA of 39.7 weeks, eight female preterm cases with a mean GA of 32.1 weeks, and eight female controls with a mean GA of 39.5 weeks were assessed by RNA-Seq. The power calculation using a false-positive rate of 0.05 (two-sided) and power of 80% to target a two-fold change indicated that six samples were sufficient to determine the significant differences between the groups. One male control and one female preterm cases were considered as outliners due to the potential contamination of other cell types and were excluded from further analysis of the differential gene expressions. The principal component analysis (PCA plot, Figure 1) indicated that the transcriptome profiles for males and females were readily distinguishable. Preterm and term birth groups also separated, suggesting significant differences in the transcriptome profiles of preterm and term placentas. Genes were considered differentially expressed, with an FDR (*q*-value) ≤ 0.05. The comparison of male SPTB vs. male term deliveries yielded a total of 724 differentially expressed genes, with 347 upregulated and 377 downregulated (Figure 2a and Supplemental Table S1) (*q* ≤ 0.05 vs. term controls; *q*-value ≤ 0.05 was considered significant). Interestingly, the difference of the transcriptome profiles between female SPTB and female term deliveries was much smaller than that of the male transcriptomes. Only 66 differentially expressed genes were identified in the comparison of female SPTB vs. term, with 28 upregulated and 38 downregulated (Figure 2b and Supplemental Table S2). Most differentially expressed genes in each comparison were unique, suggesting a different molecular basis for placenta dysfunction and possibly preterm births of males vs. females. Five differentially expressed genes were identified in both male and female SPTB compared with their term controls, including four upregulated genes, *ASB4*, *CMAS*, *KATNBL1*, and *PRR9*, and one downregulated gene, *SLC28A1* (Supplemental Table S3). Differences in the expression between SPTB and term placentas of these genes are likely due to gestational age effects and unlikely to be associated with preterm births.

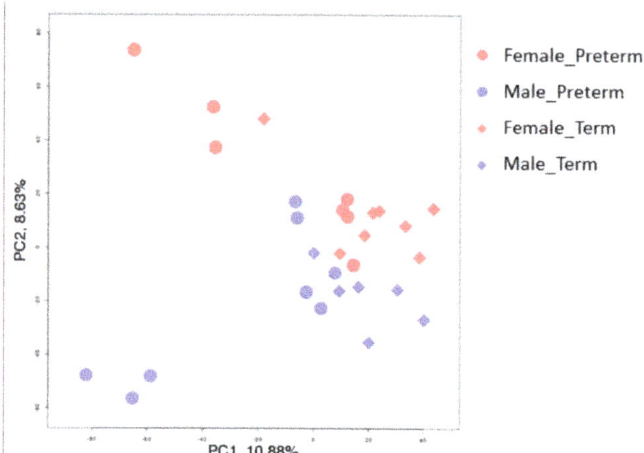


**Figure 1.** Principal component analysis (PCA plot) of the placental transcriptomes. The PCA plot revealed a significant separation between the male and female placentas, as well as the placentas from preterm and term births.

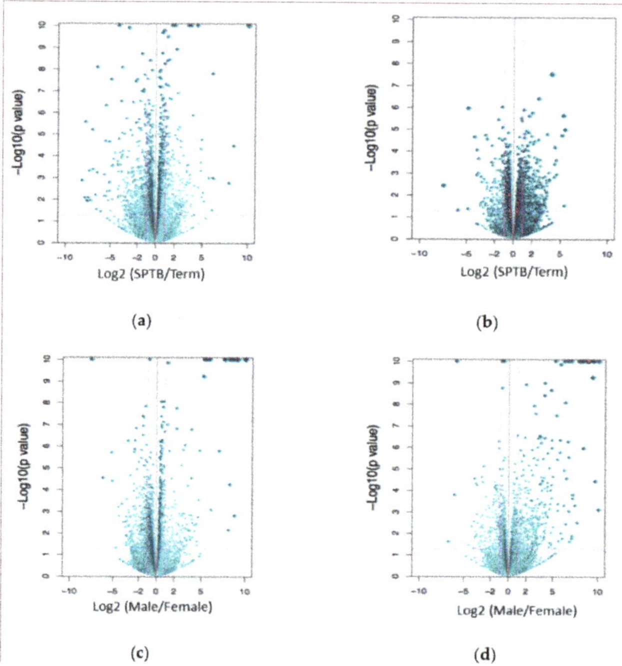


**Figure 2.** Volcano plots identifying differentially expressed genes with an FDR (*q*-value) < 0.05. Male SPTB placentas compared with term placentas (**a**). Female SPTB placentas compared with term placentas (**b**). Male term placentas compared with female term placentas (**c**). Male SPTB placentas compared with female SPTB placentas (**d**).

To determine the possible effect of gestational age (GA) on the differences in the gene expressions, we took two approaches. First, we compared our SPTB transcriptomes with the human placenta studies conducted by Eidem et al. [18] and Brockway et al. [19], who identified 37 and 170 GA-specific candidate genes, respectively. We also compared the SPTB transcriptomes with our proteomics data on four placentas from term deliveries and four placentas from elective second trimester terminations (Supplemental Table S4). In the proteomic dataset, 4711 proteins were identified, and 953 proteins were differentially expressed in term compared to second trimester placentas (proteins associated with blood were eliminated) (Supplemental Table S5). Together, we identified 54 GA-specific candidate genes in male SPTB placental transcriptome and five candidate genes in female SPTB placental transcriptome (Supplemental Table S6). These GA-specific candidate genes were removed from the datasets prior to the ingenuity pathway analysis.

### 2.3. Differences in the Transcriptome Profiles between Male and Female Placentas

Comparing male with female transcriptomes from term placentas, 319 differentially expressed genes were identified, with 177 upregulated and 142 downregulated in male compared to female placentas (Figure 2c and Supplemental Table S7). Thirty-nine differentially expressed genes were either X- or Y-chromosome-linked. In the comparison of male and female SPTB transcriptomes, in addition to 36 sex chromosome-associated genes, 144 differentially expressed genes were identified, with 105 upregulated and 39 downregulated in male compared to female STPB placentas (Figure 2d and Supplemental Table S8). These genes regulated at least 20 canonical pathways (Supplemental Table S9). Only 29 differentially expressed genes overlapped between the STPB and term groups when comparing male with female placentas. The expression of six genes, *ICAM2*, *AADACL3*, *RNR1*, *RNR2*, *MTNR1B*, and *HIST1H3H*, that were not associated with sex chromosomes showed clear sex differences (Table 2). The expression of *ICAM2*, *MTNR1B*, and *HIST1H3H* were lower in male placentas from both preterm and term births compared with female placentas. In contrast, the expression levels of *AADACL3*, *RNR1*, and *RNR2* were higher in male placentas.

**Table 2.** Overlay of differentially expressed genes comparing male and female placentas at both SPTB and term births.

| Gene | Gene Names | SPTB_Male_vs._Female LogFC | FDR | Term_Male_vs._Female LogFC | FDR |
|---|---|---|---|---|---|
| *RNR2* | l-rRNA | 2.83 | $9.30 \times 10^{-6}$ | 2.20 | $9.10 \times 10^{-5}$ |
| *AADACL3* | arylacetamide deacetylase like 3 | 5.72 | $7.50 \times 10^{-4}$ | 3.14 | $3.80 \times 10^{-4}$ |
| *RNR1* | s-rRNA | 2.55 | $3.50 \times 10^{-3}$ | 1.95 | $1.50 \times 10^{-2}$ |
| *HIST1H3H* | histone cluster 1 H3 family member h | −1.39 | $1.40 \times 10^{-2}$ | −0.70 | $4.90 \times 10^{-2}$ |
| *MTNR1B* | melatonin receptor 1B | −1.40 | $1.60 \times 10^{-2}$ | −2.02 | $2.90 \times 10^{-2}$ |
| *ICAM2* | intercellular adhesion molecule 2 | −0.75 | $1.60 \times 10^{-2}$ | −0.76 | $3.80 \times 10^{-2}$ |

### 2.4. IPA Identifies Multiple Pathways That Are Altered in Male SPTB Placentas

To identify the molecular pathways that may contribute to placenta dysfunction and, possibly, SPTB, an ingenuity pathway analysis (IPA) was used to map differentially expressed genes into functional networks. An IPA analysis of 670 non-GA-associated differentially expressed genes in male SPTB placentas revealed 65 canonical pathways that were altered in preterm births. As predicted by the activation z-score, the top canonical pathways activated in male SPTB included the NRF2-mediated oxidative stress response, xenobiotic metabolism pathways, and estrogen receptor signaling (Table 3). The top pathways inhibited in male SPTB were VDR/RXR activation and the antiproliferative role of the transducer of ERBB2 (TOB) in T-cell signaling (Table 3).

**Table 3.** Top canonical pathways altered in male SPTB placentas.

| Ingenuity Canonical Pathways | *p*-Value | z-Score |
| --- | --- | --- |
| STAT3 Pathway | $6.46 \times 10^{-5}$ | 0.63 |
| Glucocorticoid Receptor Signaling | $8.71 \times 10^{-5}$ | —— |
| Adenine and Adenosine Salvage III | $5.25 \times 10^{-4}$ | —— |
| LPS/IL-1 Mediated Inhibition of RXR Function | $5.50 \times 10^{-4}$ | −0.38 |
| Wnt/Ca+ pathway | $5.50 \times 10^{-4}$ | 0.00 |
| Guanine and Guanosine Salvage I | $9.12 \times 10^{-4}$ | —— |
| Inhibition of Matrix Metalloproteases | $1.07 \times 10^{-3}$ | 0.82 |
| Fibrosis / Stellate Cell Activation | $1.70 \times 10^{-3}$ | —— |
| Glutathione-mediated Detoxification | $2.57 \times 10^{-3}$ | —— |
| Myo-inositol Biosynthesis | $8.71 \times 10^{-3}$ | —— |
| TGF-β Signaling | $8.91 \times 10^{-3}$ | −0.82 |
| Estrogen Receptor Signaling | $1.17 \times 10^{-2}$ | 1.89 |
| Urea Cycle | $1.26 \times 10^{-2}$ | —— |
| NRF2-mediated Oxidative Stress Response | $1.29 \times 10^{-2}$ | 1.34 |
| TR/RXR Activation | $1.38 \times 10^{-2}$ | —— |
| PXR/RXR Activation | $1.38 \times 10^{-2}$ | —— |
| PPARα/RXRα Activation | $1.41 \times 10^{-2}$ | 1.26 |
| Regulation of the Epithelial-Mesenchymal Transition Pathway | $1.45 \times 10^{-2}$ | —— |
| Pyridoxal 5′-phosphate Salvage Pathway | $1.48 \times 10^{-2}$ | 0.45 |
| Purine Ribonucleosides Degradation to Ribose-1-phosphate | $1.74 \times 10^{-2}$ | —— |
| Epithelial Adherens Junction Signaling | $1.78 \times 10^{-2}$ | —— |
| Role of JAK2 in Hormone-like Cytokine Signaling | $1.91 \times 10^{-2}$ | —— |
| Histidine Degradation III | $2.29 \times 10^{-2}$ | —— |
| Sphingomyelin Metabolism | $2.29 \times 10^{-2}$ | —— |
| Superoxide Radicals Degradation | $2.29 \times 10^{-2}$ | —— |
| PI3K Signaling | $2.51 \times 10^{-2}$ | 0.33 |
| Sulfite Oxidation IV | $3.02 \times 10^{-2}$ | —— |
| Xenobiotic Metabolism CAR Signaling Pathway | $3.02 \times 10^{-2}$ | 2.71 |
| Integrin Signaling | $3.02 \times 10^{-2}$ | 0.30 |
| Antiproliferative Role of TOB in T Cell Signaling | $3.02 \times 10^{-2}$ | −2.00 |
| Xanthine and Xanthosine Salvage | $3.02 \times 10^{-2}$ | —— |
| VDR/RXR Activation | $3.09 \times 10^{-2}$ | −1.34 |
| Superpathway of D-myo-inositol (1,4,5)-trisphosphate Metabolism | $3.16 \times 10^{-2}$ | —— |
| Xenobiotic Metabolism PXR Signaling Pathway | $3.31 \times 10^{-2}$ | 2.11 |
| BEX2 Signaling Pathway | $3.31 \times 10^{-2}$ | 1.63 |
| Gap Junction Signaling | $3.98 \times 10^{-2}$ | —— |
| Xenobiotic Metabolism AHR Signaling Pathway | $4.47 \times 10^{-2}$ | 2.45 |
| cAMP-mediated signaling | $4.68 \times 10^{-2}$ | 1.51 |

In addition to the identification of altered canonical pathways, IPA disease and a biological function analysis also revealed more than 65 genes that were differentially expressed in male SPTB placentas, such as *RBP4, VEGFA, EREG, PLA2G2A, CRH, PRL,* and *LEP,* which regulated the metabolic processes, particularly lipid and fatty acid metabolism (Supplemental Table S10). This was consistent with our previous metabolomics study that showed a marked elevation of multiple acylcarnitine species and significantly decreased the fatty acid oxidation in SPTB placentas [10].

### 2.4.1. Inflammatory Signaling and Oxidative Stress Pathways Are Activated

Many environmental exposures, including infections, during pregnancy increase the production of mediators of oxidative stress and abnormal metabolism, which may lead to spontaneous preterm births [20]. Several inflammatory signaling and detoxification pathways were altered in male SPTB placentas, including xenobiotic metabolism pathways, the NRF2-mediated oxidative stress response, glutathione-mediated detoxification, superoxide radical degradation, transforming growth factor-β (TGF-β) signaling, antiproliferative role of TOB in T-cell signaling, glucocorticoid receptor signaling, and PI3K signaling (Table 3). Many of these pathways were predicted to be activated in male SPTB placentas, consistent with the presence of inflammation and oxidative stress.

Xenobiotic metabolizing enzymes and transporters play critical roles in the metabolism, elimination, and detoxification of harmful xenobiotics and toxic endogenous compounds in the placenta via nuclear receptors, including the constitutive active receptor (CAR), pregnane X receptor (PXR), and aryl hydrocarbon receptor (AHR) [21]. These three nuclear receptor superfamilies were predicted to be activated in male SPTB placentas (z-score = 2.71, 2.11, and 2.45, respectively). Twelve genes comprising these pathways were altered in male SPTB placentas, including *GSTA1, GSTM5, MGST1,* and *CHST2* (Supplemental Table S11).

Consistent with previous studies showing that oxidative stress is associated with SPTB [22,23], glutathione-mediated detoxification and the nuclear factor erythroid 2-like 2 (NRF2)-mediated oxidative stress response were both disrupted in male SPTB placentas. Eleven genes comprising these pathways were altered in male SPTB placentas, including *CAT, PRKCA,* and *SOD3* (Supplemental Table S11).

The IPA Tox analysis, which links gene expression to clinical pathology endpoints, further identified the genes directly contributing to mitochondrial dysfunction in male SPTB placentas, including *CAT, BTG2, BCL2,* and *GSTA1* (Supplemental Table S12). These genes regulate the processes such as mitochondrial transmembrane potential, permeability transition of mitochondria, depolarization of mitochondria, and reactive oxygen species detoxification.

Both TGF-β and TOB play critical roles in maintaining a normal pregnancy, and low levels of TGF-β are associated with an increased risk of preterm birth [24–26]. Both TGF-β and TOB signaling were predicted to be inhibited in male SPTB placentas (z-score = −0.82 and −2.00, respectively). *TGFB1, TGFBR1, INHBA, INHA,* and *CCNE1* were examples of differentially expressed genes comprising these pathways (Supplemental Table S11). Further, underscoring TGF-β1′s importance in SPTB placentas, its downstream signal transducers SMAD3 and SMAD4 were all predicted as inhibited regulators in male SPTB placentas. They regulate the differential expression of more than 90 genes in our dataset (Figure 3a).

Interestingly, glucocorticoid receptor signaling was altered in male, but not female, SPTB placentas. Glucocorticoids activate a number of physiologic pathways in the placenta and are also critical for fetal organ development and survival during pregnancy and parturition [27]. Previous studies have shown that term female placentas have a higher glucocorticoid receptor expression compared to term male placentas [28]. Twenty-seven genes comprising this pathway were differentially expressed, including *FGG, IL5RA, KRT24, KRT5,* and *PRL* (Supplemental Table S11).

Phosphoinositide 3-kinase (PI3K) signaling modulates the immune system during pregnancy [29]. The disruption of PI3K signaling leads to an unbalanced adaptation of the maternal innate immune system to gestation and increases the fetal mortality [30]. We found that PI3K signaling was activated in male SPTB placentas, including the differentially expressed genes *CD180, PIK3AP1,* and *PLCD3* (Supplemental Table S11).

Cumulatively, these data suggest that the key pathways regulating oxidative stress and the ability of the placentas to handle toxins are altered in preterm birth placentas.

### 2.4.2. Metabolic Pathways Are Altered in SPTB Placentas

The IPA analysis identified multiple metabolic canonical pathways that were altered in male SPTB placentas, including amino acids, sphingomyelins, vitamin B6, purines, and myo-inositol (Table 3). Pathways regulating histidine, sulfite oxidation, and urea cycle metabolism were disrupted, including the differentially expressed genes *CPS1, ARG2,* and

*MTHFD1* (Supplemental Table S13). Sulfite oxidation, catalyzed by mitochondrial inter-membrane space enzyme sulfite oxidase, is associated with oxidative stress in the placenta.

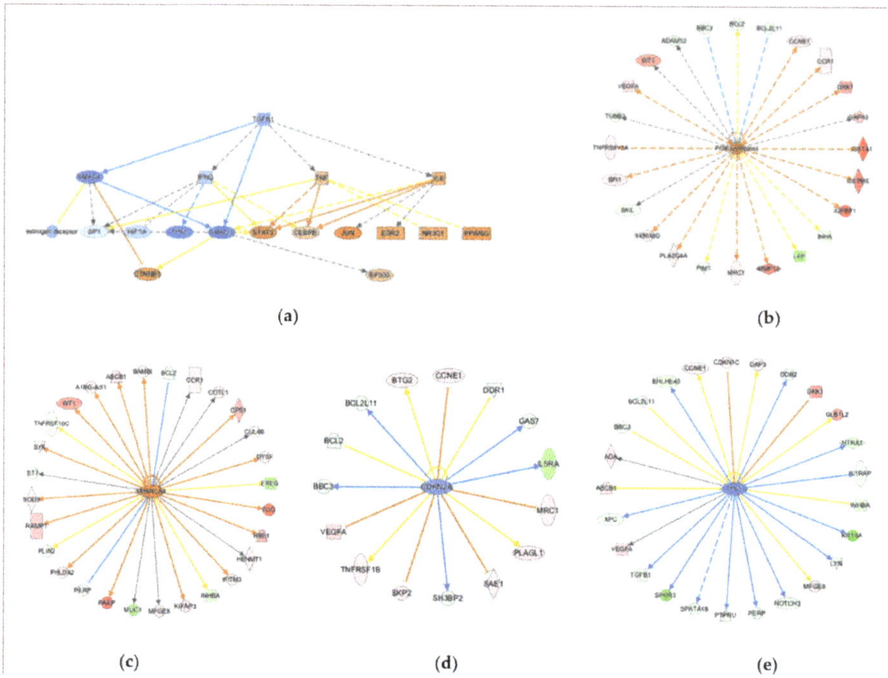

**Figure 3.** Ingenuity Pathway Analysis® (IPA)-annotated mechanistic network or differentially expressed genes regulated by critical upstream regulators. Mechanistic network regulated by TGFB1, SMAD3, and SMAD4 (**a**). Differentially expressed genes regulated by the PI3K complex (**b**), SMARCA4 (**c**), CDKN2A (**d**), and TP63 (**e**). Orange-filled and blue-filled shapes indicate predicted activation and inhibition, respectively; red-filled and green-filled shapes indicate increased and decreased expressions, respectively; orange-red lines indicate activation; blue lines indicate inhibition; yellow lines indicate findings inconsistent with the state of downstream activity; grey lines indicate that the effect was not predicted.

Sphingomyelin synthase (*SGMS1*) and sphingomyelin phosphodiesterase 2 (*SMPD2*), two genes involved in sphingomyelin metabolism, were downregulated in male SPTB placentas (Supplemental Table S13). Sphingomyelins are plasma membrane components, as well as signaling sphingolipids. An altered distribution of sphingomyelin and other sphingolipid species has been shown to play an important role in preeclampsia [31–33]. Sphingomyelin can also be degraded into phosphocholine and ceramide via SMPD2 [34]. Reduced levels of SMPD2 result in decreased levels of ceramides which is associated with impaired trophoblast syncytialization [35]. Ceramides also act as lipid secondary messengers and influence oxidative stress via regulating the expression and activity of antioxidant enzyme manganese-dependent superoxide dismutase (MnSOD) [36].

Pyridoxal 5′-phosphate (PLP), an active form of vitamin B6, acts as a coenzyme in the metabolism of amino acids, lipids, carbohydrates, and one-carbon units. PLP also functions as an antioxidant to prevent free radical generation and lipid peroxidation, modulates mitochondrial function, and regulates the immune system [37,38]. Five genes regulating PLP metabolism were differentially expressed in male SPTB placentas, including *DAPK1* and *PNPO* (Supplemental Table S13). These metabolic changes were consistent with our metabolomics study in SPTB placentas that metabolites of amino acids and sphingolipids were altered in the SPTB placentas [10]. Furthermore, purine metabolism is altered in male SPTB placentas, which was also observed in the placentas from an intrauterine inflammation preterm birth mouse model [39].

Several signaling pathways regulating general energy metabolism were altered in male SPTB placenta, including c-AMP-mediated signaling, signal transducer and activator of the transcription 3 (STAT3) pathway, Janus kinase 2 (JAK2) signaling, and PI3K signaling. c-AMP signaling is important for the differentiation and function of trophoblasts and placentas and is the major route to trigger trophoblast fusion [40]. It also interacts with protein kinase A (PKA) and MAPK signaling and plays a critical role in glucose and lipid metabolism [41]. c-AMP signaling was predicted to be activated in male SPTB placentas (z-score = 1.51). Twelve genes regulating the c-AMP signaling pathway were differentially expressed, including *ADCY1*, *PDE6H*, *ADORA2B*, and *CNGA1* (Supplemental Table S13). Not only is PI3K signaling key in immune system functions, it is also a key regulator of glucose and lipid metabolism and oxidative stress through modulating mitochondrial respiratory chain activity, oxidative phosphorylation, and mitochondrial integrity [42–44]. Via PI3K signaling, the placenta can fine-tune the supply of maternal nutrient resources to the fetus [45]. Nine genes comprising this pathway were differentially expressed (Supplemental Table S13). The PI3K complex was also predicted as an activated upstream regulator with a z-score of 2.35. It regulates expression changes of 24 genes in male SPTB placentas (Figure 3b).

These results suggest that the pathways that regulate key metabolic functions of the placenta, including fatty acids and glucose, are altered in SPTB placenta.

### 2.4.3. Retinoids, Vitamin D, and PPAR Signaling Is Disrupted

Nuclear receptors are a superfamily of transcription factors that can bind to DNA directly and regulate the gene expression upon binding to their ligands. The IPA analysis revealed that multiple nuclear receptor signaling pathways were altered in male SPTB placentas, including retinoid X receptors (RXR), peroxisome proliferator-activated receptors (PPARs), vitamin D receptor (VDR), thyroid hormone receptor (TR), and PXR (Table 3). A total of 23 genes comprising these pathways were differentially expressed, including *IGFBP1*, *IL1RL1*, *WT1*, *ADCY1*, *TGFB1*, and *GSTA1* (Supplemental Table S14). In our study, VDR/RXR activation was predicted to be inhibited in male SPTB placentas with a z-score of −1.34, and PPARα/RXRα activation was predicted to be activated with a z-score = 1.26.

### 2.4.4. Extracellular Matrix and Cell Adhesion Are Disrupted

The extracellular matrix (ECM) is important for the architecture of placental stroma, which supports trophoblasts and provides the environment for a healthy pregnancy. Placentas from pregnancies complicated by preeclampsia exhibit peri-villous coagulation and villous fibrosis, resulting from the overproduction of ECM in the connective tissue [46,47]. Multiple pathways regulating the extracellular matrix were altered in male SPTB placentas, including the Wnt/Ca+ pathway, fibrosis/stellate cell activation, the inhibition of matrix metalloproteases, and the epithelial–mesenchymal transitional pathway (Table 3). Thirty-two genes comprising these pathways were differentially expressed, including *PDE6H*, *COL11A2*, *COL24A1*, *LEP*, *IL1RL1*, *PROK1*, and *MMP12* (Supplemental Table S15).

Cell–cell and cell–ECM communication are important in coordinating proliferation and differentiation during placenta development. Cell adhesion molecules, including transmembrane receptor integrins, can facilitate cell–cell and cell–ECM adhesion [48,49]. Pathways regulating cell adhesion were altered in male SPTB placentas, including epithelial adherens junction signaling, integrin signaling, and gap junction signaling (Table 3). Twenty-five genes comprising these pathways were differentially expressed, including *NOTCH3*, *ITGAD*, *ITGB6*, *MYLK3*, *ADCY1*, and *HTR2B* (Supplemental Table S15).

### 2.4.5. Estrogen Receptor Signaling Is Activated

Interestingly, estrogen receptor signaling was predicted to be activated in male SPTB placentas with a z-score of 1.89 but was unaltered in female SPTB placentas (Table 3). Eighteen genes comprising this pathway were differentially expressed, including *ADCY1, LEP, MMP12, PROK1,* and *VEGFA* (Supplemental Table S16). Estrogen receptor signaling plays a critical role in trophoblast differentiation, placental function, and fetal development and modulates placenta and fetal communication [50–52]. Estrogen also regulates placental angiogenesis via modulating the expression of VEGF, angiopoietin-1, and angiopoietin-2 [53]. Estrogen receptor alpha and beta expression in the placenta are altered in preeclampsia and intrauterine growth restriction [54].

### 2.5. Canonical Pathways Altered in Female SPTB Placentas

An IPA analysis of 61 non-GA-associated differentially expressed genes in female SPTB placentas revealed nine canonical pathways that were altered in preterm births (Table 4). Most of these are critical pathways regulating the metabolism and nutrient sensing, including PKA signaling, insulin-like growth factor 1 (IGF-1) signaling, G-protein-coupled receptor (GPCR)-mediated nutrient sensing, $\alpha$-adrenergic signaling, and cholecystokinin/gastrin-mediated signaling. PKA pathway activation plays a major role in steroidogenic gene regulation in human placentas [55]. PKA is also located in mitochondria and regulates mitochondrial protein phosphorylation [56]. IGF-1 plays a critical role in fetal and placenta growth and development [57,58]. Interestingly, maternal IGF-1 and IGF1R polymorphisms are associated with preterm birth [59] and the expression of IGF binding proteins (IGFBPs) are altered in placentas from idiopathic spontaneous preterm births [21]. $\alpha$1-Adrenergic signaling stimulates the placenta blood flow, and dysregulation of this pathway has been implicated in placenta ischemia-induced hypertension [60]. Finally, cholecystokinin is one of the most highly upregulated genes in early placentas from women who later developed preeclampsia compared with women who experienced a normal pregnancy; however, its role in the placenta has not been investigated [61]. *CHD5, GH1, PRKAR1A, HIST1H1A, PYGM, YWHAZ,* and *CCK* were the differentially expressed genes regulating these pathways (Supplemental Table S17).

**Table 4.** Top canonical pathways altered in female SPTB placentas.

| Ingenuity Canonical Pathways | $p$-Value | z-Score |
|---|---|---|
| PPAR$\alpha$/RXR$\alpha$ Activation | $1.38 \times 10^{-2}$ | —— |
| Protein Kinase A Signaling | $2.09 \times 10^{-2}$ | 1.00 |
| $\alpha$-Adrenergic Signaling | $2.63 \times 10^{-2}$ | —— |
| NER Pathway | $3.02 \times 10^{-2}$ | —— |
| IGF$-$1 Signaling | $3.09 \times 10^{-2}$ | —— |
| GPCR-Mediated Nutrient Sensing in Enteroendocrine Cells | $3.55 \times 10^{-2}$ | —— |
| Cholecystokinin/Gastrin-mediated Signaling | $3.98 \times 10^{-2}$ | —— |
| Inhibition of ARE-Mediated mRNA Degradation Pathway | $4.17 \times 10^{-2}$ | —— |
| Nitric Oxide Signaling | $4.68 \times 10^{-2}$ | —— |

The IPA disease and biological function analysis also revealed that 17 genes in female SPTB placentas that regulate lipid, protein, and carbohydrate metabolism, including *CCK, GH1, APOC3, SULT1E1, TGM3,* and *ASB4* (Supplemental Table S18), were differentially expressed.

Overall, our data indicate that nutrient sensing and lipid, protein, and carbohydrate metabolism are disrupted in female SPTB placentas.

### 2.6. Upstream Regulators and Regulatory Networks Regulating Nutrient-Sensing, Metabolic, and Mitochondrial Function Are Altered in SPTB Placenta

The ingenuity pathway analysis can identify upstream regulators that mediate changes in the gene expression. The top upstream regulators identified in male SPTB placenta are listed in Table 5. The activated regulators in male SPTB placentas include SMARCA4, RAF1, and JUN. The important inhibited regulators include PHB2, $\alpha$-catenin, TP63, and CDKN2A. Many of these upstream regulators are involved in glucose and lipid metabolism, nutrient-sensing, and mitochondrial function. SMARCA4 is part of the large ATP-dependent chromatin remodeling complex SNF/SWI. It regulates the transcription of many genes for fatty acid and lipid biosynthesis [62,63]. In addition, SMARCA4 also plays a role in trophoblast stem cell renewal and placenta development [64]. SMARCA4 was predicted as an activated regulator (z-score = 3.20) and modulated the expression changes of 28 genes in male SPTB placentas (Figure 3c). Oncoprotein c-Jun (JUN) regulates the cell cycle and apoptosis, and the expression is altered in the placenta in pregnancies complicated by preeclampsia [65]. JUN was predicted as an activated regulator (z-score = 2.21) and regulates expression changes in 31genes in male SPTB placentas. Cyclin-dependent kinase inhibitor 2A (CDKN2A) regulates cell senescence and was predicted as an inhibited upstream regulator, and the expression of 15 of its downstream targets was disrupted (Figure 3d). Attenuation of the senescence program occurs in IUGR human placentas, and a knockdown of the CDKN2A expression results in functional and morphological abnormalities in murine placenta [66]. Tumor protein p63 (TP63) was predicted as an inhibited regulator (z-score = −2.01), modulating the expression changes of 25 genes in male SPTB placentas (Figure 3e). It regulates cytotrophoblast differentiation and fusion [67].

**Table 5.** Top upstream regulators altered in male SPTB placentas.

| Regulators | $p$-Value | Activation z-Score | # Genes Regulated |
| --- | --- | --- | --- |
| SMARCA4 | $1.66 \times 10^{-2}$ | 3.20 | 28 |
| RAF1 | $5.01 \times 10^{-4}$ | 2.40 | 16 |
| JUN | $1.83 \times 10^{-4}$ | 2.21 | 31 |
| IL33 | $1.85 \times 10^{-3}$ | 2.20 | 16 |
| IL13 | $1.00 \times 10^{-6}$ | 2.19 | 33 |
| SOX7 | $1.06 \times 10^{-3}$ | 2.12 | 6 |
| LIF | $1.31 \times 10^{-3}$ | 2.07 | 15 |
| CDKN2A | $4.44 \times 10^{-2}$ | −1.33 | 15 |
| PDX1 | $1.81 \times 10^{-2}$ | −1.71 | 11 |
| TP63 | $4.78 \times 10^{-3}$ | −2.01 | 25 |
| GATA1 | $4.93 \times 10^{-5}$ | −2.03 | 21 |
| Alpha catenin | $1.36 \times 10^{-3}$ | −2.07 | 10 |
| ROCK2 | $4.76 \times 10^{-3}$ | −2.24 | 6 |
| PHB2 | $1.27 \times 10^{-4}$ | −2.24 | 4 |

# Genes Regulated: number of genes regulated.

In female SPTB placentas, the important upstream regulators identified by IPA are listed in Table 6. Insulin (INS) was predicted as an inhibited upstream regulator in female SPTB placentas with a z-score = −1.09. Phosphatase and tensin homolog (PTEN), an inhibitor of PI3K signaling, was also predicted to be inhibited with a z-score of −1, suggesting that PI3K signaling was also activated in female SPTB placentas. Vascular endothelial growth factors (VEGFs) were predicted as inhibited upstream regulators with a z-score = −1.09. They are important for angiogenesis and vascular remodeling, a process critical for placental function and healthy pregnancies.

The IPA analysis can expand predictions and identify the potential novel master regulators responsible for the changes in the gene expressions. LGALS1, NOX1, DNAJA3, PPP2R2A, PPP2CA, and BRCA1 were a few examples of master regulators identified in male SPTB placentas (Table 7). Galectin 1 (LGALS1) is a β-galactoside-binding protein and regulates maternal–fetal immune tolerance and maintaining a normal pregnancy [68–70]. The dysregulation of LGALS1 expression is associated with preeclampsia [71], and interestingly, maternal serum LGALS1 levels are significantly higher in pregnancies with premature ruptures of the membranes [72]. LGALS1 was predicted as an activated master regulator in male SPTB placenta and interacts with 15 upstream regulators to regulate the gene expression. NADPH oxidase 1 (NOX1), a member of the NADPH oxidase family, was predicted as an activated master regulator in male SPTB placenta. NADPH oxidase is the major source of superoxide in placentas and plays a role in early placental development [73,74]. However, the overexpression of NOX1 increases oxidative stress and is associated with preeclampsia [75]. DnaJ heat shock protein family member A3 (DNAJA3) is a mitochondrial protein regulating protein folding, degradation, and complex assembly. It plays a critical role in maintaining the mitochondrial membrane potential and mitochondrial DNA integrity; however, its role in the placenta is unknown [76]. DNAJA3 was predicted to be inhibited in SPTB placentas (z-score = $-3.16$) and interact with 11 upstream regulators to regulate the gene expression. Both PPP2R2A, a regulatory subunit of protein phosphatase 2 (PP2A), and PPP2CA, the catalytic subunit of PP2A, were predicted as inhibited master regulators in male SPTB placentas with a z-score of $-3.03$ and $-1.51$, respectively. PP2A acts as a negative regulator of cell growth and division and controls energy metabolism and redox homeostasis via modulating AMP kinase (AMPK) and PI3K-AKT-mTOR signaling [77]. AMPK is a master metabolic regulator controlling glucose sensing and uptake, lipid metabolism, glycogen, cholesterol, and protein synthesis, and the induction of mitochondrial biogenesis [78,79]. mTOR regulates energy-sensing pathways and functions as an important placental growth signaling the sensor to regulate trophoblast proliferation [80–83]. Furthermore, PP2A can control genome integrity by coupling the metabolic processes with DNA damage responses [84]. The breast cancer type 1 susceptibility protein (BRCA1), a tumor suppressor, regulates the cell cycle and DNA damage repair in the placenta [85]. It also functions as a regulator of glucose and lipid metabolism, as well as mitochondrial respiration [86–88]. BRCA1 was predicted as an inhibited master regulator in the male SPTB placentas (z-score = $-2.08$) and interacts with 27 upstream regulators to modulate the gene expression.

**Table 6.** Top upstream regulators altered in female SPTB placentas.

| Regulators | $p$-Value | Activation z-Score | # Genes Regulated |
|---|---|---|---|
| HNRNPK | $2.94 \times 10^{-4}$ | —— | 3 |
| INS | $1.53 \times 10^{-3}$ | $-1.09$ | 4 |
| AKT1 | $8.10 \times 10^{-3}$ | —— | 4 |
| ZBTB16 | $1.61 \times 10^{-2}$ | —— | 3 |
| PTEN | $2.73 \times 10^{-2}$ | $-1.00$ | 5 |
| IL15 | $3.11 \times 10^{-2}$ | —— | 4 |
| ONECUT1 | $4.04 \times 10^{-2}$ | —— | 3 |
| STAT5B | $4.19 \times 10^{-2}$ | —— | 3 |
| VEGF | $4.75 \times 10^{-2}$ | $-1.09$ | 4 |

# Genes Regulated: number of genes regulated.

**Table 7.** Top master regulators altered in male SPTB placentas.

| Master Regulators | $p$-Value | Activation z-Score | # Connected Regulators |
|---|---|---|---|
| MYB | $8.33 \times 10^{-10}$ | 3.76 | 6 |
| TBK1 | $1.09 \times 10^{-10}$ | 3.14 | 20 |
| LGALS1 | $5.96 \times 10^{-10}$ | 3.03 | 15 |
| GAB2 | $7.01 \times 10^{-10}$ | 2.71 | 23 |
| NOX1 | $2.89 \times 10^{-11}$ | 2.68 | 14 |
| MAP3K8 | $5.54 \times 10^{-11}$ | 2.32 | 34 |
| PPP2CA | $4.05 \times 10^{-9}$ | $-1.51$ | 34 |
| BRCA1 | $8.78 \times 10^{-10}$ | $-2.08$ | 27 |
| CCHCR1 | $1.75 \times 10^{-11}$ | $-2.51$ | 29 |
| RBP1 | $4.24 \times 10^{-10}$ | $-2.84$ | 4 |
| PPP2R2A | $1.87 \times 10^{-8}$ | $-3.03$ | 12 |
| DNAJA3 | $1.07 \times 10^{-9}$ | $-3.16$ | 11 |
| MEN1 | $1.16 \times 10^{-8}$ | $-3.25$ | 8 |

# Connected Regulators: number of connected regulators.

NFAT and CERK are master regulators that were identified in female SPTB placentas (Table 8). Nuclear factors of activated T cells (NFAT) is a family of transcription factors that activate cytokine production and also positively regulate placental FLT-1 and sFlt-1 e15a, the secretion of sFlt-1, and the inflammatory cytokine expression [89]. NFAT is thought to be involved in the pathophysiology of preeclampsia. In addition, NFAT may act as a regulator for the parturition and induction of labor [90]. NFAT was predicted as an activated master regulator and interacted with 23 upstream regulators to regulate the gene expression. Ceramide kinase (CERK) converts ceramide to ceramide-1-phosphate. CERK was predicted as an inhibited master regulator in female SPTB placentas and interacting with 13 upstream regulators to regulate the gene expression.

**Table 8.** Top master regulators altered in female SPTB placentas.

| Master Regulators | *p*-Value | Activation z-Score | # Connected Regulators |
|---|---|---|---|
| Fe2+ | $1.97 \times 10^{-2}$ | 2.67 | 16 |
| NFAT (family) | $4.89 \times 10^{-3}$ | 2.40 | 23 |
| CAMKK2 | $1.60 \times 10^{-2}$ | 2.12 | 5 |
| BLVRA | $7.49 \times 10^{-3}$ | 2.07 | 23 |
| MAPK13 | $5.11 \times 10^{-4}$ | 2.04 | 26 |
| TRERF1 | $3.82 \times 10^{-2}$ | $-2.00$ | 2 |
| CERK | $2.16 \times 10^{-2}$ | $-2.50$ | 13 |

# Connected Regulators: number of connected regulators.

In summary, these results demonstrate that upstream regulators and master regulators important for the nutrient-sensing and metabolism are altered in SPTB placentas.

### 2.7. Interactome Network Analysis of the Transcriptome and Metabolome in SPTB Placentas

To further investigate the association of differentially expressed genes and significantly changed metabolites, identified in our previous metabolomics study [10]; an interactome network model (Figure 4) integrating the transcriptome and metabolome was generated that was connected via protein–protein or protein–metabolite interactions. Since the placentas from both sexes were used in our metabolomics study [10], we used the transcriptomes from the combined sexes (Supplemental Table S19) for the interactome network analysis. The analysis of this interactome network confirmed that several metabolic processes are altered in SPTB placentas. These modules included the genes and/or metabolite interactions that were associated with fatty acid metabolism, cholesterol biosynthesis, steroid hormone metabolism, glycolysis and gluconeogenesis, pentose phosphate pathway, TCA cycle, amino acid metabolism, purine and pyrimidine metabolism, glycosphingolipid metabolism, glycerophospholipid metabolism, phosphatidylinositol phosphate metabolism, vitamin metabolism, and xenobiotics metabolism (Table 9). The differentially expressed genes and significantly altered levels of metabolites associated within these modules in our datasets are also listed in Table 9.

**Table 9.** Metabolic pathways identified from the interactome network.

| Metabolic Pathways Enriched within the Interactome Network | Number of Gene Changes (Inferred and Non-Inferred) | Gene Changes within Dataset | Metabolite Changes within Dataset | Number of Metabolites Changes (Inferred and Non-Inferred) |
|---|---|---|---|---|
| Androgen and estrogen biosynthesis and metabolism | 98 | CYP3A5, CYP2C18, MGST1, AKR1C2, SULT1E1 | 3beta-Hydroxyandrost-5-en-17-one 3-sulfate, Androst-5-ene-3beta,17beta-diol, Estrone | 64 |

**Table 9.** *Cont.*

| Metabolic Pathways Enriched within the Interactome Network | Number of Gene Changes (Inferred and Non-Inferred) | Gene Changes within Dataset | Metabolite Changes within Dataset | Number of Metabolites Changes (Inferred and Non-Inferred) |
|---|---|---|---|---|
| Arachidonic acid metabolism | 114 | HADHA, HADHB, CYP2C18, CYP3A5, PLA2G2A, PLA2G4A, ECHS1, MGST1, CYP4F3 | 15(S)-HETE, 5(S)-HETE | 63 |
| C21-steroid hormone bioshnthesis and metabolism | 50 | CYP2C18, CYP3A5, SULT1E1 | Cholesterol, Cortisone | 45 |
| Fatty acid beta-oxidation and metabolism | 57 | CPT2, HADHA, HADHB, ECHS1 | L-Palmitoylcarnitine | 184 |
| Fat-soluble vitamin metabolism | 65 | ALDH1A1, HADHA, HADHB, ECHS1, CYP4F3 | No metabolites with significant difference | 49 |
| Fructose, galactose, and aminosugars metabolism | 62 | AKR1B1, NPL | N-Acetyl-D-glucosamine 1-phosphate, L-Glutamine, CMP-N-acetylneuraminate, N-Acetylneuraminate, N-Acetyl-D-glucosamine 6-phosphate | 65 |
| Glycerophospholipid metabolism | 113 | ADHFE1, AKR1B1, PLA2G2A, PLA2G4A, LIPA, ALDH7A1, CEL | L-Serine, sn-Glycerol 3-phosphate, Glycerone phosphate, Cholesterol, Ethanolamine phosphate, Choline phosphate | 62 |
| Glycine, serine, alanine, and threonine metabolism | 61 | DLD, BHMT2, TARS, GATM, CKMT1B | Glycine, L-Serine, L-Methionine, L-Threonine, Betaine aldehyde, Guanidinoacetate | 59 |
| Glycolysis and gluconeogenesis | 80 | ADHFE1, DLD, LDHB, ALDH7A1, PCK2 | Phosphoenolpyruvate, Glycerone phosphate, Acetyl phosphate, 2-Phospho-D-glycerate | 34 |
| Glycosphingolipid metabolism | 83 | GALC, ASAH1 | CMP, L-Serine, CMP-N-acetylneuraminate, Ethanolamine phosphate | 116 |
| Histidine and lysine metabolism | 69 | SUV39H2, ALDH7A1 | L-2-Aminoadipate, L-Glutamate, Glycine, L-Lysine, beta-Alanine, Carnosine | 50 |
| Leukotriene metabolism | 115 | ADHFE1, HADHA, HADHB, CYP2C18, CYP3A5, ECHS1, MGST1, ALDH7A1, CYP4F3 | Glycine | 83 |

**Table 9.** *Cont.*

| Metabolic Pathways Enriched within the Interactome Network | Number of Gene Changes (Inferred and Non-Inferred) | Gene Changes within Dataset | Metabolite Changes within Dataset | Number of Metabolites Changes (Inferred and Non-Inferred) |
|---|---|---|---|---|
| Linoleate metabolism | 80 | CYP2C18, CYP3A5, PLA2G2A, PLA2G4A | No metabolites with significant difference | 16 |
| Methionine and cysteine metabolism | 51 | MAT2A, LDHB, CBS | N-Acetylmethionine, L-Serine, L-Methionine, 3-Sulfino-L-alanine | 41 |
| Pentose phosphate pathway | 30 | SAT1, PGD | D-Sedoheptulose 7-phosphate, Glycerone phosphate | 27 |
| Phosphatidylinositol phosphate metabolism | 98 | MINPP1, PLCL2, OCRL, SYNJ2 | myo-Inositol, Ethanolamine phosphate | 45 |
| Prostaglandin formation | 50 | MGST1 | Prostaglandin E2 | 53 |
| Purine metabolsim | 274 | ATP8A2, ATP5J, ATP5O, POLD2, HPRT1, ATP2B1, AMPD2, POLS | L-Aspartate, L-Glutamine, Urate | 65 |
| Pyrimidine metabolism | 119 | POLD2, UPB1, CAD, POLS | L-Aspartate, CMP, L-Glutamine, beta-Alanine | 45 |
| Squalene and cholesterol biosynthesis | 26 | SC4MOL, HMGCR, IDI1 | Cholesterol | 31 |
| TCA cycle | 26 | DLD, IDH1, PCK2 | Phosphoenolpyruvate | 20 |
| Tryptophan and tyrosine metabolism | 169 | FAH, ADHFE1, HADHA, HADHB, CYP2C18, CYP3A5, ECHS1, MGST1, CAT, ALDH7A1 | Adenosine 3′,5′-bisphosphate, L-Phenylalanine, L-Tyrosine, Indole-3-acetate | 157 |
| Urea cycle and metabolism of arginine, proline, glutamate, aspartate, and asparagine | 124 | EPRS, GATM, CPS1, SAT1, ARG2, MGST1, ALDH7A1 | L-Glutamate, Glycine, L-Lysine, L-Aspartate, L-Glutamine, beta-Alanine, L-Proline, L-Asparagine, 4-Aminobutanoate, trans-4-Hydroxy-L-proline, 5-Oxoproline, N1-Acetylspermine, O-Acetylcarnitine, N-Acetylputrescine, 4-Acetamidobutanoate, N1,N12-Diacetylspermine | 108 |
| Valine, leucine, and isoleucine degradation | 54 | HADHA, HADHB, ECHS1, ALDH7A1 | L-Valine, L-Isoleucine, L-Leucine | 43 |

Table 9. Cont.

| Metabolic Pathways Enriched within the Interactome Network | Number of Gene Changes (Inferred and Non-Inferred) | Gene Changes within Dataset | Metabolite Changes within Dataset | Number of Metabolites Changes (Inferred and Non-Inferred) |
|---|---|---|---|---|
| Vitamin B metabolism | 76 | NMNAT1, PNPO, MTHFD1, NNMT, MTHFD1L, COASY | Pantothenate, FAD, L-Glutamate, Glycine, L-Lysine, L-Glutamine, L-Serine, Nicotinamide, 1-Methylnicotinamide, Pyridine-2,3-dicarboxylate | 70 |
| Xenobiotics metabolism | 85 | ADHFE1, CYP2C18, CYP3A5, AKR1C2, MGST1 | No metabolites with significant difference | 72 |

Genes or metabolites shown in red were upregulated in SPTB placentas, whereas those in green were downregulated.

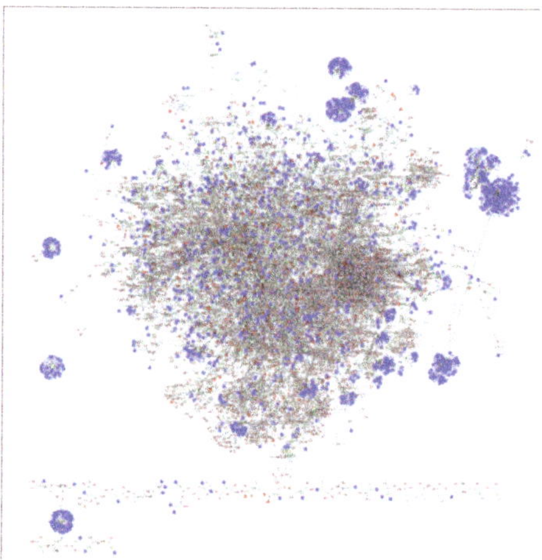

**Figure 4.** Visual representation of the interactome model. Interaction network of integrated transcriptome and metabolome was analyzed using MetScape 3.1. Dark blue circles represent differentially expressed genes in the placenta dataset; light blue circles represent inferred gene interactions; dark red circles represent significantly changed metabolites in the placenta dataset; light red circles represent inferred metabolite interactions; grey lines represent protein–protein or protein–metabolite interactions.

## 3. Discussion

In this study, we demonstrated marked changes in the expression of genes in SPTB placentas involving key pathways regulating mitochondria function, inflammation, amino acid and lipid metabolism, extracellular matrix, and detoxification. Importantly, we also show that there are marked differences in the placental transcriptome in SPTB between males and females, suggesting that there may be differences between males and females in the mechanisms by which a placenta dysfunction contributes to SPTB.

Male fetuses have a higher incidence of many pregnancy complications, including preterm births [15–17]. Preterm males also have increased morbidity and mortality after births [91]. Although the underlying mechanisms are unclear, a more proinflammatory intrauterine milieu at lower gestational ages may account for the increased incidence and/or make the male fetus more susceptible to an abnormal intrauterine milieu. In the current study, 670 differentially expressed genes were identified in male SPTB compared to term placentas. To our surprise, only 61 differentially expressed genes were found in female SPTB placentas, supporting the observations of fetal sex differences and the lower susceptibility to spontaneous preterm birth in female fetuses. The limited changes of the transcriptomes in female SPTB placentas compared with term placentas may result in a survival advantage for females and adaptive responses for a suboptimal milieu, as previously reported [92,93].

We identified multiple metabolic pathways that were altered in the SPTB placenta in our previous metabolomics study [10]. Levels of sphingolipids, steroids, amino acids, and metabolites involved in fatty acid oxidation, such as acylcarnitines, were significantly different between SPTB and term placentas. Acylcarnitines, the major metabolites increased in SPTB placentas, are intermediate oxidative lipids and are associated with proinflammatory signaling and mitochondrial dysfunction [94,95]. Multiple elevated 2-hydroxy long-chain fatty acids in SPTB placentas are potent uncouplers of oxidative phosphorylation and can impair energy homeostasis and induce the mitochondria permeability transition pore. The current study identifies the molecular basis for these changes as multiple genes, and the pathways controlling these processes were substantially altered. The IPA analysis of the transcriptome data demonstrated that many pathways and upstream regulators regulating inflammation, mitochondrial function, redox status and signaling, and energy metabolism and homeostasis were significantly altered in SPTB placentas. The alteration of these pathways suggests a fundamental disruption of mitochondria metabolism, as well as the initiation of a proinflammatory milieu in SPTB, and an activation of oxidative response/detoxification pathways may reflect an adaptive response, which ultimately fails, resulting in SPTB. In fact, our current findings were similar to the observations in our previous study with an intrauterine inflammation preterm birth mouse model [39], supporting that mitochondria dysfunction, abnormal fatty acid, and inflammation play major roles in SPTB even in the absence of overt infections.

Our finding that glucocorticoid receptor signaling was altered in male, but not female, SPTB placentas was intriguing. Glucocorticoids are critical for implantation, fetal organ development, and survival during pregnancy and parturition [27]. However, excess glucocorticoid exposure suppresses the immune system and has adverse effects on placental proliferation, angiogenesis, and glucose transport [28,96,97]. The observation that a higher glucocorticoid receptor expression in term female placentas compared to term male placentas suggests the lower glucocorticoid exposure of female fetuses during pregnancy and an enhanced immune response, which may contribute to the increased survival rate of female fetuses in an aberrant intrauterine milieu [28]. The effects of glucocorticoids on the placentas and fetuses also show a sex-specific manner. Glucocorticoid exposure increases oxidative stress in the male placentas but not the female placentas [98]. Glucocorticoids also decrease the adrenal activity in preterm males but not females [99], which may partially account for the poor perinatal outcomes of preterm males.

Other pathways that were altered in SPTB placentas include retinoids, vitamin D metabolism/signaling, thyroid hormone, and PPARs. Retinoids are lipophilic molecules and metabolites of Vitamin A (all-trans-retinol). They play important roles in regulating the energy metabolism and function as critical regulators during embryogenesis and promote the differentiation of trophoblast stem cells [100–102]. The actions of retinoids are mediated through retinoic acid receptors (RARs) and RXRs [103]. RXRs are common heterodimer partners for multiple nuclear receptors, such as PPARs, VDR, TR, and PXR [104]. Vitamin D plays a critical role in pregnancy in additional to its classical role in calcium/phosphate homeostasis and bone metabolism. It regulates he decidualization and implantation, hormone secretion, and placental immune response and defended the infections [105,106]. Vitamin D also has a potent antioxidant effect to prevent protein oxidation, lipid peroxidation, DNA damage, and maintaining a normal mitochondrial function [107]. The deficiency of vitamin D is associated with impaired fetal growth, preeclampsia, and gestational diabetes [105]. Indeed, epidemiologic studies also provide the evidence linking vitamin D insufficiency with preterm births [108]. The thyroid hormone (TH) plays a critical role in regulating the metabolic processes for normal growth. It can regulate gene expression directly, as well as crosstalk with PPAR and the liver X receptor (LXR), and modulate glucose, lipid, and cholesterol metabolism [109]. TH is important for the healthy pregnancy modulating for cell proliferation and differentiation, metabolism, and formation and functioning of the placenta [110]. It may play an important role in fine-tuning inflammation in placentas in both term and preterm labors [111]. The dysregulation of maternal thyroid hormone signaling is associated with preeclampsia, miscarriage, and intrauterine growth restriction [110]. In addition to drug transport, PXR/RXR heterodimer regulates the homeostasis/metabolism of glucose, lipid, steroids, bile acids, retinoic acid, and bone minerals [112]. PPARs modulate the inflammatory responses, cell proliferation, and cell division. They also play a central role in placental angiogenesis [113–117]. PPARs also exert the antioxidant effects and are critically important to early placental development [118,119].

Multiple pathways regulating the extracellular matrix and cell adhesion were disrupted in SPTB placentas. ECM is important for the architecture of placental stroma and supports a healthy pregnancy. Of note, ECM also plays an important role in the nutrient uptake, redox status, energy metabolism, and mitochondrial function. The ECM also regulates glucose transport, glycolysis, lipid metabolism, and the TCA cycle [120–122]. AMPK, the master metabolic regulator, also regulates integrin activity and extracellular matrix assembly [123]. ECM remodeling can modulate mitochondrial structure, dynamics, and function [124].

Integrated interactome modeling provides greater confidence in the signaling and pathways identified in the transcriptome and metabolome individually. Alterations of the lipid metabolism in SPTB placentas were identified in all three analyses of the transcriptome, metabolome, and interactome, which further supports that aberrant fatty acid metabolism may be causal to preterm birth. Interactome modeling also showed that metabolism, for almost all amino acids, glucose, steroid hormones, purines, pyrimidines, and vitamins, were altered in SPTB placentas. Collectively, our results strongly suggest that alterations and/or deficits in metabolic pathways cause placental insufficiency, ultimately resulting in SPTB.

A major limitation in spontaneous preterm birth research is the lack of human gestational controls. Eidem et al. used Rhesus macaque as gestational age controls in their study and identified 267 differentially expressed genes between preterm and term human placentas, including 29 SPTB-specific candidate genes [18]. These "SPTB" genes are enriched for functions in the metabolism, immunity, inflammation, and cell signaling, which are consistent with the results in our current study. Brockway et al. used infections related to preterm birth as a gestational age control and identified 170 SPTB-specific genes [19]. Similarly, these genes are also enriched for pathways in insulin-like growth factor (IGF) signaling, cytokine signaling, immune system, and signal transduction. While it is not a perfect control, we used second trimester placenta as a gestational age control in the current study. We also adapted the approach of Eidem et al. [18] and demonstrated that at least the differences of acylcarnitine metabolites, the major changes in the metabolome, between the Rhesus macaque preterm and term placentas, are not caused by the difference in gestational ages [10].

A well-functioning placenta plays a crucial role in normal pregnancy. Our current study has identified alterations in novel pathways and upstream regulators that may play an important role in the maintenance of normal bioenergetic metabolism and provides new insights into the underlying mechanisms of SPTB. Larger studies in preterm birth will be necessary to determine whether these findings can be generalized beyond the African American population that was studied and the possible population disparities.

## 4. Materials and Methods

### 4.1. Clinical Characteristics

Placenta samples from Black women (self-identified race) in the current study were selected from the larger Cellular Injury and Preterm Birth (CRIB 821376, NCT02441335) study at the University of Pennsylvania. CRIB enrollment criteria included women aged 18–45 years with singleton pregnancies admitted to the hospital with either spontaneous labor (defined as regular contractions and cervical dilation) or the premature rupture of membranes (PROM) occurring between 20 0/7 and 36 6/7 weeks of gestational age (preterm) or at 38 to 41 weeks of gestational age (term). The CRIB exclusion criteria included multiple gestations, fetal chromosomal abnormalities, major fetal anomalies, intrauterine fetal demise, intrauterine growth restriction, clinical chorioamnionitis, induction of labor, elective cesarean delivery, gestational diabetes, and gestational hypertension or preeclampsia. The CRIB study was approved by the Institutional Review Board at the University of Pennsylvania (protocol #821376), and patients were enrolled after written informed consent.

Second trimester placenta samples from Black women (self-identified race) were selected from the *"Trophoblast cells Isolation from Second and first Trimester placenta"* (TrISecT) study at the University of Pennsylvania and were utilized as gestational age controls in the current study. The enrollment criteria included women aged 18–45 years receiving care at the hospital due to the elective termination of a singleton pregnancy prior to 23 6/7 weeks of gestational age. The TrISecT exclusion criteria included multiple gestation, aneuploidy, and fetal congenital anomalies. The TrISecT study was approved by the Institutional Review Board at the University of Pennsylvania (protocol #827072), and patients were enrolled after written informed consent.

### 4.2. Total RNA Isolation and RNA-Seq Library Preparation

Placenta samples were collected from mid-placenta near the cord insertion on the fetal side and flash-frozen at the time of delivery (within 10 min) and stored at $-80\ °C$ prior to RNA extraction. Total RNA was extracted using TRIzol® Reagent (Invitrogen, Waltham, MA, USA), followed by Qiagen RNeasy® Mini Columns (Qiagen, Germantown, MD, USA) following the manufacturer's instructions. RNA integrity numbers greater than 7 were used for RNA-Seq Studies. RNA-Seq libraries were generated, using the Agilent SureSelect strand-specific RNA library preparation kit (Agilent, Santa Clara, CA, USA).

### 4.3. RNA-Seq and Gene Expression Analysis

RNA-Seq libraries were paired-end sequenced to 100 bp on an Illumina Hi-Seq platform in CAG Sequencing Core at the Children's Hospital of Philadelphia. RNA-seq data in .fastq files were aligned to the reference human genome (hg38) and transcriptome using the STAR (https://github.com/alexdobin/STAR, accessed on 10 June 2021) program in 2-pass mode. The alignment results were saved as indexed .bam files. Aligned reads in .bam files were loaded into R and mapped to known genes. Read pairs uniquely mapped to the sense strand of that transcription were counted to obtain a gene-level read count matrix. Differential gene expression was tested by DESeq2. Differential gene expression was evaluated by the fold change, DESeq2 *p*-value, and corresponding false discovery rate (FDR). Sequence data were deposited in NCBI's Gene Expression Omnibus and are accessible through the GEO Series accession number GSE174415. A functional analysis was conducted using QIAGEN's Ingenuity® Pathway Analysis (IPA®) (Qiagen, Germantown, MD, USA). Core analyses were performed on genes with FDR (*q*-value) $\leq 0.05$.

### 4.4. Sample Preparation for Proteomics

Frozen placenta samples from normal term births, and the second trimester ($n = 4$ for each group) were sent to the Proteomics Core Facility at the Children's Hospital of Philadelphia for protein hydrolysis, followed by peptide separation, and analyzed by LC-MS/MS on a QExactive HF mass spectrometer (Thermo Fisher Scientific, San Jose, CA, USA) coupled with an Ultimate 3000. A label-free approach was chosen for its adaptability to include new samples when needed, as well as to avoid the possible errors while the labeling techniques were applied.

*4.5. Protein Sequence Database Search and Proteomics Analysis*

MS/MS raw files were searched against a human protein sequence database, including isoforms from the UniProt Knowledgebase (taxonomy:10090 AND keyword: "Complete proteome (KW-0181)"), using MaxQuant [125] version 1.6.1.0 with the following parameters: fixed modifications, carbamidomethyl (C); decoy mode, revert; MS/MS tolerance, FTMS 20 ppm; False Discovery Rate (FDR) for both peptides and proteins of 0.01; minimum peptide length of 7; modifications included protein quantification, acetyl (protein N-term), and oxidation (M); peptides used for protein quantification, razor, and unique. iBAQ values were used for protein quantification.

Perseus (1.6.1.1) was used for proteomic data processing and statistical analysis. Protein groups containing matches to decoy database or contaminants were discarded. The data were $Log_2$-transformed and normalized by subtracting the median for each sample. Student's *t*-test was employed to identify differentially expressed proteins. Benjamini–Hochberg correction was applied to obtain an FDR.

*4.6. Integrated Network Analysis of the Transcriptome and Metabolome*

Differentially expressed genes and significantly changed metabolites identified in our previous study [10] were analyzed using MetScape3.1 [126] in Cytoscape (v3.8.0). The interactome networks were generated based on known protein–protein and protein–metabolite interactions. The metabolic pathways that were associated with protein–metabolite interactions were mapped onto each network.

**Supplementary Materials:** The following are available online at https://www.mdpi.com/article/10.3390/ijms22157899/s1: Table S1: Differentially expressed genes in male SPTB placentas. Table S2: Differentially expressed genes in female SPTB placentas. Table S3: Differentially expressed genes in both male and female SPTB placentas. Table S4: Demographics of the proteomics study. Table S5: Differentially expressed proteins comparing term births and 2nd trimester placentas. Table S6: Differentially expressed gestational age-specific candidate genes. Table S7: Differentially expressed genes comparing male with female term placentas. Table S8: Differentially expressed genes comparing male with female SPTB placentas. Table S9: Canonical pathways altered comparing male with female SPTB placentas. Table S10: Metabolic processes disrupted in male SPTB placentas. Table S11: Inflammatory signaling and detoxification pathways disrupted in male SPTB placentas. Table S12: Genes regulating mitochondrial function disrupted in male SPTB placentas. Table S13: Metabolic pathways and pathways modulating the energy metabolisms disrupted in male SPTB placentas. Table S14: Nuclear receptor signaling pathways disrupted in male SPTB placentas. Table S15: Pathways regulating extracellular matrix and cell adhesion disrupted in male SPTB placentas. Table S16: Estrogen receptor signaling disrupted in male SPTB placentas. Table S17: Genes regulating energy metabolisms disrupted in female SPTB placentas. Table S18: Metabolic processes disrupted in female SPTB placentas. Table S19: Differentially expressed genes in SPTB placentas combined.

**Author Contributions:** Conceptualization, Y.-C.L., R.A.S., M.J.F. and S.P.; methodology, Y.-C.L., Y.C., E.P. and L.S.; validation, Y.-C.L. and L.S.; formal analysis, Z.Z.; data curation, Y.-C.L., Z.Z., M.J.F., H.I. and R.A.S.; writing—original draft preparation, Y.-C.L.; writing—review and editing, R.A.S.; supervision, R.A.S.; project administration, Y.-C.L., M.J.F., R.A.S.; and funding acquisition, R.A.S. All authors have read and agreed to the published version of the manuscript.

**Funding:** This research was funded by the March of Dimes Prematurity Research Center at the University of Pennsylvania (R.A.S.).

**Institutional Review Board Statement:** The CRIB study was conducted according to the guidelines and approved by the Institutional Review Board at the University of Pennsylvania (protocol #821376). The TrISecT study was approved by the Institutional Review Board at the University of Pennsylvania (protocol #827072).

**Informed Consent Statement:** Patients were enrolled after written informed consent.

**Data Availability Statement:** Sequence data have been deposited in NCBI's Gene Expression Omnibus and are accessible through GEO Series accession number GSE174415.

Conflicts of Interest: The authors declare that the research was conducted in the absence of any commercial or financial relationships that could be construed as a potential conflict of interests.

## References


1. March of Dimes. Mission statement. Available online: http://www.marchofdimes.org/mission/mission.aspx (accessed on 27 October 2015).
2. Romero, R.; Dey, S.K.; Fisher, S.J. Preterm labor: One syndrome, many causes. *Science* **2014**, *345*, 760–765. [CrossRef] [PubMed]
3. Manuck, T.A.; Esplin, M.S.; Biggio, J.; Bukowski, R.; Parry, S.; Zhang, H.; Huang, H.; Varner, M.W.; Andrews, W.; Saade, G.; et al. The phenotype of spontaneous preterm birth: Application of a clinical phenotyping tool. *Am. J. Obstet. Gynecol.* **2015**, *212*, 487.e1–487.e11. [CrossRef] [PubMed]
4. Kourtis, A.P.; Read, J.S.; Jamieson, D.J. Pregnancy and infection. *N. Engl. J. Med.* **2014**, *371*, 1077. [CrossRef] [PubMed]
5. do Imperio, G.E.; Bloise, E.; Javam, M.; Lye, P.; Constantinof, A.; Dunk, C.; Dos Reis, F.M.; Lye, S.J.; Gibb, W.; Ortiga-Carvalho, T.M.; et al. Chorioamnionitis Induces a Specific Signature of Placental ABC Transporters Associated with an Increase of miR-331-5p in the Human Preterm Placenta. *Cell. Physiol. Biochem.* **2018**, *45*, 591–604. [CrossRef]
6. Morgan, T.K. Role of the Placenta in Preterm Birth: A Review. *Am. J. Perinatol.* **2016**, *33*, 258–266. [CrossRef]
7. Sultana, Z.; Maiti, K.; Aitken, J.; Morris, J.; Dedman, L.; Smith, R. Oxidative stress, placental ageing-related pathologies and adverse pregnancy outcomes. *Am. J. Reprod. Immunol.* **2017**, *77*. [CrossRef]
8. Battaglia, F.C. New concepts in fetal and placental amino acid metabolism. *J. Anim. Sci.* **1992**, *70*, 3258–3263. [CrossRef]
9. Shekhawat, P.; Bennett, M.J.; Sadovsky, Y.; Nelson, D.M.; Rakheja, D.; Strauss, A.W. Human placenta metabolizes fatty acids: Implications for fetal fatty acid oxidation disorders and maternal liver diseases. *Am. J. Physiol. Endocrinol. Metab.* **2003**, *284*, E1098–E1105. [CrossRef]
10. Elshenawy, S.; Pinney, S.E.; Stuart, T.; Doulias, P.T.; Zura, G.; Parry, S.; Elovitz, M.A.; Bennett, M.J.; Bansal, A.; Strauss, J.F.; et al. The Metabolomic Signature of the Placenta in Spontaneous Preterm Birth. *Int. J. Mol. Sci.* **2020**, *21*, 1043. [CrossRef]
11. Di Renzo, G.C.; Rosati, A.; Sarti, R.D.; Cruciani, L.; Cutuli, A.M. Does fetal sex affect pregnancy outcome? *Gend. Med.* **2007**, *4*, 19–30. [CrossRef]
12. Al-Qaraghouli, M.; Fang, Y.M.V. Effect of Fetal Sex on Maternal and Obstetric Outcomes. *Front. Pediatr.* **2017**, *5*, 144. [CrossRef]
13. Lorente-Pozo, S.; Parra-Llorca, A.; Torres, B.; Torres-Cuevas, I.; Nuñez-Ramiro, A.; Cernada, M.; García-Robles, A.; Vento, M. Influence of Sex on Gestational Complications, Fetal-to-Neonatal Transition, and Postnatal Adaptation. *Front. Pediatr.* **2018**, *6*, 63. [CrossRef]
14. Broere-Brown, Z.A.; Adank, M.C.; Benschop, L.; Tielemans, M.; Muka, T.; Gonçalves, R.; Bramer, W.M.; Schoufour, J.D.; Voortman, T.; Steegers, E.A.P.; et al. Fetal sex and maternal pregnancy outcomes: A systematic review and meta-analysis. *Biol. Sex. Differ.* **2020**, *11*, 26. [CrossRef]
15. McGregor, J.A.; Leff, M.; Orleans, M.; Baron, A. Fetal gender differences in preterm birth: Findings in a North American cohort. *Am. J. Perinatol.* **1992**, *9*, 43–48. [CrossRef]
16. Challis, J.; Newnham, J.; Petraglia, F.; Yeganegi, M.; Bocking, A. Fetal sex and preterm birth. *Placenta* **2013**, *34*, 95–99. [CrossRef]
17. Peelen, M.J.; Kazemier, B.M.; Ravelli, A.C.; De Groot, C.J.; Van Der Post, J.A.; Mol, B.W.; Hajenius, P.J.; Kok, M. Impact of fetal gender on the risk of preterm birth, a national cohort study. *Acta Obstet. Gynecol. Scand.* **2016**, *95*, 1034–1041. [CrossRef]
18. Eidem, H.R.; Rinker, D.C.; Ackerman, W.E.; Buhimschi, I.A.; Buhimschi, C.S.; Dunn-Fletcher, C.; Kallapur, S.G.; Pavličev, M.; Muglia, L.J.; Abbot, P.; et al. Comparing human and macaque placental transcriptomes to disentangle preterm birth pathology from gestational age effects. *Placenta* **2016**, *41*, 74–82. [CrossRef]
19. Brockway, H.M.; Kallapur, S.G.; Buhimschi, I.A.; Buhimschi, C.S.; Ackerman, W.E.; Muglia, L.J.; Jones, H.N. Unique transcriptomic landscapes identified in idiopathic spontaneous and infection related preterm births compared to normal term births. *PLoS ONE* **2019**, *14*, e0225062. [CrossRef]
20. Goldenberg, R.L.; Culhane, J.F.; Iams, J.D.; Romero, R. Epidemiology and causes of preterm birth. *Lancet* **2008**, *371*, 75–84. [CrossRef]
21. Xu, C.; Li, C.Y.; Kong, A.N. Induction of phase I, II and III drug metabolism/transport by xenobiotics. *Arch. Pharm. Res.* **2005**, *28*, 249–268. [CrossRef]
22. Aouache, R.; Biquard, L.; Vaiman, D.; Miralles, F. Oxidative Stress in Preeclampsia and Placental Diseases. *Int. J. Mol. Sci.* **2018**, *19*, 1496. [CrossRef]
23. Moore, T.A.; Ahmad, I.M.; Zimmerman, M.C. Oxidative Stress and Preterm Birth: An Integrative Review. *Biol. Res. Nurs.* **2018**, *20*, 497–512. [CrossRef]
24. Bry, K.; Hallman, M. Transforming growth factor-beta 2 prevents preterm delivery induced by interleukin-1 alpha and tumor necrosis factor-alpha in the rabbit. *Am. J. Obstet. Gynecol.* **1993**, *168*, 1318–1322. [CrossRef]
25. Blackburn, S. Cytokines in the perinatal and neonatal periods: Selected aspects. *J. Perinat. Neonatal Nurs.* **2008**, *22*, 187–190. [CrossRef]
26. Pereira, T.B.; Thomaz, E.B.; Nascimento, F.R.; Santos, A.P.; Batista, R.L.; Bettiol, H.; Cavalli, R.e.C.; Barbieri, M.A.; Silva, A.A. Regulatory Cytokine Expression and Preterm Birth: Case-Control Study Nested in a Cohort. *PLoS ONE* **2016**, *11*, e0158380. [CrossRef]

27. Chida, D.; Miyoshi, K.; Sato, T.; Yoda, T.; Kikusui, T.; Iwakura, Y. The role of glucocorticoids in pregnancy, parturition, lactation, and nurturing in melanocortin receptor 2-deficient mice. *Endocrinology* **2011**, *152*, 1652–1660. [CrossRef]
28. Dickinson, H.; O'Connell, B.A.; Walker, D.W.; Moritz, K.M. Sex-Specific Effects of Prenatal Glucocorticoids on Placental Development. In *Glucocorticoids: New Recognition of Our Familiar Friend*; Xiaoxiao, Q., Ed.; IntechOpen Limited: London, UK, 2012. [CrossRef]
29. Okkenhaug, K. Signaling by the phosphoinositide 3-kinase family in immune cells. *Annu. Rev. Immunol.* **2013**, *31*, 675–704. [CrossRef] [PubMed]
30. Kieckbusch, J.; Balmas, E.; Hawkes, D.A.; Colucci, F. Disrupted PI3K p110δ Signaling Dysregulates Maternal Immune Cells and Increases Fetal Mortality In Mice. *Cell Rep.* **2015**, *13*, 2817–2828. [CrossRef]
31. Yamazaki, K.; Masaki, N.; Kohmura-Kobayashi, Y.; Yaguchi, C.; Hayasaka, T.; Itoh, H.; Setou, M.; Kanayama, N. Decrease in Sphingomyelin (d18:1/16:0) in Stem Villi and Phosphatidylcholine (16:0/20:4) in Terminal Villi of Human Term Placentas with Pathohistological Maternal Malperfusion. *PLoS ONE* **2015**, *10*, e0142609. [CrossRef]
32. Del Gaudio, I.; Sasset, L.; Lorenzo, A.D.; Wadsack, C. Sphingolipid Signature of Human Feto-Placental Vasculature in Preeclampsia. *Int. J. Mol. Sci.* **2020**, *21*, 1019. [CrossRef]
33. Ermini, L.; Ausman, J.; Melland-Smith, M.; Yeganeh, B.; Rolfo, A.; Litvack, M.L.; Todros, T.; Letarte, M.; Post, M.; Caniggia, I. A Single Sphingomyelin Species Promotes Exosomal Release of Endoglin into the Maternal Circulation in Preeclampsia. *Sci. Rep.* **2017**, *7*, 12172. [CrossRef] [PubMed]
34. Testi, R. Sphingomyelin breakdown and cell fate. *Trends Biochem. Sci.* **1996**, *21*, 468–471. [CrossRef]
35. Singh, A.T.; Dharmarajan, A.; Aye, I.L.; Keelan, J.A. Ceramide biosynthesis and metabolism in trophoblast syncytialization. *Mol. Cell. Endocrinol.* **2012**, *362*, 48–59. [CrossRef]
36. Pahan, K.; Dobashi, K.; Ghosh, B.; Singh, I. Induction of the manganese superoxide dismutase gene by sphingomyelinase and ceramide. *J. Neurochem.* **1999**, *73*, 513–520. [CrossRef]
37. Kannan, K.; Jain, S.K. Effect of vitamin B6 on oxygen radicals, mitochondrial membrane potential, and lipid peroxidation in H2O2-treated U937 monocytes. *Free Radic. Biol. Med.* **2004**, *36*, 423–428. [CrossRef]
38. Ueland, P.M.; McCann, A.; Midttun, Ø.; Ulvik, A. Inflammation, vitamin B6 and related pathways. *Mol. Aspects Med.* **2017**, *53*, 10–27. [CrossRef]
39. Lien, Y.C.; Zhang, Z.; Barila, G.; Green-Brown, A.; Elovitz, M.A.; Simmons, R.A. Intrauterine Inflammation Alters the Transcriptome and Metabolome in Placenta. *Front. Physiol.* **2020**, *11*, 592689. [CrossRef] [PubMed]
40. Strauss, J.F.; Kido, S.; Sayegh, R.; Sakuragi, N.; Gåfvels, M.E. The cAMP signalling system and human trophoblast function. *Placenta* **1992**, *13*, 389–403. [CrossRef]
41. Ravnskjaer, K.; Madiraju, A.; Montminy, M. Role of the cAMP Pathway in Glucose and Lipid Metabolism. *Handb. Exp. Pharmacol.* **2016**, *233*, 29–49. [CrossRef]
42. Bijur, G.N.; Jope, R.S. Rapid accumulation of Akt in mitochondria following phosphatidylinositol 3-kinase activation. *J. Neurochem.* **2003**, *87*, 1427–1435. [CrossRef]
43. Cheng, Z.; Tseng, Y.; White, M.F. Insulin signaling meets mitochondria in metabolism. *Trends Endocrinol. Metab.* **2010**, *21*, 589–598. [CrossRef]
44. Goo, C.K.; Lim, H.Y.; Ho, Q.S.; Too, H.P.; Clement, M.V.; Wong, K.P. PTEN/Akt signaling controls mitochondrial respiratory capacity through 4E-BP1. *PLoS ONE* **2012**, *7*, e45806. [CrossRef]
45. Sferruzzi-Perri, A.N.; López-Tello, J.; Fowden, A.L.; Constancia, M. Maternal and fetal genomes interplay through phosphoinositol 3-kinase(PI3K)-p110α signaling to modify placental resource allocation. *Proc. Natl. Acad. Sci. USA* **2016**, *113*, 11255–11260. [CrossRef]
46. Devisme, L.; Merlot, B.; Ego, A.; Houfflin-Debarge, V.; Deruelle, P.; Subtil, D. A case-control study of placental lesions associated with pre-eclampsia. *Int. J. Gynaecol. Obstet.* **2013**, *120*, 165–168. [CrossRef]
47. Ducray, J.F.; Naicker, T.; Moodley, J. Pilot study of comparative placental morphometry in pre-eclamptic and normotensive pregnancies suggests possible maladaptations of the fetal component of the placenta. *Eur. J. Obstet. Gynecol. Reprod. Biol.* **2011**, *156*, 29–34. [CrossRef]
48. González-Amaro, R.; Sánchez-Madrid, F. Cell adhesion molecules: Selectins and integrins. *Crit. Rev. Immunol.* **1999**, *19*, 389–429.
49. Hynes, R.O. Integrins: Bidirectional, allosteric signaling machines. *Cell* **2002**, *110*, 673–687. [CrossRef]
50. Bukovsky, A.; Cekanova, M.; Caudle, M.R.; Wimalasena, J.; Foster, J.S.; Henley, D.C.; Elder, R.F. Expression and localization of estrogen receptor-alpha protein in normal and abnormal term placentae and stimulation of trophoblast differentiation by estradiol. *Reprod. Biol. Endocrinol.* **2003**, *1*, 13. [CrossRef]
51. Pepe, G.J.; Albrecht, E.D. Regulation of functional differentiation of the placental villous syncytiotrophoblast by estrogen during primate pregnancy. *Steroids* **1999**, *64*, 624–627. [CrossRef]
52. Albrecht, E.D.; Pepe, G.J. Central integrative role of oestrogen in modulating the communication between the placenta and fetus that results in primate fecal-placental development. *Placenta* **1999**, *20*, 129–139. [CrossRef]
53. Albrecht, E.D.; Pepe, G.J. Estrogen regulation of placental angiogenesis and fetal ovarian development during primate pregnancy. *Int. J. Dev. Biol.* **2010**, *54*, 397–408. [CrossRef] [PubMed]

54. Schiessl, B.; Mylonas, I.; Hantschmann, P.; Kuhn, C.; Schulze, S.; Kunze, S.; Friese, K.; Jeschke, U. Expression of endothelial NO synthase, inducible NO synthase, and estrogen receptors alpha and beta in placental tissue of normal, preeclamptic, and intrauterine growth-restricted pregnancies. *J. Histochem. Cytochem.* **2005**, *53*, 1441–1449. [CrossRef] [PubMed]

55. Tremblay, Y.; Beaudoin, C. Regulation of 3 beta-hydroxysteroid dehydrogenase and 17 beta-hydroxysteroid dehydrogenase messenger ribonucleic acid levels by cyclic adenosine 3′,5′-monophosphate and phorbol myristate acetate in human choriocarcinoma cells. *Mol. Endocrinol.* **1993**, *7*, 355–364. [CrossRef] [PubMed]

56. Ma, M.P.; Thomson, M. Protein Kinase A Subunit α Catalytic and A Kinase Anchoring Protein 79 in Human Placental Mitochondria. *Open Biochem. J.* **2012**, *6*, 23–30. [CrossRef]

57. Clemmons, D.R. Metabolic actions of insulin-like growth factor-I in normal physiology and diabetes. *Endocrinol. Metab. Clin. North. Am.* **2012**, *41*, 425–443. [CrossRef]

58. LeRoith, D.; Yakar, S. Mechanisms of disease: Metabolic effects of growth hormone and insulin-like growth factor 1. *Nat. Clin. Pract. Endocrinol. Metab.* **2007**, *3*, 302–310. [CrossRef]

59. He, J.R.; Lai, Y.M.; Liu, H.H.; Liu, G.J.; Li, W.D.; Fan, X.J.; Wei, X.L.; Xia, X.Y.; Kuang, Y.S.; Liu, X.D.; et al. Maternal IGF1 and IGF1R polymorphisms and the risk of spontaneous preterm birth. *J. Clin. Lab. Anal.* **2017**, *31*. [CrossRef]

60. Spradley, F.T.; Ge, Y.; Haynes, B.P.; Granger, J.P.; Anderson, C.D. Adrenergic receptor blockade attenuates placental ischemia-induced hypertension. *Physiol. Rep.* **2018**, *6*, e13814. [CrossRef]

61. Taher, S.; Borja, Y.; Cabanela, L.; Costers, V.J.; Carson-Marino, M.; Bailes, J.C.; Dhar, B.; Beckworth, M.T.; Rabaglino, M.B.; Post Uiterweer, E.D.; et al. Cholecystokinin, gastrin, cholecystokinin/gastrin receptors, and bitter taste receptor TAS2R14: Trophoblast expression and signaling. *Am. J. Physiol. Regul. Integr. Comp. Physiol.* **2019**, *316*, R628–R639. [CrossRef]

62. Wu, Q.; Madany, P.; Dobson, J.R.; Schnabl, J.M.; Sharma, S.; Smith, T.C.; van Wijnen, A.J.; Stein, J.L.; Lian, J.B.; Stein, G.S.; et al. The BRG1 chromatin remodeling enzyme links cancer cell metabolism and proliferation. *Oncotarget* **2016**, *7*, 38270–38281. [CrossRef]

63. Nickerson, J.A.; Wu, Q.; Imbalzano, A.N. Mammalian SWI/SNF Enzymes and the Epigenetics of Tumor Cell Metabolic Reprogramming. *Front. Oncol.* **2017**, *7*, 49. [CrossRef]

64. Kidder, B.L.; Palmer, S. Examination of transcriptional networks reveals an important role for TCFAP2C, SMARCA4, and EOMES in trophoblast stem cell maintenance. *Genome Res.* **2010**, *20*, 458–472. [CrossRef]

65. Wisdom, R.; Johnson, R.S.; Moore, C. c-Jun regulates cell cycle progression and apoptosis by distinct mechanisms. *EMBO J.* **1999**, *18*, 188–197. [CrossRef]

66. Gal, H.; Lysenko, M.; Stroganov, S.; Vadai, E.; Youssef, S.A.; Tzadikevitch-Geffen, K.; Rotkopf, R.; Biron-Shental, T.; de Bruin, A.; Neeman, M.; et al. Molecular pathways of senescence regulate placental structure and function. *EMBO J.* **2019**, *38*, e100849. [CrossRef]

67. Jaju Bhattad, G.; Jeyarajah, M.J.; McGill, M.G.; Dumeaux, V.; Okae, H.; Arima, T.; Lajoie, P.; Bérubé, N.G.; Renaud, S.J. Histone deacetylase 1 and 2 drive differentiation and fusion of progenitor cells in human placental trophoblasts. *Cell Death Dis.* **2020**, *11*, 311. [CrossRef]

68. Than, N.G.; Romero, R.; Erez, O.; Weckle, A.; Tarca, A.L.; Hotra, J.; Abbas, A.; Han, Y.M.; Kim, S.S.; Kusanovic, J.P.; et al. Emergence of hormonal and redox regulation of galectin-1 in placental mammals: Implication in maternal-fetal immune tolerance. *Proc. Natl. Acad. Sci. USA* **2008**, *105*, 15819–15824. [CrossRef]

69. Blois, S.M.; Ilarregui, J.M.; Tometten, M.; Garcia, M.; Orsal, A.S.; Cordo-Russo, R.; Toscano, M.A.; Bianco, G.A.; Kobelt, P.; Handjiski, B.; et al. A pivotal role for galectin-1 in fetomaternal tolerance. *Nat. Med.* **2007**, *13*, 1450–1457. [CrossRef]

70. Blois, S.M.; Verlohren, S.; Wu, G.; Clark, G.; Dell, A.; Haslam, S.M.; Barrientos, G. Role of galectin-glycan circuits in reproduction: From healthy pregnancy to preterm birth (PTB). *Semin. Immunopathol.* **2020**, *42*, 469–486. [CrossRef]

71. Freitag, N.; Tirado-González, I.; Barrientos, G.; Herse, F.; Thijssen, V.L.; Weedon-Fekjær, S.M.; Schulz, H.; Wallukat, G.; Klapp, B.F.; Nevers, T.; et al. Interfering with Gal-1-mediated angiogenesis contributes to the pathogenesis of preeclampsia. *Proc. Natl. Acad. Sci. USA* **2013**, *110*, 11451–11456. [CrossRef]

72. Kaya, B.; Turhan, U.; Sezer, S.; Kaya, S.; Dağ, İ.; Tayyar, A. Maternal serum galectin-1 and galectin-3 levels in pregnancies complicated with preterm prelabor rupture of membranes. *J. Matern. Fetal. Neonatal. Med.* **2020**, *33*, 861–868. [CrossRef]

73. Hernandez, I.; Fournier, T.; Chissey, A.; Therond, P.; Slama, A.; Beaudeux, J.L.; Zerrad-Saadi, A. NADPH oxidase is the major source of placental superoxide in early pregnancy: Association with MAPK pathway activation. *Sci. Rep.* **2019**, *9*, 13962. [CrossRef] [PubMed]

74. Bevilacqua, E.; Gomes, S.Z.; Lorenzon, A.R.; Hoshida, M.S.; Amarante-Paffaro, A.M. NADPH oxidase as an important source of reactive oxygen species at the mouse maternal-fetal interface: Putative biological roles. *Reprod. Biomed. Online* **2012**, *25*, 31–43. [CrossRef]

75. Cui, X.L.; Brockman, D.; Campos, B.; Myatt, L. Expression of NADPH oxidase isoform 1 (Nox1) in human placenta: Involvement in preeclampsia. *Placenta* **2006**, *27*, 422–431. [CrossRef]

76. Ng, A.C.; Baird, S.D.; Screaton, R.A. Essential role of TID1 in maintaining mitochondrial membrane potential homogeneity and mitochondrial DNA integrity. *Mol. Cell. Biol.* **2014**, *34*, 1427–1437. [CrossRef]

77. Joseph, B.K.; Liu, H.Y.; Francisco, J.; Pandya, D.; Donigan, M.; Gallo-Ebert, C.; Giordano, C.; Bata, A.; Nickels, J.T. Inhibition of AMP Kinase by the Protein Phosphatase 2A Heterotrimer, PP2APpp2r2d. *J. Biol. Chem.* **2015**, *290*, 10588–10598. [CrossRef]

78. Winder, W.W.; Hardie, D.G. AMP-activated protein kinase, a metabolic master switch: Possible roles in type 2 diabetes. *Am. J. Physiol.* **1999**, *277*, E1–E10. [CrossRef] [PubMed]

79. Claret, M.; Smith, M.A.; Batterham, R.L.; Selman, C.; Choudhury, A.I.; Fryer, L.G.; Clements, M.; Al-Qassab, H.; Heffron, H.; Xu, A.W.; et al. AMPK is essential for energy homeostasis regulation and glucose sensing by POMC and AgRP neurons. *J. Clin. Investig.* **2007**, *117*, 2325–2336. [CrossRef] [PubMed]

80. Tokunaga, C.; Yoshino, K.; Yonezawa, K. mTOR integrates amino acid- and energy-sensing pathways. *Biochem. Biophys. Res. Commun.* **2004**, *313*, 443–446. [CrossRef]

81. Kennedy, B.K.; Lamming, D.W. The Mechanistic Target of Rapamycin: The Grand ConducTOR of Metabolism and Aging. *Cell Metab.* **2016**, *23*, 990–1003. [CrossRef]

82. Wen, H.Y.; Abbasi, S.; Kellems, R.E.; Xia, Y. mTOR: A placental growth signaling sensor. *Placenta* **2005**, *26*, S63–S69. [CrossRef]

83. Jansson, T.; Aye, I.L.; Goberdhan, D.C. The emerging role of mTORC1 signaling in placental nutrient-sensing. *Placenta* **2012**, *33* (Suppl. 2), e23–e29. [CrossRef]

84. Ferrari, E.; Bruhn, C.; Peretti, M.; Cassani, C.; Carotenuto, W.V.; Elgendy, M.; Shubassi, G.; Lucca, C.; Bermejo, R.; Varasi, M.; et al. PP2A Controls Genome Integrity by Integrating Nutrient-Sensing and Metabolic Pathways with the DNA Damage Response. *Mol. Cell* **2017**, *67*, 266–281.e264. [CrossRef]

85. Hoch, D.; Bachbauer, M.; Pöchlauer, C.; Algaba-Chueca, F.; Tandl, V.; Novakovic, B.; Megia, A.; Gauster, M.; Saffery, R.; Glasner, A.; et al. Maternal Obesity Alters Placental Cell Cycle Regulators in the First Trimester of Human Pregnancy: New Insights for BRCA1. *Int. J. Mol. Sci.* **2020**, *21*, 468. [CrossRef]

86. Privat, M.; Radosevic-Robin, N.; Aubel, C.; Cayre, A.; Penault-Llorca, F.; Marceau, G.; Sapin, V.; Bignon, Y.J.; Morvan, D. BRCA1 induces major energetic metabolism reprogramming in breast cancer cells. *PLoS ONE* **2014**, *9*, e102438. [CrossRef]

87. Jackson, K.C.; Gidlund, E.K.; Norrbom, J.; Valencia, A.P.; Thomson, D.M.; Schuh, R.A.; Neufer, P.D.; Spangenburg, E.E. BRCA1 is a novel regulator of metabolic function in skeletal muscle. *J. Lipid Res.* **2014**, *55*, 668–680. [CrossRef]

88. Jackson, K.C.; Tarpey, M.D.; Valencia, A.P.; Iñigo, M.R.; Pratt, S.J.; Patteson, D.J.; McClung, J.M.; Lovering, R.M.; Thomson, D.M.; Spangenburg, E.E. Induced Cre-mediated knockdown of Brca1 in skeletal muscle reduces mitochondrial respiration and prevents glucose intolerance in adult mice on a high-fat diet. *FASEB J.* **2018**, *32*, 3070–3084. [CrossRef]

89. Ye, L.; Gratton, A.; Hannan, N.J.; Cannon, P.; Deo, M.; Palmer, K.R.; Tong, S.; Kaitu'u-Lino, T.J.; Brownfoot, F.C. Nuclear factor of activated T-cells (NFAT) regulates soluble fms-like tyrosine kinase-1 secretion (sFlt-1) from human placenta. *Placenta* **2016**, *48*, 110–118. [CrossRef]

90. Tabata, C.; Ogita, K.; Sato, K.; Nakamura, H.; Qing, Z.; Negoro, H.; Kumasawa, K.; Temma-Asano, K.; Tsutsui, T.; Nishimori, K.; et al. Calcineurin/NFAT pathway: A novel regulator of parturition. *Am. J. Reprod. Immunol.* **2009**, *62*, 44–50. [CrossRef]

91. Neubauer, V.; Griesmaier, E.; Ralser, E.; Kiechl-Kohlendorfer, U. The effect of sex on outcome of preterm infants—A population-based survey. *Acta Paediatr.* **2012**, *101*, 906–911. [CrossRef]

92. Clifton, V.L. Review: Sex and the human placenta: Mediating differential strategies of fetal growth and survival. *Placenta* **2010**, *31*, S33–S39. [CrossRef]

93. Alur, P. Sex Differences in Nutrition, Growth, and Metabolism in Preterm Infants. *Front. Pediatr.* **2019**, *7*, 22. [CrossRef] [PubMed]

94. Rutkowsky, J.M.; Knotts, T.A.; Ono-Moore, K.D.; McCoin, C.S.; Huang, S.; Schneider, D.; Singh, S.; Adams, S.H.; Hwang, D.H. Acylcarnitines activate proinflammatory signaling pathways. *Am. J. Physiol. Endocrinol. Metab.* **2014**, *306*, E1378–E1387. [CrossRef] [PubMed]

95. Batchuluun, B.; Al Rijjal, D.; Prentice, K.J.; Eversley, J.A.; Burdett, E.; Mohan, H.; Bhattacharjee, A.; Gunderson, E.P.; Liu, Y.; Wheeler, M.B. Elevated Medium-Chain Acylcarnitines Are Associated With Gestational Diabetes Mellitus and Early Progression to Type 2 Diabetes and Induce Pancreatic β-Cell Dysfunction. *Diabetes* **2018**, *67*, 885–897. [CrossRef] [PubMed]

96. Dhabhar, F.S.; McEwen, B.S. Acute stress enhances while chronic stress suppresses cell-mediated immunity in vivo: A potential role for leukocyte trafficking. *Brain Behav. Immun.* **1997**, *11*, 286–306. [CrossRef] [PubMed]

97. Korgun, E.T.; Ozmen, A.; Unek, G.; Mendilcioglu, I. The effects of glucocorticoids on fetal and placental development. In *Glucocorticoids: New Recognition of Our Familiar Friend*; Qian, X., Ed.; IntechOpen Limited: London, UK, 2012. [CrossRef]

98. Stark, M.J.; Hodyl, N.A.; Wright, I.M.; Clifton, V.L. Influence of sex and glucocorticoid exposure on preterm placental pro-oxidant-antioxidant balance. *Placenta* **2011**, *32*, 865–870. [CrossRef] [PubMed]

99. Stark, M.J.; Wright, I.M.; Clifton, V.L. Sex-specific alterations in placental 11beta-hydroxysteroid dehydrogenase 2 activity and early postnatal clinical course following antenatal betamethasone. *Am. J. Physiol. Regul. Integr. Comp. Physiol.* **2009**, *297*, R510–R514. [CrossRef]

100. Kam, R.K.; Deng, Y.; Chen, Y.; Zhao, H. Retinoic acid synthesis and functions in early embryonic development. *Cell Biosci.* **2012**, *2*, 11. [CrossRef]

101. Rhinn, M.; Dollé, P. Retinoic acid signalling during development. *Development* **2012**, *139*, 843–858. [CrossRef]

102. Brun, P.J.; Yang, K.J.; Lee, S.A.; Yuen, J.J.; Blaner, W.S. Retinoids: Potent regulators of metabolism. *Biofactors* **2013**, *39*, 151–163. [CrossRef]

103. Chambon, P. A decade of molecular biology of retinoic acid receptors. *FASEB J.* **1996**, *10*, 940–954. [CrossRef]

104. Chawla, A.; Repa, J.J.; Evans, R.M.; Mangelsdorf, D.J. Nuclear receptors and lipid physiology: Opening the X-files. *Science* **2001**, *294*, 1866–1870. [CrossRef]

105. Shin, J.S.; Choi, M.Y.; Longtine, M.S.; Nelson, D.M. Vitamin D effects on pregnancy and the placenta. *Placenta* **2010**, *31*, 1027–1034. [CrossRef]

106. Knabl, J.; Vattai, A.; Ye, Y.; Jueckstock, J.; Hutter, S.; Kainer, F.; Mahner, S.; Jeschke, U. Role of Placental VDR Expression and Function in Common Late Pregnancy Disorders. *Int. J. Mol. Sci.* **2017**, *18*, 2340. [CrossRef]
107. Wimalawansa, S.J. Vitamin D Deficiency: Effects on Oxidative Stress, Epigenetics, Gene Regulation, and Aging. *Biology (Basel)* **2019**, *8*, 30. [CrossRef]
108. Amegah, A.K.; Klevor, M.K.; Wagner, C.L. Maternal vitamin D insufficiency and risk of adverse pregnancy and birth outcomes: A systematic review and meta-analysis of longitudinal studies. *PLoS ONE* **2017**, *12*, e0173605. [CrossRef]
109. Mullur, R.; Liu, Y.Y.; Brent, G.A. Thyroid hormone regulation of metabolism. *Physiol. Rev.* **2014**, *94*, 355–382. [CrossRef]
110. Adu-Gyamfi, E.A.; Wang, Y.X.; Ding, Y.B. The interplay between thyroid hormones and the placenta: A comprehensive review†. *Biol. Reprod.* **2020**, *102*, 8–17. [CrossRef]
111. Chen, C.Y.; Chen, C.P.; Lin, K.H. Biological functions of thyroid hormone in placenta. *Int. J. Mol. Sci.* **2015**, *16*, 4161–4179. [CrossRef]
112. Ihunnah, C.A.; Jiang, M.; Xie, W. Nuclear receptor PXR, transcriptional circuits and metabolic relevance. *Biochim. Biophys. Acta* **2011**, *1812*, 956–963. [CrossRef]
113. Zelcer, N.; Tontonoz, P. Liver X receptors as integrators of metabolic and inflammatory signaling. *J. Clin. Investig.* **2006**, *116*, 607–614. [CrossRef]
114. Hong, C.; Tontonoz, P. Liver X receptors in lipid metabolism: Opportunities for drug discovery. *Nat. Rev. Drug Discov.* **2014**, *13*, 433–444. [CrossRef]
115. Tyagi, S.; Gupta, P.; Saini, A.S.; Kaushal, C.; Sharma, S. The peroxisome proliferator-activated receptor: A family of nuclear receptors role in various diseases. *J. Adv. Pharm. Technol. Res.* **2011**, *2*, 236–240. [CrossRef]
116. Feldman, P.L.; Lambert, M.H.; Henke, B.R. PPAR modulators and PPAR pan agonists for metabolic diseases: The next generation of drugs targeting peroxisome proliferator-activated receptors? *Curr. Top. Med. Chem.* **2008**, *8*, 728–749. [CrossRef]
117. Sundrani, D.P.; Karkhanis, A.R.; Joshi, S.R. Peroxisome Proliferator-Activated Receptors (PPAR), fatty acids and microRNAs: Implications in women delivering low birth weight babies. *Syst. Biol. Reprod. Med.* **2021**, *67*, 24–41. [CrossRef]
118. Stienstra, R.; Duval, C.; Müller, M.; Kersten, S. PPARs, Obesity, and Inflammation. *PPAR Res.* **2007**, *2007*, 95974. [CrossRef]
119. Matsuda, S.; Kobayashi, M.; Kitagishi, Y. Expression and Function of PPARs in Placenta. *PPAR Res.* **2013**, *2013*, 256508. [CrossRef]
120. Sullivan, W.J.; Mullen, P.J.; Schmid, E.W.; Flores, A.; Momcilovic, M.; Sharpley, M.S.; Jelinek, D.; Whiteley, A.E.; Maxwell, M.B.; Wilde, B.R.; et al. Extracellular Matrix Remodeling Regulates Glucose Metabolism through TXNIP Destabilization. *Cell* **2018**, *175*, 117–132.e121. [CrossRef]
121. Romani, P.; Brian, I.; Santinon, G.; Pocaterra, A.; Audano, M.; Pedretti, S.; Mathieu, S.; Forcato, M.; Bicciato, S.; Manneville, J.B.; et al. Extracellular matrix mechanical cues regulate lipid metabolism through Lipin-1 and SREBP. *Nat. Cell Biol.* **2019**, *21*, 338–347. [CrossRef]
122. Huang, T.; Jones, C.G.; Chung, J.H.; Chen, C. Microfibrous Extracellular Matrix Changes the Liver Hepatocyte Energy Metabolism via Integrins. *ACS Biomater. Sci. Eng.* **2020**, *6*, 5849–5856. [CrossRef]
123. Georgiadou, M.; Lilja, J.; Jacquemet, G.; Guzmán, C.; Rafaeva, M.; Alibert, C.; Yan, Y.; Sahgal, P.; Lerche, M.; Manneville, J.B.; et al. AMPK negatively regulates tensin-dependent integrin activity. *J. Cell Biol.* **2017**, *216*, 1107–1121. [CrossRef] [PubMed]
124. Bartolák-Suki, E.; Imsirovic, J.; Nishibori, Y.; Krishnan, R.; Suki, B. Regulation of Mitochondrial Structure and Dynamics by the Cytoskeleton and Mechanical Factors. *Int. J. Mol. Sci.* **2017**, *18*, 1812. [CrossRef] [PubMed]
125. Tyanova, S.; Temu, T.; Cox, J. The MaxQuant computational platform for mass spectrometry-based shotgun proteomics. *Nat. Protoc.* **2016**, *11*, 2301–2319. [CrossRef] [PubMed]
126. Karnovsky, A.; Weymouth, T.; Hull, T.; Tarcea, V.G.; Scardoni, G.; Laudanna, C.; Sartor, M.A.; Stringer, K.A.; Jagadish, H.V.; Burant, C.; et al. Metscape 2 bioinformatics tool for the analysis and visualization of metabolomics and gene expression data. *Bioinformatics* **2012**, *28*, 373–380. [CrossRef]

International Journal of
*Molecular Sciences*



*Article*

# Placental Transcriptome Adaptations to Maternal Nutrient Restriction in Sheep

Chelsie B. Steinhauser [1,†], Colleen A. Lambo [2,†], Katharine Askelson [1], Gregory W. Burns [3], Susanta K. Behura [4,5], Thomas E. Spencer [4], Fuller W. Bazer [1] and Michael Carey Satterfield [1,*]

[1]  Department of Animal Science, Texas A & M University, College Station, TX 77843, USA; csteinhauser@tamu.edu (C.B.S.); kbeaso@gmail.com (K.A.); fbazer@tamu.edu (F.W.B.)
[2]  Department of Veterinary Physiology and Pharmacology, Texas A & M University, College Station, TX 77843, USA; clambo@tamu.edu
[3]  Department of Obstetrics, Gynecology and Reproductive Biology, Michigan State University, Grand Rapids, MI 49503, USA; burnsgr2@msu.edu
[4]  Division of Animal Sciences, University of Missouri, Columbia, MO 65211, USA; behuras@missouri.edu (S.K.B.); spencerte@missouri.edu (T.E.S.)
[5]  Institute for Data Science and Informatics, University of Missouri, Columbia, MO 65211, USA
*  Correspondence: csatterfield@tamu.edu; Tel.: +1-979-845-6448
†  These authors contributed equally to this work.

**Abstract:** Placental development is modified in response to maternal nutrient restriction (NR), resulting in a spectrum of fetal growth rates. Pregnant sheep carrying singleton fetuses and fed either 100% ($n = 8$) or 50% (NR; $n = 28$) of their National Research Council (NRC) recommended intake from days 35–135 of pregnancy were used to elucidate placentome transcriptome alterations at both day 70 and day 135. NR fetuses were further designated into upper (NR NonSGA; $n = 7$) and lower quartiles (NR SGA; $n = 7$) based on day 135 fetal weight. At day 70 of pregnancy, there were 22 genes dysregulated between NR SGA and 100% NRC placentomes, 27 genes between NR NonSGA and 100% NRC placentomes, and 22 genes between NR SGA and NR NonSGA placentomes. These genes mediated molecular functions such as MHC class II protein binding, signaling receptor binding, and cytokine activity. Gene set enrichment analysis (GSEA) revealed significant overrepresentation of genes for natural-killer-cell-mediated cytotoxicity in NR SGA compared to 100% NRC placentomes, and alterations in nutrient utilization pathways between NR SGA and NR NonSGA placentomes at day 70. Results identify novel factors associated with impaired function in SGA placentomes and potential for placentomes from NR NonSGA pregnancies to adapt to nutritional hardship.

**Keywords:** placentome; pregnancy; nutrient restriction; gene expression

**Citation:** Steinhauser, C.B.; Lambo, C.A.; Askelson, K.; Burns, G.W.; Behura, S.K.; Spencer, T.E.; Bazer, F.W.; Satterfield, M.C. Placental Transcriptome Adaptations to Maternal Nutrient Restriction in Sheep. *Int. J. Mol. Sci.* **2021**, *22*, 7654. https://doi.org/10.3390/ijms22147654

Academic Editors: Hiten D. Mistry and Eun Lee

Received: 18 June 2021
Accepted: 15 July 2021
Published: 17 July 2021

**Publisher's Note:** MDPI stays neutral with regard to jurisdictional claims in published maps and institutional affiliations.



## 1. Introduction

In eutherian mammals, the placenta mediates the exchange of nutrients, gases, and waste products between mother and fetus. Impaired growth and function of the placenta is associated with fetal growth restriction, poor pregnancy outcomes, and susceptibility to a myriad of health-related consequences in adulthood [1–4]. Placental growth can be influenced by maternal exposure to environmental factors, such as malnutrition. An area of increasing interest is the elucidation of adaptive mechanisms by which the placenta can respond to maternal environmental insults in a compensatory manner to sustain adequate fetal growth despite, for example, maternal nutrient restriction.

Using a long-term nutritionally restricted pregnant sheep model, we recently reported that nutrient-restricted (NR) pregnant sheep that support rates of fetal growth similar to growth of fetuses in control-fed sheep exhibited increased expression of select amino acid transporters in the placenta and possessed increased amino acid availability in the fetal circulation [5]. In the same cohort of animals, microarray analyses revealed changes in the expression of genes whose functions were associated with the biological actions of nutrient sensing and transport and immune system activation [6]. While the aforementioned sheep studies highlight adaptive changes occurring within the ovine placenta to support normal fetal growth during maternal NR, they are limited in that analyses were performed during the final third of gestation, when placental growth and function had already reached its maximum [7]. This in fact highlights one of the biggest challenges of placental research: how do we assess early placental growth and function, while allowing pregnancy to progress, giving the opportunity to definitively link early placental growth, or early placental adaptations, with a late gestation fetal phenotype? The sheep serves as a unique and valuable model organism to address this dilemma.

The primary functional units of the sheep placenta are the placentomes, which are discrete regions where the maternal caruncle intimately interdigitates and syncytializes with the fetal cotyledon. A singleton pregnancy contains between 50 and 120 of these placentomes, which collectively support greater than 95% of the hematotrophic exchange between mother and fetus [8]. We recently developed a surgical technique to selectively remove a single placentome in close approximation to the fetus in mid-gestation without compromising fetal growth [9]. Using this approach, we have identified alterations in placental fatty acid transport in NR pregnancies with impaired fetal growth that correspond to changes in circulating levels of triglycerides, non-esterified fatty acids, and cholesterol in both the dam and the fetus [10]. Additionally, thyroid hormones are altered in NR pregnancies, with identifiable changes in placental thyroid-hormone-related genes and proteins during mid and late gestation [11].

Given the demonstrated potential of our surgical technique, the objective of the present study was to utilize a discovery-based approach to identify novel genes and biological processes associated with the earliest adaptations within the ovine placentome in response to maternal NR, due to total caloric restriction from day 35 to day 135 (term = day 147) of pregnancy, giving rise to either small-for-gestational-age (SGA) or normal-weight (NonSGA) fetuses in late gestation.

## 2. Results

### 2.1. Model Characteristics

Maternal, placental, and fetal weights, as well as select metabolite abundances for this study, have been published elsewhere [10–13]. Of importance, fetuses from NR dams were categorized by fetal weight at day 135 into quartiles, with the highest quartile being denoted as NR NonSGA ($n = 7$) and the lowest quartile as NR SGA ($n = 7$). Well-fed controls are denoted as 100% NRC ($n = 8$). Fetal weight was lower in the NR SGA ($3.8 \pm 0.2$ kg) compared to 100% NRC ($5.6 \pm 0.1$ kg) and NR NonSGA ($5.4 \pm 0.2$ kg; $p < 0.001$) fetuses [12]. Total placentome weight was lower in NR SGA ($307 \pm 16$ g) ewes compared to 100% NRC ($546 \pm 43$ g) and NR NonSGA ($524 \pm 36$ g; $p < 0.001$) pregnancies [10].

### 2.2. Differentially Expressed Genes

The numbers of differentially expressed genes (DEGs) between treatment groups and days of pregnancy are shown in Figure 1 (FDR $\leq 0.10$ and $p \leq 0.05$). At day 70 of pregnancy, 9 genes were upregulated and 11 downregulated in NR SGA compared to 100% NRC placentomes, while 13 genes were upregulated and 10 genes downregulated in NR NonSGA compared to 100% NRC placentomes (Figure 1A). *Cleavage stimulation factor subunit 2 tau (CSTF2T)* was downregulated in both NR SGA and NR NonSGA placentomes compared to 100% NRC placentomes. There were 8 genes upregulated and 10 genes downregulated in NR SGA compared to NR NonSGA placentomes (Figure 1A). *Myostatin (MSTN)* was downregulated in NR SGA compared to both 100% NRC and NR NonSGA placentomes (Figure 1A, Table 1). *Heterogeneous nuclear ribonucleoprotein K (HNRNPK)* was upregulated in 100% NRC, but downregulated in NR SGA placentomes compared to NR NonSGA placentomes. *Major histocompatibility complex, class II, DO beta (HLA-DOB)* and *peptidylglycine alpha-amidating monooxygenase (PAM)* were downregulated in 100% NRC placentomes compared to all NR placentomes (Figure 1A, Table 1).

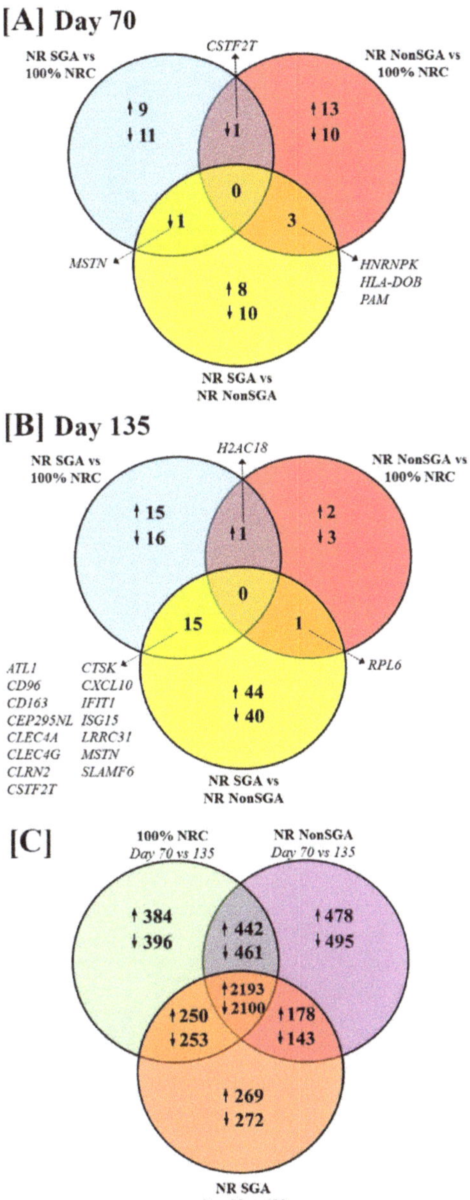


**Figure 1.** Venn diagrams depicting differentially expressed genes (DEGs) between 100% NRC ($n = 8$), NR NonSGA ($n = 7$), and NR SGA ($n = 7$) placentomes at day 70 (**A**) and day 135 (**B**) of gestation. The number of DEGs between day 70 and day 135 within each treatment group is depicted in (**C**). Genes were considered differentially expressed when FDR $\leq 0.10$ and $p \leq 0.05$.

**Table 1.** Differentially expressed genes in placentomes at day 70 of pregnancy.

| NR SGA [1] vs. 100% NRC | | | NR NonSGA vs. 100% NRC | | | NR SGA vs. NR NonSGA | | |
|---|---|---|---|---|---|---|---|---|
| **Gene** | **Log2 FC [2]** | **FDR [3]** | **Gene** | **Log2 FC** | **FDR** | **Gene** | **Log2 FC** | **FDR** |
| *MSTN* | −3.10 | 0.03 | *HNRNPK* | −6.28 | 0.01 | *HNRNPK* | −5.53 | 0.02 |
| *CSTF2T* | −2.78 | 0.08 | *UBE2D3* | −3.36 | <0.01 | *MSTN* | −3.74 | 0.03 |
| *HLA-DQA1* | −1.98 | 0.03 | *CSTF2T* | −3.17 | 0.02 | *FEM1A* | −2.35 | 0.09 |
| *SLC15A5* | −1.92 | 0.07 | *GBP4* | −2.74 | 0.01 | *EEF1A1* | −1.91 | 0.06 |
| *COL19A1* | −1.87 | 0.01 | *PPAT* | −2.23 | 0.09 | *AMER3* | −1.56 | 0.10 |
| *IL12RB2* | −1.52 | 0.01 | *OLFM1* | −1.81 | 0.09 | *HLA-DOB* | −1.28 | 0.07 |
| *GPRIN2* | −1.12 | 0.02 | *APLP1* | −1.44 | 0.01 | *MUC4* | −1.14 | 0.07 |
| *CD74* | −1.01 | <0.01 | *HLA-DQB2* | −1.38 | 0.09 | *SLC27A6* | −0.77 | 0.06 |
| *FZD2* | −0.85 | 0.05 | *HLA-DQA2* | −1.04 | 0.01 | *WNT9B* | −0.76 | 0.05 |
| *MMP9* | −0.75 | 0.10 | *NOTUM* | −0.61 | 0.10 | *JAM2* | −0.73 | 0.08 |
| *PI15* | −0.73 | <0.01 | *SH3GL3* | −0.44 | 0.01 | *APOE* | −0.67 | 0.06 |
| *LRATD1* | −0.50 | 0.08 | *MCM5* | −0.35 | 0.05 | *PAM* | −0.54 | 0.02 |
| *NFE2L3* | −0.35 | 0.08 | *GINS2* | 0.34 | 0.09 | *FAM3C* | 0.35 | 0.06 |
| *THBS1* | 0.58 | 0.10 | *TTC21B* | 0.43 | 0.10 | *ZNF462* | 0.53 | 0.04 |
| *CD163* | 0.65 | 0.06 | *PAM* | 0.50 | 0.01 | *AQP3* | 0.87 | 0.06 |
| *NDUFA4L2* | 0.76 | 0.08 | *JMY* | 0.50 | 0.09 | *S100B* | 1.04 | 0.02 |
| *PRSS12* | 0.82 | 0.01 | *MRC1* | 0.51 | 0.03 | *MYO1A* | 1.28 | 0.06 |
| *PLAC8L1* | 0.91 | 0.03 | *RNASE1* | 0.57 | 0.09 | *DOK5* | 1.31 | 0.07 |
| *CACNG3* | 1.78 | 0.08 | *FOLR3* | 0.60 | 0.01 | *EVA1A* | 1.32 | <0.01 |
| *BTNL9* | 1.92 | 0.07 | *FHL1* | 0.66 | 0.01 | *TNFRSF11B* | 2.07 | 0.06 |
| *SPATS1* | 2.31 | 0.08 | *SGCB* | 0.67 | 0.03 | *GFRA2* | 2.28 | 0.02 |
| *SLC5A12* | 2.53 | 0.08 | *NIPAL4* | 0.78 | 0.09 | *SLC10A6* | 2.77 | 0.06 |
| | | | *ZC3HAV1L* | 0.80 | 0.09 | | | |
| | | | *SVOPL* | 0.83 | 0.01 | | | |
| | | | *HLA-DOB* | 1.35 | 0.05 | | | |
| | | | *RPL6* | 1.46 | 0.01 | | | |
| | | | *PDE6B* | 2.13 | 0.01 | | | |

[1] 100% NRC: well-fed controls; NR NonSGA: nutrient-restricted normal-weight fetuses; NR SGA: nutrient-restricted low-weight fetuses. [2] Log2 FC: log2 fold change. [3] FDR: false discovery rate.

At day 135 of pregnancy, 15 genes were upregulated and 16 genes downregulated in NR SGA compared to 100% NRC placentomes, while only 2 genes were upregulated and 3 genes downregulated in NR NonSGA compared to 100% NRC placentomes (Figure 1B). *H2A clustered histone 18 (H2AC18)* was upregulated in all NR placentomes compared to 100% NRC placentomes (Figure 1B, Table 2). There were 44 genes upregulated and 40 downregulated in NR SGA compared to NR NonSGA placentomes (Figure 1B). *Ribosomal protein L6 (RPL6)* was downregulated in 100% NRC and NR SGA compared to NR NonSGA placentomes (Figure 1B, Table 2). *CD96, CD163, C-type lectin domain family 4 member A (CLEC4A), CLEC4G, CSTF2T, cathepsin K (CTSK), C-X-C motif chemokine ligand 10 (CXCL10), interferon-induced protein with tetratricopeptide repeats 1 (IFIT1), interferon-stimulated gene 15 (ISG15), leucine-rich repeat-containing 31 (LRRC31),* and *signaling lymphocyte activation molecule family member 6 (SLAMF6)* were upregulated in NR SGA compared to 100% NRC and NR NonSGA placentomes, while *glutamic pyruvic transaminase (ALT1), centrosomal protein 295 n-terminal like (CEP295NL), clarin-2 (CLRN2),* and *MSTN* were downregulated in NR SGA placentomes at day 135 of pregnancy (Figure 1B, Table 2).

The majority of genes that are differently expressed between days 70 and 135 of pregnancy are the same between 100% NRC, NR NonSGA, and NR SGA placentomes (2193 genes upregulated, 2100 genes downregulated; Figure 1C). Interestingly, there are fewer genes overall that are differentially expressed between days in the NR SGA placentomes (5658 genes) compared to NR NonSGA (6490 genes) and 100% NRC (6479 genes) placentomes (Figure 1C). Additionally, 100% NRC and NonSGA placentomes have more differentially expressed genes in common (442 upregulated, 461 downregulated) than either group does with NR SGA placentomes (250 up- and 253 downregulated, and 178 up- and 143 downregulated, respectively; Figure 1C).

All DEGs identified by treatment comparison at day 70 of pregnancy are listed in Table 1, and those from day 135 are listed in Table 2.

**Table 2.** Differentially expressed genes in placentomes at day 135 of pregnancy.

| NR SGA [1] vs. 100% NRC | | |
| --- | --- | --- |
| Gene | Log2 FC [2] | FDR [3] |
| CLRN2 | −3.82 | 0.04 |
| PIWIL1 | −3.34 | 0.09 |
| MSTN | −3.19 | 0.10 |
| NTNG1 | −2.85 | 0.06 |
| CCL11 | −2.51 | 0.09 |
| CCDC70 | −2.16 | 0.07 |
| SEMA3E | −2.06 | 0.08 |
| PPIL6 | −1.65 | 0.05 |
| CLIC6 | −1.63 | 0.02 |
| CAPN11 | −1.50 | 0.10 |
| ANPEP | −1.17 | 0.03 |
| IL1R2 | −1.11 | 0.08 |
| CASQ1 | −0.99 | 0.01 |
| SH3BP2 | −0.79 | 0.08 |
| CEP295NL | −0.78 | 0.02 |
| ETNPPL | −0.69 | 0.02 |
| CCDC80 | −0.67 | 0.02 |
| HAPLN3 | −0.64 | 0.10 |
| ATL1 | −0.51 | 0.04 |
| CHRNE | −0.43 | 0.10 |
| CTSK | 0.54 | 0.01 |
| FCGR1B | 0.73 | 0.09 |
| UMAD1 | 0.75 | 0.06 |
| CLEC4A | 0.79 | 0.10 |
| MAF | 0.83 | 0.09 |
| COLGALT2 | 0.98 | 0.01 |
| FST | 1.01 | 0.08 |
| CD163 | 1.02 | 0.01 |
| CD200R1L | 1.03 | 0.06 |
| NRN1 | 1.09 | 0.10 |
| CLEC4G | 1.11 | 0.02 |
| LRRC31 | 1.14 | 0.02 |
| ISG15 | 1.20 | 0.10 |
| IFIT1 | 1.21 | 0.01 |
| SLAMF6 | 1.21 | 0.02 |
| EPM2A | 1.29 | 0.06 |
| CD96 | 1.37 | 0.07 |
| ERVW-1 | 1.44 | 0.06 |
| P2RX2 | 1.45 | 0.04 |
| TRIML2 | 1.54 | 0.01 |
| SH2D2A | 2.01 | 0.07 |
| CXCL10 | 2.25 | 0.10 |
| ZNF623 | 3.25 | 0.01 |
| SNX10 | 3.47 | 0.07 |
| CSTF2T | 3.65 | <0.01 |
| RPL9 | 4.37 | 0.05 |
| H2AC18 | 5.56 | <0.01 |

| NR NonSGA vs. 100% NRC | | |
| --- | --- | --- |
| Gene | Log2 FC | FDR |
| CYP3A5 | −3.68 | 0.04 |
| CYP3A24 | −3.36 | 0.03 |
| ARSD | −1.43 | 0.04 |
| GLRX5 | 1.08 | 0.04 |
| RPL6 | 1.59 | <0.01 |
| DDX4 | 1.72 | 0.07 |
| H2AC18 | 5.33 | 0.03 |

| NR SGA vs. NR NonSGA | | |
| --- | --- | --- |
| Gene | Log2 FC | FDR |
| MSTN | −4.25 | <0.01 |
| KRT4 | −3.76 | 0.06 |
| CLRN2 | −3.49 | 0.08 |
| MAN2B2 | −2.73 | <0.01 |
| ANGPTL7 | −2.20 | 0.04 |
| COL25A1 | −2.08 | 0.10 |
| CCDC151 | −1.97 | 0.01 |
| RPL6 | −1.88 | <0.01 |
| LRIT1 | −1.80 | 0.09 |
| OSTN | −1.79 | 0.02 |
| MUC4 | −1.75 | 0.07 |
| TCL1B | −1.74 | <0.01 |
| CCDC86 | −1.63 | 0.09 |
| MYOM3 | −1.33 | 0.08 |
| PCYOX1 | −1.25 | 0.01 |
| ERICH4 | −1.21 | 0.10 |
| RPL30 | −1.14 | 0.06 |
| PODN | −1.11 | 0.01 |
| CES4A | −1.04 | 0.01 |
| CASQ1 | −1.02 | 0.02 |
| RLN3 | −1.00 | 0.09 |
| CCNA1 | −0.94 | 0.05 |
| HSH2D | −0.94 | 0.07 |
| CEP295NL | −0.90 | <0.01 |
| FIG4 | −0.90 | 0.04 |
| TST | −0.89 | 0.08 |
| CYP8B1 | −0.89 | 0.06 |
| SLC13A3 | −0.82 | <0.01 |
| CLCN1 | −0.77 | 0.05 |
| A2ML1 | −0.75 | 0.02 |
| ADA | −0.66 | 0.06 |
| ORAI2 | −0.65 | 0.09 |
| CRYAB | −0.64 | 0.05 |
| NINL | −0.63 | 0.06 |
| AGK | −0.61 | 0.10 |
| CALML5 | −0.61 | 0.03 |
| CLDND1 | −0.60 | 0.01 |
| ART4 | −0.58 | 0.10 |
| OTULINL | −0.55 | 0.04 |
| CRELD2 | −0.53 | 0.04 |
| FBLN1 | −0.51 | 0.04 |

| NR SGA vs. NR NonSGA Cont. | | |
| --- | --- | --- |
| Gene | Log2 FC | FDR |
| LPCAT4 | −0.48 | 0.02 |
| GSTA4 | −0.45 | 0.08 |
| IMPDH1 | −0.45 | 0.10 |
| ATL1 | −0.44 | 0.08 |
| ADAM15 | −0.42 | 0.10 |
| DGKZ | −0.40 | 0.06 |
| LAPTM5 | 0.37 | 0.04 |
| ZNFX1 | 0.39 | 0.09 |
| CLEC1A | 0.45 | 0.03 |
| SLC38A7 | 0.47 | 0.10 |
| OLA1 | 0.49 | 0.04 |
| CTSK | 0.57 | 0.01 |
| TLR2 | 0.60 | 0.02 |
| KCNJ8 | 0.61 | 0.10 |
| SERPINB2 | 0.70 | 0.10 |
| IFI44L | 0.72 | 0.09 |
| ITGB2 | 0.75 | 0.04 |
| GALNT13 | 0.78 | 0.03 |
| DPYD | 0.79 | 0.10 |
| CYTH4 | 0.79 | 0.05 |
| CD274 | 0.87 | 0.10 |
| CMTM4 | 0.89 | 0.10 |
| CLEC4A | 0.92 | 0.01 |
| IGFBP2 | 0.96 | 0.05 |
| CLEC4G | 0.98 | 0.03 |
| CD163 | 0.99 | <0.01 |
| LRRC25 | 1.02 | 0.09 |
| ITGB3 | 1.05 | 0.05 |
| EPHB1 | 1.09 | 0.10 |
| GPR182 | 1.11 | 0.09 |
| ITGAM | 1.15 | 0.01 |
| SLAMF6 | 1.19 | 0.08 |
| EVA1A | 1.26 | <0.01 |
| CD86 | 1.30 | <0.01 |
| LAD1 | 1.37 | 0.10 |
| ISG15 | 1.47 | 0.04 |
| CXCL8 | 1.47 | <0.01 |
| CXCL9 | 1.57 | 0.04 |
| CD2 | 1.57 | 0.08 |
| IFIT1 | 1.59 | <0.01 |
| BMP3 | 1.69 | 0.08 |
| CLGN | 2.06 | 0.09 |
| CXCL10 | 2.11 | 0.03 |
| CLEC4F | 2.17 | 0.09 |
| CD96 | 2.49 | <0.01 |
| SAXO1 | 3.31 | 0.08 |
| CSTF2T | 3.36 | <0.01 |
| CYP3A5 | 3.40 | 0.01 |
| RPL9 | 4.36 | 0.06 |
| CCDC148 | 4.52 | <0.01 |

[1] 100% NRC: well-fed controls; NR NonSGA: nutrient-restricted normal-weight fetuses; NR SGA: nutrient-restricted low-weight fetuses.
[2] Log2 FC: log2 fold change. [3] FDR: false discovery rate.

### 2.3. Over-Representation Analysis of GO: Molecular Functions

Genes that were differentially expressed between the three treatment groups on day 70 of pregnancy were further analyzed to determine changes in Gene Ontology (GO) molecular functions (Figure 2). Only functions with > 1 gene involved and an FDR < 0.05 are shown. While the number of DEGs was fairly low in each comparison at day 70, there were still nine functions that were differentially enriched between NR SGA and 100% NRC placentomes, including terms such as MHC class II protein binding, cytokine binding, and transmembrane signaling receptor binding (Figure 2A). MHC class II receptor activity was the only function noted in NR NonSGA compared to 100% NRC placentomes, and the heat map shows that two of the three involved genes (*major histocompatibility complex, class II, DQ alpha 2 (HLA-DQA2)* and *major histocompatibility complex, class II, DQ beta 2 (HLA-DQB2)*) were downregulated in NR NonSGA compared to 100% NRC placentomes, while *HLA-DOB* was upregulated in NR NonSGA placentomes (Figure 2B). In NR SGA compared to NR NonSGA placentomes there were eight functions noted that included terms such as receptor ligand activity, organic hydroxyl compound transporter activity, and lipid transporter activity (Figure 2C). Signaling receptor binding and cytokine activity were also functions of note identified in NR SGA versus NR NonSGA placentomes, and as they were also identified at day 135, the expression for the genes involved in those functions is shown in a heat map in Figure 2D. Genes that are upregulated in NR SGA placentomes include *family with sequence similarity 3 member C (FAM3C), S100 calcium-binding protein B (S100B),* and *osteoprotegerin (TNFRSF11B),* while *apolipoprotein E (APOE), HLA-DOB, junctional adhesion molecule 2 (JAM2), MSTN,* and *Wnt family member 9B (WNT9B)* were downregulated.

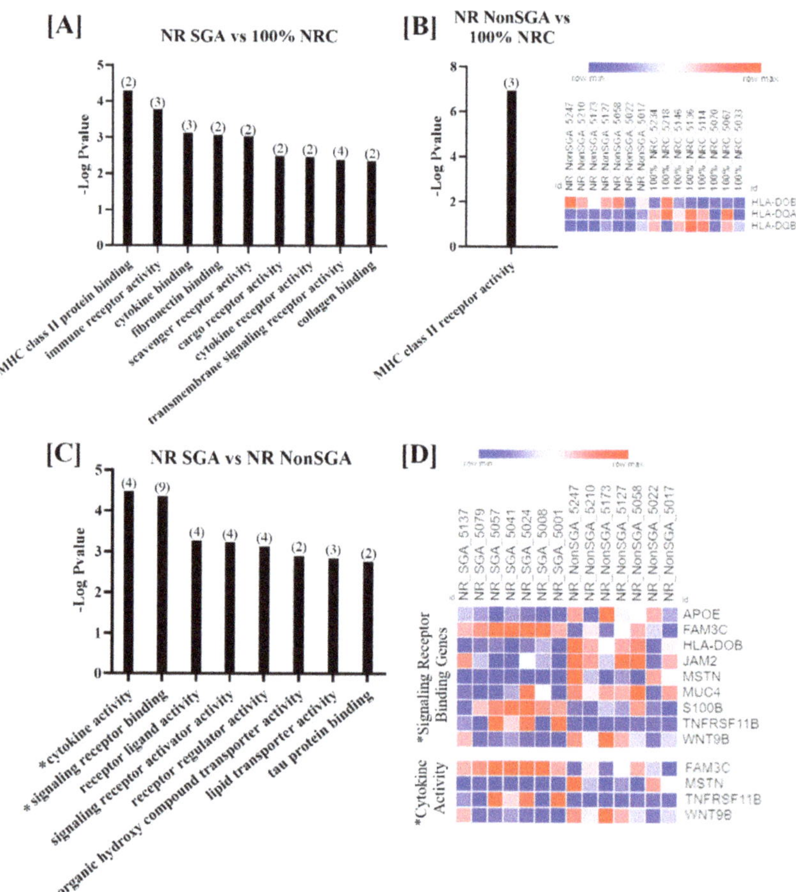

**Figure 2.** GO molecular function analyses of differentially expressed genes (DEGs) in placentomes at day 70 of pregnancy. Molecular functions differentially regulated in NR SGA compared to 100% NRC placentomes (**A**). The molecular function differentially regulated in NR NonSGA compared to 100% NRC placentomes, with a heat map of the three involved genes (**B**). Molecular functions differentially regulated in NR SGA compared to NR NonSGA placentomes (**C**), * with a heat map of genes involved in "signaling receptor binding" and "cytokine activity" (**D**). The heat maps correspond to one sample for each column and one gene for each row. All depicted GO molecular functions were significant at FDR < 0.05. Numbers in parentheses indicate the number of DEGs involved in the function.

Molecular functions that were differentially enriched between treatment groups at day 135 are shown in Figure 3. There were seven functions altered between NR SGA and 100% NRC placentomes, which included terms such as signaling receptor binding, molecular transducer activity, and polysaccharide binding (Figure 2A). Since only seven genes were differentially expressed between NR NonSGA and 100% NRC placentomes, only functions involving one or two genes were identified; those functions include hydroxylase activity for multiple molecules and oxidoreductase activity (Figure 3B). There were 15 functions with >1 gene altered in NR SGA compared to NR NonSGA placentomes, with terms such as CXCR chemokine receptor binding, amyloid-beta binding, and lipid kinase activity (Figure 3C). As at day 70, signaling receptor binding and cytokine activity were two functions with significant differences between NR SGA and NR NonSGA placentomes. The genes *MSTN* and *MUC4* continued to be downregulated in day 135 NR SGA placentomes, but none of the other genes identified in these functions at day 70 were still differentially regulated at day 135. Interestingly, multiple members of the C–X–C motif chemokine family (*CXCL8, CXCL9, CXCL10*) were upregulated in the NR SGA placentomes, as were multiple integrins (*ITGAM, ITGB2, ITGB3*; Figure 3D).

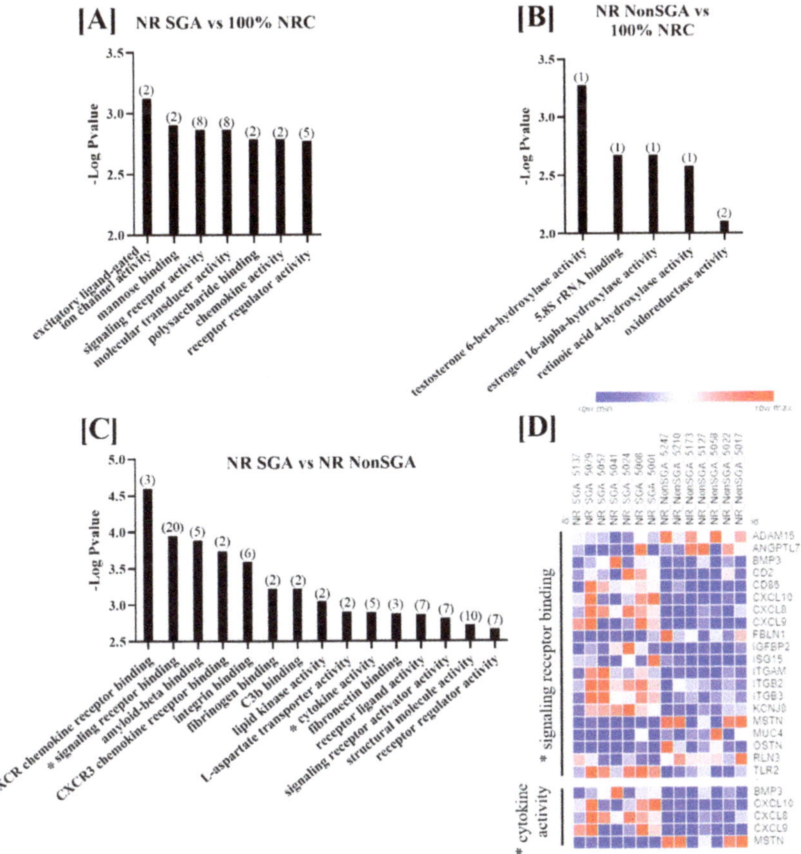

**Figure 3.** GO molecular function analyses of differentially expressed genes (DEGs) in placentomes at day 135 of pregnancy. Molecular functions differentially regulated in NR SGA compared to 100% NRC placentomes (**A**), NR NonSGA compared to 100% NRC placentomes (**B**), and NR SGA compared to NR NonSGA placentomes (**C**), * with a heat map of genes involved in "signaling receptor binding" and "cytokine activity" (**D**). The heat map corresponds to one sample for each column and one gene for each row. All depicted GO molecular functions were significant at FDR < 0.05. Numbers in parentheses indicate the number of differentially expressed genes involved in the function.

### 2.4. Transcription Factors Potentially Regulating DEGs

There is a single known transcription factor—*zinc finger protein 462 (ZNF462)*—that is differentially expressed in NR SGA compared to NR NonSGA placentomes at day 70, but little is known about which genes are specifically regulated by *ZNF462*. Therefore, the list of differentially expressed genes between NR SGA and NR NonSGA placentomes at day 70 was analyzed to determine transcription factors that are expressed in day 70 placentomes that could potentially be regulating gene expression at that time. There were nine transcription factors (*aryl hydrocarbon receptor nuclear translocator like 2 (ARNTL2), GA-binding protein transcription factor subunit alpha (GABPA), homeobox A2 (HOXA2), peroxisome proliferator-activated receptor gamma (PPARG), snail family transcriptional repressor 1 (SNAI1), SRY-box transcription factor 17 (SOX17), ZNF358, ZNF512B, ZNF740*) identified that potentially have the motifs to regulate at least 4 of the 22 identified DEGs (Figure 4). Of specific interest, *GABPA* has motifs to regulate nine of the DEGs, while all four genes with *PPARG* motifs are downregulated in SGA placentomes and all four genes with *ZNF740* motifs are upregulated in SGA placentomes (Figure 4).

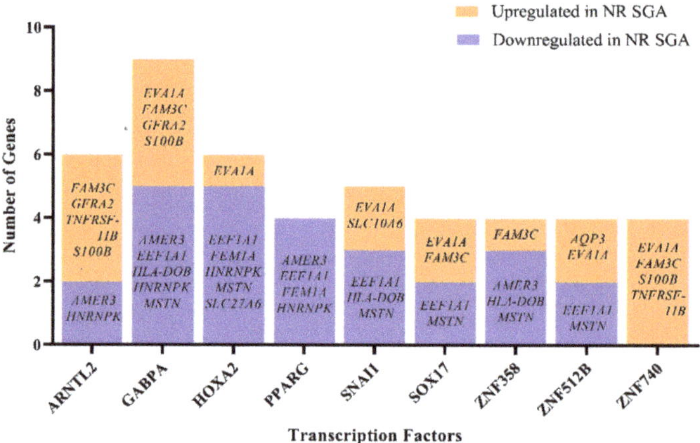


**Figure 4.** Computational analysis of transcription factors potentially regulating differentially expressed genes (DEGs) in NR SGA compared to NR NonSGA placentomes at day 70 of pregnancy. Only transcription factors that are expressed in placentomes and have ≥4 gene targets are depicted.

### 2.5. Gene Set Enrichment Analysis (GSEA) of Hallmark Pathways

Due to the relatively small number of DEGs between treatment groups, GSEA was used to further identify pathways that were specifically altered in NR SGA compared to NR NonSGA placentomes. Of note, sheep Ensembl IDs were converted to human Ensembl IDs before performing GSEA analyses due to limitations in the availability of sheep-related resources; thus, any sheep-specific genes were unable to be assessed as part of the GSEA analyses.

Nominal enrichment score (NES) values for the 50 hallmark pathways from an NR NonSGA versus 100% NRC analysis were plotted against those from a NR SGA versus 100% NRC analysis, with any point falling into a (+, −) or (−, +) quadrant being considered different between NR SGA and NR NonSGA placentomes (Figure 5). At day 70, apical surface and pancreas beta cell pathways were upregulated in NR NonSGA, but not NR SGA placentomes (Figure 5A). Additionally, pathways including fatty acid metabolism, MTORC1 signaling, IL2-STAT5 signaling, and Wnt/β-catenin signaling were upregulated in NR SGA, but not NR NonSGA placentomes (Figure 5A). At day 135, oxidative phosphorylation, reactive oxygen species, heme metabolism, and adipogenesis pathways were upregulated in NR NonSGA, but not NR SGA pathways. Pathways such as interferon alpha/gamma response, KRAS signaling, TGF-β signaling, hypoxia, and apical junctions were upregulated in NR SGA, but not NR NonSGA placentomes.

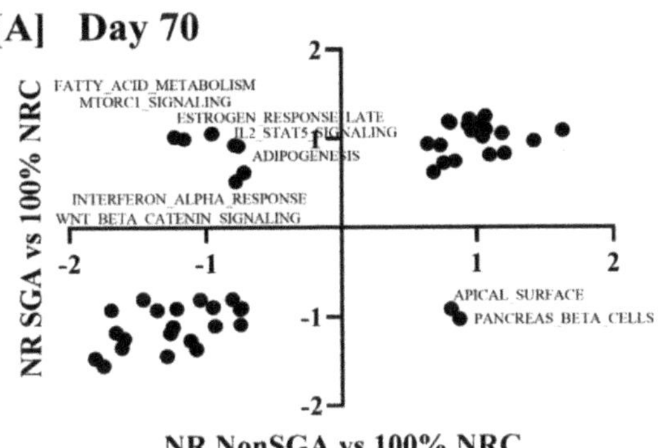


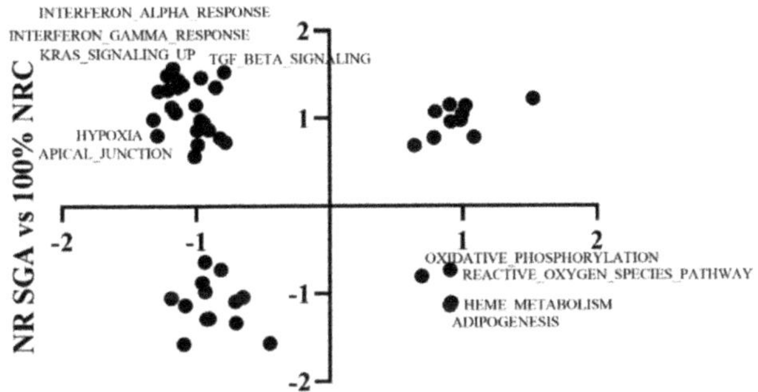


**Figure 5.** Gene set enrichment analysis of hallmark pathways in NR SGA compared to NR NonSGA placentomes at day 70 (**A**) and day 135 (**B**) of pregnancy. Nominal enrichment scores (NES) of NR NonSGA vs. 100% NRC analyses were plotted against NES values from NR SGA vs. 100% NRC analyses. Points in quadrants with (+, −) or (−, +) values were considered differentially regulated and were labeled with their pathway name.

To determine pathways that changed differently over time between 100% NRC, NR NonSGA, and NR SGA placentomes, NES values from a day 70 versus day 135 analysis of each treatment group were plotted against one another (Figure 6). Unfolded protein response, pancreas beta cells, and PI3K–AKT–MTOR signaling pathways were upregulated in 100% NRC placentomes at day 70, but not NR NonSGA, while TNF-α signaling via NF-κB and apical surface pathways were upregulated in NR NonSGA, but not 100% NRC placentomes (Figure 6A). The pancreas beta cell pathway was also upregulated in 100% NRC placentomes at day 70, but not NR SGA placentomes, while the apical surface pathway was upregulated in NR SGA, but not 100% NRC placentomes (Figure 6B). Pathways upregulated in NR SGA, but not NR NonSGA, included unfolded protein response and PI3K–AKT–MTOR signaling, while TNF-α signaling via NF-κB was upregulated in NR NonSGA, but not NR SGA placentomes (Figure 6C).

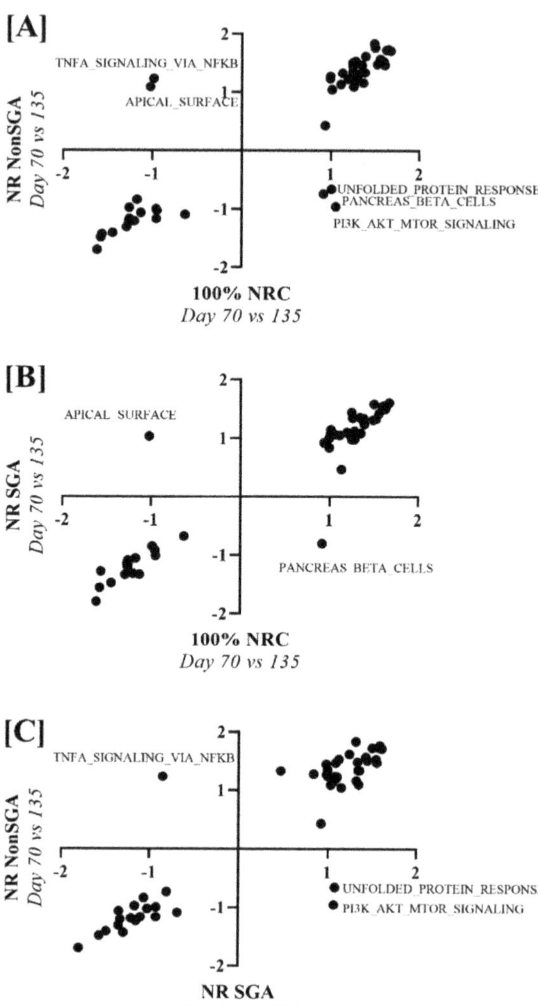


**Figure 6.** Gene set enrichment analysis of hallmark pathways between treatment groups across days of pregnancy. Nominal enrichment scores (NES) of NR NonSGA (day 70 vs. 135) analyses were plotted against NES values from 100% NRC (day 70 vs. 135) analyses (**A**). NES values of NR SGA (day 70 vs. 135) analyses were plotted against 100% NRC (day 70 vs. 135) analyses NES values (**B**). NES values of NR NonSGA (day 70 vs. 135) analyses were plotted against NR SGA (day 70 vs. 135) analyses NES values (**C**). Points in quadrants with (+, −) or (−, +) values were considered differentially regulated, and were labeled with their pathway name.

*2.6. Gene Set Enrichment Analysis (GSEA) of KEGG Pathways*

When GSEA analyses were performed between treatment groups to identify enriched KEGG pathways, only two pathways were significant at an adjusted $p$-value < 0.05. The first was the natural-killer-cell-mediated cytotoxicity pathway, which was enriched in NR SGA placentomes compared to 100% NRC placentomes at day 70 of pregnancy (Figure 7). Upregulated genes in this pathway included *major histocompatibility complex, class I G1 (HLA-G1), intercellular adhesion molecule 1-2 (ICAM1-2), UL16-binding protein 1–3 (ULBP1-3), ITGAL, ITGB2, linker for activation of T cells (LAT), protein kinase C (PKC),* and *perforin 1 (PRF1),* while downregulated genes included *CD94*, *Rac family small GTPase 1 (Rac)*, and *SHC adaptor protein 1 (Shc*; Figure 7). The second enriched pathway was β-alanine metabolism in NR SGA placentomes compared to NR NonSGA placentomes at day 70 of pregnancy (Figure 8). Upregulated genes in this pathway included *dihydropyrimidinase (DPYS)*, *upstream-binding protein 1 (UBP1)*, *glutamate decarboxylase-like 1 (GADL1)*, and *enoyl-CoA hydratase, short chain 1 (ECHS1)*, while there were no downregulated genes (Figure 8).

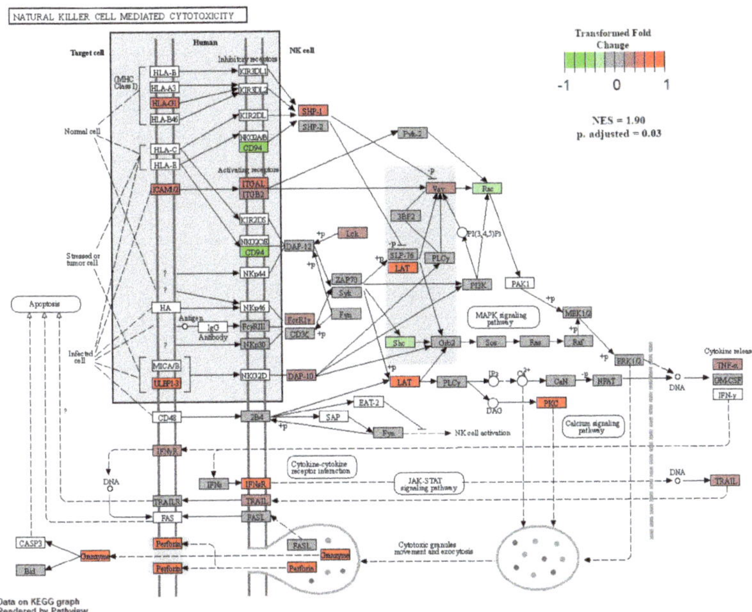

**Figure 7.** Natural-killer-cell-mediated cytotoxicity in NR SGA versus 100% NRC placentomes at day 70 of pregnancy. Gene set enrichment analysis was performed to identify KEGG pathways that were significantly enriched at adjusted *p*-values < 0.05.

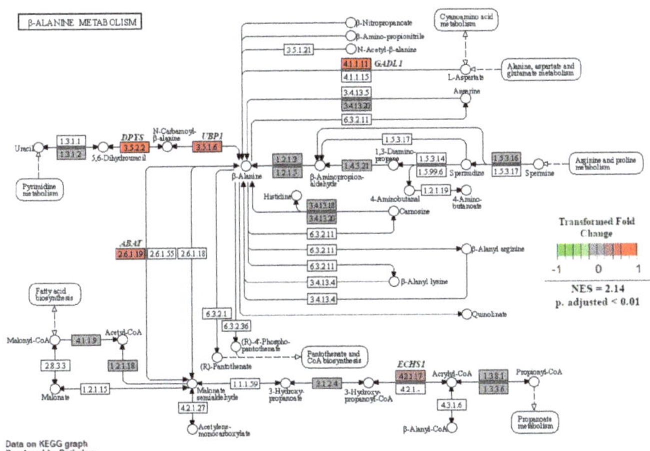

**Figure 8.** β-Alanine metabolism in NR SGA versus NR NonSGA placentomes at day 70 of pregnancy. Gene set enrichment analysis was performed to identify KEGG pathways that were significantly enriched at adjusted *p*-values < 0.05.

## 3. Discussion

Nutrient restriction during pregnancy reshapes the interaction between, and function of, the maternal, placental, and fetal compartments. Previous studies have indicated that the placenta—specifically the placentome in sheep—is not only nutrient-sensitive, but may be able to adapt to nutrient restriction as measured by the growth of the fetus [5,6]. Because placental growth precedes fetal growth, a major limitation in placental biology research—including previous studies on sheep—has been the inability to conduct a retrospective assessment of placental development responses that give rise to a spectrum of fetal growth outcomes. Using the pregnant sheep as our model organism, we developed a surgical technique to remove a single placentome during mid-gestation, and then allowed the pregnancy to proceed, which has provided the opportunity for a retrospective assessment of the early and late placentomal transcriptome based on late gestation fetal weight. Results from these analyses have revealed that a relatively low number of genes are differentially expressed between placentomes from well-fed ewes, NR ewes with NonSGA fetuses, and NR ewes with SGA fetuses—especially in comparison to the number of genes that are differentially expressed over time in the placentome—but those genes are involved in a variety of functions that lead to a broader impact later in pregnancy. Additionally, GSEA analyses have revealed a small number of unique pathways at day 70—including natural-killer-cell-mediated cell toxicity in SGA fetuses—that have interesting implications for placental development and function.

A previous study in our laboratory used a microarray analysis to identify transcriptomic changes between SGA and NonSGA placentomes from NR pregnancies at day 125 of pregnancy [6]. Major findings of that study included differentially expressed gene clusters involved in immune response, cell signaling, nutrient response, and nutrient transport [6]. These placentomes also showed histological differences, including decreased placentome volume and total maternal/fetal interface surface area in SGA pregnancies compared to NonSGA pregnancies from NR ewes [5]. The current study showed similar functional differences between SGA and NonSGA placentomes, with chemokine binding and cytokine activity altered at day 135 in addition to signaling receptor binding, nutrient transporter activity, and cell structure molecules. Indeed, certain genes–such as *CD86*, *CXCL10*, and *DPYD*—were differentially expressed in both studies, even with the small difference in timing [6]. Similar observations in late pregnancy between the two studies demonstrate that the model is reproducible, and provide confidence that DEGs identified in mid-gestation are inducing changes in placental function that lead to divergent patterns of fetal growth.

In the same cohort of animals as the current study, we previously found that triglycerides and bile acids were accumulating in allantoic fluid during mid-gestation of SGA fetuses from NR ewes, but not in allantoic fluid of NonSGA fetuses, suggesting impaired placental transport. We next found that SLC27A6 protein levels were reduced in mid-gestation placentomes from NR sheep with SGA fetuses, but not in NonSGA fetuses [9]. Interestingly, SLC27A6 is robustly expressed during placental growth, but is only modestly expressed during late gestation, highlighting a potentially novel role for this member of the fatty acid transport family in supporting early placental growth during nutritional hardship. This particular transporter family also has adaptive potential in other species, as both SLC27A2 and SLC27A6 were increased in the late-gestation baboon placenta in response to maternal nutrient restriction [14]. In addition to *SLC27A6* being differentially expressed between SGA and NonSGA placentomes from NR ewes in the current transcriptomic analysis, the molecular function of lipid kinase activity was altered, and fatty acid metabolism was noted as a pathway that was upregulated in SGA, but not NonSGA fetuses, in the GSEA analyses at day 70 of pregnancy. This connection between gene expression, protein expression, and systemic metabolic levels reinforces the value of using transcriptomic analyses to further elucidate mechanisms by which the placentome is able to adapt to nutrient restriction.

The placenta is a unique organ, in that it performs a variety of functions that would be relegated to specific organs in the adult animal. When conducting pathway analyses, this is important to consider, as the implications of certain pathways could be quite different than they would be in their intended organ. An example of this in the current study is the hallmark pathway of pancreas beta cells (M5957), which is composed of genes specifically upregulated in pancreatic beta cells. At day 70, this pathway is positively enriched in placentomes from NR pregnancies with NonSGA fetuses, but negatively enriched in NR pregnancies with SGA fetuses (Figure 5). Overall, many of the genes in the pathway that are enriched are involved in glucose and insulin regulation but, interestingly, the vitamin D receptor, *VDR*, is also enriched in this pathway in placentomes from SGA fetuses. Dysregulation of vitamin D metabolism has been implicated in intrauterine growth restriction in human pregnancies [15], and is predominantly located in the syncytial cells of the placentome during early to mid-pregnancy in sheep [16]. In addition to traditional roles in the regulation of calcium and phosphate homeostasis, vitamin D is a regulator of immune function that has been hypothesized to play an immunomodulatory role at the maternal–fetal interface [17]. Vitamin D, acting through its nuclear hormone receptor, VDR, has a large number of gene targets, and knockdown of *VDR* in mice led to increased proinflammatory cytokine and chemokine expression in placental cells, supporting a role for vitamin D in the placenta [18]. Vitamin D's status has not been assessed in this study, but is potentially of importance for future studies.

Of interest to this study is the potential interaction of vitamin D with macrophages and natural killer (NK) cells. Vitamin D can be locally activated in the placenta by macrophages, and promotes macrophage proliferation and differentiation [19,20]. At day 70 of pregnancy, *CD163*—a macrophage-specific marker whose upregulation is associated with a response to inflammation [21]—was upregulated in NR SGA placentomes compared to 100% NRC placentomes (Table 1). Additionally, *CD163* continued to be upregulated at day 135 in NR SGA placentomes compared to both NR NonSGA and 100% NRC placentomes, indicating a potential increase in either the total population of actual macrophages, or, as the cells of the placenta tend to take on the roles of other cells, the number of cells responding to perceived inflammation to help promote tissue modeling and the scavenging of apoptotic cells, as has been seen in other uterine environments [22].

NK cells can also be regulated by vitamin D, and are used as a defense mechanism against cells undergoing forms of stress [23]. By binding ligands on a target cell's surface, NK cells are activated to release cytotoxic granules onto the bound target cell in order to induce programmed cell death [24]. At day 70 of pregnancy, GSEA analysis showed an enrichment for the KEGG pathway natural-killer-cell-mediated cytotoxicity in placentomes from NR SGA pregnancies compared to those from well-fed pregnancies (Figure 7). At day 135, there was also upregulation of *CD96*—an NK cell marker—in SGA placentomes compared to placentomes from well-fed pregnancies and NR NonSGA pregnancies (Table 2). Multiple genes from a variety of steps in the cytotoxicity process were upregulated, from target cell ligands and NK cell receptors to *TNF-α*, genes involved in cytotoxic granule exocytosis, and apoptotic genes. Considering the histological changes, such as decreased surface area at the maternal–fetal interface in placentomes from SGA fetuses that were seen previously [5], the implications of increased cytotoxic activity may manifest in both histoarchitectural and functional aspects of placentome development. Of note, increased NK-cell-mediated cytotoxicity has also been associated with recurrent pregnancy loss in vitamin-D-deficient women, lending weight to the negative impact it can have on placental development [24,25].

Major histocompatibility complex (MHC) class II molecule expression serves as a mediator of T-lymphocyte response [26,27]. This is of interest because, at day 70, MHC class II protein binding was a molecular function that was altered in NR SGA compared to 100% NRC placentomes. MHC class II receptor activity was also the only molecular function that was altered in NR NonSGA compared to 100% NRC placentomes and, indeed, two of the three genes involved—*HLA-DQA2* and *HLA-DQB2*—were downregulated in the NR placentomes (Figure 2). While these two genes are considered to be nonfunctional pseudogenes in humans, they are transcribed and translated in sheep and cattle, although their exact roles are still unknown, and may be of interest to study in the context of placentome development [28].

One limitation to this study is that after mapping reads to the available sheep genome, those gene identifiers were then matched to human identifiers for further bioinformatics analyses, due to the availability of resources to analyze the human transcriptome as opposed to that of the sheep. This resulted in the loss of sheep-specific genes from the GSEA analyses that may have an important role that could be further elucidated by working through specific pathways of interest. The other area where this swap from sheep to human identifiers would need to be explored further is in the major histocompatibility complex molecules and NK cell complex molecules. While there are a number of similarities between the sheep and human genomes in these pathways, there are also documented differences that reinforce that these bioinformatics analyses are exploratory tools only, and specific mechanisms must be worked out directly [28,29].

Alterations in amino acid transport in the late-gestation placenta, and resultant changes in fetal plasma amino acid concentrations, have been well documented in NR pregnancies resulting in SGA offspring [5,30–33]. However, one of the pathways that were enriched in this study in SGA compared to NonSGA placentomes in mid-pregnancy that is unique is beta-alanine metabolism. Beta-alanine is a non-essential amino acid that is not incorporated into proteins, but serves as an intermediary that can be used for fatty acid synthesis, pyrimidine metabolism, or carnosine production [34]. The specific enzymes that are upregulated in the SGA fetuses appear to drive the production of beta-alanine from L-aspartate and uracil, with beta-alanine itself then being utilized in fatty acid biosynthesis (Figure 8). There appears to be little knowledge about a role for beta-alanine in the placenta, but its use could very well be an adaptive measure undertaken by the NR placentome to maintain function.

The GO term signaling receptor binding (GO:0005102) was identified as differential between SGA and NonSGA placentomes, and identifies genes that bind to a receptor molecule to initiate a change in cell function. At day 70, 9 of the 22 differentially expressed genes fit into this category, and especially interesting is that 4 of those—*FAM3C, MSTN, TNFRSF11B,* and *WNT9B*—were also identified in the cytokine activity molecular function (GO:0005125), which is defined as genes that interact with a receptor to control survival, growth, and differentiation. These genes have had little to no previous mechanistic association with placental development, especially in sheep, and have potential as critical regulators relative to their roles in other tissues. *FAM3C* has an insulin-independent regulatory role in hepatic gluconeogenesis and lipogenesis [35]. *MSTN*, a negative regulator of muscle development, may also be involved in cytokine production and glucose metabolism, and has been associated with preeclampsia and IUGR in human pregnancies [36,37]. *TNFRSF11B*, also known as osteoprotegerin, is a secreted factor that is a key regulator in bone metabolism, but is also pro-angiogenic, and has been associated with preeclampsia and diabetes mellitus in human pregnancies [38,39]. *WNT9B* drives mesenchymal–epithelial transitions in the urogenital system during organogenesis [40].

Signaling receptor binding was also identified at day 135 of pregnancy, and included 20 of the 100 differentially expressed genes between SGA and NonSGA placentomes. From the genes involved in signaling receptor binding, only *MSTN* and *MUC4* were differentially expressed at day 70 and day 135 of pregnancy. *MUC4* is a cell-surface membrane-bound glycoprotein that sterically masks cell surface antigens to protect cells from immune recognition—most notably in cancer cells [41]—and a specific role in the placenta has not been well defined. Multiple C–X–C motif chemokine family members (*CXCL8, CXCL9, CXCL10*) were upregulated in SGA placentomes, as was the interferon gamma response pathway, emphasizing that an inflammatory environment has been established in those placentomes late in pregnancy that was not necessarily active in mid-pregnancy, but was likely developing [42].

Identifying the regulators of differentially expressed genes—transcription factors—is a critical piece when trying to elucidate the mechanisms driving changes in placentome development. We chose to identify those transcription factors that may be of importance in differentiating between SGA and NonSGA placentomes during mid-gestation, because they potentially establish the developmental trajectory that the placentome will follow. Nine transcription factors were identified that were actually expressed in the NR placentomes that had potential to regulate at least 4 of the 22 DEGs from day 70, although none of them were expressed differently between SGA and NonSGA placentomes. Two of the transcription factors—*PPARG* and *SOX17*—have previously identified roles in the placenta, but the rest do not [43]. *PPARG* is essential for placental development, has functions in adipogenesis and inflammation, and has been previously shown to be nutrient-sensitive in the sheep placenta [44,45]. *GABPA* is a regulator of cellular energy metabolism and protein synthesis, as well as cytokine expression [46]. In addition to the 9 differential genes potentially regulated by *GABPA* at day 70 between SGA and NonSGA, there are 13 genes at day 135 potentially regulated by *GABPA*, including *carboxylesterase 4A (CES4A), claudin domain-containing 1 (CLDND1), collagen type 25 alpha 1 (COL25A1), cytohesin-4 (CYTH4), G-protein-coupled receptor 182 (GPR182), insulin-like growth factor-binding protein 2 (IGFBP2), inosine monophosphate dehydrogenase 1 (IMPDH1), ITGB2, keratin 4 (KRT4), ladinin-1 (LAD1),* and *leucine-rich repeat Ig-like transmembrane domains 1 (LRIT1). GABPA* itself is stably expressed in all placentomes at both days, but interacts with a variety of cofactors, including VDR, to achieve regulation of target genes [46,47].

## 4. Materials and Methods

### 4.1. Animal Study and Tissue Collection

Mature Hampshire ewes of similar parity, frame size, and initial body condition were fed to meet 100% of their National Research Council (NRC) [48] nutritional requirements, and served as embryo transfer recipients [5,13]. Ewes were synchronized into estrus, and a single embryo from a superovulated Hampshire donor ewe of normal body condition was transferred into the uterus of a recipient ewe on day 6 post-estrus [5]. Pregnancy was diagnosed by ultrasound on day 28 of gestation. All ewes were individually housed in pens with concrete flooring from days 28 to 135 of gestation, and fed once daily. Beginning on day 28 of gestation, body weight was measured weekly, and feed intake was adjusted based on changes in body weight. On day 35 of pregnancy, ewes were assigned randomly to either a control-fed group (100% NRC; $n = 8$) or a nutrient-restricted (NR) group (50% NRC; $n = 28$); composition of their respective diets has been published previously [49]. NR ewes were provided 50% of the total weight of feed that the control-fed group received, in order to induce a total caloric restriction equally across macromolecule groups. Vitamins and minerals were provided as recommended or in excess for all ewes.

On day 70 of pregnancy, a single placentome was surgically removed, as previously described [9]. Briefly, care was taken to remove a placentome from near the antimesometrial greater curvature of the gravid uterus and proximal to the anterior end of the amniotic membrane. The placentome was finely minced and thoroughly mixed in order to ensure representation of all cell types, before being snap-frozen in liquid nitrogen for RNA analyses. Necropsies were performed on day 135 of gestation. At this time, placentomes were dissected, weighed, and then processed, as on day 70.

Fetuses from ewes fed 100% NRC were the control group ($n = 8$). Fetuses within the NR group ($n = 28$) were segregated into quartiles based on their fetal weights. The highest (NR NonSGA; $n = 7$) and lowest (NR SGA; $n = 7$) quartiles were selected for further investigation [10,12].

### 4.2. RNA Extraction, Sequencing, and Analyses

Total RNA was extracted from snap-frozen placentomes using TRIzol reagent (Invitrogen, Carlsbad, CA, USA), according to the manufacturer's recommendations. Extracted RNA was treated with DNase I (Qiagen, Hilden, Germany) and purified using a RNeasy Mini Kit (Qiagen), before the RNA was quantified and its quality was assessed using a NanoDrop and an Agilent 2100 Bioanalyzer (Agilent Technologies, Santa Clara, CA, USA), respectively. An RNA integrity number (RIN) of >8 and a 260/230 value of >1.8 were considered acceptable. Extracted RNA was stored at −80 °C until further analyses.

Total RNA from the samples was submitted to the University of Missouri DNA Core facility (Columbia, Missouri, USA). Library construction and sequencing was conducted following the manufacturer's protocol with reagents supplied in Illumina's TruSeq Stranded mRNA sample preparation kit. Libraries were multiplexed and sequenced from both directions as 75 base pair paired-end reads on one lane on a NextSeq500 instrument. The raw sequences (FASTQ) were subjected to quality check via FastQC (http://www.bioinformatics.babraham.ac.uk/projects/fastqc/ (accessed on 16 July 2017)). The program fqtrim (https://ccb.jhu.edu/software/fqtrim/ (accessed on 16 July 2017)) was used to remove adapters, perform quality trimming (phred score > 30) by a sliding window scan (6 nucleotides), and select read lengths of 30 nucleotides or longer after trimming. The reads obtained from the quality control step were mapped to the sheep reference genome (Oar_v3.1) using Hisat2 aligner [50]. The program FeatureCounts [51] was used to quantify read counts by using the sequence alignment files of each sample. Genes with evidence of expression (counts per million; rowSum > 5) were used for model-based differential expression analysis using the edgeR robust method [52]. The false discovery rate (FDR) $\leq$ 0.10 was used as threshold for statistically significant differential expression of genes. Only protein-coding genes were included. Venn diagrams were produced using Venny 2.1 (http://bioinfogp.cnb.csic.es/tools/venny/index.html (accessed on 13 May 2021)). Over-representation analyses using DEG lists were conducted using ToppFun (http://toppgene.cchmc.org/ accessed on 18 May 2021)) with default settings [53] to identify Gene Ontology (GO) terms for the molecular function ontology (FDR < 0.05). Sheep Ensembl IDs were converted to human Ensembl IDs (genome assembly GRCh38.p13) in order to facilitate pathway analyses. Enrichment analyses were performed using gene set enrichment analysis software v4.1.0 (http://gsea-msigdb.org/ (accessed on 14 May 2021)) [54], with statistical significance set at an FDR $q$-value < 0.25. Transcription factors were identified using the Tf2DNA database (http://fiserlab.org/tf2dna_db/ (accessed on 18 May 2021)). KEGG pathway analyses were performed using the pathview package in R [55]. Data files were deposited in the National Center for Biotechnology Information (NCBI) Gene Expression Omnibus (GEO) under accession number GSE180182.

## 5. Conclusions

The results of the present study retrospectively identify novel DEGs and pathways during mid-gestation that give rise to a spectrum of fetal weight phenotypes in late gestation. Maternal nutrient restriction appears to trigger alterations in lipid metabolism, leading to a proinflammatory state that initiates a cascade of immune effects that are maintained into late gestation, specifically in those placentomes from pregnancies that produce SGA fetuses. In contrast, the placentomes from pregnancies with NonSGA fetuses are able to adapt to nutritional hardship, as evidenced by transcriptome changes in mid-pregnancy, in order to avoid the fate of the placentomes from the SGA fetuses, and become similar to the placentomes from well-fed control pregnancies in late gestation. Future studies are necessary to investigate the influence of identified key genes, such as *MSTN*, and potential systemic effectors, such as vitamin D, in order to be able to devise potential therapeutics to alleviate pregnancies resulting in small-for-gestational-age offspring.

**Author Contributions:** Conceptualization, M.C.S.; formal analysis, C.B.S., G.W.B., and S.K.B.; investigation, C.A.L., K.A., F.W.B., and M.C.S.; resources, T.E.S.; writing—original draft preparation, C.B.S.; writing—review and editing, M.C.S.; funding acquisition, M.C.S. All authors have read and agreed to the published version of the manuscript.

**Funding:** This work was supported by the Eunice Kennedy Shriver National Institute of Child Health and Human Development grant no. 1R01HD080658-01A1 (M.C.S.) from the National Institutes of Health.

**Institutional Review Board Statement:** All experimental procedures in this study were approved by, and performed in accordance with, the Texas A & M University Institutional Animal Care and Use Committee (AUP#2015-0204) and the National Institutes of Health (NIH) guidelines.

**Informed Consent Statement:** Not applicable for studies not involving humans.

**Data Availability Statement:** Data files were deposited in the National Center for Biotechnology Information (NCBI) Gene Expression Omnibus (GEO) under accession number GSE180182.

**Conflicts of Interest:** The authors declare no conflict of interest. The funders had no role in the design of the study, in the collection, analyses, or interpretation of data, in the writing of the manuscript, or in the decision to publish the results.

## References


1. DelCurto, H.; Wu, G.; Satterfield, M.C. Nutrition and reproduction: Links to epigenetics and metabolic syndrome in offspring. *Curr. Opin. Clin. Nutr. Metab. Care* **2013**, *16*, 385–391. [CrossRef]
2. Wu, G.; Bazer, F.W.; Wallace, J.M.; Spencer, T.E. Board-invited review: Intrauterine growth retardation: Implications for the animal sciences. *J. Anim. Sci.* **2006**, *84*, 2316–2337. [CrossRef] [PubMed]
3. Balasuriya, C.N.D.; Stunes, A.K.; Mosti, M.P.; Schei, B.; Indredavik, M.S.; Hals, I.K.; Evensen, K.A.I.; Syversen, U. Metabolic outcomes in adults born preterm with very low birthweight or small for gestational age at term: A cohort study. *J. Clin. Endocrinol. Metab.* **2018**, *103*, 4437–4446. [CrossRef] [PubMed]
4. Hochberg, Z.; Feil, R.; Constancia, M.; Fraga, M.; Junien, C.; Carel, J.-C.; Boileau, P.; Le Bouc, Y.; Deal, C.L.; Lillycrop, K.; et al. Child health, developmental plasticity, and epigenetic programming. *Endocr. Rev.* **2010**, *32*, 159–224. [CrossRef] [PubMed]
5. Edwards, A.K.; McKnight, S.M.; Askelson, K.; McKnight, J.; Dunlap, K.A.; Satterfield, M.C. Adaptive responses to maternal nutrient restriction alter placental transport in ewes. *Placenta* **2020**, *96*, 1–9. [CrossRef] [PubMed]
6. Edwards, A.K.; Dunlap, K.A.; Spencer, T.E.; Satterfield, M.C. Identification of pathways associated with placental adaptation to maternal nutrient restriction in sheep. *Genes* **2020**, *11*, 9. [CrossRef] [PubMed]
7. Alexander, G. Studies on the placenta of the sheep (*Ovis aries*): Placental size. *J. Reprod. Fertil.* **1964**, *7*, 289–305. [CrossRef]
8. Dunlap, K.; Brown, J.; Keith, A.; Satterfield, M. Factors controlling nutrient availability to the developing fetus in ruminants. *J. Anim. Sci. Biotechnol.* **2015**, *6*, 1–10. [CrossRef]
9. Lambo, C.A.; Edwards, A.K.; Bazer, F.W.; Dunlap, K.; Satterfield, M.C. Development of a surgical procedure for removal of a placentome from a pregnant ewe during gestation. *J. Anim. Sci. Biotechnol.* **2020**, *11*, 1–7. [CrossRef]
10. Steinhauser, C.B.; Askelson, K.; Lambo, C.A.; Hobbs, K.C.; Bazer, F.W.; Satterfield, M.C. Lipid metabolism is altered in maternal, placental, and fetal tissues of ewes with small for gestational age fetuses. *Biol. Reprod.* **2021**, *104*, 170–180. [CrossRef]
11. Steinhauser, C.; Askelson, K.; Hobbs, K.; Bazer, F.; Satterfield, M. Maternal nutrient restriction alters thyroid hormone dynamics in placentae of sheep having small for gestational age fetuses. *Domest. Anim. Endocrinol.* **2021**, *77*, 106632. [CrossRef]
12. Sandoval, C.; Lambo, C.; Beason, K.; Dunlap, K.; Satterfield, M. Effect of maternal nutrient restriction on skeletal muscle mass and associated molecular pathways in SGA and Non-SGA sheep fetuses. *Domest. Anim. Endocrinol.* **2020**, *72*, 106443. [CrossRef] [PubMed]
13. Sandoval, C.; Askelson, K.; Lambo, C.; Dunlap, K.; Satterfield, M. Effect of maternal nutrient restriction on expression of glucose transporters (SLC2A4 and SLC2A1) and insulin signaling in skeletal muscle of SGA and Non-SGA sheep fetuses. *Domest. Anim. Endocrinol.* **2021**, *74*, 106556. [CrossRef]
14. Chassen, S.S.; Ferchaud-Roucher, V.; Palmer, C.; Li, C.; Jansson, T.; Nathanielsz, P.W.; Powell, T.L. Placental fatty acid transport across late gestation in a baboon model of intrauterine growth restriction. *J. Physiol.* **2020**, *598*, 2469–2489. [CrossRef] [PubMed]
15. Chen, Y.-H.; Fu, L.; Hao, J.-H.; Yu, Z.; Zhu, P.; Wang, H.; Xu, Y.-Y.; Zhang, C.; Tao, F.-B.; Xu, D.-X. Maternal vitamin D deficiency during pregnancy elevates the risks of small for gestational age and low birth weight infants in Chinese population. *J. Clin. Endocrinol. Metab.* **2015**, *100*, 1912–1919. [CrossRef]
16. Stenhouse, C.; Halloran, K.M.; Newton, M.G.; Gaddy, D.; Suva, L.J.; Bazer, F.W. Novel mineral regulatory pathways in ovine pregnancy: II. Calcium-binding proteins, calcium transporters, and vitamin D signaling. *Biol. Reprod.* **2021**, *105*, 232–243. [CrossRef]
17. Tamblyn, J.A.; Hewison, M.; Wagner, C.L.; Bulmer, J.N.; Kilby, M.D. Immunological role of vitamin D at the maternal–fetal interface. *J. Endocrinol.* **2015**, *224*, R107–R121. [CrossRef]
18. Liu, N.Q.; Kaplan, A.; Lagishetty, V.; Ouyang, Y.B.; Ouyang, Y.; Simmons, C.F.; Equils, O.; Hewison, M. Vitamin D and the regulation of placental inflammation. *J. Immunol.* **2011**, *186*, 5968–5974. [CrossRef]
19. Schröder-Heurich, B.; Springer, C.J.P.; Von Versen-Höynck, F. Vitamin D effects on the immune system from periconception through pregnancy. *Nutrients* **2020**, *12*, 1432. [CrossRef] [PubMed]
20. Aranow, C. Vitamin D and the immune system. *J. Investig. Med.* **2011**, *59*, 881–886. [CrossRef]
21. Etzerodt, A.; Moestrup, S.K. CD163 and inflammation: Biological, diagnostic, and therapeutic aspects. *Antioxid. Redox Signal.* **2013**, *18*, 2352–2363. [CrossRef]

22. Houser, B.; Tilburgs, T.; Hill, J.; Nicotra, M.; Strominger, J. Two unique human decidual macrophage populations. *J. Immunol.* **2011**, *186*, 2633–2642. [CrossRef]

23. Balogh, G.; De Boland, A.R.; Boland, R.; Barja, P. Effect of 1,25(OH)2-vitamin D3 on the activation of natural killer cells: Role of protein kinase c and extracellular calcium. *Exp. Mol. Pathol.* **1999**, *67*, 63–74. [CrossRef]

24. Liu, Y.; Gao, S.; Zhao, Y.; Wang, H.; Pan, Q.; Shao, Q. Decidual natural killer cells: A good nanny at the maternal-fetal interface during early pregnancy. *Front. Immunol.* **2021**, *12*, 663660. [CrossRef]

25. Ota, K.; Dambaeva, S.; Han, A.R.; Beaman, K.; Gilman-Sachs, A.; Kwak-Kim, J. Vitamin D deficiency may be a risk factor for recurrent pregnancy losses by increasing cellular immunity and autoimmunity. *Hum. Reprod.* **2014**, *29*, 208–219. [CrossRef] [PubMed]

26. Mora, J.R.; Iwata, M.; Von Andrian, U.H. Vitamin effects on the immune system: Vitamins A and D take centre stage. *Nat. Rev. Immunol.* **2008**, *8*, 685–698. [CrossRef]

27. Ota, K.; Dambaeva, S.; Kim, M.W.-I.; Han, A.R.; Fukui, A.; Gilman-Sachs, A.; Beaman, K.; Kwak-Kim, J. 1,25-dihydroxy-vitamin D3 regulates NK-cell cytotoxicity, cytokine secretion, and degranulation in women with recurrent pregnancy losses. *Eur. J. Immunol.* **2015**, *45*, 3188–3199. [CrossRef] [PubMed]

28. Herrmann-Hoesing, L.M.; White, S.N.; Kappmeyer, L.S.; Herndon, D.R.; Knowles, D.P. Genomic analysis of *Ovis aries* (Ovar) MHC class IIa loci. *Immunogenetics* **2008**, *60*, 167–176. [CrossRef] [PubMed]

29. Schwartz, J.C.; Gibson, M.S.; Heimeier, D.; Koren, S.; Phillippy, A.; Bickhart, D.; Smith, T.P.L.; Medrano, J.F.; Hammond, J.A. The evolution of the natural killer complex: A comparison between mammals using new high-quality genome assemblies and targeted annotation. *Immunogenetics* **2017**, *69*, 255–269. [CrossRef]

30. Jansson, N.; Pettersson, J.; Haafiz, A.; Ericsson, A.; Palmberg, I.; Tranberg, M.; Ganapathy, V.; Powell, T.L.; Jansson, T. Down-regulation of placental transport of amino acids precedes the development of intrauterine growth restriction in rats fed a low protein diet. *J. Physiol.* **2006**, *576*, 935–946.

31. Malandro, M.S.; Beveridge, M.J.; Kilberg, M.S.; Novak, D.A. Effect of low-protein diet-induced intrauterine growth retardation on rat placental amino acid transport. *Am. J. Physiol. Physiol.* **1996**, *271*, C295–C303. [CrossRef] [PubMed]

32. Pantham, P.; Rosario, F.J.; Weintraub, S.T.; Nathanielsz, P.W.; Powell, T.L.; Li, C.; Jansson, T. Down-regulation of placental transport of amino acids precedes the development of intrauterine growth restriction in maternal nutrient restricted baboons. *Biol. Reprod.* **2016**, *95*, 98. [CrossRef] [PubMed]

33. Satterfield, M.C.; Bazer, F.W.; Spencer, T.E.; Wu, G. Sildenafil citrate treatment enhances amino acid availability in the conceptus and fetal growth in an ovine model of intrauterine growth restriction. *J. Nutr.* **2010**, *140*, 251–258. [CrossRef]

34. Tiedje, K.E.; Stevens, K.; Barnes, S.; Weaver, D. β-Alanine as a small molecule neurotransmitter. *Neurochem. Int.* **2010**, *57*, 177–188. [CrossRef]

35. Zhang, X.; Yang, W.; Wang, J.; Meng, Y.; Guan, Y.; Yang, J. FAM3 gene family: A promising therapeutical target for NAFLD and type 2 diabetes. *Metabolism* **2018**, *81*, 71–82. [CrossRef] [PubMed]

36. Peiris, H.; Mitchell, M. The expression and potential functions of placental myostatin. *Placenta* **2012**, *33*, 902–907. [CrossRef]

37. Peiris, H.; Georgiou, H.; Lappas, M.; Kaitu'U-Lino, T.; Salomón, C.; Vaswani, K.; Rice, G.E.; Mitchell, M.D. Expression of myostatin in intrauterine growth restriction and preeclampsia complicated pregnancies and alterations to cytokine production by first-trimester placental explants following myostatin treatment. *Reprod. Sci.* **2015**, *22*, 1202–1211. [CrossRef]

38. Shen, P.; Gong, Y.; Wang, T.; Chen, Y.; Jia, J.; Ni, S.; Zhou, B.; Song, Y.; Zhang, L.; Zhou, R. Expression of osteoprotegerin in placenta and its association with preeclampsia. *PLoS ONE* **2012**, *7*, e44340. [CrossRef]

39. Huang, B.; Zhu, W.; Zhao, H.; Zeng, F.; Wang, E.; Wang, H.; Chen, J.; Li, M.; Huang, C.; Ren, L.; et al. Placenta-derived osteoprotegerin is required for glucose homeostasis in gestational diabetes mellitus. *Front. Cell Dev. Biol.* **2020**, *8*, 563509. [CrossRef]

40. Carroll, T.J.; Park, J.-S.; Hayashi, S.; Majumdar, A.; McMahon, A.P. Wnt9b plays a central role in the regulation of mesenchymal to epithelial transitions underlying organogenesis of the mammalian urogenital system. *Dev. Cell* **2005**, *9*, 283–292. [CrossRef] [PubMed]

41. Bhatia, R.; Gautam, S.K.; Cannon, A.; Thompson, C.; Hall, B.R.; Aithal, A.; Banerjee, K.; Jain, M.; Solheim, J.C.; Kumar, S.; et al. Cancer-associated mucins: Role in immune modulation and metastasis. *Cancer Metastasis Rev.* **2019**, *38*, 223–236. [CrossRef]

42. Amabebe, E.; Anumba, D.O. The transmembrane G protein-coupled CXCR3 receptor-ligand system and maternal fetal allograft rejection. *Placenta* **2021**, *104*, 81–88. [CrossRef] [PubMed]

43. Igarashi, H.; Uemura, M.; Hiramatsu, R.; Hiramatsu, R.; Segami, S.; Pattarapanawan, M.; Hirate, Y.; Yoshimura, Y.; Hashimoto, H.; Higashiyama, H.; et al. Sox17 is essential for proper formation of the marginal zone of extraembryonic endoderm adjacent to a developing mouse placental disk. *Biol. Reprod.* **2018**, *99*, 578–589. [CrossRef] [PubMed]

44. Yiallourides, M.; Sébert, S.P.; Wilson, V.; Sharkey, D.; Rhind, S.M.; Symonds, M.E.; Budge, H.; Yiallourides, M.; Sébert, S.P.; Wilson, V.; et al. The differential effects of the timing of maternal nutrient restriction in the ovine placenta on glucocorticoid sensitivity, uncoupling protein 2, peroxisome proliferator-activated receptor-γ and cell proliferation. *Reproduction* **2009**, *138*, 601–608. [CrossRef] [PubMed]

45. Shalom-Barak, T.; Zhang, X.; Chu, T.; Schaiff, W.T.; Reddy, J.K.; Xu, J.; Sadovsky, Y.; Barak, Y. Placental PPARγ regulates spatiotemporally diverse genes and a unique metabolic network. *Dev. Biol.* **2012**, *372*, 143–155. [CrossRef]

46. Rosmarin, A.G.; Resendes, K.K.; Yang, Z.; McMillan, J.N.; Fleming, S.L. GA-binding protein transcription factor: A review of GABP as an integrator of intracellular signaling and protein-protein interactions. *Blood Cells Mol. Dis.* **2004**, *32*, 143–154. [CrossRef]
47. Seuter, S.; Neme, A.; Carlberg, C. ETS transcription factor family member GABPA contributes to vitamin D receptor target gene regulation. *J. Steroid Biochem. Mol. Biol.* **2018**, *177*, 46–52. [CrossRef]
48. National Research Council. *Nutrient Requirements of Small Ruminants: Sheep, Goats, Cervids, and New World Camelids*; The National Academies Press: Washington, DC, USA, 2007; p. 384.
49. Lassala, A.; Bazer, F.W.; Cudd, T.A.; Datta, S.; Keisler, D.; Satterfield, M.C.; Spencer, T.; Wu, G. Parenteral administration of L-arginine prevents fetal growth restriction in undernourished ewes. *J. Nutr.* **2010**, *140*, 1242–1248. [CrossRef]
50. Kim, D.; Langmead, B.; Salzberg, S.L. HISAT: A fast spliced aligner with low memory requirements. *Nat. Methods* **2015**, *12*, 357–360. [CrossRef]
51. Liao, Y.; Smyth, G.K.; Shi, W. FeatureCounts: An efficient general purpose program for assigning sequence reads to genomic features. *Bioinformatics* **2014**, *30*, 923–930. [CrossRef]
52. Robinson, M.D.; McCarthy, D.J.; Smyth, G.K. EdgeR: A Bioconductor package for differential expression analysis of digital gene expression data. *Bioinformatics* **2010**, *26*, 139–140. [CrossRef] [PubMed]
53. Chen, J.; Bardes, E.E.; Aronow, B.J.; Jegga, A.G. ToppGene Suite for gene list enrichment analysis and candidate gene prioritization. *Nucleic Acids Res.* **2009**, *37*, W305–W311. [CrossRef]
54. Subramanian, A.; Tamayo, P.; Mootha, V.K.; Mukherjee, S.; Ebert, B.L.; Gillette, M.A.; Paulovich, A.; Pomeroy, S.L.; Golub, T.R.; Lander, E.S.; et al. Gene set enrichment analysis: A knowledge-based approach for interpreting genome-wide expression profiles. *Proc. Natl. Acad. Sci. USA* **2005**, *102*, 15545–15550. [CrossRef] [PubMed]
55. Luo, W.; Brouwer, C. Pathview: An R/Bioconductor package for pathway-based data integration and visualization. *Bioinformatics* **2013**, *29*, 1830–1831. [CrossRef] [PubMed]

International Journal of
*Molecular Sciences*

*Article*

# Elevated Circulating and Placental SPINT2 Is Associated with Placental Dysfunction

Ciara N. Murphy [1,2,*], Susan P. Walker [1,2], Teresa M. MacDonald [1,2], Emerson Keenan [1,2], Natalie J. Hannan [1,2], Mary E. Wlodek [1,3], Jenny Myers [4], Jessica F. Briffa [3], Tania Romano [5], Alexandra Roddy Mitchell [1,2], Carole-Anne Whigham [1,2], Ping Cannon [1,2], Tuong-Vi Nguyen [1,2], Manju Kandel [1,2], Natasha Pritchard [1,2], Stephen Tong [1,2,†] and Tu'uhevaha J. Kaitu'u-Lino [1,2,†]

[1] The Department of Obstetrics and Gynaecology, Mercy Hospital for Women, The University of Melbourne, Heidelberg, VIC 3084, Australia; spwalker@unimelb.edu.au (S.P.W.); teresa.mary.macdonald@gmail.com (T.M.M.); emerson.keenan@unimelb.edu.au (E.K.); nhannan@unimelb.edu.au (N.J.H.); m.wlodek@unimelb.edu.au (M.E.W.); aroddymitchell@gmail.com (A.R.M.); drcwhigham@gmail.com (C.-A.W.); ping.cannon@unimelb.edu.au (P.C.); tuong-vi.nguyen@unimelb.edu.au (T.-V.N.); manju.kandel@unimelb.edu.au (M.K.); natashalpritchard@gmail.com (N.P.); stong@unimelb.edu.au (S.T.); t.klino@unimelb.edu.au (T.J.K.-L.)
[2] Mercy Perinatal, Mercy Hospital for Women, Heidelberg, VIC 3084, Australia
[3] The Department of Anatomy and Physiology, The University of Melbourne, VIC 3010, Australia; jessica.briffa@unimelb.edu.au
[4] Manchester Academic Health Science Centre, St Mary's Hospital, University of Manchester, Manchester M13 OJH, UK; Jenny.Myers@manchester.ac.uk
[5] The Department of Physiology, Anatomy and Microbiology, La Trobe University, Bundoora, VIC 3086, Australia; t.romano@latrobe.edu.au
[*] Correspondence: ciaram1@student.unimelb.edu.au
[†] Equal contribution.

**Citation:** Murphy, C.N.; Walker, S.P.; MacDonald, T.M.; Keenan, E.; Hannan, N.J.; Wlodek, M.E.; Myers, J.; Briffa, J.F.; Romano, T.; Roddy Mitchell, A.; et al. Elevated Circulating and Placental SPINT2 Is Associated with Placental Dysfunction. *Int. J. Mol. Sci.* **2021**, 22, 7467. https://doi.org/10.3390/ijms22147467

Academic Editors: Hiten D. Mistry and Eun Lee

Received: 21 June 2021
Accepted: 8 July 2021
Published: 12 July 2021

**Publisher's Note:** MDPI stays neutral with regard to jurisdictional claims in published maps and institutional affiliations.



**Abstract:** Biomarkers for placental dysfunction are currently lacking. We recently identified SPINT1 as a novel biomarker; SPINT2 is a functionally related placental protease inhibitor. This study aimed to characterise SPINT2 expression in placental insufficiency. Circulating SPINT2 was assessed in three prospective cohorts, collected at the following: (1) term delivery ($n$ = 227), (2) 36 weeks ($n$ = 364), and (3) 24–34 weeks' ($n$ = 294) gestation. SPINT2 was also measured in the plasma and placentas of women with established placental disease at preterm (<34 weeks) delivery. Using first-trimester human trophoblast stem cells, SPINT2 expression was assessed in hypoxia/normoxia (1% vs. 8% O2), and following inflammatory cytokine treatment (TNF$\alpha$, IL-6). Placental SPINT2 mRNA was measured in a rat model of late-gestational foetal growth restriction. At 36 weeks, circulating SPINT2 was elevated in patients who later developed preeclampsia ($p$ = 0.028; median = 2233 pg/mL vs. controls, median = 1644 pg/mL), or delivered a small-for-gestational-age infant ($p$ = 0.002; median = 2109 pg/mL vs. controls, median = 1614 pg/mL). SPINT2 was elevated in the placentas of patients who required delivery for preterm preeclampsia ($p$ = 0.025). Though inflammatory cytokines had no effect, hypoxia increased SPINT2 in cytotrophoblast stem cells, and its expression was elevated in the placental labyrinth of growth-restricted rats. These findings suggest elevated SPINT2 is associated with placental insufficiency.

**Keywords:** placental insufficiency; SPINT2/HAI-2; preeclampsia; foetal growth restriction; intrauterine growth restriction; small for gestational age

## 1. Introduction

Aberrations in placentation, particularly those amounting to restricted vascular remodelling, are associated with a constellation of obstetric complications, with significant implications for mothers and babies. This includes foetal growth restriction (FGR), in which affected foetuses fail to achieve their unique growth potential in utero, owing to inadequate uteroplacental perfusion. This confers an increased risk of perinatal morbidity and mortality upon the foetus [1]; in fact, FGR is recognised as the single largest risk factor for stillbirth [2]. Alternatively or simultaneously, placental insufficiency may manifest maternally as preeclampsia, which is characterised by persistent maternal hypertension and end organ dysfunction, arising from vascular endothelial injury [3].

Currently, there is an absence of effective and targeted treatments for both preeclampsia and FGR, and the latter, in particular, eludes precise diagnosis. Consequently, there are no interventions to rescue a poorly functioning placenta; except for iatrogenic preterm delivery, which has its own associated risks.

In the search for circulating biomarkers of placental insufficiency, SPINT1, a serine protease inhibitor, also called HGF activator inhibitor 1 (HAI-1), has been identified as a promising candidate [4]. SPINT1 is highly expressed in the placenta, where it is localised to the cell surface of villous cytotrophoblasts, and secreted in a proteolytically truncated form [5] through ectodomain shedding [6] into the maternal circulation. By virtue of its dual Kunitz domains, SPINT1 inhibits the activity of proteolytic substrates that are critical for normal placentation. Therefore, SPINT1 mediates the trophoblast secretion of degradative enzymes (serine proteinases, metalloproteinases, and collagenases), which regulate the invasion and remodelling of endometrial spiral arteries [7,8]. Inadequate or superficial remodelling may result in a suboptimal placenta, through oxidative stress-induced placental growth suppression [9], and intermittent placental perfusion, which leads to ischaemia-reperfusion injury [10]. We have recently demonstrated that SPINT1 is reduced in FGR placentas and is modulated by hypoxia; both in human placental cells and in a mouse model of FGR [4].

The validation of this serine peptidase inhibitor as an indicator of placental dysfunction justifies the investigation of its analogue SPINT2/HAI-2. To date, SPINT2 has not been thoroughly assessed in the human placenta; thus, its application in FGR diagnosis is yet to be explored. SPINT2 has a comparable tissue distribution to SPINT1 [11], and contains two extracellular inhibitory Kunitz domains, making it structurally similar to SPINT1. As with SPINT1, SPINT2 regulates matriptase [12], a transmembrane serine protease that is responsible for the degradation of the extracellular matrix components fibronectin and laminin [13]. The involvement of SPINT2 with matriptase in the placenta suggests its importance in placental development [12,14]. Indeed, in mouse placentas, SPINT2 is expressed for the duration of development in the placental labyrinth layer (the site of murine foetomaternal exchange) [12,15]. As reported in *Spint1* knockout mouse models, *Spint2*-deficient mice suffer from placental defects. However, the effects of the loss of *Spint2* extend beyond the placenta, causing embryonic lethality, unless matriptase is simultaneously ablated, in which case there are impairments to the neural tube closure [12].

Given the previous findings of dysregulated SPINT1 in FGR, it was hypothesised that the expression of SPINT2 would be similarly deranged in the placenta and maternal circulation of pregnancies that are complicated by FGR, even prior to diagnosis. Further, the expression of SPINT2 in placental trophoblasts was expected to be regulated by hypoxia and inflammation, which are signature contributors underlying placental insufficiency. Therefore, the aim of this study was to characterise the expression of SPINT2 at the mRNA and protein level in placental and plasma samples from pregnancies that have been affected by preeclampsia and/or FGR, as well as in a rodent model of placental insufficiency, and to observe any hypoxic- or inflammation-mediated changes in expression in vitro.

## 2. Results

### 2.1. SPINT2 Expression Is Deranged in Placental Dysfunction

Given that SPINT2 has not previously been analysed in the human placenta, this study first characterised its expression in pregnancies that were known to be compromised by FGR and/or preeclampsia.

In preterm FGR placentas ($n$ = 14), SPINT2 mRNA expression (Figure 1a) was highly variable and did not significantly differ from the controls ($n$ = 19). In contrast, the placentas from pregnancies that were complicated by both preeclampsia and FGR ($n$ = 20) had significantly decreased SPINT2 mRNA expression (68% of control, $p$ = 0.002), whereas those affected by preeclampsia only ($n$ = 60) had significantly increased SPINT2 mRNA levels (119% of control, $p$ = 0.03).

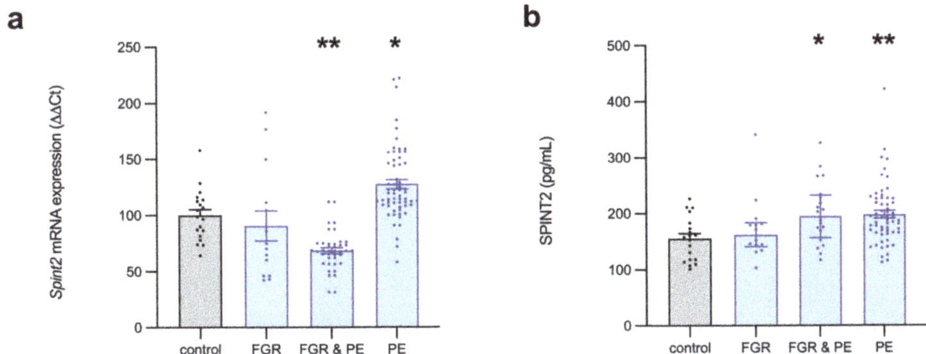

**Figure 1.** SPINT2 expression in placentas of patients with established placental disease. Compared to preterm controls, SPINT2 mRNA expression (**a**) was not altered in placentas from normotensive pregnancies affected by foetal growth restriction (FGR), but it was significantly decreased in those from pregnancies compromised by concurrent preeclampsia (PE) and FGR, and increased in preeclamptic placentas of AGA infants. SPINT2 protein expression (**b**) in these same placentas was also not changed in FGR-affected normotensive pregnancies, but was significantly elevated in all PE cases (with and without FGR). Each data point represents an individual patient sample; data are expressed as median $\pm$ IQR; * $p$ < 0.05, ** $p$ < 0.01.

The level of SPINT2 protein expression was also measured in these placentas (Figure 1b), which revealed no change in FGR placentas from normotensive pregnancies, relative to the controls. In placentas affected concurrently with preeclampsia and FGR (median = 195.4 pg/mL, IQR: 157.0–232.8 pg/mL), SPINT2 was significantly elevated ($p$ = 0.0171) compared to the controls (median = 157.6 pg/mL, IQR: 118.4–183.5 pg/mL), with a similar result observed in the placentas from pregnancies that were affected by preeclampsia alone ($p$ = 0.0042, median = 189.9 pg/mL, IQR: 165.6–226.3 pg/mL).

### 2.2. Circulating SPINT2 in FGR and/or Preeclampsia

We next sought to measure SPINT2 within the maternal circulation in prospective cohorts, prior to any diagnoses.

In maternal plasma (collected upon presentation to MHW for caesarean section at term; Figure 2a), SPINT2 was modestly elevated ($p$ = 0.0507) in those women whose infant was born small for the gestational age (SGA, birthweight < 10th centile; $n$ = 75, median = 4020 pg/mL), compared to appropriate for the gestational age (AGA, birth-weight > 10th centile) controls ($n$ = 152, median = 3407 pg/mL). Circulating SPINT2 was significantly elevated ($p$ = 0.002) in the SGA ($n$ = 128, median 2109 pg/mL, IQR 1355–3069 pg/mL) samples that were collected at 36 weeks' gestation (Figure 2b), compared to the AGA controls ($n$ = 182, median = 1614 pg/mL, IQR: 1139–2360 pg/mL). This association was lost, however, earlier in gestation, where 24- to 34-week plasma from women with underlying vascular disease (MAViS clinic samples, Figure 2c, Table S1) demonstrated no difference between the AGA control ($n$ = 179) and SGA ($n$ = 58) levels of SPINT2. In this cohort, SPINT2 did not vary across gestation (Figure 2c) in the controls, but there was a trend towards a modest increase ($p$ = 0.054) in SPINT2 across gestation in those women who were destined to birth an SGA infant ($R^2$ = 0.0645).

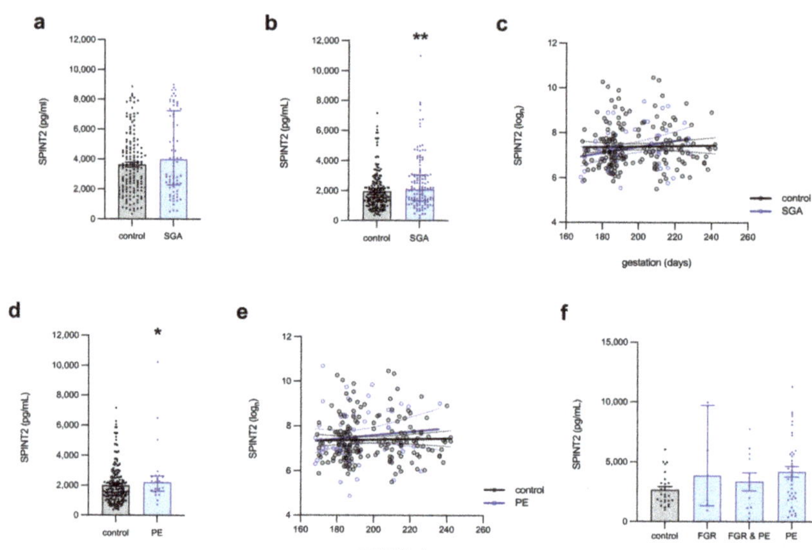

**Figure 2.** Circulating SPINT2 levels preceding preeclampsia development or birth of a small-for-gestational-age infant. In the blood of women on the day of delivery (**a**), SPINT2 protein expression was increased (approaching statistical significance, *p* = 0.051) in those who delivered an SGA infant, compared to AGA controls. This association was stronger at 36 weeks' gestation, in which there was a significant elevation of SPINT2 levels among women who later delivered a small-for-gestational-age (SGA) infant (**b**) as well as in those women who subsequently developed PE (**d**), although the significance of the latter was lost when accounting for outliers. Earlier in the pregnancy, however, at 24–34 weeks, there was no association between circulating SPINT2 and SGA (**c**) nor PE (**e**) cases in samples from women with underlying vascular disease. In this cohort, SPINT2 did not fluctuate across gestation in controls nor PE; however, there was an apparent increase in SPINT2 across gestation in those women destined to birth an SGA infant. In the plasma collected on the day of delivery from women with diagnosed placental insufficiency (**f**), circulating SPINT2 was unchanged in cases, relative to controls. Each data point represents an individual patient sample; data are expressed as median ± IQR; linear regression showing 95% confidence intervals; * *p* < 0.05, ** *p* < 0.01.

At 36 weeks' gestation (Figure 2d), circulating SPINT2 was also elevated (*p* = 0.03) in women who were destined to develop term preeclampsia (*n* = 23, median = 2233 pg/mL, IQR: 1643–2661 pg/mL), relative to the controls (*n* = 182, median 1644 pg/mL, IQR: 1218–2480 pg/mL). In the 24- to 34-week plasma from women attending the MAViS clinic (Figure 2e, Table S2), there was no difference in SPINT2 levels in those who were ultimately diagnosed with preeclampsia, relative to the controls, nor did the protein concentration change, relative to gestation, regardless of the disease status (Figure 2e).

Interestingly, there were no significant differences in circulating SPINT2 in patients delivering preterm for preeclampsia or FGR, relative to the controls (Figure 2f).

### 2.3. Hypoxic Regulation of SPINT2 in Trophoblasts

Placental insufficiency is often associated with intermittent placental hypoxia; thus, we assessed the effect of hypoxia on SPINT2 expression.

In primary trophoblasts that were isolated from term placentas, SPINT2 mRNA transcripts were significantly increased in response to hypoxia (Figure 3a; mean = 193% of control, *p* = 0.002), while secreted SPINT2 was unchanged (Figure 3b). In contrast, hypoxia had no effect on SPINT2 mRNA expression in first-trimester cytotrophoblast stem cells (Figure 3c), but SPINT2 protein secretion (Figure 3d) was significantly increased (mean = 412.8% of control, *p* = 0.008). Given SPINT2 is also likely expressed in syncytiotrophoblast, we measured expression and secretion in syncytialised first-trimester human trophoblast stem cells (hTSCs; Figure 3e), observing that oxygen tension had no effect on mRNA expression, but modestly decreased SPINT2 secretion (Figure 3f; mean = 85.6% of control, $p = 0.008$) in the syncytiotrophoblasts that were exposed to hypoxia.

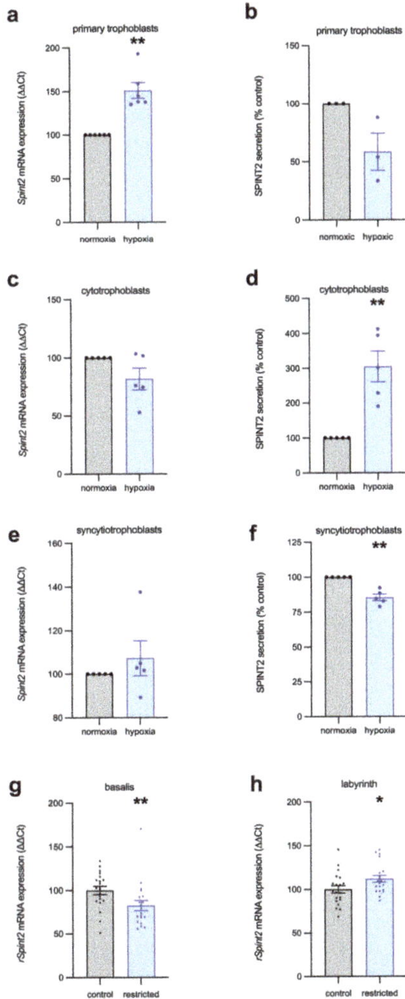

**Figure 3.** The effect of hypoxia on SPINT2 expression in placental cells. SPINT2 mRNA and protein secretion was measured in the following three types of trophoblast cultures: primary trophoblasts isolated from term placentas, and first-trimester cytotrophoblast and syncytiotrophoblast from a stem cell line. In the term primary trophoblasts (**a**), SPINT2 transcripts were significantly increased in response to hypoxia, while secreted protein levels (**b**) were not changed. Hypoxic conditions caused no alteration to the first-trimester cytotrophoblast stem cell SPINT2 mRNA (**c**), but did significantly increase the levels of SPINT2 secretion (**d**) compared to normoxic controls. The syncytialised first-trimester stem cells had no change in mRNA (**e**), although they did demonstrate decreased SPINT2 secretion (**f**). Experiments were repeated $n = 3$–5 times; data are expressed as mean $\pm$ SEM. In placentas from a rat model of uteroplacental insufficiency, changes were identified in rat SPINT2 mRNA (*rSpint2*) expression in both the basalis (**g**) and labyrinth (**h**) zones, being depressed and elevated, respectively. Each data point represents an individual rat placenta; data are expressed as median $\pm$ IQR; * $p < 0.05$, ** $p < 0.01$.

The derangement of SPINT2 expression in response to hypoxia was also demonstrated in placentas from a rat model of uteroplacental insufficiency, induced by ligating the uterine vessels, thereby impeding placental perfusion. There are distinct morphological differences (see Furukawa et al., 2011 [15]) between the rat and human placenta, thus we separated the basalis and labyrinthine layers for analysis of SPINT2. In the basalis region of the restricted placentas (Figure 3g), SPINT2 mRNA expression was significantly depressed (median = 76.9% of control, $p = 0.004$), compared to that of dams who underwent sham surgery. Interestingly, the labyrinth layer—akin to the chorionic villi (including syncytiotrophoblasts, villous cytotrophoblast, stroma and blood vessels) of the human placenta—of restricted placentas had modestly upregulated SPINT2 mRNA expression (median = 108.4% of control, $p = 0.04$).

### 2.4. SPINT2 Is Not Regulated by Inflammation

Preeclampsia is associated with placental and systemic inflammation, and we therefore assessed whether SPINT2 is influenced by pro-inflammatory stimuli. In first-trimester cytotrophoblasts (Figure 4a,c) and syncytialised trophoblasts (Figure 4e,g), we observed no significant effect on SPINT2 mRNA expression. SPINT2 secretion was also unchanged in both the cell types (Figure 4b,d,f), with only low doses of IL-6 stimulating a modest decrease ($p < 0.01$) in syncytiotrophoblasts (Figure 4g).

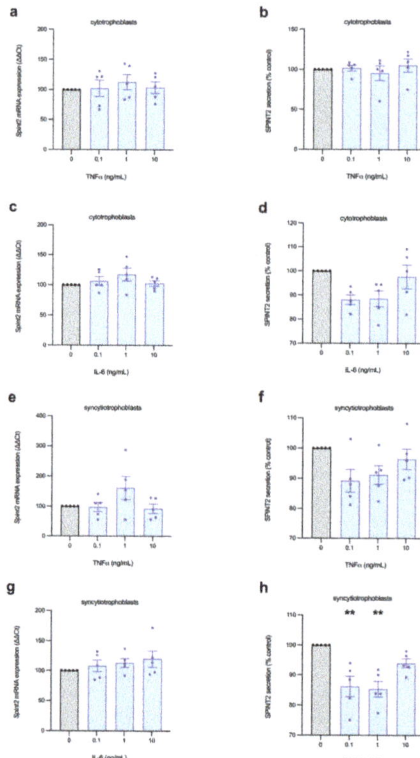

**Figure 4.** Inflammatory cytokine regulation of SPINT2 in placental cells. First-trimester cytotrophoblasts (**a–d**) and syncytiotrophoblasts (**e–h**) were treated with inflammatory cytokines, TNF or IL-6, at various doses. No significant changes were observed in SPINT2 mRNA (**a,c,e,g**) expression, nor in secreted SPINT2 (**b,d,f**); with the exception of IL-6–treated syncytiotrophoblasts (**h**), which stimulated a modest, but significant, decrease in SPINT2 secretion at lower doses. Experiments were repeated $n = 5$ times; data are expressed as mean $\pm$ SEM; ** $p < 0.01$.

## 3. Discussion

Throughout the first trimester of pregnancy, the intricate foetal–maternal interface is established through the process of placentation, which involves tightly regulated and complex pathways, the understanding of which is at present incomplete. The accumulation of anomalies in this process can lead to a dysfunctional placenta, which inadequately supplies the foetus, and can have serious consequences for the mother and baby, including preeclampsia and/or FGR. In this study, we sought to characterise the expression of SPINT2 in the placentas and maternal circulation of pregnancies that were complicated by preeclampsia and/or FGR, using three prospective cohorts, a rodent model, in trophoblasts isolated from human tissue at term, and in stem cells from the first trimester.

By measuring circulating SPINT2 levels in the cases of established disease (delivered at <34 weeks' gestation), it was apparent that SPINT2 expression is not dysregulated in FGR-like functionally related homologue SPINT1 [4]. While derangements in SPINT2 expression were apparent in the cases of placental insufficiency-mediated pregnancy complications, the measured fluctuations do not reliably reflect the disease status, making SPINT2 an overall poor biomarker candidate. Indeed, an association between circulating SPINT2 and placental insufficiency is apparent only at term (and near term, from 36 weeks onwards), with no distinction between the cases and controls in weeks 24–34, nor at preterm delivery (<34 weeks). As such, SPINT2 lacks the robust predictive potential of its relative, SPINT1 [4]. This is perhaps unsurprising, because, despite their similarities, SPINT2 is more ubiquitously expressed than SPINT1, and their encoding genes are located on different chromosomes (15 and 19, respectively). So, although they likely share a common ancestor gene, they have evolved distinctly [16]. Nevertheless, there is likely involvement of SPINT2 in placental function—given the changes in placental SPINT2 expression that have been observed.

The expression of SPINT2 mRNA in the preterm placentas was decreased in concurrent FGR and preeclampsia, but elevated where only preeclampsia is present (i.e., with AGA); whereas, SPINT2 protein expression was increased in all cases of preeclampsia (with and without FGR). It is interesting that SPINT2 mRNA and protein are both significantly elevated in preeclamptic (without FGR) placentas, whereas the placentas plagued by concurrent preeclampsia and FGR had decreased SPINT2 mRNA, but elevated protein. This is an unusual phenomenon, and the reason for this disparity has not been elucidated by the present study. It may suggest that there are post-transcriptional modifications in the preeclamptic/FGR placentas, to reduce SPINT2 protein turnover, as a compensatory mechanism for the low transcript levels; however, further study is needed to confirm this hypothesis.

In the cases of FGR, the circulating levels of SPINT2 were modestly elevated at term and at 36 weeks' gestation; however, at earlier gestations, there were no observed changes in expression. Similarly, there was elevated circulating SPINT2 at 36 weeks in those women who were destined to develop preeclampsia, while earlier gestation levels were unchanged, relative to the controls. Interestingly, SPINT2 was not altered in preterm disease, raising the possibility that derangements in circulating levels only arise in later-onset disease. Alternatively, the lack of statistical significance in the established disease plasma cohort may be attributable, in part, to having relatively few FGR samples (especially plasma; $n = 6$), and the high variability in expression for the samples that were available. A larger sample size would aid in verifying whether there are bona fide alterations in mRNA, and/or protein in the placentas and plasma of FGR-affected pregnancies.

Frequently, the dysfunctional placenta suffers from suboptimal perfusion and high levels of inflammatory factors, and, in previous research, SPINT2 has been shown to be regulated by hypoxia (in breast cancer cells) [17]. Having found SPINT2 levels to be elevated in placental insufficiency, we used in vitro methods to examine the effect of hypoxia and pro-inflammatory cytokines on the trophoblasts of the placenta, early in gestation and at term. SPINT2 expression did appear to be regulated by hypoxia in first-trimester human trophoblast stem cells (hTSCs), term primary trophoblasts, and in a rat model of restricted placental perfusion. Early in gestation (as modelled by hTSCs), SPINT2 secretion was elevated under hypoxia in cytotrophoblast stem cells, while being decreased in syncytiotrophoblasts. However, there were no changes in transcription at this early stage, with SPINT2 mRNA largely unchanged by oxygen tension. In contrast, in primary trophoblasts, isolated from placentas that were delivered at term, there was an increase in SPINT2 mRNA expression, but no change in protein secretion. This indicates that in response to hypoxic conditions early in gestation, trophoblasts alter the secretion of SPINT2; whereas, nearer to term, transcriptional changes dominate the response to hypoxia. The mechanism behind this difference is uncertain and requires further investigation. Importantly, fluctuations in SPINT2 expression, in relation to hypoxia, were also measured in rat placentas with induced uteroplacental insufficiency, with inverse changes in the basalis and labyrinth zones of the placenta. Notably, SPINT2 mRNA was upregulated in the labyrinthine region of restricted placentas, complementing the findings in term trophoblasts exposed to hypoxia.

Mouse models have previously established the importance of SPINT2 in placental development [15], and, consequently, embryonic survival. The findings presented here suggest that SPINT2 has a similar importance in human placentation, owing to its derangement in FGR and preeclamptic pregnancies.

## 4. Materials and Methods

### 4.1. Tissue and Blood Collection at Time of Preterm Delivery from Women with Established Placental Disease (Day of Delivery at <34 Weeks)

To characterise the expression of SPINT2 in the maternal circulation and placenta of FGR- and/or preeclampsia-complicated pregnancies, human specimens were obtained. All studies were approved by the Mercy Health Human Research ethics committee (R11/34).

Placental tissue samples were collected from consenting women, delivering by caesarean section at less than 34 weeks' gestation, with decision for delivery made independently by the treating obstetric team. The samples were classified as FGR ($n$ = 14), PE ($n$ = 60) or both ($n$ = 20), according to preeclampsia guidelines from ACOG 2020 [18] and FGR defined as birthweight <10th centile on local birthweight charts [19]. The control samples ($n$ = 19) were gestation-matched and obtained from women who were delivered preterm due to other complications not associated with placental insufficiency or hypertensive disorders of pregnancy, such as placenta praevia or spontaneous preterm rupture of membranes. Although the control sample comprises pregnancies with complications, gestation-matching is important when investigating proteins highly expressed in the placenta, as the pattern of expression commonly varies with advancing gestation. Samples in both the case and control group were excluded if there were congenital anomalies and/or histopathological evidence of congenital infection. Patient characteristics are detailed in Table S3.

Tissue was collected and processed within 30 min of delivery by caesarean section. Segments of tissue were dissected and washed in PBS, then samples of roughly equal size were immersed in RNAlater TM stabilisation solution (Thermo Fisher Scientific; Waltham, MA, USA) for 48 h, then snap frozen and stored at −80 °C. Subsequently, RNA or protein was extracted from tissue lysates.

Plasma samples were also collected on the day of delivery at less than 34 weeks' gestation from women delivering prematurely with FGR ($n$ = 6), PE ($n$ = 40) or both ($n$ = 11). These were compared to gestation-matched blood specimens collected from control pregnancies delivered at term ($n$ = 26). These samples were aliquoted and stored at −80 °C until future analysis. Patient characteristics of this cohort are detailed in Table S4.

### 4.2. Prospective Case-Cohorts

#### 4.2.1. Day of Delivery at Term—FLAG2

The Fetal Longitudinal Assessment of Growth 2 (FLAG2) study recruited 562 unselected women on the day of elective caesarean section at the Mercy Hospital for Women

(MHW, Melbourne, Australia). Women who were aged over 18 years with a well-dated singleton pregnancy, at $36^{+0}$–$42^{+0}$ weeks' gestation, were eligible to participate. Exclusion criteria included any suspicion of major foetal anomaly or infection; ruptured membranes; labouring women; those who had undergone cervical ripening or steroid administration before the caesarean section; and those who were positive for hepatitis B, C or HIV. A study blood sample was taken at the time of intravenous cannula placement and birthweight centile was determined using the GROW Bulk centile calculator (v8.0.4, 2019). The FLAG2 study was approved by the Mercy Health Research ethics committee (ethics approval number R11/34) and written informed consent was obtained from all participants. The total number of remaining samples used for SPINT2 analysis was 227, comprising 152 controls (appropriate for gestational age, AGA) and 75 cases (SGA). Patient characteristics are shown in Table S5.

### 4.2.2. BUMPS—36 Weeks' Gestation

The Biomarker and Ultrasound Measures for Preventable Stillbirth (BUMPS) study is a large prospective cohort collection at MHW, with samples collected from an unselected population at 28 and 36 weeks' gestation. Women were screened for eligibility and invited to participate at their oral glucose test, universally offered to non-diabetic pregnant women around 28 weeks' gestation to test for gestational diabetes mellitus. Following written informed consent, women aged over 18 years, with a singleton pregnancy and normal mid-trimester foetal morphology examination were eligible to participate. The BUMPS study was approved by the Mercy Health Research ethics committee (ethics approval number 2019-012). For this study, a case-cohort of 364 samples was selected from the first 1000 BUMPS participants, including all cases delivering an infant <10th centile (SGA; $n$ = 198) according to the GROW Bulk centile calculator (v8.0.4, 2019), all cases delivering with preeclampsia (defined according to ACOG guidelines; $n$ = 23) and a cohort of controls ($n$ = 182). Patient characteristics detailed in Table S6.

### 4.2.3. MAViS—24–34 Weeks' Gestation

SPINT2 was also measured in a high-risk cohort of patients at the Manchester Antenatal Vascular Service (the MAViS clinic; Manchester, UK). Women are referred to the clinic in early pregnancy for monitoring across gestation based on hypertensive disease, which predisposes to preeclampsia and/or FGR, allowing for longitudinal sampling between 24- and 34-weeks' gestation. The inclusion criteria for women in the MAViS study were as follows: 1. chronic hypertension (BP $\geq$ 140/90 at $\leq$20 weeks; 2. chronic hypertension requiring antihypertensive treatment from $\leq$ 20 weeks; 3. pre-gestational diabetes with evidence of vascular complications (hypertension, nephropathy); 4. history of ischaemic heart disease; and 5. previous early onset preeclampsia. A case-cohort of 294 participants was recruited between October 2011 and December 2016, with a plasma sample obtained between 24 and 34 weeks, and complete outcome data were included in the current study. These participants were selected from an overall cohort of 518 participants and included 179 control women and 115 who either delivered with preeclampsia, FGR or both. The study was granted ethics approval by the NRES Committee North West (11/NW/0426). Patient characteristics are listed in Table S7.

### *4.3. Placental Samples from a Rat Model of Placental Insufficiency*

In order to assess the in vivo expression of SPINT2, samples were obtained from a previously established rodent model of FGR. The placental deficiency in this model was induced during late gestation (at day 18 of 22), providing an in vivo model of late-onset, placental-derived FGR [20]. Uteroplacental insufficiency was induced by means of bilateral uterine vessel ligation (of both the artery and vein), to restrict the blood and nutrient supply to the foetuses. The control group underwent sham surgery mimicking this procedure, without the ligation of uterine vessels. The details of this protocol can be found in Wlodek et al. (2005) [21].

The placentas were collected, weighed and the labyrinthine layer separated from the basalis layer before being immediately frozen in liquid nitrogen, then stored at −80 °C for later analysis. RNA was extracted from both regions. The development of this model of uteroplacental insufficiency in rats was approved by the La Trobe animal ethics committee (AEC: 12–42), in accordance with the National Health and Medical Research Council's (NHMRC) Australian code for the care and use of animals for scientific purposes.

### 4.4. Human Trophoblast Stem Cells (hTSCs)

To examine SPINT2 in response to hypoxia and inflammatory stressors, first-trimester human trophoblast stem cells (hTSCs) were obtained from the RIKEN BRC through the National BioResource Project of the MEXT/AMED (Japan), as previously detailed in the manuscript from Okae et al., 2018. This cell line was isolated from first-trimester placentas under ethical approval from Tohoku University School of Medicine [22]. The cells were then cultured in specialised media, according to the optimised conditions in Okae et al., 2018 [22]. Given the localisation of SPINT2 to both cytotrophoblast and syncytiotrophoblast in the placenta [23], some cells were propagated as multipotent cytotrophoblasts, while others were directed to differentiate into the syncytiotrophoblast lineage.

### 4.5. Term Primary Cytotrophoblast Isolation

Primary cytotrophoblast cells were isolated from term placentas according to the protocol optimised by Kaitu'u-Lino et al., 2014 [24]. In summary, a segment of placenta was resected, washed, mechanically dissociated, and enzymatically digested, allowing for the collection of the isolated cells in the supernatant [24]. Cells were then cultured in preparation for subsequent analysis.

### 4.6. Simulation of Trophoblast Hypoxia

Hypoxic conditions were simulated for first-trimester hTSCs and term trophoblasts to assess the effect of inadequate oxygen perfusion during placentation and approaching term, respectively. After a 24-h incubation in 8% oxygen at 37 °C, allowing cells to adhere to the basement membrane (iMatrix-511 for hTSCs, fibronectin for primary trophoblasts), cells were incubated in different oxygen concentrations. Given the physiologically relevant oxygen tension in utero is 8%, those cells designated normoxic were incubated at 37 °C in 8% oxygen for 48 h. Hypoxia involved exposure to 1% oxygen for the same duration. The media was collected for subsequent analysis and cells lysed for RNA extraction.

### 4.7. Simulation of Placental Inflammation

Two inflammatory cytokines, tumour necrosis factor alpha (TNFα; Life Technologies, Carlsbad, CA, USA) and interleukin-6 (IL-6; In Vitro Technologies, Noble Park, VIC, Australia), were added to the media of primary trophoblasts and first-trimester hTSCs (cytotrophoblasts and syncytialised stem cells) to simulate the inflammation common to the preeclamptic placenta. These cells were incubated at 37 °C for 24 h after plating, followed by treatment with 0 ng/mL, 0.1 ng/mL, 1 ng/mL or 10 ng/mL of the recombinant cytokine, diluted with fresh media. After being cultured for a further 24 h in the treatment media, cells and media were collected.

### 4.8. Protein Extraction

Protein was isolated from placental tissue and syncytiotrophoblast, using RIPA buffer containing protease inhibitor cocktail (Sigma-Aldrich; St. Louis, MO, USA) and Halt™ phosphatase inhibitor cocktail (Thermo Fisher Scientific; Waltham, MA, USA) to lyse cells and centrifugation to pellet debris. To quantify the protein content of each sample, a Pierce™ BCA assay (Thermo Fisher Scientific) was performed according to the manufacturer's protocol. Equal protein amounts were loaded for ELISA.

*4.9. RNA Extraction*

The Genelute™ mammalian total RNA miniprep kit (Sigma-Aldrich; St Louis, MO, USA) was used to extract RNA from cultured hTSCs, primary term trophoblasts and placental tissue, as per the manufacturer's protocol.

*4.10. Reverse Transcription*

RNA extracted from samples was converted to cDNA using the Applied Biosystems™ high-capacity cDNA reverse transcription kit (Thermo Fisher Scientific; Waltham, MA, USA), following the manufacturer's guidelines. The reaction comprised 150 ng RNA solution (appropriately diluted with DEPC-treated $H_2O$). The iCycler iQ™5 (Bio-Rad, Hercules, CA, USA) protocol was run according to kit specifications, being held at 4 °C after completion until collection and storage at $-20$ °C for later PCR analysis.

*4.11. Real-Time Polymerase Chain Reaction (RT-qPCR)*

Quantitative PCR was carried out to ascertain the mRNA expression of *Spint1* and *Spint2*, relative to reference housekeeper genes. TaqMan gene expression primers (Thermo Fisher Scientific; Waltham, MA, USA) specific to the genes of interest are detailed in Table S8, including the appropriate housekeeping genes for each sample set. All PCRs were performed on the CFX384 (Bio-Rad). The average $C_t$ of sample duplicates were normalised to appropriate reference genes before being calibrated to the average $C_t$ of experimental controls, allowing the results to be expressed as percentage relative to controls.

*4.12. Enzyme Linked Immunosorbent Assays (ELISAs)*

SPINT2 protein levels were measured in maternal plasma samples, hTSC media, and placental lysates via ELISA. The large cohort analyses were analysed using an ELISA kit for SPINT2 (Sigma-Aldrich, St. Louis, MO, USA), following manufacturer's specifications. The <34 week plasma was diluted 1:12 for SPINT2.

Cellular SPINT2 in placental lysates was analysed using a SPINT2 DuoSet® ELISA (R&D Systems; Minneapolis, MN, USA). Then, 5 μg of each sample was loaded, diluted in 1% BSA in PBS according to the concentration determined by the BCA assay. Cultured hTSC media was also analysed using the R&D Systems SPINT2 kit. Media samples were undiluted, with the exception of the hypoxia studies, which were diluted 1:2 with 1% BSA in PBS.

*4.13. Statistical Analysis*

In vitro experiments were carried out in technical triplicate and repeated 3–5 times. The results of in vitro experiments were normalised to controls so data could be expressed as % control. Using GraphPad Prism 8 (GraphPad Software, Inc., San Diego, CA, USA), statistical analyses were carried out, with data first assessed for Gaussian distribution, then analysed using appropriate statistical tests. Maternal characteristics and birth outcome data (Supplementary Tables S3–S7 were compared for all women who were preeclamptic and/or delivered an SGA baby against controls using Mann–Whitney U, unpaired t, Fisher's exact or Chi-square tests. For all other data, when two groups were compared, a Student's *t*-test or Mann–Whitney U test was used according to Gaussian distribution. For more than two groups, a one-way ANOVA or Kruskal–Wallis test was used, according to Gaussian distribution, and post hoc analyses ascertained by Dunn's multiple comparisons test. Outliers were identified and accounted for using a ROUT test.

## 5. Conclusions

Unlike SPINT1, circulating SPINT2 is not consistently dysregulated in diseases of placental insufficiency—in preeclampsia or foetal growth restriction. We have shown that SPINT2 is unlikely to be a clinically useful biomarker; however, we did identify changes in placental SPINT2, which suggest it may be functionally involved in human placentation; this is a role yet to be explored.

**Supplementary Materials:** The following are available online at https://www.mdpi.com/article/10.3390/ijms22147467/s1.

**Author Contributions:** Conceptualisation, T.J.K.-L., S.T., S.P.W., C.N.M.; methodology, C.N.M., T.-V.N., P.C., M.K., N.J.H.; formal Analysis, E.K., C.N.M.; investigation, C.N.M.; resources, T.J.K.-L., N.J.H., S.T., S.P.W., M.E.W., J.F.B., T.R., J.M.; data Curation, N.P., T.M.M., A.R.M., C.-A.W., J.M.; writing—original draft preparation, C.N.M., T.J.K.-L.; writing—review and editing, all co-authors; funding acquisition, T.J.K.-L., S.T., S.P.W. All authors have read and agreed to the published version of the manuscript.

**Funding:** Funding for this work was provided by the National Health and Medical Research Council (#1065854, #1183854, #116071, #2000732), National Health and Medical Research Council Fellowships to TKL (#1159261), NJH (#1146128), ST (#1136418). The funders played no role in study design or analysis.

**Institutional Review Board Statement:** The study was conducted according to the guidelines of the Declaration of Helsinki, and approved by the Mercy Health Research ethics committee (R11/34 and 2019-012), the NRES Committee North West (11/NW/0426), and the La Trobe animal ethics committee (AEC: 12-42).

**Informed Consent Statement:** All patients who participated in this study provided written informed consent.

**Data Availability Statement:** Raw data are available upon reasonable request from the corresponding author.

**Acknowledgments:** We thank Alison Abboud, Valerie Kyritsis, Gabrielle Pell, Rachel Murdoch and Genevieve Christophers for their assistance in recruiting and characterizing participants. We thank Richard Hiscock for his assistance with sample selection for study. We also wish to thank the pathology, health information services, and prenatal clinic staff at the Mercy Hospital for Women for their assistance in conducting this research. First trimester cytotrophoblast stem cell line was obtained from the RIKEN BRC through the National BioResource Project of the MEXT/AMED, Japan.

**Conflicts of Interest:** The authors have no conflicts of interest to declare.

## References

1. Gardosi, J.; Madurasinghe, V.; Williams, M.; Malik, A.; Francis, A. Maternal and fetal risk factors for stillbirth: Population based study. *BMJ* **2013**, *346*, f108. [CrossRef]
2. Flenady, V.; Koopmans, L.; Middleton, P.; Froen, J.F.F.; Smith, G.C.; Gibbons, K.; Coory, M.; Gordon, A.; Ellwood, D.; McIntyre, H.D.; et al. Major risk factors for stillbirth in high-income countries: A systematic review and meta-analysis. *Lancet* **2011**, *377*, 1331–1340. [CrossRef]
3. Chappell, L.C.; A Cluver, C.; Kingdom, J.; Tong, S. Pre-eclampsia. *Lancet* **2021**. [CrossRef]
4. Kaitu'u-Lino, T.J.; MacDonald, T.M.; Cannon, P.; Nguyen, T.-V.; Hiscock, R.J.; Haan, N.; Myers, J.E.; Hastie, R.; Dane, K.M.; Middleton, A.L.; et al. Circulating SPINT1 is a biomarker of pregnancies with poor placental function and fetal growth re-striction. *Nat. Commun.* **2020**, *11*, 2411. [CrossRef]
5. Shimomura, T.; Denda, K.; Kitamura, A.; Kawaguchi, T.; Kito, M.; Kondo, J.; Kagaya, S.; Qin, L.; Takata, H.; Miyazawa, K.; et al. Hepatocyte growth factor activator inhibitor, a novel Kunitz-type serine protease inhibitor. *J. Biol. Chem.* **1997**, *272*, 6370–6376. [CrossRef] [PubMed]
6. Kataoka, H.; Miyata, S.; Uchinokura, S.; Itoh, H. Roles of hepatocyte growth factor (HGF) activator and HGF activator inhibitor in the pericellular activation of HGF/scatter factor. *Cancer Metastasis Rev.* **2003**, *22*, 223–236. [CrossRef]
7. McEwan, M.; Lins, R.J.; Munro, S.K.; Vincent, Z.L.; Ponnampalam, A.P.; Mitchell, M.D. Cytokine regulation during the formation of the fetal–maternal interface: Focus on cell-cell adhesion and remodelling of the extra-cellular matrix. *Cytokine Growth Factor Rev.* **2009**, *20*, 241–249. [CrossRef]
8. Cohen, M.; Meisser, A.; Bischof, P. Metalloproteinases and human placental invasiveness. *Placenta* **2006**, *27*, 783–793. [CrossRef]
9. Burton, G.J.; Jauniaux, E. Pathophysiology of placental-derived fetal growth restriction. *Am. J. Obstet. Gynecol.* **2018**, *218*, S745–S761. [CrossRef] [PubMed]
10. Burton, G.; Woods, A.; Jauniaux, E.; Kingdom, J. Rheological and physiological consequences of conversion of the maternal spiral arteries for uteroplacental blood flow during human pregnancy. *Placenta* **2009**, *30*, 473–482. [CrossRef]
11. Kawaguchi, T.; Qin, L.; Shimomura, T.; Kondo, J.; Matsumoto, K.; Denda, K.; Kitamura, N. Purification and cloning of hepatocyte growth factor activator inhibitor type 2, a Kunitz-type serine protease inhibitor. *J. Biol. Chem.* **1997**, *272*, 27558–27564. [CrossRef]

12. Szabo, R.; Hobson, J.P.; Christoph, K.; Kosa, P.; List, K.; Bugge, T.H. Regulation of cell surface protease matriptase by HAI2 is essential for placental development, neural tube closure and embryonic survival in mice. *Development* **2009**, *136*, 2653–2663. [CrossRef] [PubMed]
13. Kohama, K.; Kawaguchi, M.; Fukushima, T.; Lin, C.-Y.; Kataoka, H. Regulation of pericellular proteolysis by hepatocyte growth factor activator inhibitor type 1 (HAI-1) in trophoblast cells. *Hum. Cell* **2012**, *25*, 100–110. [CrossRef]
14. Szabo, R.; Lantsman, T.; Peters, D.E.; Bugge, T.H. Delineation of proteolytic and non-proteolytic functions of membrane-anchored serine protease Prss8/prostasin. *Development* **2016**, *143*, 2818–2828. [PubMed]
15. Furukawa, S.; Tsuji, N.; Sugiyama, A. Morphology and physiology of rat placenta for toxicological evaluation. *J. Toxicol. Pathol.* **2019**, *32*, 1–17. [CrossRef]
16. Itoh, H.; Yamauchi, M.; Kataoka, H.; Hamasuna, R.; Kitamura, N.; Koono, M. Genomic structure and chromosomal localization of the human hepatocyte growth factor activator inhibitor type 1 and 2 genes. *Eur. J. Biochem.* **2000**, *267*, 3351–3359. [CrossRef] [PubMed]
17. Generali, D.; Fox, S.B.; Berruti, A.; Moore, J.W.; Brizzi, M.P.; Patel, N.; Allevi, G.; Bonardi, S.; Aguggini, S.; Bersiga, A.; et al. Regulation of hepatocyte growth factor activator inhibitor 2 by hypoxia in breast cancer. *Clin. Cancer Res.* **2007**, *13*, 550–558. [CrossRef]
18. American College of Obstetricians and Gynecologists. Gestational Hypertension and Preeclampsia: ACOG Practice Bulletin, Number 222. *Obstet. Gynecol.* **2020**, *135*, e237–e260. [CrossRef]
19. Dobbins, T.A.; Sullivan, E.; Roberts, C.L.; Simpson, J.M. Australian national birthweight percentiles by sex and gestational age, 1998–2007. *Med. J. Aust.* **2012**, *197*, 291–294. [CrossRef] [PubMed]
20. Briffa, J.F.; O'Dowd, R.; Moritz, K.M.; Romano, T.; Jedwab, L.R.; McAinch, A.J.; Hryciw, D.H.; Wlodek, M.E. Uteroplacental insufficiency reduces rat plasma leptin concentrations and alters placental leptin transporters: Ameliorated with enhanced milk intake and nutrition. *J. Physiol.* **2017**, *595*, 3389–3407. [CrossRef] [PubMed]
21. Wlodek, M.E.; Westcott, K.T.; O'Dowd, R.; Serruto, A.; Wassef, L.; Moritz, K.M.; Moseley, J.M. Uteroplacental restriction in the rat impairs fetal growth in association with alterations in placental growth factors including PTHrP. *Am. J. Physiol. Integr. Comp. Physiol.* **2005**, *288*, R1620–R1627. [CrossRef] [PubMed]
22. Okae, H.; Toh, H.; Sato, T.; Hiura, H.; Takahashi, S.; Shirane, K.; Kabayama, Y.; Suyama, M.; Sasaki, H.; Arima, T. Derivation of human trophoblast stem cells. *Cell Stem Cell* **2018**, *22*, 50–63.e6. [CrossRef] [PubMed]
23. Walentin, K.; Hinze, C.; Schmidt-Ott, K.M. The basal chorionic trophoblast cell layer: An emerging coordinator of placenta development. *BioEssays* **2016**, *38*, 254–265. [CrossRef] [PubMed]
24. Kaitu'u-Lino, T.U.J.; Tong, S.; Beard, S.; Hastie, R.; Tuohey, L.; Brownfoot, F.; Onda, K.; Hannan, N.J. Characterization of protocols for primary trophoblast purification, optimized for functional inves-tigation of sFlt-1 and soluble endoglin. *Pregnancy Hypertens. Int. J. Women's Cardiovasc. Health* **2014**, *4*, 287–295.

International Journal of
*Molecular Sciences*

*Article*

# Placental Villous Explant Culture 2.0: Flow Culture Allows Studies Closer to the In Vivo Situation

Nadja Kupper, Elisabeth Pritz, Monika Siwetz, Jacqueline Guettler and Berthold Huppertz *

Division of Cell Biology, Histology and Embryology, Gottfried Schatz Research Center, Medical University of Graz, 8010 Graz, Austria; nadja.kupper@medunigraz.at (N.K.); elisabeth.pritz@medunigraz.at (E.P.); monika.siwetz@medunigraz.at (M.S.); jacqueline.serbin@medunigraz.at (J.G.)
* Correspondence: berthold.huppertz@medunigraz.at

**Abstract:** During pregnancy, freely floating placental villi are adapted to fluid shear stress due to placental perfusion with maternal plasma and blood. In vitro culture of placental villous explants is widely performed under static conditions, hoping the conditions may represent the in utero environment. However, static placental villous explant culture dramatically differs from the in vivo situation. Thus, we established a flow culture system for placental villous explants and compared commonly used static cultured tissue to flow cultured tissue using transmission and scanning electron microscopy, immunohistochemistry, and lactate dehydrogenase (LDH) and human chorionic gonadotropin (hCG) measurements. The data revealed a better structural and biochemical integrity of flow cultured tissue compared to static cultured tissue. Thus, this new flow system can be used to simulate the blood flow from the mother to the placenta and back in the most native-like in vitro system so far and thus can enable novel study designs.

**Keywords:** flow culture system; placenta; explant culture under flow

**Citation:** Kupper, N.; Pritz, E.; Siwetz, M.; Guettler, J.; Huppertz, B. Placental Villous Explant Culture 2.0: Flow Culture Allows Studies Closer to the In Vivo Situation. *Int. J. Mol. Sci.* **2021**, *22*, 7464. https://doi.org/10.3390/ijms22147464

Academic Editors: Hiten D. Mistry and Eun Lee

Received: 25 June 2021
Accepted: 9 July 2021
Published: 12 July 2021

**Publisher's Note:** MDPI stays neutral with regard to jurisdictional claims in published maps and institutional affiliations.

## 1. Introduction

As a fetal organ, the placenta is temporarily present during pregnancy and serves as the lungs, liver, kidney, and gut of the fetus [1]. The chorionic villi that enable exchange between mother and fetus are organized as villous trees and are freely floating in maternal plasma and blood [2]. The freely floating villi also release a bulk of substances including vesicles, hormones, and growth factors that modulate maternal and fetal physiology [1–3].

In vitro analysis of the placenta partly allows examination of its function, regulatory repertoire, and the feto–maternal interface [1,4]. Already in the 1960s, villous explant culture was a prominent approach for transport studies [4,5]. More recently, villous explant cultures were used to analyze placental hormones and factors released into the maternal circulation [6,7]. Although the cultivation of placental explants has been adapted and improved in terms of oxygen concentrations [8,9], the static culture method on the bottom of plastic wells is still the most commonly used approach [4,10]. A variety of static culturing conditions have been developed according to the study design including cultures on the bottom of a well, on a supportive mesh, in a shaking water bath, or freely floating hanging from a styrene block into the medium [4,11–13]. However, all these placental explant culture approaches are static methods with no flow around the villi and thus, all of these approaches differ dramatically from the in vivo situation.

Looking at the in vivo situation, placental villi only survive in a floating environment. Already during the first trimester of pregnancy, a first flow of maternal blood plasma traverses from invaded and plugged spiral arteries [14] through the intervillous space and back into invaded uterine veins [15,16]. After dissolution of the arterial trophoblast plugs, at the beginning of the second trimester, maternal blood enters the placenta through spiral arteries. Perfusion of the intervillous space allows maternal blood to flow around placental villi, allowing the exchange of nutrients and oxygen between mother and fetus [2,17,18].

The invasion of endoarterial trophoblasts into spiral arteries results in the dilation of the very end of the vessels [17]. This conversion is important for the subsequent blood flow into the placenta as it results in a reduced blood flow velocity into the intervillous space from 2–3 m/s inside normal arteries to 0.1 m/s within the intervillous space [2,17,18]. Hence, in vivo placental villi are used to a slight fluid shear stress from the beginning of placental perfusion, with plasma during the first trimester, which is followed by blood flow until delivery. This fits to culture experiments indicating beneficial effects of fluid shear stress on the syncytiotrophoblast [19].

Here, we argue that in vitro culture of villous explants should take place in the most functional and native way possible to get robust results representative of the in utero environment. Therefore, this study aimed at establishing a flow culture system under normal placental oxygen conditions for placental villous explants to simulate the blood flow from the mother to the placenta and back in the most native-like in vitro system so far.

## 2. Results

### 2.1. Establishment of the Flow System

The closed flow system consists of five chambers connected in series with four placental explants per chamber (Figure 1). To prevent the floating of villous explants from the chambers into the tubes, the villi were secured on the bottom of the chamber using small metal plates with needles (Figure 1D,E). To mimic the physiological oxygen concentration in the placenta during the 3rd trimester of pregnancy, $O_2$ saturation within the bioreactor was set to 8% and then verified in the medium. The analysis revealed an $O_2$ saturation of 8% within the medium after 18.9 min of circulating in the flow system at a flow rate of 1 mL/min (experimental settings are summarized in Supplemental Table S1).

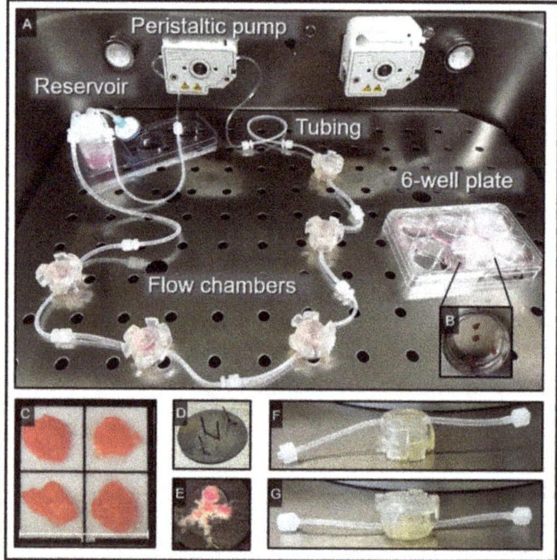


**Figure 1.** Depiction of the experimental setup. (**A**) The inside of the bioreactor is shown where temperature and gases are controlled. On the left side of the bioreactor a complete flow cycle is assembled showing the reservoir, the tubing, and the flow chambers as well as the peristaltic pump. On the right side, a six-well plate is used as a static control. (**B**) The four explants in a static well are shown. (**C**) Placental villous explants with a cross-sectional diameter of about 0.5 cm are used for the flow and the static explant culture. (**D**) To prevent sweeping away of the explants in the flow-cycle, a metal plate with needles is used to fix the placental explants. (**E**) Then, the metal plate with fixed placental villous explants is introduced into the flow chambers. (**F,G**) The flow chambers are used upside-down (**G**) to facilitate the direct exposure of explants to the stream of the medium.

### 2.2. Morphological Analysis

#### 2.2.1. β-Actin

To assess differences in tissue viability and integrity related to the different culture conditions, diverse immunohistochemistry staining protocols were applied. β-actin staining was used to display the actin cytoskeleton of the villous tissues (Figure 2A–E). In fresh tissue, β-actin staining showed the structured appearance of the cytoskeleton in placental villi, especially in villous trophoblast (Figure 2A). A qualitative analysis of β-actin staining revealed a cultivation mode as well as a time-dependent degeneration of tissues. Disintegration of the actin cytoskeleton became obvious by increased accumulation of actin microfilaments upon cultivation time (Figure 2C–E), especially in tissues cultured under static conditions (Figure 2C,E, asterisks). Since a definite answer as to whether the flow or static cultivation contributes to better tissue preservation cannot be stated based on the qualitative assessment of this staining, further immunohistochemical staining, electron microscopic analysis, as well as analysis of biochemical parameters were performed.

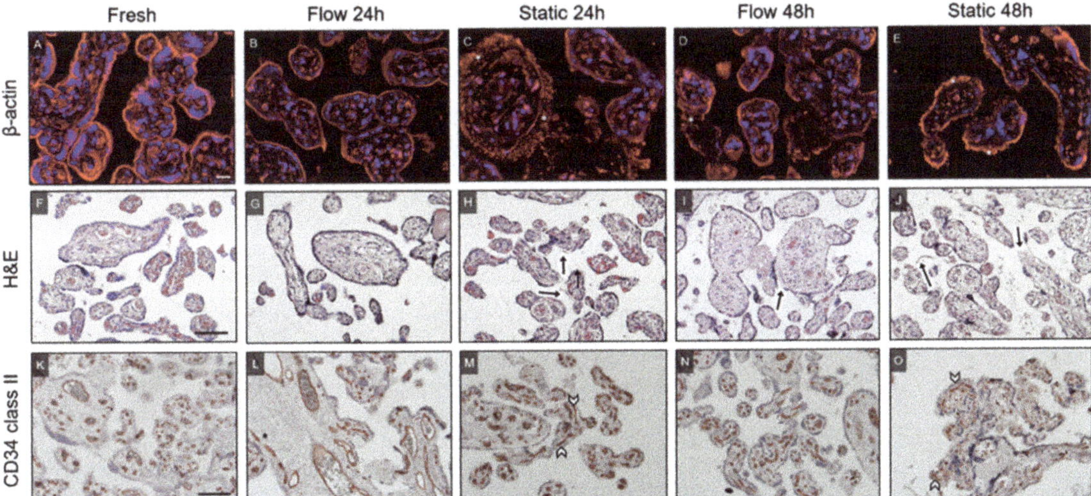


**Figure 2.** Representative immunofluorescence and (immuno-)histochemical staining of placental villous tissues. (**A–E**) The actin cytoskeleton was stained in fresh as well as flow and static cultured tissues using an anti β-actin antibody. (**A**) In fresh tissue, a structured appearance of the cytoskeleton was observed. (**B–E**) Cultivation mode and time-depended degeneration of tissue was observed in explant culture. (**C,E**) Asterisks designate increased accumulation of actin microfilaments indicating disintegration of the actin cytoskeleton in static culture. (**A–E**) Scale bar represents 20 μm. Six random spots were photographed per slide and used for analysis. (**F–J**) H&E staining of placental villi. (**F**) Freshly dissected explants show structured morphology of villi with a dense stroma and noticeable vessels and capillaries. (**F,G**) In fresh and flow cultured tissue for 24 h, the syncytiotrophoblast was attached to the stroma, indicating a healthy morphology of the explants. (**H,J**) Under static conditions, placental villi appeared partly damaged with augment parts of detached syncytiotrophoblast. (**I,J**) The stroma of flow cultured tissue for 48 h (**I**) appeared dense compared to the loose and porous appearance of the stroma in static cultured tissue for 48 h (**J**). Arrows indicate detached syncytiotrophoblast. (**F–J**) Scale bar represents 100 μm. (**K–O**) CD34 class II staining was applied to visualize feto-placental endothelial cells. (**K,L**) Fresh tissue and flow cultured tissue for 24 h show clearly defined and normally arranged endothelial cells. (**M,O**) Disrupted vessels are found in static cultured tissue for 24 h (**M**), and this collapsed appearance increased after 48 h of static culture (**O**). (**N**) After 48 h of flow culture, vessels still show structural integrity. (**M,O**) Arrowheads indicate damaged and collapsed vessels. (**K–O**) Scale bar represents 100 μm.

### 2.2.2. H&E Staining

The morphological differences were visualized, among others, using H&E staining of the villous explants (Figure 2F–J). Freshly dissected explants showed a structured morphology of villi with a dense and compact stroma embedding noticeable vessels and capillaries (Figure 2F). The syncytiotrophoblast was attached to the stroma indicating a healthy morphology of the villi prior to culture. In comparison to fresh tissue, the morphology of explants cultured for 24 h in flow or static conditions appeared partly damaged (Figure 2G,H). On a qualitative level, the morphology of flow cultured tissue seemed to show more preserved parts of villi compared to static cultured tissue. Additionally, more parts of the still attached syncytiotrophoblast seemed to appear in flow cultured tissue after 48 h compared to static cultured tissue after 48 h of cultivation (Figure 2I,J, arrows). Tissue cultured under static conditions for 48 h often showed degenerated parts of villi, which were indicated by a loose and porous appearance of the stroma and collapsed vessels (Figure 2J). In summary, a descriptive analysis of H&E staining revealed that the preserved morphology of the flow cultured tissue represents an intermediate state between freshly dissected tissue and static cultured tissue.

### 2.2.3. CD34 Class II

CD34 class II staining was used to visualize endothelial cells of vessels within villous explants (Figure 2K–O). Freshly dissected explants displayed clearly defined and normally arranged endothelial cells aiding to identify capillaries and larger blood vessels (Figure 2K). In the static cultured explants for 24 h, more damaged and disrupted vessels were found, and this collapsed appearance increased if static cultivation lasted for 48 h (Figure 2M,O, arrowheads). In comparison to the static culture, the appearance of blood vessels within the explants cultured under flow conditions represented a stage close to blood vessels in freshly dissected explants. This was indicated by the presence of still preserved structural integrity of vessel lining after 24 h and 48 h of flow cultivation (Figure 2L,N).

### 2.2.4. Cytokeratin 7

To visualize the villous trophoblast including the syncytiotrophoblast as well as villous cytotrophoblasts, cytokeratin 7 staining was applied (Figure 3A–E). Intact villi and villi with detached or disrupted syncytiotrophoblast were counted and quantified. Only a small percentage (6.8%) of damaged villi was found in freshly dissected villous explants (Figures 3A and 4A), while there was a significant increase of villi with damaged syncytiotrophoblast after 48 h of culture in both flow and static culture conditions (Figure 3B–E and Figure 4A). The highest number of damaged villi was found in static cultured explants for 48 h (43.5%) (Figure 4A).

### 2.2.5. Active Caspase 8

Levels of early apoptosis were analyzed using cleaved caspase 8 staining of placental explants cultured in flow or static conditions for 24 h or 48 h and compared to staining levels of freshly dissected explants (Figure 3F–O). Fresh tissue displayed the lowest amount of cleaved caspase 8 positive cells (0.4%) (Figure 4B). There was a 5.8 times higher amount of active caspase 8 positive cells in 48 h static cultured tissue compared to fresh tissue (Figure 4B). No differences were observed in flow cultured tissue after 24 h or 48 h of culture compared to freshly dissected explants (Figure 4B). Although not statistically significant, there seemed to be a trend toward an increase in active caspase 8-positive cells in static cultured explants between 24 and 48 h (Figure 4B). In summary, these data indicate progressive apoptosis in static cultured tissue upon cultivation time, while there is no cultivation time-dependent increase in flow cultured explants (Figure 4B).

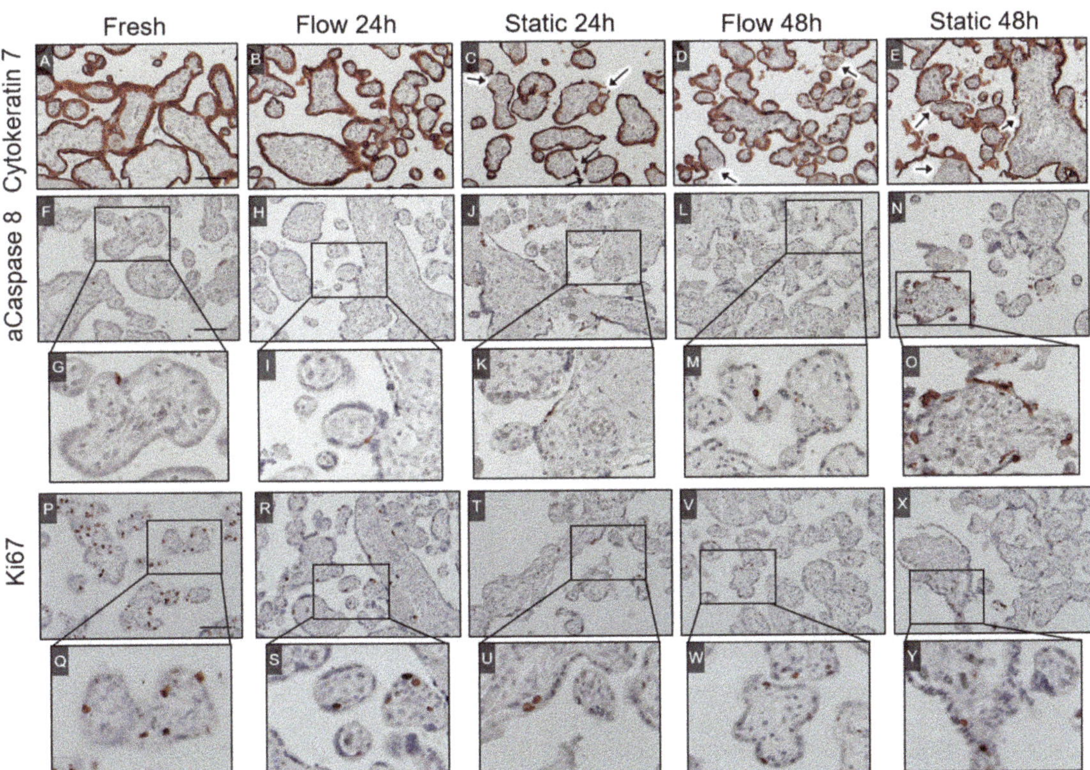

**Figure 3.** Cytokeratin 7, active caspase 8, and Ki67 staining of villous explants. (**A–E**) Cytokeratin 7 staining was used to stain the syncytiotrophoblast as well as villous cytotrophoblasts shown in red. A thick, continuous placental barrier is shown in fresh tissue (**A**) and flow cultured tissue for 24 h (**B**). After 24 h of static culture, the syncytiotrophoblast appeared thinner compared to the fresh tissue (**C**), and detached parts increased after 48 h (**E**). Arrows indicate detached syncytiotrophoblast. (**F–O**) Active caspase 8 staining was applied to stain early apoptotic cells in the villous explants, as indicated in red. (**G–O**) A zoom in for each image (**F–N**) is shown to better display cells with a positive staining for active caspase 8. (**P–Y**) Ki67 staining was applied to stain proliferating cells in the tissues, as indicated in red. (**Q–Y**) A zoom in for each image (**F–N**) is shown to better display cells with a positive staining for Ki67. (**A–E,F–N,P–X**) The scale bars represent 100 µm.

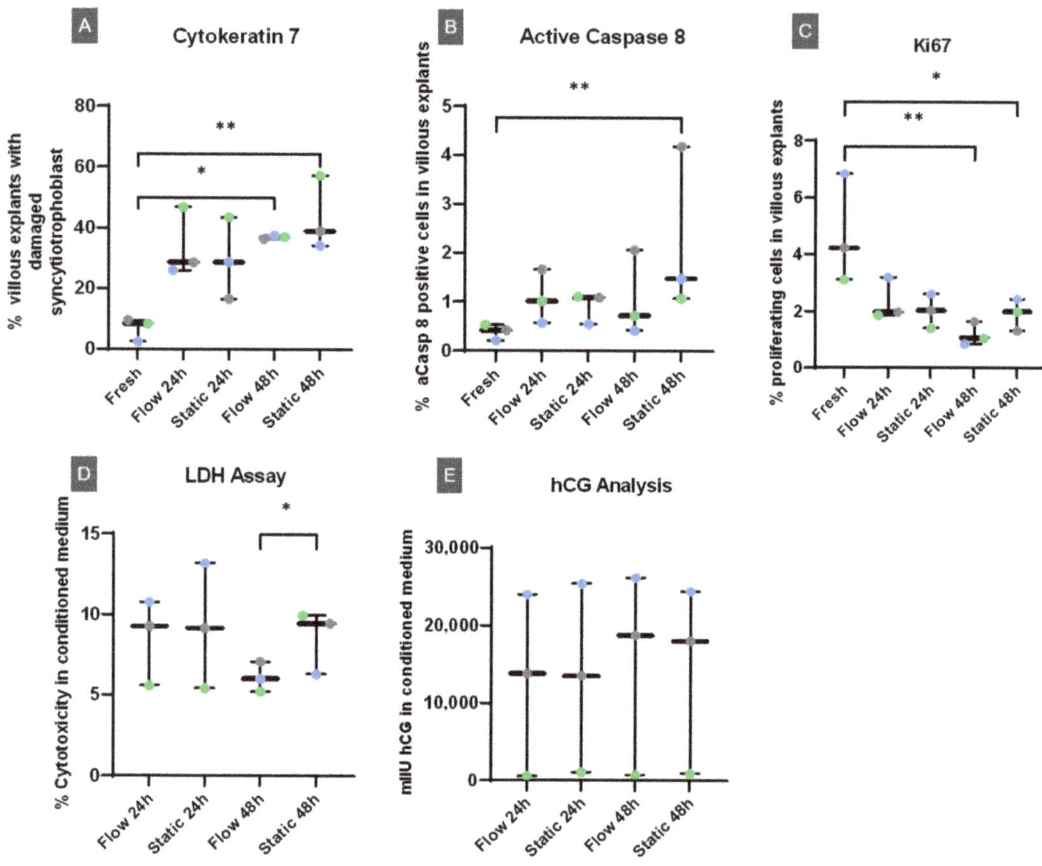


**Figure 4.** Histological and biochemical integrity and degeneration of placental tissue upon cultivation. For the quantitative analysis of each staining, twelve random spots from the explants were photographed and included for analysis. The median and all data points are shown. The same placenta is color-coded in every condition (gray, blue, green). (**A**) Cytokeratin 7 staining was used to stain villous trophoblast and assess the detachment of this layer from the villous core of villi. There was a significant increase in the amount of detached syncytiotrophoblast in cultured tissues after 48 h in both conditions. The quantification of damaged syncytiotrophoblast around villous explants was performed manually by means of counting villi with detached syncytiotrophoblast versus villi with intact syncytiotrophoblast. (**B**) The quantification of active caspase 8 was performed automatically using the software HALO and counting caspase-positive and caspase-negative cells in the whole tissue area. A small percentage of early apoptotic cells was found in fresh tissues. In tissues cultured under flow conditions, no significant increase of early apoptosis was found, while there was a time-dependent increase in early apoptosis in tissues cultured under static conditions. (**C**) Quantification of Ki67 staining was performed automatically using the software HALO and counting Ki67 positive and Ki67 negative nuclei in the whole tissue area. A high amount of proliferating cells was found in fresh tissues. Significantly decreased levels of proliferating cells were found in flow and static cultured tissue after 48 h of culture. (**D**) An LDH assay was performed to measure necrosis in the tissue cultivated under flow or static conditions. Aside from some variations, there was a significant increase in the necrotic release of LDH in static cultured tissues compared to flow cultured tissues after 48 h of culture. Three independent experiments were used for this analysis, and measurements were done in duplicates. (**E**) Endocrine function was assessed by hCG measurements in the conditioned media. There were no statistically significant differences between the samples. The same three independent experiments as for the LDH assay were used for this analysis. * $p < 0.05$, ** $p < 0.01$.

### 2.2.6. Ki67

To stain proliferating cells, an anti-Ki67 antibody was applied on the tissues used in the flow and static experiments as well as for freshly dissected explants (Figure 3P–Y). The highest level of proliferating cells was observed in fresh tissue (Figure 4C). There was a statistically significant decrease of Ki67-positive cells in villous explants between fresh tissue and 48 h cultured tissue in both culturing conditions (Figure 4C). The analysis revealed no significant differences in cell proliferation levels between static and flow cultured tissue for 24 h and 48 h (Figure 4C).

### 2.3. LDH Assay and hCG Measurement

To analyze necrosis in the cultured villous explants, an LDH assay with the conditioned media was performed (Figure 4D). A statistically significant higher cytotoxicity was observed in the media of static cultured explants for 48 h compared to cytotoxicity rates in flow cultured tissue for 48 h (Figure 4D). No further significant differences could be detected. This analysis indicates improved tissue integrity upon longer flow culture compared to static cultured tissue. HCG measurements of the conditioned media were used to analyze the endocrine function of in vitro cultured tissue (Figure 4E). The analysis revealed no statistically significant differences between the samples.

In sum, these data indicate a progressive disintegration of static cultured placental explants with time, while flow cultured tissues occasionally improved or at least maintained their status during the 48 h in vitro culture period.

### 2.4. Ultrastructural Analysis

### 2.4.1. Scanning Electron Microscopy

Scanning electron microscopy was performed to analyze the surface of placental villi in detail (Figure 5A–J). Freshly dissected explants mostly displayed a dense microvillous surface accompanied by occasional singular, small vesicular-like structures (Figure 5A,B). After 24 h of static culture, a clearly reduced number of microvilli was observed, which partially appeared shriveled (Figure 5C,D). Moreover, vesicular-like structures occasionally appeared withered and accumulated. The appearance of the tissue dramatically worsened after 48 h of static culture, as indicated by the increased presence of non-released, accumulated vesicles and the increased appearance of a stunned, disintegrated microvillous surface (Figure 5E,F). The surface of placental explants was also affected after 24 h of flow culture indicated by a slight increase of accumulated vesicular-like structures, mostly in niches between villi. In addition, the number of microvilli on the surface of villi seemed to be slightly reduced (Figure 5G,H). After 48 h of flow culture, the appearance of the tissue did not really change or even worsen compared to the appearance of the explants in the flow culture after 24 h (Figure 5I,J). In flow culture, the villi showed less accumulations of still attached vesicular-like structures, indicating a washing effect of the flowing media similar to the in vivo situation (Figure 5).

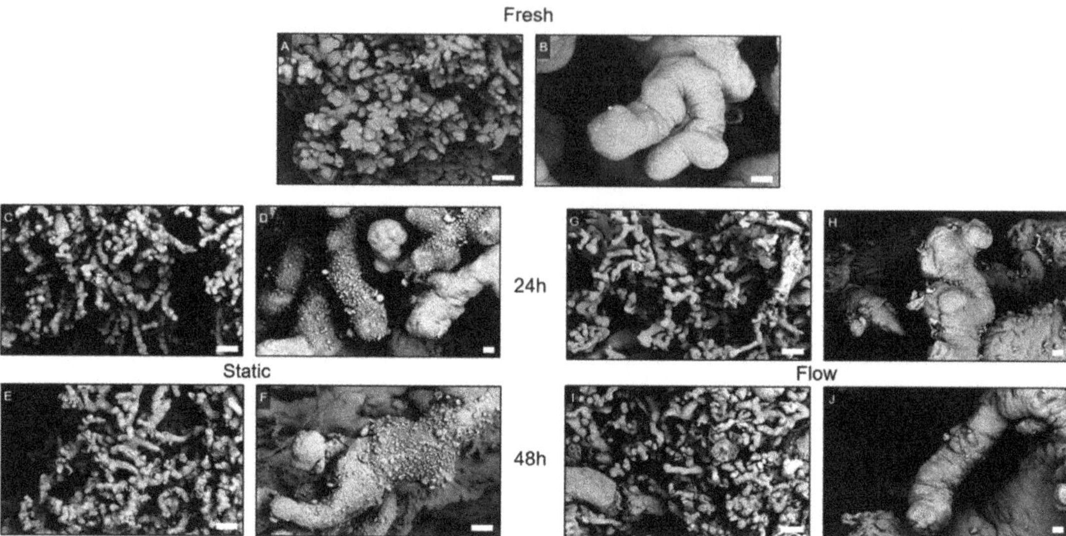



**Figure 5.** Ultrastructural analysis of representative placental tissue using scanning electron microscopy. (**A,B**) Fresh tissue showed a dense microvillous surface accompanied by singular, small vesicular-like structures. (**C,D**) After 24 h of static culture, explants revealed a reduction of microvilli and an increased accumulation of vesicular-like structures. (**E,F**) After 48 h of static culture, the structural integrity of placental explants dramatically worsened with the further increased presence of accumulated vesicles and an elevated appearance of a stunned, disintegrated microvillous surface. (**G,H**) After 24 h of flow culture, the microvillous surface was mostly preserved, while a slightly increased number of vesicular-like structures was observed. (**I,J**) After 48 h of flow culture, microvilli were still observed, and the general structural integrity did not seem to worsen between 24 h and 48 h of flow culture. Scale bar represents 100 μm (**A,C,E,G,I**), 20 μm (**B,F**), or 10 μm (**D,H,J**).

### 2.4.2. Transmission Electron Microscopy

Transmission electron microscopy was performed to get a deeper insight into the morphology of the cultured tissue (Figure 6A–E). Fresh placental explants showed a structured morphology and no evidence of intracellular vacuoles or edema (Figure 6A). The stroma appeared dense, and diverse cells and organelles were identifiable. Capillaries with erythrocytes, endothelial cells and tight junctions between endothelial cells as well as vascular smooth muscle cells were clearly visible (not shown). Villous cytotrophoblasts were noticeable underneath the syncytiotrophoblast, which was easily identifiable by the more electron-dense cytoplasm (darker gray appearance). The syncytiotrophoblast maintained its characteristic appearance of a multinucleated continuous layer with abundant microvilli on the surface and being attached to the basement membrane (Figure 6A).

After 24 h of static culture, there were obvious changes of tissue morphology. The stroma partially appeared loose and disorganized (Figure 6B). Furthermore, there was an increased appearance of lipid droplets within the cells of the explants, which were only rarely seen in fresh tissues. In concordance to H&E and cytokeratin 7 staining, the syncytiotrophoblast tended to detach from the basement membrane. Additionally, cytotrophoblasts appeared loose and vacuolarized, and nuclei appeared condensed. Moreover, the increased congestion of intravascular erythrocytes and disintegration of endothelial cells was observed (Figure 6B). The degeneration of the tissue worsened after 48 h of static culture indicated by the general vacuolated appearance (Figure 6C). Very large vacuoles within stromal cells and the degenerating syncytiotrophoblast were seen. Microvilli on the syncytial surface were either lost or appeared denuded (Figure 6C).

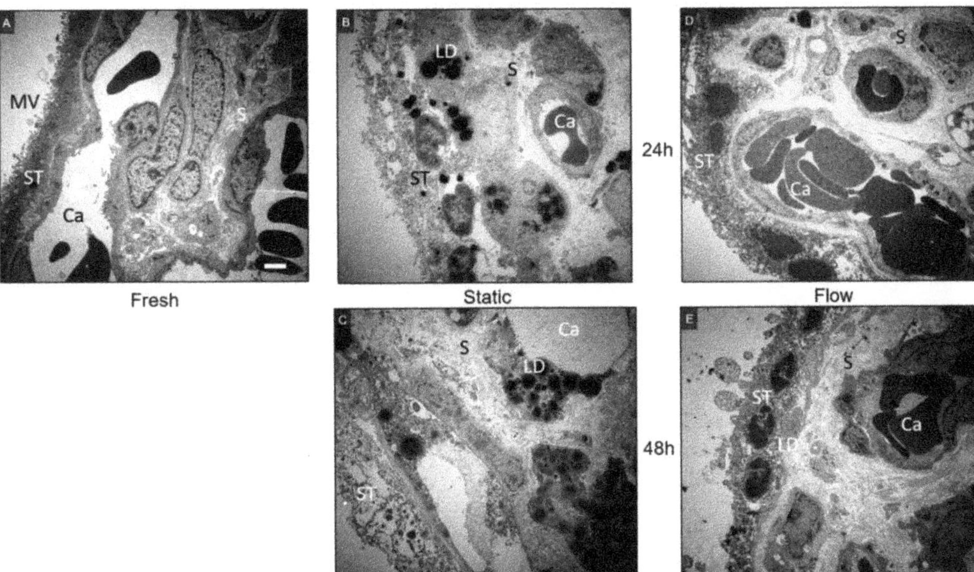

**Figure 6.** Ultrastructural analysis of representative placental tissue using transmission electron microscopy. (**A**) In fresh tissue, capillaries (Ca) with erythrocytes and endothelial cells were clearly visible. The stroma (S) appeared dense and diverse cells were identifiable. The syncytiotrophoblast (ST) was visible by its dark gray and dense appearance. This layer showed its characteristic appearance of a multinucleated continuous layer with abundant microvilli on the surface and being attached to the basement membrane. (**B**) After 24 h of static culture, the syncytiotrophoblast seemed to detach from the basal membrane and nuclei inside appeared condensed, additionally increased appearance of lipid droplets was identified. (**C**) After 48 h of static culture, accumulation of lipid droplets was found in the cells of the explants. The syncytiotrophoblast was detached from the basement membrane in great extend and appeared loose with hardly identifiable nuclei. Endothelial cells showed vast disintegration. (**D**) After 24 h of flow culture, the tissue occasionally showed parts of detached and loose syncytiotrophoblast. Capillaries and erythrocytes were still clearly visible, and the stroma mostly appeared dense. (**E**) After 48 h of flow culture, the stroma still appeared dense with clearly identifiable capillaries and endothelial cells, while a few lipid droplets appeared in stromal cells. The syncytiotrophoblast was mostly attached to the basal membrane and showed some condensed nuclei inside. Scale bar represents 2 μm. MV: Microvilli, ST: Syncytiotrophoblast, Ca: Capillary, S: Stroma, LD: Lipid droplets.

In comparison to the static cultured tissue, less lipid droplets appeared in the explants cultured under flow (Figure 6D,E). Villous explants cultured under flow for 24 h and 48 h occasionally showed parts of detached and loose syncytiotrophoblast. Erythrocytes and endothelial cells within placental vessels still preserved their normal morphology even after 48 h of flow culture (Figure 6D,E).

In summary, our data indicate that the commonly used static placental explant culture results in the disintegration of placental villous morphology. This effect is diminished through the cultivation of placental explants under flow, especially for long-term cultivation (48 h). The ultrastructural integrity of flow cultured tissue was relatively high; thus, the flow system improved tissue integrity by mimicking the in vivo situation of the blood flow from the mother to the placenta and back.

## 3. Discussion

Diverse systems are available to decipher placental function and its impact on pregnancy pathologies [4,11,13,20]. In vitro culture of isolated primary cells, transfected primary cell lines, or choriocarcinoma cell lines represent useful and simple methods [21]; however, they do not reflect the in vivo micro-environment of cells in a tissue [11]. A prominent and well-established method to look at tissues in vitro is represented by placental dual lobe perfusion [22], which enables the analysis of diverse experimental hypotheses. Although various advantages of this method are known, it is not suitable for experiments with first trimester placentas, and a complete intact organ, including intact blood vessels, is required for the perfusion of the fetal and the maternal side [11,23]. It can also only be used in short-term experiments lasting a couple of hours.

Another commonly used approach is represented by the static in vitro culture of placental explants [4,12]. Different to the in vivo situation where blood is flowing around placental villi, this static method of culturing placental explants finds the explants mostly lying in medium on the bottom of the wells and thus differs dramatically from the in vivo situation. This fact has already been discussed in various studies [11,12,24]. It is also reflected in our data shown here. Watson et al. have already stated the difficulties of placental in vitro organ culture associated with the sensitivity of the syncytiotrophoblast [24], since it represents a terminally differentiated and highly sensitive epithelium [24,25]. Therefore, we established a gentler and still simple method to cultivate placental explants under flow in contrast to the commonly used static approach.

The in vitro flow system for placental explants simulates the perfusion within the intervillous space, thus mimicking an in-utero-like environment. Explants were cultured under flow (1 mL/min, 8% $O_2$, 5% $CO_2$) for 24 h and 48 h in a closed humidified environment. The upside-down position of the chambers facilitated the direct exposure of explants to the flow of medium. In the normal upside-up position of the chambers, flow only passes on top of the explants without direct exposure of the explants to the flow of the medium.

The flow rate of 1 mL/min was used based on literature research [17,18,26–28]. So far, the in vitro culture of trophoblast under flow was performed using single cells including primary term trophoblasts, choriocarcinoma cells, as well as trophoblastic stem cells [27]. The flow rates used in these systems ranged from 2 µL/min to 5 mL/min. Hence, the flow rate used here for villous explants (1 mL/min) was chosen to be in the middle of what has been used so far. In general, maternal blood flow through the intervillous space of the placenta starts with a relatively high velocity (0.1 m/s) [17], is hypothesized to be reduced while passing the villous trees and may then speed up again when it is drained back into uterine veins. Thus, it is debatable whether it is possible to calculate the actual flow rate of blood in the intervillous space for a given placental villus, since this flow rate depends on the position of the individual villus in the intervillous space and its distance to the spiral arteries and the uterine veins.

In our study, the impact of the flow system on placental tissue viability and structural integrity was analyzed using unbiased morphological and biochemical parameters. Previous studies reported superfused placental explants at a moist cellulose filter inserted in a perfusion chamber [29–31]; however, they revealed restricted tissue viability [20]. To our knowledge, this study is the first proving the benefits of culturing tissues under flow in terms of biochemical and structural viability. We showed a well-preserved morphology of flow cultured tissue compared to static cultured tissue up to 48 h of in vitro culture. All data showed a higher tendency of tissue disintegration of static cultured tissue compared to flow cultured tissue.

The endocrine function of the explants was assessed by measuring hCG levels in conditioned media from flow and static explant cultures. In healthy pregnancies, hCG levels exponentially rise during the first weeks of gestation and peak at around 10 weeks, which is followed by a slow decline until the end of pregnancy [7,32]. It seems to play a role in placental growth by triggering the fusion and differentiation of the cytotrophoblast with the syncytiotrophoblast [33]. Notably, in our study, there was no statistically significant time-dependent and condition-dependent effect on hCG secretion. The study of Siman et al. showed that when using static conditions, hCG release from cultured explants is increased from the second day onwards [12]. The authors attributed this effect on the regeneration of the syncytiotrophoblast over time; however, they could not exclude increased damage and thus necrotic release of hCG into the media [12].

Electron microscopical imaging data provided further detailed insight into the morphology of the tissue and revealed a cultivation-time dependent degeneration, especially under static culture conditions. Although it should be noticed that transmission and scanning electron microscopical data are primarily descriptive, a progressive degeneration of the syncytiotrophoblast upon static in vitro culture of placental tissue was reported before, and it was observed in first trimester [11,34] as well as third trimester tissues [11,12]. In concordance to our data, these studies showed intracellular vacuoles in the syncytiotrophoblast after 8 h [11] and 1 day [12] in static cultured placental tissue. By contrast, in first trimester explants, Palmer et al. showed a newly formed trophoblast layer already after 48 h of static explant culture [34]. For explants from term placenta, Siman et al. indicated syncytiotrophoblast degeneration over time upon static in vitro culture followed by regeneration beginning at 7 days of in vitro culture [12].

Watson et al. stated that the general deterioration of the microvilli on the surface of static cultured tissue could be symptomatic for the degeneration of the apical membrane of the syncytiotrophoblast [24]. Due to the observed chromatin condensation and membrane blebbing, which are major characteristic features of the apoptosis process [35], the authors suggested that syncytiotrophoblast degeneration in static cultured tissue may be driven by apoptosis [12]. This notion is in line with our ultrastructural analysis of the static cultured tissue as well as confirmed with the quantification of the immunohistochemical staining for active caspase 8, which showed increased apoptosis in static cultured tissue at 48 h.

In addition, the increased appearance of lipid droplets observed after 48 h in static cultured tissue was observed in the study of Palmer et al. [34]. This further indicates increased apoptosis in static cultured tissue, since the induction of apoptosis is associated with an accumulation of cytoplasmic lipid droplets [36]. This observation could represent a trophoblastic defense mechanism protecting the placenta against lipotoxicity [37–39]. The study of Bildirici et al. showed increased lipid accumulation in placental villi from pregnancies complicated by fetal growth restriction (FGR) as well as in primary trophoblasts treated with hypoxia (<1% $O_2$), indicating hypoxia-induced diminished fatty acid oxidation. The authors supposed a link between increased lipid droplet storage in FGR and placental insufficiency [37]. This notion can be supported by our findings, showing placental tissue damage and increased lipid droplet accumulation after 48 h of static culture. In terms of flow cultured tissue, also a slight degeneration of the syncytiotrophoblast was observed after 24 h; however, the degeneration seemed to attenuate or at least remain at the same level after 48 h of flow culture.

Two new and interesting aspects of the flow cultures need to be mentioned. One is that the endothelial cells of the placental vessels within the explants only remain intact in explants cultured under flow. Although in all explants, there is no flow within the placental vessels, only those with flow on the outside of the villi show cellular integrity over the 48 h culture period, while those cultured under static conditions disintegrate quite soon. The second interesting aspect can be found on the surface of the cultured villi. Under static conditions, there is an accumulation of vesicular structures protruding from the apical membrane of the syncytiotrophoblast. By contrast, there are less such vesicular structures on the villous surface of those explants cultured under flow conditions. Such vesicles remained only in niches where very little flow is expected. Hence, it seems as if these vesicles are detached from the villous surface by means of shear stress. If this is true, then the flow culture would be an ideal culture method to mimic the in vivo situation regarding the release of vesicles from the syncytiotrophoblast into the maternal circulation.

Consequently, the analysis of the feto–maternal interface using placental explants cultured under flow conditions would enable deciphering the etiologies of different pregnancy pathologies, including preeclampsia. A growing body of literature has already elaborated the complexity of many serious and common pregnancy-related diseases such as preeclampsia, fetal growth restriction, and gestational diabetes mellitus. However, insights into the pathophysiology and diagnosis of these syndromes are still missing, which often leads to premature births with all its consequences. In addition, also affected women may have long-term health problems such as risk disposition for cardiovascular diseases [40–43]. Hence, finding specific diagnostic and curative treatments for the mother and fetus is a demanding task, requiring systemic studies of the placenta and the feto–maternal interface using innovative study designs. One of these new study designs could be provided with this flow system. Indeed, there is more to come in terms of automatic explant sampling as well as approaches for upscaling the quantity of explants per experiment. Nevertheless, to our knowledge, we are the first to cultivate placental explants under constant flow conditions. Furthermore, this approach may be used for first trimester placental explant culture as well and also to simulate diverse conditions of pregnancy by changing the variable conditions. Additionally, introducing endothelial cells directly behind the villous explants in the flow system will enable mimicking the blood flow from the placenta to the mother in the most native-like in vitro system so far.

## 4. Materials and Methods

### 4.1. Human Placental Samples

This study was approved by the ethics committee of the Medical University of Graz (31-019 ex 18/19 version 1.2). Placental tissue from 3rd trimester deliveries between weeks 34 and 40 of uncomplicated pregnancies was used for the study ($n = 3$) with written informed consent from women undergoing C-sections.

### 4.2. General Culture of Villous Explants

Immediately after delivery, samples from three areas around the central region of the placenta were dissected with a size of 2 cm$^3$. Chorionic plate, maternal decidua, and areas of visible infarcts were discarded. The remaining villous tissue was further dissected into villous explants with a wet weight of approximately 7.5 mg (about 0.5 cm cross-sectional diameter) and used for explant cultures.

Villous explants were washed with PBS (ThermoFisher Scientific, Waltham, MA, USA) and transferred into pre-warmed medium (PromoCell PC-C-22120, Heidelberg, Germany; without EGCS/h and FCS) and supplemented with 5% exosome-depleted fetal bovine serum (Gibco by Life Technologies, ThermoFisher Scientific, Waltham, MA, USA) and 1% penicillin/streptomycin (Gibco by Life Technologies, ThermoFisher Scientific, Waltham, MA, USA). The tissue was cultured at 37 °C for 24 h or 48 h in a humidified atmosphere containing 8% $O_2$ and 5% $CO_2$ under static or flow conditions using a flow bioreactor (TEB500, EBERS Medical Technology SL, Zaragoza, Spain). $O_2$ saturation in the circulating medium was verified with an external $O_2$ measurement device (PreSens, Fibox 3, Regensburg, Germany).

Flow and Static Culture of Villous Explants

The flow chambers with a dimension of 23 mm height × 37 mm diameter and 15 mm internal chamber width (Kirkstall Ltd., Quasi Vivo®, North Yorkshire, UK; Supplemental Table S2) were filled with 2 mL of PromoCell medium. The chambers were connected with tubes having a diameter of 1/16″ ID for the inlet and a diameter of 3/32″ ID for the outlet tubes. Since the inlet and outlet tubes from the chamber are positioned on the upper side of the chambers, the chambers were turned upside-down to facilitate the exposure of explants to the direct stream of medium. Four explants were transferred into each chamber and secured inside the chamber with stainless steel pins. One flow cycle consisted of five chambers connected in series with approximately 30 mg of villous tissue per chamber and a total wet weight of all explants of about 150 mg. Using the peristaltic pump system integrated into the TEB500 system and an additional pumping tube of 1.02 mm diameter (Tygon®, Bartelt, Graz, Austria), villous explants were perfused with a flow rate of 1 mL/min. The flow system was filled with a total volume of 25 mL. The specifications of the flow system are summarized in Supplemental Table S2. Depiction of the flow system is represented in Figure 1.

For the static culture, placental explants from the same placentas as used for the flow cultures were cultured in 6-well plates (NUNC, ThermoFisher Scientific, Waltham, MA, USA) filled with 4 mL of PromoCell medium and 4 villous explants per well (30 mg villous tissue per well, 150 mg villous tissue per well plate). The static explants were placed in the TEB500 flow bioreactor and cultured in the same humidified atmosphere as the flow culture explants.

### 4.3. Histology and Immunohistochemistry

Explants were fixed in formalin (4%) for up to 48 h followed by paraffin embedding using an Excelsior AS Tissue Processor (ThermoFisher Scientific, Waltham, MA, USA). Five μm sections from formalin-fixed paraffin-embedded tissues (FFPE) (Microtome Microm HM 355 S, ThermoFisher Scientific, Waltham, MA, USA) were mounted on Superfrost Plus slides (Menzel-Glaeser, Braunschweig, Germany). The sections were deparaffinized using Histolab Clear® (Histolab®, Askim, Sweden) solution and rehydrated through a graded series of ethanol. For each FFPE sample, a hematoxylin–eosin staining was performed. Antigen retrieval was performed in a microwave oven (40 min, 150 W, Miele, Guetersloh, Germany) using preheated 10 mM Tris EDTA buffer (pH 9) or 10 mM citrate buffer (pH 6).

Immunohistochemistry (IHC) was performed utilizing the UltraVision LP-Detection System HRP-Polymer (ThermoFisher Scientific, Waltham, MA, USA) according to the manufacturer's instructions. Briefly, endogenous peroxidase was blocked for 10 min with Hydrogen Peroxide Block (ThermoFisher Scientific, Waltham, MA, USA). After three washing steps (Tris-buffered saline with 0.05% Tween, TBST) slides were incubated with UltraVision Protein Block for 7 min (ThermoFisher Scientific, Waltham, MA, USA) and then incubated with the primary antibodies diluted in antibody diluent (Dako, Santa Clara, CA, USA) for 45 min: cytokeratin 7 (1:1000, OVTL 12/30, Invitrogen, Waltham, MA, USA), CD34 Class II (1:500, QBEnd-10, Dako, Santa Clara, CA, USA), cleaved caspase 8 (1:100, clone 18C8, Cell signaling, Danvers, MA, USA), or Ki67 (1:50, clone MIB-1, Dako, Santa Clara, CA, USA). Primary antibody enhancer was applied for 10 min, which was followed by incubation for 15 min with Large Volume HRP Polymer. Then, the slides were incubated with the substrate amino-ethyl carbazole (AEC substrate kit, Abcam, Cambridge, UK) for 10 min. Nuclei were counterstained with Mayer's Haemalaun for 10 min. After the staining procedure, all slides were mounted with Kaiser's Glycerin Gelatine (Merck, Darmstadt, Germany) and analyzed with a Leica DM 6000 B microscope (Wetzlar, Germany) equipped with an Olympus DP 72 Camera.

If not stated otherwise, twelve random spots per slide were photographed with 200× magnification by manual rotation of the stage using a joystick and then used for analysis. Semi-quantitative analysis was performed using the HALO software (v3.1.1076.342, indica labs, Albuquerque, NM, USA). For cleaved caspase 8 and Ki67, data are given as percentage of positive cells per total number of cells in the analyzed tissue area. A pipeline of the quantification is depicted in Supplemental Figure S1. Cytokeratin 7 was used to quantify the detachment of villous trophoblast from the villous stromal core. This was achieved by counting villi with detached or disrupted syncytiotrophoblast per total villous count.

For immunofluorescence staining, the slides were washed with PBS and incubated with UV Block (Thermo Fisher Scientific, Waltham, MA, USA). The primary antibody (anti-β-actin, 1:10,000, AC-15, Abcam, Cambridge, UK) was diluted in antibody diluent (Dako, Santa Clara, CA, USA) and incubated on the slide for 30 min. After washing steps, slides were incubated with the secondary antibody for 30 min (Alexa Fluor 555 goat-anti-mouse, 1:200, ThermoFisher Scientific, Waltham, MA, USA), while nuclei were stained with DAPI (1:1000, ThermoFisher Scientific, Waltham, MA, USA) for 5 min. Slides were dried at room temperature in the dark and mounted with ProLong™ Gold Antifade Reagent (ThermoFisher Scientific, Waltham, MA, USA). Pictures were taken with an Olympus microscope (BX3-CBH) (Hamburg, Germany) at 400× magnification. A summary of the used antibodies is shown in Supplemental Table S3.

### 4.4. Ultrastructural Analysis

The fixation and preparation of villous explants for electron microscopy was performed according to standard electron microscopy protocols. In brief, villous explants were fixed in 2% paraformaldehyde/2.5% glutardialdehyde in 0.1 M cacodylate buffer (pH 7.4) for 2 h and then transferred into 0.1 M cacodylate buffer. After post fixation in 2% osmium tetroxide in 0.1 M cacodylate buffer, samples were dehydrated in a graded series of ethanol.

#### 4.4.1. Transmission Electron Microscopy (TEM)

Samples for transmission electron microscopy were transferred into propylene oxide as an intermedium and subsequently embedded in TAAB epoxy resin (Agar Scientific, Stansted, Essex, UK). Ultrathin sections (70 nm) were cut using a Leica UC7 ultramicrotome (Leica Microsystems, Vienna, Austria) and then stained with platinum blue and lead citrate. Images were taken with 80 kV acceleration voltage using a Zeiss EM 900 transmission electron microscope (Zeiss, Oberkochen, Germany).

#### 4.4.2. Scanning Electron Microscopy (SEM)

After dehydration in a graded series of ethanol, samples for scanning electron microscopy were critically point dried (CPD 030; Bal-Tec, Balzers, Liechtenstein) and sputter coated with gold palladium (SCD 500; Bal-Tec, Balzers, Liechtenstein). Images were taken using a Zeiss Sigma 500 field emission scanning electron microscope (Zeiss, Cambridge, UK) with a back-scattered electron detector at 5 kV acceleration voltage.

### 4.5. LDH Assay and hCG Measurement

For LDH and hCG measurements, the conditioned culture media from placental villous explants were used. Prior to performing the LDH assay (LDH Cytotoxicity Detection Kit, Takara, Japan) and the hCG measurement, the medium was centrifuged at $1500 \times g$ for 10 min, and the supernatants were stored at $-80\,^{\circ}$C. The LDH assay was performed according to the manufacturer's instructions in 96-well plates (NUNC, ThermoFisher Scientific, Waltham, MA, USA). For measuring the absorbance of the samples at 492 nm with a reference wavelength at 620 nm, a Spark TM 10 M multimode microplate reader (TECAN, Maennedorf, Switzerland) was used. All samples were measured in duplicates. HCG was measured in routine immunoassay analyses at the department of Obstetrics and Gynecology at the Medical University of Graz (Dimension Xpand; Dade Behring Inc., Deerfield, IL, USA). Obtained values were corrected to the primarily used amount of medium in the experiments.

### 4.6. Statistical Analysis

Data were analyzed and visualized using GraphPad Prism 9.0.0. (San Diego, CA, USA). Unless stated otherwise, experiments were performed in triplicate. Statistical differences were calculated by ordinary one-way ANOVA using a nonparametric test (Friedman Test) without multiple comparison. A $p$-value of less than 0.05 was considered statistically significant.

**Supplementary Materials:** The following are available online at https://www.mdpi.com/article/10.3390/ijms22147464/s1.

**Author Contributions:** Conceptualization, B.H., N.K.; methodology, N.K., E.P., M.S., J.G.; software, N.K.; validation, B.H., N.K.; formal analysis, B.H., N.K.; investigation, N.K.; resources, B.H.; data curation, B.H., N.K.; writing—original draft preparation, N.K., B.H.; writing—review and editing, N.K., B.H., E.P., M.S., J.G.; visualization, N.K.; supervision, B.H.; project administration, B.H.; funding acquisition, B.H. All authors have read and agreed to the published version of the manuscript.

**Funding:** This research was funded by the Austrian Science Fund FWF (DOC 31-B26) and the Medical University of Graz, Austria, through the PhD Program Inflammatory Disorders in Pregnancy (DP-iDP).

**Institutional Review Board Statement:** This study was approved by the ethics committee of the Medical University of Graz (31-019 ex 18/19 version 1.2). Placental tissue from 3rd trimester deliveries between weeks 34 and 40 of uncomplicated pregnancies was used for the study.

**Informed Consent Statement:** Informed consent was obtained from all subjects involved in the study.

**Data Availability Statement:** The data that support the findings of this study are available from the corresponding author upon reasonable request.

**Acknowledgments:** The authors gratefully appreciate the excellent technical assistance of Sabine Maninger and the excellent support of Bettina Amtmann and Petra Winkler for tissue sampling. Open Access Funding by the Austrian Science Fund (FWF).

**Conflicts of Interest:** The authors declare no conflict of interest. The funders had no role in the design of the study; in the collection, analyses, or interpretation of data; in the writing of the manuscript, or in the decision to publish the results.

# References


1. Burton, G.J.; Jauniaux, E. What is the placenta? *Am. J. Obstet. Gynecol.* **2015**, *213*, S6.e1–S6.e4. [CrossRef]
2. Gude, N.M.; Roberts, C.T.; Kalionis, B.; King, R.G. Growth and function of the normal human placenta. *Thromb. Res.* **2004**, *114*, 397–407. [CrossRef]
3. Kupper, N.; Huppertz, B. The endogenous exposome of the pregnant mother: Placental extracellular vesicles and their effect on the maternal system. *Mol. Asp. Med.* **2021**, 100955. [CrossRef] [PubMed]
4. Miller, R.K.; Genbacev, O.; Turner, M.A.; Aplin, J.D.; Caniggia, I.; Huppertz, B. Human placental explants in culture: Approaches and assessments. *Placenta* **2005**, *26*, 439–448. [CrossRef]
5. Villee, C.A. The metabolism of the human placenta in vitro. *J. Biol. Chem.* **1953**, *205*, 113–123. [CrossRef]
6. Siwetz, M.; Blaschitz, A.; El-Heliebi, A.; Hiden, U.; Desoye, G.; Huppertz, B.; Gauster, M. TNF-α alters the inflammatory secretion profile of human first trimester placenta. *Lab. Investig.* **2016**, *96*, 428–438. [CrossRef] [PubMed]
7. Forstner, D.; Maninger, S.; Nonn, O.; Guettler, J.; Moser, G.; Leitinger, G.; Pritz, E.; Strunk, D.; Schallmoser, K.; Marsche, G.; et al. Platelet-derived factors impair placental chorionic gonadotropin beta-subunit synthesis. *J. Mol. Med.* **2020**, *98*, 193–207. [CrossRef] [PubMed]
8. Brew, O.; Sullivan, M.H.F. Oxygen and tissue culture affect placental gene expression. *Placenta* **2017**, *55*, 13–20. [CrossRef]
9. Reti, N.G.; Lappas, M.; Huppertz, B.; Riley, C.; Wlodek, M.E.; Henschke, P.; Permezel, M.; Rice, G.E. Effect of high oxygen on placental function in short-term explant cultures. *Cell Tissue Res.* **2007**, *328*, 607–616. [CrossRef]
10. Tong, M.; Chamley, L.W. Isolation and characterization of extracellular vesicles from ex vivo cultured human placental explants. *Methods Mol. Biol.* **2018**, *1710*, 117–129. [CrossRef]
11. Sooranna, S.R.; Oteng-Ntim, E.; Meah, R.; Ryder, T.A.; Bajoria, R. Characterization of human placental explants: Morphological, biochemical and physiological studies using first and third trimester placenta. *Hum. Reprod.* **1999**, *14*, 536–541. [CrossRef]
12. Simán, C.M.; Sibley, C.P.; Jones, C.J.P.; Turner, M.A.; Greenwood, S.L. The functional regeneration of syncytiotrophoblast in cultured explants of term placent. *Am. J. Physiol. Regul. Integr. Comp. Physiol.* **2001**, *280*, 1116–1122. [CrossRef]
13. Toro, A.R.; Maymó, J.L.; Ibarbalz, F.M.; Pérez, A.P.; Maskin, B.; Faletti, A.G.; Margalet, V.S.; Varone, C.L. Leptin Is an Anti-Apoptotic Effector in Placental Cells Involving p53 Downregulation. *PLoS ONE* **2014**, *9*, e99187. [CrossRef]
14. Weiss, G.; Sundl, M.; Glasner, A.; Huppertz, B.; Moser, G. The trophoblast plug during early pregnancy: A deeper insight, Histochem. *Cell Biol.* **2016**, *146*, 749–756. [CrossRef]
15. Moser, G.; Gauster, M.; Orendi, K.; Glasner, A.; Theuerkauf, R.; Huppertz, B. Endoglandular trophoblast, an alternative route of trophoblast invasion? Analysis with novel confrontation co-culture models. *Hum. Reprod.* **2010**, *25*, 1127–1136. [CrossRef] [PubMed]
16. Huppertz, B.; Weiss, G.; Moser, G. Trophoblast invasion and oxygenation of the placenta: Measurements versus presumptions. *J. Reprod. Immunol.* **2014**, *101–102*, 74–79. [CrossRef] [PubMed]
17. Burton, G.J.; Woods, A.W.; Jauniaux, E.; Kingdom, J.C.P. Rheological and Physiological Consequences of Conversion of the Maternal Spiral Arteries for Uteroplacental Blood Flow during Human Pregnancy. *Placenta* **2009**, *30*, 473–482. [CrossRef] [PubMed]
18. Wang, Y. Vascular Biology of the Placenta. Colloquium Series on Integrated Systems Physiology: From Molecule to Function to Disease. *Morgan & Claypool Life Sci.* **2010**, *2*, 1–98. [CrossRef]
19. Miura, S.; Sato, K.; Kato-Negishi, M.; Teshima, T.; Takeuchi, S. Fluid shear triggers microvilli formation via mechanosensitive activation of TRPV6. *Nat. Commun.* **2015**, *6*, 8871. [CrossRef] [PubMed]
20. Ringler, G.E.; Strauss, J.F. In Vitro Systems for the Study of Human Placental Endocrine Function. *Endocr. Rev.* **1990**, *11*, 105–123. [CrossRef]
21. Pastuschek, J.; Nonn, O.; Gutiérrez-Samudio, R.N.; Murrieta-Coxca, J.M.; Müller, J.; Sanft, J.; Huppertz, B.; Markert, U.R.; Groten, T.; Morales-Prieto, D.M. Molecular characteristics of established trophoblast-derived cell lines. *Placenta* **2021**, *108*, 122–133. [CrossRef] [PubMed]

22. Sodha, R.J.; Proegler, M.; Schneider, H. Transfer and metabolism of norepinephrine studied from maternal-to-fetal and fetal-to-maternal sides in the in vitro perfused human placental lobe. *Am. J. Obstet. Gynecol.* **1984**, *148*, 474–481. [CrossRef]
23. Hutson, J.R.; Garcia-Bournissen, F.; Davis, A.; Koren, G. The Human Placental Perfusion Model: A Systematic Review and Development of a Model to Predict In Vivo Transfer of Therapeutic Drugs. *Clin. Pharmacol. Ther.* **2011**, *90*, 67–76. [CrossRef]
24. Watson, A.L.; Palmer, M.E.; Burton, G. Human chorionic gonadotrophin release and tissue viability in placental organ culture. *Hum. Reprod.* **1995**, *10*, 2159–2164. [CrossRef]
25. Huppertz, B. IFPA Award in Placentology Lecture: Biology of the placental syncytiotrophoblast—Myths and facts. *Placenta* **2010**, *31*, S75–S81. [CrossRef] [PubMed]
26. Huppertz, B. Trophoblast differentiation, fetal growth restriction and preeclampsia. *Pregnancy Hypertens. Int. J. Women's Cardiovasc. Health* **2011**, *1*, 79–86. [CrossRef] [PubMed]
27. Brugger, B.A.; Guettler, J.; Gauster, M. Go with the Flow—Trophoblasts in Flow Culture. *Int. J. Mol. Sci.* **2020**, *21*, 4666. [CrossRef]
28. Lecarpentier, E.; Atallah, A.; Guibourdenche, J.; Hebert-Schuster, M.; Vieillefosse, S.; Chissey, A.; Haddad, B.; Pidoux, G.; Evain-Brion, D.; Barakat, A.; et al. Fluid Shear Stress Promotes Placental Growth Factor Upregulation in Human Syncytiotrophoblast Through the cAMP-PKA Signaling Pathway. *Hypertension* **2016**, *68*, 1438–1446. [CrossRef] [PubMed]
29. Barnea, E.R.; Shurtz-Swirski, R.; Kaplan, M. Factors controlling spontaneous human chorionic gonadotrophin in superfused first trimester placental explants. *Hum. Reprod.* **1992**, *7*, 1022–1026. [CrossRef]
30. Barnea, E.R.; Kaplan, M. Spontaneous, gonadotropin-releasing hormone-induced, and progesterone-inhibited pulsatile secretion of human chorionic gonadotropin in the first trimester placenta in vitro. *J. Clin. Endocrinol. Metab.* **1989**, *69*, 215–217. [CrossRef] [PubMed]
31. Lambot, N.; Lebrun, P.; Cirelli, N.; Vanbellinghen, A.; Delogne-Desnoeck, J.; Graff, G.; Meuris, S. Colloidal effect of albumin on the placental lactogen and chorionic gonadotrophin releases from human term placental explants. *Biochem. Biophys. Res. Commun.* **2004**, *315*, 342–348. [CrossRef]
32. Cole, L.A. hCG, the wonder of today's science. *Reprod. Biol. Endocrinol.* **2012**, *10*, 24. [CrossRef] [PubMed]
33. Shi, Q.J.; Lei, Z.M.; Lin, J.; Carolina, N. Novel role of human chorionic gonadotropin in differentiation of human cytotrophoblasts. *Endocrinology* **1993**, *132*, 1387–1395. [CrossRef] [PubMed]
34. Palmer, M.E.; Watson, A.L.; Burton, G.J. Morphological analysis of degeneration and regeneration of syncytiotrophoblast in first trimester placental villi during organ culture. *Hum. Reprod.* **1997**, *12*, 379–382. [CrossRef] [PubMed]
35. Huppertz, B.; Frank, H.-G.; Kingdom, J.C.P.; Reister, F.; Kaufmann, P. Villous cytotrophoblast regulation of the syncytial apoptotic cascade in the human placenta. *Histochem. Cell Biol.* **1998**, *110*, 495. [CrossRef]
36. Boren, J.; Brindle, K.M. Apoptosis-induced mitochondrial dysfunction causes cytoplasmic lipid droplet formation. *Cell Death Differ.* **2012**, *19*, 1561–1570. [CrossRef]
37. Bildirici, I.; Schaiff, W.T.; Chen, B.; Morizane, M.; Oh, S.-Y.; O'Brien, M.; Sonnenberg-Hirche, C.; Chu, T.; Barak, Y.; Nelson, D.M.; et al. PLIN2 Is Essential for Trophoblastic Lipid Droplet Accumulation and Cell Survival During Hypoxia. *Endocrinology* **2018**, *159*, 3937–3949. [CrossRef] [PubMed]
38. Young, S.G.; Zechner, R. Biochemistry and pathophysiology of intravascular and intracellular lipolysis. *Genes Dev.* **2013**, *27*, 459–484. [CrossRef]
39. Jarc, E.; Petan, T. Lipid Droplets and the Management of Cellular Stress. *Yale J. Biol. Med.* **2019**, *92*, 435–452. Available online: http://www.ncbi.nlm.nih.gov/pubmed/31543707 (accessed on 4 February 2021).
40. Li, H.; Ouyang, Y.; Sadovsky, E.; Parks, W.T.; Chu, T.; Sadovsky, Y. Unique microRNA Signals in Plasma Exosomes from Pregnancies Complicated by Preeclampsia. *Hypertension* **2020**, *75*, 762–771. [CrossRef]
41. Shahgheibi, S.; Mardani, R.; Babaei, E.; Mardani, P.; Rezaie, M.; Farhadifar, F.; Roshani, D.; Naqshbandi, M.; Jalili, A. Platelet indices and CXCL12 levels in patients with intrauterine growth restriction. *Int. J. Women's Health* **2020**, *12*, 307–312. [CrossRef] [PubMed]
42. Tallarek, A.-C.; Huppertz, B.; Stepan, H. Preeclampsia-Aetiology, Current Diagnostics and Clinical Management, New Therapy Options and Future Perspectives. *Geburtshilfe Frauenheilkd* **2012**, *72*, 1107–1116. [CrossRef] [PubMed]
43. Huppertz, B. Biology of preeclampsia: Combined actions of angiogenic factors, their receptors and placental proteins. *Biochim. Biophys. Acta Mol. Basis Dis.* **2018**, *1866*, 165349. [CrossRef] [PubMed]

International Journal of
*Molecular Sciences*

MDPI

*Article*

# Symptoms of Prenatal Depression Associated with Shorter Telomeres in Female Placenta

Isabel Garcia-Martin [1,†], Richard J. A. Penketh [2], Samantha M. Garay [1], Rhiannon E. Jones [3], Julia W. Grimstead [3], Duncan M. Baird [3] and Rosalind M. John [1,*]

1   Division of Biomedicine, Cardiff School of Biosciences, Cardiff University, Cardiff, Wales CF10 3AX, UK; isabel.garciamartin@swansea.ac.uk (I.G.-M.); GaraySM@cardiff.ac.uk (S.M.G.)
2   Department of Obstetrics and Gynaecology, University Hospital Wales, Cardiff, Wales CF14 4XW, UK; Richard.Penketh@wales.nhs.uk
3   Division of Cancer and Genetics, Cardiff School of Medicine, Cardiff University, Cardiff, Wales CF14 4XW, UK; JonesR47@cardiff.ac.uk (R.E.J.); SkinnerJW@cardiff.ac.uk (J.W.G.); BairdDM@cardiff.ac.uk (D.M.B.)
*   Correspondence: JohnRM@Cardiff.ac.uk
†   Current address: Institute of Life Science, Swansea University Medical School, Swansea, Wales, SA2 8PP, UK.

**Abstract:** Background. Depression is a common mood disorder during pregnancy impacting one in every seven women. Children exposed to prenatal depression are more likely to be born at a low birth weight and develop chronic diseases later in life. A proposed hypothesis for this relationship between early exposure to adversity and poor outcomes is accelerated aging. Telomere length has been used as a biomarker of cellular aging. We used high-resolution telomere length analysis to examine the relationship between placental telomere length distributions and maternal mood symptoms in pregnancy. Methods. This study utilised samples from the longitudinal Grown in Wales (GiW) study. Women participating in this study were recruited at their presurgical appointment prior to a term elective caesarean section (ELCS). Women completed the Edinburgh Postnatal Depression Scale (EPDS) and trait subscale of the State-Trait Anxiety Inventory (STAI). Telomere length distributions were generated using single telomere length analysis (STELA) in 109 term placenta (37–42 weeks). Multiple linear regression was performed to examine the relationship between maternally reported symptoms of depression and anxiety at term and mean placental telomere length. Results: Prenatal depression symptoms were significantly negatively associated with XpYp telomere length in female placenta (B = −0.098, $p$ = 0.026, 95% CI −0.184, −0.012). There was no association between maternal depression symptoms and telomere length in male placenta (B = 0.022, $p$ = 0.586, 95% CI −0.059, 0.103). There was no association with anxiety symptoms and telomere length for either sex. Conclusion: Maternal prenatal depression is associated with sex-specific differences in term placental telomeres. Telomere shortening in female placenta may indicate accelerated placental aging.

**Keywords:** telomere shortening; prenatal depression; placenta; sex differences

**Citation:** Garcia-Martin, I.; Penketh, R.J.A.; Garay, S.M.; Jones, R.E.; Grimstead, J.W.; Baird, D.M.; John, R.M. Symptoms of Prenatal Depression Associated with Shorter Telomeres in Female Placenta. *Int. J. Mol. Sci.* **2021**, *22*, 7458. https://doi.org/10.3390/ijms22147458

Academic Editors: Hiten D Mistry and Eun Lee

Received: 1 June 2021
Accepted: 8 July 2021
Published: 12 July 2021



## 1. Introduction

During pregnancy, women are highly vulnerable to major depression [1] and 10–15% experience at least one major depressive episode associated with the increased risk of morbidity to both mother and child [2]. Prenatal depression is co-morbid with anxiety, which can impact up to 25% of pregnancies, and both mood disorders are strongly linked to postpartum depression, all of which have a negative effect on child development in the short and longer term [3,4]. Every year in the UK > 100,000 babies are born exposed in utero to maternal depression, and considerably more are exposed to maternal anxiety. Importantly, recent surveys suggest that the rates of both depression and anxiety are increasing [5–7]. Maternal mood disorders represent a considerable clinical, financial and emotional burden to society [8]. A number of risk factors have been identified including previous history of poor mental health and socio-economic deprivation [9]. However, the biological mechanisms linking mood symptoms to adverse outcomes in children are unknown.

One biological factor linked to prenatal adversity is shortened telomere length. Telomeres are located at the ends of mammalian chromosomes and function to maintain genomic integrity [10,11]. Human telomeres consist of the hexameric DNA sequence, TTAGGG, tandemly repeated in arrays which vary in length depending on the tissue type and the age of the individual [10,12]. An individual's telomere length is set at conception through the combined contribution of paternal and maternal telomere lengths inherited through the sperm and oocyte [13,14]. Whilst telomere length is maintained by the action of the enzyme telomerase in stem cell compartments [15,16], the majority of human somatic tissues do not express sufficient telomerase and thus exhibit erosion with ongoing cell division [12]. The gradual decline in telomere length over the individual's lifespan is thought to play a role in the reduction of cellular viability via the induction of replicative senescence as a function of age [17]. Factors that accelerate or protect against telomere shortening during pregnancy may result in individuals being born whose telomere lengths differs from their chronological age [18]. Maternal age has been positively correlated with Telomere Restriction Fragment (TRF) length in newborns white blood cells [19]. In addition, placental telomere length decreases with gestational age and is influenced by parity, as multiparity appears to delay placental telomere shortening [20,21]. Maternal smoking, body mass index, stress, poor nutrition and disease status have been linked to shorter telomeres measured in cord blood DNA [22]. In adults, chronic psychosocial stress [23], major depression [24], chronic mood disorder [25] have been linked to shorter telomeres in white blood cells. Symptoms of depression reported in the first or second trimester of pregnancy have been linked to shorter cord blood telomeres in male infants but not female infants [26].

Few studies have examined telomere length in the placenta. The placenta is a fetally derived tissue that develops in concert with the fetus and is exposed to essentially the same environment [27], displaying low levels of telomerase activity and telomere shortening [28,29]. We recently reported shorter telomeres in placenta from pregnancies complicated by gestational diabetes, and this was the case only for male placenta. Here, we explored the relationship between telomere length profiles in placenta and maternally reported symptoms of depression and anxiety at term using measures of telomere length generated by high-resolution telomere length analysis (STELA).

## 2. Methods

### 2.1. Cohort

The Grown in Wales (GiW) study [6] is a longitudinal birth cohort based in South-East Wales, United Kingdom that began in September 2015 and ended recruitment in November 2016. The basis of recruitment was attendance of a pre-surgical appointment prior to an elective caesarean section (ELCS) at the University Hospital of Wales during the specified time period with the stated criteria of women aged between 18 and 45 with a singleton term pregnancy without fetal abnormalities or infectious diseases. ELCS was chosen as the mode of delivery to facilitate the collection of high-quality biological samples which included maternal serum and saliva at recruitment as well as placenta and cord blood on delivery. The major indication for ELCS was a previous caesarean. We previously reported no significant difference in late antenatal mood scores between participants with varying indications for ELCS [6]. All participants were recruited by two trained research midwives. Women were not clinically assessed for depression, anxiety or stress. A total of 355 women were originally recruited into the study, seven of whom later withdrew. Full ethical approval for the GiW study was obtained from the Wales Research Ethics Committee REC2 reference 15/WA/0004. Research was carried out in accordance with the principles of the Declaration of Helsinki as revised in 2008. Written informed consent was obtained from all the participants at recruitment.

*2.2. Materials*

2.2.1. Maternal Demographics and Birth Outcomes

Maternal lifestyle and demographics were reported by the mother in the questionnaire at recruitment which was 1–4 days prior to ELCS. Data included ethnicity, education, income, age, body mass index (BMI), pregnancy complications and whether they smoked or drank alcohol during their pregnancy. Welsh Index of Multiple Deprivation (WIMD) 2014 scores were calculated from anonymised postcodes [30]. Delivery information, fetal and placental biometry, and body mass index at initial booking was recorded from medical notes. Gestational age at the prenatal assessment was the same as gestational age at ELCS. Prevalence of obstetric risk factors are provided in Supplementary Table S1.

2.2.2. Maternal Depression and Anxiety Symptoms

Participants completed two self-reporting mood questionnaires at recruitment 1–4 days before their ELCS. Depression was measured using the Edinburgh Postnatal Depression Scale (EPDS) which comprises of ten questions each scored between zero and three, with total scores 13 and above indicating probable depression [31]. Anxiety was measured using the Trait subscale of the State Trait Anxiety Inventory (STAI) to measure general anxiety levels [32,33]. This subscale contains 20 questions scored between one and four, with final scores of 40 and above indicating high anxiety levels [32]. Missing data was addressed using participant level mean substitution for those missing <20% of data.

2.2.3. Placental Biopsies

Biopsies were collected from term placenta (37–42 weeks) by trained research midwives within two hours of delivery. For each placenta, 3–5 chorionic villous samples were taken 1 cm below the surface from the maternal side of the placenta at sites midway between the cord insertion and the lateral edge. Samples were washed three times in phosphate buffered saline to remove maternal blood and stored in RNAlater at $-80\,^{\circ}$C.

2.2.4. STELA

Placental telomere length distributions were obtained using the STELA assay as described previously [34]. Briefly, genomic DNA was extracted from approximately 40 mg of placental tissue consisting of a combined tissue sample from 3–5 separate biopsies by a standard proteinase K and phenol/chloroform protocol [35] with all samples processed in the same way by a single researcher to minimise technical variability. Diluted genomic DNA samples were PCR amplified using telorette2, teltail and the XpYp telomere specific primer XpYpE2. Amplified telomeric DNA fragments were resolved on 0.5% agarose gels and detected by Southern hybridisation using a TTAGGG repeat probe $\alpha$-$^{33}$P dCTP labelled (Perkin Elmer) before visualisation using a Typhoon FLA 9500 phosphoimager (GE Healthcare Life Sciences, United Kingdom). Telomere length distributions were determined using the ImageQuant software (GE Healthcare Life Sciences, United Kingdom). Telomere measures were generated for N = 133/337 of the placental biopsies collected. For this study, we selected 30 samples with EPDS scores $\geq$ 13 as our "depressed" group and 79 with EPDS scores < 13 as our non-depressed "low mood score" group excluding samples with a diagnosis of preeclampsia or gestational diabetes, those with gestational age < 37 weeks and those not ultimately delivering by ELCS. Two of the 109 participants did not have a STAI or an EPDS value.

*2.3. Statistical Analysis*

2.3.1. Research Question

Is telomere length affected by prenatal depression? Is there a sex-specific effect of prenatal depression on telomere length?

### 2.3.2. Hypothesis

Placentas from women exposed to prenatal depression will exhibit shorter telomeres than placentas from control women.

### 2.3.3. Statistical Steps

All statistical analysis was performed using SPSS 26.0 for Macintosh and GraphPad Prism 9. Data are expressed as means with standard deviation, or as numbers (%). Normality was assessed using Shapiro–Wilk test, histograms and normal Q–Q plots (Figure 1). Mean placental telomere length was normally distributed ($p > 0.05$) as determined by Shapiro–Wilk test. Differences between low mood score and high mood score groups (EPDS < 13 and EPDS $\geq$ 13) were assessed using $\chi2$ test or Student's *t*-test. Relationships between the main dependent variable (mean telomere length) and other confounding variables were analysed by Pearson or Spearman correlation. To assess the relationship between mean telomere length and maternal mental health scores (EPDS and STAI), unadjusted and adjusted linear regression was performed. To test the main effect of maternal perinatal mental health on telomere length, unadjusted linear regression was performed. This included as the dependent variable mean telomere length, and EPDS or STAI as the independent variable.

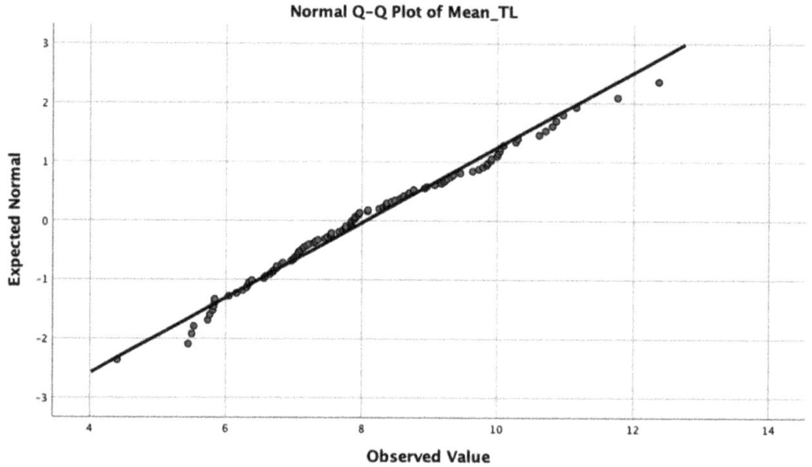


**Figure 1.** Q–Q plot of mean telomere length.

A list of relevant covariates controlled for in the linear regression included maternal age, gestational age, parity, BMI, smoking during pregnancy, alcohol during pregnancy, WIMD score and EPDS or STAI. These variables were chosen based on the established literature in the introduction and methods section. All unadjusted and adjusted linear regressions were run separately for participants who had girls or boys.

## 3. Results

### 3.1. Association between Telomere Length and Potential Confounders

As in our previous studies [34,36], this study applied STELA to measure the lengths of the XpYp telomeres located at the end of the pseudoautosomal region. Placental telomere length distributions were analysed for term placental samples obtained from 109 Grown in Wales participants delivering by ELCS excluding the pregnancy complications of gestational diabetes and preeclampsia. There was no significant relationship ($p > 0.05$) between mean telomere and the obstetric risk factors listed in Supplementary Table S1. The association between potential confounders (Table 1) including maternal age, gestational age, parity, smoking, alcohol consumption, BMI and WIMD score with telomere length was also tested. Average telomere length was correlated with WIMD score with a Spearman's rho coefficient of 0.216, $p < 0.05$. WIMD score ranks all small areas in Wales from 1 (most deprived) to 1909 (least deprived) and is therefore an indicator for socio-economic status.

**Table 1.** Association between telomere characteristics and potential confounders. Number (%) is shown; *p*-values were assessed with Pearson or Spearman rho correlation; *n* = 109 samples.

| Potential Confounder | Association with Mean Telomere Length |
|---|---|
| Maternal age | r = 0.09<br>CI = −0.030, 0.078<br>*p* = 0.37 |
| Gestation age | r = 0.15<br>CI = −0.100, 0.832<br>*p* = 0.12 |
| Parity | r = 0.09<br>CI = −0.173, 0.479<br>*p* = 0.35 |
| Smoking 14/109 (12.9%) | r = −0.00<br>CI = −0.765, 1.023<br>*p* = 0.776 |
| Alcohol 8/109 (7.3%) | r = 0.11<br>CI = −0.421, 1.85<br>*p* = 0.214 |
| BMI | r = −0.03<br>CI = −0.056, 0.093<br>*p* = 0.792 |
| WIMD | r = 0.22<br>CI = 0.000, 0.001<br>***p* = 0.027** |

BMI = body mass index; WIMD = Welsh Index of multiple deprivation.

### 3.2. Analysis of High and Low Mood Score Groups

In our previous study on GDM, we detected significantly shorter telomeres and significantly more telomeres under 5 kb in male placenta with a group size of 38 controls versus 10 GDM [36]. Using data from this study, we calculated we would need a sample size of 13 in each group ($\alpha = 0.05$ and a power of 0.8). Of the 109 sample measures analysed in this study $n = 55$ were male and $n = 54$ were female placentas. Of these, 30 were from pregnancies where women reported significant depression symptoms just prior to the ELCS with an EPDS score of $\geq$13 (high mood score), with 17 samples from male placenta and 13 samples from female placenta. Of these 30 samples, 25 women also scored above the cut-off for STAI with a score of $\geq$40, i.e., for most of these mothers there was evidence of both depression and anxiety consistent with the significant association between EPDS and STAI with a Spearman's rho coefficient of 0.877**, $p < 0.01$. The remaining 79 samples, with 38 samples from male placenta and 41 samples from female placenta, formed a non-depressed "low mood score" group. There were no differences for any of the listed characteristics between the two groups ($p > 0.05$; Table 2).

### 3.3. Multiple Linear Regression

In order to investigate the association between mean telomere length and perinatal anxiety and depression, multiple linear regression was performed. There were no significant associations when data from all participants ($n = 109$) was analysed unadjusted (Table 3). Disparate sex-specific features of prenatal vulnerability have been reported in relation to specific pregnancy conditions [26,36–39]. Data from male and female samples were therefore analysed separately. The analysis was then adjusted for maternal age, gestational age, parity, BMI, smoking during pregnancy, alcohol during pregnancy and

WIMD covariates (Figure 2 and Table 3). There was no statistical difference in mean placental telomere length between male and female placenta (mean = 7.97, SD = 1.66 versus mean = 8.13, SD = 1.47; $p$ = 0.60). EPDS scores were significantly negatively associated with shorter mean placental telomere length for female infants only ($p$ = 0.026; Figure 3A). 35% of the variance in female placental telomere length was accounted for by the EPDS score and the abovenamed covariates.

**Table 2.** Key characteristics of the study participants. Mean (SD)/range or number (%) is shown; $p$-values were assessed using independent samples $t$-test or $\chi^2$ test. Due to missing values some numbers do not add up to 100%.

| Characteristics | Low Mood Score Group: EPDS < 13 Group ($n$ = 79) | High Mood Score Group: EPDS $\geq$ 13 Group ($n$ = 30) | $p$-Value |
|---|---|---|---|
| Caucasian ethnicity | 73 (92%) | 29 (96%) | 0.52 |
| Parity: | | | |
| Primiparous | 14 (17.7%) | 7 (23.3%) | 0.82 |
| Multiparous | 65 (82.2%) | 23 (76.7%) | |
| Maternal age | 32 (5.51)/19–44 | 30 (5.68)/20–39 | 0.31 |
| ELCS | 79 (100%) | 30 (100%) | N/A |
| Birth weight (g) | 3491 (620)/2260–5080 | 3488 (499)/2460–5110 | 0.98 |
| Gestational age (weeks) | 39 (0.61)/38–41 | 38 (0.68)/37–41 | 0.12 |
| Placental weight (g) | 663 (129)/376–941 | 671 (150)/455–1060 | 0.76 |
| Fetal sex: | | | |
| Female | 38 (48%) | 17 (56%) | 0.42 |
| Male | 41 (52%) | 13 (44%) | |
| Smoking during pregnancy | 9 (11.4%) | 5 (16.7%) | 0.46 |
| Alcohol during pregnancy | 5 (6.3%) | 3 (10%) | 0.51 |

ELCS = elective caesarean delivery.

**Table 3.** Analysis of the association between mean telomere length and maternal mood scores using unadjusted and adjusted multiple linear regression models. Adjusting for maternal age, gestational age, parity, BMI, smoking during pregnancy, alcohol during pregnancy, WIMD score and EPDS/STAI. A $p$-value < 0.05 is considered statistically significant.

| Mood Scores | All | | | Male | | | Female | | |
|---|---|---|---|---|---|---|---|---|---|
| | B | 95%CI | $p$ | B | 95%CI | $p$ | B | 95%CI | $p$ |
| Unadjusted linear regressions | | | | | | | | | |
| EPDS | −0.025 | −0.079, 0.028 | 0.346 | 0.002 | −0.073, 0.078 | 0.949 | −0.063 | −0.143, 0.017 | 0.121 |
| STAI | −0.016 | −0.046, 0.015 | 0.305 | −0.013 | −0.060, 0.034 | 0.582 | −0.017 | −0.059, 0.024 | 0.399 |
| Adjusted linear regressions | | | | | | | | | |
| EPDS | −0.004 | −0.061, 0.052 | 0.884 | 0.022 | −0.059, 0.103 | 0.586 | −0.098 | −0.184, −0.012 | **0.026** |
| STAI | −0.002 | −0.033, 0.030 | 0.914 | −0.004 | −0.054, 0.046 | 0.879 | −0.016 | −0.058, 0.027 | 0.462 |

EPDS, Edinburgh Postnatal Depression Scale; STAI, Spielberger State-Trait Anxiety Inventory; WIMD, Welsh Index of Multiple Deprivation; BMI, body mass index.

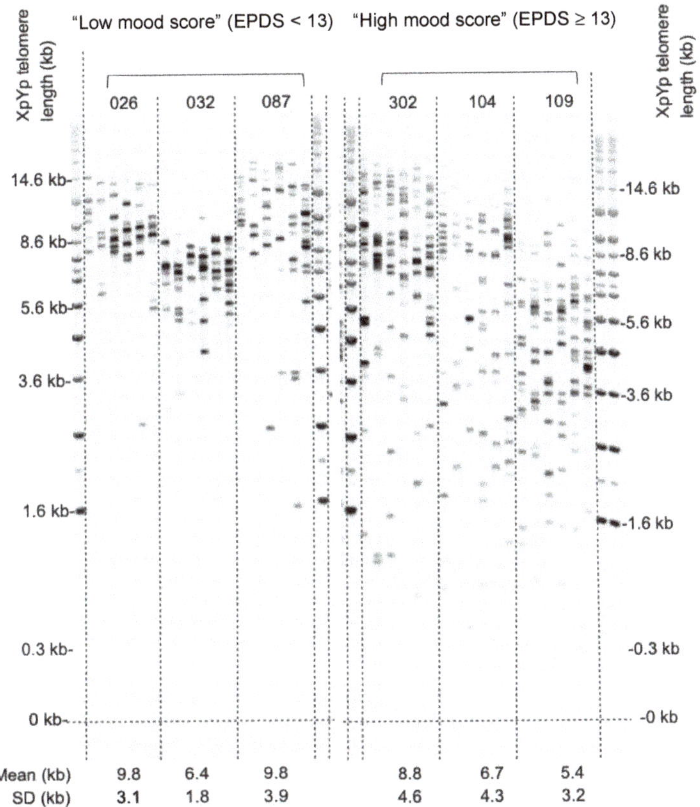



**Figure 2.** Representative STELA. Images of two autoradiographs are shown alongside molecular weight markers. The left panel shows STELA PCR reactions from three female placenta samples from the "low mood score" (EPDS < 13) group with 6 reactions run for each sample. The right panel shows comparable data for "high mood score" (EPDS ≥ 13) group. Mean telomere lengths ± standard deviation (SD) are given below the lanes for each sample. The coefficient of variation of all samples was <1 indicating low variance.

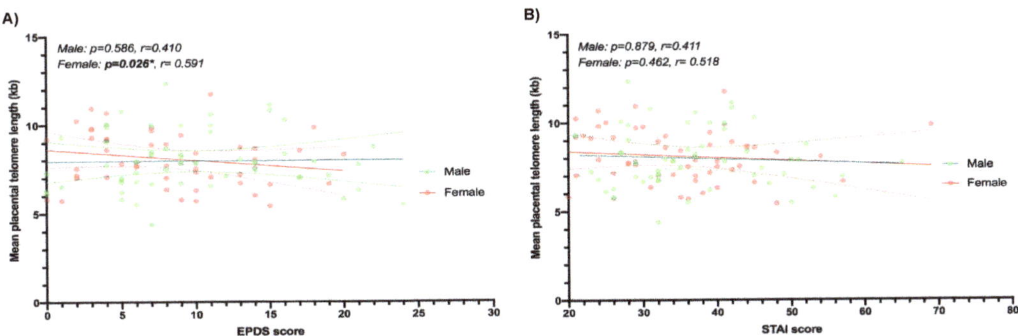



**Figure 3.** Telomere length differences in placenta according to mood scores (EPDS and STAI) for males (*n* = 55) and females (*n* = 54). The association between maternally reported depression symptoms (EPDS score) and mean placental telomere length (**A**), and the association between maternally reported anxiety symptoms (STAI score) and mean placental telomere length (**B**) are shown as linear regressions.

## 4. Discussion

Our study uncovered an association between maternally reported symptoms of depression recorded via an EPDS questionnaire completed just prior to delivery and placental telomere length. EDPS scores were significantly negatively associated with mean placental telomere length in female placenta only. This relationship remained after adjusting for factors known or suspected to influence telomere length including maternal age, gestational age, parity, smoking, alcohol, BMI and WIMD score. In contrast, STAI scores (anxiety) were not associated with placental telomere length.

The placenta is derived from the same genome as the fetus and exposed to the same factors but, due to its position between the maternal and fetal circulation, may be more impacted by adversity in pregnancy. We previously reported that term placenta show remarkable heterogeneity in placental telomere lengths [34]. We also reported that the length and distributions of telomeres were not influenced by sampling site, mode of delivery or fetal sex [34]. Using the same STELA approach, we more recently reported shorter telomeres in male placenta from pregnancies where mothers were diagnosed with gestational diabetes, but not medically treated with metformin or insulin [36]. Shorter placental telomeres have previously been associated with being born small for gestational age or growth restricted [40–43] and preeclampsia [44]. Higher maternal reporting of adverse childhood experiences has also been associated with shorter placental telomeres [45]. These data all lend support to the idea that pregnancy complications impact the maintenance of telomere integrity in the placenta which may have relevance for placental function and fetal wellbeing.

The major finding of our study was the negative association between prenatal EPDS score and telomere length in female placenta. Higher maternally reported symptoms of depression, but not anxiety, were associated with significantly shorter telomeres. In contrast, there was no association between EDPS score and telomere length in male placenta. The absence of association in male placenta was not due to a lack of statistical power as analysing all the placenta combined did not reveal an association despite twice the number of samples in the analysis ($n$ = 109 all samples v $n$ = 54 female samples v $n$ = 55 male samples). EPDS scores are used to indicate symptoms of depression with scores of 13 and above predicting an episode of clinical depression based on diagnostic criteria in the postpartum period [31,46,47]. Maternal stress during pregnancy has been associated with shorter cord blood telomeres [48–50] with one retrospective study suggesting a greater impact on females [51]. Air pollution is another adversity linked to shorter cord blood telomeres with a stronger association in girls [52]. However, in a recent study of 151 newborns, female infant cord blood telomere length was not associated with any factor tested. This study reported associations between shorter male cord blood telomeres and maternal smoking, higher body mass index, maternal sexual abuse in childhood and elevated depressive symptoms in pregnancy. Longer cord blood telomere length was associated with higher maternal educational attainment and household income in pregnancy, and greater maternal familial emotional support in childhood [26]. There are a number of differences between this study and ours, the most obvious being the nature of the samples measured (placenta versus cord blood), the technical approach (STELA versus real-time polymerase chain reaction) and the population (Welsh White versus North American White, Hispanic and Black/Haitian). In addition, Bosquet Enlow et al. measured depressive symptoms by EPDS questionnaire during the first or second trimester [26] whereas our participants completed this questionnaire just prior to term. Bosquet Enlow et al. also excluded participants endorsing drinking more than seven alcoholic drinks per week. These differences may account for our different findings. An interesting alternative possibility is that telomere length in fetus may be preserved in females exposed to stressors in pregnancy at the cost of placental telomeres while the reverse may be true in males. Recently in a new study on 146 newborns [53] reported no association between annual household income and telomere length. This study also reported a positive association between newborn relative telomere length and pregnancy-related anxiety symptoms (PRAS), depressive symptoms during pregnancy (EPDS), general anxiety symptoms (STAI), and self-reported depression prior to the current pregnancy. These findings suggested a positive and adaptive effect of maternal stress on fetal telomere biology among males, which contradicts their previous reported findings that correlated elevated depressive symptoms in pregnancy and shorter cord blood telomeres in boys. This variability in similar studies emphasises the need to dig into the influence of pre-conception factors on fetal telomere biology, as well as to define the clinical significance of shorter and longer telomere length at birth. Moreover, the simultaneity of certain factors and maternal physical and mental health issues may rescue the detrimental effect of the latter on fetal telomere biology. It can be hypothesised that compensatory factors, such as socioeconomic status, may ameliorate the negative impact of maternal depressive and anxiety symptoms on fetal telomere length.

In our study we did not find an association between shorter placental telomeres and depressive symptoms in male placenta, but we did observe an association between higher WIMD score and longer telomeres. WIMD scores are generated using postcode information and reflect the Welsh Government's official measure of relative deprivation within small areas in Wales. A number of measures feed into WIMD including income, employment, health, education, access to services, housing, community safety and physical environment. While non-deprived people can live in deprived areas, and deprived people can live in non-deprived areas, higher WIMD scores are generally indicative of lower overall levels of deprivation. Our data is consistent with the previous findings that higher maternal education and income provide protection against telomere shortening in males [26]. Moreover, a recent study by Martens et al. [54] found that higher parental socioeconomic status was associated with longer cord blood telomere length and placental telomere length in boys, but not in girls. These findings indicate that social economic factors likely play an important role in influencing cellular longevity in the exposed individuals.

*Limitations*

One limitation of our study is that we are relying on mood scores from self-reported questionnaires recorded 1–4 days before an ELCS. For our full GiW study cohort, we previously reported a highly significant association between EPDS/STAI scores and mental health history, and also between term EPDS/STAI scores and those obtained at three points after delivery for a subset of participants [6] which is reassuring. However, although self-reported questionnaires have a high acceptance by women who are thought to feel less constrained in responding as stated by the WHO (2008), a full clinical evaluation would provide further validation of our findings. Some of the factors controlled for were also self-reported including alcohol consumption, although we found that reporting for these factors was highly similar between the questionnaires and the medical notes [6]. Another limitation of this study was the nature of the cohort which had a restricted participant demographic. While minimising heterogeneity within a cohort study has value in uncovering subtle relationships with smaller cohort numbers, our findings may not be generalizable to other populations. A further limitation is that we do not have the parental telomere lengths. It is possible that our findings are an indicator of telomere length at conception and not exposures during pregnancy, although if this were the case it would be more challenging to interpret the sex-specific findings. A final limitation is that our analysis was restricted to the XpYp telomeres located at the end of the pseudoautosomal region. While this region segregates independently of sex with similar telomere distributions to autosomes [55], it is possible that our findings are confined to the telomeres we analysed. Further work is required to validate our findings in other cohorts with a greater diversity of participants and extend this analysis to other chromosomes. It will also be important to compare placental and cord blood telomere lengths from matched samples to establish the potential for inheritance in other fetally derived tissues.

## 5. Conclusions

In summary, we discovered that mothers reporting higher symptoms of depression near term gave birth to daughters with relatively shorter placental telomeres. We did not observe a similar relationship between mood symptoms and placental telomeres when mothers gave birth to sons. However, we did observe longer male placental telomeres in mothers residing in more affluent areas. These findings contribute to our understanding of the sex-specific outcomes observed for children exposed in utero to maternal depression.

**Supplementary Materials:** The following are available online at https://www.mdpi.com/article/10.3390/ijms22147458/s1.

**Author Contributions:** Conceptualization, R.M.J., D.M.B. and R.J.A.P.; methodology, D.M.B., R.E.J., J.W.G. and I.G.-M.; formal analysis, I.G.-M.; validation, S.M.G., investigation, I.G.-M.; resources, R.M.J.; data curation, I.G.-M. and R.M.J.; writing—original draft preparation, R.M.J. and I.G.-M.; writing—review and editing, R.M.J., I.G.-M. and D.M.B.; supervision, R.M.J. and D.M.B.; project administration, R.M.J.; funding acquisition, R.M.J. and R.J.A.P. All authors have read and agreed to the published version of the manuscript.

**Funding:** I.G.M. was funded by a Cardiff School of Biosciences PhD studentship. Recruitment and placental collection was supported by MRC grant MR/M013960/1. D.M.B. lab was supported by Cancer Research UK (C17199/A18246/A29202). S.M.G. was supported by MRC doctoral training grant MR/N013794/1.

**Institutional Review Board Statement:** The study was conducted according to the guidelines of the Declaration of Helsinki and ethical approval for the GiW Study cohort was obtained from the Wales Research Ethics Committee REC2, Health Research Authority, UK reference 15/WA/0004 (13 February 2015).

**Informed Consent Statement:** Written informed consent was obtained from all subjects involved in the study.

**Data Availability Statement:** For this specific study, there is a danger that sharing participant data might reveal participant identity due to the nature of the data, the specific recruitment dates and unique route of recruitment. The datasets used and analysed during the current study will therefore not be made publically available but will be available from the corresponding author on reasonable request.

**Acknowledgments:** The authors would like to thank the participants at University Hospital Wales who kindly donated their placenta as part of the Grown in Wales Study, and the research midwives involved in recruitment and sample collection.

**Conflicts of Interest:** The authors declare no conflict of interest.

## Abbreviations

| | |
|---|---|
| STELA | Single telomere length analysis |
| GiW | Grown in Wales |
| ELCS | Elective caesarean section |
| WIMD | Welsh Index of Multiple Deprivation |
| EPDS | Edinburgh Postnatal Depression Scale |
| STAI | State Trait Anxiety Index |
| BMI | Body mass index |

## References

1. Van de Velde, S.; Bracke, P.; Levecque, K. Gender differences in depression in 23 European countries. Cross-national variation in the gender gap in depression. *Soc. Sci. Med.* **2010**, *71*, 305–313. [CrossRef] [PubMed]
2. Stewart, D.E. Clinical practice. Depression during pregnancy. *N. Engl. J. Med.* **2011**, *365*, 1605–1611. [CrossRef] [PubMed]
3. Brand, S.R.; Brennan, P.A. Impact of antenatal and postpartum maternal mental illness: How are the children? *Clin Obs. Gynecol* **2009**, *52*, 441–455. [CrossRef] [PubMed]
4. Kinsella, M.T.; Monk, C. Impact of maternal stress, depression and anxiety on fetal neurobehavioral development. *Clin. Obs. Gynecol.* **2009**, *52*, 425–440. [CrossRef] [PubMed]

5.  Howard, L.; Ryan, E.; Trevillion, K.; Anderson, F.; Bick, D.; Bye, A.; Pickles, A. Accuracy of the Whooley questions and the Edinburgh Postnatal Depression Scale in identifying depression and other mental disorders in early pregnancy. *Br. J. Psychiatry* **2018**, *212*, 50–56. [CrossRef]

6.  Janssen, A.B.; Savory, K.A.; Garay, S.M.; Sumption, L.; Watkins, W.; Garcia-Martin, I.; Savory, N.A.; Ridgway, A.; Isles, A.R.; Penketh, R.; et al. Persistence of anxiety symptoms after elective caesarean delivery. *Bjpsych Open* **2018**, *4*, 354–360. [CrossRef]

7.  Pearson, R.M.; Carnegie, R.E.; Cree, C.; Rollings, C.; Rena-Jones, L.; Evans, J.; Stein, A.; Tilling, K.; Lewcock, M.; Lawlor, D.A. Prevalence of Prenatal Depression Symptoms Among 2 Generations of Pregnant Mothers: The Avon Longitudinal Study of Parents and Children. *Jama Netw. Open* **2018**, *1*, e180725. [CrossRef]

8.  Bauer, A.; Parsonage, M.; Knapp, M.; Iemmi, V.; Adelaja, B. Costs of perinatal mental health problems. *J. Affect. Disord.* **2014**, *192*, 83–90. [CrossRef]

9.  Biaggi, A.; Conroy, S.; Pawlby, S.; Pariante, C.M. Identifying the women at risk of antenatal anxiety and depression: A systematic review. *J. Affect. Disord.* **2016**, *191*, 62–77. [CrossRef]

10. Moyzis, R.K.; Buckingham, J.M.; Cram, L.S.; Dani, M.; Deaven, L.L.; Jones, M.D.; Meyne, J.; Ratliff, R.L.; Wu, J.R. A highly conserved repetitive DNA sequence, (TTAGGG)n, present at the telomeres of human chromosomes. *Proc. Natl. Acad. Sci. USA* **1988**, *85*, 6622–6626. [CrossRef]

11. De Lange, T. Shelterin: The protein complex that shapes and safeguards human telomeres. *Genes Dev.* **2005**, *19*, 2100–2110. [CrossRef]

12. Takubo, K.; Izumiyama-Shimomura, N.; Honma, N.; Sawabe, M.; Arai, T.; Kato, M.; Oshimura, M.; Nakamura, K. Telomere lengths are characteristic in each human individual. *Exp. Gerontol.* **2002**, *37*, 523–531. [CrossRef]

13. Baird, D.M.; Britt-Compton, B.; Rowson, J.; Amso, N.N.; Gregory, L.; Kipling, D. Telomere instability in the male germline. *Hum. Mol. Genet.* **2006**, *15*, 45–51. [CrossRef]

14. Baird, D.M.; Rowson, J.; Wynford-Thomas, D.; Kipling, D. Extensive allelic variation and ultrashort telomeres in senescent human cells. *Nat. Genet.* **2003**, *33*, 203–207. [CrossRef]

15. Blackburn, E.H. Switching and signaling at the telomere. *Cell* **2001**, *106*, 661–673. [CrossRef]

16. Collins, K.; Mitchell, J.R. Telomerase in the human organism. *Oncogene* **2002**, *21*, 564–579. [CrossRef]

17. Campisi, J. Senescent cells, tumor suppression, and organismal aging: Good citizens, bad neighbors. *Cell* **2005**, *120*, 513–522. [CrossRef]

18. Jylhava, J.; Pedersen, N.L.; Hagg, S. Biological Age Predictors. *EBioMedicine* **2017**, *21*, 29–36. [CrossRef] [PubMed]

19. Okuda, K.; Bardeguez, A.; Gardner, J.P.; Rodriguez, P.; Ganesh, V.; Kimura, M.; Skurnick, J.; Awad, G.; Aviv, A. Telomere length in the newborn. *Pediatr. Res.* **2002**, *52*, 377–381. [CrossRef] [PubMed]

20. Gielen, M.; Hageman, G.; Pachen, D.; Derom, C.; Vlietinck, R.; Zeegers, M.P. Placental telomere length decreases with gestational age and is influenced by parity: A study of third trimester live-born twins. *Placenta* **2014**, *35*, 791–796. [CrossRef] [PubMed]

21. Wilson, S.L.; Liu, Y.; Robinson, W.P. Placental telomere length decline with gestational age differs by sex and TERT, DNMT1, and DNMT3A DNA methylation. *Placenta* **2016**, *48*, 26–33. [CrossRef]

22. Whiteman, V.E.; Goswami, A.; Salihu, H.M. Telomere length and fetal programming: A review of recent scientific advances. *Am. J. Reprod. Immunol.* **2017**, *77*. [CrossRef] [PubMed]

23. Epel, E.S.; Blackburn, E.H.; Lin, J.; Dhabhar, F.S.; Adler, N.E.; Morrow, J.D.; Cawthon, R.M. Accelerated telomere shortening in response to life stress. *Proc. Natl. Acad. Sci. USA* **2004**, *101*, 17312–17315. [CrossRef] [PubMed]

24. Hartmann, N.; Boehner, M.; Groenen, F.; Kalb, R. Telomere length of patients with major depression is shortened but independent from therapy and severity of the disease. *Depress Anxiety* **2010**, *27*, 1111–1116. [CrossRef] [PubMed]

25. Simon, N.M.; Smoller, J.W.; McNamara, K.L.; Maser, R.S.; Zalta, A.K.; Pollack, M.H.; Nierenberg, A.A.; Fava, M.; Wong, K.K. Telomere shortening and mood disorders: Preliminary support for a chronic stress model of accelerated aging. *Biol. Psychiatry* **2006**, *60*, 432–435. [CrossRef] [PubMed]

26. Bosquet Enlow, M.; Bollati, V.; Sideridis, G.; Flom, J.D.; Hoxha, M.; Hacker, M.R.; Wright, R.J. Sex differences in effects of maternal risk and protective factors in childhood and pregnancy on newborn telomere length. *Psychoneuroendocrinology* **2018**, *95*, 74–85. [CrossRef]

27. John, R.; Hemberger, M. A placenta for life. *Reprod Biomed Online* **2012**, *25*, 5–11. [CrossRef]

28. Izutsu, T.; Kudo, T.; Sato, T.; Nishiya, I.; Ohyashiki, K.; Mori, M.; Nakagawara, K. Telomerase activity in human chorionic villi and placenta determined by TRAP and in situ TRAP assay. *Placenta* **1998**, *19*, 613–618. [CrossRef]

29. Biron-Shental, T.; Kidron, D.; Sukenik-Halevy, R.; Goldberg-Bittman, L.; Sharony, R.; Fejgin, M.D.; Amiel, A. TERC telomerase subunit gene copy number in placentas from pregnancies complicated with intrauterine growth restriction. *Early Hum. Dev.* **2011**, *87*, 73–75. [CrossRef]

30. Welsh Index of Multiple Deprivation. Available online: https://gov.wales/welsh-index-multiple-deprivation (accessed on 3 January 2021).

31. Cox, J.L.; Holden, J.M.; Sagovsky, R. Detection of postnatal depression. Development of the 10-item Edinburgh Postnatal Depression Scale. *Br. J. Psychiatry J. Ment. Sci.* **1987**, *150*, 782–786. [CrossRef] [PubMed]

32. Grant, K.A.; McMahon, C.; Austin, M.P. Maternal anxiety during the transition to parenthood: A prospective study. *J. Affect. Disord.* **2008**, *108*, 101–111. [CrossRef] [PubMed]

33. Meades, R.; Ayers, S. Anxiety measures validated in perinatal populations: A systematic review. *J. Affect. Disord.* **2011**, *133*, 1–15. [CrossRef]

34. Garcia-Martin, I.; Janssen, A.B.; Jones, R.E.; Grimstead, J.W.; Penketh, R.J.A.; Baird, D.M.; John, R.M. Telomere length heterogeneity in placenta revealed with high-resolution telomere length analysis. *Placenta* **2017**, *59*, 61–68. [CrossRef]

35. Hogan, B.; Beddington, R.; Constantini., F.; Lacy, E. *Manipulating the Mouse Embryo: A Laboratory Manual*, 2nd ed.; Cold Spring Harbor Laboratory Press: Cold Spring Harbor, NY, USA, 1994.

36. Garcia-Martin, I.; Penketh, R.J.A.; Janssen, A.B.; Jones, R.E.; Grimstead, J.; Baird, D.M.; John, R.M. Metformin and insulin treatment prevent placental telomere attrition in boys exposed to maternal diabetes. *PLoS ONE* **2018**, *13*, e0208533. [CrossRef]

37. Sumption, L.A.; Garay, S.M.; John, R.M. Low serum placental lactogen at term is associated with postnatal symptoms of depression and anxiety in women delivering female infants. *Psychoneuroendocrinology* **2020**, *116*, 104655. [CrossRef] [PubMed]

38. Dingsdale, H.; Nan, X.; Garay, S.M.; Mueller, A.; Sumption, L.A.; Chacon-Fernandez, P.; Martinez-Garay, I.; Ghevaert, C.; Barde, Y.A.; John, R.M. The placenta protects the fetal circulation from anxiety-driven elevations in maternal serum levels of brain-derived neurotrophic factor. *Transl. Psychiatry* **2021**, *11*, 62. [CrossRef]

39. Hjort, L.; Vryer, R.; Grunnet, L.G.; Burgner, D.; Olsen, S.F.; Saffery, R.; Vaag, A. Telomere length is reduced in 9- to 16-year-old girls exposed to gestational diabetes in utero. *Diabetologia* **2018**, *61*, 870–880. [CrossRef] [PubMed]

40. Davy, P.; Nagata, M.; Bullard, P.; Fogelson, N.S.; Allsopp, R. Fetal growth restriction is associated with accelerated telomere shortening and increased expression of cell senescence markers in the placenta. *Placenta* **2009**, *30*, 539–542. [CrossRef] [PubMed]

41. Biron-Shental, T.; Sukenik Halevy, R.; Goldberg-Bittman, L.; Kidron, D.; Fejgin, M.D.; Amiel, A. Telomeres are shorter in placental trophoblasts of pregnancies complicated with intrauterine growth restriction (IUGR). *Early Hum. Dev.* **2010**, *86*, 451–456. [CrossRef] [PubMed]

42. Toutain, J.; Prochazkova-Carlotti, M.; Cappellen, D.; Jarne, A.; Chevret, E.; Ferrer, J.; Idrissi, Y.; Pelluard, F.; Carles, D.; Maugey-Laulon, B.; et al. Reduced placental telomere length during pregnancies complicated by intrauterine growth restriction. *PLoS ONE* **2013**, *8*, e54013. [CrossRef]

43. Paules, C.; Dantas, A.P.; Miranda, J.; Crovetto, F.; Eixarch, E.; Rodriguez-Sureda, V.; Dominguez, C.; Casu, G.; Rovira, C.; Nadal, A.; et al. Premature placental aging in term small-for-gestational-age and growth-restricted fetuses. *Ultrasound Obs. Gynecol.* **2019**, *53*, 615–622. [CrossRef] [PubMed]

44. Biron-Shental, T.; Sukenik-Halevy, R.; Sharon, Y.; Goldberg-Bittman, L.; Kidron, D.; Fejgin, M.D.; Amiel, A. Short telomeres may play a role in placental dysfunction in preeclampsia and intrauterine growth restriction. *Am. J. Obs. Gynecol.* **2010**, *202*, 381.e1–381.e7. [CrossRef] [PubMed]

45. Jones, C.W.; Esteves, K.C.; Gray, S.A.O.; Clarke, T.N.; Callerame, K.; Theall, K.P.; Drury, S.S. The transgenerational transmission of maternal adverse childhood experiences (ACEs): Insights from placental aging and infant autonomic nervous system reactivity. *Psychoneuroendocrinology* **2019**, *106*, 20–27. [CrossRef] [PubMed]

46. Matthey, S.; Henshaw, C.; Elliott, S.; Barnett, B. Variability in use of cut-off scores and formats on the Edinburgh Postnatal Depression Scale: Implications for clinical and research practice. *Arch. Women's Ment. Health* **2006**, *9*, 309–315. [CrossRef] [PubMed]

47. Murray, L.; Carothers, A.D. The validation of the Edinburgh Post-natal Depression Scale on a community sample. *Br. J. Psychiatry J. Ment. Sci.* **1990**, *157*, 288–290. [CrossRef]

48. Entringer, S.; Epel, E.S.; Lin, J.; Buss, C.; Shahbaba, B.; Blackburn, E.H.; Simhan, H.N.; Wadhwa, P.D. Maternal psychosocial stress during pregnancy is associated with newborn leukocyte telomere length. *Am. J. Obs. Gynecol.* **2013**, *208*, 134.e1–134.e7. [CrossRef]

49. Marchetto, N.M.; Glynn, R.A.; Ferry, M.L.; Ostojic, M.; Wolff, S.M.; Yao, R.; Haussmann, M.F. Prenatal stress and newborn telomere length. *Am. J. Obs. Gynecol.* **2016**, *215*, 94. [CrossRef]

50. Send, T.S.; Gilles, M.; Codd, V.; Wolf, I.; Bardtke, S.; Streit, F.; Strohmaier, J.; Frank, J.; Schendel, D.; Sutterlin, M.W.; et al. Telomere Length in Newborns is Related to Maternal Stress During Pregnancy. *Neuropsychopharmacology* **2017**, *42*, 2407–2413. [CrossRef] [PubMed]

51. Entringer, S.; Epel, E.S.; Kumsta, R.; Lin, J.; Hellhammer, D.H.; Blackburn, E.H.; Wust, S.; Wadhwa, P.D. Stress exposure in intrauterine life is associated with shorter telomere length in young adulthood. *Proc. Natl. Acad. Sci. USA* **2011**, *108*, E513–E518. [CrossRef]

52. Rosa, M.J.; Hsu, H.L.; Just, A.C.; Brennan, K.J.; Bloomquist, T.; Kloog, I.; Pantic, I.; Mercado Garcia, A.; Wilson, A.; Coull, B.A.; et al. Association between prenatal particulate air pollution exposure and telomere length in cord blood: Effect modification by fetal sex. *Env. Res.* **2019**, *172*, 495–501. [CrossRef]

53. Bosquet Enlow, M.; Petty, C.R.; Hacker, M.R.; Burris, H.H. Maternal psychosocial functioning, obstetric health history, and newborn telomere length. *Psychoneuroendocrinology* **2021**, *123*, 105043. [CrossRef] [PubMed]

54. Martens, D.S.; Janssen, B.G.; Bijnens, E.M.; Clemente, D.B.P.; Vineis, P.; Plusquin, M.; Nawrot, T.S. Association of Parental Socioeconomic Status and Newborn Telomere Length. *Jama. Netw. Open* **2020**, *3*, e204057. [CrossRef] [PubMed]

55. Roger, L.; Jones, R.E.; Heppel, N.H.; Williams, G.T.; Sampson, J.R.; Baird, D.M. Extensive telomere erosion in the initiation of colorectal adenomas and its Assoc. with chromosomal instability. *J. Natl. Cancer Inst.* **2013**, *105*, 1202–1211. [CrossRef] [PubMed]

International Journal of
*Molecular Sciences*

*Article*

# Megalin, Proton Pump Inhibitors and the Renin–Angiotensin System in Healthy and Pre-Eclamptic Placentas

Yuan Sun [1,2,3,†], Lunbo Tan [1,3,†], Rugina I. Neuman [1], Michelle Broekhuizen [1,4], Sam Schoenmakers [5], Xifeng Lu [3] and A. H. Jan Danser [1,*]

[1]  Division of Pharmacology and Vascular Medicine, Department of Internal Medicine, Erasmus MC, 3015 CN Rotterdam, The Netherlands; sunyuan@sztu.edu.cn (Y.S.); l.tan@erasmusmc.nl (L.T.); r.neuman@erasmusmc.nl (R.I.N.); m.broekhuizen@erasmusmc.nl (M.B.)
[2]  Department of Pharmacology, College of Pharmacy, Shenzhen Technology University, Shenzhen 518118, China
[3]  Health Science Center, Department of Physiology, Shenzhen University, Shenzhen 518061, China; x.lu@szu.edu.cn
[4]  Division of Neonatology, Department of Pediatrics, Erasmus MC, 3015 CN Rotterdam, The Netherlands
[5]  Department of Obstetrics and Gynaecology, Erasmus MC, 3015 CN Rotterdam, The Netherlands; s.schoenmakers@erasmusmc.nl
*  Correspondence: a.danser@erasmusmc.nl; Tel.: +31-10-7043540
†  These authors contributed equally to this manuscript.

Citation: Sun, Y.; Tan, L.; Neuman, R.I.; Broekhuizen, M.; Schoenmakers, S.; Lu, X.; Danser, A.H.J. Megalin, Proton Pump Inhibitors and the Renin–Angiotensin System in Healthy and Pre-Eclamptic Placentas. *Int. J. Mol. Sci.* **2021**, *22*, 7407. https://doi.org/10.3390/ijms22147407

Academic Editors: Hiten D. Mistry and Eun Lee

Received: 15 June 2021
Accepted: 8 July 2021
Published: 10 July 2021

**Publisher's Note:** MDPI stays neutral with regard to jurisdictional claims in published maps and institutional affiliations.



**Abstract:** Soluble Fms-like tyrosine kinase-1 (sFlt-1) is increased in pre-eclampsia. The proton pump inhibitor (PPI) lowers sFlt-1, while angiotensin increases it. To investigate whether PPIs lower sFlt-1 by suppressing placental renin–angiotensin system (RAS) activity, we studied gene expression and protein abundance of RAS components, including megalin, a novel endocytic receptor for prorenin and renin, in placental tissue obtained from healthy pregnant women and women with early-onset pre-eclampsia. *Renin, ACE, ACE2,* and the angiotensin receptors were expressed at identical levels in healthy and pre-eclamptic placentas, while both the (pro)renin receptor and megalin were increased in the latter. Placental prorenin levels were upregulated in pre-eclamptic pregnancies. Angiotensinogen protein, but not mRNA, was detectable in placental tissue, implying that it originates from maternal blood. Ex vivo placental perfusion revealed a complete washout of angiotensinogen, while prorenin release remained constant. The PPI esomeprazole dose-dependently reduced megalin/(pro)renin receptor-mediated renin uptake in Brown Norway yolk sac epithelial cells and decreased sFlt-1 secretion from placental villous explants. Megalin inhibition blocked angiotensinogen uptake in epithelial cells. In conclusion, our data suggest that placental RAS activity depends on angiotensinogen taken up from the maternal systemic circulation. PPIs might interfere with placental (pro)renin-AGT uptake/transport, thereby reducing angiotensin formation as well as angiotensin-induced sFlt-1 synthesis.

**Keywords:** pre-eclampsia; renin–angiotensin system; megalin; proton pump inhibitors

## 1. Introduction

Pre-eclampsia is a hypertensive disorder characterized by new-onset hypertension and proteinuria occurring after 20 weeks of gestation. It is one of the major causes of worldwide maternal mortality and morbidity [1–3]. A normal pregnancy requires a ≈30% increase in extracellular fluid volume to provide an adequate blood supply for the developing uterus, placenta and fetus [4]. The renin–angiotensin system (RAS) plays a key role in this process. Estrogen stimulates hepatic angiotensinogen (AGT) expression, resulting in a 3–5-fold rise in circulating AGT [4,5]. The plasma renin concentration also rises modestly [6], and together with the rise in AGT, this results in a significant rise in plasma renin activity (PRA); thus, ensuring sufficient RAS activity to allow water and salt retention [7,8]. Remarkably, plasma prorenin levels increase much more strongly than plasma renin levels [6,9], due to prorenin release from the ovaries and, to a lesser degree, the placenta [10]. The function of this prorenin remains unclear.

In pre-eclampsia, the rise in renin and AGT in the circulation is suppressed, although sensitivity to angiotensin (Ang) II is enhanced [11]. Circulating prorenin levels in pre-eclamptic women are in the normal pregnancy range [12]. Given the local production of prorenin in the placenta, while the placenta is a key player in the development of pre-eclampsia, disturbed placental RAS activity may contribute to this phenomenon. Yet, data on prorenin levels in the uteroplacental unit of pre-eclamptic women are conflicting, with evidence for decreases, increases and no alteration [13–15]. What exactly determines the distribution of prorenin across the ovaries and uteroplacental unit, as well as its release into the blood stream, is unknown. It may require transcellular transport. For instance, amniotic fluid prorenin is of chorionic origin [16], implying that chorionic prorenin is capable of crossing the amnion membrane. A novel player in this field is megalin. Megalin, a recently discovered endocytic receptor for prorenin and renin, not only contributes to renin/prorenin reabsorption in the proximal tubule of the kidney, but simultaneously plays a role in renal Ang II generation, although how exactly this occurs is still unknown [17]. Outside the kidney, megalin is present in syncytiotrophoblast and cytotrophoblast cells in the placenta [18]. Membrane recycling of megalin is a fast process which relies on vacuolar $H^+$-ATPase (V-ATPase)-dependent endosomal acidification [19]. The latter involves the so-called (pro)renin receptor ((P)RR), which is an accessory protein of V-ATPase. Here, it is of interest to note that proton pump inhibitors (PPIs) have been observed to lower the placental release of soluble Fms-like tyrosine kinase-1 (sFlt-1), a mediator of maternal endothelial dysfunction in pre-eclampsia [20,21]. Ang II is a stimulator of sFlt-1 [22]. Although PPIs are capable of inhibiting V-ATPase [23,24], such inhibition did not contribute to their acute effects on sFlt-1 release from human trophoblast cells [25]. A further possibility is that PPIs affect sFlt-1 production indirectly, by preventing RAS component transport/uptake.

In the present study, we, therefore, set out to first quantify the expression of major RAS components (including megalin and (P)RR) and their protein abundance in healthy and pre-eclamptic placentas. We focused on early-onset pre-eclampsia (EoPE), and not late-onset pre-eclampsia, since the latter has a different pathophysiological mechanism, being more a maternal rather than a placental syndrome, and showing clear histopathological differences.

Next, we studied the release of RAS components from ex vivo-perfused placentas of healthy and pre-eclamptic pregnancies. Third, we verified whether PPIs modulate megalin-dependent renin uptake in megalin-expressing Brown Norway yolk sac epithelial cells (BN16 cells). Finally, we investigated whether PPIs regulate sFlt-1 secretion in both healthy and pre-eclamptic placental explants.

## 2. Results

Clinical characteristics of the patients are provided in Table 1. The data confirm that our PE patients fulfilled the ISSHP 2018 criteria [26].

**Table 1.** Clinical characteristics of placentas.

| Characteristic | Healthy (*n* = 25) | EoPE (*n* = 17) |
| --- | --- | --- |
| Maternal age (years) | 31 (28–36) | 30 (26–32) |
| Nulliparity/multiparity (n) | 6/19 | 11/6 * |
| Ethnicity (n = Caucasian/other) | 15/10 | 12/5 |
| Highest diastolic blood pressure (mmHg) | 80 (74–82) | 109 (100–116) * |
| Urinary protein-to-creatinine ratio | ND | 351 (92–861) * |
| Mode of delivery (n = caesarean/spontaneous) | 25/0 | 16/1 |
| Gestational age (weeks) | 39.0 (38.7–39.1) | 29.9 (28.8–31.8) * |

**Table 1.** *Cont.*

| Characteristic | Healthy (*n* = 25) | EoPE (*n* = 17) |
|---|---|---|
| Sex (n = female/male) | 7/18 | 9/8 |
| Birth weight (g) | 3465 (3175–3763) | 1125 (920–1300) * |
| Birth weight centile < 10th (n) | 0 | 9 |
| Placental weight (g) | 623 (568–729) | 295 (232–347) * |

Data are shown as n (number of cases) or median (interquartile range). * $p < 0.05$. ND, not detected; EoPE, early-onset pre-eclampsia.

### 2.1. Expression of RAS Components in Healthy and Pre-Eclamptic Placentas

Placental expression of *renin*, *AGT*, angiotensin-converting enzyme (*ACE*), *ACE2*, the Ang II type 1 and type 2 receptor (*AT1R*, *AT2R*), and *megalin* was comparable between healthy and pre-eclamptic placentas (*n* = 12 for each), while (P)RR expression was increased in the latter (Figure 1A–H). *AGT* expression was at or below the detection limit (Ct = 38) in most of the samples, even when using two additional sets of primers (Figure 1I,J). Total renin levels in healthy placental tissue amounted to 118 (range 42.5–1476) pg/g tissue, and 72 ± 2% of this was prorenin (Figure 2A–C; *n* = 16). Total renin levels were significantly ($p < 0.05$) higher in pre-eclamptic placentas (266 (range 54.5–1568) pg/g tissue), and this was due to upregulated prorenin ($p < 0.05$) rather than renin. Nevertheless, the proportion of renin in the pre-eclamptic group (29 ± 2%) was not different from that in healthy placentas. Placental AGT abundance was decreased ($p < 0.0001$) in pre-eclamptic placentas (Figure 2D,E), while the opposite was true ($p < 0.01$) for megalin protein abundance (Figure 2F,G).

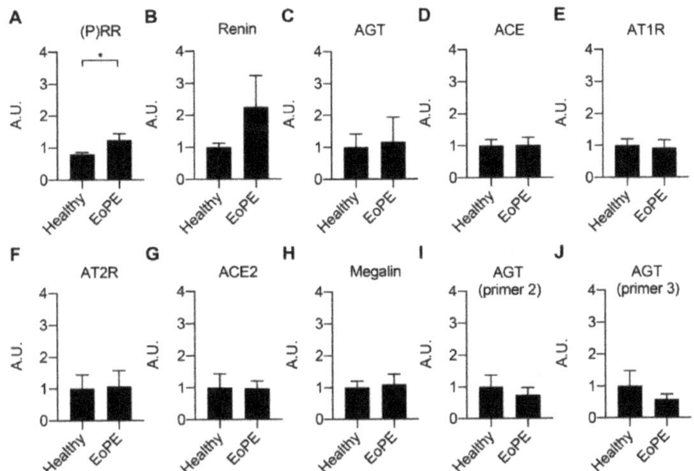

**Figure 1.** Gene expression levels of RAS components (**A–G**) and megalin (**H**) in healthy and early-onset pre-eclamptic (EoPE) placental tissue. (**I,J**): *AGT* expression was confirmed with different pairs of primers. Data are mean ± SEM of *n* = 12. * $p < 0.05$. (P)RR, (pro)renin receptor; *AGT*, angiotensinogen; *ACE*, angiotensin-converting enzyme; *AT1R*, angiotensin II type 1 receptor; *AT2R*, angiotensin II type 2 receptor; A.U., arbitrary unit.

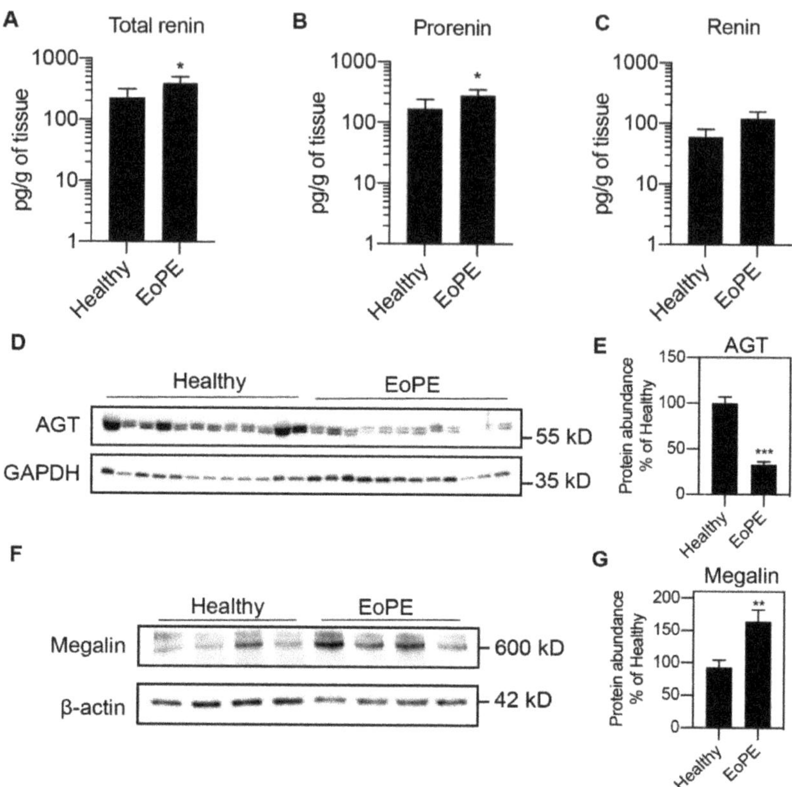

**Figure 2.** Total renin (**A**), renin (**B**) and prorenin (**C**) levels (*n* = 16) measured by immunoradiometric assay and angiotensinogen (AGT; **D,E**) and megalin (**F,G**) protein abundance (*n* = 12) measured by Western blot in healthy and early-onset pre-eclamptic (EoPE) placental tissue. Panels (**D,F**) are representative blots of AGT and megalin protein. Data were normalized versus GAPDH (**E**) or β-actin (**G**). Data are mean ± SEM. * $p < 0.05$, ** $p < 0.005$, *** $p < 0.0005$ vs. Healthy. AGT, angiotensinogen.

### 2.2. Placental Release of (Pro)renin and AGT

Prorenin release into the maternal perfusate from both healthy (*n* = 5) and pre-eclamptic (*n* = 6) placentas remained stable over the entire perfusion period. In two pre-eclamptic placenta perfusions, the perfusion had to be stopped after 90 min because of leakage. Release of prorenin tended to be higher from pre-eclamptic placentas, although significance was reached at two time points only (Figure 3A). Renin release paralleled prorenin release, albeit at >10-fold lower levels (Figure 3B). In contrast, placental AGT release from both healthy and pre-eclamptic placentas peaked initially, and then displayed a steady decline, so that the release was lowest after 180 min (Figure 3C). Neither for renin, nor for AGT, were there differences observed between healthy and pre-eclamptic placentas.

### 2.3. Megalin Internalizes AGT and PPI Decreases Renin Internalization but Not Binding in BN16 Cells

Megalin is known to be a receptor for both renin and prorenin [27]. To verify whether it also binds and internalizes AGT, we incubated BN16 cells with tagged AGT. BN16 cells accumulated AGT following a 2 h incubation period (Figure 4A). Inhibiting megalin expression greatly suppressed this accumulation, confirming that megalin also underlies AGT uptake (Figure 4B,C). We next tested whether the PPI esomeprazole regulates renin binding and internalization. Esomeprazole dose-dependently reduced renin uptake in

BN16 cells, but did not affect renin binding (Figure 4D,E). This fully paralleled the effect of bafilomycin A1 (Figure 4F,G), and illustrates that esomeprazole, such as bafilomycin A1, interferes with the megalin-mediated internalization process.

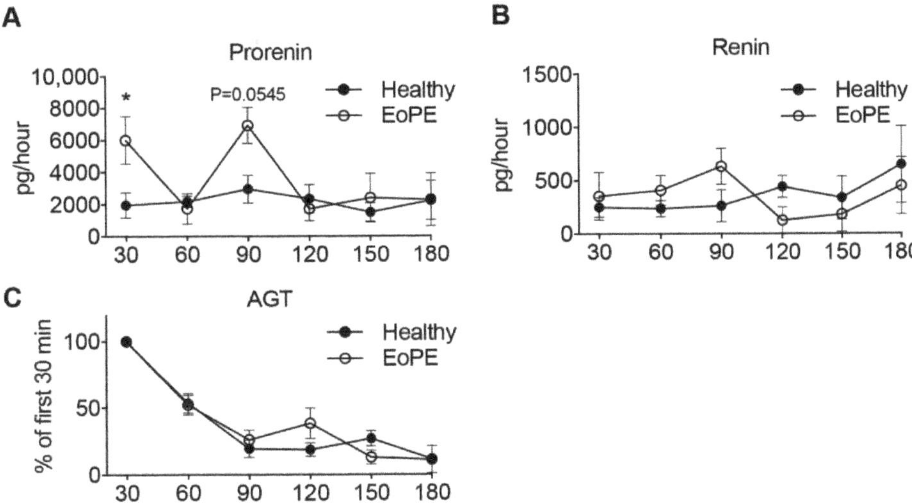

**Figure 3.** Prorenin (**A**), renin (**B**) and angiotensinogen (**C**) release from isolated perfused healthy ($n = 5$) and early-onset pre-eclamptic (EoPE, $n = 6$) cotyledons. AGT: angiotensinogen. Perfusate collection from two of the pre-eclamptic placentas could not be continued over the full 180 min due to fetal-to-maternal leakage. In such cases, only the samples prior to the occurrence of leakage were used. Data are mean ± SEM. * $p < 0.05$.

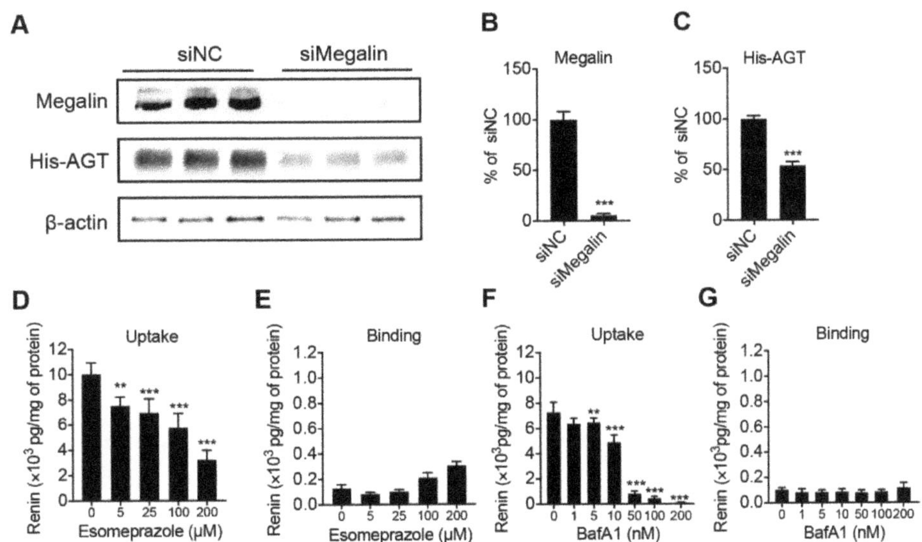

**Figure 4.** Panels (**A–C**): representative blots of His-AGT protein abundance in BN16 cells treated with negative control (siNC) or siRNA against megalin (siMegalin). Panels, (**D–G**): cell-associated renin levels after incubating BN16 cells with 100 U/L renin at 37 °C (**D,F**) or 4 °C (**E,G**) for 2 h in the presence of increasing concentrations of esomeprazole (**D,E**) or bafilomycin A1 (**F,G**). Data are mean ± SEM of $n = 5$–9. ** $p < 0.005$, *** $p < 0.0005$ vs. control.

### 2.4. PPI Reduces sFlt-1 Secretion in Human Placental Villous Explants

As expected, pre-eclamptic placenta explants release more sFlt-1 than healthy placenta explants (Figure 5A). This is in agreement with the increased Flt-1 expression in pre-eclamptic explants (Figure 5B). Esomeprazole-suppressed Flt-1 expression and sFlt-1 release particularly in pre-eclamptic explants and, as a consequence, Flt-1 expression in pre-eclamptic explants after this drug was identical to that in healthy placenta explants.

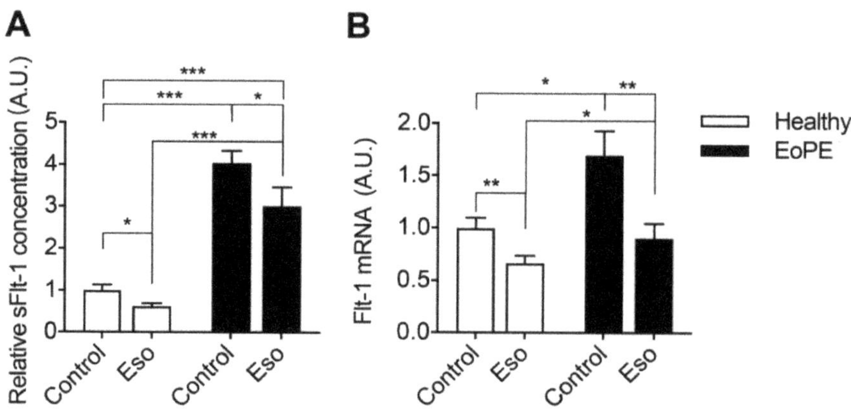


**Figure 5.** sFlt-1 levels in medium (**A**) and mRNA levels in explants (**B**) after incubating healthy or pre-eclamptic (EoPE) placental villous explants with 100 µM of esomeprazole (Eso) for 48 h. sFlt-1 level: healthy $n$ = 18, EoPE $n$ = 8; Flt-1 mRNA level: healthy $n$ = 7, EoPE $n$ = 4. * $p < 0.05$, ** $p < 0.005$, *** $p < 0.0005$ vs. 0. A.U., arbitrary unit.

## 3. Discussion

The existence of a placental RAS has been proposed decades ago [28]. Although it is believed that it may contribute to pre-eclampsia, its exact role is still unknown. This relates to the fact that we still do not fully understand how local Ang II synthesis occurs in the placenta. In the current study, we found that the mRNA levels of RAS components are comparable in healthy and pre-eclamptic placentas. This agrees with previous studies showing that *renin, AGT, ACE* and *AT2R* expression are similar in normal and pre-eclamptic decidua basalis [15,29,30]. Simultaneously, uteroplacental AT1R [30–32] and renin [32] upregulation have also been claimed to be involved in the pathogenesis of pre-eclampsia. One explanation for these conflicting observations is that different tissues were studied, ranging from endometrium and myometrium to decidual tissue. In our study, we collected tissue from the maternal side of the placenta, i.e., decidua basalis and part of chorionic villi. A second explanation is that placental RAS expression may change with increasing gestational age [32]. Thus, the 10-week difference in gestational age between the healthy and pre-eclamptic women of the current study is a confounding factor. Nevertheless, it is important to note that symptoms in early-onset pre-eclampsia (before 34 ± 0 weeks of gestation, such as in our study), develop much earlier [33], implying that changes in placental RAS expression, if having a causal effect, may be more drastic at this early gestational age.

Importantly, placental *AGT* expression in the present study was low and in most cases undetectable. Such low expression was observed earlier, when making a comparison with prolactin and renin using Northern blot analysis [34]. In situ mRNA detection via the padlock probes method also did not detect *AGT* expression in the placenta [35]. Despite the absence of significant expression, AGT protein was easily detectable in placental tissue. This implies that placental AGT must have been taken up from maternal blood. In agreement with this concept, placental AGT levels were lower in pre-eclamptic women, who are known to have greatly diminished circulating AGT levels. Moreover, perfusing the placenta with buffer resulted in a washout of AGT. Animal data similarly support a major role for AGT of maternal origin. When female mice carrying the human renin transgene were crossed with male mice carrying the human AGT transgene (both of which do not react with their mouse counterparts), the pregnant mice remained normotensive. Yet, when crossing female human AGT transgenic mice with male human renin transgenic mice, the pregnant mice did develop a pre-eclamptic phenotype, characterized by hypertension, elevated sFlt-1 levels and proteinuria [36]. Moreover, specifically inhibiting human AGT expression in the maternal liver with liver-targeted siRNA in the latter model suppressed the pre-eclamptic phenotype [37]. Taken together, these data imply that placental angiotensin generation relies on maternal AGT.

Unlike AGT release, which peaked immediately and then decreased, placental (pro)renin release remained stable over a 3 h perfusion period in healthy placentas. Since the total renin release in healthy placentas amounted to $\approx$2000 pg/h, and considering that a cotyledon wet weight amounted to $\approx$25 g, the total renin amount released during the 3 h perfusion period can be estimated to be 2500/25 $\times$ 3 = 300 pg/g. This is 2–3 times the amount detected in placental tissue. These data, therefore, strongly support the continuous synthesis and release of (pro)renin from the placenta, in full agreement with previous findings [28,38,39]. Importantly, the circulating renin levels in pre-eclamptic women are lower than those in healthy pregnant women, while their circulating prorenin levels mimic those in healthy pregnancy [12]. The present study observed 2-fold higher prorenin levels in pre-eclamptic placentas, while placental renin was identical in healthy and pre-eclamptic women. Placental (pro)renin release paralleled these latter observations. Taken together, these data suggest that placental (pro)renin synthesis occurs independently of renal (pro)renin synthesis. Yet, whether this results in independent upregulation of Ang II generation at placental tissue sites cannot be concluded, particularly since this would rely on the uptake of circulating AGT, which is lower in pre-eclampsia. Here, there might be a role for the megalin/V-ATPase pathway.

The (P)RR is an accessory protein of the V-ATPase. Its upregulation in pre-eclampsia, combined with increased megalin levels, favors increased activity of the megalin–V-ATPase pathway in this condition [17,40]. We recently reported that megalin is a novel endocytic receptor for (pro)renin in the kidney [27]. Inhibition of megalin and the (P)RR similarly suppressed endocytosis, without showing additive effects [25]. This confirms that megalin-mediated uptake relies on (P)RR/V-ATPase-dependent endosomal acidification. These data were obtained in BN16 cells, which predominantly express megalin, and no other renin/prorenin-binding receptors such as the mannose-6-phosphate receptor [27]. The latter receptor also occurs in trophoblast cells [41]. Using megalin-expressing BN16 cells, we were now able to show that the megalin–V-ATPase pathway also underlies AGT uptake. Moreover, we observed that the PPI esomeprazole, such as the V-ATPase inhibitor bafilomycin A1, inhibits megalin-mediated internalization. Given that megalin-mediated AGT uptake underlies renal angiotensin generation [17], we speculate that a similar phenomenon may occur in the placenta. Here, we emphasize that the greater sensitivity of pre-eclamptic women to Ang II [11,42] may imply that placental Ang II levels do not necessarily have to be higher than those in healthy placentas in order to exert the same or even a stronger effect. Onda et al. showed that PPIs lower sFlt-1 synthesis in placental explants (confirmed in the present study), but were unable to link this to a direct consequence of V-ATPase inhibition with bafilomycin A1 [25]. Our data now reveal a new option: PPIs might suppress placental RAS activity by interfering either with AGT uptake, with (pro)renin transport, or both. Since lower placental Ang II levels are expected to result in diminished sFlt-1 production, this concept offers a novel explanation for the observation that PPIs lower circulating sFlt-1 in pre-eclamptic women (Figure 6).

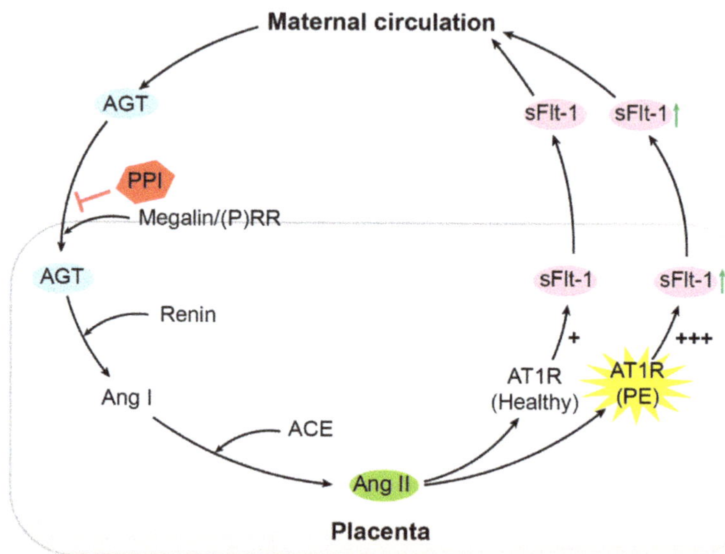



**Figure 6.** Schematic overview. AGT, angiotensinogen; ⊣, inhibition; PPI, proton pump inhibitor; (P)RR, (pro)renin receptor; Ang I, angiotensin I; ACE, angiotensin-converting enzyme; Ang II, angiotensin II; AT1R, angiotensin II type 1 receptor; PE, pre-eclampsia; sFlt-1, soluble Fms-like tyrosine kinase-1; +, normal sensitivity to Ang II; +++, increased sensitivity to Ang II.

## 4. Materials and Methods

### 4.1. Tissue Collection

The experiments were conducted using human placental tissue from women with a healthy, term pregnancy or with early-onset pre-eclampsia (<34 wk of GA) based on the ISSHP 2018 criteria [26]. All but one woman with early-onset pre-eclampsia delivered by caesarean section. The indication for caesarean section in healthy women was elective, due to either a previous caesarean section or breech position. In addition, all women gave informed consent prior to delivery. Placental tissue was collected immediately after delivery at the Erasmus Medical Center, Rotterdam, The Netherlands. Tissue sections were cut from the decidual side of the placenta and included decidua basalis and chorionic villi. They were snap frozen in liquid nitrogen and stored at −80 °C. Areas displaying necrosis, tissue damage, calcification, hematoma or tears were avoided. The study was exempted from approval by the local institutional Medical Ethics Committee according to the Dutch Medical Research with Human Subjects Law (MEC-2016-418 and MEC-2017-418).

### 4.2. Placenta Perfusion Experiments

The perfusion method used in this study was previously described by Hitzerd et al. [43]. In short, placentas were perfused at 37 °C with aerated (95% $O_2$ and 5% $CO_2$) perfusion medium consisting of Krebs–Henseleit buffer, supplemented with heparin (2500 IU/L; LEO Pharma B.V., Amsterdam, The Netherlands). The fetal circulation (closed-circuit; flow rate 6 mL/min) was established by cannulating the chorionic artery and corresponding vein of an intact cotyledon. Maternal circulation (closed-circuit; flow rate 12 mL/min) was created by placing four blunt cannulas in the intervillous space. Perfusion experiments were conducted in healthy and pre-eclamptic placentas. The healthy group consisted of placentas from uncomplicated, normotensive pregnancies in which the endothelin receptor antagonist ambrisentan (10 mg/L; Sigma-Aldrich, Schnelldorf, Germany) or the phosphodiesterase-5 inhibitor sildenafil (500 ng/mL; Pfizer Europe MA EEIG, Brussel, Belgium) was added to the maternal circulation at t = 0 [43,44]. The pre-eclamptic group consisted of placentas from early-onset pre-eclamptic pregnancies (<34 weeks) which were perfused with sildenafil or no drug. No drug-related differences in RAS component release were noted and, thus, all samples per group were combined. Samples from the maternal circulations were taken every 30 min (until 180 min) and immediately stored at $-80\,^\circ$C. To control the quality of perfusion, 100 mg/L of antipyrine (Sigma-Aldrich, St. Louis, MO, USA) and 36 mg/L of FITC-dextran (40 kDa; Sigma-Aldrich) were added to the fetal and maternal buffer, respectively. An experiment was considered successful when the fetal-to-maternal (F/M) ratio of antipyrine was >0.75 and the maternal-to-fetal (M/F) ratio of FITC-dextran was <0.03 at t = 180 min. The release of angiotensinogen, renin or prorenin was expressed per hour, correcting for the level that was present in the previous sample.

### 4.3. Placental Villous Explants

Freshly obtained placental tissue slices were cut from three different areas of each placenta and washed three times in cold phosphate-buffered saline (PBS; Lonza, Walkersville, MD, USA), after which the decidua and chorionic plate were removed. Tissue containing chorionic villi was then cut in explant blocks of 2 × 2 mm. Explants from the three different areas were combined in one well in 2 mL DMEM/F12 medium (Gibco, Thermo Fisher Scientific, Paisley, UK) containing 10% FCS (GE Healthcare, Eindhoven, The Netherlands), 1.95 g/L $NaHCO_3$ and 100 μg/mL Primocin (Invivogen, San Diego, CA, USA) in 12-well plates, and equilibrated at $37\,^\circ$C for 3 h at 8% $O_2$ and 5% $CO_2$. Thereafter, explants were transferred to a new plate and incubated with or without 100 μmol/L esomeprazole (Sigma-Aldrich) in the above medium for 24 h. sFlt-1 was measured in the medium using the human VEGFR1/Flt-1 DuoSet ELISA (R&D Systems, Minneapolis, MN, USA).

### 4.4. Placental Lysates

About 100 mg of placental tissue was homogenized with grind beads in 1 mL ice-cold buffer C (12 mmol/L $NaH_2PO_4 \cdot 2H_2O$, 86.7 mmol/L $Na_2HPO_4$, 15.9 mmol/L NaCl) and 1× complete protease inhibitors (Roche, Mannheim, Germany) at $4\,^\circ$C with a TissueLyser 24 (Shanghai Jingxin, Shanghai, China) 3 × 60 s at 60 Hz. The lysates were centrifuged at 14,000× $g$ for 10 min at $4\,^\circ$C. Then, the supernatant was transferred to a new tube and kept on ice. The pellet was resuspended in 200 μL of buffer C with protease inhibitors, and was additionally homogenized for 2 × 60 s at 60 Hz. After centrifuging at 14,000× $g$ for 10 min at $4\,^\circ$C, the second supernatant was added to the first supernatant, and the combined supernatants were stored at $-80\,^\circ$C.

### 4.5. Studies in Brown Norway Rat Yolk Sac Epithelial Cells

Brown Norway rat yolk sac epithelial cells (BN16) were cultured in Minimum Essential Media (MEM) (Gibco) supplemented with 1 × GlutaMAX (Gibco) and 10% FCS at $37\,^\circ$C in a humidified incubator with 5% $CO_2$. BN16 cells were seeded at a density of 2 × $10^5$ cells per well in 24-well plates and cultured for 48 h before carrying out experiments. To study the effect of proton pump inhibition on renin binding and internalization, BN16 cells were first washed twice with PBS, and then pre-incubated with MEM without FCS for 30 min at $37\,^\circ$C. Next, cells were washed with ice-cold PBS twice, and incubated with 300 μL MEM containing 0.2 μg/mL of recombinant human renin (a gift from Actelion Pharmaceuticals, Allschwil, Switzerland) in the presence or absence of increasing concentrations of the $H^+/K^+$ ATPase inhibitor esomeprazole (Sigma-Aldrich) or the V-ATPase inhibitor bafilomycin A1 (Sigma-Aldrich) at either $4\,^\circ$C (to quantify renin binding) or $37\,^\circ$C (to quantify renin internalization). After 2 h of incubation, the culture medium was discarded, and the cells were washed twice with ice-cold PBS containing 0.5% BSA (Sigma-Aldrich) and twice with ice-cold PBS. Then, the cells were lysed with ice-cold PBS containing 0.2% Triton X-100 (SeeVa, Heidelberg, Germany) and 1× complete protease inhibitors. Lysate was centrifuged at 14,000× $g$ for 10 min at $4\,^\circ$C to remove any cell debris and the supernatant was stored at $-80\,^\circ$C until use.

To study whether megalin regulates angiotensinogen uptake, BN16 cells were transfected with 50 µmol/L of negative control (siNC) (Invitrogen, Paisley, UK) or siRNA against megalin (siMegalin) (Invitrogen) for 48 h by using RNAi max transfection reagent (Invitrogen). Then the cells were incubated with 10 µg/mL of His-tagged recombinant rabbit angiotensinogen (Sino Biological, Hong Kong, China) for 2 h. After incubation, cells were washed once with ice-cold PBS with 0.5% BSA and twice with ice-cold PBS, and then collected for immunoblotting.

### 4.6. Measurement of (Pro)renin and Angiotensinogen

Renin was measured in the placental - and BN16 cell lysates with the Renin III (Cisbio, Gif-sur-Yvette, France) immunoradiometric assay (detection limit 2 pg/mL). Total renin was measured in the lysates with the same assay, after activating prorenin with 10 µmol/L aliskiren (allowing its detection in the assay) at 4 °C for 48 h [45]. Prorenin in the lysates was calculated by subtracting renin from total renin. Given their low levels, renin and prorenin in the placental perfusates (the latter after its conversion to renin by trypsin) were measured by the more sensitive enzyme-kinetic assay as described before (detection limit 0.05 ng angiotensin I per mL per hour) [46]. Data were converted to pg/mL given that 1 ng Ang I/mL per hour equals 2.6 pg/mL renin [47]. AGT in the perfusates was measured as the maximum quantity of Ang I that was generated during incubation with excess recombinant renin (detection limit 0.5 pmol/mL) [46].

### 4.7. Immunoblotting

BN16 cells or placental tissue pieces were homogenized in RIPA buffer (150 mmol/L NaCl, 1% Triton X-100, 0.5% sodium deoxycholate, 0.1% SDS, 50 mmol/L Tris, pH 8.0) containing 1 × complete protease inhibitors with a TissueLyser (Polytron PT2100, Littau-Lucerne, Switzerland). Lysates were cleared by centrifugation at $14,000 \times g$ for 10 min at 4 °C. The total protein concentration in supernatant was determined by BCA assay (Pierce, Waltham, MA, USA). Equal amounts of protein (30–40 µg) were loaded and separated on Precast Midi Protein Gel (Bio-Rad, Hercules, CA, USA), and transferred to PVDF membranes using semi-dry Trans-Blot Turbo Transfer system (Bio-Rad). The blots were then probed with antibodies against angiotensinogen (1:100, Abbiotec, Shenzhen, China), GAPDH (1:5000, GeneTex, Hsinchu, China), β-actin (1:5000, Merck Millipore, Darmstadt, Germany), or megalin (1:500, Biotech, Wuhan, China), and detected by using Clarity Western ECL Substrate (Bio-Rad). The intensities of bands were analyzed using ImageJ.

### 4.8. RNA Isolation and qPCR Analysis

Total RNA was extracted using the Direct-zol RNA kit (Zymo Research, Irvine, USA). One microgram total RNA was reverse transcribed using QuantiTect® Reverse Transcription Kit (Qiagen, Hilden, Germany). SYBR Green real-time quantitative PCR assays were performed on QuantStudio 7 Flex Real-Time PCR Systems (Thermo Fisher, Waltham, USA) using SYBR®Premix Ex TaqTM II kit (Qiagen, Venlo, The Netherlands). Primers used in the study are *(P)RR* (forward: 5′-TCTCAGTTCACTCCCCCTCAA-3′; reverse: 5′-GATGCTTATGACGAGACAGCAAG-3′), *renin* (forward: 5′-GCCGTCTCTACACTGCC-TGT-3′; reverse: 5′-GGAGGGTGAGTTCTGTTCCA-3′), *AGT* (forward: 5′-TCAACACCT ACGTCCACTTCC-3′; reverse: 5′-CACTGAGGTGCTGTTGTCCA-3′), *AGT* primer 2 (forward: 5′-ACCTACGTCCACTTCCAAGG-3′; reverse: 5′-GTTGTCCACCCAGAACTCCT-3′), *AGT* primer 3 (forward: 5′-ACAAGGTGGAGGGTCTCACT-3′; reverse: 5′-TGGATGGT CCGGGGAGATAG-3′), *ACE* (forward: 5′-CCAACCTCGATGTCACCAGT-3′, reverse: 5′-TCGACCCTTCCCAGAACTC-3′), *AT1R* (forward: 5′-TGACAGTCCAAAGGCTCCA-3′, reverse: 5′-TTTGATCACCTGGGTCGAAT-3′), *AT2R* (forward: 5′-GGCAACTCCACCCTTGC CACT-3′, reverse: 5′-TGCCAGAGATGTTCACAAGCCCGA-3′), *ACE2* (forward: 5′-TTCTG TCACCCGATTTTCAA-3′; reverse: 5′-TCCCAACAATCGTGAGTGC-3′), *megalin* (forward: 5′-CTGCTCCTGGCTCTCGTC-3′; reverse: 5′-CTTTGGTCCCATCACACCTC-3′), *β-actin* (forward: 5′-CTGCTCCTGGCTCTCGTC-3′; reverse: 5′-CTTTGGTCCCATCACACCTC-3′),

Pre-mRNA Processing Factor 38A (*PRPF38A*) (forward: 5′-GTTAAGGTTTGTGGGTGGCG-3′; reverse: 5′-AGCATGCGGACATACTTGAAATC-3′), DExD-Box Helicase 50 (*DDX50*) (forward: 5′-GCCTCCTGAAAGGAAATATGG-3′; reverse: 5′-AGTATCCAGTCGGAATCA TGC-3′), *Flt-1* (forward: 5′-ACAATCAGAGGTGAGCACTGCAA-3′; reverser: 5′-TCCGAG CCTGAAAGTTAGCAA-3′) (Invitrogen). *β-actin, PRPF38A* and *DDX50* were used as housekeeping genes for calculating the relative expressions of target genes, and the geometric mean of the three relative expressions were used for calculating the arbitrary unit (A.U.).

### 4.9. Statistics

Data are provided as mean $\pm$ SEM or geometric mean and range. When comparing differences between two groups, Student's *t*-test was used. One-way ANOVA or two-way ANOVA with Bonferroni post-modification was used when comparing differences between more than two groups. The Kolmogorov–Smirnov test was used to verify normal distribution. For data that were non-normally distributed, a non-parametric *t*-test (Mann–Whitney U test) was used to compare the differences between groups. Categorical variables were evaluated by Chi-square test. $p < 0.05$ was considered significant. Data below detection limit were assumed to equal the detection limit.

**Author Contributions:** Conceptualization, A.H.J.D.; formal analysis, L.T. and Y.S.; resources, R.I.N., M.B. and S.S.; perfusion experiments, R.I.N., and M.B.; writing—original draft preparation, Y.S. and X.L.; writing—review and editing, Y.S., L.T., R.I.N., M.B., S.S., X.L. and A.H.J.D. All authors have read and agreed to the published version of the manuscript.

**Funding:** This research was funded by the National Natural Science Foundation of China (grant nrs. 81800383 and 81870605), Shenzhen Key Laboratory of Metabolism and Cardiovascular Homeostasis (ZDSYS20190902092903237), and Shenzhen Municipal Science and Technology Innovation Council (grant no. JCYJ20170817093928508 and JCYJ20190808170401660).

**Institutional Review Board Statement:** The study was conducted according to the guidelines of the Declaration of Helsinki and approved by the local institutional Medical Ethics Committee according to the Dutch Medical Research with Human Subjects Law (MEC-2016-418 and MEC-2017-418).

**Informed Consent Statement:** Informed consent was obtained from all subjects involved in the study.

**Data Availability Statement:** All data supporting the reported results can be acquired from the corresponding author upon reasonable request.

**Conflicts of Interest:** The authors declare no conflict of interest.

## References

1. Mol, B.W.J.; Roberts, C.T.; Thangaratinam, S.; Magee, L.A.; de Groot, C.J.M.; Hofmeyr, G.J. Pre-eclampsia. *Lancet* **2016**, *387*, 999–1011. [CrossRef]
2. Miklus, R.M.; Elliott, C.; Snow, N. Surgical cricothyrotomy in the field: Experience of a helicopter transport team. *J. Trauma* **1989**, *29*, 506–508. [CrossRef] [PubMed]
3. Phipps, E.A.; Thadhani, R.; Benzing, T.; Karumanchi, S.A. Pre-eclampsia: Pathogenesis, novel diagnostics and therapies. *Nat. Rev. Nephrol.* **2019**, *15*, 275–289. [CrossRef]
4. Verdonk, K.; Visser, W.; Van Den Meiracker, A.H.; Danser, A.H. The renin-angiotensin-aldosterone system in pre-eclampsia: The delicate balance between good and bad. *Clin. Sci.* **2014**, *126*, 537–544. [CrossRef] [PubMed]
5. Rana, S.; Powe, C.E.; Salahuddin, S.; Verlohren, S.; Perschel, F.H.; Levine, R.J.; Lim, K.H.; Wenger, J.B.; Thadhani, R.; Karumanchi, S.A. Angiogenic factors and the risk of adverse outcomes in women with suspected preeclampsia. *Circulation* **2012**, *125*, 911–919. [CrossRef]
6. Derkx, F.H.; Alberda, A.T.; de Jong, F.H.; Zeilmaker, F.H.; Makovitz, J.W.; Schalekamp, M.A. Source of plasma prorenin in early and late pregnancy: Observations in a patient with primary ovarian failure. *J. Clin. Endocrinol. Metab.* **1987**, *65*, 349–354. [CrossRef] [PubMed]
7. Hsueh, W.A.; Luetscher, J.A.; Carlson, E.J.; Grislis, G.; Fraze, E.; McHargue, A. Changes in active and inactive renin throughout pregnancy. *J. Clin. Endocrinol. Metab.* **1982**, *54*, 1010–1016. [CrossRef]
8. Irani, R.A.; Xia, Y. Renin angiotensin signaling in normal pregnancy and preeclampsia. *Semin. Nephrol.* **2011**, *31*, 47–58. [CrossRef]
9. Sealey, J.E.; McCord, D.; Taufield, P.A.; Ales, K.A.; Druzin, M.L.; Atlas, S.A.; Laragh, J.H. Plasma prorenin in first-trimester pregnancy: Relationship to changes in human chorionic gonadotropin. *Am. J. Obstet. Gynecol.* **1985**, *153*, 514–519. [CrossRef]

10. Krop, M.; Danser, A.H. Circulating versus tissue renin-angiotensin system: On the origin of (pro)renin. *Curr. Hypertens. Rep.* **2008**, *10*, 112–118. [CrossRef]

11. Irani, R.A.; Xia, Y. The functional role of the renin-angiotensin system in pregnancy and preeclampsia. *Placenta* **2008**, *29*, 763–771. [CrossRef] [PubMed]

12. Verdonk, K.; Saleh, L.; Lankhorst, S.; Smilde, J.E.; van Ingen, M.M.; Garrelds, I.M.; Friesema, E.C.; Russcher, H.; van den Meiracker, A.H.; Visser, W.; et al. Association studies suggest a key role for endothelin-1 in the pathogenesis of preeclampsia and the accompanying renin-angiotensin-aldosterone system suppression. *Hypertension* **2015**, *65*, 1316–1323. [CrossRef] [PubMed]

13. Brar, H.S.; Kjos, S.L.; Dougherty, W.; Do, Y.S.; Tam, H.B.; Hsueh, W.A. Increased fetoplacental active renin production in pregnancy-induced hypertension. *Am. J. Obstet. Gynecol.* **1987**, *157*, 363–367. [CrossRef]

14. Kalenga, M.K.; Thomas, K.; de Gasparo, M.; De Hertogh, R. Determination of renin, angiotensin converting enzyme and angiotensin II levels in human placenta, chorion and amnion from women with pregnancy induced hypertension. *Clin. Endocrinol.* **1996**, *44*, 429–433. [CrossRef] [PubMed]

15. Herse, F.; Dechend, R.; Harsem, N.K.; Wallukat, G.; Janke, J.; Qadri, F.; Hering, L.; Muller, D.N.; Luft, F.C.; Staff, A.C. Dysregulation of the circulating and tissue-based renin-angiotensin system in preeclampsia. *Hypertension* **2007**, *49*, 604–611. [CrossRef] [PubMed]

16. Pringle, K.G.; Wang, Y.; Lumbers, E.R. The synthesis, secretion and uptake of prorenin in human amnion. *Physiol. Rep.* **2015**, *3*. [CrossRef]

17. Koizumi, M.; Ueda, K.; Niimura, F.; Nishiyama, A.; Yanagita, M.; Saito, A.; Pastan, I.; Fujita, T.; Fukagawa, M.; Matsusaka, T. Podocyte Injury Augments Intrarenal Angiotensin II Generation and Sodium Retention in a Megalin-Dependent Manner. *Hypertension* **2019**, *74*, 509–517. [CrossRef]

18. Storm, T.; Christensen, E.I.; Christensen, J.N.; Kjaergaard, T.; Uldbjerg, N.; Larsen, A.; Honore, B.; Madsen, M. Megalin Is Predominantly Observed in Vesicular Structures in First and Third Trimester Cytotrophoblasts of the Human Placenta. *J. Histochem. Cytochem.* **2016**, *64*, 769–784. [CrossRef]

19. Gleixner, E.M.; Canaud, G.; Hermle, T.; Guida, M.C.; Kretz, O.; Helmstadter, M.; Huber, T.B.; Eimer, S.; Terzi, F.; Simons, M. V-ATPase/mTOR signaling regulates megalin-mediated apical endocytosis. *Cell Rep.* **2014**, *8*, 10–19. [CrossRef]

20. Hastie, R.; Bergman, L.; Cluver, C.A.; Wikman, A.; Hannan, N.J.; Walker, S.P.; Wikstrom, A.K.; Tong, S.; Hesselman, S. Proton Pump Inhibitors and Preeclampsia Risk Among 157,720 Women. *Hypertension* **2019**, *73*, 1097–1103. [CrossRef]

21. Saleh, L.; Samantar, R.; Garrelds, I.M.; van den Meiracker, A.H.; Visser, W.; Danser, A.H.J. Low Soluble Fms-Like Tyrosine Kinase-1, Endoglin, and Endothelin-1 Levels in Women With Confirmed or Suspected Preeclampsia Using Proton Pump Inhibitors. *Hypertension* **2017**, *70*, 594–600. [CrossRef]

22. Zhou, C.C.; Ahmad, S.; Mi, T.; Xia, L.; Abbasi, S.; Hewett, P.W.; Sun, C.; Ahmed, A.; Kellems, R.E.; Xia, Y. Angiotensin II induces soluble fms-Like tyrosine kinase-1 release via calcineurin signaling pathway in pregnancy. *Circ. Res.* **2007**, *100*, 88–95. [CrossRef] [PubMed]

23. Spugnini, E.P.; Citro, G.; Fais, S. Proton pump inhibitors as anti vacuolar-ATPases drugs: A novel anticancer strategy. *J. Exp. Clin. Cancer Res.* **2010**, *29*, 44. [CrossRef] [PubMed]

24. Sabolic, I.; Brown, D.; Verbavatz, J.M.; Kleinman, J. H(+)-ATPases of renal cortical and medullary endosomes are differentially sensitive to Sch-28080 and omeprazole. *Am. J. Physiol.* **1994**, *266*, F868–F877. [CrossRef] [PubMed]

25. Onda, K.; Tong, S.; Beard, S.; Binder, N.; Muto, M.; Senadheera, S.N.; Parry, L.; Dilworth, M.; Renshall, L.; Brownfoot, F.; et al. Proton Pump Inhibitors Decrease Soluble fms-Like Tyrosine Kinase-1 and Soluble Endoglin Secretion, Decrease Hypertension, and Rescue Endothelial Dysfunction. *Hypertension* **2017**, *69*, 457–468. [CrossRef] [PubMed]

26. Brown, M.A.; Magee, L.A.; Kenny, L.C.; Karumanchi, S.A.; McCarthy, F.P.; Saito, S.; Hall, D.R.; Warren, C.E.; Adoyi, G.; Ishaku, S.; et al. Hypertensive Disorders of Pregnancy: ISSHP Classification, Diagnosis, and Management Recommendations for International Practice. *Hypertension* **2018**, *72*, 24–43. [CrossRef]

27. Sun, Y.; Goes Martini, A.; Janssen, M.J.; Garrelds, I.M.; Masereeuw, R.; Lu, X.; Danser, A.H.J. Megalin: A Novel Endocytic Receptor for Prorenin and Renin. *Hypertension* **2020**, *75*, 1242–1250. [CrossRef]

28. Poisner, A.M. Regulation of utero-placental prorenin. *Adv. Exp. Med. Biol.* **1995**, *377*, 411–426. [CrossRef]

29. Shah, D.M.; Banu, J.M.; Chirgwin, J.M.; Tekmal, R.R. Reproductive tissue renin gene expression in preeclampsia. *Hypertens. Pregnancy* **2000**, *19*, 341–351. [CrossRef]

30. Anton, L.; Brosnihan, K.B. Systemic and uteroplacental renin–angiotensin system in normal and pre-eclamptic pregnancies. *Ther. Adv. Cardiovasc. Dis.* **2008**, *2*, 349–362. [CrossRef]

31. Anton, L.; Merrill, D.C.; Neves, L.A.; Stovall, K.; Gallagher, P.E.; Diz, D.I.; Moorefield, C.; Gruver, C.; Ferrario, C.M.; Brosnihan, K.B. Activation of local chorionic villi angiotensin II levels but not angiotensin (1-7) in preeclampsia. *Hypertension* **2008**, *51*, 1066–1072. [CrossRef]

32. Anton, L.; Merrill, D.C.; Neves, L.A.; Diz, D.I.; Corthorn, J.; Valdes, G.; Stovall, K.; Gallagher, P.E.; Moorefield, C.; Gruver, C.; et al. The uterine placental bed Renin-Angiotensin system in normal and preeclamptic pregnancy. *Endocrinology* **2009**, *150*, 4316–4325. [CrossRef]

33. Burton, G.J.; Redman, C.W.; Roberts, J.M.; Moffett, A. Pre-eclampsia: Pathophysiology and clinical implications. *BMJ* **2019**, *366*, l2381. [CrossRef] [PubMed]

34. Li, C.; Ansari, R.; Yu, Z.; Shah, D. Definitive molecular evidence of renin-angiotensin system in human uterine decidual cells. *Hypertension* **2000**, *36*, 159–164. [CrossRef]

35. Nonn, O.; Fischer, C.; Geisberger, S.; El-Heliebi, A.; Kroneis, T.; Forstner, D.; Desoye, G.; Staff, A.C.; Sugulle, M.; Dechend, R.; et al. Maternal Angiotensin Increases Placental Leptin in Early Gestation via an Alternative Renin-Angiotensin System Pathway: Suggesting a Link to Preeclampsia. *Hypertension* **2021**, *77*, 1723–1736. [CrossRef]

36. Takimoto, E.; Ishida, J.; Sugiyama, F.; Horiguchi, H.; Murakami, K.; Fukamizu, A. Hypertension induced in pregnant mice by placental renin and maternal angiotensinogen. *Science* **1996**, *274*, 995–998. [CrossRef] [PubMed]

37. Haase, N.; Foster, D.J.; Cunningham, M.W.; Bercher, J.; Nguyen, T.; Shulga-Morskaya, S.; Milstein, S.; Shaikh, S.; Rollins, J.; Golic, M.; et al. RNA interference therapeutics targeting angiotensinogen ameliorate preeclamptic phenotype in rodent models. *J. Clin. Invest.* **2020**, *130*, 2928–2942. [CrossRef]

38. Lenz, T.; James, G.D.; Laragh, J.H.; Sealey, J.E. Prorenin secretion from human placenta perfused in vitro. *Am. J. Physiol.* **1991**, *260*, E876–E882. [CrossRef] [PubMed]

39. Lenz, T. Release of prorenin and placental hormones from superfused minced chorion laeve. *Acta Obstet. Gynecol. Scand.* **1997**, *76*, 903–906. [CrossRef] [PubMed]

40. Ye, F.; Wang, Y.; Wu, C.; Howatt, D.A.; Wu, C.H.; Balakrishnan, A.; Mullick, A.E.; Graham, M.J.; Danser, A.H.J.; Wang, J.; et al. Angiotensinogen and Megalin Interactions Contribute to Atherosclerosis-Brief Report. *Arterioscler. Thromb. Vasc. Biol.* **2019**, *39*, 150–155. [CrossRef] [PubMed]

41. Zygmunt, M.; McKinnon, T.; Herr, F.; Lala, P.K.; Han, V.K. HCG increases trophoblast migration in vitro via the insulin-like growth factor-II/mannose-6 phosphate receptor. *Mol. Hum. Reprod.* **2005**, *11*, 261–267. [CrossRef] [PubMed]

42. Burke, S.D.; Zsengeller, Z.K.; Khankin, E.V.; Lo, A.S.; Rajakumar, A.; DuPont, J.J.; McCurley, A.; Moss, M.E.; Zhang, D.; Clark, C.D.; et al. Soluble fms-like tyrosine kinase 1 promotes angiotensin II sensitivity in preeclampsia. *J. Clin. Invest.* **2016**, *126*, 2561–2574. [CrossRef] [PubMed]

43. Hitzerd, E.; Neuman, R.I.; Broekhuizen, M.; Simons, S.H.P.; Schoenmakers, S.; Reiss, I.K.M.; Koch, B.C.P.; van den Meiracker, A.H.; Versmissen, J.; Visser, W.; et al. Transfer and Vascular Effect of Endothelin Receptor Antagonists in the Human Placenta. *Hypertension* **2020**, *75*, 877–884. [CrossRef]

44. Hitzerd, E.; Broekhuizen, M.; Mirabito Colafella, K.M.; Glisic, M.; de Vries, R.; Koch, B.C.P.; de Raaf, M.A.; Merkus, D.; Schoenmakers, S.; Reiss, I.K.M.; et al. Placental effects and transfer of sildenafil in healthy and preeclamptic conditions. *EBioMedicine* **2019**, *45*, 447–455. [CrossRef] [PubMed]

45. Batenburg, W.W.; de Bruin, R.J.; van Gool, J.M.; Muller, D.N.; Bader, M.; Nguyen, G.; Danser, A.H. Aliskiren-binding increases the half life of renin and prorenin in rat aortic vascular smooth muscle cells. *Arterioscler. Thromb. Vasc. Biol.* **2008**, *28*, 1151–1157. [CrossRef] [PubMed]

46. Van den Heuvel, M.; Batenburg, W.W.; Jainandunsing, S.; Garrelds, I.M.; van Gool, J.M.; Feelders, R.A.; van den Meiracker, A.H.; Danser, A.H. Urinary renin, but not angiotensinogen or aldosterone, reflects the renal renin-angiotensin-aldosterone system activity and the efficacy of renin-angiotensin-aldosterone system blockade in the kidney. *J. Hypertens.* **2011**, *29*, 2147–2155. [CrossRef] [PubMed]

47. Krop, M.; Garrelds, I.M.; de Bruin, R.J.; van Gool, J.M.; Fisher, N.D.; Hollenberg, N.K.; Jan Danser, A.H. Aliskiren accumulates in Renin secretory granules and binds plasma prorenin. *Hypertension* **2008**, *52*, 1076–1083. [CrossRef]

International Journal of
*Molecular Sciences*

*Article*

# Increased Placental Cell Senescence and Oxidative Stress in Women with Pre-Eclampsia and Normotensive Post-Term Pregnancies

Paula J. Scaife [1], Amy Simpson [2], Lesia O. Kurlak [3], Louise V. Briggs [4], David S. Gardner [3], Fiona Broughton Pipkin [2], Carolyn J. P. Jones [5] and Hiten D. Mistry [6],*

1. Clinical, Metabolic and Molecular Physiology Research Group, University of Nottingham, Nottingham NG7 2RD, UK; paula.scaife@nottingham.ac.uk
2. Department of Obstetrics & Gynaecology, University of Nottingham, Nottingham NG7 2RD, UK; amy.e.simpson98@gmail.com (A.S.); Fiona.broughton_pipkin@nottingham.ac.uk (F.B.P.)
3. School of Veterinary Medicine and Science, University of Nottingham, Nottingham NG7 2RD, UK; lesia.kurlak@nottingham.ac.uk (L.O.K.); David.gardner@nottingham.ac.uk (D.S.G.)
4. School of Engineering, University of Nottingham, Nottingham NG7 2RD, UK; louise.briggs@nottingham.ac.uk
5. Maternal & Fetal Health Research Centre, Manchester Academic Health Science Centre, University of Manchester, Manchester M13 9PL, UK; carolyn.jones@manchester.ac.uk
6. Department of Women and Children's Health, School of Life Course Sciences, King's College London, London SE5 9NU, UK
* Correspondence: hiten.mistry@kcl.ac.uk

**Citation:** Scaife, P.J.; Simpson, A.; Kurlak, L.O.; Briggs, L.V.; Gardner, D.S.; Broughton Pipkin, F.; Jones, C.J.P.; Mistry, H.D. Increased Placental Cell Senescence and Oxidative Stress in Women with Pre-Eclampsia and Normotensive Post-Term Pregnancies. *Int. J. Mol. Sci.* **2021**, *22*, 7295. https://doi.org/10.3390/ijms22147295

Academic Editor: Jerome F. Strauss III

Received: 3 June 2021
Accepted: 2 July 2021
Published: 7 July 2021

**Publisher's Note:** MDPI stays neutral with regard to jurisdictional claims in published maps and institutional affiliations.



**Abstract:** Up to 11% of pregnancies extend to post-term with adverse obstetric events linked to pregnancies over 42 weeks. Oxidative stress and senescence (cells stop growing and dividing by irreversibly arresting their cell cycle and gradually ageing) can result in diminished cell function. There are no detailed studies of placental cell senescence markers across a range of gestational ages, although increased levels have been linked to pre-eclampsia before full term. This study aimed to determine placental senescence and oxidative markers across a range of gestational ages in women with uncomplicated pregnancies and those with a diagnosis of pre-eclampsia. Placentae were obtained from 37 women with uncomplicated pregnancies of 37–42 weeks and from 13 cases of pre-eclampsia of $31^{+2}$–$41^{+2}$ weeks. The expression of markers of senescence, oxidative stress, and antioxidant defence (tumour suppressor protein p16$^{INK4a}$, kinase inhibitor p21, interleukin-6 (IL-6), NADPH oxidase 4 (NOX4), glutathione peroxidases 1, 3, and 4 (GPx1, GPx3, and GPx4), placental growth factor (PlGF), and soluble fms-like tyrosine kinase-1 (sFlt-1)) genes was measured (quantitative real-time PCR). Protein abundance of p16$^{INK4a}$, IL-6, NOX4, 8-hydroxy-2′-deoxy-guanosine (8-OHdG), and PlGF was assessed by immunocytochemistry. Placental NOX4 protein was higher in post-term than term deliveries and further increased by pre-eclampsia ($p < 0.05$ for all). P21 expression was higher in post-term placentae ($p = 0.012$) and in pre-eclampsia ($p = 0.04$), compared to term. Placental P16$^{INK4a}$ protein expression was increased post-term, compared to term ($p = 0.01$). In normotensive women, gestational age at delivery was negatively associated with GPx4 and PlGF (mRNA and protein) ($p < 0.05$ for all), whereas a positive correlation was seen with placental P21, NOX4, and P16$^{INK4a}$ ($p < 0.05$ for all) expression. Markers of placental oxidative stress and senescence appear to increase as gestational age increases, with antioxidant defences diminishing concomitantly. These observations increase our understanding of placental health and may contribute to assessment of the optimal gestational age for delivery.

**Keywords:** hypertension in pregnancy; angiogenesis; endothelial function; oxidative stress; antioxidants; post-maturity; senescence

## 1. Introduction

The human placenta stops growing at ~90% of full term (~36 weeks of gestation), unlike that of other mammalian species, but the fetus continues to grow, which would presumably "stress" even a normal placenta. This feature is assumed to have evolved in parallel with upright posture and the necessary development of a very muscular uterus, delaying delivery. Ageing is a process that causes a deterioration in function at the cellular, tissue, and organ level, leading to individuals being more susceptible to disease. Short chromosomal telomeres, as well as the partial or complete insufficiency of the telomerase enzyme, have been linked to diseases caused by ageing [1]. Telomeres are protective caps made of nucleoprotein molecules located at the end of chromosomes and are necessary for protection against breaks at DNA ends, fusion of chromosome ends, and chromosome degradation [1]. Telomeres are shortened with each cell division. The rate at which this occurs is accelerated by certain stressors, such as oxidative stress [1]. Eventually, telomeres reach a dangerously short length, which initiates the process of cellular senescence, through which cells irreversibly stop growing and dividing by arresting their cell cycle and gradually ageing (becoming 'senescent') [2,3].

Prolonged pregnancy (also known as post-term pregnancy), is defined by the World Health Organisation as "the end of gestation at $\geq$42 completed weeks of gestation, measured from the first day of the last menstrual period and based on a 28 day cycle" [4]. Adverse obstetric events have been linked with pregnancies that last longer than 42 weeks, including an increased frequency of foetal death and raised risk of foetal malnutrition, intrauterine foetal hypoxia, irregular nonstress tests, respiratory distress, oligohydramnios, delivery via Caesarean section, and stillbirth [5,6]. Furthermore, long-term health problems in the child have also been associated with pregnancies extending over 42 weeks [5].

In the developed world, up to 11% of pregnancies extend to post-term. However, this figure varies significantly between different countries with further disparity between low- and middle-income countries [5]. In the UK, a woman who has not spontaneously delivered by the start of 41 weeks is offered an induction of labour. The induction is performed between $41^{+0}$ and $42^{+0}$ weeks to avoid the associated risks of prolonged pregnancy [7].

The main risk factors associated with prolonged pregnancy are genetic, the mother having been born post-term herself, previous history of prolonged pregnancy, primiparity, and obesity [8]. Moreover, a Swedish study reported that one-quarter of the risk arises from the fetal genetic background, and another quarter arises from the mother [9].

Some of the signs associated with prolonged pregnancy are reduced fetal movements, a reduced volume of amniotic fluid, and meconium-stained amniotic fluid after the membranes have ruptured [10]. A seminal study conducted by Jones and Fox in 1978 investigated the structural changes in placentae after prolonged pregnancy and revealed that most of these placentae show morphological irregularities. A possible decline in trophoblastic cell function was postulated [11].

A study by Maiti et al. concluded that placental cells start to age dramatically from 37 weeks of gestation, with oxidative stress rising as gestational age increases [12]. Other studies suggest that premature placental ageing may contribute to placental dysfunction. This could be the cause of many placenta-related pathologies, such as pre-eclampsia [13], a multisystem disorder affecting up to 5% of pregnant women. It is one of the leading causes of maternal and perinatal mortality and morbidity, especially when it occurs before 34 weeks of gestation [14]. It is characterised by de novo hypertension together with evidence of endothelial cell damage and/or significant proteinuria in the absence of urinary tract infection [15]. Pre-eclampsia is associated with shallow placentation, inadequate remodelling of the uterine vasculature, and consequent oxidative stress; antioxidant defences are inadequate, and the synthesis of a variety of angiogenic and growth factors is perturbed [16].

Senescence stressors ultimately activate the P53 and/or cyclin-dependent kinase inhibitor 2A (P16$^{INK4a}$) pathways and secondary induction of cyclin-dependent kinase inhibitor (P21) [2]. Furthermore, interleukin-6 (IL-6) depletion is correlated with reduced formation of senescence-associated heterochromatin foci (SAHF) and, thus, can also be used as a surrogate marker of senescence [17].

In summary, oxidative stress in combination with cell senescence can cause extensive tissue damage, leading to accelerated cellular ageing. We hypothesised that extended, 'post-term' pregnancies and gestational disease such as pre-eclampsia both increase oxidative stress and cell senescence in placentae, relative to healthy term controls. To investigate this, we examined the mRNA expression and protein abundance of a panel of markers of cell senescence ($P16^{INK4\alpha}$, P21, and IL-6), oxidative stress (NADPH oxidase 4 (NOX4), 8-Oxo-2′-deoxyguanosine (8-OHdg), a marker of DNA modification), antioxidant defence (glutathione peroxidases (GPx)), and placental function (PlGF and sFlt-1) in normotensive control and post-term placenta, as well as from women with pre-eclampsia.

## 2. Results

### 2.1. Participants

Baseline demographic and pregnancy outcome data are presented in Table 1. As can be seen, by definition, women who had pre-eclampsia had significantly higher blood pressures ($p < 0.05$) and significant proteinuria. They also delivered earlier, and their babies' birthweights were lower. Overall, the groups were matched for maternal age, BMI, and parity.

**Table 1.** Participant demographics and pregnancy outcome data.

| Parameter | Term 37–39$^{+6}$ ($n$ = 26) | Post-Term 41–42 ($n$ = 11) | Pre-Eclampsia ($n$ = 13) |
|---|---|---|---|
| Age at booking (year) | $31 \pm 6.0$ | $31 \pm 6.4$ | $34 \pm 3.8$ |
| BMI at booking (kg/m$^2$) | 26 [23,32] | 27 [26,31] | 30 [26,31] |
| Nulliparous, $n$ (%) | 10 (38) | 5 (45) | 8 (62) |
| Smoking, $n$ (%) | 5 (19) | 3 (27) | 1 (8) |
| Systolic blood pressure (mmHg) | $110 \pm 10.6$ [c] | $118 \pm 5.1$ [c] | $158 \pm 12.3$ [c] |
| Diastolic blood pressure (mmHg) | $77 \pm 7.8$ [c] | $75 \pm 6.0$ [c] | $97 \pm 6.8$ [c] |
| Proteinuria | - | - | 1.0 [0.4, 1.4] |
| Gestational age at delivery (weeks) | 39.4 (36.3–40.6) [a,c] | 41.3 (41.0–42) [a,b] | 37.2 (31.1–41.1) [b,c] |
| Birth weight (g) | $3533 \pm 378$ [a] | $3873 \pm 384$ [b] | $2741 \pm 1073.7$ [a,b] |
| Birthweight centile | 74 [47,94] | 86 [65,94] | 48 [26,78] |
| Placental weight (g) | $624 \pm 133.6$ | $696 \pm 92.6$ [b] | $539 \pm 182.9$ [b] |
| Baby gender, female $n$ (%) | 14 (54) | 5 (45) | 8 (62) |
| Caesarean Section $n$ (%) | 19 (73) | 11 (73) | 8 (77) |

[a][b][c] $p < 0.05$ between the respective groups. Data are presented as mean $\pm$ SD, median [IQR], or median (range) for gestational age at delivery, depending on distribution or $n$ (%). BMI: body mass index. Birthweight centiles calculated using INTERGROWTH 21 (https://intergrowth21.tghn.org/standards-tools/; accessed on 25 May 2021).

### 2.2. Gene Expression

Considering markers of oxidative stress, expression of NOX4 was present in all samples (Figure 1A). NOX4 expression was increased post-term, compared to placentae from term deliveries ($p = 0.013$). For all GPxs, no differences were observed between groups ($p > 0.05$ for all), although there was a trend towards lower GPx4 expression in both post-term and pre-eclampsia groups. When further sub-grouped by gestational age, we found significantly higher GPx4 gene expression in the 37–39$^{+0}$-week placentae (median [IQR]: 17,765 [13,366, 25,101], normalised copy number) compared to the 40–40$^{+6}$-week (8729 [2992, 13,823]; $p < 0.01$) and 41–42-week placentae (5739.1 [2160, 18,130]; $p = 0.04$), as well as in pre-eclampsia ($p = 0.04$).

All markers of senescence were expressed (Figure 1E–G) in every group. Placental P21 expression was raised in post-term deliveries ($p = 0.012$), as well as in women with pre-eclampsia ($p = 0.04$), compared to term samples. For IL-6, women with pre-eclampsia had a higher placental expression compared to both term ($p = 0.031$) and post-term ($p = 0.008$); no differences were seen between term and post-term samples ($p > 0.05$). No differences were observed between groups for P16$^{INK4\alpha}$.

We also examined expression of the placental functional markers, PlGF and sFlt-1 (Figure 1H,I). PlGF expression was lower in placentae from women with pre-eclampsia, compared to term ($p = 0.008$), but not post-term samples ($p > 0.05$); smaller differences between term and post-term expression did not reach statistical significance ($p = 0.08$). No differences were seen between groups for sFlt-1 expression ($p > 0.05$).

Lastly, when considering any impact of gestational age at delivery in only the normotensive women, negative associations were observed with placental GPx4 ($r = -0.405$; $p = 0.012$) and PlGF ($r = -0.388$; $p = 0.021$), whereas a positive correlation was seen with P21 ($r = 0.324$; $p = 0.044$).

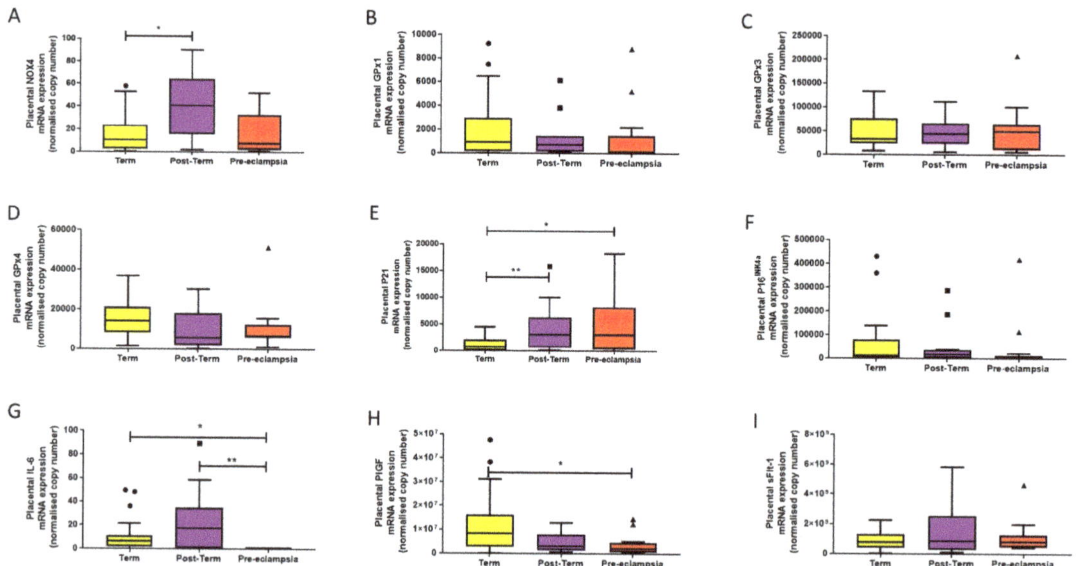

**Figure 1.** mRNA expression assessed by quantitative reverse-transcription PCR of (**A**) NADPH oxidase (NOX4), (**B–D**) glutathione peroxidases 1, 3, and 4 (GPx1, 3, and 4), respectively, (**E**) cyclin-dependent kinase inhibitor (P21), (**F**) cyclin-dependent kinase inhibitor 2A (P16$^{INK4\alpha}$), (**G**) interleukin-6 (IL-6), (**H**) placental growth factor (PlGF), and (**I**) soluble fms-like tyrosine kinase-1 (sFlt-1) in placentae from term normotensive (37–40 + 6 weeks; $n = 26$), post-term normotensive (41–42 weeks; $n = 11$), and women who had pre-eclampsia ($n = 13$). Data are presented as median [IQR]; * $p < 0.05$, ** $p < 0.005$.

### 2.3. Protein Expression

NOX4 protein expression was confirmed in all placental samples analysed, with staining localised within nuclei and syncytiotrophoblast (Figure 2A). As with gene expression, NOX4 protein expression was higher in both post-term (0.95 [0.79, 0.97]; $p = 0.017$) and pre-eclampsia samples (0.94 [0.92, 0.96]; $p < 0.0001$; Figure 2A), compared to term samples (median [IQR]: 0.80 [0.77, 0.90] positivity), with more uniform, high expression observed in pre-eclampsia ($p > 0.05$). Placental expression of 8-OHdG was localised mainly within the nuclei with some weak cytoplasmic staining (Figure 2B) and was highest in women with pre-eclampsia (median [IQR]: 0.77 [0.72, 0.83] positivity), compared to both term

(0.68 [0.61, 0.76]; $p = 0.015$) and post-term (0.70 [0.63, 0.76]; $p = 0.021$) samples (Figure 2B); similar expression was observed between term and post-term ($p > 0.05$).

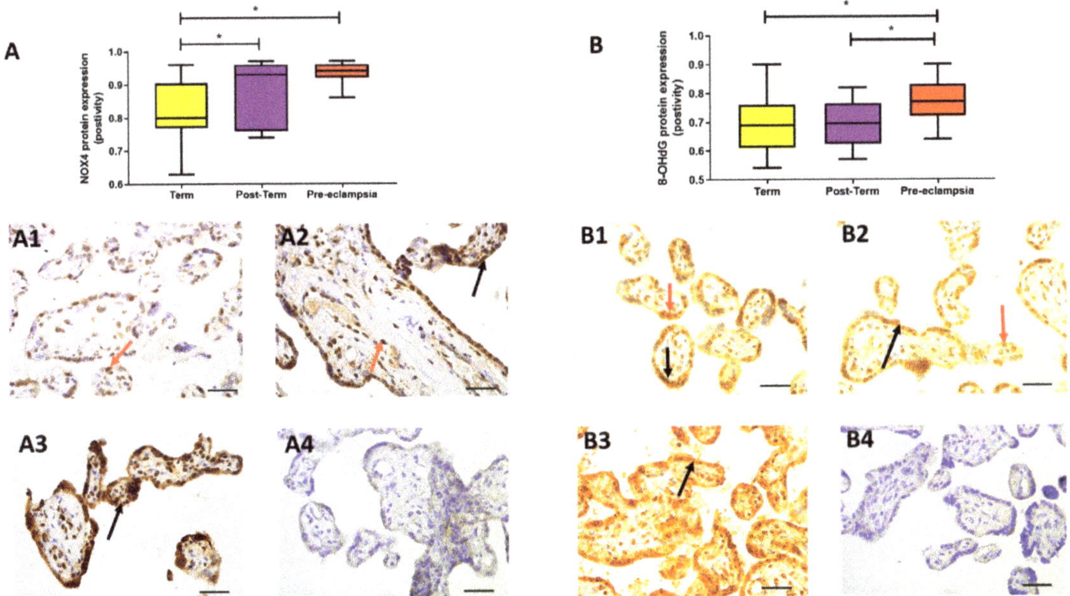

**Figure 2.** Protein expression assessed by immunohistochemistry. (**A**) NADPH oxidase (NOX4) and (**B**) 8-hydroxy-2′-deoxy-guanosine (8-OHdG) in placentae from term normotensive (37–40 + 6 weeks; $n = 26$), post-term normotensive (41–42 weeks; $n = 11$), and women who had pre-eclampsia, delivered at 31 + 2 to 41 + 2 weeks gestation ($n = 13$). Data are presented as median [IQR]; * $p < 0.05$. Photomicrographs show typical examples of immunostaining in (**A1,B1**) term, (**A2,B2**) post-term, (**A3,B3**) pre-eclampsia, and (**A4,B4**) IgG negative control. Positive protein expression appears brown and is localised mainly to the syncytiotrophoblast (black arrows) but is also evident in the nuclei (red arrows); scale bar = 100 μm.

P16$^{INK4a}$ expression differed between groups ($p = 0.036$). Expression was higher in post-term (0.22 [0.20, 0.32]) compared to term (0.13 [0.09, 0.23]; $p = 0.011$) and was found mainly in the endothelium; however, placentae from women with pre-eclampsia exhibited similar expression (0.20 [0.09, 0.40]; $p > 0.05$; Figure 3A). IL-6 localised mainly to the endothelium of the villi and overall expression was low; expression did not differ between groups ($p > 0.05$; Figure 3B).

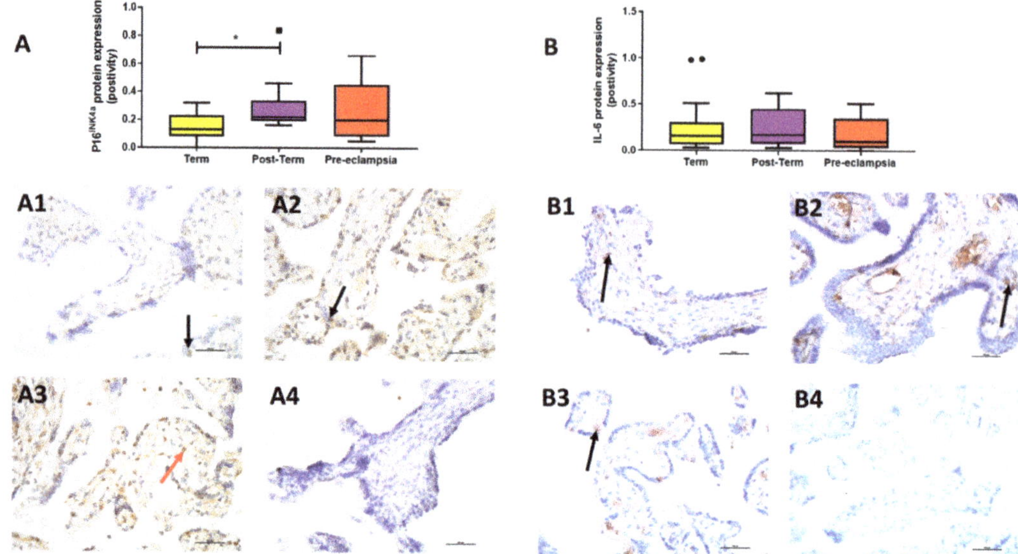

**Figure 3.** Protein expression assessed by immunohistochemistry. (**A**) P16$^{INK4a}$ and (**B**) interleukin-6 (IL-6) in placentae from term normotensive (37–40 + 6 weeks; $n$ = 26), post-term normotensive (41–42 weeks; $n$ = 11), and women who had pre-eclampsia ($n$ = 13). Data are presented as median [IQR]; * $p < 0.05$. Photomicrographs show typical examples of immunostaining in (**A1,B1**) term, (**A2,B2**) post-term, (**A3,B3**) pre-eclampsia and (**A4,B4**) IgG negative control. Positive protein expression of P16$^{INK4a}$ appears brown and is localised mainly to the syncytiotrophoblast (black arrow) with some stromal cell staining (red arrow); that of interleukin 6 is found mainly in endothelial cells (arrows); scale bar = 100 μm.

PlGF expression was weak and localised to the syncytiotrophoblast layer (Figure 4) and, although levels did not differ statistically between groups ($p > 0.05$), there was a trend towards lower expression in post-term (0.0024 [0.0013, 0.023]) and placentae from women with pre-eclampsia (0.0069 [0.0039, 0.019]), compared to term (0.0073 [0.0026, 0.045]) (Figure 4).

Again, when considering the impact of gestational age at delivery in only the normotensive women, a positive association was evident with placental P16$^{INK4a}$ expression ($r$ = 0.331; $p$ = 0.04; Figure 5A). In contrast, a negative correlation was seen with PlGF expression ($r$ = −0.367; $p$ = 0.046; Figure 5B). In addition, placental expression of P16$^{INK4a}$ and NOX4 were positively correlated ($r$ = 0.537; $p$ = 0.001; Figure 5C).

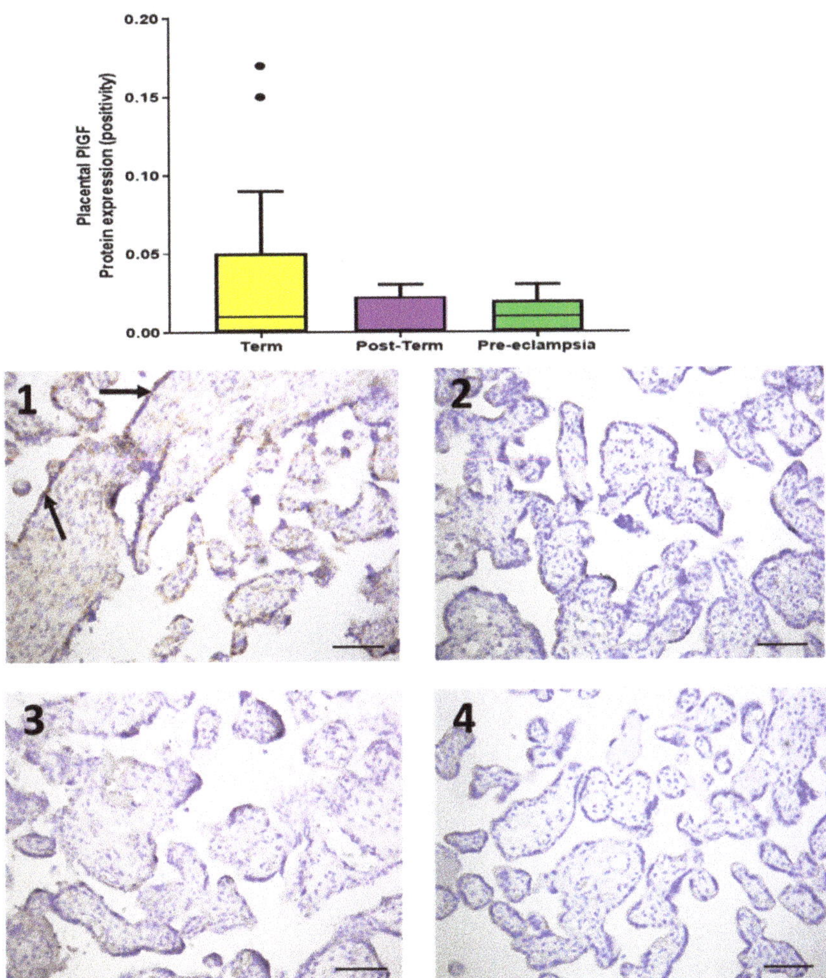

**Figure 4.** Protein expression of placental growth factor (PlGF) assessed by immunohistochemistry in placentae from term normotensive (37–40 + 6 weeks; $n = 26$), post-term normotensive (41–42 weeks; $n = 11$), and women who had pre-eclampsia ($n = 13$). Data are presented as median [IQR]. Photomicrographs show typical examples of immunostaining in (**1**) term, (**2**) post-term, (**3**) pre-eclampsia, and (**4**) IgG negative control. Positive protein expression is weak and appears brown, localised mainly to the syncytiotrophoblast (black arrows); scale bar = 100 μm.

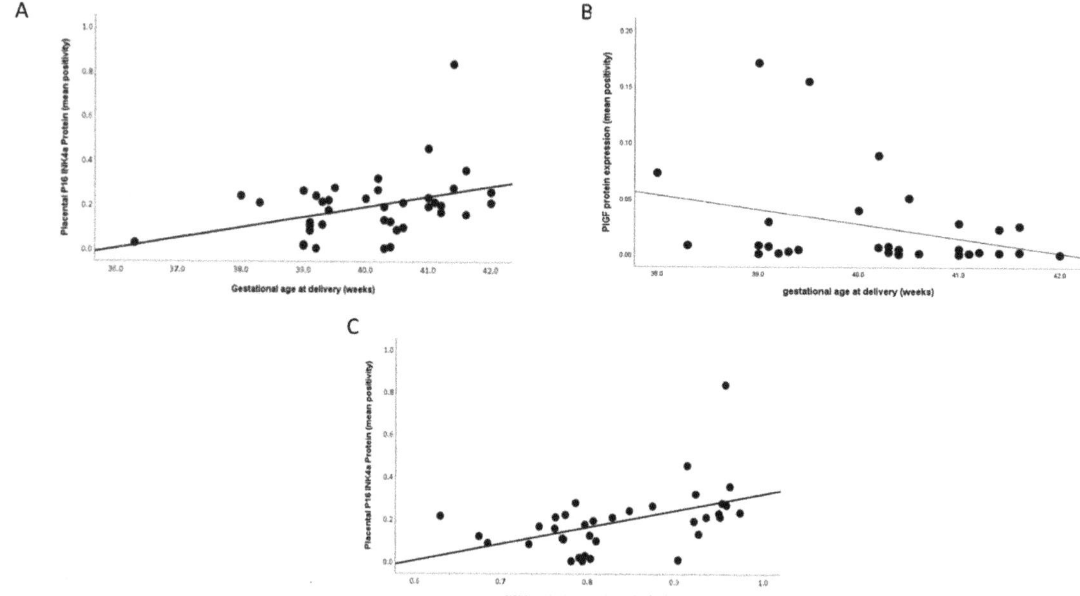

**Figure 5.** Scatter plot illustrations showing correlations between gestational age at delivery and (**A**) P16INK4a ($r = 0.331$; $p = 0.04$) and (**B**) placental growth factor (PlGF; $r = -0.367$; $p = 0.046$); (**C**) between P16[INK4a] and NADPH oxidase 4 (NOX4; $r = 0.537$; $p = 0.001$) in normotensive samples only.

## 3. Discussion

Features of placental ageing may play a role in the morbidity associated with prolonged gestation. After 39 weeks of pregnancy, the fetal death rate rises fourfold [18,19]. Maiti et al. studied placentae from three groups: term (39 weeks), late term (>41 weeks), and idiopathic stillbirths; they found increased aldehyde oxidase 1 expression (a mediator of placental ageing), DNA/RNA and lipid oxidation, lysosomes situated perinuclearly and basally as opposed to apically when localised with lysosome-associated membrane protein 2 (LAMP2), and bigger autophagosomes suggestive of inhibition of function in later-term placentae [12]. Towards the end of pregnancy and amplified during the post-term period, the foetal demands for oxygen and nutrients outstrip the placenta's ability to supply, thus leading to increased oxidative stress and subsequent deterioration of the placenta, including marked senescence [20].

We have now shown increased cell senescence and oxidative stress not only in placentae from normotensive women who delivered post term but also from women who developed pre-eclampsia, when compared to normotensive women delivering at term. The fact that increased gene and protein expression of NOX4, one of the sources of cellular reactive oxygen species (ROS), was evident in both post-term and pre-eclamptic groups implies that these placentae are under significant oxidative stress. Whilst this is well known in pre-eclampsia [21], this is the first detailed report in normotensive post-term placentae.

In this study, 8-OHdG protein expression was only raised in the placentae from women who had pre-eclampsia in concordance with others [13], reflecting oxidative DNA damage. However, one other study has shown increased 8-OHdG expression in post-term placentae, using a different technique, specifically, counting the number of positively stained nuclei [12]. Nevertheless, increased levels of oxidative stress cause damage to DNA, proteins, and lipids in placental tissue, which may manifest as another form of accelerated placental ageing. The antioxidant GPX genes also showed a trend towards reduced levels in the post-term and pre-eclamptic tissue. There were, however, no significant differences between groups in the protein expression of the cytokine IL-6, despite gene expression being raised in the pre-eclampsia group. A possible reason for this is that IL-6 is downstream of senescence, not an initiator of it [22]. Thus, placental senescence in post-term pregnancy does not activate this particular inflammatory cascade or has not been around long enough to do so.

In this study, increased protein expression of the senescence marker P16$^{INK4a}$ was observed in post-term and pre-eclamptic placental tissue. The p16/pRb and p53/p21 pathways are reported to be activated in various cell lines in response to stimuli that induce irreversible cellular senescence [23,24]. Cell fusion, an essential physiological process to establish and expand the syncytiotrophoblast, has been recognised to be a further trigger of cell senescence, and the syncytiotrophoblast features characteristics of senescent cells including the biomarker senescence-associated beta-galactosidase (SA-β-gal), together with high expression of the cyclin kinases inhibitors p16, p21, and p53 [25], which regulate cell-cycle progression at $G_1$ and S phase. It must be noted that syncytiotrophoblast senescence is a normal physiological phenomenon, which progresses as pregnancy advances [26]. However, there is increasing evidence to suggest that, when physiological senescence is accelerated, it results in placental and clinical pathology. Cindrova-Davies et al. reported placental P21 protein expression to be higher in post-term and pre-eclampsia samples [13]. Moreover, increased placental or trophoblast senescence has already been demonstrated in terms of telomere shortening, aggregation, or other measures of telomere dysfunction, both in pre-eclampsia and in normotensive women with foetal growth restriction [27–31].

Our data also demonstrated that mRNA and protein expressions of the proangiogenic growth factor PlGF were lower in placentae from women with pre-eclampsia and post-term samples, although the latter did not achieve statistical significance. It is known that pre-eclampsia is partly mediated by dysfunctional syncytiotrophoblast, and PlGF (as a marker for syncytiotrophoblast health) has appeared as a good marker for early-onset pre-eclampsia [32]. However, this is not the case in late-onset pre-eclampsia occurring towards term. Interestingly, in uncomplicated normotensive pregnancy, circulating PlGF concentrations rise steadily, peaking around 30 weeks, and then fall [32]. This suggests that the syncytiotrophoblast becomes increasingly stressed for the last 8–10 weeks of pregnancy [33] and possibly beyond. Further supporting data show that pO$_2$ measurements of maternal intervillous and umbilical venous and arterial bloods decline in the third trimester, with slowing placental growth [34]. These data collectively suggest that normotensive women also have placentae with syncytiotrophoblast stress both at term and post-term [33].

When defining 'term' pregnancy according to the American College of Obstetricians and Gynaecologists, the neonatal outcome measures on delivery vary greatly between delivery at 37$^{+0}$ and 42$^{+0}$ weeks [35]; therefore, the importance of accurately establishing the gestational age cannot be overemphasised. Nonetheless, there are many countries in which mothers are unable to access ultrasound scans and do not have verifiable measurements of gestational age, meaning that the actual rate of post-term births in such countries may be higher than that officially recorded [5].

In this study we wished both to identify whether placentae from post-term pregnancies showed evidence of increasing senescence and to determine whether placentae from pre-eclamptic women showed accelerated senescence earlier in pregnancy. Both post-term pregnancy and pre-eclampsia are associated with increased risk of sudden fetal death, presumably consequent on placental failure. This is a preliminary study; hence, we chose to investigate well-established markers of several different aspects of senescence, rather than immediately performing gene-array studies. Future studies will require the use of RNA-seq as an unbiased approach to using tissues from different gestational stages and then test known genes, as well as novel genes, to further support our initial findings. Our data demonstrate that placental oxidative stress and senescence increase, in parallel with a reduction in PlGF expression, as normotensive pregnancy progresses, while antioxidant defences diminish as gestational age increases. These features become evident earlier in gestation in women with pre-eclampsia, suggesting accelerated senescence, probably secondary to their poor antioxidant status [13]. If these markers can be detected in the circulation, they may prove useful in screening for women with more severe problems of post-term pregnancy and pre-eclampsia.

## 4. Materials and Methods

### 4.1. Cohort and Sample Collection

The study population (Table 1) consisted of two groups of normotensive women: term ($n$ = 26), with gestational age of 37–40$^{+6}$ weeks, and post-term ($n$ = 11), with gestational age of 41–42 weeks. The inclusion criteria comprised no maternal or pregnancy complications, live birth, singleton pregnancy, and delivery either vaginally or by Caesarean section. A third group consisted of women diagnosed with pre-eclampsia ($n$ = 13) and delivered between 31$^{+2}$ and 41$^{+2}$ weeks. Pre-eclampsia was defined as systolic blood pressure of $\geq$140/90 mmHg on two occasions and proteinuria $\geq$300 mg/L, 500 mg/day, or $\geq$2+ on a dipstick analysis of midstream urine after 20 weeks [36]. Detailed demographics and outcome data have previously been published [37]. The study was approved by the HRA-REC ethics committee of the University of Nottingham (REF: 15/EM/0523); written, informed consent to take part in the study was obtained from each participant.

Full-depth tissue biopsies were collected within 10 min of the placenta being delivered as previously described [37]. Samples were taken from the mid-point between the umbilical cord insertion and the periphery of the placenta, avoiding infarcts. One set of samples was snap-frozen and stored at −80 °C for RNA analysis. The other set was formalin-fixed and embedded in paraffin wax for immunohistochemistry.

### 4.2. RNA Extraction, cDNA Synthesis, and Quantitative Reverse-Transcription Polymerase Chain Reaction (RT-qPCR)

Total RNA was extracted from ~100 mg of placental tissue using QIAzol lysis reagent (Qiagen, UK) as previously described [38]. RNA (1 µg) was reverse-transcribed using the QuantiTect Reverse Transcription kit (Qiagen, UK) in a Primus96 thermocycler (Peqlab Ltd., Southampton, UK). RT-qPCR was carried out using SYBR Green chemistry (2× QuantiFast SYBR Green, Qiagen, UK) on an AB7500 Fast (Life Technologies, Cramlington, UK) using primers to *NADPH oxidase 4 (NOX4), glutathione peroxidase 1, 3, and 4 (GPX1, GPX3 GPX4), cyclin-dependent kinase inhibitor 2A (P16$^{INK4\alpha}$), cyclin-dependent kinase inhibitor (P21), interleukin-6 (IL-6), placental growth factor (PlGF),* and *soluble fms-like tyrosine kinase-1 (a splice variant of VEGF receptor 1, sFlt-1;* Table 2). Cycling conditions were as follows: a pre-PCR cycle was run for 15 min at 95 °C followed by 40 cycles of 95 °C for 10 s and 60 °C for 30 s. Abundance data for the genes of interest were expressed as normalised copy number following normalisation using GeNORM (http://medgen.ugent.be/~jvdesomp/genorm/; accessed on 3 November 2018), with stably expressed reference genes [39] *beta-2 microglobulin (B2M), tyrosine 3-monooxygenase/tryptophan 5-monooxygenase activation protein zeta (YWHAZ),* and *glyceraldehyde 3-phosphate dehydrogenase (GADPH)* (Table 2).

**Table 2.** Details of primers used.

| Gene | Accession Number | Primers | Length (bp) |
|---|---|---|---|
| *Nox4* | NM_016931.3 | 5′–TGAACTATGAGGTCAGCCTCTG–3′<br>5′–TCTCACGAATCTCCTCATGGT–3′ | 107 |
| *GPx1* | NM_201397 | 5′–CAGTCGGTGTATGCCTTCTCG–3′<br>5′–GAGGGACGCCACATTCTCG–3′ | 105 |
| *GPx3* | NM_002084 | 5′–GAGCTTGCACCATTCGGTCT–3′<br>5′–GGGTAGGAAGGATCTCTGAGTTC–3′ | 94 |
| *GPx4* | NM_001039847 | 5′–GAGGCAAGACCGAAGTAAACTAC–3′<br>5′–CCGAACTGGTTACACGGGAA–3′ | 100 |

**Table 2.** *Cont.*

| Gene | Accession Number | Primers | Length (bp) |
|---|---|---|---|
| $P16^{INK4a}$ | NM_000077.4 | 5′–CTTCGGCTGACTGGCTGG–3′<br>5′–TCATCATGACCTGGATCGGC–3′ | 129 |
| P21 | NM_078467 | 5′–TGTCCGTCAGAACCCATGC–3′<br>5′–AAAGTCGAAGTTCCATCGCTC–3′ | 139 |
| IL–6 | NM_000600 | 5′–ACTCACCTCTTCAGAACGAATTG–3′<br>5′–CCATCTTTGGAAGGTTCAGGTTG–3′ | 149 |
| PlGF | NM_001207012 | 5′–GAACGGCTCGTCAGAGGTG–3′<br>5′–ACAGTGCAGATTCTCATCGCC–3′ | 187 |
| sFlt–1 | NM_001159920 | 5′–TTTGCCTGAAATGGTGAGTAAGG–3′<br>5′–TGGTTTGCTTGAGCTGTGTTC–3′ | 117 |
| B2M | NM_004048.2 | 5′–CTTATGCACGCTTAACTATCTTAACAA–3′<br>5′–TAGGAGGGCTGGCAACTTAG–3′ | 127 |
| YWHAZ | NM_001135702.1 | 5′–ACTTTTGGTACATTGTGGCTTCAA–3′<br>5′–CCGCCAGGACAAACCAGTAT–3′ | 94 |
| GAPDH | NM_002046.3 | 5′–GGAAGCTTGTCATCAATGGAA–3′<br>5′–TGGACTCCACGACGTACTCA–3′ | 102 |

### 4.3. Immunohistochemical Staining

Placental protein expression was assessed by immunohistochemistry as previously described [38], using antibodies to NOX4, 8-Oxo-2′-deoxyguanosine (8-OHdg), $P16^{INK4\alpha}$, IL-6, and PlGF at concentrations detailed in Table 3. Immunoglobulin G (IgG) from the same host as the primary antibody was used as a negative control. All slides were assessed by the same observer, blinded to group. Quantification was performed as described previously [38], using the Positive Pixel Algorithm of Aperio ImageScope software; a visual check was also performed.

**Table 3.** Antibody details.

| Antigen | Suppler Information | Concentration (µg/mL) |
|---|---|---|
| NOX4 | Abcam, rabbit monoclonal: ab133303 | 2.18 |
| 8-0HdG | Abcam, mouse monoclonal: ab48508 | 12 |
| $p16^{INK4\alpha}$ | Abcam, rabbit polyclonal: ab108349 | 0.7 |
| IL-6 | Abcam, mouse monoclonal: ab9324 | |
| PlGF | Abcam, rabbit polyclonal: ab196666 | 10 |

### 4.4. Statistical Analysis

All tests were performed using SPSS version 26 and GraphPad Prism version 8. Summary data are presented as means $\pm$ standard deviation (SD) or median and interquartile range (IQR) as appropriate. The Kruskal–Wallis test, followed by Mann–Whitney U-test, was used for multiple group analysis. Student's *t*-tests or Mann–Whitney U-tests were applied depending on whether the data distribution was normal or skewed, as indicated by the Kolmogorov–Smirnov test. The null hypothesis was rejected when $p < 0.05$.

**Author Contributions:** Conceptualization, P.J.S., L.O.K. and H.D.M.; data curation, A.S., L.V.B., D.S.G. and H.D.M.; formal analysis, L.O.K., F.B.P., C.J.P.J. and H.D.M.; funding acquisition, H.D.M.; investigation, P.J.S., L.O.K., L.V.B., F.B.P., C.J.P.J. and H.D.M.; methodology, P.J.S., A.S., D.S.G. and H.D.M.; writing—original draft, H.D.M.; writing—review and editing, P.J.S., A.S., L.O.K., L.V.B., D.S.G., F.B.P. and C.J.P.J. All authors read and agreed to the published version of the manuscript.

**Funding:** This work was produced by HDM under the terms of a BHF Basic Science Intermediate Basic Science Fellowship (FS/15/32/31604).

**Institutional Review Board Statement:** The study was conducted according to the guidelines of the Declaration of Helsinki and approved by HRA-REC ethics committee of the University of Nottingham (REF: 15/EM/0523).

**Informed Consent Statement:** Signed, informed consent was obtained from all subjects involved in the study.

**Data Availability Statement:** Data is contained within the article and additional raw data can be obtained from the corresponding author on request.

**Acknowledgments:** We thank the women who participated in the study and the midwives/doctors whose support made this study possible. We thank Ceri Staley for help with the immunohistochemical staining of PIGF.

**Conflicts of Interest:** The authors declare no conflict of interest.

## References


1. Sultana, Z.; Maiti, K.; Aitken, J.; Morris, J.; Dedman, L.; Smith, R. Oxidative stress, placental ageing-related pathologies and adverse pregnancy outcomes. *Am. J. Reprod. Immunol.* **2017**, *77*, e12653. [CrossRef]
2. van Deursen, J.M. The role of senescent cells in ageing. *Nature* **2014**, *509*, 439–446. [CrossRef]
3. Kumari, R.; Jat, P. Mechanisms of Cellular Senescence: Cell Cycle Arrest and Senescence Associated Secretory Phenotype. *Front. Cell Dev. Biol.* **2021**, *9*, 645593. [CrossRef]
4. WHO. Recommended definitions, terminology and format for statistical tables related to the perinatal period and use of a new certificate for cause of perinatal deaths. Modifications recommended by FIGO as amended October 14, 1976. *Acta Obstet. Gynecol. Scand.* **1977**, *56*, 247–253.
5. Ayyavoo, A.; Derraik, J.G.; Hofman, P.L.; Cutfield, W.S. Postterm births: Are prolonged pregnancies too long? *J. Pediatrics* **2014**, *164*, 647–651. [CrossRef]
6. Wennerholm, U.B.; Saltvedt, S.; Wessberg, A.; Alkmark, M.; Bergh, C.; Wendel, S.B.; Fadl, H.; Jonsson, M.; Ladfors, L.; Sengpiel, V.; et al. Induction of labour at 41 weeks versus expectant management and induction of labour at 42 weeks (SWEdish Post-term Induction Study, SWEPIS): Multicentre, open label, randomised, superiority trial. *BMJ* **2019**, *367*, l6131. [CrossRef]
7. NICE. *Inducing Labour*; NICE Public Health Guidance: London, UK, 2008.
8. Galal, M.; Symonds, I.; Murray, H.; Petraglia, F.; Smith, R. Postterm pregnancy. *Facts Views Vis. Obgyn.* **2012**, *4*, 175–187.
9. Oberg, A.S.; Frisell, T.; Svensson, A.C.; Iliadou, A.N. Maternal and fetal genetic contributions to postterm birth: Familial clustering in a population-based sample of 475,429 Swedish births. *Am. J. Epidemiol.* **2013**, *177*, 531–537. [CrossRef]
10. Harding, M. Post-term Prenancy. In *Patient*; NICE: London, UK, 2016.
11. Jones, C.J.; Fox, H. Ultrastructure of the placenta in prolonged pregnancy. *J. Pathol.* **1978**, *126*, 173–179. [CrossRef]
12. Maiti, K.; Sultana, Z.; Aitken, R.J.; Morris, J.; Park, F.; Andrew, B.; Riley, S.C.; Smith, R. Evidence that fetal death is associated with placental aging. *Am. J. Obstet. Gynecol.* **2017**, *217*, 441.e1–441.e14. [CrossRef]
13. Cindrova-Davies, T.; Fogarty, N.M.E.; Jones, C.J.P.; Kingdom, J.; Burton, G.J. Evidence of oxidative stress-induced senescence in mature, post-mature and pathological human placentas. *Placenta* **2018**, *68*, 15–22. [CrossRef]
14. Steegers, E.A.; von Dadelszen, P.; Duvekot, J.J.; Pijnenborg, R. Pre-eclampsia. *Lancet* **2010**, *376*, 631–644. [CrossRef]
15. Brown, M.A.; Magee, L.A.; Kenny, L.C.; Karumanchi, S.A.; McCarthy, F.P.; Saito, S.; Hall, D.R.; Warren, C.E.; Adoyi, G.; Ishaku, S.; et al. The hypertensive disorders of pregnancy: ISSHP classification, diagnosis & management recommendations for international practice. *Pregnancy Hypertens* **2018**, *13*, 291–310. [PubMed]
16. Ives, C.W.; Sinkey, R.; Rajapreyar, I.; Tita, A.T.N.; Oparil, S. Preeclampsia-Pathophysiology and Clinical Presentations: JACC State-of-the-Art Review. *J. Am. Coll. Cardiol.* **2020**, *76*, 1690–1702. [CrossRef] [PubMed]
17. Carnero, A. Markers of Cellular Senescence. In *Cell Senescence Methods and Protocols*; Galluzzi, L., Vitala, I., Kepp, O., Kroemer, G., Eds.; Humana Press: Totowa, NJ, USA, 2013.
18. Yudkin, P.L.; Wood, L.; Redman, C.W. Risk of unexplained stillbirth at different gestational ages. *Lancet* **1987**, *1*, 1192–1194. [CrossRef]
19. Ferrari, F.; Facchinetti, F.; Saade, G.; Menon, R. Placental telomere shortening in stillbirth: A sign of premature senescence? *J. Matern. Fetal Neonatal Med.* **2016**, *29*, 1283–1288. [CrossRef]
20. Smith, R.; Maiti, K.; Aitken, R.J. Unexplained antepartum stillbirth: A consequence of placental aging? *Placenta* **2013**, *34*, 310–313. [CrossRef]
21. Williams, P.J.; Mistry, H.D.; Innes, B.A.; Bulmer, J.N.; Broughton Pipkin, F. Expression of AT1R, AT2R and AT4R and their roles in extravillous trophoblast invasion in the human. *Placenta* **2010**, *31*, 448–455. [CrossRef]
22. Hirano, T. IL-6 in inflammation, autoimmunity and cancer. *Int. Immunol.* **2021**, *33*, 127–148. [CrossRef] [PubMed]
23. Hannan, K.M.; Brandenburger, Y.; Jenkins, A.; Sharkey, K.; Cavanaugh, A.; Rothblum, L.; Moss, T.; Poortinga, G.; McArthur, G.A.; Pearson, R.B.; et al. mTOR-dependent regulation of ribosomal gene transcription requires S6K1 and is mediated by phosphorylation of the carboxy-terminal activation domain of the nucleolar transcription factor UBF. *Mol. Cell Biol.* **2003**, *23*, 8862–8877. [CrossRef]

24. Hara, K.; Yonezawa, K.; Kozlowski, M.T.; Sugimoto, T.; Andrabi, K.; Weng, Q.P.; Kasuga, M.; Nishimoto, I.; Avruch, J. Regulation of eIF-4E BP1 phosphorylation by mTOR. *J. Biol. Chem.* **1997**, *272*, 26457–26463. [CrossRef]
25. Chuprin, A.; Gal, H.; Biron-Shental, T.; Biran, A.; Amiel, A.; Rozenblatt, S.; Krizhanovsky, V. Cell fusion induced by ERVWE1 or measles virus causes cellular senescence. *Genes Dev.* **2013**, *27*, 2356–2366. [CrossRef]
26. Cox, L.S.; Redman, C. The role of cellular senescence in ageing of the placenta. *Placenta* **2017**, *52*, 139–145. [CrossRef] [PubMed]
27. Londero, A.P.; Orsaria, M.; Marzinotto, S.; Grassi, T.; Fruscalzo, A.; Calcagno, A.; Bertozzi, S.; Nardini, N.; Stella, E.; Lelle, R.J.; et al. Placental aging and oxidation damage in a tissue micro-array model: An immunohistochemistry study. *Histochem. Cell Biol.* **2016**, *146*, 191–204. [CrossRef] [PubMed]
28. Sukenik-Halevy, R.; Amiel, A.; Kidron, D.; Liberman, M.; Ganor-Paz, Y.; Biron-Shental, T. Telomere homeostasis in trophoblasts and in cord blood cells from pregnancies complicated with preeclampsia. *Am. J. Obstet Gynecol.* **2016**, *214*, 283.e1–283.e7. [CrossRef]
29. Biron-Shental, T.; Kidron, D.; Sukenik-Halevy, R.; Goldberg-Bittman, L.; Sharony, R.; Fejgin, M.D.; Amiel, A. TERC telomerase subunit gene copy number in placentas from pregnancies complicated with intrauterine growth restriction. *Early Hum. Dev.* **2011**, *87*, 73–75. [CrossRef] [PubMed]
30. Biron-Shental, T.; Sukenik-Halevy, R.; Sharon, Y.; Goldberg-Bittman, L.; Kidron, D.; Fejgin, M.D.; Amiel, A. Short telomeres may play a role in placental dysfunction in preeclampsia and intrauterine growth restriction. *Am. J. Obstet. Gynecol.* **2010**, *202*, 381.e1–381.e7. [CrossRef]
31. Biron-Shental, T.; Sukenik-Halevy, R.; Sharon, Y.; Laish, I.; Fejgin, M.D.; Amiel, A. Telomere shortening in intra uterine growth restriction placentas. *Early Hum. Dev.* **2014**, *90*, 465–469. [CrossRef]
32. Levine, R.J.; Maynard, S.E.; Qian, C.; Lim, K.H.; England, L.J.; Yu, K.F.; Schisterman, E.F.; Thadhani, R.; Sachs, B.P.; Epstein, F.H.; et al. Circulating angiogenic factors and the risk of preeclampsia. *N. Engl. J. Med.* **2004**, *350*, 672–683. [CrossRef]
33. Redman, C.W.; Staff, A.C. Preeclampsia, biomarkers, syncytiotrophoblast stress, and placental capacity. *Am. J. Obstet. Gynecol.* **2015**, *213* (Suppl. 4), S9–S11. [CrossRef]
34. Redman, C.W.; Sargent, I.L.; Staff, A.C. IFPA Senior Award Lecture: Making sense of pre-eclampsia—two placental causes of preeclampsia? *Placenta* **2014**, *35*, S20–S25. [CrossRef] [PubMed]
35. ACOG. *Definition of Term Pregnancy, Committee Opinion No. 579*; ACOG: New Orleans, LA, USA, 2013; pp. 1139–1140.
36. Brown, M.A.; Lindheimer, M.D.; de Swiet, M.; Van Assche, A.; Moutquin, J.M. The classification and diagnosis of the hypertensive disorders of pregnancy: Statement from the International Society for the Study of Hypertension in Pregnancy (ISSHP). *Hypertens Pregnancy* **2001**, *20*, IX–XIV. [CrossRef] [PubMed]
37. Mistry, H.D.; Wilson, V.; Ramsay, M.M.; Symonds, M.E.; Broughton Pipkin, F. Reduced selenium concentrations and glutathione peroxidase activity in pre-eclamptic pregnancies. *Hypertension* **2008**, *52*, 881–888. [CrossRef] [PubMed]
38. Mistry, H.D.; McCallum, L.A.; Kurlak, L.O.; Greenwood, I.A.; Broughton Pipkin, F.; Tribe, R.M. Novel expression and regulation of voltage-dependent potassium channels in placentas from women with preeclampsia. *Hypertension* **2011**, *58*, 497–504. [CrossRef]
39. Murthi, P.; Fitzpatrick, E.; Borg, A.J.; Donath, S.; Brennecke, S.P.; Kalionis, B. GAPDH, 18S rRNA and YWHAZ are suitable endogenous reference genes for relative gene expression studies in placental tissues from human idiopathic fetal growth restriction. *Placenta* **2008**, *29*, 798–801. [CrossRef] [PubMed]

International Journal of
*Molecular Sciences*

*Article*

# YB-1 is Altered in Pregnancy-Associated Disorders and Affects Trophoblast in Vitro Properties via Alternation of Multiple Molecular Traits

Violeta Stojanovska [1,*], Aneri Shah [2], Katja Woidacki [3], Florence Fischer [1], Mario Bauer [1], Jonathan A. Lindquist [2], Peter R. Mertens [2] and Ana C. Zenclussen [1,4,*]

1   Department of Environmental Immunology, Helmholtz-Centre for Environmental Research-UFZ-, 04318 Leipzig, Germany; florence.fisher@ufz.de (F.F.); mario.bauer@ufz.de (M.B.)
2   Clinic of Nephrology and Hypertension, Diabetes and Endocrinology, Otto-von-Guericke University, 39120 Magdeburg, Germany; aneri.shah@ovgu.de (A.S.); jon.lindquist@med.ovgu.de (J.A.L.); peter.mertens@med.ovgu.de (P.R.M.)
3   Medical Faculty, Otto-von-Guericke University, 39120 Magdeburg, Germany; katja.woidacki@med.ovgu.de
4   Perinatal Immunology, Saxonian Incubator for Clinical Translation, Medical Faculty, University of Leipzig, 04103 Leipzig, Germany
*   Correspondence: violeta.stojanovska@ufz.de (V.S.); ana.zenclussen@ufz.de (A.C.Z.)

**Citation:** Stojanovska, V.; Shah, A.; Woidacki, K.; Fischer, F.; Bauer, M.; Lidquist, J.A.; Mertens, P.R.; Zenclussen, A.C. YB-1 is Altered in Pregnancy-Associated Disorders and Affects Trophoblast in Vitro Properties via Alternation of Multiple Molecular Traits. *Int. J. Mol. Sci.* **2021**, *22*, 7226. https://doi.org/10.3390/ijms22137226

Academic Editors: Hiten D. Mistry and Eun Lee

Received: 4 June 2021
Accepted: 28 June 2021
Published: 5 July 2021

**Publisher's Note:** MDPI stays neutral with regard to jurisdictional claims in published maps and institutional affiliations.



**Abstract:** Cold shock Y-box binding protein-1 (YB-1) coordinates several molecular processes between the nucleus and the cytoplasm and plays a crucial role in cell function. Moreover, it is involved in cancer progression, invasion, and metastasis. As trophoblast cells share similar characteristics with cancer cells, we hypothesized that YB-1 might also be necessary for trophoblast functionality. In samples of patients with intrauterine growth restriction, YB-1 mRNA levels were decreased, while they were increased in preeclampsia and unchanged in spontaneous abortions when compared to normal pregnant controls. Studies with overexpression and downregulation of YB-1 were performed to assess the key trophoblast processes in two trophoblast cell lines HTR8/SVneo and JEG3. Overexpression of YB-1 or exposure of trophoblast cells to recombinant YB-1 caused enhanced proliferation, while knockdown of YB-1 lead to proliferative disadvantage in JEG3 or HTR8/SVneo cells. The invasion and migration properties were affected at different degrees among the trophoblast cell lines. Trophoblast expression of genes mediating migration, invasion, apoptosis, and inflammation was altered upon YB-1 downregulation. Moreover, IL-6 secretion was excessively increased in HTR8/SVneo. Ultimately, YB-1 directly binds to NF-$\kappa$B enhancer mark in HTR8/SVneo cells. Our data show that YB-1 protein is important for trophoblast cell functioning and, when downregulated, leads to trophoblast disadvantage that at least in part is mediated by NF-$\kappa$B.

**Keywords:** cold shock protein; intrauterine growth restriction; preeclampsia; placentation; apoptosis; NF-$\kappa$B

## 1. Introduction

Adequate placenta development is a prerequisite of a successful pregnancy, while inadequate placentation is a feature of many pregnancy-associated disorders, including spontaneous abortion, intrauterine growth restriction (IUGR), and preeclampsia (PE) [1,2]. Although these disorders have complex pathophysiology with incompletely understood etiology, many share similar origins, such as inadequate trophoblast proliferation and shallow trophoblast invasion [3]. The placenta arises from the trophectoderm of the developing embryo, and, as soon as the embryo-derived trophoblasts adhere to the maternal endometrium, they start to proliferate, migrate towards the basal membrane, and invade into the surrounding endometrial stroma [4]. Here, they attain to remodel the spiral arteries under tightly balanced exposure to growth factors and cytokines derived from the surrounding immune and stromal cells [5].

Interestingly, the trophoblasts share several common characteristics with malignant cells, including proliferative, migratory and invasive features [6]. Moreover, the molecular traits of the trophoblasts, e.g., gene expression and cell response to extracellular stimuli, are similar to those found in malignant cells [7]. Hence, understanding of the molecular mechanisms underlying tumor growth and invasiveness might be of essential interest in understanding the trophoblast functionality, as well.

In several cancer types, YB-1 has been reported as a promoter of cell proliferation, migration, invasion, and inflammation and as an inhibitor of apoptosis [8–11]. YB-1 is encoded by the YBX1 gene, performs pleotropic functions, and contains a highly conserved cold shock domain. This domain has an extreme affinity to bind to DNA and RNA [12], which enables YB-1 to regulate the expression of numerous genes, such as the mechanistic target of rapamycin (mTOR), vascular endothelial growth factor (VEGF), signal transducer and activator of transcription 3 (STAT3), nuclear factor 'kappa-light-chain-enhancer' of activated B-cells (NF-$\kappa$B), and major histocompatibility complex class 2 genes (MHC2) and Notch homolog 3 (NOTCH3), which are involved in cell growth and metabolism, angiogenesis, inflammation, immune system evasion, and embryo development, respectively [13–16]. In addition to the functions in carcinogenesis, YB-1 has been suggested to play a role in embryo development, as well [17]. Throughout embryogenesis, it is highly expressed in the skeletal muscle, spleen and liver, and, after birth, its expression levels quickly decrease [17]. In YBX1-/- knockout embryos, the development advances usually up to embryonic day 10.5 (E10.5), and, afterwards, they exhibit severe growth retardation, neurological and pulmonary lesions, and are embryonically lethal by E18 [18,19], which indicates that YB-1 is necessary at late developmental stages. Recently, we characterized the effects of YB-1 deficiency on placenta development in vivo [20], where trophoblast-specific YB-1 deficient mice showed reduced implantation areas and negatively affected gross placental morphometry already at E10 [20]. This shows a direct involvement of YB-1 not only in placenta growth but also in implantation processes. Knowing that YB-1 is necessary for adequate placental development, we now aim to understand the participation of YB-1 in trophoblast functionality. In that order, firstly, we checked the relevance of YB-1 mRNA expression in different pregnancy-associated disorders using patient samples. Secondly, we performed targeted overexpression and downregulation analysis of human YB-1 in two different trophoblast cell lines. Ultimately, we investigated a possible underlying connection with trophoblast specific genes of proliferation, migration, invasion, and inflammation.

## 2. Results

### 2.1. YB-1 Expression is Unaltered in Miscarriage Samples but is Impaired in IUGR and PE Syndrome

To assess whether YB-1 is affected in early pregnancy events, we analyzed the YB-1 expression in placenta samples from spontaneous abortions and normally progressing pregnancies that were legally terminated (induced abortions), which served as controls. Overview of the baseline patients characteristics is given in Table 1. Unexpectedly, there were no changes in YBX1 expression levels between the groups (Figure 1A). Next, we checked the YBX1 levels in control term placentas and placentas obtained from late pregnancy events, such as preterm birth complicated by IUGR or PE syndrome. In IUGR placentas, YBX1 expression was significantly lower in comparison to controls (Figure 1B). This was, however, not the case in the group of patients with PE syndrome, where we detected increased YBX1 expression levels (Figure 1B). To further investigate whether YB-1 can be detected in serum from pregnant women in the last trimester, we quantified the YB-1 concentrations in a next set of patients. As expected, YB-1 serum concentrations were in line with the results obtained from the gene expression analysis in the placenta, meaning that IUGR pregnancies had lower YB-1 concentrations (Figure 1C). We observed a trend towards increased YB-1 serum concentrations in PE pregnancies, but the results did not reach statistical significance compared to the controls (Figure 1C). Our data suggest that YB-1 expression is not affected in the same way in different pregnancy-associated disorders with a possible role in later stages of pregnancy.

**Table 1.** Descriptive representation of patient data. Data are presented as mean ± SD for continuous data and as % for categorical data. Spontaneous abortions were compared to induced abortions, * $p < 0.05$. IUGR and PE pregnancies were compared to control pregnancies; * $p < 0.05$; ** $p < 0.01$; **** $p < 0.0001$. PE = preeclampsia, Y = years, SC = Cesarean section.

| Patient Characteristics | Induced Abortion (n = 20) | Spontaneous Abortion (n = 20) | |
|---|---|---|---|
| Maternal age (y) | 26.79 ± 5,52 | 29.85 ± 6.84 | |
| Gestational week | 10.74 ± 1.50 | 9.69 ± 3.55* | |
| **Patient Characteristics** | **Control (n = 18)** | **IUGR (n = 12)** | **PE (n = 18)** |
| Maternal age (y) | 28.80 ± 5.25 | 28.00 ± 3.07 | 32.29 ± 8,18 |
| Gestational week | 39.65 ± 1.58 | 32.25 ± 4.09 **** | 28.97 ± 3,01 **** |
| Fetal weight (g) | 3518 ± 459.3 | 1897 ± 869.2 ** | 1017 ± 388,1 **** |
| Fetal length (cm) | 52.13 ± 2.56 | 46.17 ± 6.24 | 42.50 ± 2.12 * |
| Sex (% female) | 33.33 | 63.64 | 42.86 |
| Mode of delivery (% SC) | 40 | 63.64 | 100 |

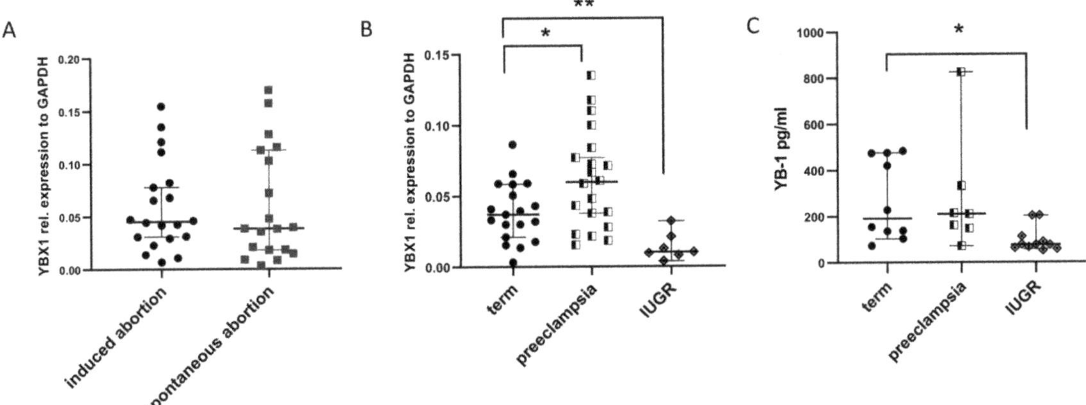

**Figure 1.** Comparison of YB-1 levels in pregnancy-associated disorders. mRNA levels of YBX1 in placenta samples from induced and spontaneous abortions (**A**), mRNA levels of YBX1 in placenta samples from term, preeclamptic, and IUGR pregnancies, at delivery (**B**), and serum YB-1 concentration comparison between term, preeclamptic, and IUGR pregnancies, at delivery (**C**). Data are presented as median with 95% confidence interval, and *p*-values were calculated with one-way ANOVA; * $p < 0.05$, ** $p < 0.01$. IUGR=intrauterine growth restriction.

### 2.2. YB-1 Enhances Trophoblast Cell Proliferation

To address whether trophoblast growth is affected by YB-1, we overexpressed YBX1 in two different trophoblast cell lines. We used the choriocarcinoma JEG3 cell line and HTR8/SVneo, a first trimester immortalized trophoblast cell line. These two cell lines have different origins and are used to assess different trophoblast characteristics [21]. Overexpression of YB-1 (Figure 2A) results in proliferative advantage at 24 h for both cell lines (Figure 2B), without affecting the number of dead cells (Supplementary Figure S1A,B). Next, we assessed whether the effect of YB-1 on trophoblast proliferation is similarly mediated by adding recombinant YB-1 (rYB-1) protein to the culture. We tested several concentrations of rYB-1, and the effects were recorded after 24 h of exposure to the recombinant protein. At 5 µg/mL, rYB-1 induced proliferation in both trophoblast cell lines, while lower concentrations of rYB-1 showed proliferative capacity only for the HTR8/SVneo cell line (Figure 2C). Collectively, these data show that YB-1 positively affects trophoblast viability and proliferation.

*2.3. Exposure to YB-1 Positively Affects Migration and Invasion of Trophoblasts*

Next, we assessed the effect of YB-1 on trophoblast migratory abilities by performing wound healing assays with HTR8/SVneo cells. rYB-1 significantly increased the trophoblast migration (Figure 3A), and, similarly, the overexpression of YBX1 in HTR8/SVneo cells leads to increased migration and complete closure of the wound in 24 h (Figure 3B). Additionally, the invasion properties of HTR8/SVneo and JEG3 cells upon exposure to rYB-1 were assessed via transwell invasion assay. In line with the previous observations, the invasion rates of HTR8/SVneo cells were around 2-fold increased after 18-h exposure to rYB-1 (Figure 3C left panel), while almost no effect was seen regarding invasion properties of JEG3 cells upon stimulation with rYB-1 (Figure 3C, right panel). Hence, YB-1 stimulated migration and invasion in HTR8/SVneo trophoblast cells but not in JEG3 choriocarcinoma cells.

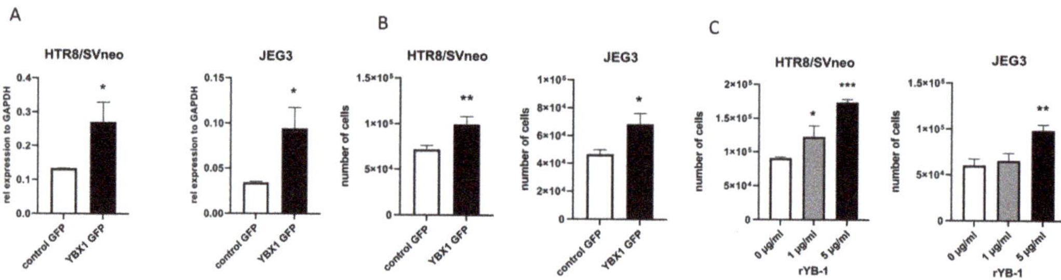

**Figure 2.** Proliferative advantage in trophoblast cell lines upon exposure to YB-1. mRNA levels of YBX1 in HTR8/SVneo and JEG3 cells 24 h after plasmid overexpression (**A**). Trophoblast cell line proliferation upon YBX1 plasmid overexpression (**B**) and after 24-h exposure to rYB-1 (**C**). The results from three independent experiments were statistically analyzed using *t*-test (**A**,**B**) or one-way ANOVA (**C**). Data are presented as mean with SD; * $p < 0.05$; ** $p < 0.01$; *** $p < 0.001$.

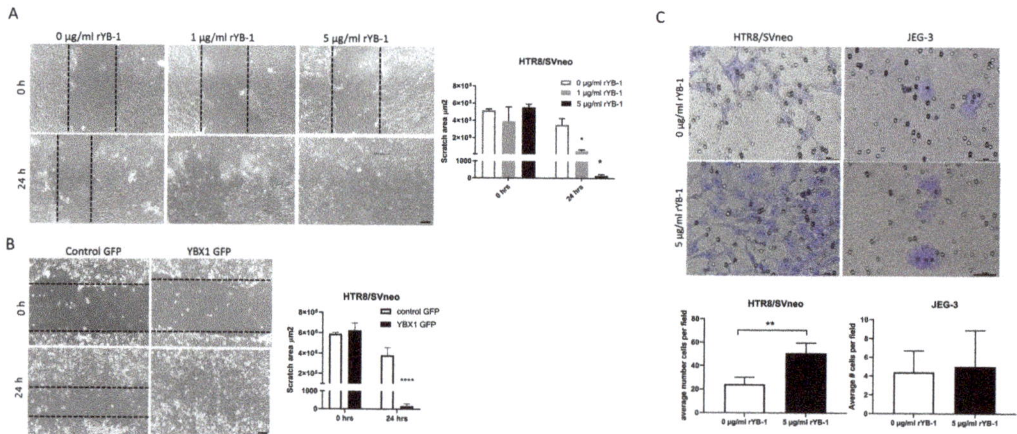

**Figure 3.** Positive effects of YB-1 on migration and invasion in HTR8/SVneo cells. Migration of HTR8/SVneo cells upon rYB-1 stimulation (**A**) and after YBX1 plasmid overexpression (**B**). Invasion rates of HTR8/SVneo and JEG3 upon 18-h rYB-1 stimulations (**C**). The results from three independent experiments were statistically analyzed using one-way ANOVA (**A**) or *t*-test (**B**,**C**). Data are presented as mean with SD; * $p < 0.05$; ** $p < 0.01$; **** $p < 0.0001$. Scale bar 50 μm.

### 2.4. Loss of YB-1 Function Reduces Proliferation and Affects Apoptosis in Trophoblasts

Given that YBX1 overexpression positively affected the trophoblast phenotype, we next investigated whether YB-1 is crucial for overall trophoblast functionality. For this, we used a control and two shRNA YB-1 lentiviral vectors (shYB-1.1 and shYB-1.2), which were used to transduce HTR8/SVneo and JEG3 cell lines. Transduction efficacy and downregulation of YB-1 was confirmed by Western Blot. In both cell lines, the controls transduced with a scramble shRNA expressed endogenous YB-1, but, upon lentiviral transduction, with both shRNA YB-1 vectors, at least a 4-fold decrease in expression was observed (Figure 4A). As indicated in Figure 4B, downregulation of YB-1 provided a proliferative disadvantage of HTR8/SVneo as the population of live cells showed a 3-fold lower numbers at 72 h and more than a 7-fold lower numbers at 96 h post-seeding of the cells (Figure 4B, upper panel). A similar trend was observed when JEG3 cells were used, where the number of live cells decreased significantly at 72 h post-seeding (Figure 4B, lower panel). Additionally, as apoptosis is an important regulator of growth in trophoblast cells, we detected the rate of apoptosis in intact HTR8/SVneo cells using a caspase 3/7 fluorescent assay. Interestingly, YB-1 downregulation resulted in a 3-fold and 6-fold increase of caspase 3/7 activated cells (Figure 4C). Furthermore, we tested the expression of genes related to apoptosis in trophoblast cells (Figure 4D). Levels of mRNA of B-cell lymphoma 2 (BCL2), caspase 3 (CASP3), and FOS Like 1 (FOSL1) were upregulated in HTR8/SVneo YB-1 downregulated cells (Figure 4D, left panel). No changes were observed for BCL2 Associated X (BAX), caspase 9 (CASP9), and galectin-3 (LGALS3) (Figure 4D, left panel). In comparison, in JEG3 YB-1 downregulated cells only, BCL2 was significantly upregulated related to controls (Figure 4D, right panel). These data indicate that, in vitro, YB-1 downregulation results in a trophoblast growth disadvantage that is further associated with alterations in apoptosis-relevant genes.

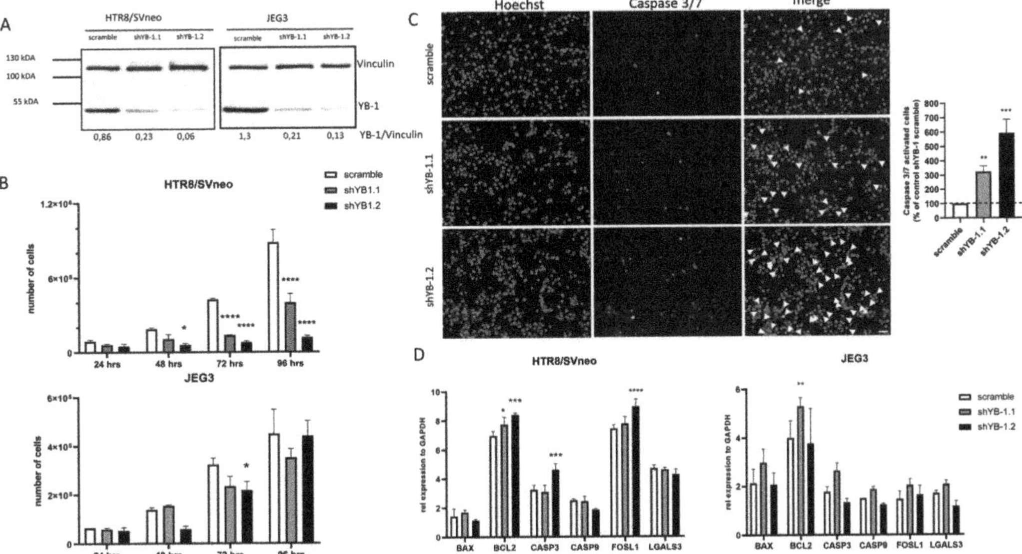



**Figure 4.** Trophoblast cell lines disadvantage upon lentiviral YB-1 downregulation. Representative Western Blot image of YB-1 protein levels 7 days upon YB-1 downregulation (**A**). Effects of YB-1 downregulation on proliferative rates (**B**), apoptosis in HTR8/SVneo cells (**C**), and apoptosis related genes expression (**D**) in trophoblast cell lines. The results from three independent experiments (except A) were statistically analyzed using one-way ANOVA. Data are presented as mean with SD; * $p < 0.05$; ** $p < 0.01$; *** $p < 0.001$; **** $p < 0.0001$. Scale bar 50 µm. BAX= BCL2-associated X, BCL2 = B-cell lymphoma 2, CASP3 = Caspase 3, CASP9 = Caspase 9, FOSL1= Fos-related antigen 1, LGALS3= Galectin-3.

*2.5. Loss of YB-1 Affects Trophoblast Functionality by Modulation of Genes Involved in Cell Migration and Invasion*

Decreased proliferation rates due to YB-1 downregulation may lead to decreased migration/invasion of trophoblast cells. In order to confirm that the impaired migratory and invasive capability of trophoblasts was not a side effect of decreased proliferation, we quantified the cell viability of YB-1 downregulated cells after 24 h in culture with an MTT assay. Both cell lines showed more than 80% cell viability compared to the controls treated with scramble shRNA (Figure 5A). Next, we evaluated the migratory properties of YB-1 downregulated HTR8/SVneo cells, and, as depicted in Figure 5B, it resulted in decreased migration in comparison to control HTR8/SVneo cells. Moreover, the invasion properties of YB-1 downregulated HTR8/SVneo and JEG3 cells were also decreased to different degrees (Figure 5C). Additionally, we analyzed the expression levels of genes involved in the promotion and inhibition of cell migration and invasion (Figure 5D). While matrix metalloproteinase 9 (MMP9) was differentially expressed in both HTR8/SVneo shYB-1 groups, the inhibitors of matrix metalloproteinases, tissue inhibitor of metalloproteinase 1 (TIMP1) and serpin family E member 1 (SERPINE1) were both significantly upregulated (Figure 5D, left panel). Additionally, there were no changes in the expression levels of macrophage migration inhibitory factor (MIF) (Figure 5D, left panel). In JEG3 cells, matrix metalloproteinase 2 (MMP2) showed disproportion in expression after YB-1 downregulation with the two shRNAs against YB-1, while MMP9, TIMP1, and MIF levels were similar to the controls (Figure 5D, right panel). Nevertheless, mRNA levels of SERPINE1 were significantly upregulated in JEG3 cells with downregulated YB-1 expression (Figure 5D, right panel). Moreover, in HTR8/SVneo YB-1 downregulated cells, NOTCH1 and NF-κB were both significantly upregulated, and signal transducer and activator of transcription 3 (STAT3) showed no changes between the groups, while NOTCH3 was downregulated (Figure 5D, left panel). As for JEG3 YB-1 downregulated cells, only NOTCH1 and STAT3 were significantly upregulated, and no changes in the expression of NOTCH3 were observed, while NF-κB showed a trend towards decreased expression (Figure 5D, right panel). Taken together, these results indicate that YB-1 downregulation promotes trophoblast dysfunctionality with effective regulation of downstream targets.

*2.6. YB-1 Mediates IL-6 Secretion and Directly Binds to NF-κB Regulatory DNA Regions*

YB-1 is implicated in inflammation and cytokine production. In the previous experiments, we showed that YB-1 downregulation altered the expression of NF-κB, NOTCH1, and NOTCH3, which are important modulators of the interleukin (IL-6) signaling pathways [22,23]. To investigate whether YB-1 downregulation also translates into changes in cytokine secretion, cell supernatants from YB-1 downregulated HTR8/SVneo and JEG3 cells were analyzed for interleukin 6 (IL-6) secretion. Interestingly, while YB-1 downregulated JEG3 cells showed similar IL-6 concentrations as the controls, HTR8/SVneo cells with downregulated YB-1 expression showed around a 40-fold increase in secreted IL-6 concentrations (Figure 6A).

Since YB-1 has translational, transcriptional, and chromatin binding abilities, we sought to investigate whether the genes that are significantly changed after YB-1 downregulation and are implicated in IL-6 secretion are functional targets of YB-1 under non-perturbed conditions. We tested the binding areas by designing CHIP-qPCR primers (Table 2) at the loci that are enriched for H3K27Ac. We chose these regions as potential binding sites for YB-1 as they indicate activation of transcription and are potential enhancer markers [24]. As a control, we confirmed that YB-1 antibody does bind to the YB-1 H3K27ac enriched area, but it does not bind to the selection position 2, which is an H3K27ac poor area (Figure 6B, lower panel). For NOTCH3, YB-1 does not bind throughout the selected H3K27ac enriched loci, which we identified as possible target marks (Figure 6C). Interestingly, we observed that YB-1 has a rudimentary binding to the chosen position 1, which lies at the exon 1 region of NF-κB in HTR/SVneo cell line (Figure 6D). Taken together, these data suggest that YB-1 might collaborate with NF-κB to induce impaired trophoblast phenotype.

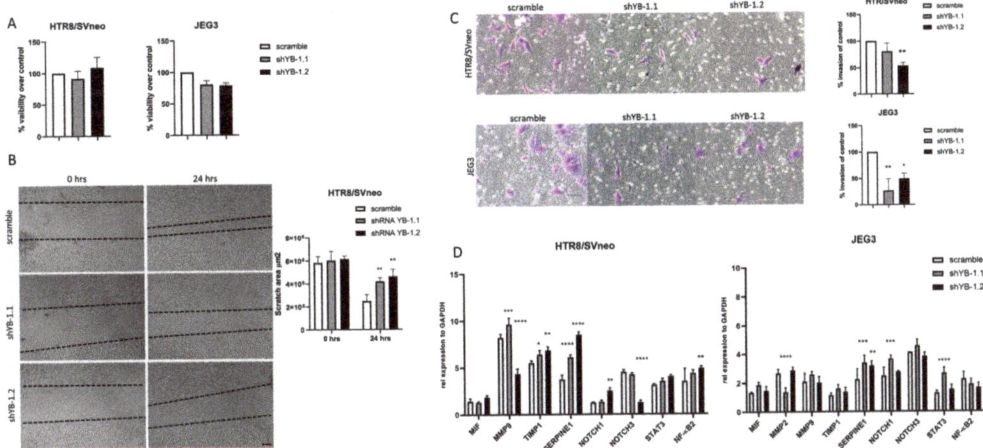

**Figure 5.** Negative effects of YB-1 on migration and invasion in HTR8/SVneo cells. YB-1 downregulation effects on trophoblast cell line viability (**A**), migration (**B**), invasion (**C**), and expression levels of genes relevant for trophoblast migration, invasion, and overall function (**D**). The results from three independent experiments were statistically analyzed using one-way ANOVA. Data are presented as mean with SD; * $p < 0.05$; ** $p < 0.01$; *** $p < 0.001$; **** $p < 0.0001$. Scale bar 50 μm. MIF = Macrophage migration inhibitory factor, MMP2 = Metalloproteinase 2, MMP9 = Metalloproteinase 9, TIMP1 = Metallopeptidase inhibitor 1, SERPINE1 = Serpin family e member 1, NOTCH1 = Notch homolog 1, NOTCH3 = Notch homolog 3, STAT3 = Signal transducer and activator of transcription 3, NF-$\kappa$B = Nuclear factor kappa-light-chain-enhancer of activated B cells.

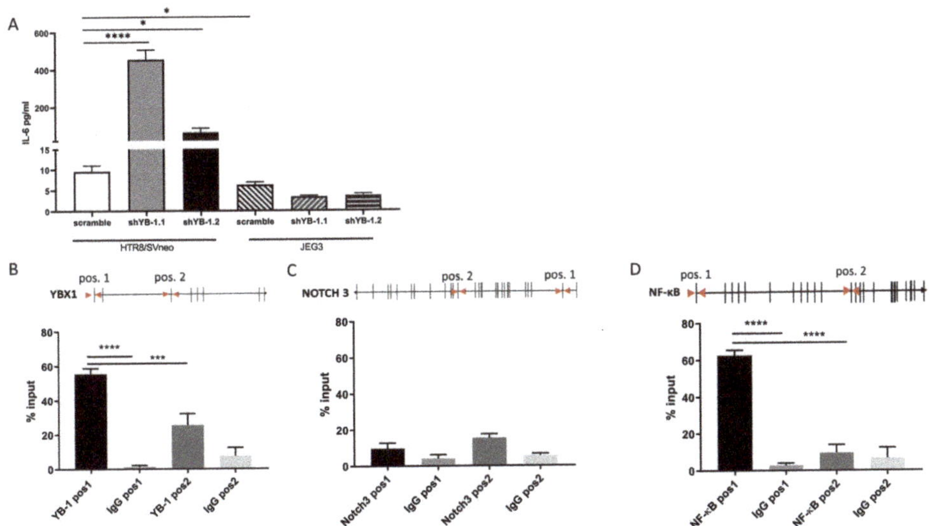

**Figure 6.** YB-1 downregulation increases IL-6 secretion and binds to NF-$\kappa$B in HTR8/SVneo cells. IL-6 was detected in cell supernatants from YB-1 downregulated trophoblast cell lines (**A**). qPCR-CHIP primer positions and YB-1 binding is presented as percent of input for YB-1 (**B**), NOTCH3 (**C**), and NF-$\kappa$B (**D**). The results from three independent experiments were statistically analyzed using one-way ANOVA. Data are presented as mean with SD; * $p < 0.05$; *** $p < 0.001$; **** $p < 0.0001$. NOTCH3 = Notch homolog 3, NF-$\kappa$B = nuclear factor kappa-light-chain-enhancer of activated B cells.

**Table 2.** qPCR-CHIP primer sequences.

| No. | Primer (Gene) Name | (Sequence 5′-3′) |
| --- | --- | --- |
| 1 | hYBX1 fw position 1 | GGGAAGCCTTTTCTTCACGG |
| 2 | hYBX1 rv position 1 | GAGTAGTCGGCCACGAAAAC |
| 3 | hYBX1 fw position 2 | GAAGCTAGGGATTGGGGTCA |
| 4 | hYBX1 rv position 2 | GCTACCGATCGAACTAGCGA |
| 5 | hNOTCH3 fw position 1 | CACAGAGGAAGTGGGTTGCT |
| 6 | hNOTCH3 rv position 1 | ATTTGCAGCCTCAGACCTCA |
| 7 | hNOTCH3 fw position 2 | ATGGGGAAACACGAGAGGTTG |
| 8 | hNOTCH3 rv position 2 | TTTGTCACTTGGGCCTGGGG |
| 9 | hNF-$\kappa$B1 (p50) fw position 1 | CCCCTCTGCCAGATCAGTATT |
| 10 | hNF-$\kappa$B1 (p50) rv position 1 | CGACTTGTGCCCAGTAAAGT |
| 11 | hNF-$\kappa$B1 (p50) fw position 2 | CTTCCTCATTCCTGCGCTAAC |
| 12 | hNF-$\kappa$B1 (p50) rv position 2 | GTAAGAGTTCCCCTCCGGTT |

## 3. Discussion

YB-1 plays an essential role in tumorigenesis by regulating cell proliferation, inflammation, migration, invasion, and apoptosis via relevant pathways [25]. During pregnancy, these processes are also crucial for the normal development of the placenta and changes in YB-1 expression might be involved in impaired placenta growth and/or functionality. We have recently shown that heterozygous YBX1 mice and mice with trophoblast-specific YBX1 deficiency display placental abnormalities with subsequent fetal growth retardation [20]. The present study demonstrates that pregnancy-associated disorders are associated with altered YB-1 concentrations in tissue and blood samples. Our experiments comprising overexpression and downregulation of YB-1 in trophoblasts show that YB-1 is an important regulator of their functionality.

Discordant YB-1 in pregnancy might lead to defective placentation and subsequent development of pregnancy-associated disorders. To date, we are the first to report results on YBX1 expression changes in patients suffering from spontaneous abortion, preterm birth, or PE syndrome. We did not observe any differences in placental YBX1 expression between subjects having a spontaneous abortion or induced termination of pregnancy. Thus, it suggests that YB-1 may not be particularly relevant in early events of pregnancy. However, in pregnancies complicated by preterm birth and intrauterine growth restriction, YB-1 transcript levels and serum concentrations were significantly decreased compared to controls. Thus, it indicates that YB-1 may have a considerable role in trophoblast physiology in the later stages of pregnancy. Surprisingly, in PE patients, YBX1 transcript levels were increased, but we could not confirm this in the few obtained matching serum samples. This discrepancy may be attributed to the slightly different pathophysiological traits underlying these disorders [26]. Preeclampsia, among defective placentation, is also characterized with angiogenic imbalance due to increased concentrations of antiangiogenic factors, such as soluble fms-like tyrosine kinase-1 (sFlt-1) and soluble endoglin (sEnd), and decreased concentrations of vascular endothelial growth factor (VEGF) and placental growth factor (PlGF) [27,28]. We speculate that this imbalance might, in turn, lead to a compensatory mechanism, such as an increase in YB-1 levels, which promotes angiogenesis [29]. Although it is known that YB-1 regulates the levels of angiogenic factors, it is yet to be elucidated whether angiogenic imbalance can trigger YB-1 upregulation. This might be the focus of subsequent studies, but it is not the main aim of the present one and will not be further discussed.

To investigate the impact of elevated YB-1 levels on the trophoblast phenotype, we used two different approaches: YBX1 overexpression and direct exposure to rYB-1 protein in the two most widely used trophoblast cell lines. The first one aims at understanding the significance of YBX1 expression in the trophoblast itself. The second one rather addresses the contribution of secreted YB-1 by other cells present at the feto-maternal interface [30]. Transient YBX1 overexpression led to an increased proliferation of both cell lines, and promotion of migration in HTR8/SVneo cells. Direct exposure to YB-1 resulted in similar outcomes, confirming the positive effect of this cold shock protein in stimulating the proliferation and migration of trophoblasts. Additionally, we tested the invasive properties of both cell lines in the presence of rYB-1, and, while no changes were registered for JEG3 cells, HTR8/SVneo cells increased their invasive activity upon YB-1 addition to the culture. The upregulation of trophoblast proliferation, migration, and invasion linked to YB-1 is in accordance with several previous studies using other cell types which demonstrated that overexpression of YBX1 acts as a potent enhancer of cellular functions [29,31]. In regard to the discrepancy we see in the affinity of YB-1 to the different trophoblast cell lines, we link this to the origin of the cell lines. The HTR8/SVneo cell line is a first trimester extravillous trophoblast cell line that was immortalized with an origin-defective simian virus 40 construct and was revealed that consists of two distinct populations; epithelial and mesenchymal [32] . Given that YB-1 is known to physically and functionally interact with the viral regulatory protein T- antigen [33], we speculate that this is the reason for the more pronounced effect in the HTR8/SVneo cell line compared to JEG3 cell line that is derived from cancerous tissue.

Normal placental development dependent not only on controlled cell proliferation and differentiation but also on programmed cell death that occurs as a rate-limiting proliferation factor [34]. Likewise, in normal pregnancies, trophoblast apoptosis increases as the pregnancy progress and in pregnancy-associated disorders, such as PE and IUGR, a greater incidence of apoptosis is observed [35]. The central players of apoptosis are the caspases, however many other genes are involved in this process [36]. Our study shows that YB-1 downregulated cells possess increased caspase activity and upregulated expression of the pro-apoptotic molecules BCL2 and FOSL1 in HTR8/SVneo cells. This clearly demonstrates that the absence of YB-1 results in excessive apoptosis in HTR8/SVneo cells via upregulation of several genes that propagate the death signal.

One essential feature for optimal placenta development is the ability of the trophoblasts first to migrate and then invade into the surrounding tissue. To date, YB-1 has been fundamentally implicated in the migration and invasion processes in several types of cancers, including breast cancer [37], lung cancer [38], melanoma [39], and spinal chordoma [40]. YB-1 downregulation leads to reduced invasive and migratory abilities of tumorigenic cells, consistent with changes in mRNA levels of genes necessary for cell proliferation and invasion, such as zinc finger protein SNAI1, NF-κB, MMP2, etc. [41]. This is also in line with our results, where we show that stable downregulation of YB-1 in trophoblast cell lines leads to a defective trophoblast phenotype that is impaired in proliferation, migration, and invasion. Furthermore, the invasive trophoblast ability is dependent on balanced protease activation and inhibition. While trophoblast invasion is mediated via MMPs, and namely MMP2 and MMP9, the inhibition is dependent on TIMP1 and SERPINE1 [42]. In our study, we show that the tissue inhibitors of matrix metalloproteinases TIMP1 and SERPINE1 were upregulated, and this may contribute to the decreased invasive profile of trophoblast cells.

Trophoblasts are known to secrete cytokines, and IL-6 is amongst the most abundantly produced factors [43]. IL-6 acts as a pro- and anti-inflammatory cytokine and modulates immune responses, angiogenesis, and trophoblast proliferation, migration, and apoptosis [44]. Moreover, cytotrophoblast cells express high IL-6 mRNA levels, which are further stimulated by IL-6 itself [45]. However, excessive production of IL-6 has been also implicated in the development of several pregnancy-associated disorders [46,47]. In the present study, we show that indeed, HTR8/SVneo cells with downregulated YB-1 secrete excessively high concentrations of IL-6. While the cytokine production is a tightly controlled process, the signaling pathways regulating IL-6 secretion are still not completely disclosed. Independently, it was reported that NOTCH3 and NF-κB could mediate the induction of IL-6 production in malignant cells [48] and in macrophages [22]. When determining the expression of NF-κB and NOTCH3 in our YB-1 downregulated HTR8/SVneo, we came to the conclusion that the expression levels were differentially affected. This provides evidence that YB-1 interacts with inflammation relevant genes that, in turn, can also perturb the trophoblast functionality.

Several studies have shown that YB-1 is required for NF-$\kappa$B activation either via IL-1$\beta$R [49] or via TNFR1 [50]. Here, we show that, in unaltered conditions, YB-1 has the affinity to directly bind to a region that encloses the first exon of the NF-$\kappa$B gene, which is also an H3K27ac enriched region as depicted in the UCSC genome browser [51]. This suggests that NF-$\kappa$B can act as a functional target of YB-1. Knowing that NF-$\kappa$B is involved in the regulation of several cellular processes, including proliferation, differentiation, apoptosis, oxidative stress, and inflammation [52], it is of no surprise that this protein is critically implicated in the placentation, as well [53]. Previously, it was shown that NF-$\kappa$B can modulate the placenta development either via regulation of factors important for trophoblast invasion, such as MMP2, MMP9, and SERPINE1 [54], or via regulation of the feto-maternal vascularization through cytokines (IL-6, IL-8) [55] and angiogenic proteins (PlGF, VEGF, sEnd) [54,56]. Hence, the detrimental roles that we observe for YB-1 in trophoblast function may be, in part, mediated by NF-$\kappa$B.

## 4. Conclusions

Our findings indicate that YB-1 acts as a potent regulator of trophoblast functionality via changes in the molecular footprint of genes involved in proliferation, apoptosis, migration, invasion, and inflammation. Moreover, we showed that YB-1 could directly bind to the NF-$\kappa$B gene, which, in turn, can shed light on the YB-1 involvement in pregnancy-associated disorders.

## 5. Materials and Methods

### 5.1. Sampling and Ethical Approval

Placental tissue was collected from induced (elective) pregnancy terminations and spontaneous abortions or shortly after birth and was processed for RNA isolation. Peripheral blood samples were obtained from healthy pregnant or pregnant patients admitted to hospital due to preterm birth or development of PE. Inclusion criteria for PE were: singleton pregnancy, hypertension (diastolic blood pressure $\geq$ 90 mmHg on at least two occasions), and proteinuria (urine dipstick > 1+ ($\geq$30 mg/dL) on at least two occasions, and protein: creatinine ratio of $\geq$0.35 or 24-h urine protein concentration $\geq$ 300 mg). For IUGR patients, the inclusion criteria were the following: singleton fetus, normotensive (120–80 mmHg), estimated fetal weight, and/or abdominal circumference <10th percentile. Exclusion criteria for all patients recruited in this study were: multiple pregnancy, presence of congenital infections and chromosomal defects, and autoimmune disorder of the mother. The clinical study was approved by the ethics board at the University of Magdeburg with reference number EK28/08. All subjects provided written informed consent, and the study was performed in accordance with the Declaration of Helsinki. Subject characteristics are presented in Table 1.

### 5.2. Cell Lines

The human choriocarcinoma cell line JEG3 (DMZO, Braunschweig, Germany) was cultured in Dulbecco's modified Eagle's medium (DMEM) (Invitrogen, Karlsruhe, Germany) supplemented with 10% fetal bovine serum (FBS, Biochrom, Berlin, Germany) and 100 nmol/L penicillin/streptomycin (Invitrogen, Karlsruhe, Germany). The immortalized human extravillous cytotrophoblast cell line HTR8/SVneo (ATCC, CRL-327) was cultured in Roswell Park Memorial Institute (RPMI) 1640 medium, supplemented with 10% fetal bovine serum (FBS, Biochrom, Berlin, Germany), 100 nmol/L penicillin/streptomycin (Invitrogen, Karlsruhe, Germany), 100 nmol/L MEM nonessential amino acids (Invitrogen, Karlsruhe, Germany), 1 mmol/L sodium pyruvate (Invitrogen, Karlsruhe, Germany), and 10 mmol/L HEPES (Invitrogen, Karlsruhe, Germany). Both cell lines were cultured at 37 °C with 5% $CO_2$ and humidified atmosphere and subcultured using 0.05% Trypsin-

EDTA (Invitrogen, Karlsruhe, Germany). The cell lines were regularly tested for absence of mycoplasma infection.

### 5.3. YB-1 Detection in Serum

Whole blood was collected in SST II Advance tubes (BD Vacutainer) and was allowed to clot for 30 min at room temperature. Afterwards, the samples were centrifuged at 2000 rpm for 10 min at 4 °C. Serum was collected, aliquoted, and stored at −80 °C until further analysis. Only samples that were not hemolyzed were used for the analysis. YB-1 serum concentrations were detected using ELISA method (Genway, Wuhan, China) following manufacturer's instructions.

### 5.4. Recombinant YB-1 Protein Harvest and Purification

Seventy percent confluent HEK293T cells were transduced with pcDNA/Flag YB-1 sequence using Ca-phosphate-DNA precipitates. After 48 h of transduction, the cells were harvested in RIPA-lysis buffer (50 nM Tris, 150 nM NaCl, 1% Nonidet P40, 0.25% sodium deoxycholate, 1 mM EDTA, 1 mM $Na_3VO_4$, 1 mM NaF), supplemented with protease inhibitors (Complete EDTA-free cocktail tablet, Roche, Mannheim, Germany), and were immunoprecipitated using anti-DYKDDDDK G1 affinity resin (Genscript, Piscataway, NJ, USA) and binding buffer (50 mM Tris-HCl pH 7.4; 1 mM EDTA, 150 mM NaCl, 1% Triton X-100). The next day, the samples were eluted with 100 µg/mL FLAG-peptide stock (Sigma-Aldrich, St. Louis, MO, USA) and dialyzed with 40% Polyethylene glycol 20,000 (Roth, Karlsruhe, Germany). Protein concentration was determined using BioRad DC protein assay (BioRad, Hercules, CA, USA).

### 5.5. YBX1 Overexpression

JEG3 and HTR8/SVneo cells were transfected using Lipofectamine 2000 (Life Technologies, Carlsbad, CA, USA), according to the manufacturer's protocol. In short, 70% confluent cells were transiently transfected with 1 µg control FuGW-eGFP or 1 µg FuGW-eGFP-YBX1 plasmid DNA. FUGW was a kind gift from David Baltimore (Addgene plasmid 14883; http://n2t.net/addgene:14883; RRID:Addgene 14883). After 24 h, the transfection efficacy was inspected by detection of GFP signal by fluorescence microscopy.

### 5.6. Lentiviral Transduction of YB-1

Downregulation of YB-1 was performed as previously described [50]. In short, the control plasmid pLKO and two different pLKO-YB-1.1 and pLKO-YB-1.2 shRNA were obtained from Sigma-Aldrich (shRNA: CCGGCCAGTTCAAGGCAGAAATATCTCGA-GATA TTTACTGCCTTGAACTGG-TTTTTG). Two micrograms of YB-1 construct, 1 µg psPAX2, and 1 µg pVSV-G with calcium phosphate precipitates was used for lentiviral transduction of human embryonic kidney HEK 293T cells. Virus containing supernatants were then added to the target cells JEG3 and HTR8/SVneo. Stably transduced cell lines were selected using puromycin (1.5 µg/mL) for 7 days. Cells were harvested 3 and 7 days after transduction to confirm the YB-1 downregulation by Western Blot analysis.

### 5.7. Protein Isolation and Western Blot Analysis

Cells were lysed in RIPA buffer (50 mM Tris-HCl, 150 mM nonidet P-40, 1 mM sodium deoxycholate, 1 mM EDTA, and 1 mM $Na_3VO_4$) containing protease inhibitor cocktail (Complete EDTA-free cocktail tablet, Roche, Mannheim Germany) at 4°C for 30 min. Protein concentrations were quantified using Bio-Rad protein assay (BioRad, Hercules, CA, USA). Only the samples that had good protein yield were used for the study. Proteins were detected using primary antibodies anti-YB-1 (Eurogentec, Liège, Belgium) and anti-vinculin (Santa Cruz, Dallas, Texas, USA) was used as a loading control. Secondary antibodies coupled to horseradish peroxidase (Southern Biotech, Birmingham, AL, USA) were used for immuno-detection. The detection was performed using Pierce ECL Western blotting substrate (Thermo Fischer Scientific, Waltham, MA, USA).

### 5.8. Functional Assays

Proliferation was assessed by the trypan blue exclusion assay using Neubauer chambers and manual quantification of cells. For the cell migration assay, 20,000 cells were plated in 24-well plates, and, after 24 h, scratch was performed with 200 µL pipette tip. Pictures were taken after 24 h, and the distances crossed were measured using an electronic grid. The mean value for the controls was set to 100%, and the data are expressed as percentages of the control value. Transwell invasion assay was conducted in 24-well plate with 8 µm pore size Transwell inserts (Corning, Durham, NC, USA). Insert membranes were precoated with 50 µl of growth factors reduced Matrigel (Corning, Bedford, MA, USA) at a concentration of 0.5 µg/mL for 1 h at 37 °C. JEG3 or HTR8/SVneo (25,000 cells/100 µL) were resuspended in serum free medium with 1 or 5 µg/mL rYB-1 added to the upper part of the Transwell chamber. In the lower part of the chamber, complete medium was added with 10% FBS as chemoattractant. Cells were incubated for 18 h and the Transwell membranes were stained with 0.2% Crystal Violet (Sigma-Aldrich, Germany). Quantification of cells on the underside of the filter were counted with brightfield microscope for average of 10 picture fields at 20× magnification total.

### 5.9. RNA Isolation and qPCR Analysis

Total RNA was isolated using TRIzol (Life Technologies, Carlsbad, CA, USA), following manufacturer's protocol. Quantity and purity of RNA were determined using Infinite F200 Nanoquant (Tecan, Grödig, Austria). The cDNA synthesis was carried out with 800 ng of total RNA by using Im Prom-II™ Reverse Transcription System (Promega, Mannheim, Germany). Real-time quantitative PCR was carried out on a LightCycler 480 System (Roche Applied Science, Mannheim, Germany) with the following cycling program: 2 min at 50 °C, 10 min at 95 °C, followed by 35 cycles of 95 °C for 15 s, 1 min at 60 °C, and 70 °C for 5 s. All reactions were performed in duplicates. Primers and UPL probes were designed and selected by the Universal Probe Library Assay Design Center (http://qpcr.probefinder.com/organism.jsp). Ct values and the expression of the reference gene GAPDH was uniform among the groups. The data was analyzed using the ddCT method.

### 5.10. IL-6 Quantification

Cell supernatants from trophoblast cell lines with downregulated YB-1 expression were collected, spun at 2000 rpm for 5 min at 4 °C, and stored at −80 °C until further analysis. IL-6 quantification was obtained with human IL-6 ELISA detection assay (R&D Systems, Minneapolis, MN, USA) according to manufacturer's instruction. Absorbance was measured using Infinite F200 microplate reader (Tecan, Grödig, Austria).

### 5.11. Caspase Activity Assay

HTR8/SVneo control and YB-1 downregulated cells were plated in 24-well plate at a density of 20,000 cells per well and incubated for 48 h under standard cell culture conditions. The caspase activity was studied using CellEvent™ Caspase-3/7detection reagent (Invitrogen, Eugene, OR, USA). After 30 min of incubation, fluorescence was observed using fluorescent microscope (KEYENCE BZ-X800, Osaka, Japan), coupled with a confocal module. Excitation and absorption wavelength were 360/40 nm and 470/40 nm, respectively. Nuclei were stained with Hoechst 33342 (Invitrogen, Eugene, OR, USA). The intensity of fluorescence was analyzed with respective KEYENCE BZ-X800 analysis software.

### 5.12. MTT Assay

One thousand cells were plated onto 96-well plates, and 3-(4,5-dimethylthiazol-2-yl)-2,5-dipheny ltetrazolium bromide (Sigma-Aldrich, St. Loius, MO USA) was added after 24 h at a final concentration of 5 mg/mL. After incubation at 37 °C and 5% $CO_2$ for 3 h, the MTT was removed and MTT formazan crystals were dissolved in 150 µL of DMSO.

Absorbance at 590 nm was determined on an automatized microtiter plate reader (BioTek Synergy HT, Watertown, MA, USA).

### 5.13. Chromatin Immunoprecipitation Assay (CHIP)

Chromatin was isolated from $1 \times 10^7$ JEG3 cells. In short, cells were cross-linked with 1% formaldehyde for 10 min and resuspended in SDS buffer (5 M NaCl, 1 M Tris-HCl, pH 8,1; 0.5 M EDTA pH 8,0; 10 M $NaN_3$, 10% SDS) with added protease inhibitor (Complete EDTA-free cocktail tablet, Roche, Mannheim, Germany), and stored at $-80\ °C$ until further analysis. After defrost and centrifugation, samples were resuspended in IP buffer (66.7 mM Tris–HCl, 100 mM NaCl, 5 mM EDTA, 0.2% $NaN_3$, 1.67% Triton-X-100, and 0.33% SDS) and sonicated 5 times for 30 s. Chromatin shearing was checked by reverse cross link reaction of 2 h on 65 °C at 850 rpm using reverse cross link buffer (Tris EDTA buffer 1x, SDS 20%, 5 M NaCl). One microgram of anti-YB-1 recombinant antibody (EP2708Y, Abcam, Cambridge, UK), or IgG from rabbit serum (I8140, Sigma-Aldrich, St. Louis, MO, USA), was used per immunoprecipitation. nProtein A Sepharose beads (GE Healthcare, Danderyd, Sweden) were used to pull-down the immune complexes. Wash of the beads and samples complexes was performed with High Salt Buffer 500 mM (1% Triton X-100, 0,1% SDS, 150 mM NaCl, 2 mM EDTA pH 8.0, 20 mM Tris-HCl) and 1x Tris-EDTA buffer. Reverse cross linking of all samples was done with reverse cross- link buffer on 65 °C at 1300 rpm for 12 h. The products were purified with QIAquick PCR purification kit (Qiagen, Hilden, Germany) following the manufacturer's instructions. The real-time PCR was performed using SYBR Green PCR Master Mix on a sequence detector (7500 Fast Real-Time PCR System; Applied Biosystems, Foster City, CA, USA) with 2.5 µL of material per point. Primer sequences are available in Table 2. The input DNA fraction corresponded to 10% of the immunoprecipitation.

### 5.14. Statistical Analysis

All results were confirmed in three independent experiments, if not otherwise stated. The patient data are presented as median and 95% confidence interval, if not otherwise stated. All other data are presented as mean with SD. Statistical data analysis was performed using GraphPad Prism software version 8.0 (GraphPad Software, San Diego, CA, USA). Differences between groups were calculated with unpaired *t*-test. For multiple comparisons, statistical difference was calculated by one-way ANOVA. When statistically significant differences were shown, post hoc analysis were performed using the Sidak test. $p < 0.05$ was considered statistically significant.

**Supplementary Materials:** The following are available online at https://www.mdpi.com/1422-006 7/22/13/7226/s1.

**Author Contributions:** V.S. and A.C.Z. conceived study concept and design; V.S., A.S., K.W. and M.B. performed and analyzed the experiments. V.S., A.S., F.F., M.B., P.R.M., J.A.L. and A.C.Z. did interpretation of the data. V.S. wrote the manuscript and A.C.Z. substantially revised the manuscript. V.S., A.S., K.W., F.F., M.B., P.M., J.A.L. and A.C.Z. reviewed and edited the manuscript. All authors have read and agreed to the published version of the manuscript.

**Funding:** This work was supported by a grant from the DFG to ACZ (ZE 526/14-1).

**Institutional Review Board Statement:** The study was conducted according to the guidelines of the Declaration of Helsinki, and approved by the ethics board of the University of Magdeburg with reference number EK28/08.

**Informed Consent Statement:** Informed consent was obtained from all subjects involved in the study.

**Acknowledgments:** The authors would like to thank Susanne Arnold, Beate Fink, Ulrike Königsmark, Lisa Wiesner, and Marcus Scharm for the excellent technical assistance and invaluable support.

**Conflicts of Interest:** The authors declare no conflict of interest.

## References


1. Romero, R.; Kusanovic, J.P.; Chaiworapongsa, T.; Hassan, S.S. Placental bed disorders in preterm labor, preterm PROM, spontaneous abortion and abruptio placentae. *Best Pr. Res. Clin. Obs. Gynaecol.* **2011**, *25*, 313–327. [CrossRef]
2. Fisher, S.J. Why is placentation abnormal in preeclampsia? *Am. J. Obs. Gynecol.* **2015**, *213*, S115–S122. [CrossRef]
3. Burton, G.J.; Jauniaux, E. Placental oxidative stress: From miscarriage to preeclampsia. *J. Soc. Gynecol. Investig.* **2004**, *11*, 342–352. [CrossRef]
4. Schumacher, A.; Zenclussen, A.C. Human Chorionic Gonadotropin-Mediated Immune Responses That Facilitate Embryo Implantation and Placentation. *Front. Immunol* **2019**, *10*, 1–14. [CrossRef]
5. Whitley, G.S.J.; Cartwright, J.E. Cellular and Molecular Regulation of Spiral Artery Remodelling: Lessons from the Cardiovascular Field. *Placenta* **2010**, *31*, 465–474. [CrossRef]
6. Mullen, C.A. Analogies between trophoblastic and malignant cells. *Am. J. Reprod. Immunol.* **1998**, *39*, 41–49. [CrossRef] [PubMed]
7. Ferretti, C.; Bruni, L.; Dangles-Marie, V.; Pecking, A.P.; Bellet, D. Molecular circuits shared by placental and cancer cells, and their implications in the proliferative, invasive and migratory capacities of trophoblasts. *Hum. Reprod. Update* **2007**, *13*, 121–141. [CrossRef]
8. Fujiwara-Okada, Y.; Matsumoto, Y.; Fukushi, J.; Setsu, N.; Matsuura, S.; Kamura, S.;Fujiwara, T.; Iida, K.;Hatano, M.; Nabeshima, A.; et al. Y-box binding protein-1 regulates cell proliferation and is associated with clinical outcomes of osteosarcoma. *Br. J. Cancer* **2013**, *108*, 836–847. [CrossRef]
9. Davies, A.H.; Barrett, I.; Pambid, M.R.; Hu, K.; Stratford, A.L.; Freeman, S.; Berquin, I.M.; Pelech, S.; Hieter, P.; Maxwell, C.; et al. YB-1 evokes susceptibility to cancer through cytokinesis failure, mitotic dysfunction and HER2 amplification. *Oncogene* **2011**, *30*, 3649–3660. [CrossRef]
10. Faury, D.; Nantel, A.; Dunn, S.E.; Guiot, M.C.; Haque, T.; Hauser, P.; Garami, M.; Bognár, L.; Hanzély, Z.; Liberski, P.P.; et al. Molecular profiling identifies prognostic subgroups of pediatric glioblastoma and shows increased YB-1 expression in tumors. *J. Clin. Oncol.* **2007**, *25*, 1196–1208. [CrossRef] [PubMed]
11. Kosnopfel, C.; Sinnberg, T.; Sauer, B.; Niessner, H.; Muenchow, A.; Fehrenbacher, B.; Schaller, M.; Mertens, P.R.; Garbe, C.; Thakur, B.K.; et al. Tumour progression stage-dependent secretion of YB-1 stimulates melanoma cell migration and invasion. *Cancers* **2020**, *12*, 1–17. [CrossRef]
12. Izumi, H.; Imamura, T.; Nagatani, G.; Ise, T.; Murakami, T.; Uramoto, H.; Torigoe, T.; Ishiguchi, H.; Yoshida, Y.; Nomoto, M.; et al.Y box-binding protein-1 binds preferentially to single-stranded nucleic acids and exhibits $3'\rightarrow5'$ exonuclease activity. *Nucleic Acids Res.* **2001**, *29*, 1200–1207. [CrossRef]
13. Lasham, A.; Print, C.G.; Woolley, A.G.; Dunn, S.E.; Braithwaite, A.W. YB-1: Oncoprotein, prognostic marker and therapeutic target? *Biochem. J.* **2013**, *449*, 11–23. [CrossRef]
14. Fujii, T.; Seki, N.; Namoto-Matsubayashi, R.; Takahashi, H.; Inoue, Y.; Toh, U.; Kage, M.; Shirouzu, K. YB-1 prevents apoptosis via the mTOR/STAT3 pathway in HER-2-overexpressing breast cancer cells. *Futur. Oncol.* **2009**, *5*, 153–156. [CrossRef] [PubMed]
15. Lindquist, J.A.; Mertens, P.R. Cold shock proteins: From cellular mechanisms to pathophysiology and disease. *Cell Commun. Signal* **2018**, *16*, 1–14. [CrossRef] [PubMed]
16. Rauen, T.; Raffetseder, U.; Frye, B.C.; Djudjaj, S.; Mühlenberg, P.J.T.; Eitner, F.; Lendahl, U.; Bernhagen, J.; Dooley, S.; Mertens, P.R. YB-1 acts as a ligand for notch-3 receptors and modulates receptor activation. *J. Biol. Chem.* **2009**, *284*, 26928–26940. [CrossRef]
17. Ito, K.; Tsutsumi, K.I.; Kuzumaki, T.; Gomez, P.F.; Otsu, K.; Ishikawa, K. A novel growth-inducible gene that encodes a protein with a conserved cold-shock domain. *Nucleic Acids Res.* **1994**, *22*, 2036–2041. [CrossRef]
18. Lu, Z.H.; Books, J.T.; Ley, T.J. YB-1 Is Important for Late-Stage Embryonic Development, Optimal Cellular Stress Responses, and the Prevention of Premature Senescence. *Mol. Cell. Biol.* **2005**, *25*, 4625–4637. [CrossRef]
19. Uchiumi, T.; Fotovati, A.; Sasaguri, T.; Shibahara, K.; Shimada, T.; Fukuda. T,; Nakamura, T.; Izumi, H.; Tsuzuki, T.; Kuwano, M.; Kohno, K. YB-1 is important for an early stage embryonic development: Neural tube formation and cell proliferation. *J. Biol. Chem.* **2006**, *281*, 40440–40449. [CrossRef]
20. Meyer, N.; Schumacher, A.; Coenen, U.; Woidacki, K.; Schmidt, H.; Lindquist, J.A.; Mertens, P.R.; Zenclussen, A.C. Y-Box Binding Protein 1 Expression in Trophoblast Cells Promotes Fetal and Placental Development. *Cells* **2020**, *9*, 1942. [CrossRef]
21. Hannan, N.J.; Paiva, P.; Dimitriadis, E.; Salamonsen, L.A. Models for study of human embryo implantation: Choice of cell lines? *Biol. Reprod.* **2010**, *82*, 235–245. [CrossRef]
22. Brasier, A.R. The nuclear factor-B-interleukin-6 signalling pathway mediating vascular inflammation. *Cardiovasc. Res.* **2010**, *86*, 211–218. [CrossRef]
23. Wongchana, W.; Palaga, T. Direct regulation of interleukin-6 expression by Notch signaling in macrophages. *Cell Mol. Immunol.* **2012**, *9*, 155–162. [CrossRef]
24. Creyghton, M.P.; Cheng, A.W.; Welstead, G.G.; Kooistra, T.; Carey, B.W.; Steine, E.J.; Hanna, J.; Lodato, M.A.; Frampton, G.M.; Sharp, P.A.; Boyer, L.A.; Young, R.A.; Jaenisch, R. Histone H3K27ac separates active from poised enhancers and predicts developmental state. *Proc. Natl. Acad. Sci. USA* **2010**, *107*, 21931–21936. [CrossRef] [PubMed]
25. Kotake, Y.; Arikawa, N.; Tahara, K.; Maru, H.; Naemura, M. Y-box binding protein 1 is involved in regulating the G2/M phase of the cell cycle. *Anticancer Res.* **2017**, *37*, 1603–1608. [CrossRef] [PubMed]

26. Villar, J.; Carroli, G.; Wojdyla, D.; Abalos, E.; Giordano, D.; Ba'aqeel, H.; Farnot, U.; Bergsjø, P.; Bakketeig, L.; Lumbiganon, P.; Campodonico, L.; Al-Mazrou, Y.; Lindheimer, M.; Kramer, M.; for the World Health Organization Antenatal Care Trial Research Group. Preeclampsia, gestational hypertension and intrauterine growth restriction, related or independent conditions? *Am. J. Obstet. Gynecol.* **2006**, *194*, 921–931. [CrossRef]
27. Maynard, S. E.; Min, J.; Merchan, J.; Lim, K.; Li, J.; Mondal, S.; Libermann, T. A; Morgon, J. P.; Sellke, F. W.; Stillman, I. E.; Epstein, F. H.; Sukhatme, V. P.; Karumanchi, S. A. Excess placental soluble fms-like tyrosine kinase 1 (sFlt1) may contribute to endothelial dysfunction, hypertension, and proteinuria in preeclampsia. *J. Clin. Invest.* **2003**, *111*, 649–658. [CrossRef]
28. Venkatesha, S.; Toporsian, M.; Lam, C.; Hanai, J.; Mammoto, T.; Kim, Y.M.; Bdolah, Y.; Lim, K.-H.; Yuan, H.-T.; Libermann, T.A; Stillman, I.E.; Roberts, D.; D'Amore, P.A.; Epstein, F. H.; Selke, F.W.; Romero, R.; SUkhatme, V.P.; Letarte, M.; Karumanchi, S.A. Soluble endoglin contributes to the pathogenesis of preeclampsia. *Nat. Med.* **2006**, *12*, 642–649. [CrossRef] [PubMed]
29. Xue, X.; Huang, J.; Yu, K.; Chen, X.; He, Y.; Qi, D.; Wu, Y. YB-1 transferred by gastric cancer exosomes promotes angiogenesis via enhancing the expression of angiogenic factors in vascular endothelial cells. *BMC Cancer* **2020**, *20*, 1–12. [CrossRef]
30. Frye, B.C.; Halfter, S.; Djudjaj, S.; Muehlenberg, P.; Weber, S.; Raffetseder, U.; En-Nia, A.; Knott, H.; Baron, J.M.; Dooley, S.; Bernhagen, J.; Mertens, P.R. Y-box protein-1 is actively secreted through a non-classical pathway and acts as an extracellular mitogen. *EMBO Rep.* **2009**, *10*, 783–789. [CrossRef]
31. Gopal, S. K.; Greening, D.W.; Mathias, R.A.; Ji, H.; Rai, A.; Chen, M.; Zhu, H.J.; Simpson, R.J. YBX1/YB-1 induces partial EMT and tumourigenicity through secretion of angiogenic factors into the extracellular microenvironment. *Oncotarget* **2015**, *6*, 13718–13730. [CrossRef]
32. Abou-Kheir, W.; Barrak, J.; Hadadeh, O.; Daoud, G. HTR-8/SVneo cell line contains a mixed population of cells. *Placenta* **2017**, *50*, 1–7. [CrossRef] [PubMed]
33. Safak, M.; Gallia, G.L.; Ansari, S.A.; Khalili, K. Physical and Functional Interaction between the Y-Box Binding Protein YB-1 and Human Polyomavirus JC Virus Large T Antigen. *J. Virol.* **1999**, *73*, 10146–10157. [CrossRef]
34. Mayhew, T.M.; Leach, L.; McGee, R.; Wan Ismail, W.; Myklebust, R.; Lammiman, M.J. Proliferation, differentiation and apoptosis in villous trophoblast at 13–41 weeks of gestation (including observations on annulate lamellae and nuclear pore complexes). *Placenta* **1999**, *20*, 407–422. [CrossRef] [PubMed]
35. Ishihara, N.; Matsuo, H.; Murakoshi, H.; Laoag-Fernandez, J.B.; Samoto, T.; Maruo, T. Increased apoptosis in the syncytiotrophoblast in human term placentas complicated by either preeclampsia or intrauterine growth retardation. *Am. J. Obstet. Gynecol.* **2002**, *186*, 158–166. [CrossRef] [PubMed]
36. Straszewski-Chavez, S.L.; Abrahams, V.M.; Mor, G. The role of apoptosis in the regulation of trophoblast survival and differentiation during pregnancy. *Endocr. Rev.* **2005**, *26*, 877–897. [CrossRef] [PubMed]
37. Lovett, D.H.; Cheng, S.; Cape, L.; Pollock, A.S.; Mertens, P.R. YB-1 alters MT1-MMP trafficking and stimulates MCF-7 breast tumor invasion and metastasis. *Biochem. Biophys. Res. Commun.* **2010**, *398*, 482–488. [CrossRef] [PubMed]
38. Zhao, S.; Wang, Y.; Guo, T.; Yu, W.; Li, J.; Tang, Z.; Yu, Z.; Zhao, L.; Zhang, Y.; Wang, Z.; Wang, P.; Li, Y.; Li, F.; Sun, Z.; Xuan, Y.; Tang, R.; Deng, W-G.; Guo, W.; Gu, C. YBX1 regulates tumor growth via CDC25a pathway in human lung adenocarcinoma. *Oncotarget* **2016**, *7*, 82139–82157. [CrossRef]
39. Jia, J.; Zheng, Y.; Wang, W.; Shao, Y.; Li, Z.; Wang, Q.; Wang, Y.; Yan, H. Antimicrobial peptide LL-37 promotes YB-1 expression, and the viability, migration and invasion of malignant melanoma cells. *Mol. Med. Rep.* **2017**, *15*, 240–248. [CrossRef]
40. Liang, C.; Ma, Y.; Yong, L.; Yang, C.; Wang, P.; Liu, X.; Zhu, B.; Zhou, H.; Liu, X.; Liu, Z. Y-box binding protein-1 promotes tumorigenesis and progression via the epidermal growth factor receptor/AKT pathway in spinal chordoma. *Cancer Sci.* **2019**, *110*, 166–179. [CrossRef]
41. Johnson, T.G.; Schelch, K.; Mehta, S.; Burgess, A.; Reid, G. Why Be One Protein When You Can Affect Many? The Multiple Roles of YB-1 in Lung Cancer and Mesothelioma. *Front. Cell Dev. Biol.* **2019**, *7*, 1–25.
42. Rahat, B.; Sharma, R.; Bagga, R.; Hamid, A.; Kaur, J. Imbalance between matrix metalloproteinases and their tissue inhibitors in preeclampsia and gestational trophoblastic diseases. *Reproduction* **2016**, *152*, 11–22. [CrossRef]
43. Siwetz, M.; Blaschitz, A.; El-Heliebi, A.; Hiden, U.; Desoye, G.; Huppertz, B.; Gauster, M. TNF-α alters the inflammatory secretion profile of human first trimester placenta. *Lab. Investig.* **2016**, *96*, 428–438. [CrossRef]
44. Goyal, P.; Brünnert, D.; Ehrhardt, J.; Bredow, M.; Piccenini, S.; Zygmunt, M. Cytokine IL-6 secretion by trophoblasts regulated via sphingosine-1-phosphate receptor 2 involving Rho/Rho-kinase and Rac1 signaling pathways. *Mol. Hum. Reprod.* **2013**, *19*, 528–538. [CrossRef]
45. Das, C.; Kumar, V.S.; Gupta, S.; Kumar, S. Network of cytokines, integrins and hormones in human trophoblast cells. *J. Reprod. Immunol.* **2002**, *53*, 257–268. [CrossRef]
46. Bowen, J.M.; Chamley, L.; Keelan, J.A.; Mitchell, M.D. Cytokines of the placenta and extra-placental membranes: Roles and regulation during human pregnancy and parturition. *Placenta* **2002**, *23*, 257–273. [CrossRef] [PubMed]
47. Zenclussen, A.C.; Kortebani, G.; Mazzolli, A.; Margni, R.; Malan Borel, I. Interleukin-6 and soluble interleukin-6 receptor serum levels in recurrent spontaneous abortion women immunized with paternal white cells. *Am. J. Reprod. Immunol.* **2000**, *44*, 22–29. [CrossRef] [PubMed]
48. Sansone, P.; Storci, G.; Tavolari, S.; Guarnieri, T.; Giovannini, C.; Taffurelli, M.; Ceccarelli, C.; Santini, D.; Paterini, P.; Marcu, K.B.; Chieco, P.; Bonafè, M. IL-6 triggers malignant features in mammospheres from human ductal breast carcinoma and normal mammary gland. *J. Clin. Invest.* **2007**, *117*, 3988–4002. [CrossRef] [PubMed]

49. Martin, M.; Hua, L.; Wang, B.; Wei, H.; Prabhu, L.; Hartley, A.V.; Jiang, G.; Liu, Y.; Lu, T. Novel serine 176 phosphorylation of YBX1 activates NF-$\kappa$B in colon cancer. *J. Biol. Chem.* **2017**, *292*, 3433–3444. [CrossRef] [PubMed]

50. Shah, A.; Plaza-Sirvent, C.; Weinert, S.; Buchbinder, J.H.; Lavrik, I.N.; Mertens, P.R.; Schmitz, I.; Lindquist, J.A. YB-1 Mediates TNF-Induced Pro-Survival Signaling by Regulating NF-$\kappa$B Activation. *Cancers* **2020**, *12*, 2188. [CrossRef]

51. Kent, W. J.; Sugnet, C.W.; Furey, T.S.; Roskin, K.M.; Pringle, T.H.; Zahler, A.M.; Haussler, A.D. The Human Genome Browser at UCSC. *Genome Res.* **2002**, *12*, 996–1006. [CrossRef]

52. Lenardo, M.J.; Baltimore, D. NF-$\kappa$B: A pleiotropic mediator of inducible and tissue-specific gene control. *Cell* **1989**, *58*, 227–229. [CrossRef]

53. Oh, S.-Y.; Hwang, J.R.; Choi, M.; Kim, Y.M.; Kim, J.S.; Suh, Y.L.; Choi, S.J.; Roh, C.R. Autophagy regulates trophoblast invasion by targeting NF-$\kappa$B activity. *Sci. Rep.* **2020**, *10*, 1–10. [CrossRef]

54. Armistead, B.; Kadam, L.; Drewlo, S.; Kohan-Ghadr, H.R. The role of NF$\kappa$B in healthy and preeclamptic placenta: Trophoblasts in the spotlight. *Int. J. Mol. Sci.* **2020**, *21*, 1775. [CrossRef] [PubMed]

55. Sakowicz, A. The role of NF$\kappa$B in the three stages of pregnancy—Implantation, maintenance, and labour: a review article. *BJOG* **2018**, *125*, 1379–1387. [CrossRef]

56. Cramer, M.; Nagy, I.; Murphy, B.J.; Gassmann, M.; Hottiger, M.O.; Georgiev, O.; Schaffner, W. NF-$\kappa$B contributes to transcription of placenta growth factor and interacts with metal responsive transcription factor-1 in hypoxic human cells. *Biol. Chem.* **2005**, *386*, 865–872. [CrossRef] [PubMed]

International Journal of
*Molecular Sciences*

*Review*

# Immunoendocrine Dysregulation during Gestational Diabetes Mellitus: The Central Role of the Placenta

Andrea Olmos-Ortiz [1], Pilar Flores-Espinosa [1], Lorenza Díaz [2], Pilar Velázquez [3], Carlos Ramírez-Isarraraz [4] and Verónica Zaga-Clavellina [5,*]

1   Departamento de Inmunobioquímica, Instituto Nacional de Perinatología Isidro Espinosa de los Reyes (INPer), Ciudad de México 11000, Mexico; nut.aolmos@gmail.com (A.O.-O.); m.pilar.flores.e@gmail.com (P.F.-E.)
2   Departamento de Biología de la Reproducción, Instituto Nacional de Ciencias Médicas y Nutrición Salvador Zubirán, Ciudad de México 14080, Mexico; lorenzadiaz@gmail.com
3   Departamento de Ginecología y Obstetricia, Hospital Ángeles México, Ciudad de México 11800, Mexico; m.pilarvs@hotmail.com
4   Clínica de Urología Ginecológica, Instituto Nacional de Perinatología Isidro Espinosa de los Reyes (INPer), Ciudad de México 11000, Mexico; drcarlos.ri@gmail.com
5   Departamento de Fisiología y Desarrollo Celular, Instituto Nacional de Perinatología Isidro Espinosa de los Reyes (INPer), Ciudad de México 11000, Mexico
*   Correspondence: v.zagaclavellina@gmail.com; Tel.: +52-55-5520-99-00 (ext. 478)

**Abstract:** Gestational Diabetes Mellitus (GDM) is a transitory metabolic condition caused by dysregulation triggered by intolerance to carbohydrates, dysfunction of beta-pancreatic and endothelial cells, and insulin resistance during pregnancy. However, this disease includes not only changes related to metabolic distress but also placental immunoendocrine adaptations, resulting in harmful effects to the mother and fetus. In this review, we focus on the placenta as an immuno-endocrine organ that can recognize and respond to the hyperglycemic environment. It synthesizes diverse chemicals that play a role in inflammation, innate defense, endocrine response, oxidative stress, and angiogenesis, all associated with different perinatal outcomes.

**Keywords:** inflammation; cytokines; adipokines; antimicrobial peptides; oxidative stress; metabolic stress; IGF-I; insulin; lactotroph hormones; angiogenesis

**Citation:** Olmos-Ortiz, A.; Flores-Espinosa, P.; Díaz, L.; Velázquez, P.; Ramírez-Isarraraz, C.; Zaga-Clavellina, V. Immunoendocrine Dysregulation during Gestational Diabetes Mellitus: The Central Role of the Placenta. *Int. J. Mol. Sci.* **2021**, 22, 8087. https://doi.org/10.3390/ijms22158087

Academic Editor: Hiten Mistry

Received: 11 June 2021
Accepted: 26 July 2021
Published: 28 July 2021

**Publisher's Note:** MDPI stays neutral with regard to jurisdictional claims in published maps and institutional affiliations.



## 1. Introduction

Gestational Diabetes Mellitus (GDM) is a transient condition characterized by carbohydrate intolerance, hyperglycemia, peripheral insulin resistance, insufficient insulin secretion or activity, endothelial dysfunction, and low-grade inflammation during pregnancy, frequently between 24 and 28 weeks of gestation [1]. Although it is a transient, GDM effects can last beyond the perinatal period and impact the health of mother and fetus in both short- and long-term [2–4].

In 2017, it was estimated that 21.3 million births (16.2%) worldwide were affected by hyperglycemia during pregnancy, with GDM contributing 86.4% of these cases [5,6]. Furthermore, an increase in the prevalence of GDM effects is expected due to the parallel increasing rate of pre-gestational obesity and excessive weight gain during pregnancy.

It is well accepted that a key event in the onset of GDM is the maternal peripheral insulin resistance. During normal pregnancy, there is a transient and physiological state of decreased insulin sensitivity, necessary to prioritize fetal glucose uptake. In response, β-cells proliferate and synthesize more insulin as a mechanism to counteract insulin resistance and favor euglycemia. However, before pregnancy, some women have a first pancreatic hit by GDM risk factors such as pre-gestational overweight, obesity, hypercaloric diet, personal or familiar antecedent of GDM, advanced maternal age, or presence of insulin resistance disorders such as polycystic ovarian syndrome [1,7]. With pregnancy, these women are exposed to a second pancreatic hit: the insulin resistance associated with early pregnancy. These two hits lead to GDM development because of inadequate compensatory changes in β-cell mass activity and proliferation due to a more pronounced insulin resistance condition, particularly during the second and third trimester of pregnancy [8,9]. This transient metabolic stress on the pancreas during pregnancy, partially explains why GDM is associated with a higher risk of post-partum development of Type 2 Diabetes Mellitus (T2DM) in the mother and fetus [10].

The American College of Obstetricians and Gynecologists (ACOG) recognizes two types of GDM. GDM Class 1 (A1GDM) patients respond to diet intervention (low glycemic index meals with low simple sugar and high fiber content) and exercise. GDM Class 2 (A2GDM) patients need pharmacologic treatment to achieve target glucose levels. For A2GDM, the first-line therapy recommended by the American Diabetes Association (ADA) is insulin, preferentially short-acting insulin (i.e., Lispro or Aspart), and long-acting insulin (i.e., Glargine or Detemir) [7,11]. However, other Societies including ACOG, German Diabetes Association, German Society of Gynecology and Obstetrics, and The Society for Maternal-Fetal Medicine recommend metformin instead [1,12]. Recent studies showed a lower risk for preeclampsia, macrosomia, neonatal hypoglycemia, and hypertensive disorders as well as better outcomes in maternal weight gain and glycemic control. No difference was observed in rates of caesarean section, neonatal respiratory distress and preterm birth compared to insulin treatment alone [13–17]. There is insufficient evidence on the long-term effects of prenatal exposure to metformin (especially because it crosses the placenta). Two recent studies showed no difference in growth and development in children of metformin-treated and insulin-treated mothers over a four-year period [18,19]. More long-term studies are needed to understand the long-term effects of metformin during pregnancy.

As it would be expected, hyperglycemia and GDM disturb placental ultrastructure and morphophysiology since the early stages of the disease. Reported placental abnormalities in GDM patients include increased placental weight, intimate glycogen deposits, increased number of syncytial knots, villous edema, and larger syncytial area and volume for favoring nutrient uptake [20–23]. Histopathologic analysis of GDM placentae indicates enhanced angiogenesis and high vasculogenesis rate evidenced by increased villous vascularity often associated with thickened immature villi capillaries and signs of placental hypoperfusion [24,25]. There has also been reported increased fibrinoid necrosis, chorangiosis, and ischemia [25,26]. Additionally, GDM syncytiotrophoblasts present an exaggerated mitochondrial dysfunction accompanied by a lower rate of glycolysis, oxidative phosphorylation, and ATP synthesis, which indicates a compromised metabolic supply and therefore placental overstress [27,28]. Lipid metabolism of the placenta is also distorted in GDM with evidence of larger lipid droplets, higher triglyceride accumulation, and fatty acid transporter expression [28,29]. This is all indicative of modification of the endocrine, immune, angiogenic, and antioxidant functions by placentae in GDM mothers. In this review, we aim to understand the placenta's role as a vital organ that acts as the interface between maternal and fetal metabolisms impacted by a pathological condition as GDM. In this scenario, the placenta as an immuno-endocrine organ is required to recognize and respond to the hyperglycemic environment by synthesizing diverse cytokines, chemokines, adipokines, antimicrobial peptides, and hormones.

## 2. Role of Placenta in the Endocrine Milieu of GDM

The endocrine system is the earliest system developing during intrauterine life. From the stage of two-blastomeres, the embryo begins to secrete the beta-fraction of the human chorionic gonadotrophin (β-hCG), which has been suggested to be a product of mRNAs previously stored in oocytes [30,31]. Later, by 6 days post-fertilization, the trophectoderm establishes its endocrine phenotype through the de novo synthesis of β-hCG. The early placenta then turns on its hormonal switch and maintains secretion of a broad panel of hormones with central activities in the maintenance of pregnancy, fetal growth, and development. These include somatostatin, placental lactogens, placental growth hormone (PlGF), gonadotropin-releasing hormone, corticotropin-releasing hormone, thyrotropin-releasing hormone, progesterone, and estradiol besides other growth factors such as the Insulin-like growth factor 1 and 2 (IGF-I and IGF-II) [32].

Inherent to its secretory phenotype, the placenta is also a target of hormonal biological activity because it expresses most receptors for these hormones and growth factors. Therefore, placental hormones act in endocrine, paracrine, and autocrine pathways in the Maternal-placental-fetal-unit (MPFU).

In the context of GDM, diverse hormonal actors take part in the beginning or progression of the hyperglycemic, insulin resistance, oxidative stress, and the meta-inflammatory state of this disease. Three years ago, Madhusmita Rout and Sajitha Lulu proposed a network of maternal and placental genes and their potential impact on the transport of nutrients from mothers with GDM to their babies [33]. Through diverse in silico tools analyzing the interaction of gene/protein/miRNA/transcription factors, they described an important dysregulation of certain placental hormones, including leptin, insulin/IGF-I and their receptors, and the placental growth hormone receptor. Recently, other hormonal candidates that might play a role in the pathophysiology of GDM have been described through comprehensive bioinformatics, gene analysis, ROC analysis, and RT-PCR, including β-hCG, oxytocin receptors, binding proteins of IGF-I, and some cytochromes involved in the steroidogenic pathway [34,35].

### 2.1. The Insulin and IGF-I/IGF-II Axis and Molecular Pathways Primarily Disturbed in GDM Placentae

Insulin resistance is one of the first alterations in the pathogenesis of GDM. Insulin acts through two known receptor isoforms: insulin receptor A (IR-A) and insulin receptor B (IR-B). These are heterodimer receptors composed of an extracellular $\alpha$-subunit that binds to insulin, and an intracellular $\beta$-subunit that binds to the insulin receptor substrate 1 (IRS-1). Both isoforms are transcripts of the same *INS* gene. However, the IR-A product lacks a sequence of 36 nucleotides in the C-terminal of the $\alpha$-subunit as a result of alternative splicing of exon 11. In contrast, IR-B is the result of the full transcription of the *INS* gene [36,37]; this explains the differences in their activities. Insulin has a higher affinity for IR-A than for IR-B. In addition, IR-A activates the Ras-Raf-mitogen-activated protein kinase (MAPK) cascade, whereas IR-B activates the phosphatidylinositol 3-kinase (PI3K) and protein kinase B (Akt) signaling.

The study of the cellular pathways activated by insulin began around 33 years ago [38]. Since then, diverse research papers have scrutinized these cellular cascades, and finally, in 1995, the role of PI3K and Akt in the glycemic control and other metabolic activities of insulin was described [39,40]. After IR-B binding to insulin, PI3K binds to tyrosine-phosphorylated IRS proteins, leading to the formation of phosphatidylinositol (3,4,5)-triphosphate (PIP3). Downstream effects of PIP3 lead to activation of 3-phosphoinositide dependent protein kinase (PDK)1 and the subsequent activation of a variety of kinases, of which Akt 1–3 is the best-studied [41]. Main metabolic activities of insulin are related to Akt phosphorylation: (i) increased glycogen synthesis by inactivation of glycogen synthase kinase-3 (GSK3) $\alpha/\beta$ and activation of glycogen synthase [39]; (ii) Decreased transcription of gluconeogenic genes in liver and autophagy genes in muscle, by phosphorylation of forkhead box (FOX) transcription factors [42]; (iii) stimulated protein synthesis and suppression of autophagy by phosphorylation of tuberous sclerosis 2 (TSC2) and the 40 kDa proline-rich Akt substrate (PRAS40) which leads to activation of mTORC1 [43,44]; (iv) increased glucose uptake by phosphorylation of TBC1 domain family member 1/Akt substrate of 160 kDa (TBC1D4/AS160) which regulates trafficking and translocation of GLUT4 cytoplasmic vesicles to the plasma membrane [45]. In line with their metabolic effects, IR-B is preferentially expressed by classical insulin-sensitive cells such as hepatocytes, adipocytes, and myocytes, which have important roles in glucose, lipid, and protein metabolism.

On the other hand, activation of IR-A induces the Grb-2/Erk 1/2 MAPK pathway related to cell growth, differentiation, and survival processes [46]. This isoform is predominantly expressed in cancer tissues, the brain, hematopoietic cells, and the placenta [36].

Although the human placenta expresses both insulin receptors, IR-A is expressed at higher levels than IR-B [47]. This differential ratio is probably related to the need for tight control of the crucial proliferative and pro-differentiation pathways during pregnancy. On the other hand, there are redundant placental pathways to help in the vital fetoplacental glucose transfer, besides insulin/IR-B mediated GLUT-4 translocation [48,49], as occurs in insulin-dependent tissues. During the first trimester of pregnancy, IR-A is expressed more in the apical membrane of syncytiotrophoblasts whereas at term IR-B is concentrated in endothelial cells of the villi microvasculature [50].

In Figure 1, we summarize the disturbed molecular pathways in GDM placentae. Immunohistochemical and blotting studies showed that GDM placentae have a lower total protein expression of IR-A, PIP3, and IRS-1 in comparison to placentae from uncomplicated pregnancies. Interestingly this occurs independently of the metabolic control of the disease which implies profound and sustained effects in placenta signaling networks [51,52]. Also, it seems that the obesogenic environment is an additional regulating factor of insulin signaling among GDM placentae, involving IRS-2, PI3K, and GLUT4, the most sensitive targets to obesity [53]. Additionally, it has also been reported that GDM placentae have a pronounced phosphorylated pattern of IR and IRS-1 proteins, accompanied by hyperphosphorylation of STAT-3, MAPK 1-3 (Erk 1/2), and Akt [54,55]. Altogether, these data suggest that GDM primes the placenta to overstimulate the insulin signaling to compensate sustained exposure to hyperglycemia. The deficit in the total protein levels of these mediators results in insufficient placental glucose uptake, and consequently in a hyperglycemic state [56,57]. Although, other authors did not observe differences in the total protein levels of placental IR and IRS-1 [54,58]. Considering these inconsistencies, we believe more studies are needed to clarify insulin signaling in GDM placentae and to understand how placental imbalance in these signaling pathways results in higher levels of inflammatory cytokines, adipokines and oxidative reactive species, insulin resistance, and vascular disorders, all of which prevail in the local placenta and peripheral tissues of GDM mothers.

Another system that can cross-react with insulin signaling is the IGF-I axis. The system consists of two ligands IGF-I and IGF-II, two receptors IGF-1R and IGF-2R, six IGF binding proteins (IGFBP 1–6), and four insulin-like growth factor binding protein-related peptides IGFBP-rP1-4 [55,59]. These growth factors mainly regulate growth and metabolism throughout the life cycle, but its activity is critical during intrauterine life for mammalian development. They provide a signal to cells to indicate that adequate nutrients are available, and therefore to enhance cellular protein synthesis, to favor hypertrophy, and to stimulate cell division [60].

Insulin and IGF-I present a relatively low homology of 34% (BLAST alignment: CAA40342.1 and AAN39451.1), but their receptors are highly homologous, around 57% (BLAST alignment: CAA28030.1 and AAA59452.1). These similarities in their structures generate promiscuous interactions between them. IGF-II binds to IR-A with an affinity close to that of insulin, but it does not bind to IR-B [61]. Additionally, random hybrids of IR-A/IR-B and hybrids of IRs/IGF-1R have been reported in the placenta [36,62,63]. IGF-1R/IR hybrids bind IGF-I and IGF-II with high affinity but bind insulin with a relatively low affinity [64].

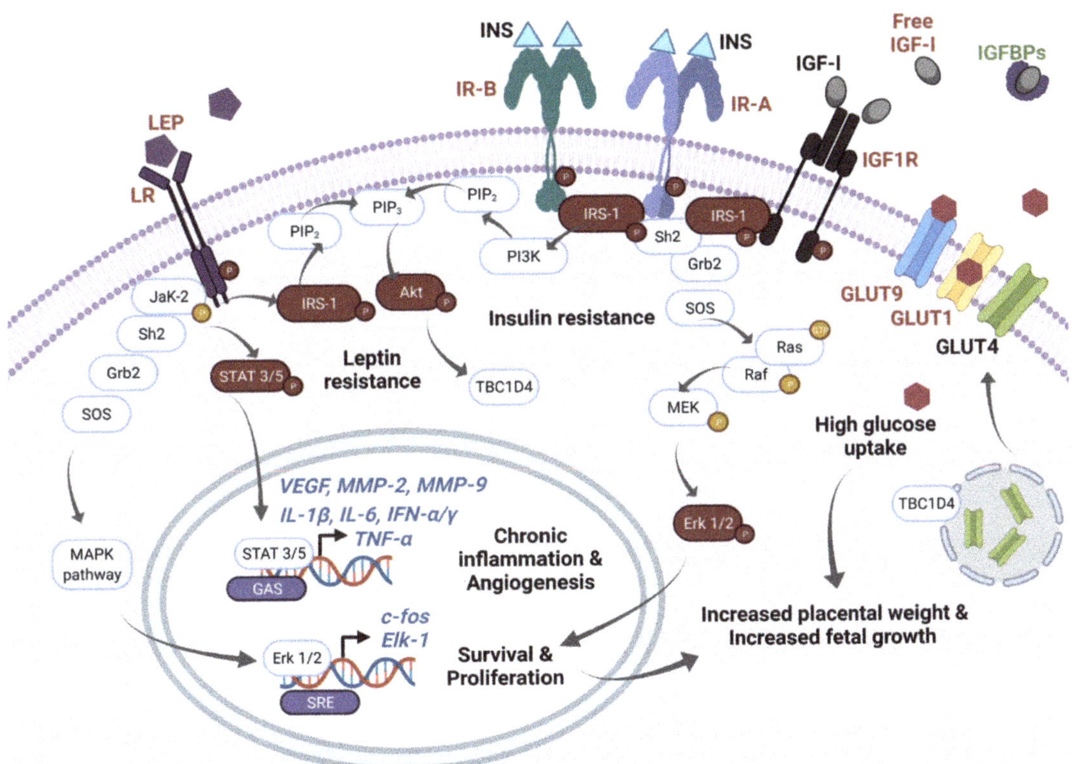


**Figure 1.** Main molecular pathways disturbed in GDM placentae. In this schematic, we highlight in red main molecules reported as overexpressed/overactivated in GDM placentae, whereas downregulation is highlighted in green. High expression of phospho-IR-A and phospho-IR-B has been reported in the GDM placenta. After insulin binding, IR-A phosphorylates IRS-1 and recruits Scheme 2. and Grb2 proteins which induce the GTPase activity of Ras, and then a GDP is exchanged by a GTP. This initiates a subsequent cascade of phosphorylations of Raf, MEK, and Erk 1/2. Finally, overexpressed phospho-Erk proteins translocate to the nucleus and recognize SRE sites favoring active transcription of c-fos and Elk-1. FOS proteins have been implicated as regulators of cell proliferation, survival, differentiation, and transformation. This pathway mediates increased placental weight and increased fetal growth in GDM. Additionally, increased levels of free IGF-I resulting from low IGFBPs serum concentrations also activate IR-A signaling as well as IGF1R and are related to proliferative effects. IGF-I excess has been also implicated in macrosomia and excessive placental growth in GDM women. On the other hand, after insulin binding to IR-B, phospho-IRS-1 activates PI3K and leads the formation of PIP3 from PIP2. Then, PIP3 activates Akt which mediates diverse metabolic effects; one of them includes GLUT4 translocation from endosomes to the cellular membrane through TBC1D4 signaling. High expression of GLUT1 and GLUT9, and probably GLUT4, mediates high glucose uptake in the GDM placenta, which can also participate in fetal and placental growth through an excess of this energetic substrate. Finally, GDM placentae present high expression of leptin and its receptor. High expression of phospho-leptin receptor recruits JaK-2 protein and activates STAT 3 or 5 proteins. Then, STATs dimerize and translocate to the nucleus and recognize GAS sites and provoke transcription of VEGF, MMP-2, MMP-9, TNF-α, IL-1α, IL-1β, IFN-α, IFN-γ, among others, which exacerbate the inflammatory placental milieu and contribute to stimulating the angiogenic process. Additionally, activation of the leptin receptor can also crosstalk with MAPK and Akt pathways. Sustained overactivation of all these pathways finally leads to clinical insulin and leptin resistance. GAS: Gamma-activated sequence. GDM: Gestational diabetes mellitus; GLUTs: Glucose transporters; IGFBPs: IGF-I binding proteins; IGF-I: Insulin-like Growth Factor 1; IGF1R: IGF-I receptor; INS: Insulin; IR-A: Insulin receptor type A; IR-B: Insulin receptor type B; LEP: Leptin; LR: Leptin receptor; SRE: Serum response elements.

Given the structural similarities between components of the IGFs/Insulin axis, it is not surprising that these hormones share signaling pathways. IGF-I binds to IGF-1R and activates two main cascades: (i) PI3K/Akt pathway via IRS-1 phosphorylation which predominantly leads to metabolic effects; (ii) Ras-Raf-MAPK pathway via SHC domain proteins which control cellular growth and differentiation [59]. On the other hand, IGF-II binds with high affinity to IGF-2R, also known as the cation-independent mannose 6 phosphate receptor. This interaction targets IGF-II for its lysosomal degradation and consequently, IGF-2R sequesters IGF-II, controlling the circulating levels of this hormone. Therefore, the biological activity of IGF-II is exclusively derived from its binding to IR-A or IGF-1R, considering that it does not bind to IR-B, as mentioned before [65].

The placenta synthesizes all components of the IGF axis from early stages at 7 weeks of gestation. IGF-I was more expressed in the second and third trimesters of pregnancy in comparison with early pregnancy, and it was expressed by practically all placental cells except syncytiotrophoblasts. Whereas IGF-II was profusely expressed by cytotrophoblasts, mesoderm core, basal plate, columnar cytotrophoblasts, amnion, and chorion. IGF-IR was expressed ubiquitously in the placenta, except in Hofbauer cells. All six IGFBPs were expressed in decidua basalis and parietalis [66].

One key insulin-like action of IGF-I is related to glucose metabolism [67]. Biomedical and clinical studies indicate that IGF-I is a hypoglycemic factor that increases glucose uptake in different kinds of cells, including euglycemic trophoblasts [68–70]. However, deeper studies are needed to confirm its metabolic effects in the GDM milieu. In contrast, IGF-II seems to present a hyperglycemic effect since overexpression of IGF-II in pancreatic β-cells results in the development of T2DM [71].

Excessive fetal growth and weight is a common complication from GDM newborn babies. Macrosomia has been explained by two central modulators: hyperglycemia and activation of the IGF-I axis. Maternal hyperglycemia increases energetic substrate availability and then stimulates excessive growth and adiposity in GDM mothers [72]. In fact, an increased concentration of glucose transporters GLUT1 and GLUT9 has been observed in GDM placentae, which favors an increased placental and fetal D-glucose uptake [49,52,73,74] (see Figure 1).

Concerning IGF-I signaling, different serum components of the pathway have been measured in GDM patients in the last 20 years. In a recent meta-analysis developed by Dr. Wang's group, in which they analyzed 12 independent studies, they found GDM was consistently associated with higher maternal IGF-I levels in mid-gestation (20–29 weeks) and late-gestation (>30 weeks), whereas serum IGF-II did not present significant changes between GDM and control mothers [60]. Interestingly, most data show significantly lower cord levels of IGFBP1, IGFBP2, IGFBP3, IGFBP-6 or IGFBP-rP1 [55,75,76], and lower maternal levels of IGFBP1 and IGFBP2 [55]. All IGFBPs bind both IGF-I and IGF-II with similar affinities (except IGFBP-6 which is essentially IGF-II specific [77]). Their metabolic effects are related to inhibition of IGF-I signaling by sequestrating it into a circulation reservoir. Consequently, diminished levels of IGFBPs and IGFBP-rPs result in higher cord blood levels of free-IGF-I in GDM patients [75]. Further, the research group of Dr. Sciacca identified an increased phosphorylation pattern of IGF-1R in placentae from metabolic uncontrolled mothers with GDM and T2DM [58]. These changes support a persistently activated IGF-I signaling in GDM placentae by increased activity of free-IGF-I (see Figure 1).

It is well known that the growth hormone (GH)-IGF-I axis is the major regulator of longitudinal growth along life. In addition, significant and positive correlations between the birth weight of newborns from GDM mothers and maternal serum IGF-I or molecules of the IGF-I signaling in GDM placentae have been described [52,78,79]. Therefore, overactivation of IGF-I signaling may be one critical factor involved in the development of macrosomia in babies from GDM mothers. Other morphologic changes in the placenta are also related to macrosomia, including broader intervillous spaces, increased terminal villus volume, a large proportion of immature villi, and a larger syncytiotrophoblast surface allowing higher amounts of glucose to cross the placenta [20,21].

One additional hypothesis in excessive fetal weight gain in GDM is related to IR/IGF-1R hybrids. A high proportion of these hybrids has been reported in skeletal muscle and adipose tissue of T2DM patients [63,80,81], and in placentae from insulin-resistant women [82]. IR/IGF-1R hybrids increase the binding sites for IGF-I and IGF-II, favoring IGF signaling, proliferation, and anabolic processes, as has been previously published in cancer models [83]. However, this hypothesis, which deserves to be further explored, has not so far been studied in the placentae of mothers with GDM.

### 2.2. The Role of Pancreatic β-Cells and Lactotroph Hormones in GDM

In a healthy pregnancy, insulin resistance increases between 50% and 60% in the third trimester, compared to the pre-pregnancy period [84]. This physiologic resistance is needed to ensure adequate delivery of glucose to a fast-growing fetus. In response, the mother needs to expand her capacity for insulin secretion which is achieved by an increase in β-pancreatic cell mass and number, finally leading to a euglycemic pregnancy. The lactotroph hormones prolactin (PRL) and human placental lactogens participate in cell-specific β-cell responses to counteract the physiological insulin resistance developed during pregnancy. These pancreatic adaptations occur before the onset of insulin resistance in pregnancy [9].

The maternal decidua is the main extra-pituitary source of PRL synthesis [85,86], although columnar trophoblasts and villous cytotrophoblasts of the placenta can also synthesize it to a lesser extent [85,87]. On the other hand, placental lactogens are exclusively synthesized during pregnancy by fetal syncytiotrophoblasts and include human placental lactogen (hPL), human chorionic somatomammotropin A and B (hCS-A and hCS-B), and the placental growth hormone (PGH). Evolutionary studies indicate that placental lactogens are closely related in their chemical structure to the human Growth Hormone (GH) as a result of three duplications and one deletion in the *GH* gene [88]. Derived from this structural homology, lactogens share with GH their binding capacity to both somatogenic and lactogenic receptors [89]. GH binds primordially to the hGH receptor and acts as a somatogen, whereas PRL and hPL bind to the prolactin receptor (PRL-R) and act as lactogens. PRL-R is a member of the cytokine receptor superfamily which presents 3 structural regions: an extracellular ligand-binding domain, a hydrophobic transmembrane domain, and an intracellular signaling domain. Multiple promoters and alternative splicing of the *PRLR* gene generate several isoforms which vary exclusively in their intracellular domains and potential recruitment of signaling mediators [90]. After ligand binding, PRL-R dimerizes which leads to the trans-phosphorylation of tyrosine residues present in Janus kinase 2 (JaK-2). This is followed by recruitment of signal transducers and transcription activators (STATs) -1, -3, or -5 which dimerize and migrate to the nucleus to enhance the expression of PRL-dependent genes [91]. During pregnancy, the binding of hPL or PRL to the long isoform of PRL-R in pancreatic β-cells activates the JaK-2/STAT-5 pathway which results in metabolic adaptations of these cells characterized for higher transcription of GLUT-2, glucokinase, insulin, survivin, cyclin D2, and Bcl6 genes [92]. GLUT-2 favors glucose uptake by β-cells, then glucose is phosphorylated by glucokinase and enters glycolysis/Krebs cycle/oxidative phosphorylation to increase ATP production. Higher ATP/ADP ratio blocks ATP-sensitive potassium channels, $K^+$ accumulation depolarizes β-cells, and voltage-gated calcium channels become activated. The resultant rise in intracellular $Ca^{2+}$ triggers insulin secretion [93]. Additionally, transcription of survivin, cyclin D2 and Bcl6 genes increases cell mitotic divisions, avoids apoptosis, and stimulates the expansion of pancreatic islets during normal pregnancy [94]. Beta-cell proliferation is also dependent on the downstream serotoninergic effect of both, PRL and hPL [95]. hPL is more potent than PRL to increase insulin secretion and β-cells proliferation, whereas GH has lower potency [96,97]. We now know that the morphologic changes in pancreatic β-cells related to pregnancy occurs largely through hPL and PRL action, but Hepatic Growth Factor (HGF), Epidermal Growth Factor (EGF), vitamin D, progestins and estrogens are also implicated [98–101] (see Figure 2).

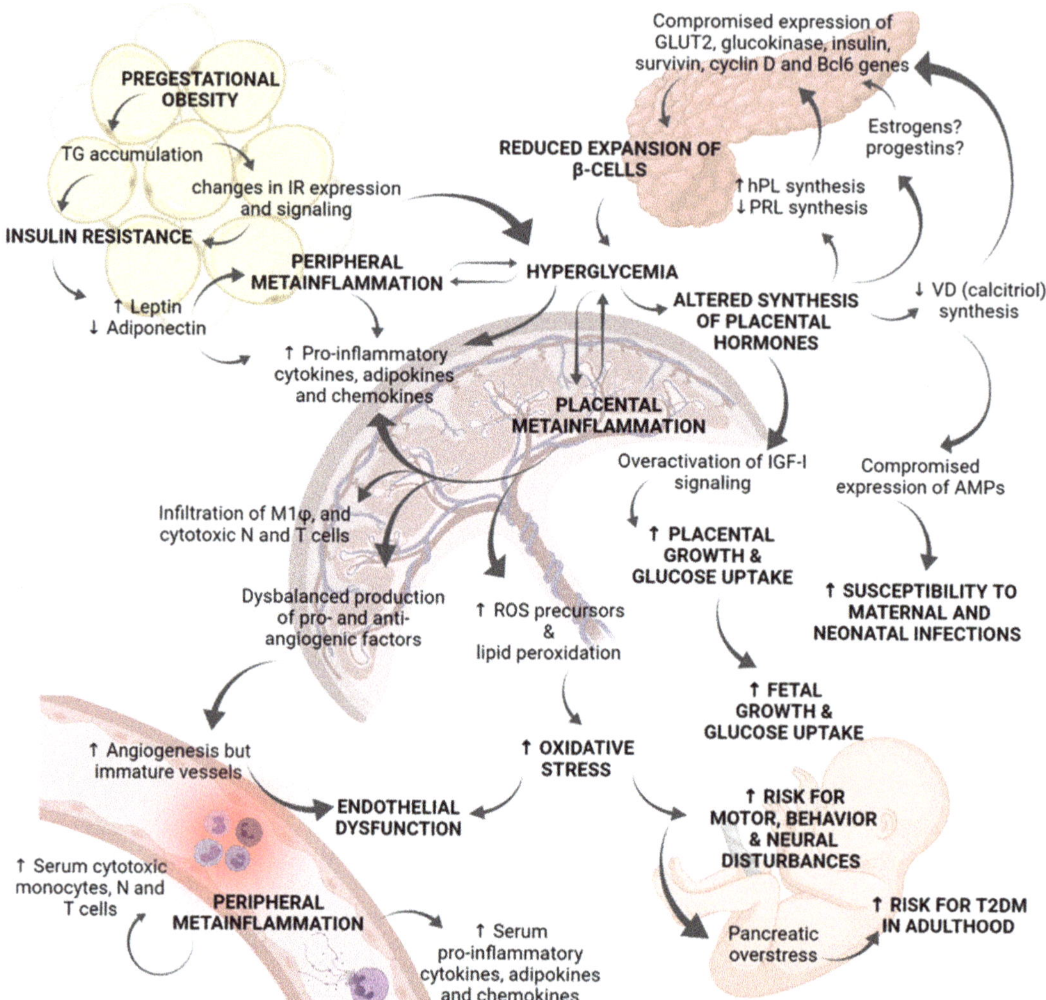



**Figure 2.** Role of placenta in the immunoendocrine dysregulations occurring in gestational diabetes mellitus. Hyperglycemia during pregnancy compromises the correct physiology of diverse organs, and particularly the placenta. Pregestational obesity is characterized by an excessive TG accumulation and changes in IR expression and signaling in adipose tissue, as well as in skeletal muscle cells. These changes in insulin-dependent tissues lead to insulin resistance. Then, adipose tissue secretes high levels of leptin and inhibits those of adiponectin. Leptin activation of JaK-2/STAT 3/5 pathway results in increased cytokine production contributing to peripheral metainflammation. Due to insulin resistance and altered IR signaling, women suffer chronic hyperglycemia. Clinical and biomedical studies indicate there is a positive regulatory loop between hyperglycemia and metainflammation, in which hyperglycemia induces placental and peripheral synthesis of pro-inflammatory cytokines, chemokines, and adipokines. Metainflammation also alters GLUT and IR expression which worsen hyperglycemia status. The synthesis of placental hormones is altered by hyperglycemia. Deficient synthesis of PRL and calcitriol (the active form of vitamin D) has been reported in GDM placentae, whereas hPL synthesis is increased. These placental hormonal changes, and probably estrogens and progestins, compromise pancreatic gene expression of GLUT2, glucokinase, insulin, survivin, cyclin D2, and Bcl6; all these genes are related to b-cells proliferation and survival. Because

of reduced b-cells expansion, hyperglycemia worsens. Additionally, deficient synthesis of calcitriol abates placental expression of antimicrobial peptides related to innate defense, increasing mother and fetus vulnerability to infections in the perinatal period. Another endocrine dysregulation is derived from IGF-I overactivation in the placenta, which explains increased placental growth and glucose uptake. The establishment of a chronic inflammatory milieu in the placenta results in: (i) Increased production of pro-inflammatory cytokines, chemokines, and adipokines; (ii) Increased placental infiltration of M1 macrophages, cytotoxic neutrophils, and T cells; (iii) Deregulated production of pro-angiogenic and anti-angiogenic factors; and (iv) Increased lipid peroxidation and synthesis of ROS precursors. Even if angiogenesis is induced, vessels in the placenta are thickened and immature. Overall, altered vessels formation, hyperglycemia, and oxidative stress induce endothelial dysfunction. Furthermore, the serum inflammatory profile is evidenced by high levels of pro-inflammatory cytokines, chemokines, and adipokines as well as a higher presence of cytotoxic monocytes, neutrophils, and T cells. Hyperglycemia and over-activation of IGF-I signaling results in increased fetal growth which may contribute to macrosomia. Finally, hyperglycemia and oxidative stress produce pancreatic overstress in the fetus, causing an increased risk for T2DM development later in life. Experimental and clinical evidence indicates that GDM fetuses present a marked neural pro-oxidative environment which may lead to neural, motor and behavior disturbances. AMPs: antimicrobial peptides; GDM: Gestational diabetes mellitus; GLUT: Glucose transporter; hPL: human placental lactogen; IGF-I: Insulin-like Growth Factor 1; IR: Insulin receptor; N: Neutrophils; PRL: Prolactin; ROS: Reactive oxygen species; T: T lymphocytes; T2DM: type 2 Diabetes mellitus; TG: triglycerides; VD: vitamin D.

Pregnancy is considered a physiologically hyperprolactinemic state, in where PRL potentiates glucose-stimulated insulin secretion and β-cell mass. However, exacerbated hyperprolactinemia, as occurs with a prolactinoma, is related to insulin resistance [102]. In GDM, blood levels of hPL are higher than in normal pregnancies and correlate with increased placental weight, macrosomia, hyperglycemia, insulin resistance, and altered values in an Oral Glucose Tolerance Test (OGTT) [103,104]. In contrast, GDM mothers present similar or even lower levels of PRL than normal pregnant women [105]. This relative contradiction seems to indicate that a delicate balance of PRL and hPL during pregnancy is needed to achieve adequate pancreatic β-cells proliferation and to avoid insulin resistance [106]. In a diabetic mouse model, low- and high- PRL treatment induced β-cell proliferation; however, low PRL levels reduced hepatic insulin resistance whereas high PRL exacerbated it and elevated apoptosis of β-cells [107]. A DNA sequencing study also supports the critical role of PRL-R in the control of glucose metabolism in GDM patients. In particular, two single nucleotide polymorphisms in the *PRLR* gene were associated with a 2-fold risk for developing GDM [103]. The study of physiological control of pancreatic β-mass expansion by lactotroph hormones still needs to be enlarged in the context of GDM and their co-interactions with estrogens/progestins hormones, inflammation, and obesity.

### 3. Hyperglycemia-Induced Metainflammation in GDM Alters Placental Immune Cells Population Favoring an Inflammatory Cytokine Signature and an Imbalance in Adipokines and Defense Peptides

The persistent exposure to hyperglycemia in GDM mothers results in a systemic response of inflammation named metainflammation [108,109]. This term defines a sustained low-grade inflammatory state characterized by an increase in serum levels of pro-inflammatory cytokines and tissue macrophage infiltration in the absence of tissue damage. Metainflammation in the context of GDM is favored by metabolic disorders such as maternal obesity or excessive weight gain, which induce inflammatory pathways leading to insulin resistance [110]. This persistent inflammatory state in GDM mothers has been associated with an increased risk for diabetes, obesity, and other poor outcomes and diseases in their offspring [111–114].

Several clinical studies in GDM patients have demonstrated alterations in a broad profile of inflammatory mediators, including lower serum levels of anti-inflammatory interleukin (IL)-10 and adiponectin, higher serum levels of pro-inflammatory TNF-α and IL-6, increased maternal serum adipokines (chemerin, leptin, omentin, visfatin, and the fatty acid-binding protein 4 FABP-4), Th1 cytokines (INF-γ, IL-2, IL-18) and chemokines

(CXCL16, IL-8) [115–120]. Intracellular signaling of these inflammatory mediators is related to a worse fetal prognosis, such as preterm delivery and premature rupture of membranes [121].

### 3.1. The Role of Immune Cells in GDM Placentae

It is well known that hyperglycemia affects innate immunity and that there is a diabetes mellitus-dependent vulnerability to infections [122]. This scenario affects the correct function of innate immune cells as well as Toll-like receptor (TLR)–dependent responses at the MPFU, including alterations in the immune functions of monocytes/macrophages, dendritic cells, NK cells, granulocytes, and Hofbauer cells.

Similar to what is seen in adipose tissue, obesity and GDM predispose to macrophage, granulocyte, and T lymphocyte infiltration into the placenta [120,123,124]. Indeed, the placenta expresses mRNA for CD68, CD14, EMR-1, and TCRa (immune cell infiltration markers), which are known to be increased in GDM placentae [123,124]. Macrophages infiltrated in GDM placentae present an M1 phenotype with a strong inflammatory response characterized by high expression of IL-6, TNF-$\alpha$, IL-1$\beta$, IL-8, and the monocyte chemoattractant protein 1 (MCP-1) [120,123].

Hofbauer cells are fetal macrophages immersed in placental villous and present an anti-inflammatory profile (M2). In a rat model of GDM, Hofbauer cells switched their M2 profile towards M1 (pro-inflammatory) and induced the oxidative stress pathway [125]. In this regard, Hofbauer cells when treated with high glucose switch their profile to an M1-type, triggering inflammatory pathways [125], similarly to macrophages in a high-glucose environment [126]. However, in human placenta from GDM women, Hofbauer cells seem to preserve their M2 phenotype despite a hyperglycemic environment [127]. M2 macrophages play a relevant role in tissue remodeling during placental development, so further studies are needed to understand the effect of GDM on these cells.

Neutrophil activity is also altered in GDM. Neutrophils seem to be significantly activated, forming a high number of neutrophil extracellular traps (NETs). Under normal circumstances, these web-like structures represent an additional mechanism of the innate immune system to protect us from invading microorganisms. However, in pathological conditions such as diabetes and cancer, platelets may also get trapped, contributing to the pathological effects of NETs, which include damage to the endothelium and thrombotic events. Notably, numerous neutrophil infiltrates have been detected in the placentae of GDM-patients [128]. In addition, elevated neutrophil-derived products including nucleosomes, neutrophil elastase, and free DNA have been found in the plasma of diabetes mellitus patients [129]. Elevated first-trimester neutrophil count has also been associated with the development of GDM and adverse pregnancy outcomes [130]. In the placenta, extracellular traps formation may be triggered by infection (e.g., by bacteria or their products) or by inflammation (e.g., preeclampsia and GDM). Notably, extracellular traps comprise a vast array of molecules with antimicrobial activity, such as elastase, cathepsin G, defensins, myeloperoxidase, hCTD, and bacterial permeability-increasing protein, which explain their bactericidal effect [131]. However, the involvement of these immune structures in some noninfectious, autoimmune, and inflammatory processes grants further studies to understand their participation in GDM pathophysiology.

Regarding NK cells, diverse comprehensive reviews have described the critical participation of NK cells, particularly decidual NK cells with CD56bright/superbright and CD16- phenotype, in pregnancy development [132,133]. However, few original articles have revised the role of these cells in GDM pathophysiology. A higher percentage of cytotoxic NK cells (CD16+ CD56dim) was observed in maternal serum of overweight GDM patients and placental extravillous tissue in comparison with euglycemic women [119,134].

### 3.2. Major Placental Cytokines in GDM

The placenta presents not only an active endocrine function, but also an important immune-modulatory action characterized by the synthesis of diverse cytokines, chemokines, and adipokines, as well as their receptors. The implication of pro-inflammatory cytokines in GDM pathology has been demonstrated in numerous studies [135–138]. Gene microarray experiments in GDM placentae showed increased expression of genes for stress-activated and inflammatory responses, with upregulation of interleukins, leptin, and TNF-$\alpha$ receptors and their downstream molecular adaptors [139].

Undoubtedly, the signaling of NF$\kappa$B is the main regulator of inflammatory pathways in normal and GDM placentae. After TNF-$\alpha$ binding with their receptor TNFR1, the adaptor protein TRADD is recruited and associated with the death domain of TNFR1. TRADD acts like platform binding for TRAF2 and RIP adaptor proteins which eventually activate the TAK1 kinase to phosphorylate and activate the IKK complex formed by the catalytic subunit IKK$\alpha$ and IKK$\beta$, and the regulatory subunit NEMO. The IKK complex phosphorylates the I$\kappa$B proteins that are constitutively bound to NF$\kappa$B, keeping this factor in the cytosol. The serine phosphorylation of I$\kappa$B proteins promotes their ubiquitination and proteolytic degradation by the proteasome, free allowing the nuclear translocation of NF$\kappa$B [140,141]. The NF$\kappa$B upregulates target genes that encoded pro-inflammatory cytokines, inducing a chronic inflammatory loop that contributes to the development of insulin resistance.

NF$\kappa$B can be activated by endogenous molecules released during tissue damage and oxidative stress, including debris from apoptotic, saturated fatty acids, heat shock proteins, advanced glycation products (AGEs) which are recognized by the TLR-4 receptor [142]. In GDM and maternal hyperglycemia there is a positive association with an increase of TLR-4 and NF$\kappa$B signaling in the placenta [143,144]. TLR-4-induction of NF$\kappa$B signaling in the placenta is an important mechanism that is altered during gestational diabetes; however, further studies are needed to elucidate the involvement of innate immunity in trophoblast functionality.

Clinical clamp assays and in vitro studies of placental perfusion demonstrate that TNF-$\alpha$ is the most significant independent predictor of insulin sensitivity in GDM patients, indicating close crosstalk between the immune and endocrine axis [145]. This immune interaction was corroborated in clinical association studies of GDM/obesity and maternal circulating levels of TNF-$\alpha$ and IL-6; this positive association remained after adjustment for total adipose mass [136,146–149]. Currently, it is known that cytokines such as TNF-$\alpha$ and IL-6 favor insulin resistance through inhibition of the insulin signaling cascade. Activation of the c-Jun N-terminal kinase (JNK) and I$\kappa$B kinase (IKK) pathway targeted by TNF-$\alpha$, increases serine phosphorylation of the IRS-1 and blocks normal tyrosine kinase activity of the IR [150]. On the other hand, IL-6 can activate the mammalian target of rapamycin (mTOR) which phosphorylates IRS-1 in serine residues, in a similar manner as TNF-$\alpha$. Another molecular signaling of insulin resistance mediated by IL-6 is related to STAT-3 activation and the consequent activation of the Suppressor of Cytokine signaling (SOCS), which inhibits tyrosine phosphorylation of IRS-1. Both serine phosphorylation and blocked tyrosine phosphorylation of IRS-1 by inflammatory mediators impaired insulin action and subsequent insulin-regulated glucose uptake [151].

IL-1$\beta$ is another important inflammatory mediator that has been shown to be increased in the placentae of obese mothers and with GDM [119,152,153]. In a first-trimester trophoblast cell line (Sw.71), high glucose levels (25 mM) increased trophoblast production of uric acid, which activated the inflammasome pathway, and positively regulated IL-1$\beta$ release [154]. In mouse models for GDM, an increase in uterine and placental IL-1$\beta$ levels and impaired glucose tolerance has been observed. In contrast, treatment with an anti-IL-1$\beta$ antibody improved glucose tolerance in GDM mice [155]. Results obtained from in vitro models of hyperglycemia support the clinical data of the inflammatory environment in the placenta of women with GDM and provide evidence of the consequences that this generates on the anatomy and functionality of the placenta. Exposure of placental trophoblast cells to high concentrations of glucose (25 and 50 mM) significantly induced the secretion of cytokines and chemokines such as TNF-$\alpha$, IL-1$\beta$, IL-6, IL-8, GRO-$\alpha$, RANTES, and G-CSF [154]. Furthermore, high glucose concentrations suppress trophoblast viability and proliferation in vitro. These effects were mediated by the increase in miR-137, which in turn decreases the expression of the protein kinase activated by AMP (PRKAA1), positively regulating the placental secretion of IL-6 [156], and upregulating Bax/Bcl-2, COX-2, and caspase 9 expression [157].

It is well known that the migration of extravillous trophoblast into the maternal decidua and their interaction with endothelial cells from uterine spiral arteries are critical processes in placental development. In this sense, in vitro studies support an active intercommunication between trophoblasts and endothelial cells. The first-trimester trophoblast exposed to hyperglycemia (25 mM) induces the release of exosomes which in turn induces the endothelial release of IL-4, IL-6, IL-8, INF-$\gamma$, and TNF-$\alpha$ [158]. Together, the placental alteration of matrix metalloproteases, pro-angiogenic, and anti-angiogenic factors, as well as the shift towards a pro-inflammatory environment can explain the peripheral endothelial dysfunction in GDM [139,159,160]. Also, this endothelial imbalance may help to explain the observed high vasculogenesis rate, and capillary immaturity in GDM placentae [20,21], as will be discussed later in this review.

### 3.3. Altered Production of Adipokines by the GDM Placenta

Adipokines are bioactive polypeptides whose dysfunction is involved in inflammation, obesity, insulin resistance, and cardiovascular diseases [161]. Although adipokines have been described as secretion products of adipose tissue, the placenta can synthesize adipokines such as chemerin, omentin-1, visfatin, leptin, and adiponectin-like adipose tissue [8,135].

Leptin acts as a regulator of satiety and energy expenditure in the central nervous system [162]. Maternal plasma leptin levels increase in the first and second trimester of pregnancy and return to pre-pregnancy levels after delivery [163]. The placental synthesis of leptin is regulated by the different placental hormones such as $\beta$-hCG and estradiol [164,165]. Several studies have shown that the secretion of leptin by the placenta exerts autocrine actions stimulating the proliferation and survival of trophoblast cells [166]. Also, leptin increases placental lipid catabolism and vasodilation, possibly increasing the availability and transport of nutrients thus favoring fetal growth [167].

Regarding GDM, most researchers reported an increase in plasma and placental leptin levels [115,168–170] while other researchers did not observe a difference compared to healthy pregnancies [171]. In vitro evidence has shown that insulin induces leptin expression in trophoblast cells [162]. It has been found that maternal hyperglycemia in GDM regulates the leptin levels in umbilical cord blood generating macrosomia in the baby and increasing their risk of obesity in the future [168].

An increase of leptin and leptin receptor expression was found in the GDM placentae [172], contributing to the increased placental weight gain observed in GDM, along with the IGF-I axis as described before (see Figure 1). The binding of leptin with its receptor activates the signaling pathways MAPK, PI3K, and JaK-STAT which are also shared by IR [54]. In GDM placentae, the basal phosphorylation of STAT-3, MAPK 1/3, and Akt are increased, causing resistance to subsequent stimulation with Leptin or Insulin in vitro, suggesting crosstalk between insulin and leptin signaling in the human placenta [54].

In explants of the placenta, leptin significantly increases the release of IL-1$\beta$, IL-6, TNF-$\alpha$, and prostaglandin E2 (PGE2) [173]. Similarly, leptin stimulates IL-6 secretion in trophoblast cells [174,175]. This increase in the production of pro-inflammatory cytokines evokes a chronic inflammatory milieu that in turn improves leptin release in the placenta, generating a vicious inflammatory loop [176].

Adiponectin is the second most studied adipokine. This peptide improves the IR signaling, makes lipid oxidation more efficient, inhibits the gluconeogenesis and the TNF-$\alpha$ signal in adipose tissue [177]. Adiponectin circulating levels increase during the first and second trimesters of normal pregnancy and later decrease post-partum [178]. Unlike leptin, low levels of adiponectin were found in maternal GDM serum compared to women with normal glucose tolerance [179,180]. It has been proposed that this decrease in serum adiponectin levels may serve as an early predictive factor for GDM development.

In terms of human placentae from pregnancies complicated with GDM, a significant downregulation of adiponectin mRNA and an upregulation of adiponectin receptor 1 (ADI-POR1) has been reported [181]. In the same study, it was observed that the cytokines INF-γ, TNF-α, and IL-6 differentially regulate the expression of adiponectin receptors (ADIPOR1 and ADIPOR2), as well as the expression and secretion of adiponectin. The researchers observed that placental adiponectin suppressed MAPK phosphorylation, particularly ERK1/2 and p38 that are essential for the onset of trophoblast differentiation, implantation, and placentation [181]. Furthermore, there is in vitro evidence that adiponectin promotes the trophoblast invasion by augmenting Matrix metalloproteinase (MMP)-2 and -9 production, and downregulating TIMP-2 mRNA expression [182]. Whereas adiponectin increases insulin sensitivity and modulates the invasion of trophoblasts, this adipokine could limit fetal and placental growth. Recent studies suggest that hypomethylation of the adiponectin gene in the placenta correlates with maternal insulin resistance and hyperglycemia, and with fetal macrosomia [183,184]. Secretion and expression of other placental adipokines have been described in women with GDM and obesity. Some studies showed an increase in many of these adipokines while others found no changes compared to women with normoglycemia [177]. More studies are needed to understand the impact of these adipokines on the development of GDM in the human.

### 3.4. Placental Innate Defense-Peptides in GDM

Regarding the innate immune system of the MPFU, the placenta, trophoblasts cells, decidual cells, stromal cells, and fetal membranes can produce several antimicrobial peptides, including human β-defensin (HBD)-1, HBD-2, HBD-3 and HBD-4, S100 proteins, human cathelicidin (hCTD) and human neutrophil peptides 1–4 (HNP 1–4) [185–191]. Moreover, the placenta also produces histones capable to neutralize certain bacterial endotoxins [192]. The main function of these defense peptides is to rapidly kill invader microorganisms; however, these multifunctional, amphipathic molecules also participate in angiogenesis, cell migration, and immune system modulation [193].

At this time, there is scarce information about the effect of diabetes and/or glucose levels on human placental antimicrobial peptides. In human amniotic epithelial cells, the high glucose culture medium is known to downregulate HBD2 production [194]. In other cell types such as macrophages, high glucose levels inhibit hCTD expression. However, in *Mycobacterium tuberculosis*-infected macrophages, hCTD levels increased as mycobacterial burden augmented, irrespective of the hyperglycemic environment [195]. Similarly, in a rat model of diabetes, lower levels of defensin BD1 have been found, but interestingly, they could be restored by insulin. On the contrary, defensin BD2 levels were found significantly higher than in non-diabetic animals, which was interpreted as the result of a glucose-dependent increased inflammatory state [196]. Likewise, in biopsies from diabetic foot ulcers, all studied defensins (HBD1-4) were overexpressed while hCTD was decreased in comparison to healthy skin. Remarkably, when an infectious agent was present, a significantly lower hCTD expression was observed. The authors concluded that the amount of antimicrobial peptide present in these diabetic tissues was not able to efficiently contain the infection [197].

Altogether, these results suggest that high blood glucose differentially regulates the expression of innate immune system components and that there seems to be an interaction between this hyperglycemic state and other assaults that may occur in diabetes, namely inflammation and infection. A particularly interesting case is that of hCTD, which is of primordial importance, especially for intracellular infections. In the human placenta, and similarly as in the cases just described, there is a differential inflammation-dependent regulation of hCTD and HBDs. Indeed, bacterial LPS endotoxin stimulated HBD2 and S100A9 mRNA levels, while significantly repressed basal and calcitriol-dependent hCTD expression [198].

## 4. Vitamin D Implications in Immune Regulation and Insulin Resistance in GDM

As discussed in the previous section, there is a major immune imbalance in GDM closely linked to the state of metainflammation. An important placental hormone with potent immune-modulating activity in pregnancy is calcitriol, the active form of Vitamin D (VD). Calcitriol shows a dual effect: anti-inflammatory (lowers inflammatory cytokines) and stimulator of the innate response (induces antimicrobial peptides expression) [189,198–200]. This suggests that the lower VD levels described in GDM [201] may worsen the inflammatory state and further decrease the innate immune response prevailing in this pathology. Specifically, hypovitaminosis-D may take its toll on placental hCTD, HBDs, and S100A9, which production depends on calcitriol transcriptional activity [189,198,202]. Moreover, maternal VD deficiency has been significantly associated with insulin resistance and a greater risk of GDM [203–205], which may be linked to the aberrant VD metabolism that takes place in the diabetic placenta [201,206]. Because of the latter, VD deficiency increases the susceptibility to maternal-fetal infections, most probably by the resultant limited immune response [207,208].

Importantly, the VD Receptor (VDR) is expressed in different organs/tissues, including the placenta and pancreas, and it is known to be involved in the regulation of glucose metabolism through modulating insulin production and secretion [101,201,209]. Indeed, VD-response elements are present in the human insulin receptor gene, which is activated by VDR-calcitriol [210,211]. Accordingly, recent reports support that calcitriol protects against insulin resistance and exacerbated inflammation at the MPFU in GDM [138,212]. Therefore, VD deficiency is associated with blood glucose and insulin alterations, partially explaining the suggested implication of VD deficiency as an independent risk factor for GDM [203].

## 5. Endovascular Changes in GDM: Endothelial Dysfunction and Overstimulation of Placental Angiogenesis

During the second trimester of pregnancy (20th week of pregnancy), the placenta receives 21% of the combined cardiac output; practically equal to the two principal organs receiving 20–25% of the cardiac output: the liver and kidneys [213–215]. This important blood influx directed to the placenta demands the formation of a highly efficient vascular network interconnecting the mother, the placenta, and the fetus. To satisfy this demand, the placenta must carry on two stages of vessel formation: vasculogenesis and angiogenesis. Vasculogenesis begins on day 21 after conception refers to the de novo formation of the primitive vascular plexus during embryonic development, via the differentiation of pluripotent hemangioblasts into endothelial cells. Then, angiogenesis takes place from day 32 until delivery, where the formation and remodeling of vascular trees occur through the proliferation and migration of endothelial cells from pre-existing blood vessels [216,217]. Both processes are essential for placental development as well as organogenesis during embryonic and fetal growth.

Physiologic control of angiogenesis is extremely complex and tightly regulated. It involves cellular interactions with the extracellular matrix, autocrine and paracrine signaling induced by hormones, cytokines, chemokines, matrix metalloproteinases (MMPs), miRNAs, as well as pro-angiogenic, anti-angiogenic factors and their receptors [218,219]. Among all these factors, VEGF (specially isoform A), known as the main inducer of angiogenesis both in vivo and in vitro [220], is actively induced by hypoxic environments through HIF-1$\alpha$ [221,222]. This pro-angiogenic molecule acts by binding to two membrane receptors: VEGFR1 (or Fms Related Receptor Tyrosine Kinase 1, Flt-1), and VEGFR2 (or Fetal Liver Kinase-1, Flk-1). Both receptors are highly expressed by terminal villous capillaries and the adjacent trophoblast [223,224].

Other important pro-angiogenic factors produced by placenta are the PlGF, which can also activate VEGFR1, neuropilin-1 (NRP-1), angiopoietins, fibroblast growth factor 2, platelet derived growth factor, and the insulin/IGF axis [225–228]. As mentioned before, HIF-1$\alpha$ is a major driver of VEGF transcription, however in vitro studies demonstrated PlGF is not regulated by this factor in the human placenta [229]. In trophoblast cells,

PlGF expression decreases in low oxygen tension conditions, whereas normoxic conditions increase PlGF contrary to what occurs in other cell types [230].

The main anti-angiogenic molecules include soluble Flt-1 (sFlt-1), which antagonizes VEGF and PlGF by sequestering them and preventing their interaction with their VEGFR-1/-2 receptors, and soluble Endoglin (sEng), a TGF-β signaling antagonist. In vitro effects of sFlt1 include vasoconstriction and endothelial dysfunction, sEng amplifies the vascular damage mediated by sFlT1 and in trophoblast cells, sEng is released in response to hypoxia, oxidative stress or inflammation [231,232].

Since the placenta is a highly vascularized organ that supplies nutrients, growth factors, oxygen, hormones, and immune mediators to the fetus, all molecules of the angiogenic axis are tightly regulated throughout pregnancy. However, as mentioned before, clinical examination of GDM placentae shows increased villous vascularity often associated with thickened immature villi capillaries [20,21,24]. Diverse experimental evidence indicate that hyperglycemia promotes angiogenesis, vasoconstriction, and a higher vessel permeability [233–238]. Therefore, hyperglycemia in GDM has been highly associated with vascular damage, endothelial dysfunction, and aberrant overstimulated placental angiogenesis. All these changes result in severe endothelial damage, leading to impairment of both the maternal and fetal vascular system [239].

The role of VEGF axis in GDM is unclear. Some experiments showed a predominant pro-angiogenic balance characterized by higher serum levels or placental content of VEGF, PlGF, VEGFR2, and HIF-1α [159,240,241]. However, other researchers observed the opposite effect of these molecules on the VEGF axis in GDM placentae or trophoblasts under hyperglycemic condition [154,242,243]. In feto-placental vessels and capillaries of GDM placentae, decreased expression of VEGF has been found [244,245]. Despite this VEGF downregulation, feto-placental endothelial cells incubated with culture media of trophoblasts isolated from GDM placentae exhibit a more pronounced vessel network formation in vitro [159,160].

Therefore, it seems that other regulatory factors besides those involved in the VEGF/VEGFR axis may be participating in placental angiogenesis and should be considered to understand vascularization and blood perfusion in GDM placentae. These factors may be derived from syncytiotrophoblasts, mesenchymal stromal cells and/or endothelial cells. In addition, other non-classical players in the control of the angiogenic process in GDM patients have been evaluated. Dr. Lana Mc-Clements found GDM placentae have a decreased expression of the anti-angiogenic protein sirtuin 1 (SIRT-1), which may help to explain the increased vascularized network observed in GDM placentae. Additionally, they cultured a trophoblast cell line (obtained from first trimester/choriocarcinoma) under conditions of hyperglycemia (25 mM) and hypoxia (2.5 and 6.5% $O_2$), and observed a downregulated expression of the anti-angiogenic proteins FK 506 binding protein-like (FKBLP) and SIRT-1 [24]. These results support the hypothesis that hyperglycemia and lower oxygen tension alter the balance of factors that regulate angiogenesis in the placenta, leading to vascular dysfunction and possibly the development of preeclampsia.

The research group of Dr. Mojgan Karimabad found increased serum levels of CXCL1 and CXCL12, chemokines with pro-angiogenic properties [246–248], in neonates from GDM mothers, whereas antiangiogenic chemokines CXCL9 and CXCL10 were downregulated, compared to neonates from healthy mothers without GDM [249]. In addition, upregulated expression and activity of MMP-2 and MMP-9 and downregulated expression of TIMP-2 has been observed in trophoblasts under hyperglycemic treatment (30 mM), favoring endothelial cell migration during angiogenesis [240]. Furthermore, the Membrane-type MMP1, a kind of MMP anchored to the cellular membrane and a key player in vascularization and angiogenesis, is also up-regulated in GDM placentae [250,251]. Interestingly, these researchers observed that this metalloproteinase is up-regulated by insulin, which signaling is overactivated in GDM placentae, as described earlier [54,55]. Also, the angiopoietin-related growth factor (AGF), another pro-angiogenic molecule, is elevated in serum of GDM mothers [252].

All considered, the changes in the angiogenic molecules profile in GDM placentae and mothers indicate an exacerbated placental vascularization, but with signs of endothelial cell dysfunction, higher vessel permeability, and compromised integrity of chorionic vasculature, like immature/injured villi capillaries. It has been hypothesized that the increased number of villi and vessels in pregnant women with hyperglycemia provides a greater surface for maternal-fetal exchange. This functional adaptation would facilitate the passage of glucose to the developing fetus, thereby contributing to explain fetal macrosomia [244,253].

Due to the human hematogenous placentation, fetal capillaries of placental villi are particularly vulnerable to any alteration in GDM maternal blood [254]. Diverse vasoactive agents are increased in GDM maternal serum, including proinflammatory cytokines, oxygen-free radicals, advanced glycation end products (AGEs), hyperglycemia, hyperinsulinemia, and hypoxia. Currently, it is well known that hyperglycemia and TNF-$\alpha$ actively stimulate plasma and endothelial production of reactive oxygen species (ROS) and toxic by-products of glycolysis, leading to the formation of AGEs [255–257] which damages macro- and especially micro-vasculature and may contribute to thrombotic and atherosclerotic events [257,258]. In addition, it has been hypothesized that this endothelial damage evidenced by increased angiogenesis rate and chorionic villous branching may conduce to increased peripheral vascular resistance. This may explain the higher maternal blood pressure and preeclampsia [24,239], that are highly frequent complications in GDM pregnancies [1,7].

## 6. GDM and Oxidative Stress

Disruption of the delicate equilibrium between antioxidants and pro-oxidants in the cellular and tissular milieu can create oxidative distress that induces accumulative damage by modifying the state of macromolecules such as proteins, lipids, and DNA [259]. A distressed intracellular and extracellular environment leads to unspecific oxidation of proteins and altered response patterns, as well as irreversible damage causing growth arrest, cell death, and inflammation [260].

Reactive Oxygen Species (ROS) are described as free radical highly reactive molecules derived from the reduction of molecular oxygen that are formed by reduction-oxidation (redox) reaction or by electronic excitation [259,260]. Some of the most known species are superoxide ($O_2^-\bullet$), hydroxide ($OH^-\bullet$), and hydrogen peroxide ($H_2O_2$) [261]. ROS have a role in cell signaling including apoptosis, gene expression, activation of cell signaling cascades, and serving as both intra- and intercellular messengers [262].

The placenta, being a "new" organ, must establish multiple adaptations to a high oxygen-rich environment. In this scenario, the intervillous space is exposed to hypoxic conditions in the first 10–12 weeks of pregnancy, when the trophoblast plug in spiral arteries is removed or disintegrated and begins the entry of complete blood into intervillous space [263,264]. When the maternal blood circulation is open through the placenta, there is a significant increase in oxidative stress in trophoblasts. The gradual opening of increasing numbers of maternal vessels allows the placental tissue to adapt to the increased oxygen tension and oxidative stress. Some degree of placental oxidative stress occurs in all pregnancies towards term, however, under normal conditions, there are local compensatory mechanisms such as the rise of antioxidant's glutathione peroxidase and catalase activity [265].

ROS generation by oxidative phosphorylation at the mitochondrial membranes may be pertinent to tissues with a high-energy demand or those containing large amounts of mitochondria such as the placenta [266]. Under normal physiological conditions, ROS play a crucial role in normal embryonic development and cell function, such as trophoblast invasion and vascular development in the placenta [265,267,268].

In the placenta, fluctuating oxygen conditions can increase ROS production, where they can act as signaling molecules. $O_2^-\bullet$ is produced via an enzymatic and non-enzymatic process that takes place mainly in mitochondria. Along a non-enzymatic process, leakage of electrons from enzymes of the mitochondrial respiratory chain takes place and reduces molecular oxygen, thereby forming $O_2^-\bullet$.

As compensatory physiological mechanisms, the enzyme manganese superoxide dismutase (MnSOD) catalyzes the diminution of $O_2^-\bullet$ and the formation of hydrogen peroxide, which is transformed to oxygen and water by glutathione peroxidase and catalase [269,270]. Likewise, MnSOD catalase and other enzymes such as glutathione peroxidase and copper/zinc superoxide dismutase (SOD1) are present in the placenta acting as antioxidants.

*ROS Changes Associated with a Hyperglycemic Environment*

Oxidative stress has been associated with multiple pathological conditions, including GDM [271]. Inflammatory modulators, such as TNF-α, have deleterious effects during gestation and especially in hyperglycemic environments by increasing the expression and activation of ROS precursors, like NADPH oxidase 4 (NOX 4) [272]. Experimental and clinical evidence support the idea that a hyperglycemic environment is associated with increased oxidative stress. It has been reported that GDM women have a decreased ability to compensate for oxidative stress, and this was associated with increased insulin resistance and reduced insulin secretion [273]. There is evidence suggesting that women with GDM produce high levels of free radicals and have compromised free-radical scavenging mechanisms. Increased ROS, together with the impaired peripheral antioxidant activity is related to the induction of congenital malformation in pre-gestational diabetic pregnancies plasma of women with GDM [274–277]. In vitro studies done in 3T3-L1 adipocytes support that oxidative stress triggers decreased GLUT-4 expression, as a result of impaired binding of nuclear proteins to the insulin-responsive element in the GLUT-4 promotor [278]. Thus, oxidative stress adds to the factors related to impaired peripheral glucose metabolic control.

In the GDM placenta, there seems to be another picture. Not only is there a reported increase in oxidative stress and lipid peroxidation compared to normal pregnant women [275,276,279], but there also appears to be a concomitant increase in antioxidant enzyme activity that compensates for it [26,275,280,281]. The activated placental Nuclear Factor Erythroid 2-Related Factor 2 (Nrf2)/Antioxidant Response Element (ARE) pathway might have led to an increased expression of antioxidant enzymes SOD1, and catalase. This may be viewed as a protective mechanism in the placenta from the further onslaught of oxidative stress [282]. Another related pathway, the Nrf2/Kelch-like ECH-associated protein 1 (Keap2) pathway, also plays a crucial role in transcriptional activation of antioxidant defense genes and restoration of redox homeostasis [282]. One more antioxidant mechanism is related to apolipoprotein D (apo D), which has been observed to increase in the villous trophoblast and adventitia tunica around the large blood vessels in placental tissue from GDM [283]. Altogether, this data suggests that placentae with GDM are more protected against oxidative damage but are more susceptible to nitrosative damage as compared to normal placentae. In Table 1, we review the main clinical results in oxidative status reported in GDM women.

**Table 1.** Main reactive oxygen and nitrogen species altered in GDM.

| Metabolite | Description | GDM Status | Reference |
|---|---|---|---|
| Protein carbonyls | Created in response to oxidative stress and are produced in the process of protein carbonylation. | Increased in plasma at 24–28 WOG | [284,285] |
| Nitrotyrosine | Superoxide $O_2^-\bullet$ is produced by the electroreduction of molecular oxygen, is highly reactive, with a short half-life. It is metabolized to hydrogen peroxide by the high activity of superoxide dismutase. In highly vascularized tissues such as the human placenta, superoxide can combine with nitric oxide to form the very reactive and damaging ROS, peroxynitrite ($ONOO^-\bullet$). $ONOO^-\bullet$ reacts with hydroxyl groups on serine, tyrosine, and threonine residues to form nitrotyrosine adducts. | Increased in the blood (24–28 WOG) | [284,286] |

**Table 1.** *Cont.*

| Metabolite | Description | GDM Status | Reference |
|---|---|---|---|
| Advanced glycation end products (AGEs) | AGEs are formed by a combination of glycation and oxidation reaction. AGEs are strongly associated with diabetic complications of pregnancy that have increased levels of glucose as well as ROS and are associated with complications of diabetes such as retinopathy, nephropathy, and neuropathy. AGEs are formed due to the non-enzymatic glycation of proteins, lipids, and nucleic acids during hyperglycemia. It has the potential to damage vasculature by modifying the substrate or utilizing AGEs and Receptor of AGE (RAGE) interaction. AGEs such as Ne-carboxymethyl lysine (CML) are produced under oxidative conditions. | AGEs are elevated in maternal serum. CML is elevated in plasma (24–30 WOG) | [287–290] |
| Oxidized low-density lipoproteins (Ox-LDL) | The formation of ox-LDL involves the oxidation of both protein and lipid components. | Elevated in maternal plasma | [291,292] |
| 4 -hydroxynonenal (4-HNE) | A product of lipid peroxidation that can modify proteins and have effects on tissue survival. | Elevated in placental and uterine tissues | [283,293] |
| Malondialdehyde (MDA) | MDA is a marker of lipid peroxidation and oxidative stress and is formed as the result of lipid peroxidation of polyunsaturated fatty acids. | Elevated maternal plasma levels at 28 WOG | [292] |
| Oxidized DNA | DNA is vulnerable to oxidation and undergoes a continuous cycle of damage and repair in living cells. The nucleoside oxidation marker 8-oxo-7,8 dihydro-2 deoxyguanosine (8OH-dG) is used as a biomarker of oxidative stress excreted in urine. | 8OH-dG level in the urine was 26% higher in women who subsequently developed GDM | [294,295] |
| Glutathione (GSH) and Glutathione peroxidase (GPX) | Under healthy conditions, cells have up to 98% of total GSH in their reduced form. GSH acts as an electron donor to reduce unstable ROS in a process that permits the formation of oxidized GSH. An altered ratio reduced GSH/oxidized GSH is an indicator of oxidative stress. While reduced GSH can directly prevent non-enzymatic oxidation of thiol groups; it also acts as a co-substrate for the antioxidant enzyme GPX. GPX can remove active peroxides, but this results in the conversion of reduced GSH to oxidized GSH. | Reduced GSH in blood at term. Diminution of GPX in blood | [296–298] |

WOG: weeks of gestation.

Finally, recent evidence suggests that the cerebral cortex and hippocampus of male and female GDM offspring present an increase in ROS and lipid peroxidation, a disruption in the glutathione status, and decreased activity of catalase and SOD1; thus, they present a marked neural pro-oxidative environment. The researchers hypothesized that cognitive behavior could be modified in an age- and sex-dependent manner [299].

## 7. Adverse Perinatal Outcomes Related to GDM

As result of the metabolic alterations that occur in the mother, and particularly in the placenta during GDM, the mother is at higher risk of complications than healthy pregnant women. Obesity and excessive weigh gain are common conditions that increase GDM development risk [300]. Obesity per se is related to increased risk of pregnancy-induced hypertension, preeclampsia, risk of venous embolism, increased need for labor induction, and cesarean sections [301]. This is particularly important because of the higher rate of overweight and obesity in women of reproductive age. However, the hallmark of GDM is maternal hyperglycemia, resulting in a broad spectrum of clinical consequences for the mother and fetus in both the short and long term. In general, the severity of complications is related to the earlier onset of GDM and correlates inversely with the degree of glycemic control [302].

For the mother, short-term obstetric complications include hypertension, preeclampsia, premature delivery, and cesarean section [303,304]. In the long-term, the mother is in a high likelihood of recurrence of GDM in successive pregnancies above 48% [305,306]. Also, a robust body of literature supports GDM hyperglycemia predisposes mothers to developing T2DM and cardiovascular diseases years after delivery [10,307–311]. Although the proportion of GDM women who develop T2DM is highly variable, a recent meta-analysis indicates a relative risk of 8-fold (95% IC: 6.5–10.6) [10]. Subsequently, this increases the chance of developing cardiovascular disease and metabolic syndrome up to threefold [308].

On the other hand, GDM is associated with the development of fetopathies. Maternal hyperglycemia during GDM generates fetal hypoxia, which can evoke birth asphyxia and fetal death. Clinical evidence supports that fetal hypoxia increases erythropoietin production, which is associated with the development of polycythemia, hyperbilirubinemia and neonatal icterus [4]. As has been described in the previous sections of the manuscript, hyperglycemia can generate fetal macrosomia, mainly in overweight or obese women before pregnancy. Macrosomia is defined as a newborn weight greater than 4000 g. Higher fetal weight is not only related to total body mass but also to an excessive fat mass/lean mass ratio, indicating higher adipose mass deposition [312,313]. An early predictor of macrosomia is increased abdominal circumference by ultrasound between weeks 20 to 24 of gestation [11]. The fetus absorbs glucose, stores it in the form of glycogen in the liver, and the excess is converted into visceral fat, increasing the fetal abdominal circumference. Moreover, fetal epicardial fat thickness has been proposed as an earlier marker (before 24 weeks) to screen GDM women [314]. Macrosomia and hyperglycemia predispose to obstetrics complications such as shoulder dystocia at birth, plexus-brachial injury, respiratory distress syndrome, as well as the need for instrumented delivery [7,315].

The fetal hyperinsulinemia induced by GDM impairs pulmonary surfactant synthesis, delaying lung maturation by 1 to 1.5 weeks, and predisposing the fetus to develop respiratory distress syndrome [316]. In the Hyperglycemia and Adverse Pregnancy Outcome (HAPO) study, a large multinational-racial and ethnically cohort was analyzed, demonstrating that high maternal glucose levels are associated with increased birth weight above the 90th percentile for gestational age, primary cesarean delivery, clinically diagnosed neonatal hypoglycemia, and cord-blood serum C-peptide level above the 90th percentile. Secondary outcomes were delivery before 37 weeks of gestation, shoulder dystocia or birth injury, need for intensive neonatal care, hyperbilirubinemia, and preeclampsia [2]. These results were supported by other studies [303,317]. In addition, adverse fetal cardiac patterns have also been evaluated in GDM neonates. Main cardiotocographic alterations include higher risk for hypoxia-related ZigZag patterns, late decelerations of the fetal heart ratio, and greater risk of fetal asphyxia [318].

GDM has a lower risk of congenital malformations because it develops after organogenesis, unlike pregestational T1DM or T2DM, where the maternal hyperglycemia acts as a teratogenic factor [319]. Although early-onset of GDM could lead to a slight increase in the rate of birth defects to 16–22 percent [320–322].

Finally, fetal programming in GDM has been studied in the context of DOHaD hypothesis (developmental origins of health and disease) to understand the long-term consequences of growing in a hyperglycemic environment during life in utero. Hyperglycemia and maternal obesity generate epigenetic changes in fetal programming, increasing the risk for obesity, T2DM, impaired glucose tolerance, insulin resistance, coronary heart disease, chronic arterial hypertension, dyslipidemia, metabolic syndrome, and some cancers in the future life of the newborn [3,323–327]. Interestingly, it has been observed that GDM effects in childhood are manifested after two years of age, such as an increase in BMI and hyperglycemia [328]. Regarding this, the HAPO follow-up study (HAPO-FUS), which included 4160 children aged 10–14 years, demonstrated that maternal hyperglycemia correlates positively with increased glucose and insulin resistance in childhood. Furthermore, hyperglycemia was associated with the increase of adiposity and metabolic distress [323]. These data confirm the hypothesis that GDM causes long-term metabolic alterations in the offspring and demonstrate the importance of timely diagnosis and treatment of this disease.

## 8. Conclusions

The functional interplay between the placenta and the maternal adipose tissue, glycemic control, and overall maternal metabolism, requires a highly controlled equilibrium that, if disturbed, may lead to GDM. Known risk factors for GDM include overweight/obesity, lack of physical activity, prediabetes, and genetic predisposition. In this pathology, maternal hyperglycemia, carbohydrate intolerance, dysfunction of beta-pancreatic and endothelial cells, as well as insulin resistance, disturb placental structure and functions. This disturbance favors a metainflammatory environment associated with increased production of inflammatory cytokines, adipokines, and oxidative reactive species, leading to an abnormal endocrine, immune and antioxidant phenotype. This maternal/placental immunoendocrine dysregulation affects the mother as well as the fetus's health in the short and long term. The severity of complications relates to an earlier onset of GDM and correlates inversely with the degree of glycemic control. Therefore, it is of utmost importance to understand the pathophysiology of GDM to develop intervention and prevention strategies, including the orientation of pregnant women to eat healthier foods and exercising. In Figure 2, we propose a multi-organ scheme integrating the role of the placenta in the immunoendocrine environment dysregulated during GDM.

**Author Contributions:** Conception of the work: V.Z.-C. and A.O.-O. All authors wrote the first draft of the manuscript. Review & editing: A.O.-O., L.D., P.F.-E. and V.Z.-C. P.V. and C.R.-I. contributed to the preparation of the clinical outcomes and the revision of the final version of the manuscript. All authors read and approved the final version of the manuscript.

**Funding:** This study was supported by the Instituto Nacional de Perinatología Isidro Espinosa de los Reyes [grant No. 2018-1-152 to A.O.-O. and 2019-1-5 to V.Z.-C.] and by CONACyT [CB-A1-S-27832 to A.O.-O.]. The funders had no role in the concept, design or writing of the manuscript; or in the decision to publish or preparation of the manuscript.

**Acknowledgments:** We thank David Wheaton and Daudi Langat for their invaluable assistance and English-style revision of the final version of this review.

**Conflicts of Interest:** The authors declare no conflict of interest.

## References

1. ACOG Practice Bulletin No. 190: Gestational Diabetes Mellitus. *Obstet. Gynecol.* **2018**, *131*, e49–e64. [CrossRef]
2. Hyperglycemia and Adverse Pregnancy Outcomes. *N. Engl. J. Med.* **2008**, *358*. [CrossRef]
3. Murray, S.R.; Reynolds, R.M. Short- and Long-Term Outcomes of Gestational Diabetes and Its Treatment on Fetal Development. *Prenat. Diagn.* **2020**, *40*. [CrossRef]
4. Bianco, M.E.; Josefson, J.L. Hyperglycemia During Pregnancy and Long-Term Offspring Outcomes. *Curr. Diabetes Rep.* **2019**, *19*. [CrossRef] [PubMed]
5. Dickens, L.T.; Thomas, C.C. Updates in Gestational Diabetes Prevalence, Treatment, and Health Policy. *Curr. Diabetes Rep.* **2019**, *19*. [CrossRef]
6. Brawerman, G.M.; Dolinsky, V.W. Therapies for Gestational Diabetes and Their Implications for Maternal and Offspring Health: Evidence from Human and Animal Studies. *Pharmacol. Res.* **2018**, *130*. [CrossRef] [PubMed]
7. McCance, D.R. *Diabetes in Pregnancy: Management from Preconception to the Postnatal Period*; NICE Guideline; NICE: Hoboken, NJ, USA, 2015; Volume 29, Available online: https://www.nice.org.uk/guidance/ng3/resources/diabetes-in-pregnancy-management-from-preconception-to-the-postnatal-period-pdf-51038446021 (accessed on 1 June 2021).
8. Plows, J.F.; Stanley, J.L.; Baker, P.N.; Reynolds, C.M.; Vickers, M.H. The Pathophysiology of Gestational Diabetes Mellitus. *Int. J. Mol. Sci.* **2018**, *19*, 3342. [CrossRef] [PubMed]
9. Moyce, B.L.; Dolinsky, V.W. Maternal β-Cell Adaptations in Pregnancy and Placental Signalling: Implications for Gestational Diabetes. *Int. J. Mol. Sci.* **2018**, *19*, 3467. [CrossRef] [PubMed]
10. Dennison, R.A.; Chen, E.S.; Green, M.E.; Legard, C.; Kotecha, D.; Farmer, G.; Sharp, S.J.; Ward, R.J.; Usher-Smith, J.A.; Griffin, S.J. The Absolute and Relative Risk of Type 2 Diabetes after Gestational Diabetes: A Systematic Review and Meta-Analysis of 129 Studies. *Diabetes Res. Clin. Pract.* **2021**, *171*. [CrossRef]

11. American Diabetes Association. Management of Diabetes in Pregnancy: Standards of Medical Care in Diabetes—2020. *Diabetes Care* **2020**, *43*. [CrossRef]
12. Schäfer-Graf, U.M.; Gembruch, U.; Kainer, F.; Groten, T.; Hummel, S.; Hösli, I.; Grieshop, M.; Kaltheuner, M.; Bührer, C.; Kautzky-Willer, A.; et al. Gestational Diabetes Mellitus (GDM)—Diagnosis, Treatment and Follow-UpGuideline of the DDG and DGGG (S3 Level, AWMF Registry Number 057/008, February 2018). *Geburtshilfe Frauenheilkd.* **2018**, *78*. [CrossRef]
13. Alqudah, A.; McKinley, M.C.; McNally, R.; Graham, U.; Watson, C.J.; Lyons, T.J.; McClements, L. Risk of Pre-Eclampsia in Women Taking Metformin: A Systematic Review and Meta-Analysis. *Diabet. Med.* **2018**, *35*. [CrossRef]
14. Martis, R.; Crowther, C.A.; Shepherd, E.; Alsweiler, J.; Downie, M.R.; Brown, J. Treatments for Women with Gestational Diabetes Mellitus: An Overview of Cochrane Systematic Reviews. *Cochrane Database Syst. Rev.* **2018**, *2018*. [CrossRef] [PubMed]
15. Feng, Y.; Yang, H. Metformin–a Potentially Effective Drug for Gestational Diabetes Mellitus: A Systematic Review and Meta-Analysis. *J. Matern. Fetal Neonatal Med.* **2017**, *30*. [CrossRef] [PubMed]
16. Wang, X.; Liu, W.; Chen, H.; Chen, Q. Comparison of Insulin, Metformin, and Glyburide on Perinatal Complications of Gestational Diabetes Mellitus: A Systematic Review and Meta-Analysis. *Gynecol. Obstet. Investig.* **2021**. [CrossRef] [PubMed]
17. Gui, J.; Liu, Q.; Feng, L. Metformin vs Insulin in the Management of Gestational Diabetes: A Meta-Analysis. *PLoS ONE* **2013**, *8*, e64585. [CrossRef]
18. Landi, S.N.; Radke, S.; Engel, S.M.; Boggess, K.; Stürmer, T.; Howe, A.S.; Funk, M.J. Association of Long-Term Child Growth and Developmental Outcomes with Metformin vs Insulin Treatment for Gestational Diabetes. *JAMA Pediatr.* **2019**, *173*. [CrossRef]
19. Wouldes, T.A.; Battin, M.; Coat, S.; Rush, E.C.; Hague, W.M.; Rowan, J.A. Neurodevelopmental Outcome at 2 Years in Offspring of Women Randomised to Metformin or Insulin Treatment for Gestational Diabetes. *Arch. Dis. Child. Fetal Neonatal Ed.* **2016**, *101*. [CrossRef]
20. Carrasco-Wong, I.; Moller, A.; Giachini, F.R.; Lima, V.V.; Toledo, F.; Stojanova, J.; Sobrevia, L.; San Martín, S. Placental Structure in Gestational Diabetes Mellitus. *Biochim. Biophys. Acta Mol. Basis Dis.* **2020**, *1866*. [CrossRef]
21. Ehlers, E.; Talton, O.O.; Schust, D.J.; Schulz, L.C. Placental Structural Abnormalities in Gestational Diabetes and When They Develop: A Scoping Review. *Placenta* **2021**. [CrossRef]
22. Lucas, M.J. Diabetes Complicating Pregnancy. *Obstet. Gynecol. Clin. N. Am.* **2001**, *28*. [CrossRef]
23. Huynh, J.; Dawson, D.; Roberts, D.; Bentley-Lewis, R. A Systematic Review of Placental Pathology in Maternal Diabetes Mellitus. *Placenta* **2015**, *36*. [CrossRef] [PubMed]
24. Alqudah, A.; Eastwood, K.A.; Jerotic, D.; Todd, N.; Hoch, D.; McNally, R.; Obradovic, D.; Dugalic, S.; Hunter, A.J.; Holmes, V.A.; et al. FKBPL and SIRT-1 Are Downregulated by Diabetes in Pregnancy Impacting on Angiogenesis and Endothelial Function. *Front. Endocrinol.* **2021**, *12*. [CrossRef] [PubMed]
25. Daskalakis, G.; Marinopoulos, S.; Krielesi, V.; Papapanagiotou, A.; Papantoniou, N.; Mesogitis, S.; Antsaklis, A. Placental Pathology in Women with Gestational Diabetes. *Acta Obstet. Gynecol. Scand.* **2008**, *87*. [CrossRef]
26. Madazli, R.; Tuten, A.; Calay, Z.; Uzun, H.; Uludag, S.; Ocak, V. The Incidence of Placental Abnormalities, Maternal and Cord Plasma Malondialdehyde and Vascular Endothelial Growth Factor Levels in Women with Gestational Diabetes Mellitus and Nondiabetic Controls. *Gynecol. Obstet. Investig.* **2008**, *65*. [CrossRef] [PubMed]
27. Fisher, J.J.; Vanderpeet, C.L.; Bartho, L.A.; McKeating, D.R.; Cuffe, J.S.M.; Holland, O.J.; Perkins, A.V. Mitochondrial Dysfunction in Placental Trophoblast Cells Experiencing Gestational Diabetes Mellitus. *J. Physiol.* **2021**, *599*. [CrossRef]
28. Valent, A.M.; Choi, H.; Kolahi, K.S.; Thornburg, K.L. Hyperglycemia and Gestational Diabetes Suppress Placental Glycolysis and Mitochondrial Function and Alter Lipid Processing. *FASEB J.* **2021**, *35*. [CrossRef] [PubMed]
29. Mishra, J.S.; Zhao, H.; Hattis, S.; Kumar, S. Elevated Glucose and Insulin Levels Decrease DHA Transfer across Human Trophoblasts via SIRT1-Dependent Mechanism. *Nutrients* **2020**, *12*, 1271. [CrossRef] [PubMed]
30. Hansis, C.; Grifo, J.A.; Tang, Y.X.; Krey, L.C. Assessment of Beta-HCG, Beta-LH MRNA and Ploidy in Individual Human Blastomeres. *Reprod. Biomed. Online* **2002**, *5*. [CrossRef]
31. Butler, S.A.; Luttoo, J.; Freire, M.O.T.; Abban, T.K.; Borrelli, P.T.A.; Iles, R.K. Human Chorionic Gonadotropin (HCG) in the Secretome of Cultured Embryos: Hyperglycosylated HCG and HCG-Free Beta Subunit Are Potential Markers for Infertility Management and Treatment. *Reprod. Sci.* **2013**, *20*. [CrossRef]
32. Feldt-Rasmussen, U.; Mathiesen, E.R. Endocrine Disorders in Pregnancy: Physiological and Hormonal Aspects of Pregnancy. *Best Pract. Res. Clin. Endocrinol. Metab.* **2011**, *25*. [CrossRef]
33. Rout, M.; Lulu, S.S. Molecular and Disease Association of Gestational Diabetes Mellitus Affected Mother and Placental Datasets Reveal a Strong Link between Insulin Growth Factor (IGF) Genes in Amino Acid Transport Pathway: A Network Biology Approach. *J. Cell. Biochem.* **2019**, *120*. [CrossRef]
34. Alur, V.; Raju, V.; Vastrad, B.; Tengli, A.; Vastrad, C.; Kotturshetti, S. Integrated Bioinformatics Analysis Reveals Novel Key Biomarkers and Potential Candidate Small Molecule Drugs in Gestational Diabetes Mellitus. *Biosci. Rep.* **2021**, *41*. [CrossRef]
35. Yang, Y.; Guo, F.; Peng, Y.; Chen, R.; Zhou, W.; Wang, H.; OuYang, J.; Yu, B.; Xu, Z. Transcriptomic Profiling of Human Placenta in Gestational Diabetes Mellitus at the Single-Cell Level. *Front. Endocrinol.* **2021**, *12*. [CrossRef]
36. Westermeier, F.; Sáez, T.; Arroyo, P.; Toledo, F.; Gutiérrez, J.; Sanhueza, C.; Pardo, F.; Leiva, A.; Sobrevia, L. Insulin Receptor Isoforms: An Integrated View Focused on Gestational Diabetes Mellitus. *Diabetes Metab. Res. Rev.* **2016**, *32*. [CrossRef]

37. Sobrevia, L.; Salsoso, R.; Fuenzalida, B.; Barros, E.; Toledo, L.; Silva, L.; Pizarro, C.; Subiabre, M.; Villalobos, R.; Araos, J.; et al. Insulin Is a Key Modulator of Fetoplacental Endothelium Metabolic Disturbances in Gestational Diabetes Mellitus. *Front. Physiol.* **2016**, *7*. [CrossRef]

38. Larner, J.; Huang, L.C.; Tang, G.; Suzuki, S.; Schwartz, C.F.W.; Romero, G.; Roulidis, Z.; Zeller, K.; Shen, T.Y.; Oswald, A.S.; et al. Insulin Mediators: Structure and Formation. *Cold Spring Harb. Symp. Quant. Biol.* **1988**, *53*. [CrossRef] [PubMed]

39. Cross, D.A.E.; Alessi, D.R.; Cohen, P.; Andjelkovich, M.; Hemmings, B.A. Inhibition of Glycogen Synthase Kinase-3 by Insulin Mediated by Protein Kinase B. *Nature* **1995**, *378*. [CrossRef]

40. Kohn, A.D.; Kovacina, K.S.; Roth, R.A. Insulin Stimulates the Kinase Activity of RAC-PK, a Pleckstrin Homology Domain Containing Ser/Thr Kinase. *EMBO J.* **1995**, *14*. [CrossRef]

41. Batista, T.M.; Haider, N.; Kahn, C.R. Defining the Underlying Defect in Insulin Action in Type 2 Diabetes. *Diabetologia* **2021**, *64*. [CrossRef] [PubMed]

42. O'Neill, B.T.; Lee, K.Y.; Klaus, K.; Softic, S.; Krumpoch, M.T.; Fentz, J.; Stanford, K.I.; Robinson, M.M.; Cai, W.; Kleinridders, A.; et al. Insulin and IGF-1 Receptors Regulate FoxO-Mediated Signaling in Muscle Proteostasis. *J. Clin. Investig.* **2016**, *126*. [CrossRef]

43. Sancak, Y.; Thoreen, C.C.; Peterson, T.R.; Lindquist, R.A.; Kang, S.A.; Spooner, E.; Carr, S.A.; Sabatini, D.M. PRAS40 Is an Insulin-Regulated Inhibitor of the MTORC1 Protein Kinase. *Mol. Cell* **2007**, *25*. [CrossRef]

44. Tee, A.R.; Fingar, D.C.; Manning, B.D.; Kwiatkowski, D.J.; Cantley, L.C.; Blenis, J. Tuberous Sclerosis Complex-1 and -2 Gene Products Function Together to Inhibit Mammalian Target of Rapamycin (MTOR)-Mediated Downstream Signaling. *Proc. Natl. Acad. Sci. USA* **2002**, *99*. [CrossRef]

45. Sano, H.; Kane, S.; Sano, E.; Mîinea, C.P.; Asara, J.M.; Lane, W.S.; Garner, C.W.; Lienhard, G.E. Insulin-Stimulated Phosphorylation of a Rab GTPase-Activating Protein Regulates GLUT4 Translocation. *J. Biol. Chem.* **2003**, *278*. [CrossRef] [PubMed]

46. Avruch, J.; Khokhlatchev, A.; Kyriakis, J.M.; Luo, Z.; Tzivion, G.; Vavvas, D.; Zhang, X.F. Ras Activation of the Raf Kinase: Tyrosine Kinase Recruitment of the MAP Kinase Cascade. *Recent Prog. Horm. Res.* **2001**, *56*. [CrossRef]

47. Sesti, G.; Tullio, A.N.; D'Alfonso, R.; Napolitano, M.L.; Marini, M.A.; Borboni, P.; Longhi, R.; Albonici, L.; Fusco, A.; Aglianò, A.M.; et al. Tissue-Specific Expression of Two Alternatively Spliced Isoforms of the Human Insulin Receptor Protein. *Acta Diabetol.* **1994**, *31*. [CrossRef] [PubMed]

48. Illsley, N.P.; Baumann, M.U. Human Placental Glucose Transport in Fetoplacental Growth and Metabolism. *Biochim. Biophys. Acta Mol. Basis Dis.* **2020**, *1866*. [CrossRef]

49. Stanirowski, P.J.; Szukiewicz, D.; Pazura-Turowska, M.; Sawicki, W.; Cendrowski, K. Placental Expression of Glucose Transporter Proteins in Pregnancies Complicated by Gestational and Pregestational Diabetes Mellitus. *Can. J. Diabetes* **2018**, *42*. [CrossRef]

50. Hiden, U.; Maier, A.; Bilban, M.; Ghaffari-Tabrizi, N.; Wadsack, C.; Lang, I.; Dohr, G.; Desoye, G. Insulin Control of Placental Gene Expression Shifts from Mother to Foetus over the Course of Pregnancy. *Diabetologia* **2006**, *49*. [CrossRef] [PubMed]

51. Alonso, A.; del Rey, C.G.; Navarro, A.; Tolivia, J.; González, C.G. Effects of Gestational Diabetes Mellitus on Proteins Implicated in Insulin Signaling in Human Placenta. *Gynecol. Endocrinol.* **2006**, *22*. [CrossRef]

52. Balachandiran, M.; Bobby, Z.; Dorairajan, G.; Gladwin, V.; Vinayagam, V.; Packirisamy, R.M. Decreased Maternal Serum Adiponectin and Increased Insulin-like Growth Factor-1 Levels along with Increased Placental Glucose Transporter-1 Expression in Gestational Diabetes Mellitus: Possible Role in Fetal Overgrowth: Regulation of Placental GLUT-1 Expression in Gestational Diabetes Mellitus. *Placenta* **2021**, *104*. [CrossRef]

53. Colomiere, M.; Permezel, M.; Riley, C.; Desoye, G.; Lappas, M. Defective Insulin Signaling in Placenta from Pregnancies Complicated by Gestational Diabetes Mellitus. *Eur. J. Endocrinol.* **2009**, *160*. [CrossRef]

54. Pérez-Pérez, A.; Guadix, P.; Maymó, J.; Dueñas, J.L.; Varone, C.; Fernández-Sánchez, M.; Sánchez-Margalet, V. Insulin and Leptin Signaling in Placenta from Gestational Diabetic Subjects. *Horm. Metab. Res.* **2015**, *48*. [CrossRef] [PubMed]

55. Gęca, T.; Kwaśniewska, A. The Influence of Gestational Diabetes Mellitus upon the Selected Parameters of the Maternal and Fetal System of Insulin-Like Growth Factors (IGF-1, IGF-2, IGFBP1-3)—A Review and a Clinical Study. *J. Clin. Med.* **2020**, *9*, 3256. [CrossRef]

56. Osmond, D.T.D.; King, R.G.; Brennecke, S.P.; Gude, N.M. Placental Glucose Transport and Utilisation Is Altered at Term in Insulin-Treated, Gestational-Diabetic Patients. *Diabetologia* **2001**, *44*. [CrossRef] [PubMed]

57. Osmond, D.T.D.; Nolan, C.J.; King, R.G.; Brennecke, S.P.; Gude, N.M. Effects of Gestational Diabetes on Human Placental Glucose Uptake, Transfer, and Utilisation. *Diabetologia* **2000**, *43*. [CrossRef]

58. Tumminia, A.; Scalisi, N.M.; Milluzzo, A.; Ettore, G.; Vigneri, R.; Sciacca, L. Maternal Diabetes Impairs Insulin and IGF-1 Receptor Expression and Signaling in Human Placenta. *Front. Endocrinol.* **2021**, *12*. [CrossRef] [PubMed]

59. Forbes, B.E.; Blyth, A.J.; Wit, J.M. Disorders of IGFs and IGF-1R Signaling Pathways. *Mol. Cell. Endocrinol.* **2020**, *518*. [CrossRef] [PubMed]

60. Wang, X.R.; Wang, W.J.; Yu, X.; Hua, X.; Ouyang, F.; Luo, Z.C. Insulin-like Growth Factor Axis Biomarkers and Gestational Diabetes Mellitus: A Systematic Review and Meta-Analysis. *Front. Endocrinol.* **2019**, *10*. [CrossRef]

61. Frasca, F.; Pandini, G.; Scalia, P.; Sciacca, L.; Mineo, R.; Costantino, A.; Goldfine, I.D.; Belfiore, A.; Vigneri, R. Insulin Receptor Isoform A, a Newly Recognized, High-Affinity Insulin-Like Growth Factor II Receptor in Fetal and Cancer Cells. *Mol. Cell. Biol.* **1999**, *19*. [CrossRef]

62. Blanquart, C.; Achi, J.; Issad, T. Characterization of IRA/IRB Hybrid Insulin Receptors Using Bioluminescence Resonance Energy Transfer. *Biochem. Pharmacol.* **2008**, *76*. [CrossRef] [PubMed]
63. Federici, M.; Porzio, O.; Zucaro, L.; Fusco, A.; Borboni, P.; Lauro, D.; Sesti, G. Distribution of Insulin/Insulin-like Growth Factor-I Hybrid Receptors in Human Tissues. *Mol. Cell. Endocrinol.* **1997**, *129*. [CrossRef]
64. Benyoucef, S.; Surinya, K.H.; Hadaschik, D.; Siddle, K. Characterization of Insulin/IGF Hybrid Receptors: Contributions of the Insulin Receptor L2 and Fn1 Domains and the Alternatively Spliced Exon 11 Sequence to Ligand Binding and Receptor Activation. *Biochem. J.* **2007**, *403*. [CrossRef] [PubMed]
65. Blyth, A.J.; Kirk, N.S.; Forbes, B.E. Understanding IGF-II Action through Insights into Receptor Binding and Activation. *Cells* **2020**, *9*, 2276. [CrossRef]
66. Han, V.K.; Bassett, N.; Walton, J.; Challis, J.R. The Expression of Insulin-like Growth Factor (IGF) and IGF-Binding Protein (IGFBP) Genes in the Human Placenta and Membranes: Evidence for IGF-IGFBP Interactions at the Feto-Maternal Interface. *J. Clin. Endocrinol. Metab.* **1996**, *81*. [CrossRef]
67. Holt, R.I.G.; Simpson, H.L.; Sönksen, P.H. The Role of the Growth Hormone-Insulin-like Growth Factor Axis in Glucose Homeostasis. *Diabet. Med.* **2003**, *20*. [CrossRef]
68. Baumann, M.U.; Schneider, H.; Malek, A.; Palta, V.; Surbek, D.V.; Sager, R.; Zamudio, S.; Illsley, N.P. Regulation of Human Trophoblast GLUT1 Glucose Transporter by Insulin-Like Growth Factor I (IGF-I). *PLoS ONE* **2014**, *9*, e106037. [CrossRef] [PubMed]
69. Scavo, L.M.; Karas, M.; Murray, M.; Leroith, D. Insulin-like Growth Factor-I Stimulates Both Cell Growth and Lipogenesis during Differentiation of Human Mesenchymal Stem Cells into Adipocytes. *J. Clin. Endocrinol. Metab.* **2004**, *89*. [CrossRef]
70. Clemmons, D.R.; Sleevi, M.; Allan, G.; Sommer, A. Effects of Combined Recombinant Insulin-like Growth Factor (IGF)-I and IGF Binding Protein-3 in Type 2 Diabetic Patients on Glycemic Control and Distribution of IGF-I and IGF-II among Serum Binding Protein Complexes. *J. Clin. Endocrinol. Metab.* **2007**, *92*. [CrossRef]
71. Devedjian, J.C.; George, M.; Casellas, A.; Pujol, A.; Visa, J.; Pelegrín, M.; Gros, L.; Bosch, F. Transgenic Mice Overexpressing Insulin-like Growth Factor-II in β Cells Develop Type 2 Diabetes. *J. Clin. Investig.* **2000**, *105*. [CrossRef]
72. Dabelea, D.; Crume, T. Maternal Environment and the Transgenerational Cycle of Obesity and Diabetes. *Diabetes* **2011**, *60*. [CrossRef] [PubMed]
73. Stanirowski, P.J.; Szukiewicz, D.; Pyzlak, M.; Abdalla, N.; Sawicki, W.; Cendrowski, K. Impact of Pre-Gestational and Gestational Diabetes Mellitus on the Expression of Glucose Transporters GLUT-1, GLUT-4 and GLUT-9 in Human Term Placenta. *Endocrine* **2017**, *55*. [CrossRef]
74. Gaither, K.; Quraishi, A.N.; Illsley, N.P. Diabetes Alters the Expression and Activity of the Human Placental GLUT1 Glucose Transporter 1. *J. Clin. Endocrinol. Metab.* **1999**, *84*. [CrossRef]
75. Borges, M.H.; Pullockaran, J.; Catalano, P.M.; Baumann, M.U.; Zamudio, S.; Illsley, N.P. Human Placental GLUT1 Glucose Transporter Expression and the Fetal Insulin-like Growth Factor Axis in Pregnancies Complicated by Diabetes. *Biochim. Biophys. Acta Mol. Basis Dis.* **2019**, *1865*. [CrossRef]
76. Lappas, M. Insulin-like Growth Factor-Binding Protein 1 and 7 Concentrations Are Lower in Obese Pregnant Women, Women with Gestational Diabetes and Their Fetuses. *J. Perinatol.* **2015**, *35*. [CrossRef]
77. Bach, L.A. Current Ideas on the Biology of IGFBP-6: More than an IGF-II Inhibitor? *Growth Horm. IGF Res.* **2016**, *30–31*. [CrossRef]
78. Shang, M.; Wen, Z. Increased Placental IGF-1/MTOR Activity in Macrosomia Born to Women with Gestational Diabetes. *Diabetes Res. Clin. Pract.* **2018**, *146*. [CrossRef]
79. Liu, K.; Wu, H.Y.; Xu, Y.H. Study on the Relationship between the Expression of IGF-1 in Umbilical Cord Blood and Abnormal Glucose Metabolism during Pregnancy. *Eur. Rev. Med. Pharmacol. Sci.* **2017**, *21*, 647–651. [PubMed]
80. Federici, M.; Zucaro, L.; Porzio, O.; Massoud, R.; Borboni, P.; Lauro, D.; Sesti, G. Increased Expression of Insulin/Insulin-like Growth Factor-1 Hybrid Receptors in Skeletal Muscle of Noninsulin-Dependent Diabetes Mellitus Subjects. *J. Clin. Investig.* **1996**, *98*. [CrossRef] [PubMed]
81. Federici, M.; Porzio, O.; Zucaro, L.; Giovannone, B.; Borboni, P.; Marini, M.A.; Lauro, D.; Sesti, G. Increased Abundance of Insulin/IGF-I Hybrid Receptors in Adipose Tissue from NIDDM Patients. *Mol. Cell. Endocrinol.* **1997**, *135*. [CrossRef]
82. Valensise, H.; Liu, Y.Y.; Federici, M.; Lauro, D.; Dell'anna, D.; Romanini, C.; Sesti, G. Increased Expression of Low-Affinity Insulin Receptor Isoform and Insulin/Insulin-like Growth Factor-1 Hybrid Receptors in Term Placenta from Insulin-Resistant Women with Gestational Hypertension. *Diabetologia* **1996**, *39*. [CrossRef]
83. Pandini, G.; Vigneri, R.; Costantino, A.; Frasca, F.; Ippolito, A.; Fujita-Yamaguchi, Y.; Siddle, K.; Goldfine, I.D.; Belfiore, A. Insulin and Insulin-like Growth Factor-I (IGF-I) Receptor Overexpression in Breast Cancers Leads to Insulin/IGF-I Hybrid Receptor Overexpression: Evidence for a Second Mechanism of IGF-I Signaling. *Clin. Cancer Res.* **1999**, *5*, 1935–1944.
84. Catalano, P.M.; Tyzbir, E.D.; Roman, N.M.; Amini, S.B.; Sims, E.A.H. Longitudinal Changes in Insulin Release and Insulin Resistance in Nonobese Pregnant Women. *Am. J. Obstet. Gynecol.* **1991**, *165*. [CrossRef]
85. Stefanoska, I.; Jovanović Krivokuća, M.; Vasilijić, S.; Ćujić, D.; Vićovac, L. Prolactin Stimulates Cell Migration and Invasion by Human Trophoblast in Vitro. *Placenta* **2013**, *34*. [CrossRef]
86. Binart, N. Prolactin and Pregnancy in Mice and Humans. *Ann. Endocrinol.* **2016**, *77*. [CrossRef] [PubMed]

87. Garzia, E.; Clauser, R.; Persani, L.; Borgato, S.; Bulfamante, G.; Avagliano, L.; Quadrelli, F.; Marconi, A.M. Prolactin and Proinflammatory Cytokine Expression at the Fetomaternal Interface in First Trimester Miscarriage. *Fertil. Steril.* **2013**, *100*. [CrossRef]

88. Cattini, P.A.; Jin, Y.; Jarmasz, J.S.; Noorjahan, N.; Bock, M.E. Obesity and Regulation of Human Placental Lactogen Production in Pregnancy. *J. Neuroendocrinol.* **2020**, *32*, e12859. [CrossRef] [PubMed]

89. Newbern, D.; Freemark, M. Placental Hormones and the Control of Maternal Metabolism and Fetal Growth. *Curr. Opin. Endocrinol. Diabetes Obes.* **2011**, *18*. [CrossRef]

90. Hu, Z.-Z.; Zhuang, L.; Meng, J.; Tsai-Morris, C.-H.; Dufau, M.L. Complex 5′ Genomic Structure of the Human Prolactin Receptor: Multiple Alternative Exons 1 and Promoter Utilization. *Endocrinology* **2002**, *143*. [CrossRef] [PubMed]

91. Abramicheva, P.A.; Smirnova, O.V. Prolactin Receptor Isoforms as the Basis of Tissue-Specific Action of Prolactin in the Norm and Pathology. *Biochemistry* **2019**, *84*. [CrossRef] [PubMed]

92. Sorenson, R.L.; Brelje, T.C. Prolactin Receptors Are Critical to the Adaptation of Islets to Pregnancy. *Endocrinology* **2009**, *150*. [CrossRef]

93. Yan, L.; Figueroa, D.J.; Austin, C.P.; Liu, Y.; Bugianesi, R.M.; Slaughter, R.S.; Kaczorowski, G.J.; Kohler, M.G. Expression of Voltage-Gated Potassium Channels in Human and Rhesus Pancreatic Islets. *Diabetes* **2004**, *53*. [CrossRef] [PubMed]

94. Huang, C.; Snider, F.; Cross, J.C. Prolactin Receptor Is Required for Normal Glucose Homeostasis and Modulation of β-Cell Mass during Pregnancy. *Endocrinology* **2009**, *150*. [CrossRef] [PubMed]

95. Kim, H.; Toyofuku, Y.; Lynn, F.C.; Chak, E.; Uchida, T.; Mizukami, H.; Fujitani, Y.; Kawamori, R.; Miyatsuka, T.; Kosaka, Y.; et al. Serotonin Regulates Pancreatic Beta Cell Mass during Pregnancy. *Nat. Med.* **2010**, *16*. [CrossRef] [PubMed]

96. Ben-Jonathan, N.; LaPensee, C.R.; LaPensee, E.W. What Can We Learn from Rodents about Prolactin in Humans? *Endocr. Rev.* **2008**, *29*. [CrossRef]

97. Brelje, T.C.; Scharp, D.W.; Lacy, P.E.; Ogren, L.; Talamantes, F.; Robertson, M.; Friesen, H.G.; Sorenson, R.L. Effect of Homologous Placental Lactogens, Prolactins, and Growth Hormones on Islet b-Cell Division and Insulin Secretion in Rat, Mouse, and Human Islets: Implication for Placental Lactogen Regulation of Islet Function during Pregnancy. *Endocrinology* **1993**, *132*. [CrossRef]

98. Ernst, S.; Demirci, C.; Valle, S.; Velazquez-Garcia, S.; Garcia-Ocaña, A. Mechanisms in the Adaptation of Maternal β-Cells during Pregnancy. *Diabetes Manag.* **2011**, *1*. [CrossRef]

99. Demirci, C.; Ernst, S.; Alvarez-Perez, J.C.; Rosa, T.; Valle, S.; Shridhar, V.; Casinelli, G.P.; Alonso, L.C.; Vasavada, R.C.; García-Ocana, A. Loss of HGF/c-Met Signaling in Pancreatic β-Cells Leads to Incomplete Maternal β-Cell Adaptation and Gestational Diabetes Mellitus. *Diabetes* **2012**, *61*. [CrossRef]

100. Hakonen, E.; Ustinov, J.; Palgi, J.; Miettinen, P.J.; Otonkoski, T. EGFR Signaling Promotes β-Cell Proliferation and Survivin Expression during Pregnancy. *PLoS ONE* **2014**, *9*, e93651. [CrossRef]

101. Sergeev, I.N.; Rhoten, W.B. 1, 25-Dihydroxyvitamin D3 Evokes Oscillations of Intracellular Calcium in a Pancreatic β-Cell Line. *Endocrinology* **1995**, *136*. [CrossRef]

102. Altuntaś, S.Ç.; Evran, M.; Sert, M.; Tetiker, T. Markers of Metabolic Syndrome in Patients with Pituitary Adenoma: A Case Series of 303 Patients. *Horm. Metab. Res.* **2019**, *51*. [CrossRef]

103. Le, T.N.; Elsea, S.H.; Romero, R.; Chaiworapongsa, T.; Francis, G.L. Prolactin Receptor Gene Polymorphisms Are Associated with Gestational Diabetes. *Genet. Test. Mol. Biomark.* **2013**, *17*. [CrossRef]

104. Ursell, W.; Brudenell, M.; Chard, T. Placental Lactogen Levels in Diabetic Pregnancy. *Br. Med. J.* **1973**, *2*, 80–82. [CrossRef]

105. Skouby, S.O.; Kühl, C.; Hornnes, P.J.; Andersen, A.N. Prolactin and Glucose Tolerance in Normal and Gestational Diabetic Pregnancy. *Obstet. Gynecol.* **1986**, *67*, 17–20.

106. Macotela, Y.; Triebel, J.; Clapp, C. Time for a New Perspective on Prolactin in Metabolism. *Trends Endocrinol. Metab.* **2020**, *31*. [CrossRef]

107. Park, S.; Kim, D.S.; Daily, J.W.; Kim, S.H. Serum Prolactin Concentrations Determine Whether They Improve or Impair β-Cell Function and Insulin Sensitivity in Diabetic Rats. *Diabetes Metab. Res. Rev.* **2011**, *27*. [CrossRef]

108. Kawaharada, R.; Masuda, H.; Chen, Z.; Blough, E.; Kohama, T.; Nakamura, A. Intrauterine Hyperglycemia-Induced Inflammatory Signalling via the Receptor for Advanced Glycation End Products in the Cardiac Muscle of the Infants of Diabetic Mother Rats. *Eur. J. Nutr.* **2018**, *57*. [CrossRef]

109. Pantham, P.; Aye, I.L.M.H.; Powell, T.L. Inflammation in Maternal Obesity and Gestational Diabetes Mellitus. *Placenta* **2015**, *36*. [CrossRef]

110. Gregor, M.F.; Hotamisligil, G.S. Inflammatory Mechanisms in Obesity. *Annu. Rev. Immunol.* **2011**, *29*. [CrossRef]

111. Yessoufou, A.; Moutairou, K. Maternal Diabetes in Pregnancy: Early and Long-Term Outcomes on the Offspring and the Concept of "Metabolic Memory". *Exp. Diabetes Res.* **2011**, *2011*. [CrossRef]

112. Westermeier, F.; Sáez, P.J.; Villalobos-Labra, R.; Sobrevia, L.; Farías-Jofré, M. Programming of Fetal Insulin Resistance in Pregnancies with Maternal Obesity by ER Stress and Inflammation. *BioMed Res. Int.* **2014**, *2014*. [CrossRef]

113. Dudele, A.; Hougaard, K.S.; Kjølby, M.; Hokland, M.; Winther, G.; Elfving, B.; Wegener, G.; Nielsen, A.L.; Larsen, A.; Nøhr, M.K.; et al. Chronic Maternal Inflammation or High-Fat-Feeding Programs Offspring Obesity in a Sex-Dependent Manner. *Int. J. Obes.* **2017**, *41*. [CrossRef]

114. Perrin, E.M.; O'Shea, T.M.; Skinner, A.C.; Bose, C.; Allred, E.N.; Fichorova, R.N.; van der Burg, J.W.; Leviton, A. Elevations of Inflammatory Proteins in Neonatal Blood Are Associated with Obesity and Overweight among 2-Year-Old Children Born Extremely Premature. *Pediatr. Res.* **2018**, *83*. [CrossRef]

115. Tsiotra, P.C.; Halvatsiotis, P.; Patsouras, K.; Maratou, E.; Salamalekis, G.; Raptis, S.A.; Dimitriadis, G.; Boutati, E. Circulating Adipokines and MRNA Expression in Adipose Tissue and the Placenta in Women with Gestational Diabetes Mellitus. *Peptides* **2018**, *101*. [CrossRef]

116. Zhang, J.; Chi, H.; Xiao, H.; Tian, X.; Wang, Y.; Yun, X.; Xu, Y. Interleukin 6 (IL-6) and Tumor Necrosis Factor α (TNF-α) Single Nucleotide Polymorphisms (SNPs), Inflammation and Metabolism in Gestational Diabetes Mellitus in Inner Mongolia. *Med. Sci. Monit.* **2017**, *23*. [CrossRef] [PubMed]

117. Angelo, A.G.S.; Neves, C.T.C.; Lobo, T.F.; Godoy, R.V.C.; Ono, É.; Mattar, R.; Daher, S. Monocyte Profile in Peripheral Blood of Gestational Diabetes Mellitus Patients. *Cytokine* **2018**, *107*. [CrossRef] [PubMed]

118. Lekva, T.; Michelsen, A.E.; Aukrust, P.; Paasche Roland, M.C.; Henriksen, T.; Bollerslev, J.; Ueland, T. CXC Chemokine Ligand 16 Is Increased in Gestational Diabetes Mellitus and Preeclampsia and Associated with Lipoproteins in Gestational Diabetes Mellitus at 5 Years Follow-Up. *Diabetes Vasc. Dis. Res.* **2017**, *14*. [CrossRef] [PubMed]

119. Hara, C.D.C.P.; França, E.L.; Fagundes, D.L.G.; de Queiroz, A.A.; Rudge, M.V.C.; Honorio-França, A.C.; Calderon, I.D.M.P. Characterization of Natural Killer Cells and Cytokines in Maternal Placenta and Fetus of Diabetic Mothers. *J. Immunol. Res.* **2016**, *2016*. [CrossRef] [PubMed]

120. Mrizak, I.; Grissa, O.; Henault, B.; Fekih, M.; Bouslema, A.; Boumaiza, I.; Zaouali, M.; Tabka, Z.; Khan, N.A. Placental Infiltration of Inflammatory Markers in Gestational Diabetic Women. *Gen. Physiol. Biophys.* **2014**, *33*. [CrossRef]

121. Tchirikov, M.; Schlabritz-Loutsevitch, N.; Maher, J.; Buchmann, J.; Naberezhnev, Y.; Winarno, A.S.; Seliger, G. Mid-Trimester Preterm Premature Rupture of Membranes (PPROM): Etiology, Diagnosis, Classification, International Recommendations of Treatment Options and Outcome. *J. Perinat. Med.* **2018**, *46*. [CrossRef]

122. Toniolo, A.; Cassani, G.; Puggioni, A.; Rossi, A.; Colombo, A.; Onodera, T.; Ferrannini, E. The Diabetes Pandemic and Associated Infections: Suggestions for Clinical Microbiology. *Rev. Med. Microbiol.* **2019**, *30*. [CrossRef] [PubMed]

123. Yu, J.; Zhou, Y.; Gui, J.; Li, A.Z.; Su, X.L.; Feng, L. Assessment of the Number and Function of Macrophages in the Placenta of Gestational Diabetes Mellitus Patients. *J. Huazhong Univ. Sci. Technol. Med. Sci.* **2013**, *33*. [CrossRef]

124. Zheng, L.; Li, C.; Qi, W.; Qiao, B.; Zhao, H.; Zhou, Y.; Lu, C. Expression of Macrophage Migration Inhibitory Factor Gene in Placenta Tissue and Its Correlation with Gestational Diabetes Mellitus. *Natl. Med. J. China* **2017**, *97*. [CrossRef]

125. Sisino, G.; Bouckenooghe, T.; Aurientis, S.; Fontaine, P.; Storme, L.; Vambergue, A. Diabetes during Pregnancy Influences Hofbauer Cells, a Subtype of Placental Macrophages, to Acquire a pro-Inflammatory Phenotype. *Biochim. Biophys. Acta Mol. Basis Dis.* **2013**, *1832*. [CrossRef] [PubMed]

126. Dasu, M.R.; Jialal, I. Free Fatty Acids in the Presence of High Glucose Amplify Monocyte Inflammation via Toll-like Receptors. *Am. J. Physiol. Endocrinol. Metab.* **2011**, *300*. [CrossRef]

127. Schliefsteiner, C.; Peinhaupt, M.; Kopp, S.; Lögl, J.; Lang-Olip, I.; Hiden, U.; Heinemann, A.; Desoye, G.; Wadsack, C. Human Placental Hofbauer Cells Maintain an Anti-Inflammatory M2 Phenotype despite the Presence of Gestational Diabetes Mellitus. *Front. Immunol.* **2017**, *8*. [CrossRef]

128. Stoikou, M.; Grimolizzi, F.; Giaglis, S.; Schäfer, G.; van Breda, S.V.; Hoesli, I.M.; Lapaire, O.; Huhn, E.A.; Hasler, P.; Rossi, S.W.; et al. Gestational Diabetes Mellitus Is Associated with Altered Neutrophil Activity. *Front. Immunol.* **2017**, *8*. [CrossRef]

129. Menegazzo, L.; Ciciliot, S.; Poncina, N.; Mazzucato, M.; Persano, M.; Bonora, B.; Albiero, M.; Vigili de Kreutzenberg, S.; Avogaro, A.; Fadini, G.P. NETosis Is Induced by High Glucose and Associated with Type 2 Diabetes. *Acta Diabetol.* **2015**, *52*. [CrossRef]

130. Sun, T.; Meng, F.; Zhao, H.; Yang, M.; Zhang, R.; Yu, Z.; Huang, X.; Ding, H.; Liu, J.; Zang, S. Elevated First-Trimester Neutrophil Count Is Closely Associated with the Development of Maternal Gestational Diabetes Mellitus and Adverse Pregnancy Outcomes. *Diabetes* **2020**, *69*. [CrossRef]

131. Vorobjeva, N.V.; Pinegin, B.V. Neutrophil Extracellular Traps: Mechanisms of Formation and Role in Health and Disease. *Biochemistry* **2014**, *79*. [CrossRef]

132. Olmos-Ortiz, A.; Flores-Espinosa, P.; Mancilla-Herrera, I.; Vega-Sánchez, R.; Díaz, L.; Zaga-Clavellina, V. Innate Immune Cells and Toll-like Receptor–Dependent Responses at the Maternal–Fetal Interface. *Int. J. Mol. Sci.* **2019**, *20*, 3654. [CrossRef]

133. Shmeleva, E.V.; Colucci, F. Maternal Natural Killer Cells at the Intersection between Reproduction and Mucosal Immunity. *Mucosal Immunol.* **2021**. [CrossRef]

134. Lobo, T.F.; Borges, C.D.M.; Mattar, R.; Gomes, C.P.; de Angelo, A.G.S.; Pendeloski, K.P.T.; Daher, S. Impaired Treg and NK Cells Profile in Overweight Women with Gestational Diabetes Mellitus. *Am. J. Reprod. Immunol.* **2018**, *79*. [CrossRef]

135. Catalano, P.M.; Huston, L.; Amini, S.B.; Kalhan, S.C. Longitudinal Changes in Glucose Metabolism during Pregnancy in Obese Women with Normal Glucose Tolerance and Gestational Diabetes Mellitus. *Am. J. Obstet. Gynecol.* **1999**, *180*. [CrossRef]

136. Atègbo, J.M.; Grissa, O.; Yessoufou, A.; Hichami, A.; Dramane, K.L.; Moutairou, K.; Miled, A.; Grissa, A.; Jerbi, M.; Tabka, Z.; et al. Modulation of Adipokines and Cytokines in Gestational Diabetes and Macrosomia. *J. Clin. Endocrinol. Metab.* **2006**, *91*. [CrossRef] [PubMed]

137. Bari, M.F.; Weickert, M.O.; Sivakumar, K.; James, S.G.; Snead, D.R.J.; Tan, B.K.; Randeva, H.S.; Bastie, C.C.; Vatish, M. Elevated Soluble CD163 in Gestational Diabetes Mellitus: Secretion from Human Placenta and Adipose Tissue. *PLoS ONE* **2014**, *9*, e101327. [CrossRef] [PubMed]

138. Lacroix, M.; Lizotte, F.; Hivert, M.-F.; Geraldes, P.; Perron, P. Calcifediol Decreases Interleukin-6 Secretion by Cultured Human Trophoblasts from GDM Pregnancies. *J. Endocr. Soc.* **2019**, *3*. [CrossRef]
139. Radaelli, T.; Varastehpour, A.; Catalano, P.; Hauguel-De Mouzon, S. Gestational Diabetes Induces Placental Genes for Chronic Stress and Inflammatory Pathways. *Diabetes* **2003**, *52*. [CrossRef] [PubMed]
140. Li, J.; Yin, Q.; Wu, H. Structural basis of signal transduction in the TNF receptor superfamily. *Adv. Immunol.* **2013**, *119*. [CrossRef]
141. Mitchell, S.; Vargas, J.; Hoffmann, A. Signaling via the NFκB System. *Wiley Interdiscip. Rev. Syst. Biol. Med.* **2016**, *8*. [CrossRef] [PubMed]
142. Lumeng, C.N. Innate Immune Activation in Obesity. *Mol. Asp. Med.* **2013**, *34*. [CrossRef]
143. Feng, H.; Su, R.; Song, Y.; Wang, C.; Lin, L.; Ma, J.; Yang, H. Positive Correlation between Enhanced Expression of TLR4/MyD88/NF-KB with Insulin Resistance in Placentae of Gestational Diabetes Mellitus. *PLoS ONE* **2016**, *11*, e0157185. [CrossRef]
144. Corrêa-Silva, S.; Alencar, A.P.; Moreli, J.B.; Borbely, A.U.; Lima, L.D.S.; Scavone, C.; Damasceno, D.C.; Rudge, M.V.C.; Bevilacqua, E.; Calderon, I.M.P. Hyperglycemia Induces Inflammatory Mediators in the Human Chorionic Villous. *Cytokine* **2018**, *111*. [CrossRef]
145. Kirwan, J.P.; Hauguel-De Mouzon, S.; Lepercq, J.; Challier, J.C.; Huston-Presley, L.; Friedman, J.E.; Kalhan, S.C.; Catalano, P.M. TNF-α Is a Predictor of Insulin Resistance in Human Pregnancy. *Diabetes* **2002**, *51*. [CrossRef]
146. Friis, C.M.; Roland, M.C.P.; Godang, K.; Ueland, T.; Tanbo, T.; Bollerslev, J.; Henriksen, T. Adiposity-Related Inflammation: Effects of Pregnancy. *Obesity* **2013**, *21*. [CrossRef]
147. Vega-Sanchez, R.; Barajas-Vega, H.A.; Rozada, G.; Espejel-Nuñez, A.; Beltran-Montoya, J.; Vadillo-Ortega, F. Association between Adiposity and Inflammatory Markers in Maternal and Fetal Blood in a Group of Mexican Pregnant Women. *Br. J. Nutr.* **2010**, *104*. [CrossRef]
148. Aye, I.L.M.H.; Lager, S.; Ramirez, V.I.; Gaccioli, F.; Dudley, D.J.; Jansson, T.; Powell, T.L. Increasing Maternal Body Mass Index Is Associated with Systemic Inflammation in the Mother and the Activation of Distinct Placental Inflammatory Pathways. *Biol. Reprod.* **2014**, *90*. [CrossRef]
149. Mohammed, A.; Aliyu, I.S. Maternal Serum Level of TNF-α in Nigerian Women with Gestational Diabetes Mellitus. *Pan Afr. Med. J.* **2018**, *31*. [CrossRef]
150. Catalano, P.M. Obesity, Insulin Resistance, and Pregnancy Outcome. *Reproduction* **2010**, *140*. [CrossRef]
151. Carey, A.L.; Febbraio, M.A. Interleukin-6 and Insulin Sensitivity: Friend or Foe? *Diabetologia* **2004**, *47*. [CrossRef]
152. Challier, J.C.; Basu, S.; Bintein, T.; Minium, J.; Hotmire, K.; Catalano, P.M.; Hauguel-de Mouzon, S. Obesity in Pregnancy Stimulates Macrophage Accumulation and Inflammation in the Placenta. *Placenta* **2008**, *29*. [CrossRef]
153. Stirm, L.; Kovářová, M.; Perschbacher, S.; Michlmaier, R.; Fritsche, L.; Siegel-Axel, D.; Schleicher, E.; Peter, A.; Pauluschke-Fröhlich, J.; Brucker, S.; et al. BMI-Independent Effects of Gestational Diabetes on Human Placenta. *J. Clin. Endocrinol. Metab.* **2018**, *103*. [CrossRef] [PubMed]
154. Han, C.S.; Herrin, M.A.; Pitruzzello, M.C.; Mulla, M.J.; Werner, E.F.; Pettker, C.M.; Flannery, C.A.; Abrahams, V.M. Glucose and Metformin Modulate Human First Trimester Trophoblast Function: A Model and Potential Therapy for Diabetes-Associated Uteroplacental Insufficiency. *Am. J. Reprod. Immunol.* **2015**, *73*. [CrossRef] [PubMed]
155. Schulze, F.; Wehner, J.; Kratschmar, D.V.; Makshana, V.; Meier, D.T.; Häuselmann, S.P.; Dalmas, E.; Thienel, C.; Dror, E.; Wiedemann, S.J.; et al. Inhibition of IL-1beta Improves Glycaemia in a Mouse Model for Gestational Diabetes. *Sci. Rep.* **2020**, *10*. [CrossRef]
156. Peng, H.Y.; Li, M.Q.; Li, H.P. High Glucose Suppresses the Viability and Proliferation of HTR-8/SVneo Cells through Regulation of the MiR-137/PRKAA1/IL-6 Axis. *Int. J. Mol. Med.* **2018**, *42*. [CrossRef]
157. Cawyer, C.; Afroze, S.H.; Drever, N.; Allen, S.; Jones, R.; Zawieja, D.C.; Kuehl, T.; Uddin, M.N. Attenuation of Hyperglycemia-Induced Apoptotic Signaling and Anti-Angiogenic Milieu in Cultured Cytotrophoblast Cells. *Hypertens. Pregnancy* **2016**, *35*. [CrossRef] [PubMed]
158. Rice, G.E.; Scholz-Romero, K.; Sweeney, E.; Peiris, H.; Kobayashi, M.; Duncombe, G.; Mitchell, M.D.; Salomon, C. The Effect of Glucose on the Release and Bioactivity of Exosomes from First Trimester Trophoblast Cells. *J. Clin. Endocrinol. Metab.* **2015**, *100*. [CrossRef]
159. Troncoso, F.; Acurio, J.; Herlitz, K.; Aguayo, C.; Bertoglia, P.; Guzman-Gutierrez, E.; Loyola, M.; Gonzalez, M.; Rezgaoui, M.; Desoye, G.; et al. Gestational Diabetes Mellitus Is Associated with Increased Pro-Migratory Activation of Vascular Endothelial Growth Factor Receptor 2 and Reduced Expression of Vascular Endothelial Growth Factor Receptor 1. *PLoS ONE* **2017**, *12*, e0182509. [CrossRef]
160. Loegl, J.; Nussbaumer, E.; Cvitic, S.; Huppertz, B.; Desoye, G.; Hiden, U. GDM Alters Paracrine Regulation of Feto-Placental Angiogenesis via the Trophoblast. *Lab. Investig.* **2017**, *97*. [CrossRef]
161. Landecho, M.F.; Tuero, C.; Valentí, V.; Bilbao, I.; de la Higuera, M.; Frühbeck, G. Relevance of Leptin and Other Adipokines in Obesity-Associated Cardiovascular Risk. *Nutrients* **2019**, *11*, 2664. [CrossRef]
162. Pérez-Pérez, A.; Maymó, J.; Gambino, Y.; Guadix, P.; Dueñas, J.L.; Varone, C.; Sánchez-Margalet, V. Insulin Enhances Leptin Expression in Human Trophoblastic Cells. *Biol. Reprod.* **2013**, *89*. [CrossRef] [PubMed]
163. Pérez-Pérez, A.; Toro, A.; Vilariño-García, T.; Maymó, J.; Guadix, P.; Dueñas, J.L.; Fernández-Sánchez, M.; Varone, C.; Sánchez-Margalet, V. Leptin Action in Normal and Pathological Pregnancies. *J. Cell. Mol. Med.* **2018**, *22*. [CrossRef]

164. Maymó, J.L.; Pérez Pérez, A.; Maskin, B.; Dueñas, J.L.; Calvo, J.C.; Sánchez Margalet, V.; Varone, C.L. The Alternative Epac/CAMP Pathway and the MAPK Pathway Mediate HCG Induction of Leptin in Placental Cells. *PLoS ONE* **2012**, *7*, e46216. [CrossRef]

165. Gambino, Y.P.; Pérez Pérez, A.; Dueñas, J.L.; Calvo, J.C.; Sánchez-Margalet, V.; Varone, C.L. Regulation of Leptin Expression by 17beta-Estradiol in Human Placental Cells Involves Membrane Associated Estrogen Receptor Alpha. *Biochim. Biophys. Acta Mol. Cell Res.* **2012**, *1823*. [CrossRef] [PubMed]

166. Magariños, M.P.; Sánchez-Margalet, V.; Kotler, M.; Calvo, J.C.; Varone, C.L. Leptin Promotes Cell Proliferation and Survival of Trophoblastic Cells. *Biol. Reprod.* **2007**, *76*. [CrossRef] [PubMed]

167. White, V.; González, E.; Capobianco, E.; Pustovrh, C.; Martínez, N.; Higa, R.; Baier, M.; Jawerbaum, A. Leptin Modulates Nitric Oxide Production and Lipid Metabolism in Human Placenta. *Reprod. Fertil. Dev.* **2006**, *18*. [CrossRef] [PubMed]

168. Côté, S.; Gagné-Ouellet, V.; Guay, S.P.; Allard, C.; Houde, A.A.; Perron, P.; Baillargeon, J.P.; Gaudet, D.; Guérin, R.; Brisson, D.; et al. PPARGC1α Gene DNA Methylation Variations in Human Placenta Mediate the Link between Maternal Hyperglycemia and Leptin Levels in Newborns. *Clin. Epigenet.* **2016**, *8*. [CrossRef] [PubMed]

169. Soheilykhah, S.; Mojibian, M.; Rahimi-Saghand, S.; Rashidi, M.; Hadinedoushan, H. Maternal Serum Leptin Concentration in Gestational Diabetes. *Taiwan. J. Obstet. Gynecol.* **2011**, *50*. [CrossRef]

170. Chen, D.; Xia, G.; Xu, P.; Dong, M. Peripartum Serum Leptin and Soluble Leptin Receptor Levels in Women with Gestational Diabetes. *Acta Obstet. Gynecol. Scand.* **2010**, *89*. [CrossRef]

171. Mokhtari, M.; Hashemi, M.; Yaghmaei, M.; Naderi, M.; Shikhzadeh, A.; Ghavami, S. Evaluation of the Serum Leptin in Normal Pregnancy and Gestational Diabetes Mellitus in Zahedan, Southeast Iran. *Arch. Gynecol. Obstet.* **2011**, *284*. [CrossRef]

172. Lepercq, J.; Cauzac, M.; Lahlou, N.; Timsit, J.; Girard, J.; Auwerx, J.; de Mouzon, S.H. Overexpression of Placental Leptin in Diabetic Pregnancy: A Critical Role for Insulin. *Diabetes* **1998**, *47*. [CrossRef]

173. Lappas, M.; Permezel, M.; Rice, G.E. Leptin and Adiponectin Stimulate the Release of Proinflammatory Cytokines and Prostaglandins from Human Placenta and Maternal Adipose Tissue via Nuclear Factor-KB, Peroxisomal Proliferator-Activated Receptor-γ and Extracellularly Regulated Kinase 1/2. *Endocrinology* **2005**, *146*. [CrossRef]

174. Cameo, P.; Bischof, P.; Calvo, J.C. Effect of Leptin on Progesterone, Human Chorionic Gonadotropin, and Interleukin-6 Secretion by Human Term Trophoblast Cells in Culture. *Biol. Reprod.* **2003**, *68*. [CrossRef] [PubMed]

175. Soh, E.B.E.; Mitchell, M.D.; Keelan, J.A. Does Leptin Exhibit Cytokine-like Properties in Tissues of Pregnancy? *Am. J. Reprod. Immunol.* **2000**, *43*. [CrossRef]

176. Miehle, K.; Stepan, H.; Fasshauer, M. Leptin, Adiponectin and Other Adipokines in Gestational Diabetes Mellitus and Pre-Eclampsia. *Clin. Endocrinol.* **2012**, *76*. [CrossRef] [PubMed]

177. De Gennaro, G.; Palla, G.; Battini, L.; Simoncini, T.; del Prato, S.; Bertolotto, A.; Bianchi, C. The Role of Adipokines in the Pathogenesis of Gestational Diabetes Mellitus. *Gynecol. Endocrinol.* **2019**, *35*. [CrossRef]

178. Retnakaran, R.; Qi, Y.; Connelly, P.W.; Sermer, M.; Hanley, A.J.; Zinman, B. Low Adiponectin Concentration during Pregnancy Predicts Postpartum Insulin Resistance, Beta Cell Dysfunction and Fasting Glycaemia. *Diabetologia* **2010**, *53*. [CrossRef] [PubMed]

179. Williams, M.A.; Qiu, C.; Muy-Rivera, M.; Vadachkoria, S.; Song, T.; Luthy, D.A. Plasma Adiponectin Concentrations in Early Pregnancy and Subsequent Risk of Gestational Diabetes Mellitus. *J. Clin. Endocrinol. Metab.* **2004**, *89*. [CrossRef]

180. Ramachandrayya, S.A.; D'Cunha, P.; Rebeiro, C. Maternal Circulating Levels of Adipocytokines and Insulin Resistance as Predictors of Gestational Diabetes Mellitus: Preliminary Findings of a Longitudinal Descriptive Study. *J. Diabetes Metab. Disord.* **2020**, *19*. [CrossRef]

181. Chen, J.; Tan, B.; Karteris, E.; Zervou, S.; Digby, J.; Hillhouse, E.W.; Vatish, M.; Randeva, H.S. Secretion of Adiponectin by Human Placenta: Differential Modulation of Adiponectin and Its Receptors by Cytokines. *Diabetologia* **2006**, *49*. [CrossRef]

182. Benaitreau, D.; dos Santos, E.; Leneveu, M.C.; Alfaidy, N.; Feige, J.J.; de Mazancourt, P.; Pecquery, R.; Dieudonné, M.N. Effects of Adiponectin on Human Trophoblast Invasion. *J. Endocrinol.* **2010**, *207*. [CrossRef] [PubMed]

183. Bouchard, L.; Hivert, M.F.; Guay, S.P.; St-Pierre, J.; Perron, P.; Brisson, D. Placental Adiponectin Gene DNA Methylation Levels Are Associated with Mothers' Blood Glucose Concentration. *Diabetes* **2012**, *61*. [CrossRef] [PubMed]

184. Dias, S.; Adam, S.; Abrahams, Y.; Rheeder, P.; Pheiffer, C. Adiponectin DNA Methylation in South African Women with Gestational Diabetes Mellitus: Effects of HIV Infection. *PLoS ONE* **2021**, *16*, e0248694. [CrossRef] [PubMed]

185. King, A.E.; Paltoo, A.; Kelly, R.W.; Sallenave, J.M.; Bocking, A.D.; Challis, J.R.G. Expression of Natural Antimicrobials by Human Placenta and Fetal Membranes. *Placenta* **2007**, *28*. [CrossRef]

186. Yarbrough, V.L.; Winkle, S.; Herbst-Kralovetz, M.M. Antimicrobial Peptides in the Female Reproductive Tract: A Critical Component of the Mucosal Immune Barrier with Physiological and Clinical Implications. *Hum. Reprod. Update* **2015**, *21*. [CrossRef]

187. Zaga-Clavellina, V.; Garcia-Lopez, G.; Flores-Espinosa, P. Evidence of in Vitro Differential Secretion of Human Beta-Defensins-1, -2, and -3 after Selective Exposure to Streptococcus Agalactiae in Human Fetal Membranes. *J. Matern. Fetal Neonatal Med.* **2012**, *25*. [CrossRef]

188. Zaga-Clavellina, V.; Martha, R.V.M.; Flores-Espinosa, P. In Vitro Secretion Profile of Pro-Inflammatory Cytokines IL-1β, TNF-α, IL-6, and of Human Beta-Defensins (HBD)-1, HBD-2, and HBD-3 from Human Chorioamniotic Membranes after Selective Stimulation with Gardnerella Vaginalis. *Am. J. Reprod. Immunol.* **2012**, *67*. [CrossRef]

189. Liu, N.; Kaplan, A.T.; Low, J.; Nguyen, L.; Liu, G.Y.; Equils, O.; Hewison, M. Vitamin D Induces Innate Antibacterial Responses in Human Trophoblasts via an Intracrine Pathway 1. *Biol. Reprod.* **2009**, *80*. [CrossRef]

190. Klaffenbach, D.; Friedrich, D.; Strick, R.; Strissel, P.L.; Beckmann, M.W.; Rascher, W.; Gessner, A.; Dötsch, J.; Meißner, U.; Schnare, M. Contribution of Different Placental Cells to the Expression and Stimulation of Antimicrobial Proteins (AMPs). *Placenta* **2011**, *32*. [CrossRef]

191. Svinarich, D.M.; Gomez, R.; Romero, R. Detection of Human Defensins in the Placenta. *Am. J. Reprod. Immunol.* **1997**, *38*. [CrossRef]

192. Kim, H.S.; Cho, J.H.; Park, H.W.; Yoon, H.; Kim, M.S.; Kim, S.C. Endotoxin-Neutralizing Antimicrobial Proteins of the Human Placenta. *J. Immunol.* **2002**, *168*. [CrossRef]

193. Oliveira-Bravo, M.; Sangiorgi, B.B.; Schiavinato, J.L.D.S.; Carvalho, J.L.; Covas, D.T.; Panepucci, R.A.; Neves, F.D.A.R.; Franco, O.L.; Pereira, R.W.; Saldanha-Araujo, F. LL-37 Boosts Immunosuppressive Function of Placenta-Derived Mesenchymal Stromal Cells. *Stem Cell Res. Ther.* **2016**, *7*. [CrossRef]

194. Szukiewicz, D.; Alkhalayla, H.; Pyzlak, M.; Szewczyk, G. High Glucose Culture Medium Downregulates Production of Human β-Defensin-2 (HBD-2) in Human Amniotic Epithelial Cells (HAEC). *FASEB J.* **2016**, *30*, 307. Available online: https://faseb.onlinelibrary.wiley.com/doi/abs/10.1096/fasebj.30.1_supplement.307.1?cited-by=yes&legid=fasebj%3B30%2F1_Supplement%2F307.1 (accessed on 1 June 2021).

195. Montoya-Rosales, A.; Castro-Garcia, P.; Torres-Juarez, F.; Enciso-Moreno, J.A.; Rivas-Santiago, B. Glucose Levels Affect LL-37 Expression in Monocyte-Derived Macrophages Altering the Mycobacterium Tuberculosis Intracellular Growth Control. *Microb. Pathog.* **2016**, *97*. [CrossRef] [PubMed]

196. Froy, O.; Hananel, A.; Chapnik, N.; Madar, Z. Differential Effect of Insulin Treatment on Decreased Levels of Beta-Defensins and Toll-like Receptors in Diabetic Rats. *Mol. Immunol.* **2007**, *44*. [CrossRef]

197. Rivas-Santiago, B.; Trujillo, V.; Montoya, A.; Gonzalez-Curiel, I.; Castañeda-Delgado, J.; Cardenas, A.; Rincon, K.; Hernandez, M.L.; Hernández-Pando, R. Expression of Antimicrobial Peptides in Diabetic Foot Ulcer. *J. Dermatol. Sci.* **2012**, *65*. [CrossRef]

198. Olmos-Ortiz, A.; García-Quiroz, J.; Avila, E.; Caldiño-Soto, F.; Halhali, A.; Larrea, F.; Díaz, L. Lipopolysaccharide and CAMP Modify Placental Calcitriol Biosynthesis Reducing Antimicrobial Peptides Gene Expression. *Am. J. Reprod. Immunol.* **2018**, *79*. [CrossRef]

199. Barrera, D.; Díaz, L.; Noyola-Martínez, N.; Halhali, A. Vitamin D and Inflammatory Cytokines in Healthy and Preeclamptic Pregnancies. *Nutrients* **2015**, *7*, 5293. [CrossRef]

200. Díaz, L.; Noyola-Martínez, N.; Barrera, D.; Hernández, G.; Avila, E.; Halhali, A.; Larrea, F. Calcitriol Inhibits TNF-α-Induced Inflammatory Cytokines in Human Trophoblasts. *J. Reprod. Immunol.* **2009**, *81*. [CrossRef] [PubMed]

201. Wang, Y.; Wang, T.; Huo, Y.; Liu, L.; Liu, S.; Yin, X.; Wang, R.; Gao, X. Placenta Expression of Vitamin D and Related Genes in Pregnant Women with Gestational Diabetes Mellitus. *J. Steroid Biochem. Mol. Biol.* **2020**, *204*. [CrossRef]

202. Walker, V.P.; Zhang, X.; Rastegar, I.; Liu, P.T.; Hollis, B.W.; Adams, J.S.; Modlin, R.L. Cord Blood Vitamin D Status Impacts Innate Immune Responses. *J. Clin. Endocrinol. Metab.* **2011**, *96*. [CrossRef]

203. Nargesi, S.; Ghorbani, A.; Shirzadpour, E.; Mohamadpour, M.; Mousavi, S.F.; Amraei, M. A Systematic Review and Meta-Analysis of the Association between Vitamin D Deficiency and Gestational Diabetes Mellitus. *Biomed. Res. Ther.* **2018**, *5*. [CrossRef]

204. Maghbooli, Z.; Hossein-Nezhad, A.; Karimi, F.; Shafaei, A.R.; Larijani, B. Correlation between Vitamin D3 Deficiency and Insulin Resistance in Pregnancy. *Diabetes Metab. Res. Rev.* **2008**, *24*. [CrossRef] [PubMed]

205. Olmos-Ortiz, A.; Avila, E.; Durand-Carbajal, M.; Díaz, L. Regulation of Calcitriol Biosynthesis and Activity: Focus on Gestational Vitamin D Deficiency and Adverse Pregnancy Outcomes. *Nutrients* **2015**, *7*, 443. [CrossRef] [PubMed]

206. Cho, G.J.; Hong, S.C.; Oh, M.J.; Kim, H.J. Vitamin D Deficiency in Gestational Diabetes Mellitus and the Role of the Placenta. *Am. J. Obstet. Gynecol.* **2013**, *209*. [CrossRef] [PubMed]

207. Akoh, C.C.; Pressman, E.K.; Whisner, C.M.; Thomas, C.; Cao, C.; Kent, T.; Cooper, E.; O'Brien, K.O. Vitamin D Mediates the Relationship between Placental Cathelicidin and Group B Streptococcus Colonization during Pregnancy. *J. Reprod. Immunol.* **2017**, *121*. [CrossRef]

208. Akoh, C.C.; Pressman, E.K.; Cooper, E.; Queenan, R.A.; Pillittere, J.; O'Brien, K.O. Low Vitamin D Is Associated with Infections and Proinflammatory Cytokines During Pregnancy. *Reprod. Sci.* **2018**, *25*. [CrossRef]

209. Norman, A.W.; Frankel, B.J.; Heldt, A.M.; Grodsky, G.M. Vitamin D Deficiency Inhibits Pancreatic Secretion of Insulin. *Science* **1980**, *209*. [CrossRef]

210. Maestro, B.; Dávila, N.; Carranza, M.C.; Calle, C. Identification of a Vitamin D Response Element in the Human Insulin Receptor Gene Promoter. *J. Steroid Biochem. Mol. Biol.* **2003**, *84*, 223–230. [CrossRef]

211. Maestro, B.; Molero, S.; Bajo, S.; Dávila, N.; Calle, C. Transcriptional Activation of the Human Insulin Receptor Gene by 1,25-Dihydroxyvitamin D3. *Cell Biochem. Funct.* **2002**, *20*, 227–232. [CrossRef]

212. Li, G.; Lin, L.; Wang, Y.L.; Yang, H. 1,25(OH)2D3 Protects Trophoblasts Against Insulin Resistance and Inflammation Via Suppressing MTOR Signaling. *Reprod. Sci.* **2019**, *26*. [CrossRef]

213. Vimpeli, T.; Huhtala, H.; Wilsgaard, T.; Acharya, G. Fetal Cardiac Output and Its Distribution to the Placenta at 11–20 Weeks of Gestation. *Ultrasound Obstet. Gynecol.* **2009**, *33*. [CrossRef]

214. Sharfuddin, A.A.; Molitoris, B.A. Pathophysiology of Acute Kidney Injury. In *Seldin and Giebisch's the Kidney*, 4th ed.; Academic Press: San Diego, CA, USA, 2008.

215. Lautt, W.W. *Hepatic Circulation Physiology and Pathophysiology*; Morgan & Claypool Life Sciences: San Rafael, CA, USA, 2009; Volume 1.

216. Zucchelli, E.; Majid, Q.A.; Foldes, G. New Artery of Knowledge: 3D Models of Angiogenesis. *Vasc. Biol.* **2020**, *1*. [CrossRef] [PubMed]
217. Kolte, D.; McClung, J.A.; Aronow, W.S. Vasculogenesis and Angiogenesis. In *Translational Research in Coronary Artery Disease: Pathophysiology to Treatment*; Academic Press: Cambridge, MA, USA, 2016.
218. Clapp, C.; Thebault, S.; Jeziorski, M.C.; Martínez De La Escalera, G. Peptide Hormone Regulation of Angiogenesis. *Physiol. Rev.* **2009**, *89*. [CrossRef] [PubMed]
219. Lamalice, L.; le Boeuf, F.; Huot, J. Endothelial Cell Migration during Angiogenesis. *Circ. Res.* **2007**, *100*, 782–794. [CrossRef]
220. Ferrara, N. Vascular Endothelial Growth Factor: Basic Science and Clinical Progress. *Endocr. Rev.* **2004**, *25*, 581–611. [CrossRef] [PubMed]
221. Levy, A.P.; Levy, N.S.; Wegner, S.; Goldberg, M.A. Transcriptional Regulation of the Rat Vascular Endothelial Growth Factor Gene by Hypoxia. *J. Biol. Chem.* **1995**, *270*. [CrossRef]
222. Gleadle, J.M.; Ratcliffe, P.J. Induction of Hypoxia-Inducible Factor-1, Erythropoietin, Vascular Endothelial Growth Factor, and Glucose Transporter-1 by Hypoxia: Evidence against a Regulatory Role for Src Kinase. *Blood* **1997**, *89*. [CrossRef]
223. Leach, L.; Babawale, M.O.; Anderson, M.; Lammiman, M. Vasculogenesis, Angiogenesis and the Molecular Organisation of Endothelial Junctions in the Early Human Placenta. *J. Vasc. Res.* **2002**, *39*. [CrossRef] [PubMed]
224. Demir, R.; Kayisli, U.A.; Cayli, S.; Huppertz, B. Sequential Steps During Vasculogenesis and Angiogenesis in the Very Early Human Placenta. *Placenta* **2006**, *27*. [CrossRef]
225. Omorphos, N.P.; Gao, C.; Tan, S.S.; Sangha, M.S. Understanding Angiogenesis and the Role of Angiogenic Growth Factors in the Vascularisation of Engineered Tissues. *Mol. Biol. Rep.* **2021**, *48*. [CrossRef]
226. Edatt, L.; Poyyakkara, A.; Raji, G.R.; Ramachandran, V.; Shankar, S.S.; Kumar, V.B.S. Role of Sirtuins in Tumor Angiogenesis. *Front. Oncol.* **2020**, *9*. [CrossRef] [PubMed]
227. Bach, L.A. Endothelial Cells and the IGF System. *J. Mol. Endocrinol.* **2015**, *54*. [CrossRef] [PubMed]
228. Slater, T.; Haywood, N.J.; Matthews, C.; Cheema, H.; Wheatcroft, S.B. Insulin-like Growth Factor Binding Proteins and Angiogenesis: From Cancer to Cardiovascular Disease. *Cytokine Growth Factor Rev.* **2019**, *46*. [CrossRef]
229. Albonici, L.; Benvenuto, M.; Focaccetti, C.; Cifaldi, L.; Miele, M.T.; Limana, F.; Manzari, V.; Bei, R. Plgf Immunological Impact during Pregnancy. *Int. J. Mol. Sci.* **2020**, *21*, 8714. [CrossRef]
230. Gobble, R.M.; Groesch, K.A.; Chang, M.; Torry, R.J.; Torry, D.S. Differential Regulation of Human PlGF Gene Expression in Trophoblast and Nontrophoblast Cells by Oxygen Tension. *Placenta* **2009**, *30*. [CrossRef] [PubMed]
231. Yinon, Y.; Nevo, O.; Xu, J.; Many, A.; Rolfo, A.; Todros, T.; Post, M.; Caniggia, I. Severe Intrauterine Growth Restriction Pregnancies Have Increased Placental Endoglin Levels: Hypoxic Regulation via Transforming Growth Factor-B3. *Am. J. Pathol.* **2008**, *172*. [CrossRef] [PubMed]
232. Pérez-Roque, L.; Núñez-Gómez, E.; Rodríguez-Barbero, A.; Bernabéu, C.; López-Novoa, J.M.; Pericacho, M. Pregnancy-Induced High Plasma Levels of Soluble Endoglin in Mice Lead to Preeclampsia Symptoms and Placental Abnormalities. *Int. J. Mol. Sci.* **2021**, *22*, 165. [CrossRef] [PubMed]
233. Ettelaie, C.; Su, S.; Li, C.; Collier, M.E.W. Tissue Factor-Containing Microparticles Released from Mesangial Cells in Response to High Glucose and AGE Induce Tube Formation in Microvascular Cells. *Microvasc. Res.* **2008**, *76*. [CrossRef]
234. Chen, Y.L.; Rosa, R.H.; Kuo, L.; Hein, T.W. Hyperglycemia Augments Endothelin-1–Induced Constriction of Human Retinal Venules. *Transl. Vis. Sci. Technol.* **2020**, *9*. [CrossRef]
235. Martin-Aragon Baudel, M.; Espinosa-Tanguma, R.; Nieves-Cintron, M.; Navedo, M.F. Purinergic Signaling During Hyperglycemia in Vascular Smooth Muscle Cells. *Front. Endocrinol.* **2020**, *11*. [CrossRef]
236. Clyne, A.M. Endothelial Response to Glucose: Dysfunction, Metabolism, and Transport. *Biochem. Soc. Trans.* **2021**, *49*. [CrossRef] [PubMed]
237. Guo, Y.; Dong, L.; Gong, A.; Zhang, J.; Jing, L.; Ding, T.; Li, P.-A.; Zhang, J.-Z. Damage to the Blood-brain Barrier and Activation of Neuroinflammation by Focal Cerebral Ischemia under Hyperglycemic Condition. *Int. J. Mol. Med.* **2021**, *48*. [CrossRef] [PubMed]
238. Babawale, M.O.; Lovat, S.; Mayhew, T.M.; Lammiman, M.J.; James, D.K.; Leach, L. Effects of Gestational Diabetes on Junctional Adhesion Molecules in Human Term Placental Vasculature. *Diabetologia* **2000**, *43*. [CrossRef] [PubMed]
239. Leach, L. Placental Vascular Dysfunction in Diabetic Pregnancies: Intimations of Fetal Cardiovascular Disease? *Microcirculation* **2011**, *18*. [CrossRef]
240. Chang, S.C.; Vivian Yang, W.C. Hyperglycemia Induces Altered Expressions of Angiogenesis Associated Molecules in the Trophoblast. *Evid. Based Complement. Altern. Med.* **2013**, *2013*. [CrossRef]
241. Li, H.P.; Chen, X.; Li, M.Q. Gestational Diabetes Induces Chronic Hypoxia Stress and Excessive Inflammatory Response in Murine Placenta. *Int. J. Clin. Exp. Pathol.* **2013**, *6*, 650–659. [PubMed]
242. Cawyer, C.R.; Horvat, D.; Leonard, D.; Allen, S.R.; Jones, R.O.; Zawieja, D.C.; Kuehl, T.J.; Uddin, M.N. Hyperglycemia Impairs Cytotrophoblast Function via Stress Signaling. *Am. J. Obstet. Gynecol.* **2014**, *211*. [CrossRef]
243. Meng, Q.; Shao, L.; Luo, X.; Mu, Y.; Xu, W.; Gao, L.; Xu, H.; Cui, Y. Expressions of VEGF-A and VEGFR-2 in Placentae from GDM Pregnancies. *Reprod. Biol. Endocrinol.* **2016**, *14*. [CrossRef] [PubMed]
244. Pietro, L.; Daher, S.; Rudge, M.V.C.; Calderon, I.M.P.; Damasceno, D.C.; Sinzato, Y.K.; Bandeira, C.; Bevilacqua, E. Vascular Endothelial Growth Factor (VEGF) and VEGF-Receptor Expression in Placenta of Hyperglycemic Pregnant Women. *Placenta* **2010**, *31*. [CrossRef] [PubMed]

245. Peng, H.Y.; Li, H.P.; Li, M.Q. High Glucose Induces Dysfunction of Human Umbilical Vein Endothelial Cells by Upregulating MiR-137 in Gestational Diabetes Mellitus. *Microvasc. Res.* **2018**, *118*. [CrossRef]

246. Korbecki, J.; Kojder, K.; Kapczuk, P.; Kupnicka, P.; Gawrońska-Szklarz, B.; Gutowska, I.; Chlubek, D.; Baranowska-Bosiacka, I. The Effect of Hypoxia on the Expression of CXC Chemokines and CXC Chemokine Receptors-A Review of Literature. *Int. J. Mol. Sci.* **2021**, *22*, 843. [CrossRef] [PubMed]

247. Bassand, K.; Metzinger, L.; Naïm, M.; Mouhoubi, N.; Haddad, O.; Assoun, V.; Zaïdi, N.; Sainte-Catherine, O.; Butt, A.; Guyot, E.; et al. MiR-126-3p Is Essential for CXCL12-Induced Angiogenesis. *J. Cell. Mol. Med.* **2021**, *25*, 6032–6045. [CrossRef] [PubMed]

248. Ma, C.; Liu, G.; Liu, W.; Xu, W.; Li, H.; Piao, S.; Sui, Y.; Feng, W. CXCL1 Stimulates Decidual Angiogenesis via the VEGF-A Pathway during the First Trimester of Pregnancy. *Mol. Cell. Biochem.* **2021**. [CrossRef]

249. Darakhshan, S.; Fatehi, A.; Hassanshahi, G.; Mahmoodi, S.; Hashemi, M.S.; Karimabad, M.N. Serum Concentration of Angiogenic (CXCL1, CXCL12) and Angiostasis (CXCL9, CXCL10) CXC Chemokines Are Differentially Altered in Normal and Gestational Diabetes Mellitus Associated Pregnancies. *J. Diabetes Metab. Disord.* **2019**, *18*. [CrossRef]

250. Hiden, U.; Lassance, L.; Ghaffari Tabrizi, N.; Miedl, H.; Tam-Amersdorfer, C.; Cetin, I.; Lang, U.; Desoye, G. Fetal Insulin and IGF-II Contribute to Gestational Diabetes Mellitus (GDM)-Associated up-Regulation of Membrane-Type Matrix Metalloproteinase 1 (MT1-MMP) in the Human Feto-Placental Endothelium. *J. Clin. Endocrinol. Metab.* **2012**, *97*. [CrossRef] [PubMed]

251. Majali-Martinez, A.; Hiden, U.; Ghaffari-Tabrizi-Wizsy, N.; Lang, U.; Desoye, G.; Dieber-Rotheneder, M. Placental Membrane-Type Metalloproteinases (MT-MMPs): Key Players in Pregnancy. *Cell Adhes. Migr.* **2016**, *10*. [CrossRef]

252. Abdullah, B.; Deveci, K.; Atilgan, R.; Kiliçli, F.; Söylemez, M.S. Serum Angiopoietin-Related Growth Factor (AGF) Levels Are Elevated in Gestational Diabetes Mellitus and Associated with Insulin Resistance. *Ginekol. Pol.* **2012**, *83*, 749–753. [PubMed]

253. Calderon, I.M.P.; Damasceno, D.C.; Amorin, R.L.; Costa, R.A.A.; Brasil, M.A.M.; Rudge, M.V.C. Morphometric Study of Placental Villi and Vessels in Women with Mild Hyperglycemia or Gestational or Overt Diabetes. *Diabetes Res. Clin. Pract.* **2007**, *78*. [CrossRef] [PubMed]

254. Baumüller, S.; Lehnen, H.; Schmitz, J.; Fimmers, R.; Müller, A.M. The Impact of Insulin Treatment on the Expression of Vascular Endothelial Cadherin and Beta-Catenin in Human Fetoplacental Vessels. *Pediatric Dev. Pathol.* **2015**, *18*. [CrossRef]

255. Rao, R.; Sen, S.; Han, B.; Ramadoss, S.; Chaudhuri, G. Gestational Diabetes, Preeclampsia and Cytokine Release: Similarities and Differences in Endothelial Cell Function. *Adv. Exp. Med. Biol.* **2014**, *814*. [CrossRef]

256. Yang, Z.; Laubach, V.E.; French, B.A.; Kron, I.L. Acute Hyperglycemia Enhances Oxidative Stress and Exacerbates Myocardial Infarction by Activating Nicotinamide Adenine Dinucleotide Phosphate Oxidase during Reperfusion. *J. Thorac. Cardiovasc. Surg.* **2009**, *137*. [CrossRef] [PubMed]

257. Giacco, F.; Brownlee, M. Oxidative Stress and Diabetic Complications. *Circ. Res.* **2010**, *107*. [CrossRef]

258. Tabit, C.E.; Chung, W.B.; Hamburg, N.M.; Vita, J.A. Endothelial Dysfunction in Diabetes Mellitus: Molecular Mechanisms and Clinical Implications. *Rev. Endocr. Metab. Disord.* **2010**, *11*. [CrossRef]

259. Sies, H.; Jones, D.P. Reactive Oxygen Species (ROS) as Pleiotropic Physiological Signalling Agents. *Nat. Rev. Mol. Cell Biol.* **2020**, *21*. [CrossRef] [PubMed]

260. Sies, H.; Berndt, C.; Jones, D.P. Oxidative Stress. *Annu. Rev. Biochem.* **2017**, *86*. [CrossRef]

261. Halliwell, B.; Gutteridge, J.M.C. Free Radicals in Biology and Medicine. *J. Free Radic. Biol. Med.* **1985**, *1*. [CrossRef]

262. Ward, J.P.T. From physiological redox signalling to oxidant stress. In *Pulmonary Vasculature Redox Signaling in Health and Disease*; Advances in Experimental Medicine and Biology; Springer: Cham, Switzerland, 2017; Volume 967. [CrossRef]

263. Burton, G.J.; Cindrova-Davies, T.; Yung, H.W.; Jauniaux, E. Oxygen and Development of the Human Placenta. *Reproduction* **2021**, *161*. [CrossRef]

264. Saghian, R.; Bogle, G.; James, J.L.; Clark, A.R. Establishment of Maternal Blood Supply to the Placenta: Insights into Plugging, Unplugging and Trophoblast Behaviour from an Agent-Based Model. *Interface Focus* **2019**, *9*. [CrossRef] [PubMed]

265. Schoots, M.H.; Gordijn, S.J.; Scherjon, S.A.; van Goor, H.; Hillebrands, J.L. Oxidative Stress in Placental Pathology. *Placenta* **2018**, *69*. [CrossRef]

266. Holland, O.J.; Hickey, A.J.R.; Alvsaker, A.; Moran, S.; Hedges, C.; Chamley, L.W.; Perkins, A.V. Changes in Mitochondrial Respiration in the Human Placenta over Gestation. *Placenta* **2017**, *57*. [CrossRef]

267. Wang, Y.; Walsh, S.W. Placental Mitochondria as a Source of Oxidative Stress in Pre-Eclampsia. *Placenta* **1998**, *19*. [CrossRef]

268. Pereira, R.D.; de Long, N.E.; Wang, R.C.; Yazdi, F.T.; Holloway, A.C.; Raha, S. Angiogenesis in the Placenta: The Role of Reactive Oxygen Species Signaling. *BioMed Res. Int.* **2015**, *2015*, 814543. [CrossRef] [PubMed]

269. Hung, T.H.; Skepper, J.N.; Burton, G.J. In Vitro Ischemia-Reperfusion Injury in Term Human Placenta as a Model for Oxidative Stress in Pathological Pregnancies. *Am. J. Pathol.* **2001**, *159*. [CrossRef]

270. Jauniaux, E.; Hempstock, J.; Greenwold, N.; Burton, G.J. Trophoblastic Oxidative Stress in Relation to Temporal and Regional Differences in Maternal Placental Blood Flow in Normal and Abnormal Early Pregnancies. *Am. J. Pathol.* **2003**, *162*. [CrossRef]

271. Cuffe, J.S.; Xu, Z.C.; Perkins, A.V. Biomarkers of Oxidative Stress in Pregnancy Complications. *Biomark. Med.* **2017**, *11*. [CrossRef] [PubMed]

272. Manea, A.; Tanase, L.I.; Raicu, M.; Simionescu, M. Transcriptional Regulation of NADPH Oxidase Isoforms, Nox1 and Nox4, by Nuclear Factor-КB in Human Aortic Smooth Muscle Cells. *Biochem. Biophys. Res. Commun.* **2010**, *396*. [CrossRef]

273. Zhu, C.; Yang, H.; Geng, Q.; Ma, Q.; Long, Y.; Zhou, C.; Chen, M. Association of Oxidative Stress Biomarkers with Gestational Diabetes Mellitus in Pregnant Women: A Case-Control Study. *PLoS ONE* **2015**, *10*, e0126490. [CrossRef]

274. Eriksson, U.J. Congenital Anomalies in Diabetic Pregnancy. *Semin. Fetal Neonatal Med.* **2009**, *14*. [CrossRef]
275. Coughlan, M.T.; Vervaart, P.P.; Permezel, M.; Georgiou, H.M.; Rice, G.E. Altered Placental Oxidative Stress Status in Gestational Diabetes Mellitus. *Placenta* **2004**, *25*. [CrossRef]
276. Lappas, M.; Permezel, M.; Rice, G.E. Release of Proinflammatory Cytokines and 8-Isoprostane from Placenta, Adipose Tissue, and Skeletal Muscle from Normal Pregnant Women and Women with Gestational Diabetes Mellitus. *J. Clin. Endocrinol. Metab.* **2004**, *89*. [CrossRef]
277. Biri, A.; Onan, A.; Devrim, E.; Babacan, F.; Kavutcu, M.; Durak, I. Oxidant Status in Maternal and Cord Plasma and Placental Tissue in Gestational Diabetes. *Placenta* **2006**, *27*. [CrossRef]
278. Pessler, D.; Rudich, A.; Bashan, N. Oxidative Stress Impairs Nuclear Proteins Binding to the Insulin Responsive Element in the GLUT4 Promoter. *Diabetologia* **2001**, *44*. [CrossRef] [PubMed]
279. Trocino, R.A.; Akazawa, S.; Ishibashi, M.; Matsumoto, K.; Matsuo, H.; Yamamoto, H.; Goto, S.; Urata, Y.; Kondo, T.; Nagataki, S. Significance of Glutathione Depletion and Oxidative Stress in Early Embryogenesis in Glucose-Induced Rat Embryo Culture. *Diabetes* **1995**, *44*. [CrossRef] [PubMed]
280. Kinalski, M.; Śledziewski, A.; Telejko, B.; Kowalska, I.; Krętowski, A.; Zarzycki, W.; Kinalska, I. Lipid Peroxidation, Antioxidant Defence and Acid-Base Status in Cord Blood at Birth: The Influence of Diabetes. *Horm. Metab. Res.* **2001**, *33*. [CrossRef] [PubMed]
281. Chaudhari, L.; Tandon, O.P.; Vaney, N.; Agarwal, N. Lipid Peroxidation and Antioxidant Enzymes in Gestational Diabetics. *Indian J. Physiol. Pharmacol.* **2003**, *47*, 441–446.
282. Manoharan, B.; Bobby, Z.; Dorairajan, G.; Jacob, S.E.; Gladwin, V.; Vinayagam, V.; Packirisamy, R.M. Increased Placental Expressions of Nuclear Factor Erythroid 2–Related Factor 2 and Antioxidant Enzymes in Gestational Diabetes: Protective Mechanisms against the Placental Oxidative Stress? *Eur. J. Obstet. Gynecol. Reprod. Biol.* **2019**, *238*. [CrossRef]
283. Navarro, A.; Alonso, A.; Garrido, P.; González, C.; González del Rey, C.; Ordoñez, C.; Tolivia, J. Increase in Placental Apolipoprotein D as an Adaptation to Human Gestational Diabetes. *Placenta* **2010**, *31*. [CrossRef]
284. Gelisgen, R.; Genc, H.; Kayali, R.; Oncul, M.; Benian, A.; Guralp, O.; Uludag, S.; Cakatay, U.; Albayrak, M.; Uzun, H. Protein Oxidation Markers in Women with and without Gestational Diabetes Mellitus: A Possible Relation with Paraoxonase Activity. *Diabetes Res. Clin. Pract.* **2011**, *94*. [CrossRef]
285. Pasternak, Y.; Biron-Shental, T.; Ohana, M.; Einbinder, Y.; Arbib, N.; Benchetrit, S.; Zitman-Gal, T. Gestational Diabetes Type 2: Variation in High-Density Lipoproteins Composition and Function. *Int. J. Mol. Sci.* **2020**, *21*, 6281. [CrossRef]
286. Kapustin, R.; Chepanov, S.; Kopteeva, E.; Arzhanova, O. Maternal Serum Nitrotyrosine, 8-Isoprostane and Total Antioxidant Capacity Levels in Pre-Gestational or Gestational Diabetes Mellitus. *Gynecol. Endocrinol.* **2020**, *36*. [CrossRef] [PubMed]
287. Bartakova, V.; Kollarova, R.; Kuricova, K.; Sebekova, K.; Belobradkova, J.; Kankova, K. Serum Carboxymethyl-Lysine, a Dominant Advanced Glycation End Product, Is Increased in Women with Gestational Diabetes Mellitus. *Biomed. Pap.* **2016**, *160*. [CrossRef]
288. Sisay, M.; Edessa, D.; Ali, T.; Mekuria, A.N.; Gebrie, A. The Relationship between Advanced Glycation End Products and Gestational Diabetes: A Systematic Review and Meta-Analysis. *PLoS ONE* **2020**, *15*, e0240382. [CrossRef]
289. Li, S.; Yang, H. Relationship between Advanced Glycation End Products and Gestational Diabetes Mellitus. *J. Matern. Fetal Neonatal Med.* **2019**, *32*. [CrossRef] [PubMed]
290. Li, H.; Dong, A.; Lv, X. Advanced Glycation End Products and Adipocytokines and Oxidative Stress in Placental Tissues of Pregnant Women with Gestational Diabetes Mellitus. *Exp. Ther. Med.* **2019**. [CrossRef]
291. Ghaneei, A.; Yassini, S.; Ghanei, M.E.; Shojaoddiny-Ardekani, A. Increased Serum Oxidized Low-Density Lipoprotein Levels in Pregnancies Complicated by Gestational Diabetes Mellitus. *Iran. J. Reprod. Med.* **2015**, *13*, 421–424. [PubMed]
292. Aydemir, B.; Baykara, O.; Cinemre, F.B.S.; Cinemre, H.; Tuten, A.; Kiziler, A.R.; Akdemir, N.; Oncul, M.; Kaya, B.; Sozer, V.; et al. LOX-1 Gene Variants and Maternal Levels of Plasma Oxidized LDL and Malondialdehyde in Patients with Gestational Diabetes Mellitus. *Arch. Gynecol. Obstet.* **2016**, *293*. [CrossRef]
293. Kopylov, A.T.; Papysheva, O.; Gribova, I.; Kotaysch, G.; Kharitonova, L.; Mayatskaya, T.; Sokerina, E.; Kaysheva, A.L.; Morozov, S.G. Molecular Pathophysiology of Diabetes Mellitus during Pregnancy with Antenatal Complications. *Sci. Rep.* **2020**, *10*. [CrossRef]
294. Qiu, C.; Hevner, K.; Abetew, D.; Enquobahrie, D.A.; Williams, M.A. Oxidative DNA Damage in Early Pregnancy and Risk of Gestational Diabetes Mellitus: A Pilot Study. *Clin. Biochem.* **2011**, *44*. [CrossRef] [PubMed]
295. Urbaniak, S.K.; Boguszewska, K.; Szewczuk, M.; Kázmierczak-Barańska, J.; Karwowski, B.T. 8-Oxo-7,8-Dihydro-2′-Deoxyguanosine (8-OxodG) and 8-Hydroxy-2′-Deoxyguanosine (8-OHdG) as a Potential Biomarker for Gestational Diabetes Mellitus (GDM) Development. *Molecules* **2020**, *25*, 202. [CrossRef]
296. Kharb, S. Low Whole Blood Glutathione Levels in Pregnancies Complicated by Preeclampsia and Diabetes. *Clin. Chim. Acta* **2000**, *294*. [CrossRef]
297. Zhang, C.; Yang, Y.; Chen, R.; Wei, Y.; Feng, Y.; Zheng, W.; Liao, H.; Zhang, Z. Aberrant Expression of Oxidative Stress Related Proteins Affects the Pregnancy Outcome of Gestational Diabetes Mellitus Patients. *Am. J. Transl. Res.* **2019**, *11*, 269–279.
298. Peuchant, E.; Brun, J.L.; Rigalleau, V.; Dubourg, L.; Thomas, M.J.; Daniel, J.Y.; Leng, J.J.; Gin, H. Oxidative and Antioxidative Status in Pregnant Women with Either Gestational or Type 1 Diabetes. *Clin. Biochem.* **2004**, *37*. [CrossRef]
299. Huerta-Cervantes, M.; Peña-Montes, D.J.; Montoya-Pérez, R.; Trujillo, X.; Huerta, M.; López-Vázquez, M.Á.; Olvera-Cortés, M.E.; Saavedra-Molina, A. Gestational Diabetes Triggers Oxidative Stress in Hippocampus and Cerebral Cortex and Cognitive Behavior Modifications in Rat Offspring: Age-and Sex-Dependent Effects. *Nutrients* **2020**, *12*, 376. [CrossRef]

300. Gujski, M.; Szukiewicz, D.; Chołuj, M.; Sawicki, W.; Bojar, I. Fetal and Placental Weight in Pre-Gestational Maternal Obesity (PGMO) vs. Excessive Gestational Weight Gain (EGWG)—A Preliminary Approach to the Perinatal Outcomes in Diet-Controlled Gestational Diabetes Mellitus. *J. Clin. Med.* **2020**, *9*, 3530. [CrossRef] [PubMed]

301. Zehravi, M.; Maqbool, M.; Ara, I. Correlation between Obesity, Gestational Diabetes Mellitus, and Pregnancy Outcomes: An Overview. *Int. J. Adolesc. Med. Health* **2021**. [CrossRef] [PubMed]

302. Ornoy, A.; Becker, M.; Weinstein-Fudim, L.; Ergaz, Z. Diabetes during Pregnancy: A Maternal Disease Complicating the Course of Pregnancy with Long-Term Deleterious Effects on the Offspring. A Clinical Review. *Int. J. Mol. Sci.* **2021**, *22*, 2965. [CrossRef] [PubMed]

303. Karasneh, R.A.; Migdady, F.H.; Alzoubi, K.H.; Al-Azzam, S.I.; Khader, Y.S.; Nusair, M.B. Trends in Maternal Characteristics, and Maternal and Neonatal Outcomes of Women with Gestational Diabetes: A Study from Jordan. *Ann. Med. Surg.* **2021**, *67*. [CrossRef]

304. Riskin, A.; Itzchaki, O.; Bader, D.; Lofe, A.; Toropine, A.; Riskin-Mashiah, S. Perinatal Outcomes in Infants of Mothers with Diabetes in Pregnancy. *Isr. Med. Assoc. J.* **2020**, *22*, 569–575. [PubMed]

305. Schwartz, N.; Nachum, Z.; Green, M.S. The Prevalence of Gestational Diabetes Mellitus Recurrence—Effect of Ethnicity and Parity: A Metaanalysis. *Am. J. Obstet. Gynecol.* **2015**, *213*. [CrossRef]

306. Liang, X.; Zheng, W.; Liu, C.; Zhang, L.; Zhang, L.; Tian, Z.; Li, G. Clinical Characteristics, Gestational Weight Gain and Pregnancy Outcomes in Women with a History of Gestational Diabetes Mellitus. *Diabetol. Metab. Syndr.* **2021**, *13*, 73. [CrossRef]

307. Herath, H.; Herath, R.; Wickremasinghe, R. Gestational Diabetes Mellitus and Risk of Type 2 Diabetes 10 Years after the Index Pregnancy in Sri Lankan Women—A Community Based Retrospective Cohort Study. *PLoS ONE* **2017**, *12*, e0179647. [CrossRef]

308. Bellamy, L.; Casas, J.P.; Hingorani, A.D.; Williams, D. Type 2 Diabetes Mellitus after Gestational Diabetes: A Systematic Review and Meta-Analysis. *Lancet* **2009**, *373*. [CrossRef]

309. Retnakaran, R. Hyperglycemia in Pregnancy and Its Implications for a Woman's Future Risk of Cardiovascular Disease. *Diabetes Res. Clin. Pract.* **2018**, *145*. [CrossRef]

310. Tobias, D.K.; Stuart, J.J.; Li, S.; Chavarro, J.; Rimm, E.B.; Rich-Edwards, J.; Hu, F.B.; Manson, J.E.; Zhang, C. Association of History of Gestational Diabetes with Long-Term Cardiovascular Disease Risk in a Large Prospective Cohort of US Women. *JAMA Intern. Med.* **2017**, *177*. [CrossRef] [PubMed]

311. Franzago, M.; Fraticelli, F.; di Nicola, M.; Bianco, F.; Marchetti, D.; Celentano, C.; Liberati, M.; de Caterina, R.; Stuppia, L.; Vitacolonna, E. Early Subclinical Atherosclerosis in Gestational Diabetes: The Predictive Role of Routine Biomarkers and Nutrigenetic Variants. *J. Diabetes Res.* **2018**, *2018*. [CrossRef] [PubMed]

312. Nobile De Santis, M.S.; Taricco, E.; Radaelli, T.; Spada, E.; Rigano, S.; Ferrazzi, E.; Milani, S.; Cetin, I. Growth of Fetal Lean Mass and Fetal Fat Mass in Gestational Diabetes. *Ultrasound Obstet. Gynecol.* **2010**, *36*. [CrossRef]

313. Elessawy, M.; Harders, C.; Kleinwechter, H.; Demandt, N.; Sheasha, G.A.; Maass, N.; Pecks, U.; Eckmann-Scholz, C. Measurement and Evaluation of Fetal Fat Layer in the Prediction of Fetal Macrosomia in Pregnancies Complicated by Gestational Diabetes. *Arch. Gynecol. Obstet.* **2017**, *296*. [CrossRef] [PubMed]

314. Aydin, S.; Fatihoglu, E. Fetal Epicardial Fat Thickness: Can It Serve as a Sonographic Screening Marker for Gestational Diabetes Mellitus? *J. Med. Ultrasound* **2020**, *28*. [CrossRef] [PubMed]

315. Blumer, I.; Hadar, E.; Hadden, D.R.; Jovanovič, L.; Mestman, J.H.; Murad, M.H.; Yogev, Y. Diabetes and Pregnancy: An Endocrine Society Clinical Practice Guideline. *J. Clin. Endocrinol. Metab.* **2013**, *98*. [CrossRef]

316. Yildiz Atar, H.; Baatz, J.E.; Ryan, R.M. Molecular Mechanisms of Maternal Diabetes Effects on Fetal and Neonatal Surfactant. *Children* **2021**, *8*, 281. [CrossRef]

317. Lee, I.L.; Barr, E.L.M.; Longmore, D.; Barzi, F.; Brown, A.D.H.; Connors, C.; Boyle, J.A.; Kirkwood, M.; Hampton, V.; Lynch, M.; et al. Cord Blood Metabolic Markers Are Strong Mediators of the Effect of Maternal Adiposity on Fetal Growth in Pregnancies across the Glucose Tolerance Spectrum: The PANDORA Study. *Diabetologia* **2020**, *63*. [CrossRef]

318. Tarvonen, M.; Hovi, P.; Sainio, S.; Vuorela, P.; Andersson, S.; Teramo, K. Intrapartal Cardiotocographic Patterns and Hypoxia-Related Perinatal Outcomes in Pregnancies Complicated by Gestational Diabetes Mellitus. *Acta Diabetol.* **2021**. [CrossRef]

319. Basu, M.; Garg, V. Maternal Hyperglycemia and Fetal Cardiac Development: Clinical Impact and Underlying Mechanisms. *Birth Defects Res.* **2018**, *110*, 1504–1516. [CrossRef] [PubMed]

320. Balsells, M.; García-Patterson, A.; Gich, I.; Corcoy, R. Major Congenital Malformations in Women with Gestational Diabetes Mellitus: A Systematic Review and Meta-Analysis. *Diabetes Metab. Res. Rev.* **2012**, *28*. [CrossRef] [PubMed]

321. Mitanchez, D. Foetal and Neonatal Complications in Gestational Diabetes: Perinatal Mortality, Congenital Malformations, Macrosomia, Shoulder Dystocia, Birth Injuries, Neonatal Complications. *Diabetes Metab.* **2010**, *36*. [CrossRef] [PubMed]

322. Martínez-Frías, M.L.; Frías, J.P.; Bermejo, E.; Rodríguez-Pinilla, E.; Prieto, L.; Frías, J.L. Pre-Gestational Maternal Body Mass Index Predicts an Increased Risk of Congenital Malformations in Infants of Mothers with Gestational Diabetes. *Diabet. Med.* **2005**, *22*. [CrossRef]

323. Lowe, W.L.; Scholtens, D.M.; Kuang, A.; Linder, B.; Lawrence, J.M.; Lebenthal, Y.; McCance, D.; Hamilton, J.; Nodzenski, M.; Talbot, O.; et al. Hyperglycemia and Adverse Pregnancy Outcome Follow-up Study (HAPO FUS): Maternal Gestational Diabetes Mellitus and Childhood Glucose Metabolism. *Diabetes Care* **2019**, *42*. [CrossRef]

324. Briana, D.D.; Germanou, K.; Boutsikou, M.; Boutsikou, T.; Athanasopoulos, N.; Marmarinos, A.; Gourgiotis, D.; Malamitsi-Puchner, A. Potential Prognostic Biomarkers of Cardiovascular Disease in Fetal Macrosomia: The Impact of Gestational Diabetes. *J. Matern. Fetal Neonatal Med.* **2018**, *31*. [CrossRef]
325. Fetita, L.S.; Sobngwi, E.; Serradas, P.; Calvo, F.; Gautier, J.F. Review: Consequences of Fetal Exposure to Maternal Diabetes in Offspring. *J. Clin. Endocrinol. Metab.* **2006**, *91*. [CrossRef]
326. Ekelund, M.; Shaat, N.; Almgren, P.; Groop, L.; Berntorp, K. Prediction of Postpartum Diabetes in Women with Gestational Diabetes Mellitus. *Diabetologia* **2010**, *53*. [CrossRef] [PubMed]
327. Magnusson, Å.; Laivuori, H.; Loft, A.; Oldereid, N.B.; Pinborg, A.; Petzold, M.; Romundstad, L.B.; Söderström-Anttila, V.; Bergh, C. The Association Between High Birth Weight and Long-Term Outcomes—Implications for Assisted Reproductive Technologies: A Systematic Review and Meta-Analysis. *Front. Pediatr.* **2021**, *9*. [CrossRef] [PubMed]
328. Crume, T.L.; Ogden, L.; Daniels, S.; Hamman, R.F.; Norris, J.M.; Dabelea, D. The Impact of in Utero Exposure to Diabetes on Childhood Body Mass Index Growth Trajectories: The EPOCH Study. *J. Pediatr.* **2011**, *158*. [CrossRef] [PubMed]

MDPI

St. Alban-Anlage 66

4052 Basel

Switzerland

Tel. +41 61 683 77 34

Fax +41 61 302 89 18

www.mdpi.com

*International Journal of Molecular Sciences* Editorial Office

E-mail: ijms@mdpi.com

www.mdpi.com/journal/ijms